T0231596

Mean Field Simulation for Monte Carlo Integration

MONOGRAPHS ON STATISTICS AND APPLIED PROBABILITY

General Editors

F. Bunea, V. Isham, N. Keiding, T. Louis, R. L. Smith, and H. Tong

1. Stochastic Population Models in Ecology and Epidemiology *M.S. Barlett* (1960)
2. Queues *D.R. Cox and W.L. Smith* (1961)
3. Monte Carlo Methods *J.M. Hammersley and D.C. Handscomb* (1964)
4. The Statistical Analysis of Series of Events *D.R. Cox and P.A.W. Lewis* (1966)
5. Population Genetics *W.J. Ewens* (1969)
6. Probability, Statistics and Time *M.S. Barlett* (1975)
7. Statistical Inference *S.D. Silvey* (1975)
8. The Analysis of Contingency Tables *B.S. Everitt* (1977)
9. Multivariate Analysis in Behavioural Research *A.E. Maxwell* (1977)
10. Stochastic Abundance Models *S. Engen* (1978)
11. Some Basic Theory for Statistical Inference *E.J.G. Pitman* (1979)
12. Point Processes *D.R. Cox and V. Isham* (1980)
13. Identification of Outliers *D.M. Hawkins* (1980)
14. Optimal Design *S.D. Silvey* (1980)
15. Finite Mixture Distributions *B.S. Everitt and D.J. Hand* (1981)
16. Classification *A.D. Gordon* (1981)
17. Distribution-Free Statistical Methods, 2nd edition *J.S. Maritz* (1995)
18. Residuals and Influence in Regression *R.D. Cook and S. Weisberg* (1982)
19. Applications of Queueing Theory, 2nd edition *G.F. Newell* (1982)
20. Risk Theory, 3rd edition *R.E. Beard, T. Pentikäinen and E. Pesonen* (1984)
21. Analysis of Survival Data *D.R. Cox and D. Oakes* (1984)
22. An Introduction to Latent Variable Models *B.S. Everitt* (1984)
23. Bandit Problems *D.A. Berry and B. Fristedt* (1985)
24. Stochastic Modelling and Control *M.H.A. Davis and R. Vinter* (1985)
25. The Statistical Analysis of Composition Data *J. Aitchison* (1986)
26. Density Estimation for Statistics and Data Analysis *B.W. Silverman* (1986)
27. Regression Analysis with Applications *G.B. Wetherill* (1986)
28. Sequential Methods in Statistics, 3rd edition *G.B. Wetherill and K.D. Glazebrook* (1986)
29. Tensor Methods in Statistics *P. McCullagh* (1987)
30. Transformation and Weighting in Regression *R.J. Carroll and D. Ruppert* (1988)
31. Asymptotic Techniques for Use in Statistics *O.E. Bandorff-Nielsen and D.R. Cox* (1989)
32. Analysis of Binary Data, 2nd edition *D.R. Cox and E.J. Snell* (1989)
33. Analysis of Infectious Disease Data *N.G. Becker* (1989)
34. Design and Analysis of Cross-Over Trials *B. Jones and M.G. Kenward* (1989)
35. Empirical Bayes Methods, 2nd edition *J.S. Maritz and T. Lwin* (1989)
36. Symmetric Multivariate and Related Distributions *K.T. Fang, S. Kotz and K.W. Ng* (1990)
37. Generalized Linear Models, 2nd edition *P. McCullagh and J.A. Nelder* (1989)
38. Cyclic and Computer Generated Designs, 2nd edition *J.A. John and E.R. Williams* (1995)
39. Analog Estimation Methods in Econometrics *C.F. Manski* (1988)
40. Subset Selection in Regression *A.J. Miller* (1990)
41. Analysis of Repeated Measures *M.J. Crowder and D.J. Hand* (1990)
42. Statistical Reasoning with Imprecise Probabilities *P. Walley* (1991)
43. Generalized Additive Models *T.J. Hastie and R.J. Tibshirani* (1990)
44. Inspection Errors for Attributes in Quality Control *N.L. Johnson, S. Kotz and X. Wu* (1991)
45. The Analysis of Contingency Tables, 2nd edition *B.S. Everitt* (1992)

Monographs on Statistics and Applied Probability 126

Mean Field Simulation for Monte Carlo Integration

Pierre Del Moral

CRC Press
Taylor & Francis Group
Boca Raton   London   New York

CRC Press is an imprint of the
Taylor & Francis Group, an **informa** business

A CHAPMAN & HALL BOOK

CRC Press
Taylor & Francis Group
6000 Broken Sound Parkway NW, Suite 300
Boca Raton, FL 33487-2742

First issued in paperback 2016

© 2013 by Taylor & Francis Group, LLC
CRC Press is an imprint of Taylor & Francis Group, an Informa business

No claim to original U.S. Government works

Version Date: 20130424

ISBN 13: 978-1-138-19873-9 (pbk)
ISBN 13: 978-1-4665-0405-9 (hbk)

This book contains information obtained from authentic and highly regarded sources. Reasonable efforts have been made to publish reliable data and information, but the author and publisher cannot assume responsibility for the validity of all materials or the consequences of their use. The authors and publishers have attempted to trace the copyright holders of all material reproduced in this publication and apologize to copyright holders if permission to publish in this form has not been obtained. If any copyright material has not been acknowledged please write and let us know so we may rectify in any future reprint.

Except as permitted under U.S. Copyright Law, no part of this book may be reprinted, reproduced, transmitted, or utilized in any form by any electronic, mechanical, or other means, now known or hereafter invented, including photocopying, microfilming, and recording, or in any information storage or retrieval system, without written permission from the publishers.

For permission to photocopy or use material electronically from this work, please access www.copyright.com (http://www.copyright.com/) or contact the Copyright Clearance Center, Inc. (CCC), 222 Rosewood Drive, Danvers, MA 01923, 978-750-8400. CCC is a not-for-profit organization that provides licenses and registration for a variety of users. For organizations that have been granted a photocopy license by the CCC, a separate system of payment has been arranged.

Trademark Notice: Product or corporate names may be trademarks or registered trademarks, and are used only for identification and explanation without intent to infringe.

Visit the Taylor & Francis Web site at
http://www.taylorandfrancis.com

and the CRC Press Web site at
http://www.crcpress.com

To Laurence, Tiffany, and Timothée

Contents

IV Theoretical aspects   269

9 Mean field Feynman-Kac models   271

10 A general class of mean field models   299

11 Empirical processes   327

Preface

Monte Carlo integration

This book deals with the theoretical foundations and the applications of mean field simulation models for Monte Carlo integration.

In the last three decades, this topic has become one of the most active contact points between pure and applied probability theory, Bayesian inference, statistical machine learning, information theory, theoretical chemistry and quantum physics, financial mathematics, signal processing, risk analysis, and several other domains in engineering and computer sciences.

The origins of Monte Carlo simulation certainly start with the seminal paper of N. Metropolis and S. Ulam in the late 1940s [435]. Inspired by the interest of his colleague S.Ulam in the poker game N. Metropolis, coined, the term "Monte Carlo Method" in reference to the "capital" of Monaco well known as the European city for gambling.

The first systematic use of Monte Carlo integration was developed by these physicists in the Manhattan Project of Los Alamos National Laboratory, to compute ground state energies of Schödinger's operators arising in thermonuclear ignition models. It is also not surprising that the development of these methods goes back to these early days of computers. For a more thorough discussion on the beginnings of the Monte Carlo method, we refer to the article by N. Metropolis [436].

As its name indicates, Monte Carlo simulation is, in the first place, one of the largest, and most important, numerical techniques for the computer simulation of mathematical models with random ingredients. Nowadays, these simulation methods are of current use in computational physics, physical chemistry, and computational biology for simulating the complex behavior of systems in high dimension. To name a few, there are turbulent fluid models, disordered physical systems, quantum models, biological processes, population dynamics, and more recently financial stock market exchanges. In engineering sciences, they are also used to simulate the complex behaviors of telecommunication networks, queuing processes, speech, audio, or image signals, as well as radar and sonar sensors.

Note that in this context, the randomness reflects different sources of model uncertainties, including unknown initial conditions, misspecified kinetic parameters, as well as the external random effects on the system. The repeated random samples of the complex system are used for estimating some *averaging* type property of some phenomenon.

Monte Carlo integration theory, including Markov chain Monte Carlo algorithms (abbreviated MCMC), sequential Monte Carlo algorithms (abbreviated SMC), and mean field interacting particle methods (abbreviated IPS) are also used to sample complex probability distributions arising in numerical probability, and in Bayesian statistics. In this context, the random samples are used for computing deterministic multidimensional *integrals*.

In other situations, stochastic algorithms are also used for solving complex *estimation problems*, including inverse type problems, global optimization models, posterior distributions calculations, nonlinear estimation problems, as well as statistical learning questions (see for instance [77, 93, 163, 274, 521]). We underline that in this situation, the randomness depends on the design of the stochastic integration algorithm, or the random search algorithm.

In the last three decades, these extremely flexible Monte Carlo algorithms have been developed in various forms mainly in applied probability, Bayesian statistics, and in computational physics. Without any doubt, the most famous MCMC algorithm is the Metropolis-Hastings model presented in the mid-1960s by N. Metropolis, A. Rosenbluth, M. Rosenbluth, A. Teller, and E. Teller in their seminal article [437].

This rather elementary stochastic technique consists in designing a reversible Markov chain, with a prescribed target invariant measure, based on sequential acceptance-rejection moves. Besides its simplicity, this stochastic technique has been used with success in a variety of application domains. The Metropolis-Hastings model is cited in *Computing in Science and Engineering* as being in the top 10 algorithms having the "greatest influence on the development and practice of science and engineering in the 20th century."

As explained by N. Metropolis and S. Ulam in the introduction of their pioneering article [435], the Monte Carlo method is, "essentially, a statistical approach to the study of differential equations, or more generally, of integro-differential equations that occur in various branches of the natural sciences."

In this connection, we emphasize that any evolution model in the space of probability measures can always be interpreted as the distributions of random states of Markov processes. This key observation is rather well known for conventional Markov processes and related linear evolution models.

More interestingly, *nonlinear* evolution models in distribution spaces can also be seen as the laws of Markov processes, but their evolution interacts in a nonlinear way with the distribution of the random states. The random states of these Markov chains are governed by a flow of complex probability distributions with often unknown analytic solutions, or sometimes too complicated to compute in a reasonable time. In this context, Monte Carlo and mean field methods offer a catalog of rather simple and cheap tools to simulate and to analyze the behavior of these complex systems.

These two observations are the stepping stones of the mean field particle theory developed in this book.

Mean field simulation

The theory of mean field interacting particle models had certainly started by the mid-1960s, with the pioneering work of H.P. McKean Jr. on Markov interpretations of a class of nonlinear parabolic partial differential equations arising in fluid mechanics [428]. We also quote an earlier article by T.E. Harris and H. Kahn [322], published in 1951, using mean field type and heuristic-like splitting techniques for estimating particle transmission energies.

Since this period until the mid-1990s, these pioneering studies have been further developed by several mathematicians; to name a few, in alphabetical order, J. Gärtner [265], C. Graham [295, 296, 297, 298, 299, 300, 301, 302, 303, 304, 305, 306], B. Jourdain [348], K. Oelschläger [450, 451, 452], Ch. Léonard [394], S. Méléard [429, 430, 431], M. Métivier [434], S. Roelly-Coppoletta [431], T. Shiga and H. Tanaka [516], and A.S. Sznitman [538]. Most of these developments were centered around solving Martingale problems related to the existence of nonlinear Markov chain models, and the description of propagation of chaos type properties of continuous time IPS models, including McKean-Vlasov type diffusion models, reaction diffusion equations, as well as generalized Boltzmann type interacting jump processes. Their traditional applications were essentially restricted to fluid mechanics, chemistry, and condensed matter theory. Some of these application domains are discussed in some detail in the series of articles [111, 112, 126, 317, 402, 128, 423]. The book [360] provides a recent review on this class of nonlinear kinetic models.

Since the mid-1990s, there has been a virtual explosion in the use of mean field IPS methods as a powerful tool in real-word applications of Monte Carlo simulation in information theory, engineering sciences, numerical physics, and statistical machine learning problems. These sophisticated population type IPS algorithms are also ideally suited to parallel and distributed environment computation [65, 269, 272, 403]. As a result, over the past few years the popularity of these computationally intensive methods has dramatically increased thanks to the availability of cheap and powerful computers. These advanced Monte Carlo integration theories offer nowadays a complementary tool, and a powerful alternative to many standard deterministic function-based projections and deterministic grid-based algorithms, often restricted to low dimensional spaces and linear evolution models.

In contrast to traditional MCMC techniques (including Gibbs sampling techniques [270], which are a particular instance of Metropolis-Hasting models), another central advantage of these mean field IPS models is the fact that their precision parameter is not related to some stationary target measure, nor of some burning time period, but only to *the size of the population*. This precision parameter is more directly related to the computational power of parallel computers on which we are running the IPS algorithms.

In the last two decades, the numerical solving of concrete and complex nonlinear filtering problems, the computation of complex posterior Bayesian distribution, as well as the numerical solving of optimization problems in evolutionary computing, has been revolutionized by this new class of mean field IPS samplers [93, 163, 161, 186, 222, 513, 547]. Nowadays, their range of application is extending from the traditional fluid mechanics modeling towards a variety of nonlinear estimation problems arising in several scientific disciplines; to name a few, with some reference pointers: Advanced signal processing and nonlinear filtering [104, 151, 161, 154, 163, 225, 222, 223, 288, 366, 364], Bayesian analysis and information theory [93, 133, 163, 165, 222, 252, 403], queueing networks [33, 296, 297, 299, 300, 301], control theory [373, 337, 338, 575], combinatorial counting and evolutionary computing [7, 163, 186, 513, 547], image processing [8, 150, 251, 476], data mining [490], molecular and polymer simulation [163, 186, 309], rare events analysis [125, 122, 163, 175, 176, 282], quantum Monte Carlo methods [16, 91, 175, 176, 326, 433, 498], as well as evolutionary algorithms and interacting agent models [82, 157, 286, 327, 513, 547].

Applications on nonlinear filtering problems arising in turbulent fluid mechanics and weather forecasting predictions can also be found in the series of articles by Ch. Baehr and his co-authors [25, 26, 28, 29, 390, 487, 537]. More recent applications of mean field IPS models to spatial point processes, and multiple object filtering theory can be found in the series of articles [101, 102, 103, 137, 221, 459, 460, 461, 510, 564, 565, 566, 567]. These spatial point processes, and related estimation problems occur in a wide variety of scientific disciplines, such as environmental models, including forestry and plant ecology modeling, as well as biology and epidemiology, seismology, materials science, astronomy, queuing theory, and many others. For a detailed discussion on these applications areas we refer the reader to the book of D. Stoyan, W. Kendall, and J. Mecke [532] and the more recent books of P.J. Diggle [218] and A. Baddeley, P. Gregori, J. Mateu, R. Stoica, and D. Stoyan [24].

The use of mean field IPS models in mathematical finance is more recent. For instance, using the rare event interpretation of particle methods, R. Carmona, J. P. Fouque, and D. Vestal proposed in [98] an interacting particle algorithm for the computation of the probabilities of simultaneous defaults in large credit portfolios. These developments for credit risk computation were then improved in the recent developments by R. Carmona and S. Crépey [95] and by the author and F. Patras in [187]. Following the particle filtering approach which is already widely used to estimate hidden Markov models, V. Genon-Catalot, T. Jeantheau, and C. Laredo [271] introduced particle methods for the estimation of stochastic volatility models.

More generally, this approach has been applied for filtering nonlinear and non-Gaussian Models by R. Casarin [106], R. Casarin, and C. Trecroci [107]. More recently, M. S. Johannes, N. G. Polson, and J.R. Stroud [345] used a similar approach for filtering latent variables such as the jump times and sizes in jump diffusion price models. Particle techniques can also be used in financial mathematics to design stochastic optimization algorithms. This version of particle schemes was used by S. Ben Hamida and R. Cont in [47] for providing a new calibration algorithm allowing for the existence of multiple global minima. Finally, in [196, 197], interacting particle methods were used to estimate backward conditional expectations for American option pricing.

For a more thorough discussion on the use of mean field methods in mathematical finance, we refer the reader to the review article [96], in the book [97].

As their name indicates, branching and interacting particle systems are of course directly related to individual based population dynamics models arising in biology and natural evolution theory. A detailed discussion on these topics can be found in the articles [56, 108, 110, 315, 313, 314, 442, 445, 446, 455], and references therein.

In this connection, I also quote the more recent and rapidly developing mean field games theory introduced in the mid-2000s by J.M. Lasry and P.L. Lions in the series of pioneering articles [381, 382, 383, 384]. In this context, fluid particles are replaced by agents or companies that interact mutually in competitive social-economic environments so that to find optimal interacting strategies w.r.t. some reward function.

Applications of game theory with multiple agents systems in biology, economics, and finance are also discussed in the more recent studies by V. Kolokoltsov and his co-authors [361, 362, 363], in the series of articles [9, 48, 83, 84, 376, 395, 449, 548], the ones by P.E. Caines, M. Huang, and R.P. Malhamé [329, 330, 331, 332, 333, 334, 335], as well as in the pioneering article by R. J. Aumann [23]. Finite difference computational methods for solving mean field games Hamilton-Jacobi nonlinear equations can also be found in the recent article by Y. Achdou, F. Camilli, and I. Capuzzo Dolcetta [4].

For a more detailed account on this new branch of games theory, I refer to the seminal Bachelier lecture notes given in 2007-2008 by P.L. Lions at the Collège de France [401], as well as the series of articles [287, 329, 426] and references therein.

A need for interdisciplinary research

As attested by the rich literature of mean field simulation theory, many researchers from different disciplines have contributed to the development of this field. However, the communication between these different scientific domains often requires substantial efforts, and often a pretty longer time frame to acquire domain specific languages so as to understand the recent developments in these fields. As a result, the communication between different scientific domains is often very limited, and many mean field simulation methods developed in some domains have been rediscovered later by others researchers from different scientific fields. For instance, mean field Feynman-Kac models are also known under a great many names and guises. In physics, engineering sciences, as well as in Bayesian statistical analysis, *the same* interacting jump mean field model is known under several lively buzzwords; to name a few: pruning [403, 549], branching selection [170, 286, 484, 569], rejuvenation [8, 134, 275, 336, 490], condensation [339], look-ahead and pilot exploration resampling [263, 403, 405], Resampled Monte Carlo and RMC-methods [556], subset simulation [20, 21, 22, 396, 399], Rao-Blackwellized particle filters [280, 444, 457], spawning [138], cloning [310, 500, 501, 502], go-with-the-winner [7, 310], resampling [324, 522, 408], rejection and weighting [403], survival of the fittest [138], splitting [121, 124, 282], bootstrapping [43, 289, 290, 409], replenish [316, 408], enrichment [54, 336, 262, 374], and many other botanical names.

On the other hand, many lines of applied research seem to be developing in a blind way, with at least no visible connections with mathematical sides of the field. As a result, in applied science literature some mean field IPS models are presented as natural heuristic-like algorithms, without a single performance analysis, nor a discussion on the robustness and the stability of the algorithms.

In the reverse angle, there exists an avenue of mathematical open research problems related to the mathematical analysis of mean field IPS models. Researchers in mathematical statistics, as well as in pure and applied probability, must be aware of ongoing research in some more applied scientific domains. I believe that interdisciplinary research is one of the key factors to develop innovative research in applied mathematics, and in more applied sciences.

It is therefore essential to have a unified and rigorous mathematical treatment so that researchers in applied mathematics, and scientists from different application domains, can learn from one field to another. This book tries to fill the gap with providing a unifying mathematical framework, and a self-contained and up-to-date treatment, on mean field simulation techniques for Monte Carlo integration. I hope it will initiate more discussions between different application domains and help theoretical probabilist researchers, applied statisticians, biologists, statistical physicists, and computer scientists to better work across their own disciplinary boundaries.

On the other hand, besides the fact that there exists an extensive number of books on simulation and conventional Monte Carlo methods, and Markov chain Monte Carlo techniques, very few studies are related to the foundations and the application of mean field simulation theory. To guide the reader we name a few reference texbooks on these conventional Monte Carlo simulations. The series of books by E.H.L. Aarts and J.H.M. Korst [1, 2], R.Y. Rubinstein [503, 504, 505, 506], and G. Pflug [471] discuss simulation and randomized techniques to combinatorial optimization, and related problems in operation research. The books by W.R. Gilks, S. Richardson, and D.J. Spiegelhalter [276], the one by B. D. Ripley [491], and the one by C.P. Robert and G. Casella [493] are more centered around conventional Monte Carlo methods, and MCMC techniques with applications in Bayesian inference.

The books by G. Fishman [256], as well as the one by S. Asmussen and

P.W. Glynn [18], discuss standard stochastic algorithms including discretization techniques of diffusions, Metropolis-Hasting techniques, and Gibbs samplers, with several applications in operation research, queueing processes, and mathematical finance. The book by P. Glasserman [284] contains more practically oriented discussions on the applications of Monte Carlo methods in mathematical finance. Rare event Monte Carlo simulation using importance sampling techniques are discussed in the book by J.A. Bucklew [85]. The book by O. Cappé, E. Moulines, and T. Rydèn [93], as well as the book [222] edited by A. Doucet, J.F.G. de Freitas, and N.J. Gordon, are centered around applications of sequential Monte Carlo type methods to Bayesian statistics, and hidden Markov chain problems.

In this connection, the present book can be considered as a continuation of the monograph [163] and lecture notes [161] edited in 2000, and the more recent survey article [161, 186], dedicated to Feynman-Kac type models. These studies also discuss continuous time models, and uniform concentration inequalities. In the present volume, we present a more recent unifying mean field theory that applies to a variety of discrete generation IPS algorithms, including McKean-Vlasov type models, branching and interacting jump processes, as well as Feynman-Kac models, and their extended versions arising in multiple object nonlinear filtering, and stochastic optimal stopping problems.

I have also collected a series of new deeper studies on general evolution equations taking values in the space of nonnegative measures, as well as new exponential concentration properties of mean field models. Researchers and applied mathematicians will also find a collection of modern stochastic perturbations techniques, based on backward semigroup methods, and second order Taylor's type expansions in distribution spaces. Last, but not least, the unifying treatment presented in this book sheds some new light on interesting links between physical, engineering, statistical, and mathematical domains which may appear disconnected at first glance. I hope that this unifying point of view will help to develop fruitfully this field further.

Use and interpretations of mean field models

I have written this book in the desire that post-graduate students, researchers in probability and statistics, as well as practitioners will use it.

To accomplish this difficult task, I have tried to illustrate the use of mean field simulation theory in a systematic and accessible way, across a wide range of illustrations presented through models with increasing complexity in a variety of application areas. The illustrations I have chosen are very often at the crossroad of several seemingly disconnected scientific disciplines, including biology, physics, engineering sciences, probability, and statistics.

In this connection, I emphasize that most of the mean field IPS algorithms

I have discussed in this book are mathematically identical, but their interpretations strongly depend on the different application domains they are thought. Furthermore, different ways of interpreting a given mean field particle technique often guide researchers' and development engineers' intuition to design and to analyze a variety of consistent mean field stochastic algorithms for solving concrete estimation problems. This variety of interpretations is one of the threads that guide the development of this book, with constant interplays between the theory and the applications.

In fluid mechanics, and computational physics, mean field particle models represent the physical evolution of different kinds of macroscopic quantities interacting with the distribution of microscopic variables. These stochastic models includes physical systems such as gases, macroscopic fluid models, and other molecular chaotic systems. One central modeling idea is often to neglect second order fluctuation terms in complex systems so that to reduce the model to a closed nonlinear evolution equation in distribution spaces (see for instance [126, 526], and references therein). The mean field limit of these particle models represents the evolution of these physical quantities. They are often described by nonlinear integro-differential equations.

In computational biology and population dynamic theory, the mathematical description of mean field genetic type adaptive populations, and related spatial branching processes, is expressed in terms of birth and death and competitive selection type processes, as well as mutation transitions of individuals, also termed particles. The state space of these evolution models depends on the application domain. In genealogical evolution models, the ancestral line of individuals evolves in the space of random trajectories. In genetic population models, individuals are encoded by strings in finite or Euclidian product spaces. Traditionally, these strings represent the chromosomes or the genotypes of the genome. The mutation transitions represent the biological random changes of individuals. The selection process is associated with fitness functions that evaluate the adaptation level of individuals. In this context, the mean field limit of the particle models is sometimes called the infinite population model. For finite state space models, these evolutions are described by deterministic dynamical systems in some simplex. In more general situations, the limiting evolution belongs to the class of measure valued equations.

In computer sciences, mean field genetic type IPS algorithms (*abbreviated GA*) are also used as random search heuristics that mimic the process of evolution to generate useful solutions to complex optimization problems. In this context, the individuals represent candidate solutions in a problem dependent state space; and the mutation transition is analogous to the biological mutation so as to increase the variability and the diversity of the population of solutions. The selection process is associated with some fitness functions that evaluate the quality of a solution w.r.t. some criteria that depend on the problem at hand. In this context, the limiting mean field model is often given by some Boltzmann-Gibbs measures associated with some fitness potential functions.

In advanced signal processing as well as in statistical machine learning theory, mean field IPS evolution models are also termed Sequential Monte Carlo samplers. As their name indicates, the aim of these methodologies is to sample from a sequence of probability distributions with an increasing complexity on general state spaces, including excursion spaces in rare event level splitting models, transition state spaces in sequential importance sampling models, and path space models in filtering and smoothing problems. In signal processing these evolution algorithms are also called particle filters. In this context, the mutation-selection transitions are often expressed in terms of a prediction step, and the updating of the particle population scheme. In this case, the limiting mean field model coincides with the evolution of conditional distributions of some random process w.r.t. some conditioning event. For linear-Gaussian models, the optimal filter is given by Gaussian conditional distributions, with conditional means and error covariance matrices computed using the celebrated Kalman filter recursions. In this context, these evolution equations can also be interpreted as McKean-Vlasov diffusion type models. The resulting mean field model coincides with the Ensemble Kalman Filters (*abbreviated EKF*) currently used in meteorological forecasting and data assimilation problems.

In physics and molecular chemistry, mean field IPS evolution models are used to simulate quantum systems to estimate ground state energies of a many-body Schödinger evolution equation. In this context, the individuals are termed walkers to avoid confusion with the physical particle based models. These walkers evolve in the set of electronic or macromolecular configurations. These evolution stochastic models belong to the class of Quantum and Diffusion Monte Carlo methods (*abbreviated QMC and DMC*). These Monte Carlo methods are designed to approximate the path space integrals in many-dimensional state spaces. Here again these mean field genetic type techniques are based on a mutation and a selection style transition. During the mutation transition, the walkers evolve randomly and independently in a potential energy landscape on particle configurations. The selection process is associated with a fitness function that reflects the particle absorption in an energy well. In this context, the limiting mean field equation can be interpreted as a normalized Schrödinger type equation. The long time behavior of these non-linear semigroups is related to top eigenvalues and ground state energies of Schrödinger's operators.

In probability theory, mean field IPS models can be interpreted into two ways. Firstly, the particle scheme can be seen as a step by step projection of the solution of an evolution equation in distribution spaces, into the space of empirical measures. More precisely, the empirical measures associated with a mean field IPS model evolve as a Markov chain in reduced finite dimensional state spaces. In contrast to conventional MCMC models based on the long time behavior of a single stochastic process, the mean field IPS Markov chain model is associated with a population of particles evolving in product state spaces. In this sense, mean field particle methods can be seen as a stochastic linearization of nonlinear equations. The second interpretation of these models relies on an original stochastic perturbation theory of nonlinear evolution equations in distribution spaces. More precisely, the local sampling transitions of the population of individuals induce some local sampling error, mainly because the transition of each individual depends on the occupation measure of the systems. In this context, the occupation measures of the IPS model evolve as the limiting equation, up to these local sampling errors.

A unifying theoretical framework

Most of the book is devoted to the applications of mean field theory to the analysis of discrete generation and nonlinear evolution equations in distribution spaces. Most of the models come from discrete time approximations of continuous time measure valued processes. Because of the importance of continuous time models in physical, finance, and biological modeling, as well as in other branches of applied mathematics, several parts of this book are concerned with the deep connections between discrete time models measure valued processes and their continuous time version, including linear and nonlinear integro-differential equations.

Our mathematical base strongly differs from the traditional convergence to equilibrium properties of conventional Markov chain Monte Carlo techniques. In addition, in contrast with traditional Monte Carlo samplers, mean field IPS models are not based on statistically independent particles, so that the law of large numbers cannot be used directly to analyze the performance of these interacting particle samplers.

The analysis of continuous time interacting particle systems developed in the last decades has been mainly centered around propagation of chaos properties, or asymptotic theorems, with very little information on the long time behavior of these particle models and on their exponential concentration properties.

To the best of my knowledge, the first studies on discrete generation mean field particle models, uniform quantitative estimates w.r.t. the time parameter, and their applications to nonlinear filtering problems and genetic type particle algorithms were started in the mid-1990s in [151, 152], as well as in the series of articles [154, 198, 169, 170]. In the last two decades, these studies have been further developed in various theoretical and more applied directions. For a detailed discussion on these developments, with a detailed bibliography, I refer to [161, 163, 186], and references therein. In this connection, I apologize in advance for any possible errors or omissions due to the lack of accurate information.

While exploring the convergence analysis of discrete generation mean field models, the reader will encounter a great variety of mathematical techniques, including nonlinear semigroup techniques in distribution spaces, interacting empirical process theory, exponential concentration inequalities, uniform convergence estimates w.r.t. the time horizon, as well as functional fluctuation theorems. These connections are remarkable in the sense that every probabilistic question related to the convergence analysis of IPS models requires to combine several mathematical techniques. For instance, the uniform exponential concentration inequalities developed in this book are the conjunction of backward nonlinear semigroup techniques, Taylor's type first order expansions in distribution spaces, $\mathbb{L}_m$-mean error estimates, and Orlicz norm analysis or Laplace estimates.

Uniform quantitative $\mathbb{L}_m$-mean error bounds, and uniform concentration estimates w.r.t. the time horizon rely on the stability properties of the limiting nonlinear semigroup. These connections between the long time behavior of mean field particle models and the stability properties of the limiting flow of measures are not new. We refer the reader to the articles [161, 163, 169, 170, 160, 182] published in the early 2000s in the context of Feynman-Kac models, and the more recent studies [166, 186] for more general classes of mean field models.

The analysis of discrete generation genetic type particle models is also deeply connected to several remarkable mathematical objects including complete ancestral trees models, historical and genealogical processes, partially observed branching processes, particle free energies, as well as backward Markov particle models. A large part of the book combines semigroup techniques with a stochastic perturbation analysis to study the convergence of interacting particle algorithms in very general contexts. I have developed, as systematically as I can, the application of this stochastic theory to applied models in numerical physics, dynamical population theory, signal processing, control systems, information theory, as well as in statistical machine learning research.

The mean field theory developed in this book also offers a rigorous and unifying framework to analyze the convergence of numerous heuristic-like algorithms currently used in physics, statistics, and engineering. It applies to any problem which can be translated in terms of nonlinear measure valued evolution equations.

At a first reading, the frustration some practitioners may get when analyzing abstract and general nonlinear evolution models in distribution state spaces will be rapidly released, since their mean field interacting particle approximations provide instantly a collection of powerful Monte Carlo simulation tools for the numerical solution of the problem at hand. Several illustrations of this mean field theory are provided in the context of McKean-Vlasov diffusion type models, Feynman-Kac distribution flows, as well as spatial branching evolution models, probability hypothesis density equations, and association tree based models arising in multiple object nonlinear filtering problems.

I also emphasize that the present book does not give a full complete treatment of the convergence analysis of mean field IPS models. To name a few missing topics, we do not discuss increasing and strong propagation of chaos properties, and we say nothing on Berry-Esseen theorems and asymptotic large deviation principles. In the context of Feynman-Kac models, the reader interested in these topics is recommended to consult the monograph [163], and references therein. To the best of our knowledge the fluctuation, and the large deviation analysis, of the intensity measure particle models and some of the extended Feynman-Kac models discussed in this book are still open research problems.

While some continuous time models are discussed in the first opening chapter, Chapter 2, and in Chapter 5, I did not hesitate to concentrate most of the exposition to discrete time models. The reasons are threefold:

Firstly, in the opening chapter I will discuss general state space Markov chain models that encapsulate without further work continuous time Markov processes by considering path spaces or excursion type continuous time models. The analysis of discrete time models only requires a smaller prerequisite on Markov chains, while the study of continuous time models would have required a different and specific knowledge on Markov infinitesimal generators and their stochastic analysis. In this opening chapter, I underline some connections between discrete generation and mean field IPS simulation and the numerical solving of nonlinear integro-differential equations. A more detailed and complete presentation of these continuous time models and their stochastic analysis techniques would have been too much digression.

The interested reader in continuous time models is recommended to consult the series of articles [295, 302, 303, 428, 429, 430, 451, 452, 538] and the more recent studies [57, 109, 295, 349, 418]. Interacting jumps processes and McKean-Vlasov diffusions are also discussed in the book [360] and the series of articles [161, 173, 176, 178, 498]. Uniform propagations of chaos for interacting Langevin type diffusions can also be found in the recent study [200]. For a discussion on sequential Monte Carlo continuous time particle models, we also refer the reader to the articles by A. Eberle, and his co-authors [237, 238, 239].

Uniform propagation of chaos for continuous time mean-field collision and jump-diffusion particle models can also be found in the more recent studies [440, 441]. The quantitative analysis developed in these studies is restricted to weak propagation of chaos estimates. It involves rather complex stochastic and functional analysis, but it also relies on backward semigroup expansions entering the stability properties of the limiting model, as the ones developed in [161, 163, 169, 170, 160, 182] in the early 2000s, and in the more recent studies [166, 186]. In this connection, an important research project is to extend the analysis developed in [440, 441] to obtain some useful quantitative uniform concentration inequalities for continuous time mean field, collision and interacting jump-diffusion models w.r.t. the time horizon.

The second reason is that, apart from some mathematical technicalities, the study of continuous time models often follows the same intuitions and the same semigroup analysis of discrete time models. On the other hand, to the best of my knowledge, various nonasymptotic convergence estimates developed in this book, such uniform concentration inequalities w.r.t. the time horizon, the convergence analysis at the level of the empirical process, as well as the analysis of backward Feynman-Kac particle models, remain nowadays open problems for continuous time mean field models.

Our third reason comes from two major practical and theoretical issues. Firstly, to get some feasible computer implementation of mean field IPS algorithms, continuous time models require introduction of additional levels of approximation. On the other hand, it is well known that at any time discretization schemes induce an additional and separate bias in the resulting discrete generation particle algorithm (see for instance [155] in the context of nonlinear filtering models).

Last, but not least, the convergence analysis of time discretization schemes is also based on different mathematical tools, with specific stochastic analysis techniques. In this connection, it is also well known that some techniques used to analyze continuous time models, such as the spectral analysis of Langevin type reversible diffusions, fail to describe the convergence properties of their discrete time versions.

A contents guide

One central theme of this book will be the applications of mean field simulation theory to the mathematical and numerical analysis of nonlinear evolution equations in distribution spaces.

To whet the appetite and to guide the different classes of users, the opening chapters, Chapter 1, and Chapter 2, provide an overview of the main topics that will be treated in greater detail in the further development of this book, with precise reference pointers to the corresponding chapters and sections.

These two chapters *should not be skipped* since they contain a detailed exposition of the mean field IPS models, and their limiting measure valued processes discussed in this book. To support this work, we also discuss some high points of the book, with a collection of selected convergence estimates, including contraction properties of nonlinear semigroups, nonasymptotic variance theorems, and uniform concentration inequalities.

Our basis is the theory of Markov processes. Discrete, and continuous time, Markov processes play a central role in the analysis of linear, and nonlinear evolution equations, taking values in the space probability measures. In this context, these evolution models in distribution spaces can always be interpreted as the distribution of the random states of a Markov process. In this interpretation, the theory of Markov processes offers a natural way to solve these measure valued equations, by simulating repeated samples of the random paths of the stochastic process. To provide a development of the area pertinent to each reader's specific interests, in the opening chapter, Chapter 1, we start with a rather detailed discussion on different classes of measure valued evolution models, and their linear, and nonlinear, Markov chain interpretation models.

In order to have a concrete basis for the further development of the book, in Chapter 2, we illustrate the mathematical models with a variety of McKean-Vlasov diffusion type models, and Feynman-Kac models, with concrete examples arising in particle physics, biology, dynamic population and branching processes, signal processing, Bayesian inference, information theory, statistical machine learning, molecular dynamics simulation, risk analysis, and rare event simulations.

In Chapter 3, we introduce the reader to Feynman-Kac models and their application domains, including spatial branching processes, particle absorption evolutions, nonlinear filtering, hidden Markov chain models, as well as sensitivity measure models arising in risk analysis, as well as in mathematical finance. We also provide some discussions on path space models, including terminal time conditioning principles and Markov chain bridge models.

It is assumed that the interpretation of mean field Feynman-Kac models strongly depends on the application area. Each interpretation is also a source of inspiration for researchers and development engineers to design stochastic particle algorithms equipped with some tuning parameters, adaptive particle criteria, and coarse grained techniques. To guide the reader's intuition, in Chapter 4, we present four equivalent particle interpretations of the Feynman-Kac mean field IPS models; namely the branching process interpretation and related genetic type algorithms (*abbreviated GA*), the sequential Monte Carlo methodology (*abbreviated SMC*), the interacting Markov chain Monte Carlo sampler (*abbreviated i-MCMC*), and the more abstract mean field interacting particle system interpretation (*abbreviated IPS*).

From the mathematical perspective, we recall that these four interpretation models are, of course, equivalent. Nevertheless, these different interpretations offer complementary perspectives that can be used to design and to analyze more sophisticated particle algorithms. These new classes of IPS models often combine branching techniques, genealogical tree samplers, forward-backward sampling, island type coarse graining models, adaptive MCMC methodologies, and adaptive selection type models [11, 186, 227, 279, 280]. More sophisticated mean field IPS algorithms include adaptive resampling rules [60, 158, 159], as well as particle MCMC strategies [11], particle SMC techniques and interacting Kalman filters [12, 132, 163, 225, 253, 444, 457], approximate Bayesian computation style methods [36, 154, 155, 156], adaptive temperature schedules [279, 280, 511], and convolution particle algorithms [497, 558].

We also emphasize that McKean-Vlasov diffusion type models can be also combined with mean field Feynman-Kac type models. This sophisticated class of mean field IPS filtering models has been used with success to solve nonlinear filtering problems arising in turbulent fluid mechanics and weather forecasting predictions [25, 26, 28, 29, 390, 487, 537]. Related, but different classes of coarse grained and branching type mean field IPS models are also discussed in the series of articles [135, 144, 161, 186].

In Chapter 5, we discuss some connections between discrete generation Feynman-Kac models and their continuous time version. We briefly recall some basic facts about Markov processes and their infinitesimal generators. For a more rigorous and detailed treatment, we refer to the reference textbooks on stochastic calculus [355, 375, 454, 488]. We also use these stochastic modeling techniques to design McKean jump type interpretations of Feynman-Kac models, and mean field particle interpretations of continuous time McKean models, in terms of infinitesimal generators.

Chapter 6 focuses on nonlinear evolution of intensity measure models arising in spatial branching processes and multiple object filtering theory. The analysis and the computation of these nonlinear equations are much more involved than the traditional single object nonlinear filtering equations. Because of their importance in practice, we have chosen to present a detailed derivation of these conditional density models. Of course these conditional density filters are not exact; they are based on some Poisson hypothesis at every updating step. The solving of the optimal filter associated with a spatial branching signal requires computing all the possible combinatorial structures between the branching particles and their noisy observations.

The second part of this book is dedicated to applications of the mean field IPS theory to two main scientific disciplines: namely, particle absorption type models arising in quantum physics and chemistry; and filtering and optimal control problems arising in advanced signal processing and stochastic optimal control. These two application areas are discussed, respectively, in Chapter 7 and in Chapter 8. Some application domains related to statistical machine learning, signal processing, and rare event simulation are also discussed in the first two opening chapters of the book, Chapter 1, and Chapter 2. The list of applications discussed in this book is by no means exhaustive. It only reflects some scientific interests of the author.

Chapter 9 is an intermediate as well as a pedagogical step towards the refined stochastic analysis of mean field particle models developed in the further development of the book, dedicated to the theoretical analysis of this class of models. We hope the rather elementary mathematical material presented in this opening chapter will help to make the more advanced stochastic analysis developed in the final part of the book more accessible to theoreticians and practitioners both. This chapter focuses on Feynman-Kac particle models. More general classes of particle models arising in fluid mechanics, spatial branching theory, and multiple object filtering literature are discussed in Chapter 10, as well as in Chapter 13.

Our main goals are to analyze in some more detail the theoretical structure of mean field interpretations of the Feynman-Kac models. This opening chapter, on the theoretical analysis of mean field models, does not pretend to have strong leanings towards applications. Our aim was just to provide some basic elements on the foundations of mean field IPS models in the context of Feynman-Kac measures. The applied value of Feynman-Kac particle models has already been illustrated in some details in Chapter 1, as well as in Chapter 7, and in Chapter 8, through a series of problems arising in rare event simulation, particle absorption models, advanced signal processing, stochastic optimization, financial mathematics, and optimal control problems.

In Chapter 10, we investigate the convergence analysis of a general and abstract class of mean field IPS model. We begin with a formal description of mean field IPS models associated with Markov-McKean interpretation models of nonlinear measure valued equations in the space of probability measures. Then, we present some rather weak regularity properties that allow development of stochastic perturbation techniques and a first order fluctuation analysis of mean field IPS models. We also present several illustrations, including Feynman-Kac models, interacting jump processes, simplified McKean models of gases, Gaussian mean field models, as well as generalized McKean-Vlasov type models with interacting jumps. Most of the material presented in this chapter is taken from the article [166]. Related approaches to the analysis of general mean field particle models can be found in the series of articles [144, 152, 168].

Chapter 11 is dedicated to the theory of empirical processes and measure concentration theory. These techniques are one of the most powerful mathematical tools to analyze the deviations of particle Monte Carlo algorithms. In the last two decades, these mathematical techniques have become some of the most important steps forward in infinite dimensional stochastic analysis and advanced machine learning techniques, as well as in the development of statistical nonasymptotic theory.

For an overview of these subjects, we refer the reader to the seminal books of D. Pollard [480], one of A.N. Van der Vaart and J.A. Wellner [559], and the remarkable articles by E. Giné [277], M. Ledoux [391], and M. Talagrand [542, 543, 544], and the article by R. Adamczak [6]. In this chapter, we have collected some more or less well known stochastic techniques for analyzing the concentration properties of empirical processes associated with independent random sequences on a measurable state space. We also present stochastic perturbation techniques for analyzing exponential concentration properties of functional empirical processes. Our stochastic analysis combines Orlicz's norm techniques, Kintchine's type inequalities, maximal inequalities, as well as Laplace-Cramèr-Chernov estimation methods.

Most of the mathematical material presented in this chapter is taken from a couple of articles [166, 186]. We also refer the reader to a couple of articles by the author, with M. Ledoux [160] and S. Tindel [199], for complementary material related to fluctuations, and Donsker's type theorems, and Berry-Esseen type inequalities for genetic type mean field models. Moderate deviations for general mean field models can also be found in the article of the author, with S. Hu and L.M. Wu [174].

Chapter 12 and Chapter 13 are concerned with the semigroup structure, and the weak regularity properties of Feynman-Kac distribution flows, and their extended version discussed in Chapter 6. Chapter 14 is dedicated to the convergence analysis of the particle density profiles associated. It covers

Feynman-Kac particle models, and their extended version discussed in Chapter 6, as well as more general classes of mean field particle models.

This chapter presents some key first order Taylor's type decompositions in distribution spaces, which are the progenitors for our other results. Further details on the origins and the applications on these expansions and their use in the bias and the fluctuation analysis of Feynman-Kac particle models can be found in [151, 161, 163, 169, 186].

In Chapter 15, we focus on genealogical tree based particle models. We present some equivalence principles that allow application without further work of most of the results presented in Chapter 14 dedicated to the stochastic analysis of particle density profiles.

The last two chapters of the book, Chapter 16 and Chapter 17, are dedicated to the convergence analysis of particle free energy models and backward particle Markov chains.

Lecture courses on advanced Monte Carlo methods

The format of the book is intended to serve three different audiences: researchers in pure and applied probability, who might be interested in specializing in mean field particle models; researchers in statistical machine learning, statistical physics, and computer sciences, who are interested in the foundations and the performance analysis of advanced mean field particle Monte Carlo methods; and post-graduate students in probability and statistics, who want to learn about mean field particle models and Monte Carlo simulation.

I assume that the reader has some familiarity with basic facts of Probability theory and Markov processes. The prerequisite texts in probability theory for understanding most of the mathematical material presented in this volume are the books by P. Billingsley [53], and the one by A. N. Shiryaev [518], as well as the book by S. N. Ethier and T. G. Kurtz [375], dedicated to continuous time Markov processes.

The material in this book can serve as a basis for different types of advanced courses in pure and applied probability. To this end, I have chosen to discuss several classes of mean field models with an increasing level of complexity, starting from the conventional McKean-Vlasov diffusion type models, and Feynman-Kac distribution flows, to more general classes of backward particle Markov models, interacting jump type models, and abstract evolution equations in the space of nonnegative measures.

The mathematical analysis developed in this book is also presented with increasing degrees of refinements, starting from simple inductive proofs of $\mathbb{L}_m$-mean error estimates, to more sophisticated functional fluctuations theorems, and exponential concentration inequalities expressed in terms of the bias and the variance of mean field approximation schemes.

To make the lecture easier, I did not try to avoid repetitions, and any chapter starts with an introduction connecting the results developed in earlier parts with the current analysis. On the other hand, with a very few exceptions, I gave my best to give a self-contained presentation with detailed proofs.

The first type of lectures, geared towards applications of mean field particle methods to Bayesian statistics, numerical physics, and signal processing, could be centered around one of the two opening chapters, and one on selected application domains, among the ones developed in the second part of the book.

The second type, geared towards applications of Feynman-Kac IPS models, could be centered around the Feynman-Kac modeling techniques presented in Chapter 3, and their four probabilistic interpretations discussed in Chapter 4.

The third type, geared towards applications of mean field particle methods to biology and dynamic population models, could also be centered around the material related to spatial branching processes and multiple object filtering problems presented in Chapter 2 and in Chapter 6.

There is also enough material to cover a course in signal processing on Kalman type filters, including quenched and annealed filters, forward-backward filters, interacting Kalman filters, and ensemble Kalman filters developed in Section 8.2 and in Section 8.3.

More theoretical types of courses could cover one selected application area and the stability properties of Feynman-Kac semigroups and their extended version developed in Chapter 13. A semester-long course could cover the stability properties of nonlinear semigroups and the theoretical aspects of mean field particle models developed in the last part of the book.

More specialized theoretical courses on advanced signal processing would cover multi-object nonlinear filtering models, the derivation of the probability hypothesis density equations, their stability properties, and data association tree models (Section 6.3 and Chapter 13).

There is also enough material to support three other sequences of mathematical courses. These lectures could cover the concentration properties of general classes of mean field particle models (Chapter 1 and Chapter 10), concentration inequalities of Feynman-Kac particle algorithms (Chapter 14, Chapter 15, Chapter 16, and Chapter 17), and the concentration analysis of extended Feynman-Kac particle schemes (Chapter 6, Section 14.7, and Section 14.8).

Acknowledgments

Some material presented in this book results from fruitful collaboration with several colleagues; to name a few, in alphabetical order: René Carmona, François Caron, Dan Crisan, Peng Hu, Pierre Jacob, Jean Jacod, Ajay Jasra, Terry Lyons, Lawrence Murray, Nadia Oudjane, Michele Pace, Philipp Protter, and more particularly Arnaud Doucet, Alice Guionnet, Laurent Miclo, Emmanuel Rio, Sumeetpal S. Singh, Ba-Ngu Vo, and Li Ming Wu.

Most of the text also proposes many new contributions to the subject. The reader will also find a series of new deeper studies of topics such as uniform concentration inequalities w.r.t. the time parameter and functional fluctuation theorems, as well as a fairly new class of application modeling areas.

This book grew out of a series of lectures given in the Computer Science and Communications Research Unit of the University of Luxembourg in February and March 2011, and in the Sino-French Summer Institute in Stochastic Modeling and Applications (CNRS-NSFC Joint Institute of Mathematics), held at the Academy of Mathematics and System Science, Beijing, on June 2011.

I would also like to express my gratitude to INRIA which gave me the opportunity to undertake this project, the Mathematical Institute of Bordeaux, and the Center of Applied Mathematics (CMAP) of the École Polytechnique of Palaiseau.

I am also grateful to the University of New South Wales of Sydney, the University of Luxembourg, the University of Warwick, the Academy of Mathematics and System Science of Beijing, Oxford University, the Commonwealth Scientific and Industrial Research Organization (CSIRO) of Sydney, and the Institute for Computational and Experimental Research in Mathematics (ICERM) of Brown University, as well as the Institut Non Linéaire of Nice Sophia Antipolis, and the CNRS where parts of this project were developed.

Last, but not least, I would like to thank John Kimmel for his precious editorial assistance, as well as for his encouragement during these last five years.

Pierre Del Moral
INRIA Bordeaux, Dec. 2012

Frequently used notation

We present some basic notation and background on stochastic analysis and integral operator theory. Most of the results with detailed proofs can be located in the book [163], dedicated to Feynman-Kac formulae and their interacting particle interpretations. Our proofs contain cross-references to this rather well known material; so the reader may wish to skip this section.

We denote, respectively, by $\mathcal{M}(E)$, $\mathcal{M}_+(E)$, $\mathcal{M}_0(E)$, $\mathcal{P}(E)$, and $\mathcal{B}_b(E)$, the set of all finite signed measures on some measurable space $(E, \mathcal{E})$, the subset of positive measures, the convex subset of measures with null mass, the set of all probability measures, and the Banach space of all bounded and measurable functions f equipped with the uniform norm $\|f\| = \mathrm{Sup}_{x \in E}|f(x)|$. We also denote by $\mathrm{Osc}(E)$, and by $\mathcal{B}_1(E)$, the set of $\mathcal{E}$-measurable functions f with oscillations $\mathrm{osc}(f) = \mathrm{Sup}_{x,y}|f(x) - f(y)| \leq 1$, and, respectively, with $\|f\| \leq 1$. Given a pair of functions $(f_1, f_2) \in (\mathcal{B}_b(E_1) \times \mathcal{B}_b(E_2))$ on some measurable state spaces $(E_1, \mathcal{E}_1)$ and $(E_2, \mathcal{E}_2)$. We denote by $(f_1 \otimes f_2) \in \mathcal{B}_b(E_1 \times E_2)$ the tensor product function defined for any $(x_1, x_2) \in (E_1 \times E_2)$ by

$$(f_1 \otimes f_2)(x_1, x_2) = f_1(x_1)f_2(x_2)$$

Given a pair of measures $(\mu_1, \mu_2) \in (\mathcal{M}(E_1) \times \mathcal{M}(E_2))$, we denote by $(\mu_1 \otimes \mu_2) \in \mathcal{M}(E_1 \times E_2)$ the tensor product measure defined for any $f \in \mathcal{B}_b(E_1 \times E_2)$ by

$$(\mu_1 \otimes \mu_2)(f) = \int \mu_1(dx_1)\mu_2(dx_2)f(x_1, x_2)$$

Sometimes, we simplify the presentation, denoting by $d\mu$ the measure $d\mu(x) = \mu(dx)$, where dx stands for an infinitesimal neighborhood of the state $x \in E$.

We shall slightly abuse the notation, denoting by 0 and 1 the zero and the unit elements in the semi-rings $(\mathbb{R}, +, \times)$ and by 0 and 1 the zero and the unit elements in the set of functions on some state space E.

The Lebesgue integral of a function $f \in \mathcal{B}_b(E)$, with respect to a measure $\mu \in \mathcal{M}(E)$, is denoted by

$$\mu(f) = \int \mu(dx)\ f(x)$$

When f is some indicator function 1_A of some $A \in \mathcal{E}$, we slightly abuse the notation, and we set

$$\mu(A) := \mu(1_A)$$

We also denote by δ_a the Dirac unit mass measure concentrated at the point $a \in E$

$$\delta_a \ : \ A \in \mathcal{E} \ \mapsto \ \delta_a(A) = 1_A(a)$$

Notice that $\delta_a(f) = f(a)$, for any $a \in E$, and any function $f \in \mathcal{B}_b(E)$. We say that a measure μ is absolutely continuous w.r.t. another measure ν on E, and we write $\mu \ll \nu$, when we have $\nu(A) = 0 \Rightarrow \mu(A) = 0$, for any $A \in \mathcal{E}$. When $\mu \ll \nu$, we denote by $d\mu/d\nu$ the Radon-Nykodim derivative function.

The total variation distance on $\mathcal{M}(E)$ is defined for any $\mu \in \mathcal{M}(E)$ by

$$\|\mu\|_{\mathrm{tv}} = \frac{1}{2} \sup_{(A,B) \in \mathcal{E}^2} (\mu(A) - \mu(B)) = \sup\{|\mu(f) - \nu(f)| \ f \in \mathrm{Osc}(E)\}$$

The Boltzmann relative entropy between a couple of measures $\mu \ll \nu$ on E is the nonnegative distance-like criteria defined by

$$\mathrm{Ent}\,(\mu|\nu) := \mu\left(\log \frac{d\mu}{d\nu}\right) \quad \text{if} \quad \mu \ll \nu$$

If $\mu \not\ll \nu$ then we set $\mathrm{Ent}\,(\mu|\nu) = \infty$.

We recall that a bounded integral operator Q from a measurable space $(E, \mathcal{E})$ into an auxiliary measurable space $(F, \mathcal{F})$ is an operator $f \mapsto Q(f)$ from $\mathcal{B}_b(F)$ into $\mathcal{B}_b(E)$ such that the functions

$$x \in E \mapsto Q(f)(x) := \int_F Q(x, dy) f(y)$$

are $\mathcal{E}$-measurable and bounded, for any $f \in \mathcal{B}_b(F)$. Depending on its action, the operator Q is alternatively called a bounded integral operator from E into F, or from $\mathcal{B}_b(F)$ into $\mathcal{B}_b(E)$.

A positive operator is a bounded integral operator Q, such that $Q(f) \geq 0$, for any $f \geq 0$. A Markov kernel, or a Markov transition, is a positive and bounded integral operator Q with $Q(1) = 1$.

A bounded integral operator Q from a measurable space $(E, \mathcal{E})$ into an auxiliary measurable space $(F, \mathcal{F})$ also generates a dual operator

$$\mu(dx) \mapsto (\mu Q)(dx) = \int \mu(dy) Q(y, dx)$$

from $\mathcal{M}(E)$ into $\mathcal{M}(F)$ defined by $(\mu Q)(f) := \mu(Q(f))$. Sometimes, with a slight abuse of notation, to the presentation we write $Q(x, A)$ instead of $Q(1_A)(x)$.

Given some measure μ on E, sometimes we write $\mu \otimes Q$, the measure on $(E \times F)$ defined by

$$[\mu \otimes Q](d(x, y)) = \mu(dx) Q(x, dy)$$

For Markov transitions Q, the function $x \mapsto Q(f)(x)$ represents the local averages of f around x. More precisely, if $Y(x)$ stands for some random variable with law $Q(x, dy)$ we have

$$Q(f)(x) = \mathbb{E}\left(f(Y(x))\right)$$

In the same way, the mapping $\mu \in \mathcal{P}(E) \mapsto \mu Q \in \mathcal{P}(F)$ represents the distributions of the random states Y associated with a given random transition $X \rightsquigarrow Y$, starting with the distribution $\mathrm{Law}(X) = \mu$, and sampling a state Y with distribution $Q(X, dy)$. More precisely, for any measurable subset $A \subset F$ we have

$$(\mu Q)(A) := \mathbb{P}\left(Y \in A\right) = \int \mathbb{P}(X \in dx) \, \mathbb{P}\left(Y \in A \,|\, X = x\right) = \int \mu(dx) \, Q(x, A)$$

In this context, $\mu \otimes Q$ represents the distribution of the couple of random variables (X, Y); that is, we have that

$$(\mu \otimes Q)(d(x, y)) = \mathbb{P}(X \in dx) \, \mathbb{P}\left(Y \in dy \,|\, X = x\right) = \mathbb{P}\left((X, Y) \in d(x, y)\right)$$

Given some bounded integral operator Q from $\mathcal{B}_b(F)$ into $\mathcal{B}_b(E)$, and some functions $(f, g) \in \mathcal{B}_b(F)^2$, sometimes we simplify notation and we denote by $Q\left([f - Q(f)] \, [g - Q(g)]\right) \in \mathcal{B}_b(E)$, and $Q\left([f - Q(f)]^2\right) \in \mathcal{B}_b(E)$ the functions

$$Q\left([f - Q(f)]^2\right)(x) \quad := \quad Q\left([f - Q(f)(x)]^2\right)(x)$$
$$Q\left([f - Q(f)] \, [g - Q(g)]\right)(x) \quad := \quad Q\left([f - Q(f)(x)] \, [g - Q(g)(x)]\right)(x)$$

for any $f, g \in \mathcal{B}_b(F)$.

When the bounded integral operator Q has a constant mass, that is, when $Q(1)(x) = Q(1)(y)$ for any $(x, y) \in E^2$, the operator $\mu \mapsto \mu Q$ maps $\mathcal{M}_0(E)$ into $\mathcal{M}_0(F)$. In this situation, we let $\beta(Q)$ be the Dobrushin coefficient of a bounded integral operator Q defined by the formula

$$\beta(Q) := \sup \, \{\mathrm{osc}(Q(f)) \, ; \quad f \in \mathrm{Osc}(F)\} \tag{0.1}$$

Many models that we consider in this book are defined in terms of collections of Markov transitions K_η from E into F, indexed by the set of probability measures $\eta \in \mathcal{M}(E)$. To avoid unnecessary repetition of technical abstract conditions, we frame the standing assumption that the mappings

$$(x, \eta) \in (E \times \mathcal{P}(E)) \mapsto K_\eta(x, A) \in \mathbb{R}$$

are measurable for any measurable subset $A \subset E$. In other words, the Markov transition $\mathcal{K}((x, \eta), dy) = K_\eta(x, dy)$ from $(E \times \mathcal{P}(E))$ into F is well defined.

Given a pair of bounded integral operators (Q_1, Q_2), we let $(Q_1 Q_2)$ are the composition operator defined by $(Q_1 Q_2)(f) = Q_1(Q_2(f))$. For time homogeneous state spaces, we denote by $Q^m = Q^{m-1}Q = QQ^{m-1}$ the m-th composition of a given bounded integral operator Q, with $m \geq 1$.

Given some bounded integral operators $Q_n(x_{n-1}, dx_n)$, from some measurable state space $(E_{n-1}, \mathcal{E}_{n-1})$ into a possibly different measurable state space $(E_n, \mathcal{E}_n)$, we denote by $Q_{p,n}$ the semigroup defined by

$$Q_{p,n} = (Q_{p+1} Q_{p+2} \ldots Q_n) = Q_{p+1} Q_{p+1,n}$$

with the convention $Q_{n,n} = Id$, the identity matrix for $p = n$.

Given a positive and bounded potential function G on E, we also denote by Ψ_G the Boltzmann-Gibbs mapping from $\mathcal{P}(E)$ into itself defined for any $\mu \in \mathcal{P}(E)$ by

$$\Psi_G(\mu)(dx) = \frac{1}{\mu(G)} \, G(x) \, \mu(dx) \tag{0.2}$$

There is no loss of generality to assume that G is a $]0, 1]$-valued function. For $[0, 1]$-valued potential functions, the transformation is only defined on measures μ s.t. $\mu(G) > 0$. Boltzmann-Gibbs mappings can be interpreted as a nonlinear Markov transport model

$$\Psi_G(\mu) = \mu S_{\mu,G} \tag{0.3}$$

for some Markov transitions $S_{\mu,G}$ from E into itself. Next we provide three different types of models of current use in the further development of the book.

Firstly, for $[0, 1]$-valued potential functions G s.t. $\mu(G) > 0$, we can choose

$$S_{\mu,G}(x, dy) := G(x) \, \delta_x(dy) + (1 - G(x)) \, \Psi_G(\mu)(dy)$$

For $[1, \infty[$-valued potential functions G s.t. $\mu(G) > 1$, we can also choose the Markov transitions

$$S_{\mu,G}(x, dy) := \frac{1}{\mu(G)} \, \delta_x(dy) + \left(1 - \frac{1}{\mu(G)}\right) \, \Psi_{(G-1)}(\mu)(dy)$$

For any bounded positive functions G, the Equation (0.3) is also met for

$$S_{\mu,G}(x, dy) := \epsilon_\mu G(x) \, \delta_x(dy) + (1 - \epsilon_\mu G(x)) \, \Psi_G(\mu)(dy) \tag{0.4}$$

for any $\epsilon_\mu \geq 0$ s.t. $\epsilon_\eta G(x) \leq 1$ for μ-almost every state x. For instance, we can choose $\epsilon_\mu = 0$, $\epsilon_\mu = 1/\|G\|$, or preferably $\epsilon_\mu = 1/\mu - ess \sup G$, where $\mu - ess \sup G$ stands for the μ-essential supremum of G.

Finally, we observe that (0.3) is always met with the transitions

$$S_{\mu,G}(x, dy) := (1 - a_{G,\mu}(x)) \, \delta_x(dy) + a_{G,\mu}(x) \, \Psi_{(G-G(x))_+}(\mu)(dy) \tag{0.5}$$

with the rejection rate

$$
\begin{aligned}
a_{G,\mu}(x) \quad &:= \quad \mu\left((G - G(x))_+\right)/\mu(G) \\
&= \quad 1 - \mu(G \wedge G(x))/\mu(G) \in [0, 1]
\end{aligned}
$$

We prove (0.3) using the decompositions below

$$\mu\left(S_{\mu,G}(f)\right) := \mu\left((1 - a_{G,\mu})\ f\right) + \frac{1}{\mu(G)}\ \int \mu(dx)\mu\left((G - G(x))_+\ f\right)$$

and

$$\frac{1}{\mu(G)}\ \int \mu(dx)\mu\left((G - G(x))_+\ f\right) = \Psi_G(\mu)(f) - \mu\left(f(1 - a_{G,\mu})\right)$$

The last assertion is proved using the following decompositions

$$\mu(Gf) - \mu(G)\mu(f)$$

$$= \int \mu(dy)\mu(dx)\left(G(y) - G(x)\right)_+\ f(y)$$

$$- \int \mu(dx)\mu(dy)\left(G(y) - G(x)\right)_+\ f(x)$$

$$= \int \mu(dx)\mu\left(f\left(G - G(x)\right)_+\right) - \mu(G) \times \mu(a_{G,\mu}f)$$

Given a Polish space E (i.e., a separable and completely metrizable topological space), we denote by $D([a, b], E)$ the set of càdlàg paths from the interval $[a, b]$ into E. The abbreviation càdlàg comes from the French description "continue à droite, limitée à gauche," and the English translation "right continuous with left limits."

We end this section with some more or less traditional notation of current use in the further development of this book.

For a Gaussian and centered random variable U, s.t. $E(U^2) = 1$, we also recall that

$$\mathbb{E}\left(U^{2m}\right) = b(2m)^{2m} := (2m)_m\ 2^{-m}$$

$$\mathbb{E}\left(|U|^{2m+1}\right) \leq b(2m+1)^{2m+1} := \frac{(2m+1)_{(m+1)}}{\sqrt{m+1/2}}\ 2^{-(m+1/2)}$$

$$(0.6)$$

with $(q + p)_p := (q + p)!/q!$. The second assertion comes from the fact that

$$\mathbb{E}\left(U^{2m+1}\right)^2 \leq \mathbb{E}\left(U^{2m}\right)\mathbb{E}\left(U^{2(m+1)}\right)$$

and therefore

$$b(2m+1)^{2(2m+1)} := \mathbb{E}\left(U^{2m}\right)\mathbb{E}\left(U^{2(m+1)}\right)$$

$$= 2^{-(2m+1)}\ (2m)_m\ (2(m+1))_{(m+1)}$$

The formula (0.6) is now a direct consequence of the following decompositions

$$(2(m+1))_{(m+1)} = \frac{(2(m+1))!}{(m+1)!} = 2\ \frac{(2m+1)!}{m!} = 2\ (2m+1)_{(m+1)}$$

and
$$(2m)_m = \frac{1}{2m+1} \frac{(2m+1)!}{m!} = \frac{1}{2m+1}(2m+1)_{(m+1)}$$

We also mention that
$$b(m) \le b(2m) \tag{0.7}$$

Indeed, for even numbers $m = 2p$ we have
$$b(m)^{2m} = b(2p)^{4p} = \mathbb{E}(U^{2p})^2 \le \mathbb{E}(U^{4p}) = b(4p)^{4p} = b(2m)^{2m}$$

and for odd numbers $m = (2p+1)$, we have
$$\begin{aligned}
b(m)^{2m} &= b(2p+1)^{2(2p+1)} = \mathbb{E}\left(U^{2p}\right) \mathbb{E}\left(U^{2(p+1)}\right) \\
&\le \mathbb{E}\left((U^{2p})^{\frac{(2p+1)}{p}}\right)^{\frac{p}{2p+1}} \mathbb{E}\left((U^{2(p+1)})^{\frac{(2p+1)}{p+1}}\right)^{\frac{p+1}{2p+1}} \\
&= \mathbb{E}\left(U^{2(2p+1)}\right) = b(2(2p+1))^{2(2p+1)} = b(2m)^{2m}
\end{aligned}$$

The gradient ∇f and the Hessian $\nabla^2 f$ of a smooth function $f : \theta = (\theta^i)_{1 \le i \le d} \in \mathbb{R}^d \mapsto f(\theta) \in \mathbb{R}$ are defined by the functions
$$\nabla f = \left(\frac{\partial f}{\partial \theta^1}, \frac{\partial f}{\partial \theta^2}, \cdots, \frac{\partial f}{\partial \theta^d} \right)$$

and
$$\nabla^2 f = \begin{pmatrix}
\frac{\partial^2 f}{\partial^2 \theta^1} & \frac{\partial^2 f}{\partial \theta^1 \partial \theta^2} & \cdots & \frac{\partial^2 f}{\partial \theta^1 \partial \theta^d} \\
\frac{\partial^2 f}{\partial \theta^2 \partial \theta^1} & \frac{\partial^2 f}{\partial \theta^2 \partial^2} & \cdots & \frac{\partial^2 f}{\partial \theta^2 \partial \theta^d} \\
\vdots & \vdots & \vdots & \\
\frac{\partial^2 f}{\partial \theta^d \partial \theta^1} & \frac{\partial^2 f}{\partial \theta^d \partial \theta^2} & \cdots & \frac{\partial^2 f}{\partial^2 \theta^d}
\end{pmatrix}$$

Given a $(d \times d')$ matrix M with random entries $M(i,j)$, we write $\mathbb{E}(M)$ the deterministic matrix with entries $\mathbb{E}(M(i,j))$. We also denote by $(.)_+$, $(.)_-$ and $\lfloor . \rfloor$, respectively, the positive, negative, and integer part operations. Sometimes we also use the notation $\mathbb{R}_+ = [0, \infty[$. The maximum and minimum operations are denoted respectively by $\vee$ and $\wedge$

$$a \vee b = \max(a,b) \quad \text{and} \quad a \wedge b = \min(a,b) \quad \text{as well as} \quad a_+ = a \vee 0$$

We also use the traditional conventions
$$\left(\sum_\emptyset, \prod_\emptyset \right) = (0,1) \quad \text{and} \quad \left(\sup_\emptyset, \inf_\emptyset \right) = (-\infty, +\infty)$$

Symbol Description

$b(m)$
Constants in Kintchine inequalities. Page xliii.

c
A cemetery state (or sometimes some finite constant). Pages 92, 165, 93, 177

$d_\nu \Psi_G, d_{p,n}^N, d_{p,n}, \boldsymbol{d}_{\boldsymbol{p,n}}^{\boldsymbol{N}}$
First order integral operators. Pages 85, 429, 513.

$D([a,b], E)$
The set of càdlàg paths from the interval $[a,b]$ into a Polish space E. Pages xliii, 5.

$E, E_n, \boldsymbol{E_n}$
Measurable state spaces. Pages 3, 5, 23, 29, 45, 50, 80, 80, 102, 103.

EKF
Ensemble Kalman Filters. Page xxvii.

$f_n, \boldsymbol{f_n}$
Bounded measurable functions on some state space. Pages xxxix, 469, 511.

$\mathcal{F}_n^N$
The σ-field generated by the random variables ξ_p, with $0 \le p \le n$. Page 456.

$G, G_n, G_n', \boldsymbol{G_n}$
Nonnegative potential functions, nonabsorption rates, fitness or likelihood functions, branching numbers. Pages xlii, 11, 24, 29, 50, 61, 73, 75, 145, 175, 203, 216, 228, 391, 469.

GA
Genetic algorithms. Page 111.

$(G_{p,n}, \mathbf{G}_{\mathbf{p,n}}, P_{p,n})$
The potential function $G_{p,n} = Q_{p,n}(1)$ and the normalized Feynman-Kac semigroup $P_{p,n} = Q_{p,n}(f)/Q_{p,n}(1)$ and their path space versions. Pages 291, 370, 470.

$\overline{G}_n, \overline{G}_{p,n}$
The normalized potential function $\overline{G}_n = G_n/\eta_n(G_n)$ and $\overline{G}_{p,n} = G_{p,n}/\eta_p(G_{p,n})$. Pages 369, 490

$g_{p,n}, \mathbf{g}_{\mathbf{p,n}}$
Constants related to $(G_{p,n}, \mathbf{G}_{\mathbf{p,n}})$. Pages 370, 470.

g_n
The constant related to G_n. Page 370.

$\mathbf{H(G,P)}, \mathbf{H}_m(\mathbf{G}, \mathbf{M})$
Regularity conditions on the pair of potential-transitions (G_n, M_n). Page 372.

IPS
Interacting Particle System models. Page 111.

i.i.d.
Abbreviation for independent and identically distributed. Page 6.

i-MCMC
Interacting Markov chain Monte Carlo models. Page 111.

Random fields models associated with the measures $\gamma_n^N, \eta_n^N, \Gamma_n^N,$ and $\mathbb{Q}_n^N$. Pages 38, 288, 323, 427, 456, 502.

$V_n, W_n^\gamma, W_n^\eta, W_n^\Gamma, W_n^\mathbb{Q}$

The limiting centered Gaussian fields associated with $V_n^N, W_n^{\gamma,N}, W_n^{\eta,N}, W_n^{\Gamma,N}, W_n^{\mathbb{Q},N}$. Pages 38, 40, 323, 488, 523.

$X_n, X_n^c, X_n^h, X_n^\theta, X_n', \mathbf{X}_n, \mathcal{X}_n$

The random states of a Markov chain at a given time n. Pages 3, 22, 80, 28, 49, 50, 64 74, 24, 97, 147.

$\beta(M)$

Dobrushin contraction parameter of a Markov transition M. Page xli.

$\xi_n^i, \boldsymbol{\xi}_n^i$

The i-th individual at time n in a particle algorithm. Pages 43, 117, 229, 278, 328, 425, 472.

$\xi_{p,n}^i$

The ancestor of level $p(\leq n)$ of a current i-th individual $\xi_{n,n}^i$ at time n in a genealogical tree model. Pages 44, 229, 246, 472.

$\tau_{k,l}(n), \overline{\tau}_{k,l}(m)$

Finite constants related to $g_{p,n}^k$ and $\beta(P_{p,n})^l$. Page 380.

$\Phi, \Phi_n, \Phi_{p,n}$

Homogeneous and time inhomogeneous measure valued transformations. Pages 11, 26, 27, 38, 43, 87, 143, 226, 263, 273, 284, 286, 291, 300, 320, 320, 371, 380, 383.

$\Psi_G(\mu)$

The Boltzmann-Gibbs transformation of a measure μ w.r.t. some potential function G. Page xxxix.

Part I

Introduction

Chapter 1

Monte Carlo and mean field models

1.1 Linear evolution equations

This opening section provides a brief overview of some traditional Monte Carlo simulation of linear evolution equations associated with discrete and continuous time Markov processes. Our discussion is restricted to direct Monte Carlo simulation methods. More refined reduction variance techniques, including importance sampling techniques, sequential Monte Carlo methodologies, and related mean field sampling techniques, are discussed in Chapter 4.

1.1.1 Markov chain models

Markov chain models are one of the simplest stochastic models developed in probability theory. These processes are characterized by the fact that their past and their future evolutions are independent, given the present value of the chain. In other words, the next state of a Markov chain only depends on its current state. For finite and ordered state space valued processes, the evolution of the chain is characterized by a sequence of stochastic matrices. These matrices represent the transitions probability of the chain between two consecutive time integers. In other instances, Markov chains are represented by random dynamical systems, defined in terms of a recursive equation relating the next state, with the current state, and some noisy random variables. In applied probability and engineering sciences, these random dynamical systems are also called nonlinear state space models. Much of probability theory is devoted to the simulation and the analysis of these stochastic processes. Reference books on this subjects are [74, 312, 352, 365, 438, 439, 468, 509, 539].

In discrete time settings, the random states of these models are defined in terms of a sequence of random variables X_n indexed by the integer time parameter $n \in \mathbb{N}$, and taking values in some measurable state spaces E_n. By the "memoryless" property of the Markov process, the evolution of the random states is characterized by the probability description of the elementary transitions $X_{n-1} \in E_{n-1} \rightsquigarrow X_n \in E_n$ between two consecutive integers; that is, we have that

$$K_n(x_{n-1}, dx_n) = \mathbb{P}\left(X_n \in dx_n \mid X_{n-1} = x_{n-1}\right) \qquad (1.1)$$

By construction, the laws of the random states $\eta_n = \text{Law}(X_n)$ of the process satisfy the linear evolution equation

$$\eta_n = \eta_{n-1} K_n \tag{1.2}$$

We further assume that we have a dedicated Monte Carlo simulation tool to draw independent random samples of the elementary transitions $X_{n-1} \rightsquigarrow X_n$. In this situation, by the law of large numbers, the distribution of the random trajectories

$$\mathbb{P}_n = \text{Law}(X_0, \ldots, X_n)$$

can be approximated by the sequence of occupation measures

$$\mathbb{P}_n^N := \frac{1}{N} \sum_{1 \leq i \leq N} \delta_{(X_0^i, \ldots, X_n^i)}$$

associated with N independent copies $(X_n^i)_{1 \leq i \leq N}$ of the stochastic process X_n, as $N \to \infty$. More formally, we have the almost sure convergence

$$\mathbb{P}_n^N(\mathbf{f_n}) \longrightarrow_{N \to \infty} \mathbb{P}_n(\mathbf{f_n}) = \mathbb{E}\left(\mathbf{f_n}(X_0, \ldots, X_n)\right)$$

for any bounded measurable function $\mathbf{f_n}$ on the path space $\boldsymbol{E_n}$ given by the product space

$$\boldsymbol{E_n} := (E_0 \times \ldots \times E_n)$$

The above convergence estimate can be made precise in various ways. Several asymptotic and nonasymptotic estimates are available, including fluctuation theorems, exponential concentration inequalities, large deviation properties, as well as empirical process convergence analysis. For instance, we have the following exponential concentration inequalities. For any measurable function $\mathbf{f_n}$ on $\boldsymbol{E_n}$ s.t. $\text{osc}(\mathbf{f_n}) \leq 1$, any $N \geq 1$, and any $x \geq 0$, the probability of the following event

$$\left| [\mathbb{P}_n^N - \mathbb{P}_n](\mathbf{f_n}) \right| \leq \frac{x + \log 2}{3N} + \sigma_n(\mathbf{f_n}) \sqrt{\frac{2(x + \log 2)}{N}} \tag{1.3}$$

is greater than $1 - e^{-x}$, with the variance parameter

$$\sigma_n^2(\mathbf{f_n}) = \mathbb{P}_n(\mathbf{f_n}^2) - \mathbb{P}_n(\mathbf{f_n})^2 \; (\leq 1)$$

A proof of this rather well known result is provided in Proposition 11.6.5. For further details on this subject, we also refer the reader to Chapter 11 dedicated to the stochastic analysis of empirical processes associated with independent random variables *(abbreviated r.v.)*.

The methodology, and the stochastic analysis developed in this book, applies to abstract Markov processes taking values in virtually arbitrary measurable state spaces. This abstract framework encapsulates a surprising wide class of random processes, including Markov chain taking values in transition spaces, as well as excursions, and historical processes taking values in path space models.

Some illustrations of these application models are discussed in the beginning of Section 1.1.2, Section 2.7.1, Section 2.7.2, as well as in Section 7.5.3. For a more thorough discussion on Markov chains on general state spaces, and their regularity properties, we also refer the reader to Section 3.1.1. Concrete examples of Markov chain models can also be found in the books by S. Asmussen, P.W. Glynn [18], O. Cappé, E. Moulines, and T. Rydèn [93], G. Fishman [256], and S.P. Meyn and R.L. Tweedie [438].

We end this section with an important observation. If we replace in (1.2) the Markov transition K_n by some positive and bounded integral operator Q_n, then we obtain an evolution equation for nonnegative measures $\gamma_n \in \mathcal{M}_+(E_n)$

$$\gamma_n = \gamma_{n-1} Q_n \qquad (1.4)$$

starting at $\gamma_0 = \eta_0 \in \mathcal{P}(E_0)$. These classes of models can be interpreted in terms of Feynman-Kac distributions. These evolution equations are, and their Markov interpretations are discussed in some detail in Section 1.4.2.

1.1.2 Continuous time models

1.1.2.1 Continuous vs discrete time models

The abstract Markov chain framework introduced in Section 1.1.1 encapsulates continuous time evolutions of nonanticipative processes, as well as càdlàg Markov processes $(X'_t)_{t \geq 0}$, taking values on some Polish state space E (i.e., a separable and completely metrizable topological space).

For instance, given some time mesh $(t_n)_{n \geq 0}$, the sequence of random variables X_n defined by

$$X_n = X'_{t_n} , \qquad X_n = (X'_t)_{t_n \leq t \leq t_{n+1}} , \qquad \text{and resp.} \quad X_n = (X'_t)_{0 \leq t \leq t_n}$$

are Markov chains taking values in $E_n = E$, $E_n = D([t_n, t_{n+1}], E)$, and resp. $E_n = D([0, t_n], E)$, where $D([a, b], E)$ stands for the set of càdlàg paths from the time interval $[a, b]$ into the set E.

Of course the Monte Carlo simulation of continuous time models $(X'_t)_{t \geq 0}$ often requires to discretize the time into small time intervals, just like a video is actually discretized into a series of discrete generation frame sequences.

For instance, let us suppose we are given an $\mathbb{R}^d$-valued Itô stochastic differential equation

$$dX'_t = a'_t(X'_t)\, dt + \sigma'_t(X'_t)\, dW_t \qquad (1.5)$$

with some initial random variable X'_0 with distribution $\eta'_0 = \text{Law}(X'_0)$. In the above display, W_t is a standard d-dimensional Wiener process, and for any $x \in \mathbb{R}^d$, $\sigma'_t(x) = (\sigma'_{t,i,j}(x))_{1 \leq i,j \leq d}$ and $a'_t(x) = (a'_{t,i}(x))_{1 \leq i \leq d}$ are, respectively, a symmetric nonnegative definite matrix and a vector with appropriate dimensions.

A Euler type discretization $X_n \simeq X'_{t_n}$ of the diffusion (1.5) is given by the Markov chain

$$X_n = X_{n-1} + a_n (X_{n-1}) \; \Delta + \sigma_n (X_{n-1}) \; \left(W_{t_n} - W_{t_{n-1}} \right) \qquad (1.6)$$

on the time mesh $(t_n)_{n \geq 0}$ s.t. $(t_n - t_{n-1}) = \Delta > 0$, with the drift and diffusion functions given by

$$(a_n, \sigma_n) = \left(a'_{t_{n-1}}, \sigma'_{t_{n-1}} \right)$$

and the initial random variable, $X_0 = X'_0$. High order or implicit time discretization schemes can also be used to cure the instability of some evolution models.

1.1.2.2   Jump-diffusion processes

Suppose we are given a jump type Markov process X'_t that evolves between jump times T_n as in (1.5). The jump times T_n are defined in terms of a sequence $(e_n)_{n \geq 1}$ of independent and identically distributed *(abbreviated i.i.d.)* exponentially distributed random variables with unit parameter. The successive jump times are defined sequentially by the following recursion

$$T_n = \inf \left\{ t \geq T_{n-1} \; : \; \int_{T_{n-1}}^{t} \lambda'_u(X'_u) \, du \geq e_n \right\} \qquad (1.7)$$

with $T_0 = 0$, and some nonnegative function λ'_u. At jump times T_n, the process at X'_{T_n-} jumps to a new location X'_{T_n} randomly chosen with distribution $P'_{T_n-}(X'_{T_n-}, dy)$. In the above discussion, we have implicitly assumed that the parameters $(a'_t, \sigma'_t, \lambda'_t)$ are sufficiently regular, so that the jump diffusion defined by (1.5) and (1.7) is well defined on the real line $\mathbb{R}_+ = [0, \infty[$.

The conditional transitions of the process X'_t are given for any $0 \leq r \leq s \leq t$, and any bounded Borel function f on $\mathbb{R}^d$ by the formulae

$$
\begin{aligned}
P'_{r,t}(f)(x) \;\; &:= \;\; \mathbb{E}\left(f(X'_t) \mid X'_r = x \right) \\
&= \;\; \mathbb{E}\left(\mathbb{E}\left(f(X'_t) \mid X'_s \right) \mid X'_r = x \right) = P'_{r,s}(P'_{s,t}(f))(x)
\end{aligned}
$$

In probability theory, the r.h.s. semigroup property is also called the Chapman-Kolmogorov equation.

A discrete time approximation of the jump-diffusion model on a time mesh t_n is given by a Markov chain X_n with elementary transitions

$$K_n(x, dz) = \int M_n(x, dy) \; J_n(y, dz) \qquad (1.8)$$

where M_n stands for the transition of the Markov chain (1.6), and J_n is the geometric jump type Markov transition defined in terms of the couple

$$(G_n, P_n) = \left(\exp\left(-\Delta \lambda'_{t_n} \right), P'_{t_n} \right)$$

by the following equation

$$J_n(y, dz) = G_n(y) \; \delta_y(dz) + (1 - G_n(y)) \; P_n(y, dz) \qquad (1.9)$$

We end this section, we an informal derivation of the infinitesimal generator of the jump-diffusion process (1.7) in terms of the Markov chain with elementary transitions K_n.

Infinitesimal generators play a central role in stochastic analysis. They provide a complete analytic description of continuous time Markov process in terms of integro-differential operators L'_t acting on some domain $\mathcal{D}$ of sufficiently regular functions f with the following formulae

$$\lim_{s \to t} \left[\frac{P'_{s,t} - Id}{t - s} \right] (f)(x) \quad = \quad \lim_{s \to t} \frac{\mathbb{E}\left((f(X'_t) - f(x)) \mid X'_s = x \right)}{t - s} = L'_t(f)(x)$$

This yields the first order expansion of the semigroup

$$P'_{s,t} = Id + (t - s) \; L'_t \; + O((t - s))$$

The regularity property of the functions depends on the dynamics of the Markov process under study. The generator of pure jump models are defined on every bounded measurable functions. The one of $\mathbb{R}^d$-valued diffusion processes is a second order differential operator, only defined on twice differentiable and bounded functions.

Taking $\Delta \simeq 0$ in (1.9), we find the approximations

$$J_n - Id = \left(1 - \exp\left(-\Delta \lambda'_{t_n} \right) \right) \; (P'_{t_n} - Id) \simeq \lambda'_{t_n} \; (P'_{t_n} - Id) \; \Delta$$

In much the same way, for any smooth functions f we have

$$\begin{aligned}
(M_n - Id)(f)(x) \quad &\simeq \quad \mathbb{E}\left(f\left(a'_{t_{n-1}}(x) \; \Delta + \sigma'_{t_{n-1}}(x) \; (W_{t_n} - W_{t_{n-1}}) \right) - f(x) \right) \\
&\simeq \quad L'_{t_{n-1}}(f)(x) \; \Delta
\end{aligned}$$

with the operator L'_t defined by the following formula

$$L'_t \quad := \quad \sum_{i=1}^{d} a'_{t,i} \; \partial_i + \frac{1}{2} \sum_{i,j=1}^{d} \left(\sigma'_t (\sigma'_t)^T \right)_{i,j} \; \partial_{i,j} \qquad (1.10)$$

A more detailed derivation of this approximation is provided in Section 1.1.2.3. Combining these couple of approximations, we find that

$$\begin{aligned}
(K_n - Id)(f) \quad &= \quad [(Id + (M_n - Id))(Id + (J_n - Id)) - Id](f) \\
&\simeq \quad (M_n - Id)(f) + (J_n - Id)(f) = L_{t_{n-1}}(f) \; \Delta \quad (1.11)
\end{aligned}$$

with the infinitesimal generator of the jump-diffusion model

$$L_t(f)(x) := L'_t(f)(x) \; + \lambda'_t(x) \int [f(y) - f(x)] \; P'_t(x, dy) \qquad (1.12)$$

For a more detailed discussion on discrete time approximations of jump processes in the context of Feynman-Kac models, we refer the reader to Chapter 5.

1.1.2.3   Integro-differential equations

Continuous time Markov processes are intimately related to intro-differential linear evolution equations. To underline the role of Monte Carlo simulation of discrete generation Markov chain models in the numerical solving of these equations, this section provides a brief description of the continuous time version of the Markov transport Equation (1.2).

Using standard stochastic calculus, we can show that the evolution equation of the laws η'_t of the random states X'_t of the jump diffusion model (1.7) is given by weak integro-differential equation

$$\frac{d}{dt}\eta'_t(f) = \eta'_t(L_t(f)) \tag{1.13}$$

for sufficiently regular functions f on E, with the integro-differential operator L_t defined in (1.12). The infinitesimal generator (1.10) of the diffusion process (1.5) is sometimes rewritten in the following form

$$L'_t(f) = a'_t \, \nabla f + \frac{1}{2} \, \mathrm{Tr}\left(\sigma'_t(\sigma'_t)^T \nabla^2 f\right)$$

with the line vector $a'_t = \left(a'_{t,i}\right)_{1 \leq i \leq d}$, the column gradient vector $\nabla f = (\partial_i f)^T_{1 \leq i \leq d}$, and the Hessian matrix $\nabla^2 f = (\partial_{i,j} f)_{1 \leq i,j \leq d}$.

For the convenience of the nonexpert readers in stochastic calculus, we provide an informal derivation of (1.13) when $\lambda'_t = 0$. Generators of jump processes are discussed in Chapter 5. Using Taylor's formula, we find that

$$
\begin{aligned}
df(X'_t) &= f(X'_t + dX'_t) - f(X'_t) \\
&= \sum_{i=1}^{d} \partial_i f(X'_t) \, dX'^i_t + \frac{1}{2} \sum_{i,j=1}^{d} \partial_{i,j} f(X'_t) \, dX'^i_t dX'^j_t
\end{aligned}
$$

Using the fact that

$$dX'^i_t dX'^j_t = \sum_{1 \leq k,l \leq d} \sigma'^i_{k,t}(X'_t) \, \sigma'^j_{l,t}(X'_t) \, dW^k_t dW^l_t$$

and applying the rules $dW^k_t \times dW^l_t = 1_{k=l} \, dt$, and $dt \times dW^i_t = 0$, we conclude that

$$dX'^i_t dX'^j_t = \sum_{1 \leq k \leq d} \sigma'^i_{k,t}(X'_t) \, \sigma'^j_{k,t}(X'_t) \, dt = \left(\sigma'_t(\sigma'_t)^T\right)_{i,j}(X'_t) \, dt$$

and therefore

$$df(X'_t) = L'_t(f)(X'_t) \, dt + dM_t(f)$$

with the Martingale term

$$dM_t(f) := \sum_{1 \leq i,j \leq d} \partial_i f(X'_t) \, \sigma'^i_{j,t}(X'_t) \, dW^j_t$$

Using the fact that $\mathbb{E}\left(M_t(f)\right) = 0$, this implies that

$$d\eta'_t(f) = d\mathbb{E}\left(f(X'_t)\right) = \mathbb{E}\left(df(X'_t)\right) = \mathbb{E}\left(L'_t(f)(X'_t)\right) dt = \eta'_t\left(L'_t(f)\right) dt$$

For a more thorough and rigorous discussion on these probabilistic models, we refer the reader to the seminal books by Daniel Revuz and Marc Yor [488], Ioannis Karatzas and Steven Shreve [355], as well as the book by Stewart Ethier and Thomas Kurtz [375], the one by Bernt Øksendal [454], and my review article with N. Hadjiconstantinou [173].

1.1.2.4 Fokker-Planck differential equations

We further assume that $P'_t(x, dy) = q_t(x, y)\, dy$ and the law of the random states $\eta'_t(dy) = p_t(y)\, dy$ have a smooth density $q_t(x, y)$, and $p_t(y)$ w.r.t. the Lebesgue measure dy on $\mathbb{R}^d$. In this situation, the equation (1.13) takes the form

$$\frac{d}{dt}\eta'_t(f) \;=\; \int f(x)\,\frac{dp_t}{dt}(x)\, dx = \int p_t(x)\, L'_t(f)(x)\, dx$$
$$+ \int p_t(x)\, \lambda'_t(x)\, q_t(x, y)\, (f(y) - f(x))\, dxdy$$

For smooth functions f with compact support, we have the integration by part formulae

$$\int p_t(x)\, L'_t(f)(x)\, dx = \int L'^\star_t(p_t)(x)\, f(x)\, dx$$

with the dual differential operator

$$L'^\star_t(p_t) = -\sum_{i=1}^{d} \partial_i\left(a'_{t,i}\, p_t\right) + \frac{1}{2}\sum_{i,j=1}^{d} \partial_{i,j}\left(\left(\sigma'_t(\sigma'_t)^T\right)_{i,j}\, p_t\right)$$

On the other hand, we have that

$$\int p_t(x)\, \lambda'_t(x)\, q_t(x, y)\, (f(y) - f(x))\, dxdy$$
$$= \int f(x)\,\left[\left(\int p_t(y)\, \lambda'_t(y)\, q_t(y, x)\, dy\right) - p_t(x)\, \lambda'_t(x)\right] dx$$

from which we conclude that

$$\int f(x)\,\frac{dp_t}{dt}(x)\, dx = \int f(x)\, L^\star_t(p_t)(x)\, dx$$

with

$$L^\star_t(p_t)(x) = L'^\star_t(p_t)(x) + \left[\left(\int p_t(y)\, \lambda'_t(y)\, q_t(y, x)\, dy\right) - p_t(x)\, \lambda'_t(x)\right]$$

Since this equation is valid for any smooth functions f, we conclude that

$$\frac{dp_t}{dt}(x) = L_t'^{\star}(p_t)(x) + \left(\int p_t(y) \ \lambda_t'(y) \ q_t(y, x) \ dy \right) - p_t(x) \ \lambda_t'(x) \qquad (1.14)$$

Inversely, any equation of the form (1.14) can be interpreted as the probability densities of the random states of a jump diffusion model of the form (1.7).

The Fokker-Planck type integro-differential equation (1.14) is sometimes rewritten in the following form

$$\frac{dp_t}{dt}(x) + \text{div}\left(a_t' \ p_t\right) - \frac{1}{2}\nabla^2 : \left(\sigma_t'(\sigma_t')^T p_t\right) - \Theta_t(p_t) = 0$$

with the operators

$$\text{div}\left(a_t' \ p_t\right) \quad := \quad \sum_{i=1}^{d} \partial_i \left(a_{t,i}' \ p_t\right)$$

$$\nabla^2 : \left(\sigma_t'(\sigma_t')^T p_t\right) \quad = \quad \sum_{i,j=1}^{d} \partial_{i,j} \left(\left(\sigma_t'(\sigma_t')^T\right)_{i,j} \ p_t\right)$$

$$\Theta_t(p_t) \quad := \quad \int p_t(y) \ \lambda_t'(y) \ [q_t(y, x) \ dy - \delta_x(dy)]$$

1.1.2.5   Monte Carlo approximations

Monte Carlo methods allow to reduce the numerical solving of integro-differential equations of the form (1.13) and (1.14) to the stochastic simulation of the jump diffusion model (1.7) presented in Section 1.1.2.1. To be more precise, we let $\mathbb{P}_t$ be the distribution of the jump-diffusion model $X_t' := (X_s')_{0 \le s \le t}$ defined in (1.7), on some time interval $[0, t]$.

Given N independent copies $X_t'^i := (X_s'^i)_{0 \le s \le t}$ of X_t', by the law of large numbers, we have the almost sure convergence

$$\mathbb{P}_t^N(\mathbf{f_t}) := \frac{1}{N} \sum_{i=1}^{N} \mathbf{f_t}(X_t'^i) \longrightarrow_{N \to \infty} \mathbb{P}_t(\mathbf{f_t}) = \mathbb{E}\left(\mathbf{f_t}(X_t')\right)$$

for any bounded measurable function $\mathbf{f_t}$ on the path space $\mathbf{E_t} := D([0, t], \mathbb{R}^d)$. In particular, in the context of (1.14) we have the almost sure convergence

$$\eta_t'^N(f) := \frac{1}{N} \sum_{i=1}^{N} f(X_t'^i) \longrightarrow_{N \to \infty} \eta_t'(f) = \int f(x) \ p_t(x) \ dx$$

for any bounded Borel function f on $\mathbb{R}^d$.

Besides its mathematical elegance, this type of assertion is not related to a concrete Monte Carlo simulation technique, since most of the time the jump diffusion process cannot be sampled perfectly. On the other hand, the fact that the empirical measure has the unbiasedness property $\mathbb{E}\left(\eta_t^{\prime N}(f)\right) = \eta_t'(f)$ doesn't give any information on the performance of the Monte Carlo methods. The best way to quantify the performance of the Monte Carlo model is to analyze the concentration properties of the the occupation measures $\eta_t^{\prime N}$ around their limiting values η_t'.

In this context, the discrete generation model (1.2), associated with the Markov K_n defined in (1.8), can be seen as an approximation of the jump type Markov process defined in (1.7). Therefore, we can use the Monte Carlo simulation method described in Section 1.1.1 to approximate the continuous time integro-differential Equation (1.13).

We also refer to the articles by E. Pardoux and D. Talay [469], V. Bally and D. Talay [32], as well as the seminal articles by P.E. Kloeden, E. Platen, and their co-authors [367, 368, 369, 370], for a detailed convergence analysis of time discretization schemes.

1.2 Nonlinear McKean evolutions

This section is concerned with nonlinear evolution models in the space of probability measures. These models describe the evolution of the laws of the random states of McKean type Markov chain models, interacting with the distribution flow of the random states. More general nonlinear models in the space of nonnegative measures are discussed in Section 2.4.1.

1.2.1 Discrete time models

In Section 1.1.1 and Section 1.1.2, we have presented a class of Markov chain models, with a linear evolution of the distributions of the random states. In more general situations, the law of the random states $\eta_n = \text{Law}(X_n)$ of the stochastic process satisfies a nonlinear evolution equation of the following form

$$\eta_n = \Phi_n(\eta_{n-1}) \qquad (1.15)$$

for some nonnecessarily linear one step mapping Φ_n from $\mathcal{P}(E_{n-1})$ into $\mathcal{P}(E_n)$. Notice that the Markov chain model discussed in (1.2) is associated with the linear mapping

$$\Phi_n \; : \; \eta \in \mathcal{P}(E_{n-1}) \; \mapsto \; \Phi_n(\eta) = \eta K_n \in \mathcal{P}(E_n)$$

Inversely, *any* evolution equation of the form (1.15) can be interpreted as the evolution of the distributions of the random states of a Markov chain taking values in some state spaces E_n. More precisely, there always exists some collection of Markov transition $K_{n,\eta}$ indexed by the time parameter $n \in \mathbb{N}$, and the set of probability measures η on E_{n-1}, such that

$$\eta_n = \eta_{n-1} K_{n,\eta_{n-1}} \tag{1.16}$$

For instance, we can take $K_{n,\eta}(x_{n-1}, .) = \Phi_n(\eta)$. In this situation, the random states X_n form a sequence of independent random variables with distributions η_n satisfying the evolution Equation (1.15). Examples of nonlinear one step mappings Φ_n on distribution spaces are discussed in Section 3.1.3 and in Section 3.1.4 dedicated, respectively, to Boltzmann-Gibbs transformations, and Feynman-Kac operators. We also refer the reader to Section 8.2.3, for applications of these models in the context of linear Gaussian filtering models, and their mean field ensemble Kalman filters.

By construction, the random states X_n of the system can again be interpreted as the distribution of "memoryless" Markov processes with elementary transitions

$$K_{n,\eta_{n-1}}(x_{n-1}, dx_n) = \mathbb{P}\left(X_n \in dx_n \mid X_{n-1} = x_{n-1}\right)$$

that depend on the distribution $\eta_{n-1} = \mathrm{Law}(X_{n-1})$ of the current state X_{n-1} of the system. These stochastic processes are called the McKean interpretations of the nonlinear equation in distribution space (1.15). The distributions of the random trajectories $(X_p)_{0 \le p \le n}$ of the chain are given by the McKean distributions

$$\mathbb{P}\left((X_0, \ldots, X_n) \in d(x_0, \ldots, x_n)\right) = \eta_0(dx_0) \prod_{1 \le p \le n} K_{p,\eta_{p-1}}(x_{p-1}, dx_p) \tag{1.17}$$

The existence of an overall distribution $\mathbb{P}$ on the set of trajectories $\Omega := \prod_{n \ge 0} E_n$ of the McKean-Markov chain is granted by the Ionescu-Tulcea's theorem (cf. for instance the theorem on page 249, in the textbook by A. Shiryaev [518]).

We end this section with an important observation. If we replace in (1.15) the mappings Φ_n by some one step mappings Ξ_n from $\mathcal{M}_+(E_{n-1})$ into $\mathcal{M}_+(E_n)$, then we obtain an evolution equation in the space of all nonnegative measures $\gamma_n \in \mathcal{M}_+(E_n)$

$$\gamma_n = \Xi_n(\gamma_{n-1}) \tag{1.18}$$

starting at $\gamma_0 = \eta_0 \in \mathcal{P}(E_0)$.

These evolution equations and their Markov interpretations are discussed in some detail in Section 2.4.1. Illustrations of these abstract models in the context of multiple object filtering problems are also provided in Section 2.4.2. Further details on the mathematical and numerical analysis of these nonlinear semigroups are provided in Chapter 6, dedicated to nonlinear evolutions of intensity measures.

1.2.2 Continuous time models

To underline the role of mean field simulation in the numerical solving of nonlinear integro-differential equations, we provide a description of some continuous time versions of the Markov transport Equation (1.16). The discrete time approximation of these models are presented in section 1.3.

1.2.2.1 Nonlinear jump-diffusion models

We consider the general jump-diffusion processes (1.7) discussed in the end of Section 1.1.1. Continuous time McKean processes X_t' are defined as in (1.5) and (1.7) by replacing the parameters $(a_t', \sigma_t', \lambda_t', P_t')$ by some collection of functional parameters

$$(a_t'(.,\eta_t'), \sigma_t'(.,\eta_t'), \lambda_{t,,\eta_t'}', P_{t,\eta_t'}')$$

that depend on the distribution flow $\eta_t' = \text{Law}(X_t')$ of the random states X_t'. To avoid unnecessary technical discussion, it is implicitly assumed that the model parameters are sufficiently regular, so that the corresponding nonlinear jump diffusion process is well defined on the real line $\mathbb{R}_+$.

In this situation, the corresponding integro-differential Equation (1.13) takes the following form

$$\frac{d}{dt}\eta_t'(f) = \eta_t'\left(L_{t,\eta_t'}(f)\right) \tag{1.19}$$

with the collection of integro-differential operators, indexed by the time parameter t, and the set of probability distributions η on $\mathbb{R}^d$, given by

$$L_{t,\eta}(f)(x) = L_{t,\eta}'(f)(x) + \lambda_{t;\eta}'(x) \int [f(y) - f(x)] \; P_{t,\eta}'(x,dy)$$

$$L_{t,\eta}' := \sum_{i=1}^{d} a_{t,i}'(x,\eta) \; \partial_i + \frac{1}{2} \sum_{i,j=1}^{d} (\sigma_t'(x,\eta)(\sigma_t'(x,\eta))^T)_{i,j} \; \partial_{i,j} \tag{1.20}$$

We notice that the second order differential operator $L_{t,\eta}'$ defined in (1.20) is the infinitesimal generator of the nonlinear diffusion process

$$dX_t' = a_t'(X_t', \eta_t') \; dt + \sigma_t'(X_t', \eta_t') \; dW_t \tag{1.21}$$

The corresponding nonlinear diffusion jump process X_t' evolves between jumps times T_n as in (1.21). As in (1.7), the jump times T_n are defined by the recursion

$$T_n = \inf\left\{t \geq T_{n-1} : \int_{T_{n-1}}^{t} \lambda_{u,\eta_u'}'(X_u') \; du \geq e_n\right\} \tag{1.22}$$

At jump times T_n, the process at X_{T_n-}' jumps to a new location X_{T_n}' randomly chosen with a distribution $P_{T_n-,\eta_{T_n-}'}'(X_{T_n-}',dy)$ that depends on the distribution η_{T_n-}' of the random state X_{T_n-}' of the process before the jump.

1.2.2.2   Nonlinear integro-differential equations

We further assume that the jump transition $P'_{t,\eta'_t}(x, dy) = q_{t,\eta'_t}(x, y) \, dy$ and the law of the random states $\eta'_t(dy) = p_t(y) \, dy$ of the jump-diffusion process (1.21) have a smooth density $q_{t,\eta'_t}(x, y)$, and $p_t(y)$ w.r.t. the Lebesgue measure dy on $\mathbb{R}^d$. To simplify the presentation, with a slight abuse of notation we set $\left(a'_t(., p_t), \sigma'_t(., p_t), q_{t,p_t}, \lambda'_{t,p_t}, L'_{t,p_t}\right)$ instead of $\left(a'_t(., \eta'_t), \sigma'_t(., \eta'_t), q_{t,\eta'_t}, \lambda'_{t,\eta'_t}, L'_{t,\eta'_t}\right)$.

In this notation, using the same arguments as the ones we used in Section 1.1.2.3, the equation (1.19) takes the form

$$
\begin{aligned}
\frac{\partial}{\partial t} \eta'_t(f) &= \int f(x) \, \frac{\partial p_t}{\partial t}(x) \, dx \\
&= \int p_t(x) \, L'_{t,p_t}(f)(x) \, dx \\
&\quad + \int p_t(x) \, \lambda'_{t,p_t}(x) \, q_{t,p_t}(x, y) \, (f(y) - f(x)) \, dx dy
\end{aligned}
$$

from which we conclude that

$$
\frac{\partial p_t}{\partial t}(x) = L'^{\star}_{t,p_t}(p_t)(x) + \left(\int p_t(y) \, \lambda'_{t,p_t}(y) \, q_{t,p_t}(y, x) \, dy \right) - p_t(x) \, \lambda'_t(x) \quad (1.23)
$$

with the dual differential operator

$$
L'^{\star}_{t,p_t}(p_t)
$$

$$
= -\sum_{i=1}^d \partial_i \left(a'_{t,i}(., p_t) \, p_t \right) + \frac{1}{2} \sum_{i,j=1}^d \partial_{i,j} \left(\left((\sigma'_t(., p_t)(\sigma'_t(., p_t))^T \right)_{i,j} \, p_t \right)
$$

Inversely, any equation of the form (1.23) can be interpreted as the probability densities of the random states of a jump diffusion model of the form (1.22).

The mean field particle methods developed in Section 1.2.2.5 and in Section 1.5 allow to reduce the numerical solving of nonlinear integro-differential equations of the form (1.23) to the stochastic simulation of the jump-diffusion process (1.21). The nonlinear equations (1.19) and (1.23), and their probabilistic interpretations were introduced in the early 1950s in [351] by M. Kac to analyze the Vlasov kinetic equation of plasma [378]. These lines of ideas were further developed by H. P. McKean in [428].

For a more thorough and rigorous discussion on these nonlinear models, and their mean field particle interpretations we refer the reader to the series of articles by C. Graham [302, 303, 304], and the ones by C. Graham and S. Méléard [295, 297, 299, 300, 301].

1.2.2.3   Feynman-Kac distribution flows

When the functions $a'_{t,i}(x, \eta) = a'_{t,i}(x)$ and $\sigma'_t(x, \eta) = \sigma'_t(x)$ depend on the parameter η, the operator $L'_{t,\eta} = L'_t$ defined in (1.20) coincides with the infinitesimal generator of the diffusion process defined in (1.5). We also consider the parameters

$$\lambda'_{t;\eta} := V_t(x) \quad \text{and} \quad P'_{t,\eta}(x, dy) := \eta(dy) \tag{1.24}$$

In this situation, the Equation (1.19) is given by the following quadratic evolution model

$$\frac{d}{dt}\eta'_t(f) = \eta'_t(L'_t(f)) + \eta'_t(V_t)\eta'_t(f) - \eta'_t(fV_t) = \eta'_t\left(L^V_{t,\eta'_t}(f)\right) \tag{1.25}$$

with the "normalized" Schrödinger operator

$$L^V_{t,\eta} := L'_t - (V_t - \eta_t(V))$$

In this particular situation, it is well known that the solution of this equation is given by the Feynman-Kac model

$$\eta'_t(f) = \gamma'_t(f)/\gamma'_t(1) \quad \text{with} \quad \gamma'_t(f) = \mathbb{E}\left(f(X'_t)\, \exp\left[-\int_0^t V_s(X'_s)ds\right]\right)$$

These continuous time models are discussed in [161, 176, 178, 179, 185], as well as in [295, 428, 429, 430, 431], and in Chapter 5 in the present book.

Returning to the McKean interpretation models discussed in Section 1.2.2.1, the flow of measures $\eta'_t = \text{Law}(X'_t)$ can be interpreted as the distributions of the random states X'_t of a jump type Markov process.

Between the jumps, X'_t follows the diffusion Equation (1.5). At jump times T_n, occurring with the stochastic rate $V_t(X'_t)$, the process $X'_{T_n-} \rightsquigarrow X'_{T_n}$ jumps to a new location, randomly chosen with the distribution $P'_{T_n-,\eta_{T_n}-}(x, dy) := \eta'_{T_n-}(dy)$.

1.2.2.4 Feynman-Kac models and parabolic equations

Feynman-Kac models are commonly used in computer sciences and applied mathematics to solve certain classes of parabolic differential equations in terms of stochastic processes. Inversely, for reasonably large dimension problems we can alternatively approximate Feynman-Kac semigroups using deterministic grid type methods. Because of their importance in practice, I have chosen to briefly describe some Feynman-Kac interpretations of these partial differential equations.

We denote by $P'_{s,t}$ the semigroup associated with the generator L'_t and we assume that the following limiting commutation properties are satisfied:

$$\frac{P'_{s,t+\epsilon} - P'_{s,t}}{\epsilon} = P'_{s,t}\left(\frac{P'_{t,t+\epsilon} - Id}{\epsilon}\right) \rightarrow_{\epsilon\downarrow 0} \frac{d}{dt}P'_{s,t} = P'_{s,t}L'_t$$

$$\frac{P'_{s-\epsilon,t} - P'_{s,t}}{-\epsilon} = \left(\frac{P'_{s-\epsilon,s} - Id}{-\epsilon}\right)P'_{s,t} \rightarrow_{\epsilon\downarrow 0} \frac{d}{ds}P'_{s,t} = -L'_s P'_{s,t}$$

for any $s \leq t$. We also consider the Feynman-Kac semigroup

$$
\begin{aligned}
Q'_{s,t}(f)(x) &= \mathbb{E}\left(f(X'_t)\; e^{-\int_s^t V_t(X'_r)dr}\; |X'_s = x\right) \\
&= \mathbb{E}\left(f(X'_t)\left(1 - \int_s^t V_r(X'_r)\; e^{-\int_s^r V_\tau(X'_\tau)d\tau}\right)\; |X'_s = x\right) \\
&= P'_{s,t}(f)(x) - \int_s^t Q'_{s,r}(V_r P'_{r,t}(f))(x)\; dr
\end{aligned}
$$

By construction, we have

$$
\begin{aligned}
\frac{d}{dt}Q'_{s,t}(f) &= P'_{s,t}(L'_t(f)) - Q'_{s,t}(V_t f) - \int_s^t Q'_{s,r}(V_r P'_{r,t}(L'_t(f)))\; dr \\
&= Q'_{s,t}(L'^V_t(f)) \quad \text{with} \quad L'^V_t(f) = L'_t(f) - V_t f
\end{aligned}
$$

and

$$
\frac{Q'_{s-\epsilon,t} - Q'_{s,t}}{-\epsilon} = \left(\frac{Q'_{s-\epsilon,s} - Id}{-\epsilon}\right)Q'_{s,t} \;\to_{\epsilon \downarrow 0}\; \frac{d}{ds}Q'_{s,t} = -L'^V_s Q'_{s,t} \qquad (1.26)
$$

This shows that the function $u(s, x) := Q'_{s,t}(f)(x)$ is solution of the Cauchy problem

$$
\begin{cases}
\dfrac{du}{ds} + L'_s u = V_s u & \forall s \leq t \\[2mm]
u(t, x) = f(x)
\end{cases}
$$

Using previous calculations, given a couple of bounded functions $\left(f_t^{(1)}, f_t^{(2)}\right)$ it is also immediate to check that the function

$$
v(s, x) := Q'_{s,t}(f_t^{(1)})(x) + \int_s^t Q'_{s,r}(f_r^{(2)})(x)\; dr \qquad s \leq t
$$

satisfies the equation

$$
\frac{dv}{ds}(s, x) = -L'^V_s Q'_{s,t}(f_t^{(1)})(x) - \int_s^t L'^V_s Q'_{s,r}(f_r^{(2)})(x)\; dr - f_s^{(2)}
$$

from which we conclude that

$$
\begin{cases}
\dfrac{dv}{ds} + L'_s v = V_s v + f_s^{(2)} & \forall s \leq t \\[2mm]
v(t, x) = f_t^{(1)}(x)
\end{cases}
$$

Inversely, we can show that the solutions of the above equations are given by the Feynman-Kac models defined above.

1.2.2.5 Mean field interacting particle models

The continuous time mean field particle model associated with the integro-differential equations discussed in Section 1.2.2.1, Section 1.2.2.2, and Section 1.2.2.3, is a Markov chain $\xi_t^{(N)} := (\xi_t^{(N,i)})_{1 \le i \le N}$ on the product state space $(\mathbb{R}^d)^N$, with infinitesimal generator defined, for sufficiently regular functions F on $(\mathbb{R}^d)^N$, by the following formulae

$$\mathcal{L}_t(F)(x^1, \ldots, x^N) := \sum_{1 \le i \le N} L_{t,m(x)}^{(i)}(F)(x^1, \ldots, x^i, \ldots, x^N) \qquad (1.27)$$

In the above display, $m(x) := \frac{1}{N} \sum_{1 \le i \le N} \delta_{x^i}$ stands for the occupation measure of the population $x = (x^i)_{1 \le i \le N} \in E^N$; and for any $\eta \in \mathcal{P}(\mathbb{R}^d)$ $L_{t,\eta}^{(i)}$ stands for the operator $L_{t,\eta}$ defined in (1.19), acting on the function $x^i \mapsto F(x^1, \ldots, x^i, \ldots, x^N)$. In other words, every individual $\xi_t^{(N,i)}$ is a jump-diffusion model defined as in (1.21) and (1.22) by replacing the unknown measures η_t' by their particle approximations $\eta_t'^N = \frac{1}{N} \sum_{j=1}^N \delta_{\xi_t^{(N,j)}}$.

Using (1.24), the generator of the Feynman-Kac model (1.20) is given for any sufficiently regular function F by the formula

$$L_{t,m(x)}^{(i)}(F)(x^1, \ldots, x^N)$$

$$:= \sum_{1 \le i \le N} L_t'^{(i)}(F)(x^1, \ldots, x^i, \ldots, x^N) + \sum_{1 \le i \le N} V(x^i)$$

$$\times \int \left[F(x^1, \ldots, x^{i-1}, y, x^{i+1}, \ldots, x^N) - F(x^1, \ldots, x^i, \ldots, x^N) \right] \, m(x)(dy)$$

Between jumps, the particles evolve independently as the diffusion process defined in (1.5). At rate V, the particles jump to a new location randomly selected in the current population.

Further details on these continuous time particle models are provided in Section 5.4. As for the linear models discussed in Section 1.1.2.3, continuous time mean field models are far from being related to a concrete Monte Carlo simulation technique, since most of the time the resulting interacting jump diffusion processes cannot be sampled perfectly. One natural way to solve this problem is to consider the time discretization schemes presented in Section 1.3, and the discrete generation mean field models discussed in Section 1.5.

On the other hand, as in the linear case discussed in the end of Section 1.1.2.5, the fact that $\lim_{N \uparrow \infty} \mathbb{E}\left(\eta_t'^N(f)\right) = \eta_t'(f)$, or any related propagation of chaos estimates, doesn't give any very precise information on the performance of the mean field particle methods.

More precisely, these types of weak convergence results state that the average values of the occupation measures, and the distributions of any finite particle blocks behaves as the limiting mean field measures, and blocks of independent particles. The best way to quantify the performance of the Monte Carlo model is to analyze the concentration properties of the occupation measures $\eta_t'^N$ around their limiting values η_t'.

1.3   Time discretization schemes

1.3.1   Euler type approximations

Using the same reasoning as in the end of Section 1.1.2, a discrete time approximation on a time mesh t_n of the jump-diffusion model discussed in Section 1.2.2.1 is given by a Markov chain X_n with elementary transitions

$$K_{n+1,\eta_n}(x,dz) = \int M_{n+1,\eta_n}(x,dy)\, J_{n+1,\bar{\eta}_n}(y,dz) \qquad (1.28)$$

with the probability measures $\bar{\eta}_n$ given by

$$\bar{\eta}_n = \eta_n \qquad \text{or} \qquad \bar{\eta}_n := \eta_n M_{n+1,\eta_n}$$

In the above displayed integral formula, M_{n+1,η_n} stands for the transition of the Markov chain

$$X_{n+1} = X_n + \mathbf{a_{n+1}}(X_n, \eta_n)\, \Delta + \boldsymbol{\sigma}_{n+1}(X_n, \eta_n)\, \left(W_{t_{n+1}} - W_{t_n}\right)$$

with $(\mathbf{a_{n+1}}, \boldsymbol{\sigma}_{n+1}) = \left(a'_{t_n}, \sigma'_{t_n}\right)$, and $\eta_n = \text{Law}(X_n)$. The geometric jump type Markov transition $J_{n+1,\eta}$ is now defined in terms of the parameters

$$(G_{n+1,\eta}, P_{n+1,\eta}) = \left(\exp\left(-\Delta\lambda_{t_{n+1},\eta}\right), P'_{t_{n+1},\eta}\right)$$

by the following equation

$$J_{n+1,\eta}(y,dz) = G_{n+1,\eta}(y)\, \delta_y(dz) + (1 - G_{n+1,\eta}(y))\, P_{n+1,\eta}(y,dz)$$

In this situation, the discrete generation model (1.16), associated with the Markov transitions (1.28), can be seen as an approximation of the jump type Markov process with infinitesimal generator (1.19). Therefore, using the same line of arguments as the ones given in Section 1.1.2.2, the distribution flow η_n of the random states of the discrete generation process given in (1.16) approximates the solution η'_{t_n} of the continuous time integro-differential Equation (1.19). More precisely, taking $\Delta \simeq 0$ in (1.28), we have the approximation

$$\begin{aligned}
J_{n+1,\bar{\eta}_n} - Id \quad &= \left(1 - \exp\left(-\Delta\lambda_{t_{n+1},\bar{\eta}_n}\right)\right)\, \left(P'_{t_{n+1},\bar{\eta}_n} - Id\right) \\
&\simeq \quad \lambda_{t_{n+1},\bar{\eta}_n}\, \left(P'_{t_{n+1},\bar{\eta}_n} - Id\right) \simeq \lambda_{t_{n+1},\eta_n}\, \left(P'_{t_{n+1},\eta_n} - Id\right)\, \Delta
\end{aligned}$$

and for any smooth functions f we also have that

$$(M_{n+1,\eta_n} - Id)(f)(x)$$

$$\simeq \mathbb{E}\left(f\left(x + a'_{t_n}(x,\eta_n)\,\Delta + \sigma'_{t_n}(x,\eta_n)\,\left(W_{t_{n+1}} - W_{t_n}\right)\right) - f(x)\right)$$

$$\simeq L'_{t_n,\eta_n}(f)(x)\,\Delta$$

with the second order differential operator $L'_{t,\eta}$ defined in (1.20). As in (1.11), we conclude that

$$K_{n+1,\eta_n} \simeq Id + L_{t_n,\eta_n} \ \Delta \simeq Id + L_{t_n,\eta'_{t_n}} \tag{1.29}$$

with the infinitesimal generator $L_{t,\eta}$ defined in (1.20).

If we choose $\bar{\eta}_n = \eta_n M_{n+1,\eta_n}$ in (1.28), then the elementary transitions of the McKean-Markov chain $X_n \rightsquigarrow X_{n+1}$ are decomposed into two steps

$$X_n \xrightarrow{\ M_{n+1,\eta_n}\ } \overline{X}_n \xrightarrow{\ J_{n+1,\bar{\eta}_n}\ } X_{n+1} \quad \text{with} \quad \text{Law}\left(\overline{X}_n\right) = \bar{\eta}_n$$

If we set $X_{n+1/2} := \overline{X}_n$, and $\eta_{n+1/2} := \bar{\eta}_n$, then the corresponding discrete generation McKean model is defined in terms of the slightly different equations

$$\eta_n \rightsquigarrow \eta_{n+1/2} = \eta_n K_{n+1/2,\eta_n} \rightsquigarrow \eta_{n+1} = \eta_{n+1/2} K_{n+1,\eta_{n+1/2}} \tag{1.30}$$

with the McKean transitions

$$\left(K_{n+1/2,\eta}, K_{n+1,\eta}\right) = \left(M_{n+1,\eta}, J_{n+1,\eta}\right)$$

Up to a change of time, the discrete evolution model described above has exactly the same form as the one discussed in (1.16). To the best of our knowledge, the refined analysis of the couple of discrete time approximation schemes discussed above has not been covered in the literature of McKean diffusions with jumps.

1.3.2 A jump-type Langevin model

The choice of the time discretization schemes discussed in Section 1.3 is extremely important in mean field IPS methodologies. For instance, a simple Euler type discretization model may fail to transfer the desired regularity properties of the continuous Feynman-Kac model to the discrete time one.

We illustrate this assertion with a discussion on an overdamped Langevin diffusion, on an energy landscape associated with a given energy function $V \in \mathcal{C}^2(\mathbb{R}^d, \mathbb{R}_+)$ on $E = \mathbb{R}^d$, for some $d \geq 1$. This model is defined by the following diffusion equation

$$dX'_t = -\beta \ \nabla V(X'_t) + \sqrt{2} \ dB_t \tag{1.31}$$

where ∇V denotes the gradient of V, β an inverse temperature parameter, and B_t a standard Brownian motion on $\mathbb{R}^d$ (also called a Wiener process). The infinitesimal generator associated with this continuous time process is given by the second order differential operator

$$L'_\beta = -\beta \ \nabla V \cdot \nabla + \Delta$$

Under some regularity conditions on V, the diffusion X'_t is geometrically ergodic with an invariant measure given by

$$d\pi_\beta = \frac{1}{\mathcal{Z}_\beta} \, e^{-\beta V} \, d\lambda \tag{1.32}$$

where λ stands for the Lebesgue measure on $\mathbb{R}^d$, and $\mathcal{Z}_\beta$ is a normalizing constant. When the inverse temperature parameter β_t depends on the time parameter t, the time inhomogeneous diffusion X'_t has a time inhomogeneous generator L'_{β_t}.

We further assume that $\pi_{\beta_0} = \mathrm{Law}(X'_0)$, and we set $\beta'_t := \frac{d\beta_t}{dt}$. By construction, we have

$$\frac{d}{dt}\pi_{\beta_t}(f) = \beta'_t \, (\pi_{\beta_t}(V)\pi_{\beta_t}(f) - \pi_{\beta_t}(fV)) \quad \text{and} \quad \pi_{\beta_t}L'_{\beta_t} = 0$$

Using these observations, we readily check that π_{β_t} satisfies the Feynman-Kac evolution equation defined as in (1.25) by replacing V_t by $\beta'_t V$. More formally, we have that

$$\frac{d}{dt}\pi_{\beta_t}(f) = \pi_{\beta_t}\left(L'_{\beta_t}(f)\right) + \beta'_t \left(\pi_{\beta_t}(V)\pi_{\beta_t}(f) - \pi_{\beta_t}(fV)\right)$$

from which we conclude that

$$\pi_{\beta_t}(f) = \gamma_t(f)/\gamma_t(1) \quad \text{with} \quad \gamma_t(f) := \mathbb{E}\left(f(X'_t) \, \exp\left(-\int_0^t \beta'_s \, V(X'_s)ds\right)\right)$$

In summary, we have described a McKean interpretation of Boltzmann-Gibbs measures (1.32) associated with some nondecreasing inverse cooling schedule. Returning to the McKean interpretations of the Feynman-Kac models discussed in Section 1.2.2.3, the flow of measures

$$\eta_t := \pi_{\beta_t} = \mathrm{Law}(X_t)$$

can be interpreted as the distributions of the random states X_t of a jump type Markov process. Between the jumps, X_t follows the Langevin diffusion Equation (1.31). At jump times T_n, with the stochastic rate $\beta'_t V_t(X_t)$, the process $X_{T_n-} \rightsquigarrow X_{T_n}$ jumps to a new site randomly chosen with the distribution $\eta_{T_n-}(dy)$. The discrete time version of these Feynman-Kac interpretation models is discussed in Section 4.3.1, dedicated to interacting MCMC models.

A more detailed analysis of these continuous time sequential Monte Carlo models can be found in the articles by M. Rousset and his co-authors [392, 498, 499], as well as in the articles by A. Eberle and C. Marinelli [237, 238, 239]. More general continuous time Feymann-Kac models are also discussed in the series of articles [161, 173, 176, 178, 179], as well as in the more recent study [171].

As usual, in the continuous time framework, to get some feasible solution we need to introduce a time discretization scheme. To this end, we let W_{n+1} be a sequence of centered and reduced Gaussian variables on $\mathbb{R}^d$. Firstly, starting from some random state X_n, we propose a random state Y_{n+1} using the conventional Euler scheme

$$Y_{n+1} = X_n - \beta \ \nabla V(X_n)/m + \sqrt{2/m} \ W_{n+1} \qquad (1.33)$$

Then, we accept this state $X_{n+1} = Y_{n+1}$ with probability

$$1 \wedge \left(e^{-\beta(V(Y_{n+1})-V(X_n))} \times \frac{p_m(Y_{n+1}, X_n)}{p_m(X_n, Y_{n+1})} \right)$$

Otherwise, we stay in the same location $X_{n+1} = X_n$.

In the above display, the function p_m stands for the probability density of the Euler scheme proposition

$$p_m(x, y) = \frac{1}{(4\pi/m)^{d/2}} \ \exp\left(-\frac{m}{4} \ \|y - x + \beta \ \nabla V(x)/m\|^2 \right)$$

The resulting Markov chain model X_n is often referred to as the Metropolis-adjusted Langevin algorithm (*abbreviated (MALA)*). One of the main advantages of the above construction is that the Markov chain X_n is reversible w.r.t. to π_β, and it has the same fixed point π_β as the continuous time model. Without the acceptance-rejection rate, the Markov chain (1.33) reduces to the standard Euler approximation of the Langevin diffusion model (1.31). In this situation, the Markov chain may even fail to be ergodic, when the vector field ∇V is not globally Lipschitz [494, 421]. We refer the reader to [69, 281, 494, 495, 568] for further details on the stochastic analysis of these Langevin diffusion models. We also mention that the Euler scheme diverge in many situations, even for uniformly convex functions V. At the cost of some additional computational effort, a better idea is to replace (1.33) by the implicit backward Euler scheme given by

$$Y_{n+1} + \beta \ \nabla V(Y_{n+1})/m = X_n + \sqrt{2/m} \ W_{n+1}$$

When the inverse temperature parameter β_{t_n} depends on the time parameter t_n, the transition $X_n \rightsquigarrow X_{n+1}$ is defined as above by replacing β by β_{t_n}. Furthermore, if we assume that $\pi_{\beta_0} = \text{Law}(X_0)$, then we also have that

$$\pi_{\beta_{t_n}}(f) \propto \mathbb{E}\left(f(X_n) \ \exp\left(-\sum_{0 \leq p < n} (\beta_{t_{p+1}} - \beta_{t_p}) \ V(X_p) \right) \right)$$

For a more thorough discussion on the stochastic analysis of these discrete generation sequential Monte Carlo models, we refer the reader to [165, 186], as well as to Section 2.5.2, Section 2.6, Section 4.3.1, and Section 8.6 in the present book.

1.4　Illustrative examples

In this section, we illustrate the abstract nonlinear Markov chain models presented in (1.16), with two important examples arising in statistical physics and applied probability. In Section 1.4.1, we discuss stochastic kinetic models and McKean-Vlasov diffusion type models arising in fluid mechanics and condensed matter theory. Section 1.4.2 is concerned with Feynman-Kac models.

A more detailed discussion on the variety of application domains of these Feynman-Kac models is provided in Section 2.5, as well as in Chapter 7 and in Chapter 8.

We also refer the reader to the series of articles [97, 161, 163, 186], and references therein.

1.4.1　Stochastic kinetic models

In computational physics, and more particularly in fluid mechanics, nonlinear stochastic kinetic models represent the evolution of complex stochastic systems with so many particles that they can be treated as a continuum. In this interpretation, a single particle interacts with the distribution of the whole population of individuals.

Another central idea, commonly used in this stochastic modeling, is to express the evolution of macroscopic quantities in terms of averages w.r.t. the distribution of microscopic variables. In kinetic theory, these stochastic models are used to represent the evolutions of various physical systems, such as gases, macroscopic fluid models, and molecular chaotic systems.

The foundations of kinetic theory were laid down by J.C. Maxwell [423, 424, 425]. Further details on these models can be found in the series of articles [111, 112, 113, 128, 294, 402, 554] and in the more probabilistic treatments [295, 428, 429, 430, 538, 545].

More recently, these Vlasov type equations have also come to play an important role in mathematical biology, as well as in socio-economics and stochastic games theory.

In biology, they are used to model animal competitions as well as bacteria dynamics with group interactions, or complex cell motions [44, 45, 442, 477]. In stochastic games theory, McKean-Vlasov models represent competing multiple class agents weakly coupled w.r.t. their dynamics and their performance [82, 148, 287, 311, 329, 426, 401, 381, 382, 383, 384].

We quote recent applications of Markov-McKean models in filtering problems arising in turbulent fluid mechanics and weather forecasting prediction developed by Ch. Baehr and his co-authors in a series of articles [25, 26, 28, 29, 390, 487, 537].

The time evolution of these stochastic kinetic models can be defined in terms of discrete generation nonlinear Markov chains discussed in Section 1.2.1, or in terms of the integro-differential equations presented in Section 1.2.2. In this connection, we also quote the series of articles [13, 61, 62, 63, 372] in the late 1990s on the rate of convergence of some time discretization schemes of McKean-Vlasov diffusions.

1.4.1.1 McKean-Vlasov type models

Prototypes of discrete generation and nonlinear Markov-McKean models are given by McKean-Vlasov-Fokker-Planck diffusion type models arising in fluid mechanics, as well as in mean field game theory. These models are also used in advanced signal processing, and more particularly in forecasting data assimilation problems. We refer the reader to Section 8.2.3, dedicated to a McKean diffusion type interpretation of the Kalman filter, and their mean field ensemble Kalman filters.

In dimension $d = 1$, these nonhomogeneous Markov models are given by an $\mathbb{R}$-valued stochastic process defined by the recursive equation

$$X_n - X_{n-1} = \mathbf{a_n}(X_{n-1}, \eta_{n-1}) + \boldsymbol{\sigma_n}(X_{n-1}, \eta_{n-1})\, W_n \qquad (1.34)$$

with $\eta_{n-1} := \mathrm{Law}(X_{n-1})$. In the above displayed formula, X_0 is a r.v. $(W_n)_{n\geq 0}$ is a collection of i.i.d. centered Gaussian random variables with unit variance, and the drift and diffusion functions are defined by

$$\mathbf{a_n}(X_{n-1}, \eta_{n-1}) = \int a_n(X_{n-1}, x_{n-1})\, \eta_{n-1}(dx_{n-1})$$

$$\boldsymbol{\sigma_n}(X_{n-1}, \eta_{n-1}) = \int \sigma_n(X_{n-1}, x_{n-1})\, \eta_{n-1}(dx_{n-1}) \qquad (1.35)$$

for some regular mappings a_n and σ_n. Whenever $\sigma_n(x, y) \geq \epsilon$, for some $\epsilon > 0$, the laws of the random states $\eta_n = \mathrm{Law}(X_n)$ satisfy the evolution Equation (1.16), with the McKean transitions given by

$$K_{n,\eta}(x, dy) = \frac{1}{\sqrt{2\pi\sigma_n^2(x, \eta)}}\, \exp\left\{-\frac{1}{2}\left(\frac{(y - x) - \mathbf{a_n}(x, \eta)}{\sigma_n(x, \eta)}\right)^2\right\} dy$$

1.4.1.2 Nonlinear state space models

A more general class of McKean-Markov chain models on some measurable state spaces E_n are given by the recursive formulae

$$X_n = F_n(X_{n-1}, \eta_{n-1}, W_n) \quad \text{with} \quad \eta_{n-1} := \mathrm{Law}(X_{n-1}) \qquad (1.36)$$

In the above display, W_n is a collection of independent, (and independent of $(X_p)_{0\leq p<n})$ r.v. taking values in some state space $\mathcal{W}_n$, and F_n is some measurable mapping from $(E_{n-1} \times \mathcal{P}(E_{n-1}) \times \mathcal{W}_n)$ into E_n.

Here again, we easily check that the laws of the random states $\eta_n = \text{Law}(X_n)$ satisfy the evolution Equation (1.16), with the McKean transitions

$$K_{n,\eta}(f)(x) = \mathbb{E}\left[f(F_n(x, \eta, W_n))\right]$$

We refer the reader to Section 10.4.3 and Section 10.4.6 in the present book, for a detailed discussion on McKean model of gases and McKean-Vlasov type models with interacting jumps. More general discrete time generation Markov-McKean models of the form (1.36) are discussed in the series of articles [144, 149, 152, 168, 161, 198].

1.4.2   Feynman-Kac models

Feynman-Kac models are one of the central application domains of the mean field particle theory developed in this book. One of their fascinations today is their use in modeling a very large class of physical, biological, and engineering problems.

These modern stochastic modeling techniques arise in a variety of application domains, including particle physics, biology, and branching processes, nonlinear filtering, and hidden Markov chain problems (*abbreviated HMM*), Bayesian inference, financial mathematics, and many others.

For a more thorough discussion of these application models, with a list of reference pointers, we refer the reader to the preface of this book, as well as to Chapter 3, dedicated to Feynman-Kac models, and four selected probabilistic interpretations. A complementary exposition, with different applications domains, can also be found in the monograph [163], as well as in a couple of lecture notes [161, 186].

The origins of Feynman-Kac models started with the work of Richard Phillips Feynman who provided in 1942 an heuristic connection between the Schrödinger equation and the path integration theory of Norbert Wiener. These lines of investigations were pursued and amplified by the mathematician Mark Kac in the early 1950s.

The central idea was to express the semigroup of a quantum particle evolving in a potential in terms of a functional path-integral formula. Intuitively speaking, Feynman-Kac measures enter the effects of the potential in the distribution of the random trajectories of the particles.

From a more probabilistic point of view, Feynman-Kac measures represent the distributions $\mathbb{P}_n$ of the random paths $(X_0, \ldots, X_n)$ of a given reference Markov process X_n, taking values in some state spaces E_n, and weighted by a collection of nonnegative potential functions G_n, up to some normalizing constant $\mathcal{Z}_n$. In this sense, these models are natural mathematical extensions of the traditional change of probability measures, commonly used in importance sampling. More formally, Feynman-Kac models are defined by the formulae

$$d\mathbb{Q}_n := \frac{1}{\mathcal{Z}_n}\left\{\prod_{0 \leq p < n} G_p(X_p)\right\} d\mathbb{P}_n \qquad (1.37)$$

Notice that measures $\mathbb{Q}_n$ are well defined, as soon as $\mathcal{Z}_n \neq 0$. We also denote by Γ_n the unnormalized measures defined by

$$d\Gamma_n = \mathcal{Z}_n \times d\mathbb{Q}_n = \left\{ \prod_{0 \leq p < n} G_p(X_p) \right\} d\mathbb{P}_n \qquad (1.38)$$

Unless otherwise stated, to avoid unnecessary repetition of technical abstract conditions, we frame the standing assumption that the potential functions G_n are upper and lower bounded by some positive constant. This condition is clearly met for continuous potential functions on compact metric spaces. It can be easily relaxed using standard cutoff techniques (such as those presented in Section 8.7.3.2, in the context of optimal stopping problems), or using the change of measures techniques presented in Section 2.5 and Section 7.2.2 in the research monograph [163].

We also emphasize that Feynman-Kac measures can always be interpreted as conditional distributions of a given stochastic process w.r.t. a sequence of conditioning events. For a more detailed discussion on these conditional distribution models, we refer the reader to Section 3.3.2 and Chapter 7, dedicated to a particle absorption interpretations of general and abstract Feynman-Kac models.

Some illustrations of Feynman-Kac conditional distributions are also provided in Section 1.4.2.2, including Markov chain restrictions, self-avoiding walks models, and particle absorption models.

Another series of applications areas related to theoretical chemistry, quantum physics, population dynamic models, signal processing and computer sciences, with some reference pointers, is provided in Section 2.2 and Section 2.3, as well as in Chapter 7 dedicated to particle motions in absorbing medium.

A more formal illustration of Feynman-Kac modeling in statistical machine learning theory is provided in Section 2.5, including nonlinear filtering, hidden Markov chain models, Bayesian inference, and stochastic optimization models. Applications related to risk analysis and rare event simulation are discussed in Section 2.7. We also refer the reader to Chapter 8 for a more thorough discussion on application areas related to signal processing and control system theory.

Let us pause for a while and discuss the different nature of Feynman-Kac models and the kinetic models discussed in Section 1.4.1.

Nonlinear kinetic models are complex McKean-Markov models, which might exhibit different types of long time behavior. Nevertheless, their mean field IPS interpretations discussed further in the development of Section 1.5 are defined in terms of elementary Markov chain models in product spaces.

In addition, the application domains of this class of McKean models are often restricted to the mathematical modeling of turbulent fluid, or physical evolutions of gases.

In contrast with these nonlinear state space models, the class of Feynman-Kac distributions has an exceptionally rich mathematical structure that has considerably influenced several research directions in mathematical physics, stochastic process theory, and many other scientific disciplines.

To name a few, in Hidden Markov chain problems, the Feynman-Kac normalizing constants are related to the likelihood functions of a sequence of observations. In physics, the same normalizing constants are related to ground states and free energy computations. In rare event simulation, these parameters also represent the probability of critical events. In filtering theory, the backward Markov interpretation of Feynman-Kac models is used for smoothing the signal trajectories. In mathematical finance, the same backward models are used to compute Greek derivatives and sensitivity measures. These probabilistic models are discussed in more detail in Section 1.4.3.

Depending whether the reference stochastic process is defined in continuous time or in discrete time, the flows of Feynman-Kac distributions satisfy nonlinear evolution equations of the form (1.15) or an integro-differential equation of the form (1.25). The nonlinear semigroups of discrete generation Feynman-Kac flows are presented in some detail in Section 1.4.2.1.

1.4.2.1   McKean-Markov semigroups

We consider the n-th marginal (γ_n, η_n) of the Feynman-Kac measures $(\Gamma_n, \mathbb{Q}_n)$ defined in (1.37) and (1.38). By construction, we have

$$\eta_n(f_n) = \gamma_n(f_n)/\gamma_n(1) \quad \text{with} \quad \gamma_n(f_n) := \mathbb{E}\left(f_n(X_n) \prod_{0 \leq p < n} G_p(X_p) \right)$$
(1.39)

for any $f_n \in \mathcal{B}_b(E_n)$. We let M_n be the Markov transitions of the reference Markov chain X_n. In this notation, the flow of measures η_n satisfies a nonlinear evolution equation of the form (1.15), with the one step mappings

$$\Phi_{n+1}(\eta) = \Psi_{G_n}(\eta)M_{n+1}$$
(1.40)

In the above displayed formula, $\Psi_{G_n}(\eta)$ stands for the Boltzmann-Gibbs measures defined in (0.2). We prove this claim using the fact that the unnormalized distributions γ_n defined in (1.39) satisfy the following *linear* evolution equation

$$\gamma_{n+1} = \gamma_n Q_{n+1}$$
(1.41)

with the integral operators

$$Q_{n+1}(x_n, dx_{n+1}) = G_n(x_n) \, M_{n+1}(x_n, dx_{n+1})$$

A detailed proof of (1.29) and (1.30) is provided in Section 3.2.2
From these observations, for any $p \leq n$ we find that

$$\gamma_n = \gamma_p Q_{p,n}$$
(1.42)

with the semigroup $Q_{p,n}$ defined for any $(x_p, f_n) \in (E_p \times \mathcal{B}_b(E_n))$ by

$$Q_{p,n}(f_n)(x_p) := \mathbb{E}\left(f_n(X_n) \prod_{p \le q < n} G_q(X_q) \mid X_p = x_p \right)$$

Extended Feynman-Kac semigroups associated with integral operators $Q_{n+1} := Q_{n+1,\gamma_n}$ that depend on the measures γ_n are discussed in Section 2.4.1. Now, we come to McKean interpretations of the flow of normalized measures η_n. Here again, the choice of the McKean transitions $K_{n+1,\eta}$ satisfying the compatibility condition

$$\eta K_{n+1,\eta} = \Psi_{G_n}(\eta) M_{n+1}$$

is far from being unique. For instance, for $[0,1]$-valued potential functions G_n, and any $\epsilon \in [0,1]$, we can choose

$$K_{n+1,\eta}(x, dy) = \epsilon G_n(x) M_{n+1}(x, dy) + (1 - \epsilon G_n(x)) \Psi_{G_n}(\eta) M_{n+1}(dy) \quad (1.43)$$

A detailed discussion on these evolution equations is provided in Section 3.2.2 and in Section 3.4. Section 4.4.5 also contains different choices of McKean transitions that satisfy the compatibility condition presented above. The continuous time versions of these Feynman-Kac models and their McKean interpretations are discussed in Chapter 5.

The structure and the stability properties of the semigroup

$$\forall p \le n \qquad \eta_n = \Phi_{p,n}(\eta_p) \quad \text{with} \quad \Phi_{p,n} = \Phi_n \circ \Phi_{n-1} \circ \ldots \circ \Phi_{p+1}$$

associated with these Feynman-Kac models are developed in some detail in Chapter 12; see also [163] for a more detailed discussion on the contraction properties of these nonlinear semigroups.

The stability properties of these semigroups ensure that the distribution flow η_n forgets its initial condition η_0. For instance, in Section 12.2.1 we shall present some sufficient conditions on the parameters (G_n, M_n) under which we have the following exponential forgetting property

$$\beta_{p,n} := \sup_{\mu,\nu} \|\Phi_{p,n}(\mu) - \Phi_{p,n}(\nu)\|_{\text{tv}} \le a \, \exp\left(-b(n-p)\right) \quad (1.44)$$

for any $p \le n$, for some positive and finite parameters (a, b).

As we shall see in the further development of Chapter 14, this regularity property also ensures that the local errors induced by some local approximation scheme will not propagate w.r.t. the time parameter. In other words, these critical robustness and stability properties allow us to obtain uniform estimates for any numerical approximation scheme (based on local approximations) w.r.t. the time horizon.

1.4.2.2 Some conditional distributions

In this section, we illustrate the abstract Feynman-Kac models defined in (1.37), with a series of typical and more concrete examples of conditional distributions.

- If we choose indicator potential functions $G_n = 1_{A_n}$, then we have

$$
\begin{aligned}
\mathcal{Z}_n &= \mathbb{P}\left(X_p \in A_p, \ \forall 0 \leq p < n\right) \\
\mathbb{Q}_n &= \text{Law}\left((X_0, \ldots, X_n) \mid X_p \in A_p, \ \forall 0 \leq p < n\right)
\end{aligned}
$$

- We assume that $X_n = (X_0', \ldots, X_n')$ is the historical process associated with a random walk evolving in a d-dimensional lattice $E = \mathbb{Z}^d$. In this situation, if we set $G_n(X_n) = 1_{\mathbb{Z}^d - \{X_0', \ldots, X_{n-1}'\}}$, then we find that

$$
\begin{aligned}
\mathcal{Z}_n &= \mathbb{P}\left(X_p' \neq X_q', \ \forall 0 \leq p < q < n\right) \\
\mathbb{Q}_n &= \text{Law}\left((X_0', \ldots, X_n') \mid X_p' \neq X_q', \ \forall 0 \leq p < q < n\right)
\end{aligned}
$$

- We let X_n^c be a particle Markov chain model evolving in an absorbing medium with an absorption rate $1 - G_n$ taking values in $[0, 1]$. If we denote by T the absorption time of the particle, then we have

$$
\mathcal{Z}_n = \mathbb{P}\left(T \geq n\right) \quad \text{and} \quad \mathbb{Q}_n = \text{Law}\left((X_0^c, \ldots, X_n^c) \mid T \geq n\right)
$$

For a more thorough discussion on these models, we refer the reader to Section 3.3.2 and Chapter 7 dedicated to particle absorption models.

- We consider a time homogeneous particle absorption model with a free exploration μ-reversible Markov transition M and a positive nonabsorption rate function G. Whenever they exist, we let λ be the largest eigenvalue of the integral operator

$$
Q(x, dy) = G(x)M(x, dy)
$$

on $\mathbb{L}_2(\mu)$, and $h(x)$ be a positive eigenvector $Q(h) = \lambda h$. In this situation, the limiting nonabsorption exponential rate is given by

$$
\lim_{n \to \infty} \frac{1}{n} \log \mathcal{Z}_n = \lim_{n \to \infty} \frac{1}{n} \log \mathbb{P}\left(T \geq n\right) = \log \lambda
$$

In addition, the density of the quasi-invariant limiting distribution w.r.t. the measure μ is given by

$$
\lim_{n \to \infty} \mathbb{P}\left(X_n^c \in dx \mid T > n\right) \propto h(x) \ \mu(dx)
$$

Moreover, the Feynman-Kac model on path space coincides with the distribution $\mathbb{P}_n^h$ of the random paths $(X_0^h, \ldots, X_n^h)$ of the Doob h-process with Markov transitions $M^h(x, dy) \propto M(x, dy)h(y)$; more precisely, we have that

$$
d\mathbb{Q}_n = \frac{1}{\mathbb{E}(h^{-1}(X_n^h))} \ h^{-1}(X_n^h) \ d\mathbb{P}_n^h
$$

These particle absorption models are discussed in more detail in Section 3.3.2 and in Chapter 7.

1.4.3 Some structural properties

As we already mentioned in the opening introduction of Section 1.4.2, Feynman-Kac models reveal a surprisingly and extremely rich mathematical structure. These structural properties are often used to design and to analyze sophisticated and powerful particle approximation models. We illustrate this assertion, with four important properties.

1.4.3.1 Historical processes

The first structural property is related to the historical process $X_n = (X'_0, \ldots, X'_n)$ of a Markov chain X'_n evolving in some measurable state spaces E'_n. We let η_n be the n-th marginal distributions of the Feynman-Kac measures $\mathbb{Q}_n$ associated with the historical Markov chain model X_n, and with a collection of potential functions $G_n(X_n) = G'_n(X'_n)$ that only depend on the terminal state X'_n.

By construction, it is readily checked that η_n are again defined by the Feynman-Kac measure on path space

$$d\eta_n := \frac{1}{\mathcal{Z}'_n} \left\{ \prod_{0 \leq p < n} G'_p(X'_p) \right\} d\mathbb{P}'_n \tag{1.45}$$

where $\mathbb{P}'_n$ is the distribution of the random variables $X_p = (X'_0, \ldots, X'_p)$, with $p \leq n$. From the pure mathematical viewpoint, these marginal distributions have the same form as the Feynman-Kac measures $\mathbb{Q}_n$. For more detail on these historical processes, we refer to Section 3.4.

1.4.3.2 Continuous time models

In continuation to the remarks we made in the beginning of Section 1.1.2, discrete time Feynman-Kac models also encapsulate without further work continuous time models. More precisely, let us consider a continuous time Markov process $(X'_t)_{t \geq 0}$ taking values in some Polish state space E, and a sequence of measurable and bounded functions

$$V \; : \; (t, x) \in (\mathbb{R}_+ \times E) \mapsto V_t(x) \in \mathbb{R}$$

As in Section 1.1.2, we consider a time mesh $(t_n)_{n \geq 0}$, with $t_0 = 0$, and we let $E_n = D([t_n, t_{n+1}], E)$ be the set of càdlàg paths from the interval $[t_n, t_{n+1}]$ into E. We consider the Feynman-Kac model (1.37) associated with the sequence of random variables X_n, and the potential functions G_n on E_n, defined by

$$X_n = (X'_t)_{t_n \leq t \leq t_{n+1}} \quad \text{and} \quad G_n(X_n) = \exp\left(\int_{t_n}^{t_{n+1}} V_t(X'_t) \, dt \right) \tag{1.46}$$

By construction, it is readily checked that

$$d\mathbb{Q}_n = \frac{1}{\mathcal{Z}_n} \exp\left\{ \int_0^{t_n} V_t(X'_t) \, dt \right\} d\mathbb{P}'_{t_{n+1}}$$

where $\mathbb{P}'_{t_{n+1}}$ is the distribution of $(X'_t)_{0 \le t \le t_{n+1}}$ on $D([0, t_{n+1}], E)$.

1.4.3.3   Unnormalized models

Besides the interesting equivalence principles discussed in Section 1.4.3.1 and Section 1.4.3.2, we underline that the normalizing constants $\mathcal{Z}_n$, as well as the measures $\mathbb{Q}_n$, can be represented in terms of the flow of marginal measures $(\eta_p)_{0 \le p \le n}$. More precisely, we have the easily checked multiplicative formulae

$$\mathcal{Z}_n = \mathbb{E}\left(\prod_{0 \le p < n} G_p(X_p) \right) = \prod_{0 \le p < n} \eta_p(G_p) \tag{1.47}$$

More generally, using this formula the unnormalized measures $\gamma_n := \mathcal{Z}_n \times \eta_n$ can also be rewritten in the following form

$$\gamma_n(f_n) = \eta_n(f_n) \times \prod_{0 \le p < n} \eta_p(G_p) \tag{1.48}$$

for any $f_n \in \mathcal{B}_n(f_n)$. A detailed proof of these formulae can be found in Section 3.2.2.

1.4.3.4   Backward Markov chain models

The description of $\mathbb{Q}_n$ in terms of $(\eta_p)_{0 \le p \le n}$ relies on some additional regularity property. We further assume that the Markov transitions M_n are absolutely continuous with respect to some measures λ_n on E_n, and for any $(x, y) \in (E_{n-1} \times E_n)$ we have

$$H_n(x, y) := \frac{dM_n(x, \cdot)}{d\lambda_n}(y) > 0 \tag{1.49}$$

In this situation, the Feynman-Kac measures $\mathbb{Q}_n$ can be expressed in terms of the flow of measures η_n with the following backward formula

$$\mathbb{Q}_n = \eta_n \otimes \mathbb{Q}_{n|n} \tag{1.50}$$

with the Markov transition $\mathbb{Q}_{n|n}$ from E_n into $\mathbf{E_{n-1}} := \prod_{0 \le p < n} E_p$ defined by

$$\mathbb{Q}_{n|n}(x_n, d(x_0, \dots, x_{n-1})) := \prod_{q=1}^{n} \mathbb{M}_{q, \eta_{q-1}}(x_q, dx_{q-1})$$

In the above display, $\mathbb{M}_{n+1, \eta_n}$ stands for the collection of Markov transitions defined by

$$\mathbb{M}_{n+1, \eta_n}(x, dy) = \frac{\eta_n(dy) \ G_n(y) \ H_{n+1}(y, x)}{\eta_n \left(G_n H_{n+1}(\cdot, x) \right)} \tag{1.51}$$

If we take the unit potential functions $G_n = 1$, the backward formula (1.50) reduces to the conventional backward representation of conditional distribution of the random paths $(X_0, \dots, X_{n-1})$ given the terminal time X_n; that is we have that

$$\mathbb{P}\left((X_0, X_1, \ldots, X_{n-1}) \in d(x_0, x_1, \ldots, x_{n-1}) \mid X_n = x_n\right)$$

$$= \mathbb{M}_{n, \eta_{n-1}}(x_n, dx_{n-1}) \cdots \mathbb{M}_{2, \eta_1}(x_2, dx_1) \mathbb{M}_{1, \eta_0}(x_1, dx_0)$$

To the best of our knowledge, these forward-backward representations of Feynman-Kac measures were introduced by Ruslan L. Stratonovitch in the early 1960s in the context of nonlinear filtering [533]. The proof and the applications of these Backward Markov interpretations are provided in Section 3.4.1 and in Section 3.4.2. For a more thorough discussion on these backward Markov chain models, and their application in advanced signal processing, and hidden Markov chain problems, we also refer the reader to [220] and to a series of articles of the author with A. Doucet and S.S. Singh [193, 194, 195].

1.5 Mean field particle methods

This section is concerned with mean field particle interpretations of the McKean models presented in (1.16). We also illustrate these particle models in the context of the McKean-Vlasov diffusions and the Feynman-Kac distribution flows, discussed in Section 1.4.1 and in Section 1.4.2. The continuous time version of these mean field models is discussed in some detail in Chapter 5, Section 5.4. These mean field models provide a nice probabilistic interpretation of the nonlinear jump-diffusion models (1.21) and (1.22) presented in Section 1.2.2.1, but generally with little practical value.

Indeed, to get some computer implementation of these mean field particle models, we need to consider some time discretization schemes. In this context, the discrete generation mean field model is of the same form as the one of the nonlinear discrete time McKean models discussed in (1.16). The convergence analysis of these discrete time models is discussed in Chapter 5, in the context of Feynman-Kac models.

1.5.1 Interacting particle systems

With the exception of some very special cases, the evolution Equations (1.15) and their continuous time versions (1.19) cannot be solved explicitly. For instance, in the context of the Feynman-Kac models presented in (1.40), it can be shown that the measures η_n, or their continuous time versions discussed in Section 5, cannot be represented in a closed form on some finite dimensional state-space. A proof of this result can be found in the article by M. Chaleyat-Maurel and D. Michel [127], in the context of nonlinear filtering problems. In addition, conventional harmonic type approximation schemes, or related linearization style techniques, such as the extended Kalman filter, often provide poor estimation results for highly nonlinear models.

On the other hand, more traditional numerical techniques for solving evolution equations of the form (1.16) using deterministic type grid approximations require extensive calculations, and they rarely cope with nonlinear distribution flows with strongly varying probability masses, or with high dimensional problems. In contrast with these conventional numerical techniques, the mean field IPS technology presented in this section provides an accurate stochastic grid approximation scheme, equipped with an interacting mechanism tracking the probability mass variations of the distribution flow. In other words, these advanced Monte Carlo methods take advantage of the nonlinearities of the model, to drive the particle populations in regions with high probability mass. We also mention that the stochastic perturbation techniques presented in the further development of Section 2.1 show that mean field models can also be interpreted as a natural stochastic linearization of nonlinear evolution equations in distribution spaces.

To design a mean field IPS model associated with a given collection of McKean transitions $K_{n,\eta}$, we further assume that we have a dedicated Monte Carlo simulation tool to draw independent random samples of the elementary transitions

$$K_{n,m(y)}(x_{n-1}, dx_n) \qquad (1.52)$$

for any empirical measure associated with N given states $(y^i)_{1 \le i \le N} \in E_{n-1}^N$ given by the formula

$$m(y) := \frac{1}{N} \sum_{1 \le i \le N} \delta_{y^i}$$

When the transitions (1.52) are easy to simulate, we design an N-particle approximation model by evolving a Markov chain $\xi_n := (\xi_n^i)_{1 \le i \le N} \in E_n^N$ with elementary transitions

$$\mathbb{P}\left(\xi_{n+1}^{(N)} \in dx_{n+1} \mid \xi_n\right) := \prod_{1 \le i \le N} K_{n+1, \eta_n^N}(\xi_n^{(N,i)}, dx_{n+1}^i) \qquad (1.53)$$

where $dx_{n+1} = dx_{n+1}^1 \times \ldots \times dx_{n+1}^N$ stands for an infinitesimal neighborhood of a point $x_{n+1} = (x_{n+1}^i)_{1 \le i \le N} \in E_{n+1}^N$, and η_n^N stands fo the empirical measures defined by

$$\eta_n^N := m\left(\xi_n^{(N)}\right) = \frac{1}{N} \sum_{i=1}^{N} \delta_{\xi_n^{(N,i)}}$$

The initial system $\xi_0 = (\xi_0^{(N,i)})_{1 \le i \le N}$ consists of N i.i.d. random variables with common law η_0.

To simplify the presentation, with a slight abuse of notation, when there is no possible confusion we often suppress the index parameter $(.)^{(N)}$ so that we write ξ_n and ξ_n^i instead of $\xi_n^{(N,i)}$ and $\xi_n^{(N,i)}$.

The N-particle model ξ_n approximates the target probability distributions η_n by the occupation measures η_n^N of a large cloud of interacting particles; that is, we have that

$$\eta_n^N \longrightarrow_{N\to\infty} \eta_n$$

More is true, we can also prove that the occupation measure of the random trajectories of the particles converges, as $N \to \infty$, to the McKean measures defined in (1.17); that is, we have that

$$\frac{1}{N} \sum_{i=1}^{N} \delta_{(\xi_0^i,\ldots,\xi_n^i)}(d(x_0,\ldots,x_n)) \longrightarrow_{N\to\infty} \eta_0(dx_0) \prod_{1\leq p\leq n} K_{p,\eta_{p-1}}(x_{p-1},dx_p)$$

$$(1.54)$$

In the context of Feynman-Kac models, the analysis of the McKean particle measures described above is discussed in some detail in the book [163]. To our knowledge, their analysis for general discrete generation mean field models is still an open research subject.

One of the central topics of this book relies on the convergence analysis of their particle density profiles η_n^N. We refer the reader to Section 2.1 and Section 2.2, for a stochastic perturbation analysis of these models, and the description of some exponential concentration inequalities. A more thorough study of the convergence analysis of these models can be found in Chapter 14. More general mean field IPS interpretations of nonlinear models in the space of nonnegative measures are discussed in Section 2.4.1. The next sections present some examples of models satisfying the effective sampling condition (1.52).

We end this section with a discussion on the connections between these discrete generation mean field models and their continuous time version discussed in Section 1.2.2.5. For time homogeneous state spaces $E_n = E = \mathbb{R}^d$, the Markov transition $\mathcal{M}_{n+1}$ of the discrete generation mean field model (1.53) on E^N is defined for any sufficiently regular function F on E^N by

$$\mathcal{M}_{n+1}(F)(x) := \int_{E^N} \left(\prod_{1\leq i\leq N} K_{n+1,m(x)}(x,dy^i) \right) F(y^1,\ldots,y^N)$$

When the collection of transitions $K_{n+1,\eta}$ are given by (1.28), using the approximation formula (1.29) with $\Delta \simeq 0$ we find that

$$[\mathcal{M}_{n+1} - Id]\, F(x)$$

$$\simeq \prod_{1\leq i\leq N} \left(Id + L_{t_n,m(x)}\, \Delta \right)^{(i)} F(x^1,\ldots,x^i,\ldots,x^N) - F(x)$$

$$\simeq \sum_{1\leq i\leq N} L_{t_n,m(x)}^{(i)}(F)(x^1,\ldots,x^i,\ldots,x^N)\, \Delta = \mathcal{L}_{t_n}(F)(x)\, \Delta$$

In the above display, $\mathcal{L}_t$ and $L_{t,\eta}$ stands for the generators defined in (5.16) and (1.19), and $\left(Id + L_{t_n,m(x)}\, \Delta \right)^{(i)}$ stands for the operator $\left(Id + L_{t_n,m(x)}\, \Delta \right)$

acting on the function $x^i \mapsto F(x^1, \ldots, x^i, \ldots, x^N)$. This shows that the discrete generation mean field model can be interpreted as an Euler type approximation on a time mesh t_n of the continuous time particle model introduced in Section 1.2.2.5.

1.5.2 McKean-Vlasov diffusion models

The effective sampling condition stated in (1.52) is clearly satisfied for the McKean-Vlasov models (1.24) introduced in Section 1.4.1.1. In this case (1.52) is the distribution of the random variable

$$x_{n-1} + \frac{1}{N} \sum_{1 \leq i \leq N} a_n(x_{n-1}, y^i) + \frac{1}{N} \sum_{1 \leq i \leq N} \sigma_n(x_{n-1}, y^i) \, W_n \qquad (1.55)$$

In this situation, using the formula (1.55) we find that the N-particle model defined in (1.53) can be rewritten in the following form

$$\begin{aligned}
\xi_{n+1}^i &:= \xi_n^i + \mathbf{a}_{n+1}(\xi_n^i, \eta_n^N) + \boldsymbol{\sigma}_{n+1}(\xi_n^i, \eta_n^N) \, W_{n+1}^i \\
&= \xi_n^i + \frac{1}{N} \sum_{1 \leq j \leq N} a_{n+1}(\xi_n^i, \xi_n^j) + \frac{1}{N} \sum_{1 \leq j \leq N} \sigma_{n+1}(\xi_n^i, \xi_n^j) \, W_{n+1}^i
\end{aligned}$$

In the above displayed formulae, $\eta_n^N = \frac{1}{N} \sum_{i=1}^N \delta_{\xi_n^i}$ is the occupation measure of the population at time n; and $(W_{n+1}^i)_{1 \leq i \leq N}$ stands for N independent copies of the random variables W_{n+1}. Illustrations of these Gaussian mean field models in the context of filtering problems are discussed in some detail in Section 8.2.3, dedicated to ensemble Kalman filters.

1.5.3 General state space models

The effective sampling condition (1.52) is also met for the general state space models presented in (1.36), as soon as we have a dedicated Monte Carlo technique to sample the random variables $F_n(x_{n-1}, m(y), W_n)$. In this case (1.52) is given by

$$K_{n,m(y)}(x_{n-1}, dx_n) = \mathbb{P}\left(F_n(x_{n-1}, m(y), W_n) \in dx_n\right)$$

In this situation, we find that the N-particle model defined in (1.53) can be rewritten in the following form

$$\xi_{n+1}^i = F_{n+1}(\xi_n^i, \eta_n^N, W_{n+1}^i) \quad \text{with} \quad \eta_n^N = \frac{1}{N} \sum_{i=1}^N \delta_{\xi_n^i} \qquad (1.56)$$

where $(W_{n+1}^i)_{1 \leq i \leq N}$ stands for N independent copies of the r.v. W_{n+1}.

1.5.4 Interacting jump-diffusion models

We return to the Euler type approximation models discussed in Section 1.3.1. In this context, if we choose $\bar{\eta}_n = \eta_n$ in (1.28), then the mean field IPS model $\xi_n = (\xi_n^i)_{1 \leq i \leq N}$ is defined by sampling N random transitions

$$\xi_n^i \rightsquigarrow \xi_{n+1}^i \sim \left(M_{n+1,\eta_n^N} J_{n+1,\eta_n^N}\right)(\xi_n^i, .) \quad \text{with} \quad \eta_n^N = \frac{1}{N} \sum_{i=1}^N \delta_{\xi_n^i}$$

If we choose $\bar{\eta}_n := \eta_n M_{n+1,\eta_n}$ in (1.28), then the mean field IPS model $\xi_n = (\xi_n^i)_{1 \leq i \leq N}$ associated with the distribution flow (1.30) is now defined in terms of the two step transitions

$$\xi_n^i \rightsquigarrow \xi_{n+1/2}^i \sim M_{n+1,\eta_n^N}(\xi_n^i, .) \rightsquigarrow \xi_{n+1}^i \sim J_{n+1,\eta_{n+1/2}^N}(\xi_{n+1/2}^i, .)$$

with the occupation measures at the intermediate time steps defined by

$$\eta_n^N = \frac{1}{N} \sum_{i=1}^N \delta_{\xi_n^i} \quad \text{and} \quad \eta_{n+1/2}^N = \frac{1}{N} \sum_{i=1}^N \delta_{\xi_{n+1/2}^i}$$

1.5.5 Feynman-Kac particle models

Condition (1.52) is also met for the Feynman-Kac model discussed in (1.43), as soon as the potential functions $G_n(x)$ can be evaluated at any state x, and as soon as we can draw independent random samples of the elementary transitions M_n. In this situation, the McKean transitions (1.52) are given by the following formulae

$$K_{n,m(y)}(x_{n-1}, dx_n)$$

$$= \epsilon G_{n-1}(x_{n-1}) \, M_n(x_{n-1}, dx_n) + (1 - \epsilon G_{n-1}(x_{n-1})) \, \Psi_{G_{n-1}}(m(y)) \, M_n(dx_n) \tag{1.57}$$

with the weighted empirical measure

$$\Psi_{G_{n-1}}(m(y)) \, M_n := \sum_{1 \leq i \leq N} \frac{G_{n-1}(y^i)}{\sum_{1 \leq j \leq N} G_{n-1}(y^j)} \, M_n(y^i, .)$$

In the above display, $\Psi_{G_{n-1}}$ stands for the Boltzmann-Gibbs transformations defined in (0.2).

In this situation, we use formula (1.57) to check that the N-particle model defined in (1.53) can be interpreted in terms of an interacting jump type genetic type particle model with a two step selection-mutation transition.

During the selection stage, each particle ξ_{n-1}^i evaluates its potential value $G_{n-1}(\xi_{n-1}^i)$. With a probability $\epsilon G_{n-1}(\xi_{n-1}^i)$ it remains in the same location.

Otherwise, it jumps to a fresh new location ξ_{n-1}^j randomly chosen with a probability proportional to $G_{n-1}(\xi_{n-1}^j)$. During the mutation stage, each particle evolves randomly according to the Markov transition M_n.

From the statistical, or from the stochastic, point of view, these interacting particle systems can be interpreted as a sophisticated acceptance-rejection sampling technique, equipped with an interacting recycling mechanism. This mean field stochastic algorithm can also be interpreted as a population of individuals mimicking natural evolution mechanisms. During a mutation stage, the particles evolve independently of one another, according to the same probability transitions M_n. During the selection stage, particles with small relative values are killed, while the ones with high relative values are multiplied.

For a more thorough discussion on the interpretations of Feynman-Kac mean field models we refer the reader to Chapter 4. A more detailed presentation of these genetic type particle models is provided in Section 4.1.1, including pseudocode block diagrams. We also refer the reader to Section 4.4.5 for a more detailed presentation of the mean field particle model described above.

Chapter 2

Theory and applications

2.1 A stochastic perturbation analysis

In this chapter, we give a brief exposition of the modern mathematical theory that is useful for the analysis of the asymptotic behavior of mean field particle models. We briefly discuss an original stochastic perturbation analysis of mean field particle models in terms of local sampling error propagations and backward semigroup techniques. We also present some exponential concentration inequalities that apply to general McKean particle models.

In order that our presentation is both broad and has some coherence, we have chosen to concentrate on three main results: nonasymptotic $\mathbb{L}_m$-mean error bounds, exponential concentration inequalities, and central limit theorems for particle occupation measures. Readers interested in propagation of chaos estimates for finite particle blocks, large deviation principles, and Berry-Esseen type estimates are recommended to consult the companion monograph [163], as well as the series of articles [178, 183, 188, 189].

This choice is dictated, at least in part, by the fact that we want to make the propaganda of a rather general stochastic perturbation technique developed in a joint work of the author with L. Miclo [161], A. Guionnet [169, 170], J. Jacod and Ph. Protter [154, 155, 156] for analyzing the performance of IPS Monte Carlo algorithms. The central idea is to use backward semigroup expansions to express any global error quantity in terms of the local sampling errors induced by the mean field simulation.

2.1.1 A backward semigroup expansion

In this section, we present a brief introduction to a stochastic perturbation analysis of an abstract class of nonlinear semigroups in distribution spaces. A more detailed description is provided in Section 14.1.

The stochastic perturbation model associated with discrete mean field models presented in Section 1.5 is defined in terms of a sequence of centered random fields V_n^N defined by the following equation

$$\eta_n^N = \Phi_n(\eta_{n-1}^N) + \frac{1}{\sqrt{N}}\, V_n^N \tag{2.1}$$

In the above display, η_n^N stands for the particle density profiles associated with a mean field IPS model (1.53). We recall that this particle model depends on the McKean interpretation model (1.16) of a flow of measures defined by the evolution Equation (1.15). It is essential to observe that the perturbation Equation (2.1) only differs from the limiting evolution (1.15) by the Monte Carlo precision parameter $1/\sqrt{N}$ and the centered random fields V_n^N. In addition, under rather weak regularity conditions, we can prove that the sequence of random fields V_n^N converge in law, as $N \uparrow \infty$, to a sequence of independent centered Gaussian fields V_n, with a covariance functional given for any functions $(f, g) \in \mathcal{B}_b(E_n)$ by the formula

$$\mathbb{E}(V_n(f)V_n(g)) = \eta_{n-1}\left(K_{n,\eta_{n-1}}\left[f - K_{n,\eta_{n-1}}(f)\right]\left[g - K_{n,\eta_{n-1}}(g)\right]\right)$$

A proof of this theorem for general IPS models is provided in Section 10.6. For a further discussion on these local sampling random field models, we refer to Section 4.4.4.

Further on in this section, we denote by $\Phi_{p,n}$, with $p \leq n$, the semigroup of the limiting evolution equation $\eta_n = \Phi_{p,n}(\eta_p)$ defined in (1.15). To analyze the propagation properties of the local sampling errors V_n^N, *up to a second order remainder measure*, we assume that the one step mappings Φ_n governing the Equation (1.15) have a first order decomposition

$$\Phi_n(\eta) - \Phi_n(\mu) \simeq (\eta - \mu)d_\mu\Phi_n \qquad (2.2)$$

for some first order integral operator $d_\mu\Phi_n$ from $\mathcal{B}(E_n)$ into $\mathcal{B}(E_{n-1})$, s.t. $d_\mu\Phi_n(1) = 0$.

More formally, we assume that the mappings $\Phi_n : \mathcal{P}(E_{n-1}) \to \mathcal{P}(E_n)$ are Gâteaux differentiable at any $\mu \in \mathcal{P}(E_{n-1})$ in any direction $(\eta - \mu) \in \mathcal{M}_0(E_{n-1})$, for some $\eta \in \mathcal{P}(E_{n-1})$; in the sense that

$$\lim_{\epsilon \downarrow 0}\left\|\frac{1}{\epsilon}\left[\Phi_n(\mu + \epsilon(\eta - \mu)) - \Phi_n(\mu)\right] - (\eta - \mu)d_\mu\Phi_n\right\|_{\text{tv}} = 0 \qquad (2.3)$$

More details on this first order regularity property (2.2) are provided in Section 10.2. By construction, using the semigroup property, we also have the first order decompositions

$$\Phi_{p,n}(\eta) - \Phi_{p,n}(\mu) \simeq (\eta - \mu)d_\mu\Phi_{p,n}$$

for some first order integral operator $d_\mu\Phi_{p,n}$ from $\mathcal{B}_b(E_n)$ into $\mathcal{B}_b(E_p)$, s.t. $d_\mu\Phi_{p,n}(1) = 0$. In other words, the mapping $\Phi_{p,n} : \mathcal{P}(E_p) \to \mathcal{P}(E_n)$ is Gâteaux differentiable at any $\mu \in \mathcal{P}(E_p)$ in any direction $(\eta - \mu) \in \mathcal{M}_0(E_p)$, with $\eta \in \mathcal{P}(E_p)$; in the sense that

$$\lim_{\epsilon \downarrow 0}\left\|\frac{1}{\epsilon}\left[\Phi_{p,n}(\mu + \epsilon(\eta - \mu)) - \Phi_{p,n}(\mu)\right] - (\eta - \mu)d_\mu\Phi_{p,n}\right\|_{\text{tv}} = 0$$

The semigroup properties of the integral operators $d_\mu \Phi_{p,n}$ are deduced from the the chain rule

$$\forall 0 \leq p \leq q \leq n \quad \Phi_{p,n} = \Phi_{q,n} \circ \Phi_{p,q} \implies d_\mu \Phi_{p,n} = d_\mu \Phi_{p,q} \; d_{\Phi_{p,q}(\mu)} \Phi_{q,n}$$

We further assume that the one step mappings Φ_n, and their first order derivatives $\mu \in \mathcal{P}(E_{n-1}) \mapsto d_\mu \Phi_n \in \mathcal{L}(\mathcal{B}_b(E_n), \mathcal{B}_b(E_{n-1}))$ are continuous, where $\mathcal{L}(\mathcal{B}_b(E_n), \mathcal{B}_b(E_{n-1}))$ stands for the set of linear mappings from $\mathcal{B}_b(E_n)$ into $\mathcal{B}_b(E_{n-1})$.

In this situation, $\Phi_{p,n}$, and their first order derivatives $\mu \in \mathcal{P}(E_p) \mapsto d_\mu \Phi_n \in \mathcal{L}(\mathcal{B}_b(E_n), \mathcal{B}_b(E_p))$ are continuous, and we have

$$\frac{1}{\epsilon}\left[\Phi_{p,n}\left(\mu + \epsilon(\eta - \mu) \right) - \Phi_{p,n}(\mu) \right] = \int_0^1 (\eta - \mu) d_{\mu + \epsilon t(\eta - \mu)} \Phi_{p,n} \; dt$$

where the r.h.s. integral is the Gelfand-Pettis weak sense integral [78].

The stability properties of the semigroup $\Phi_{p,n}$ are encapsulated into the following contraction parameter

$$\beta_{p,n} := \sup_{\eta \in \mathcal{P}(E_p)} \beta(d_\eta \Phi_{p,n}) \tag{2.4}$$

In the above display, $\beta(Q)$ stands for the Dobrushin contraction coefficient of the bounded integral operator Q defined in (0.1). Using the pivotal backward semigroup decomposition formula

$$\begin{aligned}
\eta_n^N - \eta_n &= \sum_{q=0}^n \left[\Phi_{q,n}(\eta_q^N) - \Phi_{q,n}(\Phi_q(\eta_{q-1}^N)) \right] \tag{2.5} \\
&= \sum_{q=0}^n \left[\Phi_{q,n}\left(\Phi_q(\eta_{q-1}^N) + \frac{1}{\sqrt{N}} V_q^N \right) - \Phi_{q,n}\left(\Phi_q(\eta_{q-1}^N) \right) \right]
\end{aligned}$$

with the convention $\Phi_0(\eta_{-1}^N) = \eta_0$ for $p = 0$, we find that

$$\begin{aligned}
W_n^{\eta,N} &:= \sqrt{N} \left[\eta_n^N - \eta_n \right] \\
&= \sum_{q=0}^n \int_0^1 V_q^N d_{\Phi_q(\eta_{q-1}^N) + \frac{t}{\sqrt{N}} V_q^N} \Phi_{p,n} \; dt \simeq \sum_{q=0}^n V_q^N d_{\Phi_q(\eta_{q-1}^N)} \Phi_{p,n}
\end{aligned}$$

Of course, the r.h.s. approximation requires some additional regularity conditions on the mapping $\mu \mapsto d_\mu \Phi_{p,n}$. We refer to Chapter 10 for a more detailed discussion on these Taylor's type expansions. Using this first order decomposition we readily prove the following convergence in law

$$W_n^{\eta,N} \simeq \sum_{q=0}^n V_p^N d_{p,n} \to_{N \uparrow \infty} W_n^\eta = \sum_{p=0}^n V_p d_{p,n}.$$

In the above displayed formula $d_{p,n} = d_{\Phi_p(\eta_{p-1})}\Phi_{p,n} = d_{\eta_p}\Phi_{p,n}$. This implies that

$$\mathbb{E}(W_n^\eta(f_n)^2) = \sum_{p=0}^n \mathbb{E}\left((V_p\,[d_{p,n}(f_n)])^2\right) \le \sum_{p=0}^n \sigma_p^2\,\beta_{p,n}^2 := \overline{\sigma}_n^2 \qquad (2.6)$$

with the uniform local variance parameters:

$$\sigma_n^2 := \sup_{f_n \in \mathrm{Osc}(E_n)} \sup_{\mu \in \mathcal{P}(E_{n-1})} \left|\mu\left(K_{n,\mu}\,[f_n - K_{n,\mu}(f_n)]^2\right)\right| \qquad (2.7)$$

From this discussion, we find that

$$\sup_{n \ge 0} \sum_{p=0}^n \sigma_p^2\,\beta_{p,n}^2 < \infty \implies \sup_{n \ge 0} \mathbb{E}(W_n^\eta(f_n)^2) < \infty$$

2.1.2   Exponential concentration inequalities

Using the backward semigroup expansion presented 2.1.1, we prove the following concentration inequalities. For any $N \ge 1$, $n \ge 0$, $f_n \in \mathrm{Osc}(E_n)$, and any $x \ge 0$ the probability of the following event is greater than $1 - e^{-x}$

$$\left|[\eta_n^N - \eta_n](f_n)\right| \le \frac{r_n}{N}\left(1 + (x + \sqrt{x})\right) + \overline{\sigma}_n\left(\sup_{0 \le p \le n} \beta_{p,n}\right)\sqrt{\frac{2(x + \log 2)}{N}}$$

with some finite constant r_n that depends on the bias of the particle approximation model, and with the variance parameters $\overline{\sigma}_n$ introduced in the r.h.s. of (2.6). A detailed proof of these estimates is provided in Section 14.6.

The analysis of these types of concentration inequalities is one of the central themes of this book. As well as their mathematical elegance, these results have great pragmatic value. The variance parameter σ_n^2 defined in (2.7) can be computed explicitly, or at least estimated easily, for different choices of McKean transition models. In the same vein, the first order expansions (2.2) can also be computed explicitly, or at least estimated easily, using a first order Taylor expansion of the one step mappings Φ_n associated with the limiting equation. In addition, the integral operators $d_{p,n} = d_{\eta_p}\Phi_{p,n}$ coincide with the semigroup associated with the operator $d_n = d_{\eta_{n-1}}\Phi_n$.

On the other hand, these inequalities allow entering the stability properties of the limiting nonlinear semigroup into the robustness and the convergence properties of the mean field IPS models, giving a powerful weapon to derive uniform concentration inequalities w.r.t. the time horizon.

The Dobrushin contraction coefficient $\beta_{p,n}$ defined in (2.4) is directly related to the contraction properties of the nonlinear semigroup $\Phi_{p,n}$. The estimation of these coefficients depends on the problem under study. Some illustrations related to Feynman-Kac semigroups are provided in Section 12.2.

The stochastic perturbation analysis discussed above is developed in some detail in the article [166], as well as in part IV of the present book dedicated to the stochastic analysis of general IPS models. Chapter 14 is mainly concerned with the stochastic analysis of the particle density profiles associated with Feynman-Kac models. Chapter 10 is dedicated to the analysis of a general class of mean field particle models, including McKean-Vlasov type models and Feynman-Kac particle models. First order decomposition theorems for general mean field models are also developed in Section 14.6.

We emphasize that by using some key decompositions, the stochastic perturbation technique developed in this book applies without further work to analyze the convergence of any approximating scheme, as soon as we have some quantitative information on the fluctuations of the corresponding local errors.

For the general class of $\mathbb{R}^d$-Markov-McKean models introduced in (1.36), under some regularity conditions on the drift functions, the convergence of the mean field IPS models (1.56) can alternatively be analyzed using more conventional coupling techniques, combined with the well known Gronwald's Lemma. For instance, extending the arguments presented in Section 3.3.2 in [152], for any $m \geq 1$ and $n \geq 0$ we readily prove the $\mathbb{L}_m$-mean error estimates

$$\mathbb{E}\left(\left\|\mathcal{E}_n^{(N,1)} - X_n\right\|^m\right)^{1/m} \leq b(m)\ c(n)/\sqrt{N} \tag{2.8}$$

with a Markov chain X_n defined as in (1.36) with $E_n = \mathbb{R}^d$, for some $d \geq 1$, $W_n = W_n^1$ and starting at the same location $X_0 = \xi_0^1$. In the above display, the parameters $b(m)$ are given in (0.6), and $c(n) < \infty$ stands for some finite constant. A detailed proof of these estimates is housed in Section 10.5. The analysis of the convergence properties of the ensemble Kalman filters presented in Section 8.2.3 can also be developed extending these coupling techniques [388]. For some particular classes of mean field models, we can obtain uniform variance estimates w.r.t. the time parameter, combining Gronwald's Lemma with some regularity of the system [419]. In terms of Wasserstein metric, the estimate (2.8) implies that

$$W_m\left(\mathbb{P}_n^{(N,1)}, \eta_n\right) := \inf \mathbb{E}\left(\left\|X_n^{(N)} - X_n\right\|^m\right)^{1/m} \leq b(m)\ c(n)/\sqrt{N}$$

In the above display, the infimum is taken over all joint distributions of the r.v. $X_n^{(N)}$ and X_n, with marginals $\mathbb{P}_n^{(N,1)} = \mathrm{Law}\left(\mathcal{E}_n^{(N,1)}\right)$, and $\eta_n = \mathrm{Law}(X_n)$.

To get one step further, we let $\overline{\eta}_n^N = \frac{1}{N}\sum_{i=1}^N \delta_{X_n^i}$ be the occupation measure associated with N *independent* copies $(X_n^i)_{1 \leq i \leq N}$ of the nonlinear Markov chain X_n defined as in (1.36), associated with the random perturbations W_n^i and starting at the locations $X_0^i = \xi_0^i$.

For any Lipschitz function f on $\mathbb{R}^d$ with Lipschitz constant $l(f) \leq 1$, and oscillations $\mathrm{osc}(f) \leq 1$, we have the estimates

$$
\begin{aligned}
\left| \eta_n^N(f) - \eta_n(f) \right| &= \left| \eta_n^N(f) - \overline{\eta}_n^N(f) \right| + \left| \overline{\eta}_n^N(f) - \eta_n(f) \right| \\
&\leq l(f) \frac{1}{N} \sum_{i=1}^{N} \left\| \xi_n^i - X_n^i \right\| + \left| \overline{\eta}_n^N(f) - \eta_n(f) \right|
\end{aligned}
$$

Using the $\mathbb{L}_m$-mean error bound (2.8), we readily check that

$$
\sqrt{N} \, \mathbb{E}\left(\left| \eta_n^N(f) - \eta_n(f) \right|^m \right)^{1/m} \leq 2b(m)c(n)
$$

for some finite constant $c(n) < \infty$. Later on in this book, we shall see that these $\mathbb{L}_m$-mean error type estimates directly imply, without further work, exponential concentration inequalities. More precisely, the comparison Lemma (cf. (11.18) in Lemma 11.3.1), the maximal Olicz's norm inequality stated in Lemma 11.3.6, combined with the definition (0.6) of the parameters $b(m)$, formulae (11.5), and (11.21) yield the following exponential concentration inequality.

For any $x \geq 0$, and any $N \geq 1$, the probability any of the following events is greater than $1 - e^{-x}$

$$
\begin{aligned}
\sqrt{N} \, \left| \eta_n^N(f) - \eta_n(f) \right| &\leq 4\sqrt{2/3} \, c(n) \, \sqrt{x + \log 2} \\
\sqrt{N} \sup_{0 \leq p \leq n} \left| \eta_p^N(f) - \eta_p(f) \right| &\leq 4\sqrt{2/3} \, c(n) \, \sqrt{6(x + \log 2) \log(9 + n)}
\end{aligned}
$$

2.2 Feynman-Kac particle models

This section is dedicated to mean field Feynman-Kac models. We provide a brief overview of some nonasymptotic results developed in this book, including concentration inequalities for particle density profiles, genealogical tree occupation measures, particle free energy models, and particle backward Markov chain models. These different classes of mean field models are particle interpretations of the path space models and the unnormalized disributions discussed in Section 1.4.3. In the next sections, we exploit the Feynman-Kac mathematical structures presented in Section 1.4.3 to design several types of particle approximation models.

In subsequent pages of this section, to avoid unnecessary repetition of technical abstract conditions on (G_n, M_n), we frame the standing assumption that the potential functions G_n are chosen so that $\sup_{x,y}(G_n(x) \, G_n(y)) = g < \infty$, for some finite constant g that does not depend on the time parameter n. We refer the reader to page 25 for a discussion on this condition and some strategies to relax this rather strong condition. We also assume that the stability property (1.44) is satisfied, and we have $\sigma = \sup_n \sigma_n < \infty$, with the uniform variance parameters σ_n associated with some McKean interpretation model and defined in (2.7).

We also denote by $b(m)$ the parameters defined in (0.6), and (c, c_1, c_2) stand for some finite constants, whose values may vary from line to line, but they do not depend on the time horizon n. In addition, the constants $c_1 > c_2$ depend on the bias and, respectively, on the variance of the particle approximation model under study.

2.2.1 Particle density profiles

We recall that occupation measures $\eta_n^N = \frac{1}{N} \sum_{i=1}^N \delta_{\xi_n^i}$ of the N-particle model converge, as $N \uparrow \infty$, to the flow of Feynman-Kac measures η_n introduced on page 26. More precisely, we have the uniform bias and variance estimates

$$\sup_{n \geq 0} \left\{ \left| \mathbb{E}\left(\eta_n^N(f)\right) - \eta_n(f) \right| \vee \mathbb{E}\left(\left[\eta_n^N(f) - \eta_n(f)\right]^2 \right) \right\} \leq \sigma^2 \, c/N$$

for some finite constant c whose values do not depend on the time parameter n. The proofs of these estimates are provided in Section 14.3.2.

In addition, for any $N \geq 1$, $n \geq 0$, $f_n \in \mathrm{Osc}(E_n)$, and any $x \geq 0$ the probability of each of the following event is greater than $1 - e^{-x}$

$$\left| [\eta_n^N - \eta_n](f_n) \right| \leq \frac{c_1}{N} \left(1 + (x + \sqrt{x}) \right) + c_2 \, \sigma \, \sqrt{\frac{x}{N}}$$

$$\sup_{0 \leq p \leq n} \left| [\eta_p^N - \eta_p](f_p) \right| \leq c \, \sqrt{x \log{(n + e)}/N}$$

For time homogeneous Feynman-Kac models $(E_n, G_n, M_n) = (E, G, M)$, the regularity condition (1.44) ensures that the one step Feynman-Kac mapping $\eta \mapsto \Phi(\eta) = \Psi_G(\eta)M$ has a unique fixed point measure

$$\eta_\infty = \Phi(\eta_\infty) \tag{2.9}$$

In this context, we readily prove the following uniform concentration inequality. For any $N \geq 1$, $n \geq 0$, $f \in \mathrm{Osc}(E)$, and any $x \geq 0$ the probability of the following event is greater than $1 - e^{-x}$

$$\left| [\eta_n^N - \eta_\infty](f) \right| \leq a \, e^{-bn} + \frac{c_1}{N} \left(1 + (x + \sqrt{x}) \right) + c_2 \, \sigma \, \sqrt{\frac{x}{N}}$$

With the parameters (a,b) defined in (1.33).

A detailed proof of these estimates is provided in Section 14.5.1. The extension of these concentration inequalities at the level of the empirical processes associated with some collection of functions is developed in Section 14.5.2. For a more thorough discussion of the quasi-invariant measure η_∞, and related Doob's h-processes, we also refer the reader to Section 7.6.

2.2.2   Concentration of interacting processes

In this section we illustrate the interacting empirical process analysis developed in Section 14.5.2, with a concentration inequality for particle repartition functions. We also discuss some direct consequences related to particle quantile estimates. For any $x = (x_i)_{1 \leq i \leq d}$, and any cell $(-\infty, x] = \prod_{i=1}^d (-\infty, x_i] \subset E_n = \mathbb{R}^d$, we let

$$F_n(x) = \eta_n \left(1_{(-\infty, x]} \right) \quad \text{and} \quad F_n^N(x) = \eta_n^N \left(1_{(-\infty, x]} \right)$$

In Corollary 14.5.7, we shall prove that for any $x \geq 0$, $n \geq 0$, and any population size $N \geq 1$, the probability of the following event

$$\sqrt{N} \, \left\| F_n^N - F_n \right\| \leq c \, \sqrt{d \, (x + 1)}$$

is greater than $1 - e^{-x}$. This concentration inequality ensures that the sequence of particle repartition function F_n^N converges, as $N \uparrow \infty$, to F_n, almost surely for the uniform norm. For $d = 1$, we let F_n^- be the generalized inverse on $[0, 1]$ of the function F_n; that is, we have that

$$F_n^-(u) := \inf \{ x \in \mathbb{R} \; : \; F_n(x) \geq u \}$$

We let $F_n^-(u) = q_{n,u}$ be the quantile of some order $u \in [0, 1]$, and we denote by ζ_n^i the order particle statistic associated with the particle system ξ_n^i at time n; that is, we have that

$$\zeta_n^1 := \xi_n^{\tau(1)} \leq \zeta_n^2 := \xi_n^{\tau(2)} \leq \ldots \leq \zeta_n^N := \xi_n^{\tau(N)}$$

for some random permutation τ of the set $\{1, \ldots, N\}$. We also denote by $q_{n,u}^N := \zeta_n^{1 + \lfloor Nu \rfloor}$ the u-particle quantile. By construction, we have that

$$
\begin{aligned}
\left| F_n \left(q_{n,u}^N \right) - F_n(q_{n,u}) \right| &\leq \left| F_n \left(q_{n,u}^N \right) - F_n^N(q_{n,u}^N) \right| + \left| F_n^N(q_{n,u}^N) - u \right| \\
&\leq \left\| F_n^N - F_n \right\| + ((1 + \lfloor Nu \rfloor)/N - u) \\
&\leq \left\| F_n^N - F_n \right\| + 1/N
\end{aligned}
$$

This clearly implies that $q_{n,u}^N$ converge almost surely to $q_{n,u}$, as N tends to ∞. In addition, for any $x \geq 0$, $n \geq 0$, and any population size $N \geq 1$, the probability of the following event

$$\left| F_n \left(q_{n,u}^N \right) - u \right| \leq c \, \sqrt{d \, (x + 1)/N} + 1/N$$

is greater than $1 - e^{-x}$.

2.2.3   Genealogical tree particle measures

Using the structural stability property (1.45) of the n-th time marginal measures of Feynman-Kac models associated with a reference historical process discussed in Section 1.4.3.1, we define the genealogical tree evolution model of a genetic type Feynman-Kac model. Equivalence principles between Feynman-Kac n-th time marginal measures and path space models are discussed in more details in Section 3.4. In this context, using the estimates described in Section 2.2.1, we can show, without further work, that the N-occupation measures of the genealogical tree model converge, as $N \uparrow \infty$, to the complete Feynman-Kac distribution on path space defined in (1.37).

More formally, the ancestral line of the i-th individual in the mean field Feynman-Kac population at time n is denoted by

$$\boldsymbol{\xi}_n^i := \left(\xi_{0,n}^i, \xi_{1,n}^i, \ldots, \xi_{n,n}^i\right) \in \boldsymbol{E}_n := (E_0 \times \ldots \times E_n)$$

and the occupation measure of the genealogical tree is given by

$$\eta_n^N := \frac{1}{N} \sum_{i=1}^{N} \delta_{\boldsymbol{\xi}_n^i} \longrightarrow_{N \uparrow \infty} \mathbb{Q}_n \tag{2.10}$$

Further details on the convergence of these measures can be found in Section 4.1.2 and Section 15.1.1. For instance, in Section 15.2.2, we prove that for any $m \geq 2$, $N \geq 1$, $n \geq 0$, and any function $\boldsymbol{f}_n \in \mathrm{Osc}(\boldsymbol{E}_n)$, we have

$$\mathbb{E}\left(\left|[\eta_n^N - \mathbb{Q}_n](\boldsymbol{f}_n)\right|^m\right)^{1/m}$$

$$\leq 2b(2m)^2 \, \frac{(n+1)}{N} \, c_1 + 2(m-1) \, \sqrt{\frac{(n+1)}{N}} \, c_2$$

In Section 15.5, we also derive the following concentration inequalities. For any $n \geq 0$, any $\boldsymbol{f}_n \in \mathrm{Osc}(\boldsymbol{E}_n)$, and any $N \geq 1$, the probability of the event

$$\left|[\eta_n^N - \mathbb{Q}_n]\,(\boldsymbol{f}_n)\right| \leq c_1 \, \frac{n+1}{N} \, (1 + (x + \sqrt{x})) + c_2 \, \sqrt{\frac{n+1}{N}} \, \sqrt{x}$$

is greater than $1 - e^{-x}$, for any $x \geq 0$.

2.2.4 Weak propagation of chaos expansions

In this short section, we discuss the bias properties of mean field IPS models. As the reader has certainly noticed, mean field IPS algorithms are not exact sampling techniques. They induce biased estimates, as soon as we replace in (1.53) the limiting measures by their particle approximations.

Nevertheless, the magnitude of the bias can be reduced drastically, at any desired order. More precisely, in a couple of articles [188, 189], we have derived Taylor's type expansions for the law of q-blocks of path particles

$$\mathbb{P}_n^{(N,q)} := \mathrm{Law}(\boldsymbol{\xi}_n^1, \ldots, \boldsymbol{\xi}_n^q)$$

with some finite block size $q \leq N$. More precisely, for any $m \geq 1$ we have that

$$\mathbb{P}_n^{(N,q)} = \mathbb{Q}_n^{\otimes q} + \sum_{1 \leq l \leq m} \frac{1}{N^l} \, d_l \mathbb{P}_n^{(q)} + \mathrm{O}\left(\frac{1}{N^{m+1}}\right)$$

for some signed distributions $d_l \mathbb{P}_n^{(q)} \in \mathcal{M}(E_n^q)$, with $l \geq 1$, expressed in terms of coalescent trees. Using the traditional Romberg-Richardson interpolation, we design a mean field IPS model with the desired order of approximation. More formally, for any $l \geq 1$, we have

$$\sum_{1 \leq m \leq l} \frac{(-1)^{l-m}}{m!} \frac{m^l}{(l-m)!} \mathbb{P}_n^{(mN,q)} = \mathbb{Q}_n^{\otimes q} + \mathrm{O}(1/N^l)$$

In particular, for $q = 1$, and for any $l \geq 1$ and $f_n \in \mathcal{B}_b(E_n)$ we have

$$\mathbb{E}\left(\eta_n^{(l,N)}(f_n)\right) = \mathbb{Q}_n(f_n) + \mathrm{O}(N^{-l})$$

with the particle measures

$$\eta_n^{(l,N)} := \sum_{1 \leq m \leq l} \frac{(-1)^{l-m}}{m!} \frac{m^l}{(l-m)!} \eta_n^{mN}$$

For more thorough discussions on the propagation properties of mean field type particle models, we refer the reader to the series of articles [177, 178, 295, 429, 430, 538], as well as to Chapter 8 in the monograph [163].

To the best of our knowledge, Taylor's type expansions for McKean-Vlasov diffusion models and generalized Boltzmann type models are still open research subjects.

2.2.5 Particle free energy models

Mimicking the product formula (1.47), the normalizing constants and the unnormalized distributions (1.48) in the Feynman-Kac model (1.37) can be computed using the following estimates

$$\mathcal{Z}_n^N := \prod_{0 \leq p < n} \eta_p^N(G_p) \longrightarrow_{N \to \infty} \mathcal{Z}_n = \prod_{0 \leq p < n} \eta_p(G_p) \qquad (2.11)$$

as well as

$$\gamma_n^N(f_n) := \eta_n^N(f_n) \times \prod_{0 \leq p < n} \eta_p^N(G_p)$$

For path space models, and genealogical tree based measures η_n^N on E_n, for any $f_n \in \mathcal{B}_b(E_n)$, we also set

$$\gamma_n^N(f_n) := \eta_n^N(f_n) \times \left\{ \prod_{0 \leq p < n} \eta_p^N(G_p) \right\}$$

In this context, we have the following important unbiasedness property

$$\mathbb{E}\left(\gamma_n^N(f_n)\right) = \Gamma_n(f_n) \quad \left(\overset{f_n=1}{\Longrightarrow} \mathbb{E}\left(\mathcal{Z}_n^N\right) = \mathcal{Z}_n \right) \qquad (2.12)$$

with the unnormalized Feynman-Kac measure Γ_n defined in (1.38).

A more detailed discussion on these unnormalized particle measures is provided in Section 4.1.4. Further details on the stochastic analysis of these unnormalized particle models are provided in Chapter 16, including $\mathbb{L}_m$-mean error estimates, fluctuation theorems, and exponential concentration inequalities. For instance, we prove in Section 16.5 the following $\mathbb{L}_m$-mean error estimates

$$\mathbb{E}\left(\left[\frac{\mathcal{Z}_n^N}{\mathcal{Z}_n} - 1\right]^m\right)^{1/m} \leq a_1(m) \left(\frac{n}{N}\right)^{1/2} \left(1 + \frac{1}{n} \left(\frac{n}{N}\right)^{m/2} a_2(m)\right)^{n/m}$$

for some finite constants $a_1(m)$ and $a_2(m)$ whose values only depend on the parameter $m \geq 1$. In Section 16.6, we also derive a series of exponential concentration inequalities. For instance, for any $N \geq 1$, and any $n \geq 0$, the probability of any of the following events is greater than $1 - e^{-x}$, for any $x \geq 0$

$$\left|\frac{1}{n}\log \mathcal{Z}_n^N - \frac{1}{n}\log \mathcal{Z}_n\right| \leq \frac{c_1}{N}\left(1 + (x + \sqrt{x})\right) + c_2\, \sigma\, \sqrt{\frac{x}{N}}$$

For time homogeneous models $(E_n, G_n, M_n) = (E, G, M)$ with a reversible Markov transition M w.r.t. to some reference probability measure μ the quantity $\frac{1}{n}\log \mathcal{Z}_n$ is closely related to the log of the largest eigenvalue λ of the operator $Q(x, dy) = G(x)M(x, dy)$ on $\mathbb{L}_2(\mu)$. This eigenvalue coincides with the average potential value w.r.t. the fixed point measure η_∞ discussed in (2.9); that is, we have that

$$\lambda = \eta_\infty(G)$$

We refer to Section 7.6 for a more thorough discussion on these models.

Under some regularity conditions, we also prove that for any $N \geq 1$, and any $n \geq 0$, the probability of any of the following events

$$\left|\frac{1}{n}\log \mathcal{Z}_n^N - \frac{1}{n}\log \lambda\right| \leq \frac{c}{n} + \frac{c_1}{N}\left(1 + (x + \sqrt{x})\right) + c_2\, \sigma\, \sqrt{\frac{x}{N}}$$

is greater than $1 - e^{-x}$, for any $x \geq 0$..

2.2.6 Backward particle Markov chain models

Mimicking the backward Markov chain formula (1.50), we design a N-particle approximation of the measures $(\mathbb{Q}_n, \Gamma_n)$ based on the complete ancestral tree

$$\mathbb{Q}_n^N = \eta_n^N \otimes \mathbb{Q}_{n|n}^N \tag{2.13}$$

with the Markov transitions from E_n into $\boldsymbol{E_{n-1}} = \prod_{0 \leq p < n} E_p$, given by

$$\mathbb{Q}_{n|n}^N(x_n, d(x_0, \ldots, x_{n-1})) := \prod_{q=1}^{n} \mathbb{M}_{q, \eta_{q-1}^N}(x_q, dx_{q-1})$$

and

$$\Gamma_n^N := \mathcal{Z}_n^N \times \mathbb{Q}_n^N$$

For any $\boldsymbol{f_n} \in \mathcal{B}_b(\boldsymbol{E_n})$, we quote the important unbiasedness property

$$\mathbb{E}\left(\Gamma_n^N(\boldsymbol{f_n})\right) = \Gamma_n(\boldsymbol{f_n}) \tag{2.14}$$

with the unnormalized Feynman-Kac measure Γ_n defined in (1.38).

We refer to Chapter 17 for the stochastic analysis of these backward particle models. To give a sample of the convergence estimates developed in this chapter we restrict our attention to functional integration of normalized additive functionals $\overline{\boldsymbol{f}}_{\boldsymbol{n}}$ associated with some collection of functions $f_n \in \mathrm{Osc}(E_n)$, and given by

$$\overline{\boldsymbol{f}}_{\boldsymbol{n}}(x_0, \ldots, x_n) = \frac{1}{n+1} \sum_{p=0}^{n} f_p(x_p)$$

In Section 17.6, we prove the following uniform unbiasedness property

$$\sup_{n \geq 0} \left| \mathbb{E}\left(\mathbb{Q}_n^N(\overline{\boldsymbol{f}}_{\boldsymbol{n}})\right) - \mathbb{Q}_n(\overline{\boldsymbol{f}}_{\boldsymbol{n}}) \right| \leq c/N$$

$$\mathbb{E}\left(\left|\left[\mathbb{Q}_n^N - \mathbb{Q}_n\right](\overline{\boldsymbol{f}}_{\boldsymbol{n}})\right|^m\right)^{1/m} \leq \left[\frac{(m-1)b(m)}{\sqrt{N(n+1)}} + \frac{b(2m)^2}{N}\right] c$$

for any $m \geq 2$. Working a little harder, in Section 17.9 we also prove that the probability of the events

$$\left|\left[\mathbb{Q}_n^N - \mathbb{Q}_n\right](\overline{\boldsymbol{f}}_{\boldsymbol{n}})\right|$$

$$\leq \frac{1}{N}\left(c_1\left(1 + (x + \sqrt{x})\right) + c_2\frac{x}{(n+1)}\right) + c_2\,\sigma\,\sqrt{\frac{x}{(n+1)N}}$$

is greater than $1 - e^{-x}$, for any $x \geq 0$, $n \geq 0$, and any $N \geq 1$.

In addition, under some regularity conditions (cf. Section 17.3, Corollary 17.3.7), for any $x \geq 0$, $N \geq 1$, $n \geq 0$, and any $\mathbf{f_n} \in \mathcal{B}_b(\mathbf{E_{n-1}})$ the probability of the event

$$\int \left|\mathbb{Q}_{n|n}^N(\mathbf{f_n})(x_n) - \mathbb{Q}_{n|n}(\mathbf{f_n})(x_n)\right| \lambda_n(dx_n) \leq c(n)\,\sqrt{x/N} \tag{2.15}$$

is greater than $1 - e^{-x}$, for some finite constant $c(n) < \infty$.

The extension of these concentration inequalities at the level of the empirical processes associated with some collection of functions is developed in Section 17.9.2.

2.3 Extended Feynman-Kac models

The Feynman-Kac models presented in (1.37) and (1.30) can be extended in various ways. In the further development of this section we discuss three important extensions.

The first one is concerned with vector d-dimensional Feynman-Kac measures. In Section 3.3.6, we shall use these noncommutative Feynman-Kac models to estimate the gradient of Markov semigroups in terms of Feynman-Kac particle models.

The second one is related to a class of quenched and annealed models arising in a variety of application domains, including particle absorption models, nonlinear filtering problems, and fixed parameter estimations problems arising in hidden Markov chain models. This series of application domains is discussed in Section 7.4 and Section 8.3.1, as well as in Section 8.4.

The third one is concerned with a simple model of intensity measure evolutions arising in spatial branching processes, equipped with spontaneous birth rates. We refer the reader to Chapter 6, Section 6.1, for a more detailed discussion on these models.

2.3.1 Noncommutative models

We consider a Markov chain X_n taking values in some measurable state spaces E_n. We equip the space $\mathbb{R}^d$ with some norm $\|.\|$, and we let $\mathbb{S}^{d-1} \subset \mathbb{R}^d$ be the unit sphere associated with this norm. We consider a collection of potential functions taking values in the space of matrices

$$G_n \ : \ x \in E_n \mapsto G_n(x) \in \mathbb{R}^{d \times d} \quad \text{such that} \quad \|G_n(x).u\| > 0$$

for any $u \in \mathbb{S}^{d-1}$, and any $x \in E_n$. We let $\mathcal{B}_b(E_n, \mathbb{R}^d)$ be Banach space of all bounded measurable function f_n from E_n into $\mathbb{R}^d$. We consider the d-dimensional vector Feynman-Kac measures γ_n defined for any $f_n \in \mathcal{B}_b(E_n, \mathbb{R}^d)$, and any $u_0 \in \mathbb{S}^{d-1}$ by

$$\gamma_n(f_n).u_0 := \mathbb{E}\left(f_n(X_n). \left[\prod_{0 \leq p < n} G_p(X_p) \right] .u_0 \right)$$

with the directed product of noncommutative matrices

$$\prod_{0 \leq p < n} G_p(X_p) := G_{n-1}(X_{n-1}) \dots G_1(X_1) G_0(X_0)$$

One natural way to turn these vector measure models into the Feynman-Kac models presented in (1.39) is to consider the random walk on the sphere $\mathbb{S}^{d-1}$ defined by the recursion

$$U_{n+1} := G_n(X_n).U_n / \|G_n(X_n).U_n\|$$

with the initial condition $U_0 = u_0$. In this situation, we have

$$\left\| \left[\prod_{0 \leq p < n} G_p(X_p) \right] . u_0 \right\| = \prod_{0 \leq p < n} G_p(X_p)$$

with the Markov chain $\boldsymbol{X_n}$, and the potential functions $\boldsymbol{G_n}$ defined by

$$\boldsymbol{X_n} := (X_n, U_n) \in \boldsymbol{E_n} := (E_n \times \mathbb{S}^{d-1}) \quad \text{and} \quad \boldsymbol{G_n(X_n)} := \|G_n(X_n).U_n\|$$

In this notation, we readily check that

$$\gamma_n(f_n).u_0 := \mathbb{E}\left(f_n(X_n) \prod_{0 \leq p < n} G_p(X_p) \right) := \boldsymbol{\gamma_n(f_n)}$$

with the function $\boldsymbol{f_n} \in \mathcal{B}_b(\boldsymbol{E_n})$ defined by $\boldsymbol{f_n(X_n)} := f_n(X_n).U_n$.

2.3.2   Quenched and annealed models

We consider Markov chain X_n taking values in some product state spaces

$$X_n := (X_n^1, X_n^2) \in E_n := (E_n^1 \times E_n^2)$$

We let M_n be the elementary transitions of the chain X_n from E_{n-1} into E_n, and we consider a sequence of $[0,1]$-valued potential functions G_n on E_n, and we set $G_{n,X_n^1}(X_n^2) = G_n(X_n^1, X_n^2)$. We also set $\boldsymbol{E_n} := \prod_{0 \leq p \leq n} E_p$, and

$$\forall i \in \{1,2\} \quad \boldsymbol{X_n^i} := (X_0^i, \ldots, X_n^i) \in \boldsymbol{E_n^i} := \prod_{0 \leq p \leq n} E_p^i$$

We assume that $X^1 := (X_n^1)_{n \geq 0}$ is a Markov chain; and given some realization of this chain, X_n^2 is a Markov chain with elementary transitions M_{n,X_n^1} and initial distribution η_{0,X_0^1}. We let

$$(\mathbb{Q}_{n,X^1}, \Gamma_{n,X^1}, \gamma_{n,X^1}, \eta_{n,X^1}), \quad \text{and resp.} \quad (\mathbb{Q}_n, \Gamma_n, \gamma_n, \eta_n)$$

be the quenched, and resp. the annealed, Feynman-Kac models on $\boldsymbol{E_n^2}$, and resp. $\boldsymbol{E_n}$, associated with $(G_{n,X_n^1}, M_{n,X_n^1})$, and resp. (G_n, M_n).

Our next objective is to express the annealed models in terms of the quenched models. To this end, we consider the Markov chain

$$\mathcal{X}_n := (X_n^1, \eta_{n,X^1}) \in \mathcal{E}_n := (E_n^1 \times \mathcal{P}(E_n^2))$$

and the potential functions

$$\mathcal{G}_n(\mathcal{X}_n) = \eta_{n,X^1}(G_{n,X_n^1})$$

We also consider the historical process

$$\boldsymbol{X_n} = (X_0, \ldots, X_n) \in \boldsymbol{\mathcal{E}_n} = (\mathcal{E}_0 \times \ldots \times \mathcal{E}_n)$$

In this notation, using the multiplicative formulae (1.47), for any $i \in \{1,2\}$ we have the following Feymnan-Kac formulae

$$\mathbb{E}\left(f_n^i(X_n^i) \prod_{0 \le p < n} G_p(X_p)\right) = \mathbb{E}\left(\mathcal{F}_n^i(\boldsymbol{X_n}) \prod_{0 \le p < n} \mathcal{G}_p(\boldsymbol{X_p})\right) \qquad (2.16)$$

for any $f_n^i \in \mathcal{B}_b(E_n^i)$, with the functions $\mathcal{F}_n^i \in \mathcal{B}_b(\boldsymbol{\mathcal{E}_n})$ defined below

$$\mathcal{F}_n^1(\boldsymbol{X_n}) = f_n^1(X_n^1) \quad \text{and} \quad \mathcal{F}_n^2(\boldsymbol{X_n}) = \mathbb{Q}_{n,X^1}(f_n^2)$$

A more thorough discussion on these Feynman-Kac representations, including the proof of (2.16), is provided in Section 7.4 dedicated to particle motions in absorbing medium.

Using the unbiasedness properties (2.11) and (2.14) of the unnormalized distributions $\left(\gamma_{n,X^1}^N, \Gamma_{n,X^1}^N\right)$ associated with the mean field IPS approximation of the Feynman-Kac measures $(\gamma_{n,X^1}, \Gamma_{n,X^1})$, the above formulae are valid if we replace $(\boldsymbol{X_n}, \mathcal{G}_n)$ by

$$\boldsymbol{X_n^N} := \left(X_n^1, \eta_{n,X^1}^N\right) \in \left(E_n^1 \times \mathcal{P}(E_n^2)\right) \quad \text{and} \quad \mathcal{G}_n(\boldsymbol{X_n^N}) = \eta_{n,X^1}^N(G_{n,X_n^1})$$

and the measures $\mathbb{Q}_{n,X^1}$ by the backward particle model $\mathbb{Q}_{n,X^1}^N$ discussed in Section 2.2.6, or alternatively, by the genealogical tree based particle measures $\eta_{n,\boldsymbol{X^1}}^N$ presented in Section 2.2.3.

Formulae (2.16) allows computing the annealed Feynman-Kac models in terms of genetic type mean field IPS models with mutation transitions dictated by the reference processes $\boldsymbol{X_n}$, resp. $\boldsymbol{X_n^N}$, and a selection fitness function $\mathcal{G}_n(\boldsymbol{X_n})$, resp. $\mathcal{G}_n(\boldsymbol{X_n^N})$.

A more detailed discussion on these particle models is provided in Section 7.5. We also refer the reader to Section 2.6 and Section 12.6 in the monograph [163].

2.3.3 Branching intensity distributions

Another important generalization of the Feynman-Kac model discussed in (1.41) is given by evolution equations in the space of nonnegative measures of the following form

$$\gamma_{n+1} = \gamma_n Q_{n+1} + \mu_{n+1} \qquad (2.17)$$

In the above display, μ_n stands for a given sequence of nonnegative measures on E_n, and Q_{n+1} is the nonnegative integral operator defined in (1.41), in terms of a nonnegative potential function G_n on E_n, and some Markov transition M_{n+1}, from E_n into E_{n+1}. As usual, we also let η_n be the normalized distributions defined for any $f_n \in \mathcal{B}_b(E_n)$ by $\eta_n(f_n) = \gamma_n(f_n)/\gamma_n(1)$.

As mentioned in the Introduction, these models are often used to describe the evolution of the intensity measures of spatial branching models, equipped with spontaneous branchings. We refer the reader to Chapter 6 for the detailed description of these branching models and their applications in multiple objects filtering problems. The mean field particle interpretation of these evolution equations is briefly discussed in Section 2.3.3.3. For a more detailed presentation, we refer to Section 6.1.5.

The construction and the analysis of these IPS models are slightly more involved than the ones of conventional Feynman-Kac models, but they may be understood via the following evolution equations

$$
\left\{
\begin{array}{rcl}
\gamma_{n+1}(1) & = & \gamma_n(1)\, \eta_n(G_n) + \mu_{n+1}(1) \\[2mm]
\eta_{n+1} & = & \alpha_n\,(\gamma_n(1), \eta_n)\ \Psi_{G_n}(\eta_n) M_{n+1} + (1 - \alpha_n\,(\gamma_n(1), \eta_n))\ \overline{\mu}_{n+1}
\end{array}
\right.
\tag{2.18}
$$

with the collection of $[0,1]$-parameters $\alpha_n\,(m, \eta)$ defined below

$$
\alpha_n\,(m, \eta) = \frac{m\,\eta(G_n)}{m\,\eta(G_n) + \mu_{n+1}(1)}
$$

In the above displayed formula, $\Psi_{G_n}(\eta_n)$ stands for the Boltzmann-Gibbs measures defined in (0.2). A detailed proof of these evolution equations is provided in Section 6.1.3.

2.3.3.1　Continuous time models

Spatial branching models on some measurable state space E' can be interpreted as a Markov process evolving in the state space $E = \cup_{d \geq 0} E'^d$, where d stands for the number of individual in the population. For $d = 0$, we use the convention $E'^0 = \{c\}$, where c stands for a cemetery state.

We let $(V_t^{(1)}, V_t^{(2)})$ be a collection of bounded positive functions on E', μ_t a flow of positive measures on E', and L_t the infinitesimal generator of a continuous Markov process Y_t on E'. For any $x = (x^1, \ldots, x^d) \in E$ we set $m(x) = \frac{1}{d} \sum_{i=1}^d \delta_{x^i}$ and $\underline{m}(x) = \sum_{i=1}^d \delta_{x^i}$. For $d = 0$, we use the convention $m(x) = \delta_c = \underline{m}(x)$. We also assume that $\mu_t(1) > 0$, and we set $\overline{\mu}_t = \mu_t/\mu_t(1)$ the normalized probability measure on E'.

We consider a time inhomogeneous spatial branching model $(X_t := X_t^i)_{1 \leq i \leq N_t}$ taking values in E, with an infinitesimal generators $\mathcal{L}_t := \sum_{0 \leq i \leq 3} \mathcal{L}_t^{(i)}$. The first generator $\mathcal{L}_t^{(1)}$ represents the evolution of the branching process between the birth and death jumps. It is defined for any sufficiently smooth function F on E, and for any $x = (x^1, \ldots, x^d) \in E$ by

$$
\mathcal{L}_t^{(0)}(F)(x) = \sum_{i=1}^d L_t^{(i)}(F)(x^1, \ldots, x^i, \ldots, x^d)
$$

where $L_t^{(i)}$ stands for the operator L_t acting on the function $x^i \mapsto$

$F(x^1, \ldots, x^i, \ldots, x^d)$. The second one represent the death of the particles at rate $V_t^{(1)}$. It is defined by

$$\mathcal{L}_t^{(1)}(F)(x) = \sum_{i=1}^{d} V_t^{(1)}(x^i) \left[F(\check{\theta}_i(x)) - F(x) \right]$$

with the mappings $\check{\theta}_i(x) := (x^1, \ldots, x^{i-1}, x^{i+1}, \ldots, x^d)$. The third one represents the birth and the duplication of new particle according to its $V_t^{(2)}$-fitness value

$$\mathcal{L}_t^{(2)}(F)(x) = \sum_{i=1}^{d} \int [F(x, u) - F(x)] \ V_t^{(2)}(u) \ m(x)(du)$$

The last generator represents spontaneous births at rate $\mu_t(1)$ in the population, and it is given by the formula

$$\mathcal{L}_t^{(3)}(F)(x) = \mu_t(1) \int [F(x, u) - F(x)] \ \overline{\mu}_t(du)$$

If we consider the test functions $F(x) = \underline{m}(x)(f)$, then we find that

$$\mathcal{L}_t(F)(x) = \underline{m}(x)(L_t(f)) + \underline{m}(x)(V_t f) + \mu_t(f) \quad \text{with} \quad V_t = (V_t^{(2)} - V_t^{(1)})$$

This shows that the density measures γ_t on E' defined by the first moment of the spatial branching model

$$\gamma_t(f) := \mathbb{E}\left(\mathcal{X}_t(f)\right) \quad \text{with the occupation measures} \quad \mathcal{X}_t := \sum_{i=1}^{N_t} \delta_{X_t^i}$$

satisfy the evolution equation

$$\frac{d}{dt} \gamma_t(f) = \gamma_t(L_t^V(f)) + \mu_t(f) \quad \text{with the Schrödinger operator} \quad L_t^V = L_t + V_t$$

More general classes of branching infinitesimal generators satisfying the above property can be found in [181]. To get one step further, we consider the Feynman-Kac semigroup discussed in section 1.2.2.3, and defined for any f on E' and $s \le t$, by the following formula

$$Q_{s,t}(f)(x) = \mathbb{E}\left(f(Y_t) \exp\left(\int_s^t V_t(Y_r) dr\right) \mid Y_s = x\right)$$

Using (1.26), we find that

$$\frac{d}{ds} \gamma_s(Q_{s,t}(f)) = \mu_s(Q_{s,t}(f))$$

from which we conclude that

$$\gamma_t(f) = \gamma_s Q_{s,t}(f) + \int_s^t \mu_r(Q_{r,t}(f)) \, dr$$

A discrete time approximation of these models on a time mesh t_n is given by the formulae

$$\gamma_{t_n}(f) = \gamma_{t_{n-1}} Q_{t_{n-1},t_n}(f) + \mu_{t_{n-1}}(f)\,(t_n - t_{n-1}) + \mathrm{O}\left((t_n - t_{n-1})^2\right)$$

The resulting discrete time model has the same form as the discrete generation model (2.17) discussed in Section 2.3.3.

2.3.3.2   Stability properties

We let $\Lambda_{p,n} = \left(\Lambda_{p,n}^1, \Lambda_{p,n}^2\right)$, with $p \le n$, be the nonlinear semigroup associated with the evolution models (2.18); that is, we have that

$$
\begin{aligned}
(\gamma_n(1), \eta_n) &:= \Lambda_{p,n}(\gamma_p(1), \eta_p)\\
&= \left(\Lambda_{p,n}^1(\gamma_p(1), \eta_p), \Lambda_{p,n}^2(\gamma_p(1), \eta_p)\right)
\end{aligned}
$$

The stability properties of these semigroups are discussed in Section 13.1. In Section 13.1.2, we provide several weak Lipschitz type estimates, in terms of the Feynman-Kac semigroup associated with the operators Q_n defined in (1.42). In Section 13.1.3, we also discuss the stability properties of the three typical scenarios $G_n = 1$, $\sup_{E_n} G_n < 1$, and $\inf_{E_n} G_n > 1$. We provide sufficient conditions on the Markov transitions M_n under which we have the following contraction inequalities

$$\left\| \Lambda_{p,n}^2(m, \eta') - \Lambda_{p,n}^2(m, \eta) \right\|_{tv} \le a\, e^{-b(n-p)}$$

for some positive and finite parameters (a, b).

2.3.3.3   Mean field models

To describe with some precision the mean field particle interpretation of these models, we consider a collection of Markov transitions S_{n,η_n} satisfying the compatibility condition

$$\eta_n S_{n,\eta_n} = \Psi_{G_n}(\eta_n) \tag{2.19}$$

Several examples of such transitions are provided in (0.3). In this situation, we can rewrite the evolution equation of the normalized measures as follows:

$$\eta_{n+1} = \eta_n K_{n+1,(\gamma_n(1),\eta_n)} \quad \text{with} \quad K_{n+1,(m,\eta)} = S_{n,\eta} M_{n+1,(m,\eta)}$$

and the Markov transitions $M_{n+1,(m,\eta)}$ from E_n into E_{n+1} defined by

$$M_{n+1,(m,\eta)}(x, .) = \alpha_n\,(m, \eta)\, M_{n+1}(x, .) + (1 - \alpha_n\,(m, \eta))\,\overline{\mu}_{n+1}$$

The mean field particle interpretation of these evolution equations is the Markov chain $(\gamma_n^N(1), \xi_n) \in \left(\mathbb{R}_+ \times E_n^N\right)$ defined by

$$
\left\{
\begin{aligned}
\gamma_{n+1}^N(1) &= \gamma_n^N(1)\,\eta_n^N(G_n) + \mu_{n+1}(1)\\[2mm]
\mathbb{P}\left(\xi_{n+1} \in dx \mid \xi_n, \gamma_n^N(1)\right) &= \prod_{i=1}^{N} K_{n+1,(\gamma_n^N(1),\eta_n^N)}(\xi_n^i, dx^i)
\end{aligned}
\right. \tag{2.20}
$$

with the pair of occupation measures $\left(\gamma_n^N, \eta_n^N\right)$ defined by

$$\eta_n^N := \frac{1}{N} \sum_{i=1}^{N} \delta_{\xi_n^i} \quad \text{and} \quad \gamma_n^N(f_n) := \gamma_n^N(1) \times \eta_n^N(f_n)$$

for any $f_n \in \mathcal{B}_b(E_n)$. The stochastic analysis of these IPS models is provided in Section 14.7. We mention that the unnormalized particle measures γ_n^N satisfy the unbiasedness property $\mathbb{E}\left(\gamma_n^N(f_n)\right) = \gamma_n(f_n)$. Furthermore, under some regularity properties on the parameters (G_n, M_n) we prove the following nonasymptotic variance estimates

$$\mathbb{E}\left(\left[\frac{\gamma_n^N(1)}{\gamma_n(1)} - 1\right]^2\right) \leq c \; \frac{n+1}{N-1} \; \left(1 + \frac{c}{(N-1)}\right)^n$$

for any $N > 1$, and $n \geq 0$, with some finite constant $c < \infty$, whose values do not depend on the time parameter. Section 14.7.4 also discusses several exponential concentration inequalities. For instance, in the three scenarios discussed in Section 2.3.3.2, we prove the following result:

For any $n \geq 0$, $N \geq 1$, any collection of functions $f_n \in \mathrm{Osc}(E_n)$, and any $x \geq 0$, the probability of any of the following events

$$\left|[\eta_n^N - \eta_n](f_n)\right| \leq c \; \sqrt{(x+1)/N}$$

and

$$\sup_{0 \leq p \leq n} \left|[\eta_p^N - \eta_p](f_p)\right| \leq c \; \sqrt{(x+1)\log{(n+e)}/N}$$

is greater than $1 - e^{-x}$, for some finite constant $c < \infty$, whose values do not depend on the time parameter (cf. Theorem 14.7.7 and Corollary 14.7.8).

2.4   Nonlinear intensity measure equations

There are strong connections between Feynman-Kac models and spatial branching population models. For instance, in Section 4.1 we shall discuss a natural genetic type dynamic population interpretation of the Feynman-Kac models (1.37). More general spatial branching interpretation models are also presented in Section 6.1.1. In Section 2.3.3, we also discuss the evolution of intensity measures associated with spatial branching models, equipped with spontaneous branchings.

This section is concerned with a general and abstract class of nonlinear intensity measures. We also refer to Section 6.3 for applications of these models in multiple objects nonlinear filtering problems. A more thorough discussion on these branching models is provided in Chapter 6, dedicated to nonlinear evolutions of intensity measures.

2.4.1   Flows of positive measures

We consider a sequence of measurable state spaces E_n and some (non-necessarily linear) one step mapping Ξ_n from $\mathcal{M}_+(E_{n-1})$ into $\mathcal{M}_+(E_n)$. We associate with these objects the evolution equations

$$\gamma_n = \Xi_n(\gamma_{n-1})$$

starting from some initial nonnegative measure $\gamma_0 \in \mathcal{M}_+(E_0)$. These abstract models are natural extensions of the $\mathcal{P}(E_n)$-valued equations discussed in Section 1.2.1, to evolutions equations in the space of nonnegative measures. We observe that these evolutions equations can always be rewritten in the following form

$$\gamma_n = \gamma_{n-1}Q_{n,\gamma_{n-1}} \tag{2.21}$$

for some collection of bounded integral operators $Q_{n,\gamma}$ from $\mathcal{B}_b(E_n)$ into $\mathcal{B}_b(E_{n-1})$, indexed by the time parameter $n \geq 1$ and the set of measures $\gamma \in \mathcal{M}_+(E_n)$. As in (1.16), the choice of these operators is not unique; for instance, we can take $Q_{n,\gamma_{n-1}}(x, .) \propto \Xi_n(\gamma_{n-1})$.

In this interpretation, the evolution Equation (2.21) can also be interpreted as natural extensions of the Feynman-Kac model presented in (1.41), and in (2.17). As usual, we let $\eta_n \in \mathcal{P}(E_n)$ be the normalized distributions given by

$$\eta_n(f) = \gamma_n(f)/\gamma_n(1) \quad \text{and we set} \quad G_{n,\gamma_n} := Q_{n+1,\gamma_n}(1)$$

for any $f \in \mathcal{B}_b(E_n)$. In Section 6.2.1, we shall see that the mass-probability process $(\gamma_n(1), \eta_n)$ satisfies the following measure valued equations

$$\begin{cases} \gamma_{n+1}(1) &= \eta_n(G_{n,\gamma_n})\,\gamma_n(1) \\[2mm] \eta_{n+1} &= \Psi_{G_{n,\gamma_n}}(\eta_n)\,M_{n+1,\gamma_n} \end{cases} \tag{2.22}$$

with the Markov transitions $M_{n,\gamma}$ defined for any $f \in \mathcal{B}_b(E_n)$ and $\gamma \in \mathcal{M}_+(E_{n-1})$ by the following equation

$$M_{n,\gamma}(f) := Q_{n,\gamma}(f)/Q_{n,\gamma}(1)$$

In the above displayed formula, Ψ_G stands for the Boltzmann-Gibbs transformations associated with a given potential function G defined in (0.2).

The next section, Section 2.4.2, illustrates these abstract models in the context of multi-object filtering problems. The stability properties of these models are discussed in Section 2.4.3, and their mean field approximations are developed in Section 2.4.4.

2.4.2   Multi-object nonlinear filtering problems

The measure valued process introduced in (2.22) arises in the modeling of spatial branching processes, and more particularly in multiple objects filtering problems. In this context, we are given a random-measure valued signal $\mathcal{X}_n := \sum_{i=1}^{N_n^X} \delta_{X_n^i}$ associated with a spatial branching process, with a random number N_n^X of individuals $(X_n^i)_{1 \leq i \leq N_n^X}$. The intensity measures of this branching process satisfy an evolution equation of the form (2.17). For a more precise discussion on spatial branching process and their intensity measures, we refer the reader to Section 6.1.2.

At every time step, every random state $X_n^i = x$ generates an observation Y_n^i on some measurable state space E_n^Y, with a distribution of the following form $d_n(x)g_n(x,y)\,\lambda_n(dy)$. The function d_n is a $[0,1]$-valued function that represents the detection probability of the targets; and the function $y \mapsto g_n(x,y)$ stands for the density of a given probability measure on E_n^Y, w.r.t. some reference measure λ_n. By construction, with a probability $1 - d_n(x)$, the state $X_n^i = x$ is not detectable, and it goes in a virtual cemetery state.

The full spatial observation point process $\mathcal{Y}_n := \sum_{i=1}^{N_n^Y} \delta_{Y_n^i}$ consists of the collection of random observations Y_n^i of detectable targets, plus some noisy random observations $\mathcal{Y}_n'$ unrelated to the signal. It is commonly assumed that these noisy clutter random measures $\mathcal{Y}_n'$ are given by a collection of spatial Poisson processes, with intensity measures $h_n(y)\,\lambda_n(dy)$ on E_n^Y.

A more detailed presentation of this complex nonlinear filtering problem is provided in Section 6.3.1. We let $(\gamma_n, \widehat{\gamma}_n) \in \mathcal{M}_+(E_n)^2$ be the conditional intensity measures defined, for any $f_n \in \mathcal{B}_b(E_n)$, by the equations

$$
\begin{aligned}
\gamma_n(f_n) &:= \mathbb{E}\left(\mathcal{X}_n(f_n) \mid \mathcal{Y}_0, \ldots, \mathcal{Y}_{n-1}\right) \\
\widehat{\gamma}_n(f_n) &:= \mathbb{E}\left(\mathcal{X}_n(f_n) \mid \mathcal{Y}_0, \ldots, \mathcal{Y}_{n-1}, \mathcal{Y}_n\right)
\end{aligned}
$$

Under some Poisson type approximation, we prove that these distribution flows satisfy the filtering equation

$$
\begin{aligned}
\gamma_{n+1} &:= \widehat{\gamma}_n Q_n + \mu_n \\
\widehat{\gamma}_n(f) &:= \gamma_n((1 - d_n)f_n) + \int \mathcal{Y}_n(dy)\,(1 - \beta_{\gamma_n}(y))\,\Psi_{d_n g_n(y, \cdot)}(\gamma_n)(f_n)
\end{aligned}
$$

with the $[0,1]$-valued functions

$$
y \mapsto \beta_{n,\gamma_n}(y) := h_n(y)/\left[h_n(y) + \gamma_n(d_n g_n(\cdot, y))\right]
$$

For a detailed derivation of these nonlinear multiple-object filtering equations, we refer the reader to Section 6.3 and Section 6.3.2.

These nonlinear filtering equations clearly fit into the abstract framework presented in (2.21). In signal processing, and more particularly in multiple targets tracking literature, these equations are called the Probability Hypothesis Density equations (*abbreviated (PHD)*) (see for instance [137, 101, 102, 103, 221, 510, 411], and references therein).

Since its inception by Mahler [411] in 2003, the PHD filter has attracted substantial interest to date. The development of numerical solutions for the

PHD filter [101, 102, 103, 221] and the seminal articles [564, 565, 566, 567] have opened the door to numerous novel extensions and applications.

The reader who wishes to know more details about these models and specific applications in signal processing is recommended to consult the pioneering series of papers by R. Mahler [292, 293, 412, 413], and the articles by R. Mahler [410, 411], B.T. Vo, B.N. Vo, and A. Cantoni [564, 565], and the more recent article by S.S. Singh, B.N. Vo, A. Baddeley, and S. Zuyev [510].

For a single fully detectable target, without spawning and spontaneous birth rate, these equations coincide with the traditional optimal filtering equations, discussed in Section 2.5.1, as well as in Chapter 8, dedicated to the applications of mean field theory in advanced signal processing, and optimal control theory.

In more general situations, the stability analysis as well as the numerical solving of these evolution equations are much more involved than the ones of the traditional single-target filtering. They involve complex computations of conditional distributions of spatial branching signal processes with respect to noisy and partial observations delivered by spatial point sensors.

To the best of our knowledge, the first rigorous studies on the stability properties of these models, including uniform estimates for mean field approximations w.r.t. the time parameters, and the analysis of interacting Kalman type models and particle filter-based association measures, are given in a couple of articles [102, 103]. We also refer to [459, 460, 461], for more thorough discussions on the numerical aspects of these models.

2.4.3   Stability properties

Before designing any type of numerical approximation scheme, one crucial problem is to analyze the stability properties of the PHD filtering equations, discussed in Section 2.4.2.

There are two main reasons for this:

The first one comes from the fact that the initial conditions of the targets are usually unknown. In this case, it is clearly essential to check whether or not the PHD filtering equation corrects by itself any erroneous initial condition.

The second main reason comes from numerical analysis, and the desire to design robust and stable algorithms. Indeed, it is more or less well known that this type of regularity condition is critical to ensure that the local errors, induced by some local approximation scheme, will not propagate w.r.t. the time parameter.

Section 13.2 is dedicated to the stability properties of the nonlinear semigroup

$$
\begin{aligned}
(\gamma_n(1), \eta_n) \quad &:= \quad \Lambda_{p,n}(\gamma_p(1), \eta_p) \\
&= \quad \left(\Lambda^1_{p,n}(\gamma_p(1), \eta_p), \Lambda^2_{p,n}(\gamma_p(1), \eta_p)\right) \qquad p \leq n
\end{aligned}
$$

associated with the abstract evolution Equations (2.21).

In Section 13.3.2, we provide some sufficient conditions under which we have for any $p \leq n$, $u, u' \in I_p$, $\eta, \eta' \in \mathcal{P}(E_p)$, and $f_n \in \mathrm{Osc}_1(E_n)$ the following weak Lipschitz's type contraction inequalities

$$\left| \Lambda^1_{p,n}(u', \eta') - \Lambda^1_{p,n}(u, \eta) \right|$$

$$\leq c^{1,1} \, e^{-\lambda(n-p)} \, |u - u'| + c^{1,2} \, e^{-\lambda(n-p)} \int \, |[\eta - \eta'](\varphi)| \, \Sigma^1_{p,n,u',\eta'}(d\varphi)$$

and

$$\left| \Lambda^2_{p,n}(u', \eta')(f_n) - \Lambda^2_{p,n}(u, \eta)(f_n) \right]$$

$$\leq c^{2,1} \, e^{-\lambda(n-p)} \, |u - u'| + c^{2,2} \, e^{-\lambda(n-p)} \int \, |[\eta - \eta'](\varphi)| \, \Sigma^2_{p,n,u'\eta'}(f, d\varphi)$$

for some exponential rate $\lambda > 0$, some finite constants $c^{i,j} < \infty$, with $i, j \in \{1, 2\}$, some probability measures $\Sigma^1_{p,n,u',\eta'}$ and $\Sigma^2_{p,n,m'\eta'}(f, .)$ on $\mathrm{Osc}(E_p)$. For a more precise description of these parameters and the proof of these contraction inequalities, we refer the reader to Theorem 13.3.3 on page 409.

In the context of multiple target tracking problems, these regularity conditions ensure that the PHD is exponentially stable for small clutter intensities, sufficiently high detection probability, and high spontaneous birth rates.

2.4.4   Mean field models

To describe with some precision the mean field particle interpretation of these models, we consider a collection of Markov transitions S_{n,η_n} satisfying the compatibility condition

$$\eta_n S_{n,\gamma_n} = \Psi_{G_{n,\gamma_n}}(\eta_n)$$

Several examples of such transitions are provided in (0.3). In this situation, we can rewrite the evolution equation of the normalized measures as follows:

$$\eta_{n+1} = \eta_n K_{n+1,\gamma_n} \quad \text{with} \quad K_{n+1,\gamma_n} = S_{n,\gamma_n} M_{n+1,\gamma_n}$$

The mean field particle interpretation of the evolution Equations (2.21) is the Markov chain $(\gamma_n^N(1), \xi_n) \in (\mathbb{R}_+ \times E_n^N)$ defined by

$$\begin{cases} \gamma_{n+1}^N(1) & = \; \gamma_n^N(1) \, \eta_n^N(G_{n,\gamma_n^N(1)\eta_n^N}) \\[2em] \mathbb{P}\left(\xi_{n+1} \in dx \mid \xi_{n,\gamma_n^N(1)} \right) & = \; \prod_{i=1}^{N} K_{n+1,\gamma_n^N(1)\eta_n^N}(\xi_n^i, dx^i) \end{cases} \qquad (2.23)$$

The stochastic analysis of these IPS models is provided in Section 14.8. For instance, under some stability conditions we prove the following result:

For any $n \geq 0$, $N \geq 1$, any collection of functions $f_n \in \text{Osc}(E_n)$, and any $x \geq 0$ the probability of any of the following events

$$\left|[\eta_n^N - \eta_n](f_n)\right| \vee \left|\gamma_n^N(1) - \gamma_n(1)\right| \leq c \sqrt{(x+1)/N}$$

$$\sup_{0 \leq p \leq n} \left[\left|[\eta_p^N - \eta_p](f_p)\right| \vee \left|\gamma_p^N(1) - \gamma_p(1)\right|\right] \leq c \sqrt{(x+1)\log(n+e)/N}$$

is greater than $1 - e^{-x}$, for some finite constant $c < \infty$ whose values do not depend on the time parameter (cf. Corollary 14.8.4).

2.5   Statistical machine learning models

Statistical machine learning theory merges applied probability theory and statistics with engineering and computer sciences. This new branch of numerical probability theory is mainly concerned with the development of stochastic algorithms. These algorithms are mainly used to estimate unknown parameters, or random processes, using partial and noisy observations delivered by some physical sensors, or given by some statistical databases. The central ideas are to integrate the observation sequences to estimate unknown properties of data, or to predict the behavior of the underlying physical random process.

Nonlinear filtering theory, as well as Bayesian statistics, provides a natural mathematical framework to formulate these estimation problems, in terms of conditional distributions of an unknown parameter, or a given stochastic process w.r.t. a sequence of observations.

In the further development of Section 2.5.1, Section 2.5.2.2, and Section 2.5.2.3, we shall show that these conditional probabilities can be encapsulated into the Feynman-Kac models discussed in Section 1.4.2. Further details on these topics can be found in the series of books and articles [30, 50, 51, 154, 155, 161, 163, 344], as well as in Chapter 8, Section 8.1, and Section 8.6, in the present book, dedicated to filtering problems and fixed parameter estimation in HMM models.

Another subject of growing interest in statistical machine learning literature, and the scientific computing community, concerns calibration problems and related uncertainty propagations in numerical codes.

Modern computers are now capable of simulating very complex physical and engineering systems. Nevertheless, all the mathematical models are far from being certain and error-free. In some instances, the physical environment of the system under study is often too complex to formalize perfectly. In other instances, the different physical scales of some physical phenomenon, such as turbulent fluid models, are extremely difficult to capture with high precision. In this context, the randomness is used to cure the imperfections of the mathematical models.

From another perspective, the reliability, and the accuracy of computational approximation models, relies on complex calibration processes, often combined with the random dispersion analysis of the inputs, and the many other sources of randomness entering into the system evolution. In this context, given some reference physical observation, we would like to calibrate the model parameters, so that the outputs simulated by some numerical code coincide with this reference data, or at least behave as much as possible as the "real model" associated with these physical observations.

Last, but not least, given a successfully calibrated model, one might be also interested in computing the probability that the random outputs belong to some critical event; that is, to find the law of the unknown input parameters, including the random perturbation sources, leading to such critical events.

In Section 2.5.2, we provide an interpretation of these estimation problems in terms of Boltzmann-Gibbs measures. We also present a natural Feynman-Kac interpretation of these measures that allows applying mean field type MCMC techniques.

2.5.1 Nonlinear filtering

The nonlinear filtering problem consists in computing the conditional distributions of a Markov signal process, given some noisy and partial observations. More formally, let us suppose that

$$Y_n = h_n(X_n) + V_n \tag{2.24}$$

is a partial observation process on $\mathbb{R}^d$ of some Markov chain X_n, where V_n is a sequence of i.i.d. random variables with density g_n on $\mathbb{R}^d$. We let $p_n(y_n \mid x_n)$ be the conditional density of the random variable Y_n given the random state $X_n = x_n$. In this notation, if we set

$$G_n(x_n) := p_n(y_n \mid x_n) = g_n(y_n - h_n(x_n))$$

in the Feynman-Kac model (1.37), then we have

$$\mathbb{Q}_n = \text{Law}\left((X_0, \ldots, X_n) \mid Y_p = y_p, \ \forall 0 \le p < n\right)$$

Furthermore, the density $q_n(y_0, \ldots, y_n)$ of the random sequence of observations $(Y_0, \ldots, Y_n)$ is given by the normalizing constant

$$\mathcal{Z}_n = q_n(y_0, \ldots, y_n)$$

We recall that the flow of n-th time marginals η_n satisfy the nonlinear evolution equation

$$\eta_{n+1} := \Psi_{G_n}(\eta_n)M_{n+1}$$

Remark 2.5.1 *Inversely, we can show that any solution of the evolution equation given above coincides with the n-th time marginals of the Feynman-Kac measures $\mathbb{Q}_n$ defined in (1.37). This observation is essential to solve hidden Markov chain problems and stochastic optimization models.*

In filtering literature, the measure η_n is called the optimal predictor, while $\widehat{\eta}_n = \Psi_{G_n}(\eta_n)$ is called the optimal filter. The evolution equation described above is decomposed into two steps. The first one $\eta_n \rightsquigarrow \widehat{\eta}_n$ is called the updating stage, and the second one $\widehat{\eta}_n \rightsquigarrow \eta_{n+1}$ the prediction transition.

Linear-Gaussian filtering models are discussed in some detail in Section 8.2. Forward-backward Kalman filters are developed in Section 8.2.1 and in Section 8.2.2. McKean interpretations of Kalman filters and their mean field ensemble Kalman filters are developed in Section 8.2.3. Interacting Kalman filters are discussed in Section 8.3.

2.5.2   Boltzmann-Gibbs measures

Boltzmann-Gibbs measures are one of the most central mathematical models of classical statistical physics. In this context, the main problem is to deduce macroscopic equilibrium behaviors of thermodynamic physical systems, from complex disordered microscopic interacting structures.

As their name indicates, Boltzmann-Gibbs measures were introduced independently in the early 1900s by Ludwig Boltzmann and Josiah Willard Gibbs in their seminal studies on statistical entropy theory and micro-macro canonical ensemble theory [58, 273]. In the late 1960s, Roland Lvovich Dobrushin [219] and Oscar E. Lanford and David Ruelle [379] and also developed a new theory to design probability measures on finite product spaces, by specification systems of conditional distributions w.r.t. the complement of finite volume measures, with prescribed boundary conditions.

The two prototypes of physical systems are particles in liquid-vapor models of real gases, interacting via Van der Waals forces, and atoms' configurations and their magnetic moments in crystal lattices of ferromagnetic metals (iron, colbalt, nickel) in thermal equilibrium. In this context, adjacent atoms and their electronic configurations tend to have the same angular moment (i.e., the same spins).

In Section 2.5.2.4, we shall discuss the application of our framework through two simplified disordered models; namely, the traditional Ising model and the more sophisticated Sherrington-Kirkpatrick model.

In subsequent pages of this section, we present a natural Feynman-Kac formulation of Boltzmann-Gibbs measures that allows applying the mean field theory presented in Section 1.4.2.1 and in Section 1.5. We illustrate these results, with four typical examples related to restriction models and uncertainty propagations, hidden Markov chain models and their particle approximations, and stochastic optimization models.

For a more thorough discussion on these models, and their interacting MCMC simulation, we refer the reader to Section 4.3.

From the pure probabilistic point of view, Boltzmann-Gibbs measures μ_n are defined in terms of some reference measure λ on some abstract measurable state space, say E, weighted by some product of potential functions $h_p : E \mapsto [0, \infty[$, with $p \leq n$. More formally, these measures are defined, up to some normalizing constant $\mathcal{Z}_n$, by the following formulae

$$\mu_n(dx) = \frac{1}{\mathcal{Z}_n} \left\{ \prod_{0 \leq p \leq n} h_p(x) \right\} \lambda(dx) \tag{2.25}$$

We further assume that we have a dedicated MCMC elementary transition M_n, with the target measure $\mu_n = \mu_n M_n$, at any time $n \geq 0$. It is now readily checked that

$$\left. \begin{array}{rcl} \mu_n & = & \mu_n M_n \\ \mu_n & = & \Psi_{h_n}(\mu_{n-1}) \end{array} \right\} \implies \mu_n = \mu_n M_n = \Psi_{h_n}(\mu_{n-1}) M_n \tag{2.26}$$

In other words, if we set $G_n = h_{n+1}$, in the Feynman-Kac model (1.37), then we have $\mu_n = \eta_n$. In other words, the Boltzmann-Gibbs measures μ_n can also be interpreted as the n-th time marginals of the Feynman-Kac measures $\mathbb{Q}_n$, with the potential functions $G_n = h_{n+1}$ and the reference Markov chain X_n, associated with the Markov chain Monte Carlo elementary transition M_n. It is also easily checked that

$$\mathcal{Z}_n = \mu_{n-1}(h_n) \times \mathcal{Z}_{n-1} \Rightarrow \mathcal{Z}_n/\mathcal{Z}_0 = \mathbb{E}\left(\prod_{0 \leq p \leq n} G_p(X_p) \right) \tag{2.27}$$

where X_n stands for a Markov chain with initial distribution μ_0 and elementary transition probabilities M_n.

The mean field particle model associated with these Feynman-Kac representations can be interpreted as interacting MCMC models. These adaptive MCMC algorithms are discussed in some detail in Section 4.3.

2.5.2.1 Uncertainty propagation models

The Boltzmann-Gibbs measures (2.25) associated with the indicator functions $h_n := 1_{A_n}$, with nonincreasing sequence of subsets $A_n \subset E$, take the following simple form

$$\mu_n(dx) = \frac{1}{\mathcal{Z}_n} 1_{A_n}(x) \lambda(dx) \quad \text{and} \quad \mathcal{Z}_n := \lambda(A_n) \tag{2.28}$$

These measure restriction models arise in a variety of application domains, including in rare event simulation and stochastic optimization models. In this section, we illustrate these models in the context of calibration problems and uncertainty propagations in complex numerical codes (or in numerical metamodels). These problems are often formulated in terms of a classical input-output transformation $I \mapsto O = C(I)$. The inputs I have some distribution λ. They represent the sources of randomness, some tuning parameters, or unknown kinetic parameters of the code. The output variable O can be interpreted as the outputs of a numerical approximation of some partial differential equation representing a given physical, chemical, or some biological phenomenon.

The prototype of questions arising in practice is the following: We are given a decreasing sequence of unlikely critical domains, say $\mathcal{O}_n$, in the space of the outputs, and we want to estimate both the probability that the outputs fall into these sets, as well as the distribution of the inputs leading to these critical events. More formally, if we set $A_n := C^{-1}(\mathcal{O}_n)$ these probabilistic objects coincide with the normalizing constants $\mathcal{Z}_n$, and the conditional distributions μ_n introduced in (2.28).

In the example discussed above, we can choose in (2.26) the MCMC transition

$$M_n(x, dy) = K(x, dy)\, 1_{A_n}(y) + (1 - K(x, A_n))\, \delta_x(dy)$$

for any λ-reversible Markov transition K. In engineering and scientific computing literature, the mean field models associated with these restriction models are also called subset simulation algorithms [20, 21, 22, 396, 399].

2.5.2.2   Hidden Markov chains

Hidden Markov models are particular instances of Bayesian dynamic networks and nonlinear filtering problems. Their foundations and their applications in statistics go back to the seminal work of Baum and his co-authors, in the mid-1960s [38, 39, 40, 41, 42]. Since this period, HMM models are standard Bayesian models, with applications in a variety of domains, including speech and signal processing, econometrics and financial mathematics, bioinformatics, and many others. A more detailed discussion on these models can be found in the series of books [50, 51, 93, 276, 493].

These statistical estimation problems are defined in terms of a Markov chain model (X_n^θ, Y_n^θ) that depend on the realization of some random variable Θ, with distribution λ, on some state space. The state X_n^θ is itself a Markov chain. The process $Y_n^\theta \in \mathbb{R}^d$ represents the noisy and partial observations of the random state variables X_n^θ. The central problem is to estimate the parameter Θ, given the observation sequence. Whenever it exists, we let

$$h_n(\theta) := p_n(y_n \mid \theta,\, (y_0, \dots, y_{n-1}))$$

be the density of the random variable Y_n given a realization of the random sequence $(\Theta, (Y_0, \dots, Y_{n-1})) := (\theta, (y_0, \dots, y_{n-1}))$. In this notation, the conditional density $q_n((y_0, \dots, y_n) \mid \theta)$ of the random sequence of observations $(Y_0, \dots, Y_n)$ given $\Theta = \theta$ is given by the product likelihood formula

$$\left\{ \prod_{0 \leq p \leq n} h_p(\theta) \right\} = q_n((y_0, \dots, y_n) \mid \theta)$$

We conclude that the Boltzmann-Gibbs measures (2.25) associated with the functions h_n represent the desired conditional distributions

$$\mu_n = \text{Law}\,(\Theta \mid (Y_0, \ldots, Y_n) = (y_0, \ldots, y_n)) \tag{2.29}$$

Frequentist statisticians often use stochastic gradient type techniques, to compute the parameter that maximizes the likelihood functions $\theta \mapsto q_n((y_0, \ldots, y_n) \mid \theta)$. Some of these stochastic optimization techniques, including Expected Maximization algorithms, and stochastic gradient estimates, are discussed in Section 8.6.3, and in Section 8.6.4.

It is rather well known that one the main drawback of these gradient techniques is that they often fail to obtain a global extremum of multimodal likelihood functions. In addition, they provide no information on the variability of the unknown parameter given the observations. Last, but not least, the maximum of a given probability density function may be located in some region with extremely low probability mass.

For all of these reasons, Bayesian statisticians prefer to use MCMC style methods to approximate the conditional distribution (2.29) by a cloud of samples. The main drawback of this methodology is that the estimation strongly depends on the choice of the prior distribution λ. In addition, these Bayesian techniques are much more time consuming then the gradient style variational techniques discussed above. The main reason comes from the fact that the performance of any MCMC model strongly depends on its mixing properties.

In this connection, the mean field simulation of the Boltzmann-Gibbs interpretations (2.25) of the conditional distributions (2.29) offer new population style algorithms that can be implemented in parallel computers. The precision parameter of these mean field schemes is not related to some MCMC burning period, but to the size of the mean field model, and the computational power of the parallel computers on which we implement the algorithms.

2.5.2.3 Particle approximation models

In the above example, we have implicitly assumed that the likelihood functions

$$h_n \;:\; \theta \mapsto h_n(\theta) := p_n(y_n \mid \theta,\, (y_0, \ldots, y_{n-1}))$$

are explicitly known, or at least their values can be easily computed at any given state θ in a reasonable time. This condition is met for conditionally linear-Gaussian models (cf. for instance Section 8.6.1), but in more general situations we need to resort to another level of approximation.

To this end, we return to the filtering model discussed in (2.24). We further assume that the signal-observation Markov chain (X_n^θ, Y_n^θ) depends on the realization of some random variable Θ, with distribution ν on some state space. In this situation, the conditional likelihood function $G_{\theta,n}$ of the observation Y_n^θ and the Markov transition $M_{\theta,n}$ of the chain X_n^θ, also depend on the realization of the random variable $\Theta = \theta$.

We let $P(\theta, d\xi)$ be the distribution of the mean field Feynman-Kac model

$(\xi_{\theta,n})_{n \geq 0}$ associated with the potential functions $G_{\theta,n}$ and the Markov transitions $M_{\theta,n}$. A more detailed description of this model is given in Section 1.5.5, and Section 2.2. We also refer the reader to Section 4.4.2, and Section 4.4.5, for a more thorough discussion on these particle models.

We denote by λ the distribution of the random sequence

$$\boldsymbol{X} := (\Theta, (\xi_{\Theta,n})_{n \geq 0})$$

that is, we have that $\lambda(d\boldsymbol{x}) = \lambda(d(\theta,\xi)) = \nu(d\theta) \ P(\theta,d\xi)$, where $d\boldsymbol{x} = d(\theta,\xi) = d\theta \times d\xi$ stands for an infinitesimal neighborhood of the point $\boldsymbol{x} = (\theta,\xi)$. We also consider the functions h_n defined for any $\boldsymbol{x} = (\theta,\xi)$ by

$$\boldsymbol{h_n}(\boldsymbol{x}) := \frac{1}{N} \sum_{1 \leq i \leq N} G_{\theta,n}(\xi_{\theta,n}^i)$$

Using (2.11) and (2.24), for any fixed θ, we have the *unbiased* estimates

$$\prod_{0 \leq p \leq n} h_p(\theta,\xi) \longrightarrow_{N \to \infty} q_n((y_0,\dots,y_n) \mid \theta)$$

The unbiased property ensures that the θ-marginals of the measures

$$\boldsymbol{\mu_n}(d\boldsymbol{x}) = \frac{1}{Z_n} \left\{ \prod_{0 \leq p \leq n} h_p(\boldsymbol{x}) \right\} \lambda(d\boldsymbol{x}) \tag{2.30}$$

coincide with the conditional distributions of Θ given the sequence of observations $(Y_0,\dots,Y_n) = (y_0,\dots,y_n)$.

In summary, we have transformed Boltzmann-Gibbs measures (2.29) with unknown likelihood functions h_n on the parameter space, into more complex Boltzmann-Gibbs measures (2.30), *but involving computable functions $\boldsymbol{h_n}$*, on an extended state space. The new state variables are the mean field model associated with a given value of the state parameter. In statistical literature, these auxiliary variables are sometimes called the latent variables, in reference to the fact that they are inferred by the mathematical model.

The mean field simulation of the Boltzmann-Gibbs measures (2.30) are described in some details in Section 4.3.6, as well as in Section 8.6. In Bayesian statistical literature these models are also called SMC^2 algorithms [134], in reference to the fact that there are two different levels of mean field particle simulations. The first one to define the Boltzmann-Gibbs measures (2.30) in terms of mean field particle models associated with the different values of the parameter. The second one comes from the mean field simulation of the flow of Boltzmann-Gibbs measures (2.30), using the MCMC transitions. In this connection, we refer the reader to the seminal article by C. Andrieu, A. Doucet, and R. Holenstein [11], presenting an MCMC methodology in a statistical framework, to sample general target distributions of the form (2.30).

2.5.2.4 Stochastic optimization

We consider a nondecreasing sequence of inverse temperature parameters β_n, with $\beta_0 = 0$, and a given energy type function V, on some state space E, equipped with some reference distribution λ. If we set

$$h_n(x) = \exp\left(-\left(\beta_n - \beta_{n-1}\right) V(x)\right)$$

in (2.25), then we find that

$$\mu_n(dx) = \frac{1}{\mathcal{Z}_n} \exp\left(-\beta_n V(x)\right) \lambda(dx) \quad \text{with} \quad \mathcal{Z}_n = \int \lambda(dx) \exp\left(-\beta_n V(x)\right)$$

$$(2.31)$$

We mention that the measures μ_n concentrate on the global minimum values of the potential function V, as $n \to \infty$, as soon as $\beta_n \uparrow \infty$. For instance, for finite state spaces E equipped with the counting measure $\lambda(x) = 1/\text{Card}(E)$, if we set $\mathcal{V}_\star = \{x \in E \,:\, V(x) = V_\star\}$, with $V_\star = \min_y V(y)$, then we have

$$\sum_{y \in E} e^{-\beta_n[V(y)-V_\star]} = \text{Card}\left(\mathcal{V}_\star\right) + \sum_{y \not\ni \mathcal{V}_\star} e^{-\beta_n[V(y)-V_\star]} \to_{n\uparrow\infty} \text{Card}\left(\mathcal{V}_\star\right)$$

This implies that μ_n converges to the uniform measure on the set $\mathcal{V}_\star$; that is, we have that

$$\mu_n(x) = \frac{e^{-\beta_n[V(x)-V_\star]}}{\sum_{y \in E} e^{-\beta_n[V(y)-V_\star]}} \to_{n\uparrow\infty} \mu_\infty(x) := \frac{1}{\text{Card}\left(\mathcal{V}_\star\right)} \, 1_{\mathcal{V}_\star}(x)$$

In this interpretation, the sampling of the Boltzmann-Gibbs measures μ_n is asympotically equivalent to that of sampling uniformly a global minimum value of V.

A prototype model of Boltzmann-Gibbs measures arising in the analysis of disordered systems is the spin glasses Ising model. In this context, each state $x \in E = \{-1, 1\}^d$ represents a configuration of spins $+1$ or -1 on some points in the lattice. The dependencies between the points are represented by some coupling intensity function $\theta \,:\, (i,j) \in \{1,\ldots,d\}^2 \mapsto \theta_{i,j}$. The energy of a given spin configuration x is given by the function

$$V(\theta, x) := \sum_{1 \leq i \leq j \leq d} \theta_{i,j} \, x(i) \, x(j) + h \sum_{i=1}^{d} x(i)$$

The behavior of the physical system at inverse temperature β_n is given by the Boltzmann-Gibbs measures (2.31), with the counting measure λ on $E = \{-1, 1\}^d$. In this context, the normalizing constant $\mathcal{Z}_n$ is often called the partition function. Disordered models correspond to random mappings Θ. For instance in the Sherrington-Kirkpatrick model introduced in 1975 in their seminal article [514], $\Theta_{i,j}$ are assumed to be i.i.d. centered Gaussian random variables. Directed polymer models arising in statistical physics are defined in much the same way. For instance, the micro-state of a system consists of d particles $x_i = (p_i, r_i)$, with a momentum vector p_i and a position coordinate $r_i = (r_i^1, r_i^2, r_i^3)$, with $1 \leq i \leq d$. The energy of the system is given by some function

$$V(x) := \sum_{i=1}^{d} \left(\frac{1}{2m} \|p_i\|^2 + mgr_i^1 \right)$$

where m represents the mass of the particle, r_i^1 its height, and g the gravitation constant. The probability distribution of the physical system at inverse temperature β_n is again given by the Boltzmann-Gibbs measures (2.31), with the Lebesgue measure λ.

For a more thorough discussion on these models, we refer the reader to [141, 202, 205], and references therein. We also refer the reader to Section 4.3.4 and Section 4.3.5 for a more detailed analysis of annealed Boltzmann-Gibbs distributions and their mean field IPS sampling.

2.6   Molecular dynamics simulation

2.6.1   Hamiltonian and Langevin equations

Molecular dynamics simulation is concerned with the analysis of the fluctuations, and the conformal changes of proteins and nucleic acids in biological molecules. The central problem is to understand the macroscopic properties of a molecule through the simulation of a microscopic system of atomic interacting particles in a given force field model.

More formally, we consider the microscopic evolution of a many-body system formed by k atomic particles in the Euclidian space $E = \mathbb{R}^3$ with possibly k different masses $m = (m_i)_{1 \leq i \leq k}$. Their spatial positions, and their velocities, are denoted by the letters $q = (q_i)_{1 \leq i \leq k}$, and $p = (p_i)_{1 \leq i \leq k}$. These particles move under the influence of some external forces $F_i(q)$ according to Newton's second law

$$m_i \frac{d^2 q_i}{dt^2} = F_i(q) \tag{2.32}$$

The velocity vector $p_i = m_i \frac{dq_i}{dt}$ is called the particle momenta of the system, and the couple $x = (q, p)$ is called the phase vector.

We further assume that the force field is conservative, in the sense that

$$F(q) = -\nabla_q V(q) = \left(-\frac{\partial V}{\partial q_i}(q) \right)_{1 \leq i \leq k}$$

for some interparticle potential function $V : E^k \to \mathbb{R}$. In this situation, we can reformulate the evolution equations (2.32) in terms of the Hamiltonian or energy functional

$$H(q, p) = \sum_{i=1}^{k} \frac{p_i^2}{2m_i} + V(q_1, \ldots, q_k)$$

with the following equations

$$\begin{cases} \dfrac{dq_i}{dt} &= \dfrac{p_i}{m_i} = \dfrac{\partial H}{\partial p_i}(q, p) \\ \dfrac{dp_i}{dt} &= F_i(q) = -\dfrac{\partial V}{\partial q_i}(q) = -\dfrac{\partial H}{\partial q_i}(q, p) \end{cases}$$

We notice that these evolution equations are time reversible, in the sense that they have the same form if we consider the time transformation $\tau(t) = -t$. In other words, the microscopic physics doesn't depends on the time flow direction. We also notice the conservation property

$$\frac{d}{dt} H(q, p) = \sum_{i=1}^{k} \left[\frac{\partial H}{\partial q_i}(q, p) \frac{dq_i}{dt} + \frac{\partial H}{\partial p_i}(q, p) \frac{dp_i}{dt} \right] = 0 \qquad (2.33)$$

Solid and liquid states of rare-gas elements with closed shell configurations only involves particles interacting with weak van de Waals bonds in terms of the pair-potential function

$$V(q_1, \ldots, q_k) = \sum_{1 \le i < j \le k} V_{LJ}(\|q_j - q_i\|)$$

with the Lennard Jones potential functions

$$V_{LJ}(r) = 4\epsilon \left[\left(\frac{\tau}{r} \right)^{12} - \left(\frac{\tau}{r} \right)^6 \right]$$

The parameter ϵ represents the depth of the potential well, and τ the finite distance at which the interaction potential becomes null. Notice that $\inf_r V_{LJ}(r) = V_{LJ}(2^{1/6}\tau) = -\epsilon$, so that for $r \ge 2^{1/6}\tau$ the potential is attractive, and repulsive for $r \le 2^{1/6}\tau$. The term $(\tau/r)^{1}2$ describes the short range Pauli repulsion forces due to overlapping electron orbitals, while the term $(\tau/r)^6$ represents the attraction and the van der Waals dispersion forces at long range distances. The repulsion term has no real theoretical foundations; it is sometimes replaced by the Buckingham exponential-6 potential $\exp(-r/\tau)$. To avoid the degeneracy of the Lennard Jones potential at short range distances, we often use cut-off techniques. For instance, we can replace $V_{LJ}(r)$ by $\overline{V}_{LJ}(r) = (V_{LJ}(r) - V_{LJ}(r_c)) \, 1_{r < r_c}$ or by $\overline{V}_{LJ}(r) = (V_{LJ}(r) - V_{LJ}(r_c) - V'_{LJ}(r_c)(r - r_c)) \, 1_{r < r_c}$ for some well chosen cut-off radius r_c.

We associate with the Hamiltonian function, the canonical measures on the phase space

$$\mu_\beta(dx) = \frac{1}{\mathcal{Z}_\beta} e^{-\beta H(x)} \, dx \qquad (2.34)$$

where $\mathcal{Z}_\beta$ is a normalizing constant, and $dx = dqdp$ stands for the Lebesgue measure on $\mathbb{R}^{3k+3k}$, and $x = (q, p)$ stands for a given point in the phase space. We also consider the q-marginal measures

$$\overline{\mu}_\beta(dq) = \frac{1}{\overline{\mathcal{Z}}_\beta} \, e^{-\beta V(q)} \, dq$$

where $\overline{\mathcal{Z}}_\beta$ is a normalizing constant, and dq stands for the the Lebesgue measure on the position space $\mathbb{R}^{3k}$.

The Boltzmann-Gibbs measures μ_β, and respectively $\overline{\mu}_\beta$, can be interpreted as the invariant measure of the Langevin type stochastic dynamics

$$\begin{cases} dq_i &=& \beta \dfrac{\partial H}{\partial p_i}(q, p) \, dt \\[2mm] dp_i &=& -\beta \dfrac{\partial H}{\partial q_i}(q, p) \, dt - \sigma^2 \dfrac{p_i}{m_i} dt + \sigma\sqrt{2} \, dW_t^i \end{cases} \tag{2.35}$$

and respectively

$$dq_i = -\beta \frac{\partial V}{\partial q_i}(q) \, dt + \sqrt{2} \, dW_t^i \tag{2.36}$$

where $(W^i)_{1 \leq i \leq k}$ stands for k independent standard Brownian motion on $\mathbb{R}^3$. As mentioned in [499], the additional external Brownian forces represent the fluctuations of the many-body system, balanced by dissipative and viscous damping forces.

We check these claims using the infinitesimal generators of the diffusion processes (2.35) and (2.36), given respectively for any smooth function on $\mathbb{R}^{3k+3k}$ by

$$L_\beta = \beta \sum_{i=1}^k \left[\frac{\partial H}{\partial p_i} \frac{\partial}{\partial q_i} - \left(\frac{\partial H}{\partial q_i} + \sigma^2 \frac{\partial H}{\partial p_i} \right) \frac{\partial}{\partial p_i} \right] + \sigma^2 \sum_{i=1}^k \frac{\partial^2}{\partial p_i^2}$$

and for any smooth function g on $\mathbb{R}^{3k}$ by

$$\overline{L}_\beta(g) = -\beta \sum_{i=1}^k \frac{\partial V}{\partial q_i} \frac{\partial g}{\partial q_i} + \sum_{i=1}^k \frac{\partial^2 g}{\partial q_i^2} = e^{\beta V} \sum_{i=1}^k \frac{\partial}{\partial q_i} \left(e^{-\beta V} \frac{\partial g}{\partial q_i} \right)$$

In this stochastic framework, the conservation properties (2.33) take the following form.

Lemma 2.6.1 *For any $\beta \in \mathbb{R}$, we have*

$$\mu_\beta L_\beta = 0 \quad and \quad \overline{\mu}_\beta \overline{L}_\beta = 0$$

In addition $\overline{\mu}_\beta$ is $\overline{L}_\beta$-reversible, in the sense that for any smooth couple of functions (g, h) with compact support on $\mathbb{R}^{3k}$ we have

$$\overline{\mu}_\beta \left(g \, \overline{L}_\beta(h) \right) = \overline{\mu}_\beta \left(\overline{L}_\beta(g) \, h \right)$$

Proof:

By a simple integration by part formula, for any smooth function f with compact support on $\mathbb{R}^{3k+3k}$ we check that

$$\int e^{-\beta H(x)} \, L_\beta(f)(x) \, dx$$

$$= -\beta \sum_{i=1}^k \int f(x) \frac{\partial}{\partial q_i} \left(e^{-\beta H} \frac{\partial H}{\partial p_i} \right)(x) \, dx$$

$$+\beta \sum_{i=1}^k \int f(x) \frac{\partial}{\partial p_i} \left(e^{-\beta H} \left(\frac{\partial H}{\partial q_i} + \sigma^2 \frac{\partial H}{\partial p_i} \right) \right)(x) \, dx$$

$$+\sigma^2 \sum_{i=1}^k \int f(x) \frac{\partial^2}{\partial p_i^2} \left(e^{-\beta H} \right)(x) \, dx$$

This implies that

$$\mu_\beta \left(L_\beta(f) \right)$$

$$= \sum_{i=1}^k \mu_\beta \left\{ f \left[\left(\beta^2 \frac{\partial H}{\partial p_i} \frac{\partial H}{\partial q_i} - \beta \frac{\partial^2 H}{\partial q_i \partial p_i} \right) - \beta^2 \frac{\partial H}{\partial p_i} \left(\frac{\partial H}{\partial q_i} + \sigma^2 \frac{\partial H}{\partial p_i} \right) \right] \right\}$$

$$+ \sum_{i=1}^k \mu_\beta \left\{ f \left[\beta \left(\frac{\partial^2 H}{\partial q_i \partial p_i} + \sigma^2 \frac{\partial^2 H}{\partial p_i^2} \right) - \sigma^2 \beta \frac{\partial^2 H}{\partial p_i^2} + \sigma^2 \beta^2 \left(\frac{\partial H}{\partial p_i} \right)^2 \right] \right\} = 0$$

In much the same way, for any smooth functions (g, h) with compact support on $\mathbb{R}^{3k}$ we find that

$$\int e^{-\beta V(q)} g(q) \, \overline{L}_\beta(h)(q) \, dq = \sum_{i=1}^k \int g(q) \frac{\partial}{\partial q_i} \left(e^{-\beta V} \frac{\partial h}{\partial q_i} \right)(q) \, dq$$

$$= -\sum_{i=1}^k \int e^{-\beta V(q)} \frac{\partial g}{\partial q_i}(q) \frac{\partial h}{\partial q_i}(q) \, dq$$

$$= \sum_{i=1}^k \int h(q) \frac{\partial}{\partial q_i} \left(e^{-\beta V} \frac{\partial g}{\partial q_i} \right)(q) \, dq$$

This clearly ends the proof of the lemma. ∎

2.6.2 Feynman-Kac path integration models

We let $X_t = (q_t, p_t)$ be the solution of the hypo-elliptic Langevin dynamics (2.35) in the phase space E^{3k+3k}. We consider a C^1 inverse cooling schedule $\beta : t \in \mathbb{R}_+ \mapsto \beta_t$, and we set $\beta'_t := \frac{d\beta_t}{dt}$. Using the arguments presented in Section 1.3.2, we find that

$$\mu_{\beta_t}(f) = \eta_t(f) := \gamma_t(f)/\gamma_t(1)$$

with the unnormalized Feynman-Kac measures

$$\gamma_t(f) := \mathbb{E}\left(f(X_t) \, \exp\left(-\int_0^t \beta_s' \, H(X_s)ds\right)\right)$$

Using the fact that

$$\frac{d}{dt}\log \gamma_t(1) = -\beta_t' \, \frac{\gamma_t(H)}{\gamma_t(1)} = -\beta_t' \, \eta_t(H) = -\beta_t' \, \mu_{\beta_t}(H) = \frac{d}{dt}\log \mathcal{Z}_{\beta_t}$$

we conclude that

$$\begin{aligned}
\gamma_t(1) &= \mathbb{E}\left(\exp\left(-\int_0^t \beta_s' \, H(X_s)ds\right)\right) \\
&= \exp\left(-\int_0^t \beta_s' \, \eta_s(H) \, ds\right) = \mathcal{Z}_{\beta_t}/\mathcal{Z}_{\beta_0}
\end{aligned}$$

This formula is known as the Jarzinsky equality [129, 130, 342, 343]. In statistical physics, the weight functions

$$\mathcal{W}_t(X) = \int_0^t \beta_s' \, H(X_s)ds$$

represent the out of equilibrium virtual work of the system on the time horizon t. The discrete time version of these Feynman-Kac formulae are discussed in Section 3.4.3. In much the same way, if we consider the solution $X_t = q_t$ of the ellipic overdamped Langevin dynamics (2.36) in the phase space E^{3k}, we find that

$$\overline{\mu}_{\beta_t}(f) = \eta_t(f) := \gamma_t(f)/\gamma_t(1)$$

with the unnormalized Feynman-Kac measures

$$\gamma_t(f) := \mathbb{E}\left(f(X_t) \, \exp\left(-\int_0^t \beta_s' \, V(X_s)ds\right)\right)$$

In this situation, we have the free energy formulae

$$\gamma_t(1) = \mathbb{E}\left(e^{-\int_0^t \beta_s' \, V(X_s)ds}\right) = e^{-\int_0^t \beta_s' \, \eta_s(V) \, ds} = \overline{\mathcal{Z}}_{\beta_t}/\overline{\mathcal{Z}}_{\beta_0}$$

These models are particular instances of the Feynman-Kac models discussed in Section 1.2.2.3. Their continuous time mean field models are defined in Section 1.2.2.5, as well as in Chapter 5. We can alternatively use the discrete time embedding models discussed in Section 1.4.3.2, or their discrete time approximations presented in Section 1.3, and in Section 1.5.5. In molecular dynamics simulation literature, the interacting particles in the mean field model are called walkers or replica, and the test functions f are often called observables.

For a more thorough discussion on these molecular dynamics models, the ergodic properties of the Langevin diffusions (2.35), and (2.36), including their mean field particle interpretations, we refer the reader to the series of articles [92, 161, 176, 178, 179, 180, 392, 498, 499], and the book by T. Lelièvre, M. Rousset, and G. Stoltz [393].

2.7 Risk analysis and rare event simulation

The analysis of rare events arises in various scientific areas including physics, biology, engineering science, and financial mathematics.

For instance, in nuclear physics, we might be interested in computing the probability that some radiation escapes from some containment before being absorbed by some obstacle. In biology, these rare events may be related to extinction probabilities of some population evolution model. In engineering sciences, these critical events are often related to a catastrophic failure, such as a buffer overflows in communication networks. Finally, in financial mathematics, they arise in the analysis of portfolio credit risk models. In this context, the critical events represent ruin processes, or credit payment default probabilities. Importance sampling techniques are perhaps one of the most widely used alternative to crude Monte Carlo simulation of unlikely events. The idea is to generate samples from a different judiciously chosen distribution, rather than from the distribution of interest. These statistical techniques have two main drawbacks. Very often, the twisted distribution cannot be chosen as we would like, since we need to have a dedicated technique to sample random variables w.r.t. these measures. On the other hand, these importance sampling techniques are intrusive in the sense that we need to twist the reference random process, so that to produce unphysical trajectories. In Section 2.7.1, we present a nonintrusive mean field IPS technique for the simulation of importance sampling distributions without altering the nature of the reference process. Further details on these models, including applications in fiber optics communication and financial risk analysis, can also be found in a couple of articles [98, 171, 172].

Section 2.7.2 is dedicated to mean field multilevel simulation. These techniques are often termed multilevel splitting particle methods or sequential Monte Carlo samplers in the literature on rare event simulation, Further detail on these IPS models can be found in the review article [184], as well as in the series of articles [124, 125, 346].

The final section, Section 2.7.3, is concerned with the mean field IPS computation of Dirichlet problems with boundary conditions. These problems arise in a variety of application areas of physics, including fluid mechnanics and plasma dynamics, as well as in optics and traffic engineering. For a detailed discussion on these problems in the context of elliptic-hypebolic equations of Keldysh type we refer the reader to the monograph [456].

2.7.1 Importance sampling and twisted measures

Computing the probability of some events of the form $\{V_n(X_n) \geq a\}$, for some energy like function V_n and some threshold a, is often performed using the importance sampling distribution of the state variable X_n with some multiplicative Boltzmann weight function $\exp(\beta V_n(X_n))$, associated with some inverse temperature parameter β. These twisted measures can be described by a Feynman-Kac model in transition space by setting

$$G_n(X_{n-1}, X_n) = \exp\{\beta[V_n(X_n) - V_{n-1}(X_{n-1})]\}$$

For instance, it is easily checked that

$$\mathbb{P}(V_n(X_n) \geq a) = \mathbb{E}\left(\mathbf{f_n}(\mathbf{X_n}) \prod_{0 \leq p < n} G_p(\mathbf{X}_p)\right)$$

with the function $\mathbf{f}_n(\mathbf{X}_n) = 1_{V_n(X_n) \geq a} \, e^{-\beta V_n(X_n)}$, and the potential function and the reference Markov chain

$$\mathbf{X}_n = (X_n, X_{n+1}) \quad \text{and} \quad G_n(\mathbf{X}_n) = \exp\{\beta(V_{n+1}(X_{n+1}) - V_n(X_n))\}$$

We let $\mathbb{Q}_n$ be the Feynman-Kac model (1.37) associated with the reference Markov chain $\mathbf{X}_n$ and the potential function G_n. In the same vein, we have the Feynman-Kac formulae

$$\mathbb{E}(\varphi_n(X_0, \ldots, X_n) \mid V_n(X_n) \geq a) = \mathbb{Q}_n(F_{n,\varphi_n})/\mathbb{Q}_n(F_{n,1})$$

with the function $F_{n,\varphi_n}(X_0, \ldots, X_n) = \varphi_n(X_0, \ldots, X_n) \, 1_{V_n(X_n) \geq a} \, e^{-\beta V_n(X_n)}$. The mean field IPS simulation of these Feynman-Kac distributions is defined in Section 1.5 and in Section 2.2.

2.7.2   Multilevel splitting simulation

We consider some Markov chain $(X'_n)_{n \geq 0}$ taking values in some finite state space E'. We assume that the chain X'_n starts in some given subset $X'_0 \in A \subset E'$ with a given distribution ν_0. We also let (B, C) be a pair of subsets (B, C) such that $A \cap C = \emptyset = B \cap C$. We also assume that the triplet (A, B, C) is chosen so that for any initial state $x \in A$ the chain X'_n hits one of the sets B, or C in finite time.

We let T_A be the entrance time of X' into a given subset A. One would like to estimate the probability that the chain hits B before C

$$\mathbb{P}(T_{B \cup C} < T_C) = \mathbb{P}(X'_{T_{B \cup C}} \in B) = \mathbb{E}(1_B(X'_{T_{B \cup C}}))$$

and the law of the excursions given the fact that it reached B before C

$$\text{Law}(X'_t \, ; \, 0 \leq t \leq T_{B \cup C} \mid T_{B \cup C} < T_C)$$

Of course we have implicitly assumed that $\mathbb{P}(T_{B \cup C} < T_C) > 0$ so that the conditional distributions are well defined. During its excursion from A to B, the process eventually visits a decreasing sequence of level sets $(B_n)_{n=0,\ldots,m}$

$$A = B_0 \supset B_1 \supset \ldots \supset B_m = B \tag{2.37}$$

This decomposition reflects the successive gateways the stochastic process needs to cross before entering into the relevant rare event.

To simplify the presentation, we slightly abuse the notation, and we write T_n instead of $T_{B_n \cup C}$. In this simplified notation, to capture the behavior of X between the different levels we introduce the excursion-valued Markov chain

$$X_n = (T_n, (X'_t \; ; \; T_{n-1} \leq t \leq T_n)) \in E = \cup_{p \leq q}(\{q\} \times (E')^{(q-p+1)}) \quad (2.38)$$

Under our assumptions, these entrance times are finite and

$$(T_{B \cup C} < T_C) = (T_m < T_C) = \bigcap_{1 \leq p \leq m} (T_p < T_C)$$

To check whether or not the n-th excursion has reached the desired n-th level, we consider the potential functions on E defined for each $n \in \{0, \ldots, m\}$ and $x = (x_q)_{p \leq q \leq r} \in (E')^{(r-p+1)}$, by $G_n(r, x) = 1_{B_n}(x_r)$. In this notation we have for each $n \leq m$

$$(T_n < T_C) = \bigcap_{1 \leq p \leq n}(T_p < T_C) = \bigcap_{1 \leq p \leq n}(G_p(X_p) = 1)$$

$$(X_0, \ldots, X_n)$$

$$= ((0, X'_0), (T_1, (X'_t \; ; \; 0 \leq t \leq T_1)), \ldots, (T_n, (X'_t \; ; \; T_{n-1} \leq t \leq T_n)))$$

If we write $[X'_t \; ; \; 0 \leq t \leq T_n]$ instead of $(X_0, \ldots, X_n)$, the sequence of excursions of X' between the levels, then for any $n \leq m$ and any function f_n on the product space E^{n+1} we have the Feynman-Kac formulae

$$\mathbb{E}_{\nu_0}\left(f_n(X_0, \ldots, X_n) \prod_{p=1}^{n} G_p(X_p)\right) = \mathbb{E}_{\nu_0}\left(f_n([X'_t \; ; \; 0 \leq t \leq T_n]) \, 1_{T_n < T_C}\right)$$

$$(2.39)$$

The mean field IPS simulation of these Feynman-Kac distributions is defined in Section 1.5 and in Section 2.2.

2.7.3 Dirichlet problems with hard boundary conditions

We consider the same excursion model discussed in Section 2.7.2 but we replace the potential function by the function G_n on E defined for any $n \in \{0, \ldots, m\}$, $0 \leq p \leq r$, and $x = (x_q)_{p \leq q \leq r} \in (E')^{(r-p+1)}$ by

$$G_n(r, x) = 1_{B_n}(x_r) \prod_{p < q \leq r} G'(x_q) \quad (2.40)$$

with some nonnegative functions G' on the finite set E'. In this situation, the r.h.s. expectation in (2.39) is given by

$$\mathbb{E}_{\nu_0}\left(f_n([X'_t \; ; \; 0 \leq t \leq T_n]) \, 1_{B_n}(X'_{T_n}) \prod_{p=1}^{T_n} G'(X'_p)\right)$$

We recall that these Feynman-Kac formulae can be computed using the mean field IPS simulation described in Section 1.5 and in Section 2.2. Next, we examine the excursion-valued models (2.37) when $A = E' - C$, and $(B_n \cap C) = \emptyset$. For $\nu_0 = \delta_x$, with $x \in A$, $n = m$, and any function f on E', the above expectations are given by the following function

$$h(x) = \mathbb{E}_x \left(f(X'_T) 1_B(X'_T) \prod_{p=1}^{T} G'(X'_p) \right)$$

with the first time $T(= T_m)$ the process X' hits the domain $D := (B \cup C)$. Using the fact that $x \in D \Rightarrow T = 0$, we extend the function h on C by setting $h(x) = f(x) 1_B(x)$. By a conditioning argument, for any $x \notin D$, we have

$$h(x) = \mathbb{E}_x \left(G'(X'_1) \, \mathbb{E}_{X'_1} \left(f(X'_T) 1_B(X'_T) \prod_{p=1}^{T} G'_p(X'_p) \right) \right) = \mathbb{E}_x \left(G'(X'_1) h(X'_1) \right)$$

From the above discussion, if $M'(x, y)$ is the Markov transition of the chain X', then we see that the function h satisfies the following Dirichlet problem with hard boundary conditions

$$\begin{cases} M'(G'h)(x) &= h(x) & \text{for} \quad x \notin D \\ h(x) &= f(x) 1_B(x) & \text{for} \quad x \in D \end{cases}$$

For a more thorough discussion on the Dirichlet problem for more general models, we refer the reader to the book [163].

Part II

Feynman-Kac models

Chapter 3

Discrete time Feynman-Kac models

3.1 A brief treatise on evolution operators

The mean field Feynman-Kac models presented in Section 1.5.5 can be interpreted as genetic type particle models, with mutation-selection transitions.

The mutation process is defined in terms of abstract Markov transitions on measurable state spaces that may depend on the time parameter. This rather abstract framework encapsulates random excursion models between level subsets, Markov bridges, and other classes of historical type processes. Some illustrations of these path space models are presented in Section 1.1.2. We also refer to the construction of continuous models (1.46) presented in Section 1.4.3, and the discussion of historical processes given in Section 1.4.3.1.

The selection transitions of genetic type IPS models can be interpreted as *a universal acceptance-rejection* sampling technique, equipped with an interacting recycling mechanism, associated with potential functions. These rejection style simulation transitions can be interpreted in various ways, depending on the application model they represent. For instance, in the context of particle absorption models, the potential functions are interpreted in terms of killing and absorption rates. In evolutionary computing, the fitness functions can be thought as performance weight functions at the level of the individuals. From the pure mathematical point of view, these recycling mechanisms are related to a change of probability mass distribution at the level of populations. The mass variation is expressed in terms of a nonlinear Boltzmann-Gibbs transformation on the set of distributions. In Bayesian statistics literature, these transformations are also called the Bayes' rules.

In the rest of the section, we provide an introduction to these two evolution operators. We have chosen to describe in some detail these operators on some abstract state spaces. This abstract framework permits greater flexibility and generality, with very little extra effort, in the study of the regularity properties of Feynman-Kac type semigroups, as well as in the convergence analysis of mean field IPS models.

3.1.1   Markov operators

This section provides a brief treatise on Markov processes on general measurable state spaces. We also present some contraction properties of Markov semigroups. These results are used in Chapter 12, dedicated to the analysis of Feynman-Kac semigroups, and their stability properties.

We let $(E_n, \mathcal{E}_n)$ be a collection of measurable state spaces. An E_n-valued Markov chain X_n is defined by a collection M_n of Markov transitions from E_{n-1} into E_n, and some initial distribution η_0 on E_0.

By construction, we have

$$\mathbb{P}\left((X_0, \ldots, X_n) \in d(x_0, \ldots, x_n)\right) = \eta_0(dx_0) \prod_{p=1}^{n} M_p(x_{p-1}, dx_p)$$

Given $X_p = x_p$ for some $p < n$, the law of the random state X_n is given by

$$\mathbb{P}\left(X_n \in dx_n \mid X_p = x_p\right) := M_{p,n}(x_0, dx_p)$$

with the semigroup $M_{p,n}$ of integral operators defined by

$$M_{p,n} = M_{p+1} \ldots M_{n-1} M_n$$

with the convention $M_{n,n} = Id$, the identity operator. In the same way, the distribution η_n of X_n satisfies the integral formula $\eta_n = \eta_p M_{p,n}$.

Definition 3.1.1 *The historical process associated with some reference Markov chain X_n is defined by the sequence of random paths*

$$\mathbf{X}_n = (X_0, \ldots, X_n) \in \mathbf{E_n} := (E_0 \times \ldots \times E_n)$$

Notice that the Markov transitions of the chain $\mathbf{X}_n$ is given for any

$$\boldsymbol{y_n} = ((y_0, \ldots, y_{n-1}), y_n) = (\boldsymbol{y_{n-1}}, y_n) \in \mathbf{E_n} = (\mathbf{E_{n-1}} \times E_n)$$

and any $\boldsymbol{x_{n-1}} \in \mathbf{E_{n-1}}$ by the following formulae

$$\mathbf{M}_n(\boldsymbol{x_{n-1}}, d\boldsymbol{y_n}) = \delta_{\boldsymbol{x_{n-1}}}(d\boldsymbol{y_{n-1}}) \, M_n(y_{n-1}, dx_n) \qquad (3.1)$$

We recall that the Dobrushin ergodic coefficient $\beta(M)$ of a Markov transition M from E_1 into E_2, introduced on page xli, is the norm of the operator M from $\mathcal{M}_0(E_1)$ into $\mathcal{M}_0(E_2)$; that is, we have the equivalent formulations

$$\beta(M) \quad = \quad \sup \|M(x, .) - M(y, .)\|_{\mathrm{tv}} = \sup_{\mu \in \mathcal{M}_0(E_1)} \|\mu M\|_{\mathrm{tv}} / \|\mu\|_{\mathrm{tv}}$$

with the first supremum taken over all $(x, y) \in E_1^2$. We also have the contraction inequalities

$$\|\mu M - \nu M\|_{\mathrm{tv}} \leq \beta(M) \, \|\mu - \nu\|_{\mathrm{tv}} \quad \text{and} \quad \mathrm{osc}(M(f)) \leq \beta(M) \, \mathrm{osc}(f) \qquad (3.2)$$

A detailed proof of these well known formulae can be found in [163]. Several rather crude estimates can be underlined. For instance, we have

$$(\forall x, y, z \in E_1 \quad M(x, dz) \geq \epsilon\, M(y, dz)) \Rightarrow \beta(M) \leq (1 - \epsilon)$$

In the same vein, we have $\beta(M) \leq (1 - \epsilon)$ as soon as

$$\forall x, y \in E_1 \quad M(x, dy) \geq \epsilon\, \mu(dy) \quad \text{for some} \quad \mu \in \mathcal{P}(E_2) \tag{3.3}$$

The last assertion comes from the fact that the Markov transition

$$M_\mu(x, dy) = \frac{1}{1 - \epsilon} \left[M(x, dy) - \epsilon \mu(dy) \right]$$

is such that

$$[M(x, dz) - M(y, dz)] = (1 - \epsilon)\ [M_\mu(x, dz) - M_\mu(y, dz)]$$

Given a pair of Markov transitions M_1 from E_1 into E_2, and M_2 from E_2 into E_3 , and any $f_3 \in \mathrm{Osc}(E_3)$ we have

$$\mathrm{osc}(M_1 M_2(f_3)) \leq \beta(M_2)\, \mathrm{osc}(M_1(f_2)) \leq \beta(M_1)\beta(M_2)\, \mathrm{osc}(f_1)$$

with

$$f_2 = M_2(f_3)/\beta(M_2) \in \mathrm{Osc}(E_2) \quad \text{and} \quad f_1 = M_1(f_2)/\beta(M_1) \in \mathrm{Osc}(E_1)$$

This clearly implies that $\beta(M_1 M_2) \leq \beta(M_1)\beta(M_2)$. Iterating this argument, for any collection of Markov transitions M_n such that $\beta(M_n) \leq (1 - \epsilon)$ we have the quantitative contraction estimate

$$\beta(M_1 M_2 \ldots M_n) \leq \prod_{1 \leq p \leq n} \beta(M_p) \leq (1 - \epsilon)^n$$

We end this section with an interesting contraction property of a Markov transition

$$M_G(x, dy) = \frac{M(x, dy)G(y)}{M(G)(x)} = \Psi_G(\delta_x M)(dy) \tag{3.4}$$

associated with a potential function G, with

$$g = \sup_{x,y} G(x)/G(y) < \infty \tag{3.5}$$

It is easily checked that

$$\begin{aligned} |M_G(f)(x) - M_G(f)(y)| &= |\Psi_G(\delta_x M)(f) - \Psi_G(\delta_y M)(f)| \\ &\leq g \, \|\delta_x M - \delta_y M\|_{\mathrm{tv}} \end{aligned}$$

from which we conclude that

$$\beta\,(M_G) \leq g\,\beta\,(M) \tag{3.6}$$

Further details on the regularity properties of the transformations Ψ_G are provided in Section 3.1.3.

3.1.2   Markov chain Monte Carlo models

In this section, we provide a brief introduction to MCMC models. We present the traditional Metropolis-Hasting algorithm and Gibbs sampling principles.

Markov chain Monte Carlo algorithms are rather standard stochastic simulation methods for sampling from a given target distribution, say μ on some measurable state space E. These algorithms are particular instances of the Markov chain models presented in Section 3.1.1. Because of their importance in practice, but also to illustrate the abstract Markov chain models introduced in Section 3.1.1, we provide a brief discussion on these Markov chain Monte Carlo algorithms. For a detailed discussion on MCMC models, and their stochastic analysis, we refer the reader to the review articles by P. Diaconis [206, 211, 216], and references therein.

The central idea behind MCMC methodologies is to design a judicious Markov transition $P(x, dy)$, with nice stability properties, that has the desired probability measure $\mu = \mu P$ as its invariant measure. After a rather large number of runs, and when the chain is sufficiently stable, the ergodic theorem tells us that the occupation measures of the random states X_n of the chain with Markov transition P approximate μ.

The Metropolis-Hastings algorithm is the most famous MCMC model of current use in practice. For a detailed discussion on this model, we refer the reader to the pioneering article by N. Metropolis, A. Rosenbluth, M. Rosenbluth, A. Teller, and E. Teller [437], the more recent review article by N. Metropolis [436], and the series of articles by P. Diaconis [212, 213, 215].

The mathematical analysis of this Markov chain model is also well developed. We refer the reader to the series of seminal articles by P. Diaconis and his co-authors [207, 208, 209, 210, 214, 216]. These works reveal fascinating connections between the design and the performance analysis of MCMC models with powerful pure and applied mathematical techniques, ranging from representation theory, micro-local analysis, log-Sobolev inequalities, and spectral analysis. Besides the fact that these techniques provide very sharp rates of convergence, it is clearly, of course, out of the scope of this book to review these methods. In this section, we content ourselves with presenting one of the simplest ways to analyze the convergence of an MCMC algorithm.

Firstly, we choose a Markov transition K on some state space E such that $K^m(n, dy) \geq \epsilon \nu(dy)$, for some $m \geq 1$ $\epsilon > 0$, and some $\nu \in P(E)$. For instance, this condition is met for aperiodic and irreducible Markov transitions K on some finite state space E. It is also satisfied for absolutely continuous Markov transitions on compact spaces, as well as for bi-Laplace transitions and Gaussian transitions with constant drift outside some compact domain [161, 163, 170]. This ensures that $\beta(K^m) \leq (1 - \epsilon)$.

We consider the probability measures

$$(\mu \otimes K)_0 (d(x, y)) \quad := \quad \mu(dx) K(x, dy) := (\mu \otimes K)_1 (d(y, x))$$

We assume that $(\mu \otimes K)_1 \ll (\mu \otimes K)_0$, and we set

$$G := d \, (\mu \otimes K)_1 \, / d \, (\mu \otimes K)_0$$

The Metropolis-Hastings model is a Markov chain with μ-reversible acceptance-rejection style transitions of the following form

$$P(x, dy) = K(x, dy) \, a(x, y) + \left(1 - \int K(x, dz) \, a(x, z)\right) \delta_x(dy)$$

To guarantee the reversibility property, we often choose one of the following acceptance rates

$$a = G/(1 + G) \qquad \text{or} \qquad a = 1 \wedge G \tag{3.7}$$

When the proposal transition $K(x, .) = \nu$ is given by some probability measure ν, that does not depend on the current state x, the resulting MCMC sampler is sometimes called an independent Metropolis-Hastings model.

Under our assumptions we have

$$P^m(x, dy) \geq \epsilon \, a_\star^m \, \nu(dy) \quad \text{with} \quad a_\star := \inf_{(x,y) \in E^2} a(x, y)$$

This implies that

$$\beta(P^m) \leq (1 - \epsilon \, a_\star^m)$$

We end this section with a discussion on the Gibbs sampler associated with some measure μ defined on some product state space $E = (E_1 \times E_2)$. We assume that the following disintegration property is satisfied

$$\mu = \mu_1 \otimes L_{1,2} = \mu_2 \otimes L_{2,1}$$

with the first and second marginals, μ_1 and μ_2, and the corresponding conditional probability measures $L_{1,2}$ and $L_{2,1}$.

In this situation, one natural choice of proposal transition is

$$K = K_1 K_2$$

with the proposal transitions given for any $i \in \{1, 2\}$ by

$$K_i((x^i, x^j), d(y^i, y^j)) := \delta_{x^i}(dy^i) L_{i,j}(y^i, dy^j)$$

By construction, the transitions K_i are reversible w.r.t. the measure μ, so that the Metropolis-Hasting model with proposal transition has unit acceptance rate. The resulting Markov chain model is often called the Gibbs sampler. These constructions can be extended to product state spaces of any dimension.

3.1.3　Boltzmann-Gibbs transformations

This section is concerned with Boltzmann-Gibbs transformations on the space of probability measures. These transformations are essential to describe the evolution equation of infinite population models. In the context of mean field Feynman-Kac models, the Boltzmann-Gibbs transformation is the limiting probability mass transformation associated with the selection operator. These limiting evolution models are described in some detail in Section 4.1.3.

In the present section, we briefly explore the regularity properties of these measure valued transformations, including Taylor's type first order decompositions. This approach will be particularly useful to analyze the stability properties of Feynman-Kac nonlinear semigroups, as well as the convergence properties of their mean field approximations.

Definition 3.1.2 *Given a positive and bounded potential function G on E, we denote by Ψ_G the Boltzmann-Gibbs mapping from $\mathcal{P}(E)$ into itself, defined for any $\mu \in \mathcal{P}(E)$ by*

$$\Psi_G(\mu)(dx) = \frac{1}{\mu(G)}\ G(x)\ \mu(dx) \tag{3.8}$$

There is no loss of generality to assume that G is a $]0,1]$-valued function. For $[0,1]$-valued potential functions, the transformation is only defined on measures μ s.t. $\mu(G) > 0$. To avoid unnecessary repetition of technical abstract conditions, and unless otherwise stated, in the further development of this section we frame the standing assumption that G is chosen so that

$$g := \sup_{x,y}\left(G(x)/G(y)\right) < \infty \tag{3.9}$$

To begin with, we start with a rather strong Lipschitz type estimate.

Lemma 3.1.3 *For any pair of measures μ and ν, and any bounded positive function G, we have*

$$\|\Psi_G(\mu) - \Psi_G(\nu)\|_{\mathrm{tv}} \leq \frac{\|G\|}{\mu(G) \vee \nu(G)}\ \|\mu - \nu\|_{\mathrm{tv}} \tag{3.10}$$

Proof:
There is no loss of generality to assume that G is a $]0,1]$-valued function. We prove (3.10) using the fact that the mapping Ψ_G can be expressed in the following form

$$\Psi_G(\mu) = \mu S_\mu \tag{3.11}$$

with the Markov transitions

$$S_\mu(x, dy) = G(x)\ \delta_x(dy) + (1 - G(x))\ \Psi_G(\mu)(dy) \tag{3.12}$$

On the other hand, we notice that

$$\Psi_G(\mu) - \Psi_G(\nu) = (\mu - \nu)S_\mu + \nu(S_\mu - S_\nu)$$
$$\nu(S_\mu - S_\nu) = (1 - \nu(G))\left[\Psi_G(\mu) - \Psi_G(\nu)\right]$$

from which we find the formula

$$\Psi_G(\mu) - \Psi_G(\nu) = \frac{1}{\nu(G)}\ (\mu - \nu)S_\mu \qquad (3.13)$$

In addition, using (3.3) we have

$$S_\mu(x, .) \geq (1 - \|G\|)\ \Psi_G(\mu) \implies \beta(S_\mu) \leq \|G\|$$

The end of the proof of (3.10) is now clear. This ends the proof of the lemma. ∎

The second part of this section is concerned with the weak regularity properties of the Boltzmann-Gibbs transformations. Our Taylor's type first order expansions will be expressed in terms of the first order integral operators defined below.

Definition 3.1.4 *For any positive potential function G, and any measures ν on E, we denote by $d_\nu\Psi_G$, and $d'_\nu\Psi_G$, the integral operators from $\mathcal{B}_b(E)$ into itself defined for any $f \in \mathcal{B}_b(E)$ by*

$$d_\nu\Psi_G(f) \quad := \quad G_\nu\ (f - \Psi_G(\nu)(f)) \quad and \quad G_\nu := G/\nu(G)$$

$$d'_\nu\Psi_G(f) \quad := \quad (\nu(G)/\|G\|)\ d_\nu\Psi_G(f)$$
$$= \quad G'\ (f - \Psi_G(\nu)(f)) \quad and \quad G' := G/\|G\|\ (\leq 1) \quad (3.14)$$

By construction, for any $(x, f) \in (E \times \mathcal{B}_b(E))$ we have

$$d_\nu\Psi_G(f)(x) \quad := \quad \int d_\nu\Psi_G\ (x, dy)\ f(y) = G_\nu(x)\ (f(x) - \Psi_G(\nu)(f))$$

or equivalently

$$d_\nu\Psi_G(x, dy) = G_\nu(x)\ (\delta_x - \Psi_G(\nu))\ (dy)$$

We readily check that

$$\|d'_\nu\Psi_G(f)\| \leq \mathrm{osc}(f) \quad and \quad \|d_\nu\Psi_G(f)\| \leq g\ \mathrm{osc}(f) \qquad (3.15)$$

Lemma 3.1.5 *For any positive potential function G, any measures μ and ν, and any function $f \in \mathrm{Osc}(E)$, we have the first order decompositions*

$$[\Psi_G(\mu) - \Psi_G(\nu)]\ (f)$$

$$= \frac{1}{\mu(G_\nu)}\ (\mu - \nu)d_\nu\Psi_G(f) = \frac{1}{\mu(G')}\ (\mu - \nu)d'_\nu\Psi_G(f) \qquad (3.16)$$

$$= (\mu - \nu)d_\nu\Psi_G(f) - \frac{1}{\mu(G_\nu)}\ (\mu - \nu)^{\otimes 2}\ [G_\nu \otimes d_\nu\Psi_G(f)]$$

In addition, we have the estimates

$$|[\Psi_G(\mu) - \Psi_G(\nu)](f)| \le g \left[|(\mu - \nu)d_\nu\Psi_G(f)| \wedge |(\mu - \nu)d'_\nu\Psi_G(f)|\right]$$

$$\|d_\mu\Psi_G(f) - d_\nu\Psi_G(f)\| \le g \left(|(\nu - \mu)(G_\nu)| + |(\nu - \mu)d_\nu\Psi_G(f)|\right) \quad (3.17)$$

Proof:
We prove the first assertion using the decompositions

$$\frac{1}{\mu(G_\nu)} (\mu - \nu)(d_\nu\Psi_G(f))$$

$$= \left(1 - \frac{1}{\mu(G_\nu)} (\mu - \nu)(G_\nu)\right) (\mu - \nu)(d_\nu\Psi_G(f))$$

and

$$|[\Psi_G(\mu) - \Psi_G(\nu)](f)| \le \frac{1}{\mu(G')} |(\mu - \nu)(d'_\nu\Psi_G(f))|$$
$$\le g |(\mu - \nu)(d'_\nu\Psi_G(f))|$$

On the other hand, we have the decomposition

$$d_\mu\Psi_G(f) - d_\nu\Psi_G(f)$$

$$= G_\mu \left\{(\nu - \mu)(G_\nu) (f - \Psi_G(\mu)(f)) + (\nu - \mu)(d_\nu\Psi_G(f))\right\}$$

The proof of the second assertion is a direct consequence of this decomposition. This ends the proof of the lemma. ∎

As in (2.3), the integral operators $d_\nu\Psi_G$ introduced in definition 3.1.4, and the decompositions (3.16) can be interpreted as Gâteaux derivatives, and weak Taylor's type expansions of the first order. For instance, for any measure $\mu \in \mathcal{P}(E)$, if we set $\rho = (\mu - \nu) \in \mathcal{M}_0(E)$ then for any $f \in \mathcal{B}_b(E)$ we have

$$\lim_{\epsilon \to 0} \frac{1}{\epsilon} [\Psi_G(\nu + \epsilon\rho) - \Psi_G(\nu)](f) = \rho(d_\nu\Psi_G(f)) = \frac{\partial}{\partial\epsilon}\Psi_G(\nu+\epsilon\rho)(f)_{|\epsilon=0} \quad (3.18)$$

as well as

$$\lim_{\epsilon \downarrow 0} \left\|\frac{1}{\epsilon} [\Psi_G(\mu + \epsilon\rho) - \Psi_G(\mu)] - \rho \, d_\mu\Psi_G\right\|_{tv} = 0$$

Lemma 3.1.6 *For any couple of positive potential functions G_1 and G_2, and for any $\nu \in \mathcal{P}(E)$, we have the composition differential rule*

$$d_\nu (\Psi_{G_1} \circ \Psi_{G_2}) = d_\nu\Psi_{G_2} \, d_{\Psi_{G_2}(\nu)}\Psi_{G_1}$$

Proof:

Notice that for any $\mu \in \mathcal{P}(E)$ and $\epsilon \in]0, 1[$, we have

$$\Psi_{G_1}\left(\Psi_{G_2}(\nu) + \epsilon \left[\frac{1}{\epsilon}\left(\Psi_{G_2}(\nu + \epsilon\rho) - \Psi_{G_2}(\nu)\right)\right]\right) = \Psi_{G_1 G_2}(\nu + \epsilon\rho)$$

with the null mass measure $\rho = (\mu - \nu) \in \mathcal{M}_0(E)$.

Using this formula, we prove that

$$\frac{1}{\epsilon}\left(\Psi_{G_1}\left(\Psi_{G_2}(\nu) + \epsilon \left[\frac{1}{\epsilon}\left(\Psi_{G_2}(\nu + \epsilon\rho) - \Psi_{G_2}(\nu)\right)\right]\right) - \Psi_{G_1}\left(\Psi_{G_2}(\nu)\right)\right)$$

converge, as $\epsilon \to 0$ to the measure

$$\rho \, d_\nu \Psi_{G_2} \, d_{\Psi_{G_2}(\nu)} \Psi_{G_1} = \rho \, d_\nu \left(\Psi_{G_1} \circ \Psi_{G_2}\right)$$

We end the proof taking $\rho = \delta_x$. This ends the proof of the lemma. ∎

3.1.4 Feynman-Kac transformations

Feynman-Kac transformations (1.40) are expressed in terms of the Markov models discussed in Section 3.1.1 and the Boltzmann-Gibbs transformations presented in Section 3.1.3. As we shall see in the further development of Section 4.1.3, these transformations represent the one step evolution of infinite population genetic models.

In this section, we follow the same route as the one taken in Section 3.1.3. We explore the regularity properties of these measure valued transformations, including Taylor's type first order decomposition. This approach is also useful to analyze the stability properties of Feynman-Kac nonlinear semigroups, as well as the convergence properties of their mean field approximations.

Further on in this section, we let G be a positive and bounded potential function on E s.t. (3.9), and we denote by M some Markov transition from E, into some possibly different measurable state space $(F, \mathcal{F})$.

Definition 3.1.7 *We denote by Φ the one step Feynman-Kac mapping from $\mathcal{P}(E)$ into $\mathcal{P}(F)$ defined for any $\mu \in \mathcal{P}(E)$ by*

$$\Phi(\mu) = \Psi_G(\mu)M$$

with the Boltzmann-Gibbs transformation Ψ_G defined in (3.8).

It is instructive to observe that the transformation

$$\widehat{\Phi}(\eta) := \Psi_G(\eta M) = \Psi_{\widehat{G}}(\eta)\widehat{M} \tag{3.19}$$

is equivalent to the one defined above, with the potential function $\widehat{G}$, and the Markov transition $\widehat{M}$ given by

$$\widehat{G} := M(G) \quad \text{and} \quad \widehat{M}(f) := M(Gf)/M(G)$$

In addition, using (3.13) we have the decomposition

$$\Phi(\mu) - \Phi(\nu) = \frac{1}{\nu(G)} \, (\mu - \nu) S_\mu M \qquad (3.20)$$

with the collection of Markov transitions S_μ defined in (3.12). From the previous discussion, we also find the following Lipschitz estimates

$$\|\Phi(\mu) - \Phi(\nu)\|_{\mathrm{tv}} \leq \frac{\|G\|}{\mu(G) \vee \nu(G)} \, \beta(M) \, \|\mu - \nu\|_{\mathrm{tv}} \qquad (3.21)$$

Alternatively, we deduce (3.21) from the couple of inequalities (3.2) and (3.10).

We end this section with the analysis of the weak regularity properties of Feynman-Kac transformations.

Definition 3.1.8 *We denote by $d_\nu \Phi$ and $d'_\nu \Phi$ the integral operators from $\mathcal{B}_b(F)$ into $\mathcal{B}_b(E)$ defined by*

$$d_\nu \Phi := d_\nu \Psi_G M \quad and \quad d'_\nu \Phi := d'_\nu \Psi_G M \qquad (3.22)$$

By construction, for any $f \in \mathcal{B}_b(F)$ and $x \in E$, we have

$$d_\nu \Phi(f)(x) \quad := \quad G_\nu(x) \, [M(f)(x) - \Psi_G(\nu) M(f)]$$

Using (3.15), we readily check that

$$\|d'_\nu \Phi(f)\| \leq \beta(M) \, \mathrm{osc}(f) \quad and \quad \|d_\nu \Phi(f)\| \vee \|d'_\nu \Phi(f)\| \leq g \, \beta(M) \, \mathrm{osc}(f)$$

with the parameter g defined in (3.9). Using the same lines of arguments as the ones we used in the proof of Lemma 3.1.6, we prove the following technical lemma.

Lemma 3.1.9 *We consider some positive potential functions G_1, resp. G_2, on some measurable state spaces E_1, and resp. E_2. We also consider a Markov transition M_1, resp. M_2, from E_1 into E_2, resp. from E_2 into E_3. We let Φ_1, resp. Φ_2, be the Feynman-Kac transformations from $\mathcal{P}(E_1)$ into $\mathcal{P}(E_2)$, resp. from $\mathcal{P}(E_2)$ into $\mathcal{P}(E_3)$, defined for any $(\nu_1, \nu_2) \in (\mathcal{P}(E_1) \times \mathcal{P}(E_2))$ by*

$$\Phi_1(\nu_1) = \Psi_{G_1}(\nu_1) M_1 \quad and \quad \Phi_2(\nu_2) = \Psi_{G_2}(\nu_2) M_2$$

In this situation, for any $\nu \in \mathcal{P}(E_1)$ we have the composition differential rule

$$d_\nu \, (\Phi_2 \circ \Phi_1) = d_\nu \Phi_1 \, d_{\Phi_1(\nu)} \Phi_2$$

We end this section with a direct consequence of Lemma 3.1.5.

Lemma 3.1.10 *For any couple of measures $(\mu, \nu) \in \mathcal{P}(E)^2$, and any function $f \in \mathrm{Osc}(F)$, we have the first order decomposition*

$$[\Phi(\mu) - \Phi(\nu)] \, (f)$$

$$= (\mu - \nu) d_\nu \Phi(f) - \frac{1}{\mu(G_\nu)} \, (\mu - \nu)^{\otimes 2} \, [G_\nu \otimes d_\nu \Phi(f)] \qquad (3.23)$$

In addition, we have the estimates

$$|[\Phi(\mu) - \Phi(\nu)](f)| \leq g \left[|(\mu - \nu)d_\nu\Phi(f)| \wedge |(\mu - \nu)d'_\nu\Phi(f)|\right]$$

$$\|d_\mu\Phi(f) - d_\nu\Phi(f)\| \leq g \left[|(\nu - \mu)(G_\nu)| + |(\nu - \mu)d_\nu\Phi(f)|\right] \quad (3.24)$$

As in (2.3) and (3.18), the integral operators $d_\nu\Phi$ introduced in definition 3.1.8, and the decompositions (3.23) can be interpreted in terms of Gâteaux derivatives, and weak Taylor's type expansions of the first order. For instance, for any couple of probability measures $(\mu, \eta) \in \mathcal{P}(E)^2$ we have

$$\lim_{\epsilon \downarrow 0} \left\| \frac{1}{\epsilon} \left[\Phi(\mu + \epsilon(\eta - \mu)) - \Phi(\mu)\right] - (\eta - \mu)d_\mu\Phi \right\|_{tv} = 0$$

3.2 Feynman-Kac models

3.2.1 Path integration measures

In this section, we present an alternative description of the Feynman-Kac measures defined in Section 1.4.2.

Suppose we are given a sequence of distributions $\mathbb{Q}_n$ with increasing complexity on some product state spaces $\mathbf{E_n} = \prod_{p=0}^{n} E_p$, in terms of a proportional recursion equation of the following form

$$\mathbb{Q}_n(d(x_0, \ldots, x_n)) \propto \mathbb{Q}_{n-1}(d(x_0, \ldots, x_{n-1})) \times Q_n(x_{n-1}, dx_n)$$

In the above display, $Q_n(x_{n-1}, dx_n)$ stands for some positive integral operator from E_{n-1} into E_n, and $\mathbb{Q}_0 = \eta_0$, some initial probability measure on E_0.

Each probability measure $\mathbb{Q}_n$ is only known pointwise on $\mathbf{E_n}$, up to a normalizing constant, say $\mathcal{Z}_n$. By construction, we clearly have that

$$\mathbb{Q}_n(d(x_0, \ldots, x_n)) := \frac{1}{\mathcal{Z}_n} \eta_0(dx_0) \prod_{p=1}^{n} Q_p(x_{p-1}, dx_p) \quad (3.25)$$

for some normalizing constants $\mathcal{Z}_n$. We illustrate this rather abstract formulation with the Feynman-Kac measure $\mathbb{Q}_n$ defined in (1.37). In this case, we have

$$\mathbb{Q}_n(d(x_0, \ldots, x_n)) = \frac{\mathcal{Z}_{n-1}}{\mathcal{Z}_n} \times \mathbb{Q}_{n-1}(d(x_0, \ldots, x_{n-1})) Q_n(x_{n-1}, dx_n)$$

$$\propto \mathbb{Q}_{n-1}(d(x_0, \ldots, x_{n-1})) \times Q_n(x_{n-1}, dx_n) \quad (3.26)$$

with the integral operator

$$Q_n(x_{n-1}, dx_n) := G_{n-1}(x_{n-1}) \, M_n(x_{n-1}, dx_n) \qquad (3.27)$$

In the reverse angle, any model (3.25) has the form (3.27) with

$$G_{n-1}(x_{n-1}) = Q_n(1)(x_{n-1}) \quad \text{and} \quad M_n(x_{n-1}, dx_n) = \frac{Q_n(x_{n-1}, dx_n)}{Q_n(1)(x_{n-1})}$$

From the pure mathematical point of view, this shows that the measures (3.25) have the same form as the Feynman-Kac measures defined in (1.37).

3.2.2 Evolution equations

In this section, we describe in some detail the semigroups of the n-th terminal time marginals (η_n, γ_n) of the Feynman-Kac measures $(\mathbb{Q}_n, \Gamma_n)$ defined in Section 1.4.2.

By a simple conditioning argument, for any $f_n \in \mathcal{B}_b(E_n)$, we have

$$
\begin{aligned}
\gamma_n(f_n) &= \mathbb{E}\left(G_{n-1}(X_{n-1}) \, \mathbb{E}\left(f_n(X_n) \mid X_{n-1} \right) \prod_{0 \leq p < (n-1)} G_p(X_p) \right) \\
&= \mathbb{E}\left(Q_n(f_n)(X_{n-1}) \prod_{0 \leq p < (n-1)} G_p(X_p) \right) = \gamma_{n-1}(Q_n(f_n))
\end{aligned}
$$

with the positive integral operator Q_n defined in (3.27). It is now a simple exercise to check that for any $f_{n+1} \in \mathcal{B}_b(E_{n+1})$, we have that

$$\eta_{n+1}(f_{n+1}) = \frac{\gamma_n Q_{n+1}(f_{n+1})}{\gamma_n Q_{n+1}(1)} = \frac{\eta_n Q_{n+1}(f_{n+1})}{\eta_n Q_{n+1}(1)} = \frac{\eta_n(G_n M_{n+1}(f_{n+1}))}{\eta_n(G_n)}$$

from which we conclude that

$$\eta_{n+1} = \Phi_{n+1}(\eta_n) := \Psi_{G_n}(\eta_n) M_{n+1} \qquad (3.28)$$

with the Boltzmann-Gibbs transformation Ψ_{G_n} defined in (0.3). In addition, the normalizing constants $\mathcal{Z}_n$ can be expressed in terms of the flow of marginal measures η_p, from the origin $p = 0$ up to the current time n.

Lemma 3.2.1 *For any $n \geq 0$, the normalizing constants $\mathcal{Z}_n$ in (1.37) are given by*

$$\mathcal{Z}_n := \gamma_n(1) = \mathbb{E}\left(\prod_{0 \leq p < n} G_p(X_p) \right) = \prod_{0 \leq p < n} \eta_p(G_p) \qquad (3.29)$$

This multiplicative formula is easily checked using the induction formulae

$$\gamma_{n+1}(1) = \gamma_n(Q_{n+1}(1)) = \gamma_n(G_n) = \eta_n(G_n)\,\gamma_n(1)$$

The abstract formulae discussed above are more general than it may appear. For instance, they can be used to analyze without further work path space models, including historical processes or transition space models, as well as finite excursion models and distribution state space models. These functional models also encapsulate quenched Feynman-Kac models and Brownian type bridges, as well as linear Gaussian Markov chains conditioned on starting and terminal end points.

For a more thorough discussion on these path space models, we refer the reader to Section 2.4, Section 2.6, Chapters 11-12 in the monograph [163], as well as to the Section 3.4.3, Chapter 7, and Chapter 8 of the present book.

3.3 Some illustrations

In this section we present a series of motivating illustrations of the abstract Feynman-Kac models presented in Section 1.4.2 and in Section 3.2.

The first one is concerned with a simple class of spatial branching processes. We show that their intensity measures are given by a Feynman-Kac distribution flow. In this connection, we also refer the reader to Section 4.1.1, for a discussion on genetic type branching models. More sophisticated spatial branching models, equipped with survival probabilities and spontaneous birth rates, are discussed in Section 2.3.3, as well as in Section 6.2 and Section 6.3.2, in the context of multiple objects filtering problems.

The second application domain is related to particle motions in an absorbing, and possibly random, medium. In this context, the distribution of the random trajectories of a nonabsorbed particle is given by a Feynman-Kac measure in path space. For a more thorough discussion on these models we refer the reader to Chapter 7 dedicated to particle absorption models, including quenched, annealed, and mean field particle absorption models.

Section 3.3.4 is concerned with the Feynman-Kac representation of nonlinear filtering problems dicussed in Section 2.5.1. In this context, we recall that the selection-mutation transition of the mean field IPS model is often referred to as the updating-prediction transition of the optimal filter. Section 3.3.5 is dedicated to fixed parameter estimation problems arising in hidden Markov chain models. For a detailed discussion on these couple of nonlinear estimation problems, we refer the reader to Chapter 8, dedicated to applications of Feynman-Kac models in signal processing and stochastic control theory.

Section 3.3.6 is concerned with a Feynman-Kac interpretation of the gradient of Markov semigroups. These models can be encapsulated into the noncommutative Feynman-Kac framework discussed in Section 2.3.1. In Section 3.3.7, we review some basic foundations of Malliavin Greeks derivatives. We provide a natural Feynman-Kac representation that allows designing mean field IPS approximation models.

3.3.1 Spatial branching processes

We denote by $\mathbf{E} = \cup_{p \geq 0} E^p$ the state space of a branching process with individuals taking values on some measurable state space E. The integer $p \geq 0$ represents the size of the population. For $p = 0$ we use the convention $E^0 := \{c\}$, where c stands for an auxiliary cemetery state.

We let $M_n(x, dy)$, with $n \geq 1$, be a sequence of Markov transitions from E into inself, and by $(g_n^i(x))_{i \geq 1, x \in E, n \geq 0}$ we denote a collection of integer number-valued random variables with uniformly finite first moments. We further assume that for any $x \in E$, and any $n \geq 0$, $(g_n^i(x))_{i \geq 1}$ are identically distributed, and we set

$$G_n(x) := \mathbb{E}(g_n^i(x))$$

Our branching process is defined as follows. We start at some point x_0 with a single particle, that is $p_0 = 1$ and $\zeta_0 = \zeta_0^1 = x_0 \in E^{p_0} = E$. This particle branches into $\widehat{p}_0$ offsprings $\widehat{\zeta}_0 = (\widehat{\zeta}_0^1, \ldots, \widehat{\zeta}_0^{\widehat{p}_0}) \in E^{\widehat{p}_0}$, with $\widehat{p}_0 = g_0^1(\zeta_0^1)$.

Each of these individuals explores randomly the state space E, according to the transition M_1. At the end of this mutation step, we have a population of $p_1 = \widehat{p}_0$ particles $\zeta_1^i \in E$ with distribution $M_1(\widehat{\zeta}_0^i, .)$, $i = 1, \ldots, p_1$. Then each of these particles ζ_1^i branches into $g_1^1(\zeta_1^i)$ offsprings. At the end of this transition, we have $\widehat{p}_1$ particles $\widehat{\zeta}_1 = (\widehat{\zeta}_1^1, \ldots, \widehat{\zeta}_1^{\widehat{p}_1}) \in E^{\widehat{p}_1}$, with $\widehat{p}_1 = \sum_{i=1}^{p_1} g_1^i(\zeta_1^i)$.

Then, each of these individuals explores randomly the state space E, according to the transition M_2, and so on.

Whenever the system dies, $\widehat{p}_n = 0$ at a given time n, we set $\widehat{\zeta}_q = \zeta_{q+1} = c$, and $\widehat{p}_q = p_{q+1} = 0$, for any $q \geq n$.

By construction, we have $p_{n+1} = \widehat{p}_n$, and $\sum_{i=1}^{\widehat{p}_n} f(\widehat{\zeta}_n^i) = \sum_{i=1}^{p_n} g_n^i(\zeta_n^i) f(\zeta_n^i)$, for any function $f \in \mathcal{B}_b(E)$. If we consider the random measures

$$\mathcal{X}_n = \sum_{i=1}^{p_n} \delta_{\zeta_n^i} \quad \text{and} \quad \widehat{\mathcal{X}}_n = \sum_{i=1}^{\widehat{p}_n} \delta_{\widehat{\zeta}_n^i}$$

then, we find that

$$\mathbb{E}(\widehat{\mathcal{X}}_n(f) \mid \zeta_n) = \mathcal{X}_n(G_n \, f) \quad \text{and} \quad \mathbb{E}(\mathcal{X}_{n+1}(f) \mid \widehat{\zeta}_n) = \widehat{\mathcal{X}}_n(M_n(f))$$

This clearly implies that

$$\mathbb{E}(\mathcal{X}_{n+1}(f) \mid \zeta_n) = \mathcal{X}_n(G_n M_{n+1}(f))$$

We readily conclude that the first moments of the branching distributions $\mathcal{X}_n$ are given by the Feynman-Kac model

$$\mathbb{E}(\mathcal{X}_n(f)) = \mathbb{E}_{x_0}\left(f(X_n) \prod_{0 \leq k < n} G_k(X_k) \right) := \gamma_n(f)$$

In the above display, X_n stands for the Markov chain on E with Markov transitions M_n. In this interpretation, the mean number of individuals in the current population is given by $\mathbb{E}(\mathcal{X}_n(1)) = \gamma_n(1)$. These measures have exactly the same form as the ones presented in Section 3.2. Therefore, the first moment of a spatial branching process (including the mean number of individuals) can be approximated using a mean field IPS model *with fixed population size*, using the mean birth numbers as selection fitness functions. In probability theory, the stochastic process $\mathcal{X}_n$ is called a Branching Markov chain. The long time behavior of these branching models, their connections with particle absorption models, and their applications in physics and biology is a rapidly developing subject in probability theory. We refer the reader to the series of articles [17, 18, 19, 52, 73, 341, 358, 359, 448], the more recent studies [5, 35, 49, 201, 203, 204, 321, 515], and references therein.

3.3.2   Particle absorption models

We consider the absorbed Markov chain model defined by the following synthetic diagram

$$X_n^c \in E_n^c := E_n \cup \{c\} \xrightarrow{\;absorption\;\sim(1-G_n)\;} \widehat{X}_n^c \xrightarrow{\;exploration\;\sim M_{n+1}\;} X_{n+1}^c \quad (3.30)$$

The chain X_n^c starts at some initial state X_0^c randomly chosen with distribution η_0. During the absorption stage, we set $\widehat{X}_n^c = X_n^c$ with probability $G_n(X_n^c)$; otherwise we put the particle in an auxiliary cemetery state $\widehat{X}_n^c = c$. When the particle $\widehat{X}_n^c$ is still alive (that is, if we have $\widehat{X}_n^c \in E_n$), it performs an elementary move $\widehat{X}_n^c \rightsquigarrow X_{n+1}^c$ according to the Markov transition M_{n+1}. Otherwise, the particle is absorbed and we set $X_p^c = \widehat{X}_p^c = c$, for any $p > n$.

If we let T be the first time $\widehat{X}_n^c = c$, then we have the Feynman-Kac representation formulae

$$\mathbb{Q}_n = \mathrm{Law}((X_0^c, \dots, X_n^c) \mid T \geq n) \quad \text{and} \quad \mathcal{Z}_n = \mathbb{P}(T \geq n)$$

3.3.3   Quenched and annealed absorption models

We consider a particle model X_n^c evolving in a random absorbing medium that depends on some random parameter Θ, with distribution μ on some state space S. For any given $\Theta = \theta$, we denote by $M_{\eta,n}$ the Markov transition of the particle, and $G_{n,\eta}$ the absorption potential associated with this value of the parameter. We also let $\mathbb{Q}_{\theta,n}$ be the Feynman-Kac measure (1.37) associated with the Markov transitions $M_{\eta,n}$ and the potential functions $G_{\theta,n}$. In this notation, we have

$$\mathbb{Q}_{\theta,n} = \mathrm{Law}((X_0^c, \dots, X_n^c) \mid \Theta = \theta, \ T \geq n)$$

By (3.29), the normalizing constant $\mathcal{Z}_n(\theta)$ of the measures $\mathbb{Q}_{\theta,n}$ is expressed in terms of the flow of p-th time marginals $(\eta_{\theta,p})_{p<n}$, with the following formulae

$$\mathcal{Z}_n(\theta) = \mathbb{P}\left(T \geq n \mid \Theta = \theta\right) = \prod_{0 \leq p < n} \eta_{\theta,p}(G_{\theta,p})$$

from which we conclude that

$$\mathbb{P}\left(\Theta \in d\theta \mid T \geq n\right) = \frac{1}{\mathcal{Z}_n} \, \mathcal{Z}_n(\theta) \, \nu(d\theta)$$

For a more thorough discussion on these quenched and annealed Feynman-Kac models, we refer the reader to Chapter 7. An illustration of the conditional distributions discussed above in the context of HMM models is also provided in Section 3.3.5.

3.3.4   Nonlinear filtering problems

The filtering problem is defined by a reference Markov chain model X_n and a sequence of partial and noisy observations Y_n. The pair process (X_n, Y_n) forms a Markov chain on some product space $(E_n^X \times E_n^Y)$, with elementary transitions

$$\mathbb{P}\left((X_n, Y_n) \in d(x,y) \mid (X_{n-1}, Y_{n-1})\right) = M_n(X_{n-1}, dx) \, g_n(x, y) \, \lambda_n(dy)$$
$$(3.31)$$

for some positive likelihood function g_n, some reference measure λ_n on E_n^Y, and with the Markov transitions M_n of X_n. We fix the sequence of observations $Y_n = y_n$, and we set

$$G_n(x_n) = p_n(y_n|x_n) = g_n(x_n, y_n)$$

the likelihood function of an observation $Y_n = y_n$ w.r.t. a given signal state $X_n = x_n$. In this situation, we find that

$$\mathbb{Q}_n = \mathrm{Law}((X_0, \ldots, X_n) \mid \forall 0 \leq p < n \; Y_p = y_p) \text{ and } \mathcal{Z}_{n+1} = p(y_0, \ldots, y_n)$$

A more rigorous description of these filtering models, and the abusive Bayesian notation we have used, is provided in Section 8.1.1.

3.3.5   Hidden Markov chain problems

We consider a pair signal-observation filtering model (X, Y) that depend on some random variable Θ, with distribution μ on some state space S. For any given $\Theta = \theta$, we let $M_{\theta,n}$ be the transitions of the chain X_n, and we set

$$G_{\theta,n}(x_n) = p_n(y_n|x_n, \theta)$$

the likelihood function of a given observation $Y_n = y_n$ w.r.t. a given signal state $X_n = x_n$ and the value of the parameter $\Theta = \theta$. We let $\mathbb{Q}_{\theta,n}$ be the

Feynman-Kac measure (1.37) associated with the Markov transitions $M_{\theta,n}$ and the potential functions $G_{\theta,n}$. Arguing as above, given a realization of the parameter $\Theta = \theta$, we have

$$\mathbb{Q}_{\theta,n} = \text{Law}((X_0, \ldots, X_n) \mid \forall 0 \le p < n \quad Y_p = y_p, \ \Theta = \theta)$$

In this case, the n-th time marginal of $\mathbb{Q}_{\theta,n}$ is given by

$$\eta_{\theta,n} = \text{Law}(X_n \mid \forall 0 \le p < n \quad Y_p = y_p, \ \Theta = \theta)$$

Using the multiplicative formula (3.29), we also prove that

$$\mathcal{Z}_{n+1}(\theta) = p(y_0, \ldots, y_n \mid \theta) = \prod_{0 \le p \le n} h_p(\theta)$$

with

$$
\begin{aligned}
h_p(\theta) &= p(y_p \mid y_0, \ldots, y_{p-1}, \theta) \\
&= \int p(y_p \mid x_p, \theta) \, dp(x_p \mid \theta, y_0, \ldots, y_{p-1}) = \eta_{\theta,p}(G_{\theta,p})
\end{aligned}
$$

We conclude that

$$\mathbb{P}(\Theta \in d\theta \mid \forall 0 \le p \le n \quad Y_p = y_p) = \frac{1}{\mathcal{Z}_n} \, \mathcal{Z}_n(\theta) \, \mu(d\theta)$$

with the normalizing constants

$$\mathcal{Z}_n := \int \mathcal{Z}_n(\theta) \, \mu(d\theta) = p(y_0, \ldots, y_{n-1})$$

A more rigorous description of these filtering models, and the abusive Bayesian notation we have used, is provided in Section 8.6.

In some instances, such as in conditionally linear Gaussian models, the normalizing constants $\mathcal{Z}_n(\theta)$ can be computed explicitly, and we can use a Metropolis-Hasting style Markov chain Monte Carlo method to sample the target measures μ_n. We can also turn this MCMC scheme into an interacting Markov chain Monte Carlo algorithm. For a detailed discussion on these mean field MCMC models, we refer the reader to Section 2.5.2.2, Section 4.3.6, Section 8.6.1, and Section 8.6.2.

3.3.6 Gradient of Markov semigroups

We consider a Markov chain model given by an iterated $\mathbb{R}^d$-valued random process

$$X_{n+1} := F_n(X_n) = (F_n \circ F_{n-1} \circ \cdots \circ F_0)(X_0) \tag{3.32}$$

starting at some random state X_0, with a sequence of random smooth functions of the form

$$F_n(x) = \mathcal{F}_n(x, W_n) \tag{3.33}$$

In the above display, W_n is a collection of independent r.v. taking values in $\mathbb{R}^{d'}$, for some $d' \geq 1$; and $\mathcal{F}_n$ are some smooth functions, from $\mathbb{R}^{d+d'}$ into $\mathbb{R}^d$.

The semigroup of the Markov chain X_n is the expectation operator defined for any regular function f_n and any state x by

$$P_{n+1}(f_{n+1})(x) := \mathbb{E}\left(f_{n+1}(X_{n+1}) \mid X_0 = x\right) = \mathbb{E}\left(f(X_{n+1}(x))\right)$$

with the random flows $(X_n(x))_{n\geq 0}$ defined for any $n \geq 0$ by the following equation

$$X_{n+1}(x) = F_n(X_n(x)) \quad \text{with the initial condition } X_0(x) = x$$

By construction, for any $1 \leq i, j \leq d$ we have the first variational equation

$$\frac{\partial X_{n+1}^i}{\partial x^j}(x) = \sum_{1\leq k\leq d} \frac{\partial F_n^i}{\partial x^k}(X_n(x)) \frac{\partial X_n^k}{\partial x^j}(x) \tag{3.34}$$

On the other hand, we have that

$$\frac{\partial P_{n+1}(f)}{\partial x^j}(x) = \mathbb{E}\left(\sum_{1\leq i\leq d} \frac{\partial f}{\partial x^i}(X_{n+1}(x)) \frac{\partial X_{n+1}^i}{\partial x^j}(x)\right) \tag{3.35}$$

We denote by $V_n = (V_n^{(i,j)})_{1\leq i,j\leq d}$ and $A_n = (A_n^{(i,j)})_{1\leq i,j\leq d}$ the random $(d \times d)$-matrices with the i-th line and j-th column entries

$$V_n^{(i,j)}(x) = \frac{\partial X_n^i}{\partial x^j}(x)$$

$$A_n^{(i,j)}(x) = \frac{\partial F_n^i}{\partial x^j}(x) = \frac{\partial \mathcal{F}_n^i(., W_n)}{\partial x^j}(x) := \mathcal{A}_n^{(i,j)}(x, W_n)$$

We mention that the matrix V_n coincides with the Jacobian matrix $J(X_n)(x) = \frac{\partial X_n}{\partial x}(x)$ of the function $X_n : x \in \mathbb{R}^d \mapsto X_n(x) \in \mathbb{R}^d$.

In this notation, the Equation (3.34) can be rewritten in terms of the following random matrix formulae

$$V_{n+1}(x) = A_n(X_n(x)) V_n(x) := \prod_{p=0}^{n} A_p(X_p(x)) \tag{3.36}$$

In the above display, $\prod_{p=0}^{n} A_p$ stands for the noncommutative product of the random matrices A_p, taken in the order $A_n, A_{n-1}, \ldots, A_0$.

In the same way, the Equation (3.35) can be rewritten as

$$\nabla P_{n+1}(f_{n+1})(x) = \mathbb{E}\left(\nabla f_{n+1}(X_{n+1}) V_{n+1} \mid X_0 = x\right) \tag{3.37}$$

with $V_{n+1} := \prod_{0\leq p\leq n} A_p(X_p)$. For instance, for one dimensional models of the form

$$X_{n+1} = \mathcal{F}_n(X_n, W_n) = X_n + b(X_n) \ \Delta + \sigma(X_n) \ \sqrt{\Delta} \ W_n , \tag{3.38}$$

with some $\Delta > 0$, and some sequence of independent and centered Gaussian random variables W_n, it is readily checked that

$$A_n(x) = A_n(x, W_n) = \left(1 + \frac{\partial b}{\partial x}(x) \, \Delta + \frac{\partial \sigma}{\partial x}(x) \, \sqrt{\Delta} \, W_n\right)$$

and therefore

$$
\begin{aligned}
V_{n+1}(x) \quad &= \quad \prod_{p=0}^{n} \left(1 + \frac{\partial b}{\partial x}(X_p) \, \Delta + \frac{\partial \sigma}{\partial x}(X_p) \, \sqrt{\Delta} \, W_p\right) \\
&\simeq_{\Delta \downarrow 0} \quad \exp \sum_{0 \le p \le n} \left(\frac{\partial b}{\partial x}(X_p) \, \Delta + \frac{\partial \sigma}{\partial x}(X_p) \, \sqrt{\Delta} \, W_p\right)
\end{aligned}
$$

In this context (3.37) has the same form as the Feynman-Kac models (1.39).

In the multidimensional case, we proceed as follows. We equip the space $\mathbb{R}^d$ with some norm $\|.\|$. We assume that for any state U_0 in the unit sphere $\mathcal{S}^{d-1} \subset \mathbb{R}^d$, we have

$$\|V_{n+1} \, U_0\| > 0$$

In this situation, we have the multiplicative formulae

$$\nabla f_{n+1}(X_{n+1}) \, V_{n+1} \, U_0 = [\nabla f_{n+1}(X_{n+1}) \, U_{n+1}] \prod_{0 \le p \le n} \|A_p(X_p) \, U_p\|$$

with the well defined $\mathcal{S}^{d-1}$-valued Markov chain defined by

$$U_{n+1} = A_n(X_n)U_n / \|A_n(X_n)U_n\| \ (\Longleftrightarrow U_{n+1} = V_{n+1} \, U_0 \, / \, \|V_{n+1} \, U_0\| \ )$$

If we choose $U_0 = u_0$, then we obtain the following Feynman-Kac interpretation of the gradient of a semigroup

$$\nabla P_{n+1}(f_{n+1})(x) \, u_0 = \mathbb{E}\left(F_{n+1}(\mathcal{X}_{n+1}) \prod_{0 \le p \le n} \mathcal{G}_p(\mathcal{X}_p)\right) \qquad (3.39)$$

In the above display, $\mathcal{X}_n$ is the Markov chain sequence $\mathcal{X}_n := (X_n, U_n, W_n)$, starting at (x, u_0, W_0); and the functions F_{n+1} and $\mathcal{G}_n$ are defined by

$$F_{n+1}(x, u, w) := \nabla f_{n+1}(x) \, u \quad \text{and} \quad \mathcal{G}_n(x, u, w) := \|A_n(x, w) \, u\|$$

We quote the interpolation formula

$$
\begin{aligned}
P_{n+1}(f)(y) - P_{n+1}(f)(x) \quad &= \quad \int_0^1 \nabla P_{n+1}(f_{n+1})(ty + (1-t)x) \, (y-x)' \, dt \\
&= \quad \mathbb{E}\left(\nabla f(X_{n+1}^{(x,y)}) \left[\prod_{0 \le p \le n} A_p(X_p^{(x,y)})\right] (y-x)'\right)
\end{aligned}
$$

where $X_n^{(x,y)}$ stands for Markov chain (3.32) starting at some random state $X_0^{(x,y)}$ uniformly chosen in the line segment between x and y.

We end this section with some rather crude upper bound that can be estimated, uniformly w.r.t. the time parameter, under appropriate regularity conditions on the reduced Markov chain model (X_n, W_n). To this end, firstly we notice that

$$\mathcal{G}_n(x, u, w) := \|\mathcal{A}_n(x, w)\, u\| \leq \; \begin{aligned} G_n(x, w) &:= \|\mathcal{A}_n(x, w)\| \\ &:= \sup_{u \in \mathcal{S}^{d-1}} \|\mathcal{A}_n(x, w)\, u\| \end{aligned}$$

This implies that

$$\|\nabla P_{n+1}(f_{n+1})(x)\| := \sup_{1 \leq i \leq d} \left| \frac{\partial}{\partial x^i} P_{n+1}(f_{n+1})(x) \right|$$

$$\leq \|F_{n+1}\| \times \mathbb{E} \left(\prod_{0 \leq p \leq n} G_p(X_p, W_p) \right)$$

The r.h.s. functional expectation in the above equation can be approximated using the mean field particle approximations of the multiplicative Feyman-Kac formulae (3.29), with the reference Markov chain (X_n, W_n), and the potential functions G_n.

We end this section with a Feynman-Kac model associated with extremal Lyapunov trajectories. Firstly, in terms of the Jacobian matrices $J_n(X_n)$, we notice that

$$\prod_{0 \leq p \leq n} \mathcal{G}_p(\mathcal{X}_p) = \prod_{0 \leq p \leq n} \|A_p(X_p)\, U_p\| = \prod_{0 \leq p \leq n} \frac{\|J(X_{p+1})\, u_0\|}{\|J(X_p)\, u_0\|} = \|J(X_{n+1})\, u_0\|$$

Replacing vector norms $\|J(X_n)\, u_0\|$ by matrix norms $\|J(X_n)\|$, and setting

$$\boldsymbol{X_n} = (X_n, X_{n+1}) \quad \text{and} \quad \boldsymbol{G_n(X_n)} = \|\mathrm{Jac}\,(X_{n+1})\|^{\alpha} / \|\mathrm{Jac}\,(X_n)\|^{\alpha}$$

for some $\alpha \in \mathbb{R}$, the Feynman-Kac model described above reduces to the Lyapunov weighted dynamics model

$$d\boldsymbol{Q_n} = \frac{1}{\mathcal{Z}_n} \|\mathrm{Jac}\,(X_n)\|^{\alpha} \; d\boldsymbol{P_n}$$

where $\boldsymbol{P_n}$ stands for the law of the random trajectories $(\boldsymbol{X_0}, \ldots, \boldsymbol{X_n})$. When $\alpha < 0$, the measure $\boldsymbol{Q_n}$ favors low Lyapunov trajectories, while for $\alpha > 0$, the $\boldsymbol{Q_n}$ favors high Lyapunov trajectories. This Feynman-Kac model coincides with the rare event stochastic models developed in [171], and presented in Section 2.7.1. The mean field particle models associated with these Feynman-Kac formula have been used in the series of articles [377, 540, 541, 546] to sample atypical rare event trajectories in nonlinear stochastic processes.

We notice that all these Feynman-Kac models we have discussed have exactly the same form as the matrix models discussed in Section 2.3.1. In addition, using the same analysis as above, we easily design mean field estimates of $\mathbb{E}\left(\|\mathcal{Y}_n\|^q\right)$, for any reasonably large $q \geq 0$, just replacing the potential functions $\|Au\|$ by the functions $\|Au\|^q$. We also mention that in physics literature, the mean field IPS simulation models are sometimes called "Resampled Monte Carlo methods" [556].

3.3.7 Malliavin Greeks derivatives

We return to the one dimensional model (3.38). For nonsmooth functions f_{n+1} we can use the following Gaussian regularization kernel

$$P_{n+1,\epsilon}(f_{n+1})(x) := \mathbb{E}\left(f_{n+1}(X_{n+1}(x) + \epsilon Y)\right) \qquad (3.40)$$

for some auxiliary Gaussian variable, independent of the process X_n. From the statistical viewpoint, this approximation procedure is interpreted as a Gaussian kernel density estimation of the distribution of $X_{n+1}(x)$. Combining (3.37) with (3.40), we end up with the following approximation formula

$$\frac{\partial}{\partial x} P_{n+1,\epsilon}(f_{n+1})(x)$$

$$= \mathbb{E}\left(\epsilon^{-1}\left[f_{n+1}(X_{n+1}(x) + \epsilon Y) - f_{n+1}(X_{n+1}(x))\right] \ Y \ V_{n+1}(x)\right)$$

From a more probabilistic point of view, the Gaussian regularization formula (3.40) can also be interpreted as the addition of an additional Gaussian move in the evolution of the chain $X_{n+1}(x)$. This suggests that we can alternatively use the last transition to regularize the semigroup

$$P_{n+1}(f_{n+1})(x) = \mathbb{E}\left(\mathbb{E}\left(f_{n+1}(X_{n+1}(x))\big|X_n(x)\right)\right)$$

$$= \mathbb{E}\left[\int f_{n+1}(x_{n+1}) \ \mathbb{P}\left(X_{n+1}(x) \in dx_{n+1}\,|X_n(x)\right)\right]$$

Letting $H_{n+1}(x_n, x_{n+1})$ be the density of the Markov transition $X_n \rightsquigarrow X_{n+1}$ w.r.t. the Lebesgue measure, we find that

$$P_{n+1}(f_{n+1})(x) = \mathbb{E}\left[\int f_{n+1}(x_{n+1}) \ H_{n+1}(X_n(x), x_{n+1}) \ dx_{n+1}\right]$$

Arguing as above we find that

$$\frac{\partial}{\partial x} P_{n+1}(f_{n+1})(x) = \mathbb{E}\left(f_{n+1}(X_{n+1}(x)) \ dH_{n+1}(X_n(x), X_{n+1}(x)) \ V_n(x)\right)$$

with the weight function

$$
\begin{aligned}
dH_{n+1}(x_n, x_{n+1}) &= \frac{\partial}{\partial x_n} \log H_{n+1}(x_n, x_{n+1}) \\
&= \left(\left(\frac{(x_{n+1} - x_n) - b(x_n)\Delta}{\sigma(x_n)\sqrt{\Delta}} \right)^2 - 1 \right) \frac{\partial}{\partial x} \log \sigma(x_n) \\
&\quad + \left(\frac{(x_{n+1} - x_n) - b(x_n)\Delta}{\sigma(x_n)\sqrt{\Delta}} \right) \frac{1 + \frac{\partial b}{\partial x}(x_n)\Delta}{\sigma(x_n)\sqrt{\Delta}}
\end{aligned}
$$

In the context of financial mathematics, these formulae, and the corresponding weighted conventional Monte Carlo approximations, have been recently proposed by N. Chen and P. Glasserman [131]. This framework is an alternative to the Malliavin Greeks derivative calculus introduced in the pioneering articles by E. Fournié, J.M. Lasry, J. Lebuchoux, P.L. Lions, and N. Touzi [260, 261].

In this connection, we briefly recall some foundations of Malliavin derivatives. We let $P_{s,t}$ be the semigroup associated with the diffusion stochastic equation (1.5), with $d = 1$; that is, we have that

$$
P_{s,t}(f)(X_s) = \mathbb{E}\left(f(X_t) \mid X_s \right)
$$

Using elementary backward calculations, for any $0 \le s \le t$ we find that

$$
P_{s,t}(f)(X_s) = P_{0,t}(f)(X_0) + \int_0^s \frac{\partial P_{r,t}(f)}{\partial x}(X_r)\, \sigma(X_r)\, dW_r
$$

If we set $s = t$ in the above equation, then we find that

$$
\begin{aligned}
&\mathbb{E}\left[f(X_t(x)) \int_0^t \frac{\partial X_s}{\partial x}(x)\, \sigma^{-1}(X_s(x)))\, dW_s \right] \\
&= \mathbb{E}\left[\int_0^t \frac{\partial P_{s,t}(f)}{\partial x}(X_s(x))\, \frac{\partial X_s}{\partial x}(x)\, ds \right]
\end{aligned}
\tag{3.41}
$$

as soon as σ is a regular positive function. Recalling that

$$
\frac{\partial}{\partial x} P_{0,t}(f)(x) = \frac{\partial}{\partial x} \mathbb{E}\left[P_{s,t}(f)(X_s(x)) \right] = \mathbb{E}\left[\frac{\partial P_{s,t}(f)}{\partial x}(X_s(x))\, \frac{\partial X_s}{\partial x}(x) \right]
$$

and using (3.41) we find that

$$
\mathbb{E}\left[f(X_t(x)) \int_0^t \frac{\partial X_s}{\partial x}(x)\, \sigma^{-1}(X_s(x)))\, dW_s \right] = t\, \frac{\partial}{\partial x} P_{0,t}(f)(x)
$$

This yields the Malliavin formulation of the semigroup derivatives

$$
\frac{\partial}{\partial x} P_{0,t}(f)(x) = \mathbb{E}\left[f(X_t(x))\, \frac{1}{t} \int_0^t \sigma^{-1}(X_s(x))\, \frac{\partial X_s}{\partial x}(x)\, dW_s \right]
$$

A more rigorous derivation of the above equations is provided in [260, 261].
Using an Euler type time discretization model

$$X_{(n+1)\Delta} - X_{n\Delta} = b(X_{n\Delta}) \Delta + \sigma(X_{n\Delta}) \sqrt{\Delta} Y_n \qquad (3.42)$$

with a sequence of independent and centered Gaussian random variables Y_n,
we have the Feynman-Kac approximation model

$$\frac{\partial}{\partial x} P_{0,(n+1)\Delta}(f)(x) \simeq_{\Delta \downarrow 0} \frac{1}{(n+1)\sqrt{\Delta}} \sum_{0 \le p \le n} \mathbb{E}\left(f(X_{(n+1)\Delta}(x)) \mathcal{Z}_p(x)\right)$$

$$(3.43)$$

with the random weight function

$$\mathcal{Z}_p(x) := \varphi(X_{p\Delta}(x), Y_p) \prod_{0 \le q < p} G_q(X_{q\Delta}(x), Y_q)$$

$$\varphi(x, y) = \sigma^{-1}(x) y \quad \text{and} \quad G_q(x, y) = 1 + \frac{\partial b}{\partial x}(x) \Delta + \frac{\partial \sigma}{\partial x}(x) \sqrt{\Delta} y$$

The ratio $1/\sqrt{\Delta}$ in the r.h.s. of (3.43) may induce degenerative numerical
estimates. One way *to remove* this term in the numerical scheme is to use the
following formula

$$\mathbb{E}\left(f(X_{(n+1)\Delta}(x)) \mathcal{Z}_p(x)\right) = \mathbb{E}\left(\Upsilon_{p+1,n+1}(f)[X_{p\Delta}(x), Y_p] \times \mathcal{Z}_p(x)\right)$$

with the function

$$\Upsilon_{p+1,n+1}(f)[x, y] = P_{(p+1)\Delta,(n+1)\Delta}(f)\left(x + b(x)\Delta + \sigma(x)\sqrt{\Delta}y\right)$$
$$- P_{(p+1)\Delta,(n+1)\Delta}(f)(x + b(x)\Delta)$$

Under some appropriate regularity conditions, we notice that

$$\Upsilon_{p+1,n+1}(f)[x, y]$$

$$\simeq_{\Delta \downarrow 0} P_{p\Delta,(n+1)\Delta}(f)\left(x + b(x)\Delta + \sigma(x)\sqrt{\Delta}y\right) - P_{p\Delta,(n+1)\Delta}(f)(x + b(x)\Delta)$$

$$\simeq_{\Delta \downarrow 0} \frac{\partial P_{p\Delta,(n+1)\Delta}(f)}{\partial x}(x) \, \sigma(x)\sqrt{\Delta} \, y$$

This implies that the gradient of the Markov semigroup is approximated by

$$\frac{\partial}{\partial x} P_{0,(n+1)\Delta}(f)(x) \simeq_{\Delta \downarrow 0} \frac{1}{(n+1)} \sum_{0 \le p \le n} U_{p,n}(f)(x)$$

with the Feynman-Kac formulae

$$U_{p,n}(f)(x) := \mathbb{E}\left(\frac{\partial P_{p\Delta,(n+1)\Delta}(f)}{\partial x}(X_{p\Delta}(x)) \, Y_p^2 \prod_{0 \le q < p} G_q(X_{q\Delta}(x), Y_q)\right)$$

3.4   Historical processes

This section is dedicated to Feynman-Kac models associated with historical processes. We discuss in more detail the different structural stability properties presented in Section 1.4.3.

3.4.1   Feynman-Kac path integrals

This first section is concerned with the first structural stability property discussed in Section 1.4.3.1. This property concerns the Feynman-Kac distribution flows presented in Section 3.2.2. We show that the time marginal measures coincide with the path space measures (1.37), when we consider historical reference processes and potential functions only depending on the terminal time random state. We shall use this property to show that the genealogical tree based models presented in Section 4.1.2 coincide with the standard mean field IPS model in path spaces. Further details on these equivalence principles are also provided in Section 12.2.3.3.

We consider the historical process $\mathbf{X_n} = (X_0, \ldots, X_n)$ associated with some reference E_n-valued Markov chain X_n. We also let $\mathbf{G_n}$ be a sequence of positive and bounded potential functions on $\mathbf{E_n} := (E_0 \times \ldots \times E_n)$, whose values only depend on the final state of the paths; that is, we have that

$$\mathbf{G_n} \;:\; \mathbf{x_n} = (x_0, \ldots, x_n) \in \mathbf{E_n} \mapsto \mathbf{G_n}(\mathbf{x_n}) = G_n(x_n) \qquad (3.44)$$

for some positive and bounded potential function G_n on E_n.

We let (γ_n, η_n) be the Feynman-Kac model associated with the pair $(\mathbf{G_n}, \mathbf{M_n})$ on the path spaces $\mathbf{E_n}$. By construction, for any function $\mathbf{f_n}$ on $\mathbf{E_n}$, we have

$$
\begin{aligned}
\gamma_n(\mathbf{f_n}) &= \mathbb{E}\left(\mathbf{f_n}(\mathbf{X_n}) \prod_{0 \le p < n} \mathbf{G_p}(\mathbf{X_p}) \right) \\
&= \mathbb{E}\left(\mathbf{f_n}(X_0, \ldots, X_n) \prod_{0 \le p < n} G_p(X_p) \right)
\end{aligned}
$$

from which we conclude that

$$(\gamma_n, \eta_n) = (\Gamma_n, \mathbb{Q}_n) \qquad (3.45)$$

where $(\Gamma_n, \mathbb{Q}_n)$ are the unnormalized and the normalized Feynman-Kac measures on path space associated with (G_n, M_n) and defined in (1.37).

The end of this section is concerned with the proof of the backward Markov interpretation models presented in Section 1.4.3.4. We prove the formula (1.50) using the decomposition

$$\mathbb{Q}_n(d(x_0,\ldots,x_n)) = \frac{\mathcal{Z}_{n-1}}{\mathcal{Z}_n} \, \mathbb{Q}_{n-1}(d(x_0,\ldots,x_{n-1})) \, Q_n(x_{n-1},dx_n)$$

We prove the following formulae

$$\eta_n(dx_n) = \frac{\mathcal{Z}_{n-1}}{\mathcal{Z}_n} \, \eta_{n-1} Q_n(dx_n) \tag{3.46}$$

$$= \frac{\mathcal{Z}_{n-1}}{\mathcal{Z}_n} \, \eta_{n-1} \left(G_{n-1} H_n(.,x_n) \right) \, \lambda_n(dx_n) \tag{3.47}$$

and $\mathcal{Z}_n/\mathcal{Z}_{n-1} = \eta_{n-1} Q_n(1) = \eta_{n-1}(G_{n-1})$. Recalling that $\mathcal{Z}_0 = 1$, this clearly implies that

$$\prod_{p=1}^{n} \frac{d\eta_{p-1} Q_p}{d\eta_p}(x_p) = \prod_{p=1}^{n} \frac{\mathcal{Z}_p}{\mathcal{Z}_{p-1}} = \mathcal{Z}_n$$

Using these couple of observations, we readily prove the desired backward decomposition formula.

3.4.2 Time reversal Markov models

The future and the past of Markov chains are statistically independent. This symmetry property is preserved running backward in time the chain at equilibrium. Nevertheless the Markov transitions of the time reversal chain differ from the original model. This section is concerned with the description of the time reversal Markov chain model and its mean field IPS approximation model. This framework allows using the complete genealogical tree model to design a mean field IPS approximation of the Feynman-Kac measures in path space. These algorithms are presented in Section 2.2.6, and they are discussed in more detail in the further development of Section 4.1.6 and Chapter 17.

We consider a Markov chain X_n on some measurable state spaces E_n with initial distribution $\eta_0 = \text{law}(X_0)$ and Markov transitions M_n. We let $\mathbb{P}_n$ be the law of the historical process

$$\boldsymbol{X_n} := (X_0,\ldots,X_n) \in \boldsymbol{E_n} := (E_0 \times \ldots \times E_n)$$

and we denote by $\eta_n = \text{Law}(\boldsymbol{X_n})$, the n-th time marginal of $\mathbb{P}_n$. We further assume that the Markov transitions M_n are absolutely continuous with respect to some measures λ_n on E_n, and for any $(x,y) \in (E_{n-1} \times E_n)$ we have

$$H_n(x,y) := \frac{dM_n(x,\cdot)}{d\lambda_n}(y) > 0 \tag{3.48}$$

In this situation, using the backward formulae (1.50), we find that

$$\mathbb{P}_n = \eta_n \otimes \mathbb{P}_{n|n} \tag{3.49}$$

with the conditional distribution of $(X_0, \ldots, X_{n-1})$, given $X_n = x_n$, given by

$$\mathbb{P}_{n|n}(d(x_0, \ldots, x_{n-1}) \mid x_n) = \prod_{p=1}^{n} \mathbb{M}_{p, \eta_{p-1}}(x_p, dx_{p-1})$$

In the above display, $\mathbb{M}_{n, \eta_{n-1}}$ stands for the time reversal Markov transitions

$$\mathbb{M}_{n+1, \eta_n}(x_{n+1}, dx_n) \propto \eta_n(dx_n) H_n(x_n, x_{n+1})$$

We end this section with an illustration of these formulae for time homogeneous models $(E_n, H_n, \lambda_n, M_n) = (E, H, \lambda, M)$, with an invariant measure $\mu = \mu M$. In this situation, if we choose $\eta_0 = \mu$, then we have $\eta_n = \mu$, and therefore

$$\mathbb{M}_{n+1, \eta_n}(y, dx) = \frac{\mu(dx) H(x, y)}{\mu(H(., y))} := \mathbb{M}_\mu(y, dx)$$

Under appropriate regularity conditions on the function H, this shows that $\mathbb{M}_\mu$ is the dual operator of M from $\mathbb{L}_2(\mu)$ into itself; and we have the following reversibility property

$$\mu(dx) M(x, dy) = (\mu M)(dy) \mathbb{M}_\mu(y, dx) = \mu(dy) \mathbb{M}_\mu(y, dx) \Rightarrow \mu \mathbb{M}_\mu = \mu$$

3.4.3   Terminal time conditioning

This section is dedicated to Feynman-Kac representations of the distributions of the random trajectory of a Markov chain with fixed terminal conditions.

We use the same Markov models, and the same notation, as the ones defined in Section 3.4.2. We let π_n be a collection of probability measures on some measurable state spaces E_n. We fix a given time horizon n, and we denote by $X_p^{(n)}$ the E_{n-p}-valued Markov chain sequence defined by

$$\mathbb{P}\left((X_0^{(n)}, \ldots, X_n^{(n)}) \in d(x_0, \ldots, x_n)\right) = \pi_n(dx_0) L_n(x_0, dx_1) \cdots L_1(x_{n-1}, dx_n)$$

where L_n stands for a sequence of Markov transitions from E_n into E_{n-1}.

Our objective is to compute the conditional distributions of this chain, given a fixed terminal condition. To this end, we further assume that (π_n, M_n, L_n) are chosen so that

$$(\pi_n \otimes L_n)_1 \ll (\pi_{n-1} \otimes M_n)_0 \tag{3.50}$$

with the measure $(\pi_{n-1} \otimes M_n)_0 = \pi_{n-1} \otimes M_n$ and

$$(\pi_n \otimes L_n)_1(d(x, y)) := \pi_n(dy) L_n(y, dx)$$

Whenever it exists, we consider the Metropolis-Hasting ratio

$$G_{n-1} := d(\pi_n \otimes L_n)_1 / d(\pi_{n-1} \otimes M_n)_0$$

The following time reversal formulae allow computing the conditional distributions of Markov chains with fixed terminal condition

$$\mathbb{P}\left(\left(X_0^{(n)}, X_1^{(n)}, \ldots, X_{n-1}^{(n)}\right) \in d(x_n, x_{n-1}, \ldots, x_1) \mid X_n^{(n)} = x_0\right)$$

$$\propto \left\{\prod_{1 \le p \le n} M_p(x_{p-1}, dx_p)\right\} \times \left\{\prod_{p=0}^{n-1} G_p(x_p, x_{p+1})\right\} \tag{3.51}$$

We prove this assertion using the decomposition

$$\mathbb{P}\left(\left(X_0^{(n)}, X_1^{(n)}, \ldots, X_{n-1}^{(n)}\right) \in d(x_n, x_{n-1}, \ldots, x_1) \mid X_n^{(n)} = x_0\right)$$

$$= \frac{d\pi_0}{d(\pi_n L_n \ldots L_1)}(x_0) \left\{\prod_{p=1}^{n} \frac{\pi_p(dx_p) L_p(x_p, dx_{p-1})}{\pi_{p-1}(dx_{p-1}) M_p(x_{p-1}, dx_p)} M_p(x_{p-1}, dx_p)\right\}$$

In conclusion, using (3.51) we prove the following proposition.

Proposition 3.4.1 *For any function* $\boldsymbol{f_n} \in \mathcal{B}_b(\boldsymbol{E_n})$, *we have the following Feynman-Kac representations*

$$\mathbb{E}\left(\boldsymbol{f_n}\left(X_n^{(n)}, \ldots, X_0^{(n)}\right)\right) = \mathbb{E}\left(\boldsymbol{f_n}\left(X_0, \ldots, X_n\right) \left\{\prod_{p=0}^{n-1} G_p(X_p, X_{p+1})\right\}\right)$$

In addition, for any $x_0 \in E_0$ *we have*

$$\mathbb{E}\left(\boldsymbol{f_n}\left(X_n^{(n)}, \ldots, X_0^{(n)}\right) \middle| X_n^{(n)} = x_0\right)$$

$$\propto \mathbb{E}\left(\boldsymbol{f_n}\left(X_0, \ldots, X_n\right) \left\{\prod_{p=0}^{n-1} G_p(X_p, X_{p+1})\right\} \middle| X_0 = x_0\right)$$

In the above displayed formulae X_n *stands for a Markov chain with initial distribution* π_0 *and elementary Markov transitions* M_n.

For a more detailed analysis of this class of models, with more rigorous derivations of these conditional expectations we refer the reader to the article [164], and to Chapter 5 in the monograph [163].

We end this section with some comments on these models. Firstly, if we choose $\pi_n = \eta_n = \eta_{n-1} M_n$ and $L_n = \mathbb{M}_{n, \eta_{n-1}}$, then we have

$$\pi_n(dx_n) L_n(x_n, dx_{n-1}) = (\eta_{n-1} M_n)(dx_n) \mathbb{M}_{n, \eta_{n-1}}(x_n, dx_{n-1})$$
$$= \eta_{n-1}(dx_{n-1}) M_n(x_{n-1}, dx_n)$$

In this situation, we have $G_{n-1} = 1$. On the other hand, if we choose the homogeneous model $(\pi_n, L_n, M_n) = (\pi, L, L)$, then the function

$$G_n = G = d(\pi \otimes L)_1 / d(\pi \otimes L)_0$$

has the same form as the traditional Metropolis-Hastings acceptance-rejection ratio discussed in Section 3.1.2.

Finally, if we choose some π_n-reversible Markov transitions $M_n = L_n$ on some state space $E_n = E$, then we find that

$$\mathbb{P}\left(\left(X_0^{(n)}, X_1^{(n)}, \ldots, X_n^{(n)}\right) \in d(x_n, x_{n-1}, \ldots, x_0)\right)$$

$$= \left\{\prod_{0 \le p < n} \frac{d\pi_{p+1}}{d\pi_p}(x_p)\right\} \times \mathbb{P}\left((X_0, \ldots, X_n) \in d(x_0, \ldots, x_n)\right)$$

If consider the Boltzmann-Gibbs measures (2.25) discussed in Section 2.5.2

$$\pi_n(dx) = \frac{1}{\mathcal{Z}_n}\left\{\prod_{0 \le p \le n} h_p(x)\right\} \lambda(dx)$$

then we have that

$$\left\{\prod_{0 \le p < n} \frac{d\pi_{p+1}}{d\pi_p}(x_p)\right\} = \frac{\mathcal{Z}_0}{\mathcal{Z}_n} \prod_{0 \le p < n} h_{p+1}(x_p)$$

from which we conclude that

$$\frac{\mathcal{Z}_n}{\mathcal{Z}_0}\,\mathbb{P}\left(\left(X_0^{(n)}, X_1^{(n)}, \ldots, X_n^{(n)}\right) \in d(x_n, x_{n-1}, \ldots, x_0)\right)$$

$$= \mathbb{P}\left((X_0, \ldots, X_n) \in d(x_0, \ldots, x_n)\right) \times \prod_{0 \le p < n} h_{p+1}(x_p) \tag{3.52}$$

If we integrate these equations w.r.t. the random trajectories, then we find that

$$\mathbb{E}\left(\prod_{0 \le p < n} h_{p+1}(X_p)\right) = \mathcal{Z}_n/\mathcal{Z}_0 \tag{3.53}$$

We mention that the above formula coincides with (2.27).

In statistical mechanics, and more particularly in molecular dynamics literature, the continuous time version of the above time reversal formula is sometimes called the Crooks-Jarzinsky formula [129, 130, 146, 342, 343]. In this context, formula (3.53) relates the free energy ratio $\mathcal{Z}_n/\mathcal{Z}_0$ between two states connected to the work $W\left((X_p)_{p<n}\right) := -\sum_{0 \le p < n} \log h_{p+1}(X_p)$ done on the system starting from the equilibrium π_0 and evolving with the Markov transitions M_n. For a more thorough discussion on these models, we refer the reader to the book by T. Lelièvre, M. Rousset, and G. Stoltz [393] (cf. for instance, Section 4.2, Theorem 4.10, on page 288).

3.4.4 Markov bridge models

Markov bridges are particular instances of the reference Markov chains of the abstract Feynman-Kac model (1.37). Depending on the choice of the potential functions in (1.37), these Markov bridge models can be associated with several application domains, including filtering problems or rare event analysis of bridge processes.

We assume that the elementary Markov transitions M_n of the chain X_n satisfy the regularity condition (1.49) for some density functions H_n and some reference measure λ_n. In this situation, the semigroup Markov transitions $M_{p,n+1} = M_{p+1} M_{p+2} \ldots M_{n+1}$ are absolutely continuous with respect to the measure λ_{n+1}, for any $0 \leq p \leq n$, and we have

$$M_{p,n+1}(x_p, dx_{n+1}) = H_{p,n+1}(x_p, x_{n+1}) \, \lambda_{n+1}(dx_{n+1})$$

with the density function

$$H_{p,n+1}(x_p, x_{n+1}) = M_{p,n}\left(H_{n+1}(\cdot, x_{n+1})\right)(x_p)$$

Thanks to these regularity conditions, we readily check that the path distributions of Markov bridges starting at x_0 and ending at x_{n+1} at the final time horizon $(n+1)$ are given by

$$\mathbb{B}_{(0,x_0),(n+1,x_{n+1})}\left(d(x_1, \ldots, x_n)\right)$$

$$:= \mathbb{P}\left((X_1, \ldots, X_n) \in d(x_1, \ldots, x_n) \mid X_0 = x_0, \ X_{n+1} = x_{n+1}\right)$$

$$= \prod_{1 \leq p \leq n} M_p(x_{p-1}, dx_p) \, \frac{dM_{p,n+1}(x_p, \cdot)}{dM_{p-1,n+1}(x_{p-1}, \cdot)}(x_{n+1})$$

$$= \prod_{1 \leq p \leq n} \frac{M_p(x_{p-1}, dx_p) \, H_{p,n+1}(x_p, x_{n+1})}{M_p\left(H_{p,n+1}(\cdot, x_{n+1})\right)(x_{p-1})}$$

Using some abusive (but less notationally consuming) Bayesian notation, we can rewrite these formulae as follows

$$p((x_1, \ldots, x_n) \mid (x_0, x_{n+1}))$$

$$= \frac{p(x_{n+1}|x_n)}{p(x_{n+1}|x_{n-1})} \, p(x_n|x_{n-1}) \cdots \frac{p(x_{n+1}|x_p)}{p(x_{n+1}|x_{p-1})} \, p(x_p|x_{p-1}) \cdots$$

$$\cdots \frac{p(x_{n+1}|x_1)}{p(x_{n+1}|x_0)} \, p(x_1|x_0)$$

with

$$\frac{dM_p(x_{p-1}, \cdot)}{d\lambda_p}(x_p) = p(x_p|x_{p-1}) \quad \text{and} \quad H_{p,n+1}(x_p, x_{n+1}) = p(x_{n+1}|x_p)$$

For linear-Gaussian models, the Markov bridge transitions

$$\frac{M_p(x_{p-1}, dx_p) \, H_{p,n+1}(x_p, x_{n+1})}{M_p \left(H_{p,n+1}(\cdot, x_{n+1}) \right)(x_{p-1})} = \frac{p(x_{n+1}|x_p)}{p(x_{n+1}|x_{p-1})} \, p(x_p|x_{p-1}) \lambda_p(dx_p)$$

can be computed explicitly using the traditional regression formula, or equivalently by the updating step of the Kalman filter presented in Section 8.2.

3.5   Feynman-Kac sensitivity measures

3.5.1   Introduction

This section is dedicated to the design of Feynman-Kac sensitivity measures. The analysis of these quantities arises in a variety of application domains. To name a few, in nonlinear filtering, they are used to estimate filter derivatives, as well as the gradient of log-likelihood functions in hidden Markov chain models [193, 194, 195, 220]. In financial mathematics, they are also used to compute option price type sensitivities [79, 96, 97, 260]. In this context, sensitivity measures allow the traders to determine how sensitive the values of options are to small changes in the set of parameters on which they depend. These parameters include the initial price of assets, the volatility parameter, or the risk free rates.

In financial mathematics literature, these risk measures are often named "Greeks" mainly because they are denoted by Greek letters: the delta is the first derivative of the option value w.r.t. to the underlying price. This quantity can be computed using the mean field simulation schemes associated with the Feynman-Kac representations of the gradients of Markov semigroups developed in Section 3.3.6. We can also use the Feynman-Kac functional representations of the Malliavin Greeks derivatives presented in Section 3.3.7. The gamma measures the second order sensitivity to the price. The vega is not a Greek letter, but it measures the first order sensitivity w.r.t. the volatility parameter discussed in (3.58) in Section 3.5.3. The theta is the derivative of the option value w.r.t. the final time to expiry of the option. Finally, the rho is the first derivative of the option value with respect to the risk free applicable interest rate presented in (3.59) in Section 3.5.3.

3.5.2   Some derivation formulae

We let $\theta \in \mathbb{R}^d$ be some parameter that may represent some kinetic type parameters related to the free evolution model or to adaptive potential functions. We assume that the free evolution model $X_n^{(\theta)}$ associated to some value of the parameter θ is given by a one-step probability transition of the form

$$M_{\theta,n}(x, dx') := \mathbb{P}\left(X_n^{(\theta)} \in dx' | X_{n-1}^{(\theta)} = x\right) = H_{\theta,n}(x, x') \, \lambda_n(dx')$$

for some positive density functions $H_{(\theta,n)}$ and some reference probability measures λ_n. To simplify the presentation, we assume that the initial distribution $\eta_0 = \lambda_0$.

We also consider a collection of functions $G_{\theta,n}$ that depend on θ. We also assume that the gradient, and the Hessian of the logarithms of these functions, w.r.t. the parameter θ, are well defined.

We let $(\Gamma_{\theta,n}, \mathbb{Q}_{\theta,n})$ be the Feynman-Kac measure associated with a given value of θ, and defined for any bounded measurable function $\mathbf{f_n}$ on the path space $\mathbf{E_n} = (E_0 \times \ldots \times E_n)$ by $\mathbb{Q}_{\theta,n}(\mathbf{f_n}) = \Gamma_{\theta,n}(\mathbf{f_n})/\Gamma_{\theta,n}(1)$, with

$$\Gamma_{\theta,n}(\mathbf{f_n}) = \mathbb{E}\left(\mathbf{f_n}(X_0^{(\theta)}, \ldots, X_n^{(\theta)}) \prod_{0 \leq p < n} G_{\theta,p}\left(X_p^{(\theta)}\right)\right) \qquad (3.54)$$

We also denote by $(\gamma_{\theta,n}, \eta_{\theta,n})$, the n-th time marginal measures of $(\Gamma_{\theta,n}, \mathbb{Q}_{\theta,n})$.

We observe that

$$\Gamma_{\theta,n}(\mathbf{f_n}) = \boldsymbol{\lambda_n}\left(\mathbf{f_n} \ \exp\left(\mathbb{L}_{\theta,n}\right) \ \right) \quad \text{with} \quad \boldsymbol{\lambda_n} = \otimes_{0 \leq p \leq n} \lambda_p$$

and the additive functional

$$\mathbb{L}_{\theta,n}(x_0, \ldots, x_n) := \sum_{p=1}^{n} \log\left(G_{\theta,p-1}(x_{p-1}) H_{\theta,p}(x_{p-1}, x_p)\right) \qquad (3.55)$$

By using simple derivation calculations, we prove that the first order derivative of the option value w.r.t. θ is given by

$$\begin{aligned} \nabla\Gamma_{\theta,n}(\mathbf{f_n}) &= \Gamma_{\theta,n}(\mathbf{f_n}\Lambda_{\theta,n}) \quad \text{with} \quad \Lambda_{\theta,n} := \nabla\mathbb{L}_{\theta,n} \qquad &(3.56)\\ \nabla^2\Gamma_{\theta,n}(\mathbf{f_n}) &= \Gamma_{\theta,n}\left[\mathbf{f_n}(\Lambda_{\theta,n})'\Lambda_{\theta,n} + \mathbf{f_n}\nabla^2\mathbb{L}_{\theta,n}\right] \qquad &(3.57) \end{aligned}$$

3.5.3 Some illustrations

Next, we illustrate the above discussion with the sensitivity to changes in the diffusion coefficient of the stochastic Equation (1.6), with $d = 1$. We consider a Markov chain $X_n^{(\theta)}$ that satisfies the following equation

$$X_n^{(\theta)} - X_{n-1}^{(\theta)} = b\left(X_{n-1}^{(\theta)}\right)\Delta + \left[\sigma\left(X_{n-1}^{(\theta)}\right) + \theta\,\sigma'\left(X_{n-1}^{(\theta)}\right)\right]\left(W_{t_n} - W_{t_{n-1}}\right)$$

for some σ' s.t. $\sigma + \theta\,\sigma' > 0$ for any $\theta \in [0,1]$. In this situation, we have

$$\frac{\partial}{\partial\theta}\sum_{p=1}^{n}\log\left(H_{\theta,p}(x_{p-1}, x_p)\right)$$

$$= \sum_{p=1}^{n}\frac{\sigma'(x_{p-1})}{\sigma(x_{p-1}) + \theta\sigma'(x_{p-1})}\left[\left(\frac{(x_p - x_{p-1}) - b(x_{p-1})\Delta}{(\sigma(x_{p-1}) + \theta\sigma'(x_{p-1}))\sqrt{\Delta}}\right)^2 - 1\right] \qquad (3.58)$$

To consider sensitivity to changes in the drift of the stochastic Equation (1.6), with $d = 1$, we assume that $X_n^{(\theta)}$ satisfies equation

$$X_n^{(\theta)} - X_{n-1}^{(\theta)} = \left[b\left(X_{n-1}^{(\theta)}\right) + \theta b'\left(X_{n-1}^{(\theta)}\right) \right] \Delta + \sigma \left(X_{n-1}^{(\theta)}\right) \left(W_{t_n} - W_{t_{n-1}}\right)$$

for some function b'. In this situation, we have

$$\frac{\partial}{\partial \theta} \sum_{p=1}^{n} \log \left(H_{\theta,p}(x_{p-1}, x_p)\right)$$

$$= \sum_{p=1}^{n} \left[(x_p - x_{p-1}) - [b(x_{p-1}) + \theta b'(x_{p-1})] \Delta \right] \times b'(x_{p-1})/\sigma^2(x_{p-1})$$

Now, suppose that changes in potential energy functions are given by $\log G_n = [V_n + \theta V'_n]$, for some nonnegative functions V_n and V'_n. In this situation, we have that

$$\frac{\partial}{\partial \theta} \sum_{0 \le p < n} \log \left(G_{\theta,p}(x_p)\right) = - \sum_{0 \le p < n} V'_p(x_p) \tag{3.59}$$

We illustrate these models with a Feynman-Kac model associated with a reference Markov chain $X_n^{(\theta)} = X_n$, whose values do not depend on θ

$$\gamma_{\theta,n}(f_n) = \mathbb{E} \left(f_n(X_n) \, \exp \left\{ - \sum_{0 \le q < n} V_{\theta,q}(X_q) \right\} \right)$$

for some smooth functions $\theta \mapsto V_{\theta,n}$. Then, using the backward Markov chain model we have

$$\nabla \gamma_{\theta,n}(f_n) = - \sum_{0 \le p < n} \gamma_{\theta,n} \left(f_n \, \mathbb{M}_{n,\eta_{\theta,n-1}} \cdots \mathbb{M}_{p+1,\eta_{\theta,p}} \left(\nabla V_{\theta,p}\right) \right)$$

Chapter 4

Four equivalent particle interpretations

4.1 Spatial branching models

4.1.1 Genetic particle models

One of the most natural interpretations of Feynman-Kac particle models relies on branching process theory, interacting particle systems, and adaptive population dynamics. In this context, mean field IPS models are often interpreted as genetic type algorithms (*abbreviated GA*) equipped with mutation-selection transitions, on some possibly different state spaces.

As their names indicate, in biology these stochastic models represent the genetic type evolutions of individuals, organisms, or phenotypes. For instance, in gene analysis, each population of individuals represents a chromosome, and each individual particle represents a gene. In this setting, the fitness potential function is usually time-homogeneous. This function represents the quality, as well as the adaptation potential value, of the set of genes in a given chromosome [327]. These particle algorithms can also be used in dynamical population analysis, to model changes in the structure of populations in time and in space.

From the pure computational viewpoint, these stochastic models allow encoding the natural evolution mechanisms in a computer. In contrast with the MCMC technology discussed in Section 2.5.2.4 and Section 3.1.2, in evolutionary computing literature, genetic algorithms are used as a meta-heuristic type scheme. These GA models, and their variants equipped with crossover mechanisms, or with more or less different selection style techniques, are used to solve optimization problems on some complex solution state spaces, without any reference to some target probability measure.

The central idea behind any evolutionary algorithm is to use some kind of universal Darwinism as a natural stochastic search of complex solution state spaces. The mutation process mimics the natural random evolution and the variability of organisms or individuals, while the selection pressure is related to the viability, or the fertility, of the organisms. In this interpretation, the selection fitness can also be thought of as the probability that an individual reproduces in a favorable environment.

The algorithm starts with a pool of individuals that represent the initial candidate solutions. We evaluate the fitness of the population members. During the selection process, the individuals with higher fitness are more likely to duplicate into a given number of offsprings. During the mutation process, we increase the variability of the population with local random moves of the offsprings. The desired solution is expected to be discovered gradually, by iterating these evolution mechanisms, up until some termination criteria are met.

Of course, one expects to find the optimal solutions when the number of individuals in the random search population tends to infinity. In the next sections, Section 4.1.2 and Section 4.1.3, we shall see that it is *not exactly the case*, even when the computational power and the number of samples tend to infinity. More precisely, in Section 4.1.2, we shall see that the complete genealogical tree structure of these GA algorithms converge, as the computational power tends to infinity, to a fully *deterministic* Feynman-Kac probability measure on a path space. Section 4.1.3 also provides an intuitive derivation of the evolution equation of the infinite population model, in terms of Feynman-Kac distribution flows.

We notice that the basic genetic algorithm pseudocode is given by the following remarkably simple block diagram.

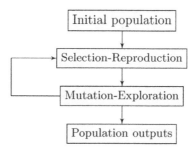

Next, we provide a more formal description of the GA algorithm.

The genetic evolution starts with a pool of N candidate possible solutions $\xi_0 := (\xi_0^1, \ldots, \xi_0^N)$, randomly chosen w.r.t. some distribution η_0, on some initial state space, say E_0. The coordinates ξ_0^i also termed particle, individuals, organisms, or phenotypes depend on the interpretation model they represent. The genetic evolution of the individuals is broken into two separate steps: the selection and the mutation transition.

During the selection-reproduction stage, multiple individuals in the current population
$$\xi_n := (\xi_n^1, \ldots, \xi_n^N) \in E_n^N$$
at time $n \in \mathbb{N}$ are stochastically selected, based on some problem dependent fitness function $G_n : E_n \to [0, \infty[$. These functions measure the quality of a solution in a given solution state space E_n, at a given time step n. More practically, we choose a random proportion B_n^i of an existing solution ξ_n^i in

the current population with a mean value proportional to its fitness $G_n(\xi_n^i)$ to breed a brand new generation of "improved" solutions

$$\widehat{\xi}_n := (\widehat{\xi}_n^1, \dots, \widehat{\xi}_n^N) \in E_n^N$$

For instance, for every index i, with a probability $\epsilon_n G_n(\xi_n^i)$, we set $\widehat{\xi}_n^i = \xi_n^i$; otherwise we replace ξ_n^i with a new individual $\widehat{\xi}_n^i = \xi_n^j$ randomly chosen from the whole population with a probability proportional to $G_n(\xi_n^j)$. The parameter $\epsilon_n \geq 0$ is a tuning parameter that must satisfy the constraint $\epsilon_n G_n(\xi_n^i) \leq 1$, for every $1 \leq i \leq N$. Many other branching strategies can be found in [144, 161, 163].

To describe with some conciseness the mean field interpretation of the selection mechanism, we let $\eta_n^N = \frac{1}{N} \sum_{1 \leq i \leq N} \delta_{\xi_n^i}$ be the occupation measure of the system ξ_n at time n. The central idea of the selection transition is to approximate the weighted measure

$$\Psi_{G_n}(\eta_n^N) := \sum_{1 \leq i \leq N} \frac{G_n(\xi_n^i)}{\sum_{1 \leq j \leq N} G_n(\xi_n^j)} \delta_{\xi_n^i}$$

by the empirical measure

$$\widehat{\eta}_n^N := \frac{1}{N} \sum_{1 \leq i \leq N} \delta_{\widehat{\xi}_n^i} = \frac{1}{N} \sum_{1 \leq i \leq N} B_n^i \, \delta_{\xi_n^i}$$

for some random integer numbers B_n^i. In the above display, Ψ_{G_n} stands for the Boltzmann-Gibbs transformation defined in (0.2).

During the mutation step, every selected individual $\widehat{\xi}_n^i$ mutates to a new solution $\xi_{n+1}^i = x$, randomly chosen with a distribution $M_{n+1}(\widehat{\xi}_n^i, dx)$ on a possibly different solution state space E_{n+1}.

Most of the time, the mutation operator generating new candidate solutions is given by some random function $F_{n+1}(., W_{n+1})$, where W_{n+1} is a random variable taking values in some state space $\mathcal{W}_{n+1}$, and F_{n+1} is a measurable mapping from $(E_n \times \mathcal{W}_{n+1})$ into E_{n+1}. In this case, we have

$$M_{n+1}(\widehat{\xi}_n^i, dx) = \mathbb{P}\left(\xi_{n+1}^i \in dx \mid \widehat{\xi}_n^i\right) = \mathbb{P}\left(F_{n+1}\left(\widehat{\xi}_n^i, W_{n+1}\right) \in dx \mid \widehat{\xi}_n^i\right)$$

In this situation, we sample N random copies $(W_{n+1}^i)_{1 \leq i \leq N}$ of W_n, and for any $1 \leq i \leq N$, we set

$$\xi_{n+1}^i = F_{n+1}\left(\widehat{\xi}_n^i, W_{n+1}^i\right)$$

In summary, the genetic algorithm $\xi_n := (\xi_n^i)_{1 \leq i \leq N}$ is a simple Markov model combining a mutation transition and a selection transition

$$\xi_n := (\xi_n^i)_{1 \leq i \leq N} \xrightarrow{\text{selection}} \widehat{\xi}_n := (\widehat{\xi}_n^i)_{1 \leq i \leq N} \xrightarrow{\text{mutation}} \xi_{n+1} \qquad (4.1)$$

The pseudocode of this genetic algorithm is described below.

Algorithm 4.1: Genetic algorithm pseudocode

<u>Initialization</u>
Fix some population size parameter $N \geq 1$.
Sample N r.v. $\xi_0 := (\xi_0^i)_{1 \leq i \leq N}$ with some given law η_0.

for $k = 1$ to n do
{For each time step k}

   <u>Selection</u>
   for $i = 1$ to N do
   {For each particle i}
$$
\widehat{\xi}_{k-1}^i := \begin{cases} \xi_{k-1}^i, & \text{with probability } \epsilon_{k-1} G_{k-1}(\xi_{k-1}^i) \\ \widehat{\xi}_{k-1}^i, & \text{a r.v. with law } \sum_{i=1}^{N} \dfrac{G_{k-1}(\xi_{k-1}^i)}{\sum_{j=1}^{N} G_{k-1}(\xi_{k-1}^j)} \, \delta_{\xi_{k-1}^i}, & \text{otherwise.} \end{cases}
$$
   end for

   <u>Mutation</u>
   for $i = 1$ to N do
   {For each particle}
$$\xi_k^i := F_k(\widehat{\xi}_{k-1}^i, W_k^i),$$
   end for
end for

The parameter ϵ_{k-1} is a tuning parameter. For instance, we can choose

$$\epsilon_{k-1} = 1 / \max_{1 \leq j \leq N} G_{k-1}(\xi_{k-1}^j)$$

If we set $\epsilon_{k-1} = 0$, then we remove the acceptance probability, so that the selection transition coincides with the classical proportional or roulette selection. For $[0, 1]$-valued fitness functions G_{k-1}, we can choose $\epsilon_{k-1} = 1$ so that $\widehat{\xi}_{k-1}^i = \xi_{k-1}^i$ with probability $G_{k-1}(\xi_{k-1}^i)$.

This generational random process is repeated until some desired termination condition has been reached.

The question of why these genetic algorithms often succeed at generating high fitness solutions of complex practical problems is not really well understood in evolutionary computing literature.

One crucial comment is that the size of the population N should be a precision parameter, so that in some sense we solve the problem at hand when N tends to ∞. In other words, when the computational resources $N \to \infty$, the genetic search model should increase its ability to find the desired solution. Some convergence results for fixed, but sufficiently large, population sizes are developed in the series of articles [115, 116, 117, 118, 119, 120, 162].

4.1.2   Genealogical tree measures

One way to understand the performance of the GA model is to analyze the genealogical structure of the individuals in a large population model. More precisely, if we interpret the updating-selection transition as a birth and death process, then arises the important notion of *the ancestral line* of an individual. More precisely, when a particle $\widehat{\xi}_{n-1}^i \longrightarrow \xi_n^i$ evolves to a new location ξ_n^i, we can interpret $\widehat{\xi}_{n-1}^i$ as the parent of ξ_n^i. Looking backwards in time and recalling that the particle $\widehat{\xi}_{n-1}^i$ has selected a site ξ_{n-1}^j in the configuration at time $(n-1)$, we can interpret this site ξ^j_{n-1} as the parent of $\widehat{\xi}^i_{n-1}$ and therefore as the ancestor denoted $\xi^i_{n-1,n}$ at level $(n-1)$ of ξ^i_n. Running backwards in time we may trace the whole ancestral line

$$\xi^i_{0,n} \longleftarrow \xi^i_{1,n} \longleftarrow \cdots \longleftarrow \xi^i_{n-1,n} \longleftarrow \xi^i_{n,n} = \xi^i_n \qquad (4.2)$$

An illustration of the genealogical tree model associated with $N = 3$ particles and a time horizon $n = 5$ is given below

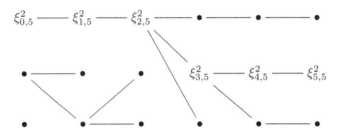

Each parent node has a certain random number of children or offspring.

One could expect that these genealogical tree models exhibit different asymptotic behaviors, depending on their sampling, or on the problem at hand. This is not the case. In terms of proportions, and probability measures, we do not have a lot of variability. The sequence of random occupation measures of the genealogical tree $\eta^N_n := N^{-1} \sum_{1 \leq i \leq N} \delta_{\xi^i_n}$, with the ancestral lines

$$\xi^i_n := (\xi^i_{0,n}, \xi^i_{1,n}, \ldots, \xi^i_{n,n}) \in \mathbf{E_n} := (E_0 \times E_0 \times \ldots \times E_n)$$

become more and more deterministic, as the size of the population $N \uparrow \infty$. More precisely, for any $\mathbf{f_n} \in \mathcal{B}_b(\mathbf{E_n})$, we have the almost sure convergence

$$\lim_{N \to \infty} \eta^N_n(\mathbf{f_n}) = \mathbb{Q}_n(\mathbf{f_n}) \qquad (4.3)$$

with the Feynman-Kac probability measure $\mathbb{Q}_n$ defined in (1.37).

In particular, using the fact that all the ancestral trajectories have the same law, we have that

$$\mathbb{E}\left(\eta^N_n(\mathbf{f_n})\right) = \mathbb{E}\left(\mathbf{f_n}(\xi^1_{0,n}, \xi^1_{1,n}, \ldots, \xi^1_{n,n})\right)$$

Using (4.3), we conclude that

$$\lim_{N \to \infty} \mathbb{E}\left(\mathbf{f_n}(\xi^1_{0,n}, \xi^1_{1,n}, \ldots, \xi^1_{n,n})\right) = \mathbb{Q}_n(\mathbf{f_n})$$

In other words, we have

$$\lim_{N \to \infty} \text{Law}\left(\xi^1_{0,n}, \xi^1_{1,n}, \ldots, \xi^1_{n,n}\right) = \lim_{N \to \infty} \text{Law}(\boldsymbol{\xi}^1_n) = \mathbb{Q}_n$$

For a more detailed discussion on these convergence properties, we refer the reader to Section 2.2, in the opening chapter; as well as to Section 15.2.2

and Section 15.5, for more refined $\mathbb{L}_m$-mean error estimates and exponential concentration inequalities.

We end this section with some comments on the mean field approximation of the extremum values of the log-likelihood of the Feynman-Kac measures $\mathbb{Q}_n$, on Euclidian state spaces.

We consider a Feynman-Kac model $\mathbb{Q}_n$, associated with some fitness function G_n and some Markov transitions M_n on $E_n = \mathbb{R}^d$, for some $d \geq 1$. We further assume that $G_n = \exp U_n$, for some bounded function U_n, and the Markov transitions M_n are absolutely continuous with respect to the Lebesgue measure λ on $\mathbb{R}^d$, and for any $(x, y) \in (E_{n-1} \times E_n)$, and for any $n \geq 1$, we have

$$\log \frac{dM_n(x,)}{d\lambda}(x) := W_n(x, y)$$

with some bounded function W_n. We also assume that $\eta_0 \ll \lambda$, with some bounded function $W_0 := \log \frac{d\eta_0}{d\lambda}$.

In this situation, we have $\mathbb{Q}_n \ll \boldsymbol{\lambda_n}$, with $\boldsymbol{\lambda_n} := \lambda^{\otimes(n+1)}$. In addition, we also have the Boltzmann-Gibbs formulae

$$d\mathbb{Q}_n = \frac{1}{\mathscr{Z}_n} \exp\left(\boldsymbol{V_n}\right) \, d\boldsymbol{\lambda_n}$$

with the log-likelihood $\boldsymbol{V_n}$ defined for any $\boldsymbol{x_n} = (x_0, \ldots, x_n) \in \boldsymbol{E_n}$ by

$$\boldsymbol{V_n}(\boldsymbol{x_n}) := \sum_{p=0}^{n} [W_p(x_{p-1}, x_p) + U_{p-1}(x_{p-1})]$$

In the above display, we have used the conventions $U_{-1} = 0$, the null function, and $W_0(x_{-1}, x_0) = W_0(x_0)$, when $p = 0$. In Section 15.3 (cf. Theorem 15.3.3), we shall prove the convergence in probability

$$\lim_{N \uparrow \infty} \max_{1 \leq i \leq N} \boldsymbol{V_n}(\xi_{0,n}^i, \ldots, \xi_{n,n}^i) = \boldsymbol{\lambda_n} - essup \boldsymbol{V_n}$$

This implies that the trajectory with "high probability mass" can be computed using the ancestral lines of the genealogical tree model. Applications of this result in the context of filtering problems, and stochastic optimal control, are discussed in Section 8.1.3 and in Section 8.3.4.

4.1.3   Infinite population models

In evolutionary computing literature, infinite genetic population models are often used in finite type state spaces to understand the behavior of the population in terms of a well defined deterministic dynamical systems in unit simplex state spaces [398, 484, 569, 573]. Under some regularity properties of the mutation transitions, the correspondence between the long time behavior of the finite, and the infinite, population models is now rather well understood, using the uniform convergence estimates presented in the opening chapter (see also one of the first rigorous studies in this fields [161, 169, 170], the research monograph [163], and the references therein). We have already mentioned that the fixed points of these infinite population semigroups are directly connected to the top eigenvalues and the corresponding eigenfunctions of Schrödinger's type integral operators. We shall return to these questions in Section 7.6, and Section 7.6.2, dedicated to Doob's h-processes, and quasi-invariant measures.

In this short section, we show that the evolution equation of the infinite population model is defined in terms of a nonlinear Feynman-Kac semigroup.

As direct consequence of (4.3), we first observe that the occupation measures of the population of individuals at time n

$$\eta_n^N = \frac{1}{N} \sum_{i=1}^{N} \delta_{\xi_{n,n}^i} = \frac{1}{N} \sum_{i=1}^{N} \delta_{\xi_n^i}$$

converge, as $N \uparrow \infty$, to the n-th time marginal η_n of the measure $\mathbb{Q}_n$. In genetic algorithm literature, the sequence of measures η_n is called the infinite population model.

Remark 4.1.1 *In the reverse angle, the solution η_n of an evolution equation of the form (3.28) is given by a Feynman-Kac model with potential functions G_n and a reference Markov chain with elementary transitions M_n.*

We end this section, with a local mean field interpretation of the GA model.

Firstly, we observe that the fluctuations of the empirical measures $\widehat{\eta}_n^N$ around the weighted measure $\Psi_{G_n}(\eta_n^N)$ are encapsulated into the sequence of local sampling random fields

$$V_n^{N,select} := \sqrt{N} \left[\widehat{\eta}_n^N - \Psi_{G_n}(\eta_n^N) \right] \quad \Longleftrightarrow \quad \widehat{\eta}_n^N = \Psi_{G_n}(\eta_n^N) + \frac{1}{\sqrt{N}} V_n^{N,select}$$

As shown in [144], the minimal requirement to get some interesting nonasymptotic variance estimate is to check the following local conditions

$$\mathbb{E}\left(V_n^{N,select}(f_n)\right) = 0 \quad \text{and} \quad \sup_{N \geq 1} \mathbb{E}\left(V_n^{N,select}(f_n)^2\right) < \infty \qquad (4.4)$$

for any bounded measurable function f_n on E_n. In the same vein, the fluctuations of the empirical measures η_{n+1}^N around the measure $\widehat{\eta}_n^N M_{n+1}$ are encapsulated into the sequence of local sampling random fields

$$V_{n+1}^{N,mut} := \sqrt{N} \left[\eta_{n+1}^N - \widehat{\eta}_n^N M_{n+1} \right] \quad \Longleftrightarrow \quad \eta_{n+1}^N = \widehat{\eta}_n^N M_{n+1} + \frac{1}{\sqrt{N}} V_{n+1}^{N,mut}$$

In this context, it is easily checked that

$$\mathbb{E}\left(V_{n+1}^{N,mut}(f_{n+1})\right) = 0 \quad \text{and} \quad \sup_{N \geq 1} \mathbb{E}\left(\left[V_{n+1}^{N,mut}(f_{n+1})\right]^2\right) < \infty$$

for any bounded measurable function f_{n+1} on E_{n+1}.

$$
\begin{aligned}
\eta_{n+1}^N &= \widehat{\eta}_n^N M_{n+1} + \frac{1}{\sqrt{N}} V_{n+1}^{N,mut} \\
&= \Psi_{G_n}(\eta_n^N) M_{n+1} + \frac{1}{\sqrt{N}} \left(V_{n+1}^{N,mut} + V_n^{N,select} M_{n+1} \right)
\end{aligned}
$$

We conclude that

$$
\eta_{n+1}^N = \Phi_{n+1}(\eta_n^N) + \frac{1}{\sqrt{N}} V_{n+1}^N \tag{4.5}
$$

with the one step Feyman-Kac transformations Φ_{n+1}, and the local sampling random fields V_{n+1}^N, defined for any $\eta \in \mathcal{P}(E_n)$, and any $N \geq 1$, by

$$
\Phi_n(\eta) = \Psi_{G_n}(\eta) M_{n+1} \quad \text{and} \quad V_{n+1}^N = V_{n+1}^{N,mut} + V_n^{N,select} M_{n+1}
$$

Roughly speaking, the local perturbation formula (4.5) indicates that

$$
\eta_{n+1}^N = \Phi_{n+1}(\eta_n^N) + \frac{1}{\sqrt{N}} V_{n+1}^N \longrightarrow_{N\uparrow\infty} \eta_{n+1} = \Phi_{n+1}(\eta_n)
$$

A more refined asymptotic analysis of these local sampling random fields models is provided in Section 10.6.4.

4.1.4 Particle normalizing constants

Surprisingly, the analysis of the unnormalized infinite population models has not received much attention in evolutionary computing literature. More interestingly, it appears that the average of the unnormalized finite population model coincides with its limiting measure. This key observation has been proved in the mid-1990s in the article [151], in the context of nonlinear filtering applications.

In this section, we provide a brief description of these unnormalized particle measures and their limiting quantities. We also discuss some important consequences of the unbiasedness properties of these particle measures.

We recall that the normalizing constants $\mathcal{Z}_n$ of the Feynman-Kac measure (1.37) are given by the formula

$$
\mathcal{Z}_n = \gamma_n(1) = \mathbb{E}\left(\prod_{0 \leq p < n} G_p(X_p) \right)
$$

In Section 3.2.2, we saw that these constants are expressed in terms of the flow of marginal measures η_p, from the origin $p = 0$, up to the current time n, with the following product formulae:

$$
\mathcal{Z}_n := \gamma_n(1) = \mathbb{E}\left(\prod_{0 \leq p < n} G_p(X_p) \right) = \prod_{0 \leq p < n} \eta_p(G_p) \tag{4.6}
$$

More generally, for any $f_n \in \mathcal{B}_b(E_n)$, we have

$$\gamma_n(f_n) = \eta_n(f_n) \times \mathcal{Z}_n = \eta_n(f_n) \times \prod_{0 \leq p < n} \eta_p(G_p) \tag{4.7}$$

Mimicking these product formulae, we set

$$\gamma_n^N(f_n) := \eta_n^N(f_n) \times \prod_{0 \leq p < n} \eta_p^N(G_p) \quad \text{and} \quad \mathcal{Z}_n^N := \gamma_n^N(1) = \prod_{0 \leq p < n} \eta_p^N(G_p) \tag{4.8}$$

These rather complex particle measures satisfy the unbiasedness property:

$$\mathbb{E}\left(\eta_n^N(f_n) \prod_{0 \leq p < n} \eta_p^N(G_p)\right) = \mathbb{E}\left(f_n(X_n) \prod_{0 \leq p < n} G_p(X_p)\right) \tag{4.9}$$

A proof of this result can be found in Section 14.7.3. Furthermore, for any bounded measurable function f_n, we have the almost sure convergence result

$$\lim_{N \to \infty} \gamma_n^N(f_n) = \gamma_n(f_n)$$

For a detailed discussion on the convergence of these particle unnormalized measures, we refer the reader to Section 2.2.5 and to Chapter 16.

The unnormalized measures γ_n^N associated with the genealogical tree occupation measures η_n^N defined in Section 4.1.2 are defined for some bounded function f_n on the product space $\mathbf{E_n} := (E_0 \times \ldots \times E_n)$ by

$$\gamma_n^N(f_n) := \eta_n^N(f_n) \times \prod_{0 \leq p < n} \eta_p^N(G_p)$$

In the above display, η_p^N stands for the p-th time marginals of the measures η_p^N, for any $p \geq 0$. These unnormalized measures on path space share the same convergence properties as the measures γ_n^N introduced in (4.8).

For instance, combining the equivalence principles developed in Chapter 15 with the unbiasedness property 4.9 we prove that

$$\mathbb{E}\left(\eta_n^N(f_n) \prod_{0 \leq p < n} \eta_p^N(G_p)\right) = \mathbb{E}\left(f_n(X_0, \ldots, X_n) \prod_{0 \leq p < n} G_p(X_p)\right) \tag{4.10}$$

We end this section with particle approximation of the Feynman-Kac sensitivity measures introduced in Section 3.5.2.

We denote by $(\gamma_{\theta,n}^N, \eta_{\theta,n}^N)$ the genealogical tree based particle measures associated with the Feynman-Kac measures $(\Gamma_{\theta,n}, \mathbb{Q}_{\theta,n})$ indexed by the parameter θ and defined in (3.54).

Under appropriate regularity conditions, the sensitivity measures

$\nabla\Gamma_{\theta,n}(\mathbf{f_n})$ and $\nabla^2\Gamma_{\theta,n}(\mathbf{f_n})$ defined in (3.56) and (3.57) can be approximated by the unbiased particle models

$$\nabla_N\Gamma_{\theta,n}(\mathbf{f_n}) \quad := \quad \gamma_{\theta,n}^N(\mathbf{f_n}\Lambda_{\theta,n})$$

$$\nabla_N^2\Gamma_{\theta,n}(\mathbf{f_n}) \quad = \quad \gamma_{\theta,n}^N\left[\mathbf{f_n}(\nabla\mathbb{L}_{\theta,n})'(\nabla\mathbb{L}_{\theta,n}) + \mathbf{f_n}\nabla^2\mathbb{L}_{\theta,n}\right] \qquad (4.11)$$

with the additive functional $\Lambda_{\theta,n}$ defined in (3.55).

4.1.5   Island particle models

This section focuses on island genetic type particle models. These coarse grained parallel procedures are very popular in genetic algorithms literature. Further details on this topic can be found in the articles [422, 520, 572], and the references therein. These techniques are commonly used in scientific computing research to speed up interacting genetic search algorithms.

In this section, we provide a mean field Feyman-Kac representation of these island population models. We consider the Markov chain $\mathcal{X}_n$ and the potential function $\mathcal{G}_n$ defined by

$$\mathcal{X}_n := \xi_n \in \mathcal{E}_n := E_n^N \quad \text{and} \quad \mathcal{G}_n(\mathcal{X}_n) := \frac{1}{N}\sum_{i=1}^N G_n(\xi_n^i)$$

Rewriting (4.9) in terms of these mathematical objects, for any $n \geq 0$, and any function $f_n \in \mathcal{B}_b(E_n)$, we find that

$$\mathbb{E}\left(f_n(X_n)\prod_{0\leq p<n}G_p(X_p)\right) = \mathbb{E}\left(\mathcal{F}_n(\mathcal{X}_n)\prod_{0\leq p<n}\mathcal{G}_p(\mathcal{X}_p)\right) \qquad (4.12)$$

with the function $\mathcal{F}_n$ on $\mathcal{E}_n$ defined by

$$\mathcal{F}_n(\mathcal{X}_n) := \frac{1}{N}\sum_{i=1}^N f_n(\xi_n^i)$$

We let $\mathcal{M}_n$ be the Markov transitions of the chain $\mathcal{X}_n$ from $\mathcal{E}_{n-1}$ onto $\mathcal{E}_n$. We denote by $\mathcal{Q}_n$ the Feynman-Kac measures associated with the pair $(\mathcal{G}_n, \mathcal{M}_n)$. We recall that we can approximate $\mathcal{Q}_n$ using an $\mathcal{E}_n^{N'}$-valued genetic type particle model

$$\zeta_n := (\zeta_n^i)_{1\leq i\leq N'} \xrightarrow{\ \ selection\ \ } \widehat{\zeta}_n := (\widehat{\zeta}_n^i)_{1\leq i\leq N'} \xrightarrow{\ \ mutation\ \ } \zeta_{n+1}$$

with mutation transitions $\mathcal{M}_n$ and selection fitness functions $\mathcal{G}_n$. It is important to notice that the particles in this model can be interpreted as islands with N individuals

$$\zeta_n^i := \left(\zeta_n^{i,j}\right)_{1\leq j\leq N} \quad \text{and} \quad \widehat{\zeta}_n^i := \left(\widehat{\zeta}_n^{i,j}\right)_{1\leq j\leq N} \in \mathcal{E}_n = E_n^N$$

In situation, we run in parallel several N-genetic type interacting particle algorithms. Each of them evolves in one of the N' islands. At a geometric stochastic rate, the populations of N individuals between the N' islands interact according to a selective migration process dictated by the fitness functions $\mathcal{G}_n$. By construction, we notice that the island selection mechanism is expressed in terms of the averaged fitness of the N individuals in a given island.

Formula (4.12) shows that this N'-island particle model is mathematically equivalent to the Feynman-Kac genetic type particle model discussed in Section 4.1.1 and in Section 4.1.2. The performance analysis of this coarse grained type particle model in terms of the number of individuals in each island has been stated in the recent article [227].

4.1.6 Complete ancestral tree models

4.1.6.1 Backward Markov models

In contrast to the genealogical tree models discussed in Section 4.1.2, the complete ancestral tree incorporates all the ancestral lineages of the individuals during their evolutions. More formally, the complete ancestral tree model is defined by the set of all the population of individuals $\xi_p = (\xi_p^i)_{1 \leq i \leq N}$, from the origin $p = 0$ up to a given final time horizon $p = n$.

An illustration of the complete ancestral tree model associated with $N = 3$ particles, and a time horizon $n = 4$, is given by the following diagram.

In contrast with the genealogical tree models introduced in Section 4.1.2, all the ancestral levels in the complete tree are filled, with the whole populations at every time step.

In this section, we discuss in more detail the backward Markov interpretation model presented in Section 2.2.6. As its name indicates, this model allows to read sequentially, and backward in time, the ancestral lines of the individuals with stochastic matrices. We further assume that the Markov transitions M_n satisfy the regularity condition (1.49) stated in Section 1.4.3.4. We recall from (2.13) that the backward particle approximations of the measures $\mathbb{Q}_n$ are given by the sequence of particle measures

$$\mathbb{Q}_n^N = \eta_n^N \otimes \mathbb{Q}_{n|n}^N \qquad (4.13)$$

with the Markov transitions from E_n into $\boldsymbol{E}_{n-1} = \prod_{0 \le p < n} E_p$, given by

$$\mathbb{Q}_{n|n}^N(x_n, d(x_0, \dots, x_{n-1})) := \prod_{q=1}^{n} \mathbb{M}_{q, \eta_{q-1}^N}(x_q, dx_{q-1})$$

For a detailed discussion on the convergence of these backward particle measures, we refer the reader to Section 2.2.6 and to Chapter 17.

This backward particle model has many special features. In the next section, Section 4.1.6.2, we present a matrix formulation of this particle model.

We also discuss two applications of this framework. In Section 4.1.6.3 we present an online estimation of the stability properties of regular Feynman-Kac semigroups. The last section, Section 4.1.6.4, is concerned with the computation of the sensitivity measures discussed in Section 3.5.2.

4.1.6.2 Random matrix formulation

Notice that the computation of integrals w.r.t. the particle measures $\mathbb{Q}_n^N$ are reduced to summations over the particle locations ξ_n^i. It is therefore natural to identify a population of individuals $(\xi_n^1, \dots, \xi_n^N)$ at time n to the ordered set of indexes $\{1, \dots, N\}$. In this framework, the occupation measures and the functions are identified with the following line and column vectors

$$\eta_n^N := \left[\frac{1}{N}, \dots, \frac{1}{N} \right] \quad \text{and} \quad f_n := \begin{pmatrix} f_n(\xi_n^1) \\ \vdots \\ f_n(\xi_n^N) \end{pmatrix}$$

and the matrices $\mathbb{M}_{n, \eta_{n-1}^N}$ by the $(N \times N)$ matrices

$$\mathbb{M}_{n, \eta_{n-1}^N} := \begin{pmatrix} \mathbb{M}_{n, \eta_{n-1}^N}(\xi_n^1, \xi_{n-1}^1) & \cdots & \mathbb{M}_{n, \eta_{n-1}^N}(\xi_n^1, \xi_{n-1}^N) \\ \vdots & \vdots & \vdots \\ \mathbb{M}_{n, \eta_{n-1}^N}(\xi_n^N, \xi_{n-1}^1) & \cdots & \mathbb{M}_{n, \eta_{n-1}^N}(\xi_n^N, \xi_{n-1}^N) \end{pmatrix} \qquad (4.14)$$

with the (i, j)-entries

$$\mathbb{M}_{n, \eta_{n-1}^N}(\xi_n^i, \xi_{n-1}^j) = \frac{G_{n-1}(\xi_{n-1}^j) H_n(\xi_{n-1}^j, \xi_n^i)}{\sum_{k=1}^{N} G_{n-1}(\xi_{n-1}^k) H_n(\xi_{n-1}^k, \xi_n^i)}$$

For instance, the $\mathbb{Q}_n$-integration of normalized additive linear functionals of the form

$$\mathbf{f_n}(x_0, \ldots, x_n) = \frac{1}{n+1} \sum_{0 \leq p \leq n} f_p(x_p) \qquad (4.15)$$

is given by the particle matrix approximation model

$$\mathbb{Q}_n^N(\mathbf{f_n}) = \frac{1}{n+1} \sum_{0 \leq p \leq n} \eta_n^N \mathbb{M}_{n,\eta_{n-1}^N} \mathbb{M}_{n-1,\eta_{n-2}^N} \cdots \mathbb{M}_{p+1,\eta_p^N}(f_p)$$

This Markov interpretation allows computing complex Feynman-Kac path integrals using simple random matrix operations on finite sets. Roughly speaking, this methodology allows reducing Feynman-Kac path integration problems on general state spaces to Markov path integration on *finite state spaces*, with cardinality N.

Nevertheless, the computational cost N^2 of these particle random matrix models can be prohibitive in some applications. In this case, we can replace the full matrix averaging technique on the finite sets of N individuals, by some judicious sampling approximation scheme. In this connection, we quote a rejection sampling method, recently proposed in [220].

4.1.6.3 Stability properties

This section is concerned with an interesting application of the backward particle models presented in Section 1.4.3.4, and Section 4.1.6.1, to the estimation of the stability properties of Feynman-Kac distribution flows. We further assume that the Markov transitions M_n of the Feynman-Kac model satisfy the regularity condition (1.49) stated in Section 1.4.3.4.

We let $(\mathbb{Q}'_n, \Gamma'_n, \eta'_n, \gamma'_n, \mathcal{Z}'_n)$ be the Feynman-Kac models defined as $(\mathbb{Q}_n, \Gamma_n, \eta_n, \gamma_n, \mathcal{Z}_n)$, by replacing the initial measure η_0 by some probability measure $\eta'_0 \ll \eta_0 \in \mathcal{P}(E_0)$. Under our assumptions, for any initial distributions $\eta'_0 \ll \eta_0 \in \mathcal{P}(E_0)$, we have $\eta'_n \ll \eta_n \in \mathcal{P}(E_n)$. For any $n \geq 0$, we set $h_n := d\eta'_n/d\eta_n$. In this notation, we have the formula

$$h_n = \mathbb{M}_{n,\eta_{n-1}} \cdots \mathbb{M}_{1,\eta_0}(h_0)/\eta_n(\mathbb{M}_{n,\eta_{n-1}} \cdots \mathbb{M}_{1,\eta_0}(h_0))$$

A detailed proof of this formula is provided in Section 12.4.2 (cf. Lemma 12.4.7).

Replacing in the above display the flow of measures $(\eta_p)_{0 \leq p \leq n}$ by the N-particle density profiles $(\eta_p^N)_{0 \leq p \leq n}$ we obtain a new particle approximation model h_n^N of the functions h_n

$$h_n^N = \mathbb{M}_{n,\eta_{n-1}^N} \cdots \mathbb{M}_{1,\eta_0^N}(h_0)/\eta_n^N(\mathbb{M}_{n,\eta_{n-1}^N} \cdots \mathbb{M}_{1,\eta_0^N}(h_0))$$

By construction, we also have the following recursive equations

$$h_n^N = \mathbb{M}_{n,\eta_{n-1}^N}(h_{n-1}^N)/\eta_n^N(\mathbb{M}_{n,\eta_{n-1}^N}(h_{n-1}^N)) \quad \text{with} \quad h_0^N = h_0$$

We can estimate the convergence of these models using the exponential concentration inequalities (2.15) presented in the opening chapter.

4.1.6.4   Particle sensitivity measures

In this short section we present unbiased particle approximation of the Feynman-Kac sensitivity measures introduced in Section 3.5.

We denote by $\mathcal{Z}_{\theta,n}^N$ the N-particle approximation of the normalizing constant $\mathcal{Z}_{\theta,n} := \Gamma_{\theta,n}(1)$. We also consider the N-particle backward measures associated with some value of the parameter θ and defined in (4.13)

$$\mathbb{Q}_{\theta,n}^N = \eta_n^N \otimes \mathbb{Q}_{n|n}^N \quad \text{and} \quad \Gamma_{\theta,n}^N := \mathcal{Z}_{\theta,n}^N \times \mathbb{Q}_{\theta,n}^N$$

Under appropriate regularity conditions, the sensitivity measures $\nabla \Gamma_{\theta,n}(\mathbf{f_n})$ and $\nabla^2 \Gamma_{\theta,n}(\mathbf{f_n})$ defined in (3.56) and (3.57) can be approximated by the *unbiased* particle models

$$
\begin{aligned}
\nabla_N \Gamma_{\theta,n}(\mathbf{f_n}) &:= \Gamma_{\theta,n}^N(\mathbf{f_n}\Lambda_{\theta,n}) \\
\nabla_N^2 \Gamma_{\theta,n}(\mathbf{f_n}) &= \Gamma_{\theta,n}^N\left[\mathbf{f_n}(\nabla\mathbb{L}_{\theta,n})'(\nabla\mathbb{L}_{\theta,n}) + \mathbf{f_n}\nabla^2\mathbb{L}_{\theta,n}\right] \quad (4.16)
\end{aligned}
$$

with the additive functional $\Lambda_{\theta,n}$ defined in (3.55).

4.2   Sequential Monte Carlo methodology

4.2.1   Importance sampling and resampling models

The second interpretation of Feynman-Kac IPS models relies on advanced sequential Monte Carlo methodologies (*abbreviated SMC*), used in Bayesian statistics and in advanced stochastic simulation theory.

Sequential Monte Carlo methods, also termed Sampling Importance Resampling (*abbreviated SIR*) [324, 519, 522], as well as particle filters [151, 288, 366], refer to a class of Monte Carlo methods that combines sequentially importance sampling with resampling steps; other related branching rules are discussed in [144, 161, 163]. These sophisticated Monte Carlo techniques can be interpreted as an adaptive version of the more traditional sequential Importance sampling methods (*abbreviated SIS*).

Through the last two decades, this SMC technology has been applied to various problems arising in Bayesian statistics [133, 165, 252, 403], risk analysis and rare event simulation [125, 122, 163, 175, 176, 282], as well as in advanced signal processing [104, 151, 161, 154, 163, 225, 222, 223, 288, 366, 364]. We also refer to [94, 142, 190, 223, 550] and the books [93, 163, 222] for more detailed reviews on SMC methods, with rather detailed bibliographical references.

Importance sampling is one of the most popular variance reduction techniques that can be used in Monte Carlo sampling of complex distributions, such as those arising in rare event analysis [87, 491, 523, 525]. We also refer to the series of article by P. Dupuis and his co-authors [228, 229, 230, 231,

232, 233]. The central idea is to find a judicious importance sampling distribution, also termed the twisted or the biased measure, which encourages important state space regions that depend on the average quantity we want to estimate. The resulting estimation depends on some weights functions that are proportional to the Radon-Nikodym derivative between the reference physical distribution and the twisted measure we have chosen. We can use biased weighted averages with weights that sum to one, or unbiased estimates with weights that do not sum to one, depending if we know the Radon-Nikodym exactly, or up to some normalizing constant.

Of course, when the target distribution is given by the law of the random trajectories of some Markov process, the importance sampling distributions are expressed in terms of the distributions of the trajectories of some different twisted Markov chain model.

The performance of these importance sampling techniques strongly depends on the choice of the twisted measures. In some instances, the variance of the twisted estimates can be much more larger than the one associated with the crude Monte Carlo method. These troubles often reflect the degeneracy of the weight functions for large time horizon problems. In general, the twisted distributions of the importance sampling Markov chain model become orthogonal to the reference distributions, when the time horizon tends to infinity. More precisely, the random Radon-Nikodym weight functions tend to ∞, or to 0. In practice, after a few time steps, one of the weights in the biased importance sampling estimator tends to 1, and the other ones tends to 0. A more thorough discussion of these degeneracy problems, and related SMC methodologies, is given in Section 12.3.3 and in the research monograph [163].

Another drawback of importance sampling techniques comes from the fact that they need to replace the original process by some twisted stochastic processes. These importance methods break down as soon as the physical system is too complex to run in unrealistic regimes. For a more detailed discussion on these problems, and related SMC methodologies, we refer the reader to Section 2.7.1 in the opening chapter of the present book.

Sequential Monte Carlo methodologies solve these physical problems. In addition, these mean field simulation techniques are intended to cure the degeneracy of the Radon-Nikodym weights functions. In scientific computing literature, and Bayesian statistic articles, this key solution is often presented as the "central Resampling step," mainly because we remove the random weights after the resampling step. Roughly speaking, at every time step, or periodically, we resample the individuals according to the weighted discrete distribution of the twisted population. At the end of this resampling stage, we replace their (unnormalized) weights by 1 over the number of samples.

This resampling stage can be interpreted as a spatial branching rule [144, 161, 163], as a bootstrap simulation [43, 277, 289, 290, 409], or an acceptance-rejection technique equipped with an adaptive recycling mechanism [186, 163, 198]. These statistical and probabilistic interpretations are clearly directly related to the spatial branching models, and the genetic type IPS schemes, we have discussed in some detail in Section 4.1.1.

Of course, these "resampling" strategies introduce locally in time additional Monte Carlo sampling errors (cf. Section 4.1.3), but these errors are compensated in some sense by the stabilization of the weights. The understanding on how the local errors propagate w.r.t. the time horizon is one of the central questions developed in the third part of this book, dedicated to the theoretical analysis of mean field IPS models. We also refer the reader to Section 2.1 in the opening chapter, dedicated to stochastic perturbation techniques and presenting some uniform convergence estimates.

The rest of this section is organized as follows:

In Section 4.2.2, we briefly present some key importance sampling formulae for Markov chain models. Section 4.2.3 is dedicated to conventional importance sampling Monte Carlo methods based on weighted i.i.d. copies of the twisted Markov chain model. In Section 4.2.4, we review some basic sequential importance sampling and resampling principles of SMC methodologies. In Section 4.2.5, we show that these models are mathematically equivalent to the mean field IPS simulation of Feynman-Kac models in transition state spaces. In Section 4.2.5.2, we present a genetic type formulation of the SMC algorithms discussed in Section 4.2.4. In the last section, Section 4.2.5.3, we present some extended adaptive SMC models.

4.2.2   Importance sampling formulae

This section gives a brief presentation of some change of measures used in SIS and SMC methodologies. Suppose we are given a sequence of target distributions $\mathbb{Q}_n$ with increasing sampling complexity on some product state space models $\mathbf{E_n} := (E_0 \times \ldots \times E_n)$, in terms of a proportional recursion

$$\mathbb{Q}_n(d(x_0, \ldots, x_n)) \propto \mathbb{Q}_{n-1}(d(x_0, \ldots, x_{n-1})) \times Q_n(x_{n-1}, dx_n) \qquad (4.17)$$

for some bounded positive integral operators $Q_n(x_{n-1}, dx_n)$ from $\mathcal{B}_b(E_n)$ into $\mathcal{B}_b(E_{n-1})$. As shown in Section 3.2, these distributions coincide with the Feynman-Kac models (1.37). We denote by η_n the n-th time marginal of $\mathbb{Q}_n$.

Sequential Monte Carlo techniques, including importance sampling algorithms and related Metropolis-Hasting Markov chain Monte Carlo schemes, are based on choosing a reference Markov chain model X_n that is easy to generate. Practically, we choose an initial distribution μ_0 and one step transitions

$$\mathbb{P}(X_n \in dx_n | X_{n-1} = x_{n-1}) = M_n(x_{n-1}, dx_n)$$

such that $Q_{n+1}(x_n, .) \ll M_{n+1}(x_n, .)$, for η_n-almost every $x_n \in E_n$, and for any $n \geq 0$. In this situation, the Radon-Nikodym derivatives between these integral operators are well defined, and we set

$$\mathbf{G_n}(x_n, x_{n+1}) = \frac{dQ_{n+1}(x_n, .)}{dM_{n+1}(x_n, .)}(x_{n+1}) \qquad (4.18)$$

Notice that the law of $(X_0, \ldots, X_n)$ is given by

$$\mathbb{P}_n(d(x_0, x_1, \ldots, x_n)) = \mu_0(dx_0) M_1(x_0, dx_1) \ldots M_n(x_{n-1}, dx_n)$$

In this notation, we have the change of measure formula given by the Boltzmann-Gibbs transformation

$$\mathbb{Q}_n = \Psi_{W_n}(\mathbb{P}_n) \tag{4.19}$$

with the importance weight functions given by the formulae

$$
\begin{aligned}
W_n(x_0, \ldots, x_n) &:= \frac{d\eta_0}{d\mu_0}(x_0) \prod_{0 \le p < n} \mathbf{G_p}(x_p, x_{p+1}) \\
&= W_{n-1}(x_0, \ldots, x_{n-1}) \times \mathbf{G_{n-1}}(x_{n-1}, x_n) \tag{4.20}
\end{aligned}
$$

For the Feynman-Kac measure $\mathbb{Q}_n$ defined in (1.37), if we choose $\mu_0 = \eta_0$ and M_n the elementary transition of the reference Markov chain X_n, then we have

$$\mathbf{G_n}(x_n, x_{n+1}) = G_n(x_n) \Longrightarrow W_n(x_0, \ldots, x_n) = \prod_{0 \le p < n} G_p(x_p)$$

4.2.3 Traditional Monte Carlo methods

Importance sampling techniques are based on the law of the large numbers for independent and identically distributed random samples. We sample N independent copies $\mathbf{X_n^i} := (X_0^i, \ldots, X_n^i)$, of the chain X_n, and we set

$$\mathbb{P}_n^N = \frac{1}{N} \sum_{i=1}^N \delta_{\mathbf{X_n^i}}.$$

By construction, for any function $\mathbf{f_n}$ on $\mathbf{E_n}$, we have the unbiased Monte Carlo estimates: $\lim_{N \to \infty} \mathbb{P}_n^N(W_n \, \mathbf{f_n}) = \mathbb{P}_n(W_n \, \mathbf{f_n})$, from which we conclude that $\lim_{N \to \infty} \Psi_{W_n}(\mathbb{P}_n^N)(\mathbf{f_n}) = \mathbb{Q}_n(\mathbf{f_n})$.

Another strategy is to run an independent Metropolis-Hasting algorithm $(X_0^i, \ldots, X_n^i)_{i \ge 1}$, with proposal transition $\mathbb{P}_n$ and target measure $\mathbb{Q}_n$. A transition $(X_0^i, \ldots, X_n^i) = x \rightsquigarrow (X_0^{i+1}, \ldots, X_n^{i+1}) = y$ is accepted with probability

$$\min\left(1, \frac{d(\mathbb{Q}_n \otimes \mathbb{P}_n)_1}{d(\mathbb{Q}_n \otimes \mathbb{P}_n)_0}(x, y)\right) = \min\left(1, \frac{W_n(y)}{W_n(x)}\right)$$

with the measures

$$
\begin{aligned}
(\mathbb{Q}_n \otimes \mathbb{P}_n)_0(d(x, y)) &= \mathbb{Q}_n(dx) \, \mathbb{P}_n(dy) \\
(\mathbb{Q}_n \otimes \mathbb{P}_n)_1(d(x, y)) &= \mathbb{P}_n(dx) \, \mathbb{Q}_n(dy)
\end{aligned}
$$

By the ergodic theorem, we have the following estimate

$$\lim_{N \to \infty} \frac{1}{N} \sum_{i=1}^N f_n(X_0^i, \ldots, X_n^i) = \mathbb{Q}_n(f_n)$$

Notice that the performance of the above estimate depends on the stability properties of the Metropolis-Hasting-Markov chain. In contrast with the importance sampling technique described in Section 4.2.3, the random samples $(X_0^i, \ldots, X_n^i)_{i \geq 1}$ are no more independent, but often repeated by the acceptance-rejection scheme.

4.2.4 Sequential Monte Carlo methods

Most of the SMC technology is based on rewriting (4.18) with the importance sampling transition M_{n+1} in terms of the target distributions

$$
\begin{aligned}
\mathbf{G_n}(x_n, x_{n+1}) \quad &\propto \quad \frac{\text{Target at time } (n+1)}{\text{Target at time } (n) \times \text{Twisted transition}} \\
&\propto \quad \frac{\mathbb{Q}_{n+1}(d(x_0, \ldots, x_{n+1}))}{\mathbb{Q}_n(d(x_0, \ldots, x_n)) \times M_{n+1}(x_n, dx_{n+1})} \\
&:= \quad \frac{dQ_{n+1}(x_n, \cdot)}{dM_{n+1}(x_n, \cdot)}(x_{n+1})
\end{aligned}
\tag{4.21}
$$

The SMC algorithm amounts to sampling in parallel a given number, say N, of twisted transitions M_{n+1}. Then we compute the importance weights $\mathbf{G_n}$ of each sample and we "resample" according to the weighted distribution.

A more synthetic picture of the SMC importance sampling algorithm is provided below.

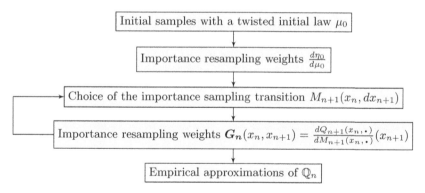

More formally, suppose we have an N-empirical approximation of the desired target measure $\mathbb{Q}_n$ at time n

$$
\widehat{\boldsymbol{\eta}}_n^N := \frac{1}{N} \sum_{i=1}^{N} \delta_{\widehat{\xi}_n^i} \simeq_{N \uparrow \infty} \mathbb{Q}_n
$$

based on some N random samples $\widehat{\boldsymbol{\xi}}_n^i := \left(\widehat{\xi}_{0,n}^i, \widehat{\xi}_{1,n}^i, \ldots, \widehat{\xi}_{n,n}^i \right) \in \boldsymbol{E}_n$. From every end point, say $\widehat{\xi}_{n,n}^i$, we sample an elementary transition $\widehat{\xi}_{n,n}^i \rightsquigarrow$

$\xi_{n+1}^i \sim M_{n+1}(\widehat{\xi}_n^i, dx_{n+1})$, and we set

$$\xi_{n+1}^i := \left(\xi_{0,n+1}^i, \xi_{1,n+1}^i, \ldots, \xi_{n+1,n+1}^i\right) = \left(\widehat{\boldsymbol{\xi}}_n^i, \xi_{n+1}^i\right) \in \boldsymbol{E}_{n+1} = (\boldsymbol{E}_n \times E_{n+1})$$

In this notation, the locally twisted measure $\mathbb{Q}_n(d(x_0, \ldots, x_n)) \times M_{n+1}(x_n, dx_{n+1})$ is clearly approximated by the N-empirical measure

$$\boldsymbol{\eta}_{n+1}^N := \frac{1}{N} \sum_{i=1}^N \delta_{\boldsymbol{\xi}_{n+1}^i}$$

Using (4.21), this implies that the target measure $\mathbb{Q}_{n+1}$ at time $(n+1)$ is approximated by the weighted measures

$$\sum_{i=1}^N \frac{\boldsymbol{G_n}(\xi_{n,n+1}^i, \xi_{n+1,n+1}^i)}{\sum_{j=1}^N \boldsymbol{G_n}(\xi_{n,n+1}^j, \xi_{n+1,n+1}^j)} \, \delta_{\boldsymbol{\xi}_{n+1}^i} \quad \simeq_{N\uparrow\infty} \quad \mathbb{Q}_{n+1} \qquad (4.22)$$

Then, we design an N-empirical approximation of the desired target measure $\mathbb{Q}_{n+1}$ at time $(n+1)$

$$\widehat{\boldsymbol{\eta}}_{n+1}^N := \frac{1}{N} \sum_{i=1}^N \delta_{\widehat{\boldsymbol{\xi}}_{n+1}^i} \quad \simeq_{N\uparrow\infty} \quad \mathbb{Q}_{n+1}$$

by "resampling" N random variables $\widehat{\boldsymbol{\xi}}_{n+1}^i$, with the weighted distribution (4.22), and so on.

Letting $\widehat{\eta}_n$ be the n-th time marginal of $\mathbb{Q}_n$, a more refined pseudocode of the SMC sampler with the marginal target measures $\widehat{\eta}_n \otimes M_{n+1}$ including an acceptance rate that is given in the following diagram.

Algorithm 4.2: SMC pseudocode

Initialization
Fix some population size parameter $N \geq 1$.
Sample N particles $\xi_0 := (\xi_0^i)_{1 \leq i \leq N}$, with $\xi_0^i = \left(\zeta_0^i, x_1^i\right) \in (E_0 \times E_1)$, with law $\eta_0(dx_0)M_1(x_0, dx_1)$.

for $k = 1$ to n do
  {For each time step k}

  Selection
  for $i = 1$ to N do
    {For each particle i}
$\widehat{\xi}_{k-1}^i$
$= \left(\widehat{\zeta}_{k-1}^i, \widehat{x}_k^i\right) \in (E_{k-1} \times E_k)$

$$:= \begin{cases} \xi_{k-1}^i = \left(\zeta_{k-1}^i, x_k^i\right), & \text{with probability } \epsilon_{k-1}\boldsymbol{G_{k-1}}(\xi_{k-1}^i)) \\ \widetilde{\xi}_{k-1}^i, \text{ a r.v. with } \text{law} \sum_{i=1}^N \dfrac{\boldsymbol{G_{k-1}}(\xi_{k-1}^i)}{\sum_{j=1}^N \boldsymbol{G_{k-1}}(\xi_{k-1}^j)} \, \delta_{\xi_{k-1}^i}, & \text{otherwise.} \end{cases}$$

  end for

  Mutation
  for $i = 1$ to N do
    {For each particle i}

    $\xi_k^i := \left(\zeta_k^i, x_{k+1}^i\right)$,

    with $\zeta_k^i = \widehat{x}_k^i$ and x_{k+1}^i a r.v. with law $M_{k+1}(\widehat{x}_k^i, dx_{k+1})$
  end for
end for

Using the same notation used in the pseudocode description we have the convergence

$$\frac{1}{N} \sum_{i=1}^{N} \delta_{(\zeta_n^i, \chi_{n+1}^i)} \simeq_{N\uparrow\infty} \widehat{\eta}_n \otimes M_{n+1} \tag{4.23}$$

This implies that

$$\frac{1}{N} \sum_{i=1}^{N} \delta_{(\widehat{\zeta}_n^i, \widehat{\chi}_{n+1}^i)} \simeq_{N\uparrow\infty} \Psi_{\mathbf{G_n}} \left(\frac{1}{N} \sum_{i=1}^{N} \delta_{(\zeta_n^i, \chi_{n+1}^i)} \right)$$

$$\simeq_{N\uparrow\infty} \Psi_{\mathbf{G_n}} \left(\widehat{\eta}_n \otimes M_{n+1} \right) \propto \widehat{\eta}_n \otimes Q_{n+1}$$

To estimate the normalizing constants $\mathcal{Z}_{n+1}$ of the measures $\mathbb{Q}_{n+1}$, we notice that

$$(\widehat{\eta}_n \otimes M_{n+1})(\mathbf{G_n}) = \mathcal{Z}_{n+1}/\mathcal{Z}_n \implies \mathcal{Z}_{n+1} = \mathcal{Z}_n \times (\widehat{\eta}_n \otimes M_{n+1})(\mathbf{G_n})$$

Using this observation, we design the following online *unbiased* estimates

$$\begin{aligned}
\mathcal{Z}_{n+1}^N &= \mathcal{Z}_n^N \times \frac{1}{N} \sum_{1 \leq i \leq N} \mathbf{G_n}(\zeta_n^i, \chi_{n+1}^i) \\
&= \mathcal{Z}_n^N \times \frac{1}{N} \sum_{1 \leq i \leq N} \frac{dQ_{n+1}(\zeta_n^i, \cdot)}{dM_{n+1}(\zeta_n^i, \cdot)}(\chi_{n+1}^i) \tag{4.24}
\end{aligned}$$

with the initial value $\mathcal{Z}_0^N = 1$.

Of course this brief presentation of the sequential Monte Carlo methodology does not answer the question on how the local importance sampling errors propagate w.r.t. the time parameter.

In the next section, we provide a Feynman-Kac representation that allows applying, without further work, all the stochastic analysis tools we have developed for mean field Feynman-Kac models.

An illustration of these twisted SMC models in the context of filtering problems is discussed in the introduction of Section 8.2. In this context, the target distributions $\mathbb{Q}_n$ represent the conditional distribution of the paths of a given signal w.r.t. partial and noisy observations, and the twisted transitions are given by Gaussian transitions associated with Kalman-type filters.

4.2.5 Feynman-Kac formulations

4.2.5.1 Twisted path space measures

In this section, we provide an alternative and equivalent Feynman-Kac description of the SMC sampler discussed in Section 4.2.4. Firstly, we notice that the sequential Monte Carlo strategy is based on the Feynman-Kac representation of any sequence of measures given by a recursion of the form

$$\mathbb{Q}_n(d(x_0, \ldots, x_n)) \propto \mathbb{Q}_{n-1}(d(x_0, \ldots, x_{n-1})) \times Q_n(x_{n-1}, dx_n)$$

in terms of some importance sampling distribution of the form

$$
\begin{aligned}
\mathbb{P}_n(d(x_0, x_1, \ldots, x_n)) &= \mathbb{P}((X_0, \ldots, X_n) \in d(x_0, \ldots, x_n)) \\
&= \mu_0(dx_0) M_1(x_0, dx_1) \ldots M_n(x_{n-1}, dx_n)
\end{aligned}
$$

If we choose $\mu_0 = \eta_0$ then the change of measure (4.19) has the following form

$$
\mathbb{Q}_n(d(x_0, x_1, \ldots, x_n)) = \frac{1}{Z_n} \left\{ \prod_{0 \le p < n} \mathbf{G_p}(x_p, x_{p+1}) \right\} \mathbb{P}_n(d(x_0, x_1, \ldots, x_n))
$$

$$(4.25)$$

with the potential function

$$
\mathbf{G_p}(x_p, x_{p+1}) = \frac{dQ_{p+1}(x_p, \cdot)}{dM_{p+1}(x_p, \cdot)}(x_{p+1})
$$

In this situation, for any bounded measurable function $\mathbf{f_n}$ on the product state space $\mathbf{E_n} := (E_0 \times \ldots \times E_n)$ we have the functional formula

$$
\mathbb{Q}_n(\mathbf{f_n}) = \frac{\mathbb{E}\left(\mathbf{f_n}(X_0, \ldots, X_n) \prod_{0 \le p < n} \mathbf{G_p}(X_p, X_{p+1})\right)}{\mathbb{E}\left(\prod_{0 \le p < n} \mathbf{G_p}(X_p, X_{p+1})\right)}
$$

In the above formula, we can replace the potential functions $\mathbf{G_p}$ by any functions proportional to $\mathbf{G_p}$.

Even if they look different from the Feynman-Kac model introduced in (1.37), from the pure mathematical point of view all of these measures on path space are equivalent. To be more precise, we consider the Markov chain on the transition space defined by

$$
\mathbf{X_n} := (X_n, X_{n+1}) \in \mathbf{E_n} = (E_n \times E_{n+1})
$$

In this notation, we have the following representation formulae

$$
\mathbb{Q}_n(\mathbf{f_n}) \propto \mathbb{E}\left(\mathbf{f_n}(\mathbf{X_0}, \ldots, \mathbf{X_n}) \prod_{0 \le p < n} \mathbf{G_p}(\mathbf{X_p})\right)
$$

with the functions

$$
\mathbf{f_n}(\mathbf{X_0}, \ldots, \mathbf{X_n}) = f_n(X_0, \ldots, X_n) \Longrightarrow \mathbb{Q}_n(f_n) = \mathbf{Q_n}(\mathbf{f_n}) \tag{4.26}
$$

In the above display, $\mathbf{Q_n}$ stands for the Feynman-Kac measure associated with the Markov chain $\mathbf{X_n}$, on the transition space $\mathbf{E_n} = (E_n \times E_{n+1})$, and the potential functions $\mathbf{G_n}$.

The measures $\mathbf{Q_n}$ clearly coincide with the ones defined in (1.37), with an $\mathbf{E_n}$-valued reference Markov chain $\mathbf{X_n}$, with elementary transitions $\mathbf{M_n}$, from $\mathbf{E_{n-1}}$ into $\mathbf{E_n}$, defined for any $\mathbf{x_{n-1}} = (x_{n-1}, x_n) \in \mathbf{E_{n-1}}$, and any $\mathbf{y_n} = (y_n, y_{n+1}) \in \mathbf{E_n}$, by

$$\mathbf{M_n}\left(\mathbf{x_{n-1}}, d\mathbf{y_n}\right) \quad = \quad \delta_{x_n}(dy_n)\, M_{n+1}(x_n, dx_{n+1}) \qquad (4.27)$$

In this notation, we observe that the n-th time marginal $\boldsymbol{\mu_n}$ of $\mathbf{Q_n}$ satisfies the evolution equation

$$\boldsymbol{\mu_n} = \Psi_{\mathbf{G_{n-1}}}(\boldsymbol{\mu_{n-1}})\mathbf{M_n} \qquad (4.28)$$

4.2.5.2   A mean field genetic simulation

The genetic type IPS particle interpretation of the Feynman-Kac models discussed in Section 4.2.5 is presented in Section 4.1.1. The particle algorithm consists in evolving an N particle with a mutation transition $\mathbf{M_n}$ and the selection fitness function $\mathbf{G_n}$. In this case, the population of particles after the mutation transition is defined by

$$\xi_n = \left(\xi_n^1, \dots, \xi_n^N\right) \quad \text{with} \quad \xi_n^i = \left(\zeta_n^i, \chi_{n+1}^i\right) \in \mathbf{E_n} = (E_n \times E_{n+1})$$

for every $1 \leq i \leq N$, while the population of selected individuals is given by

$$\widehat{\xi}_n = \left(\widehat{\xi}_n^1, \dots, \widehat{\xi}_n^N\right) \quad \text{with} \quad \widehat{\xi}_n^i = \left(\widehat{\zeta}_n^i, \widehat{\chi}_{n+1}^i\right) \in \mathbf{E_n} = (E_n \times E_{n+1})$$

for every $1 \leq i \leq N$.

This genetic type algorithm clearly coincides with the SMC sampler presented in Algorithm 4.2 on page 129, which in turns coincides with the genetic algorithm pseudocode described on page 114, when we consider the mutation transition $\mathbf{M_n}$ and the selection fitness function $\mathbf{G_n}$.

The normalizing constants $\mathbf{Z_n} = \mathbb{E}\left(\prod_{0 \leq p < n} \mathbf{G_p}(\mathbf{X_p})\right)$, of the measures $\mathbf{Q_n}$, are estimated online using the product formulae (3.29). In this situation, using (4.26) the unbiased N-particle approximation of the normalizing constant $\mathcal{Z}_n = \mathbf{Z_n}$ of the initial measure $\mathbb{Q}_n$ is given by

$$\mathcal{Z}_n^N := \prod_{0 \leq p < n} \eta_p^N(\mathbf{G_p}) \quad \text{with} \quad \eta_p^N := \frac{1}{N} \sum_{1 \leq i \leq N} \delta_{\left(\zeta_p^i, \chi_{p+1}^i\right)}$$

We readily check these this formulae coincide with the ones given in (4.24).

4.2.5.3   Adaptive importance sampling models

By construction the Feynman-Kac measures $\mathbb{Q}_n$ do not depend on the choice of twisted Markov transitions M_n. Thus, the choice of the importance sampling strategy is very flexible.

For instance, we can choose the twisted transitions M_{n+1} online at every time step. We can choose the $(n+1)$-th twisted Markov transition M_{n+1} depending on the particle approximations (4.23) of the measures $\widehat{\eta}_{n-1} \otimes M_n$

at time n, without altering the importance sampling formulae (4.26). In this connection, we observe that the measures μ_n defined in (4.28) are given by

$$\mu_{n-1} = \widehat{\eta}_{n-1} \otimes M_n$$

Thus, if we choose a Markov transition M_{n+1} that depends on μ_{n-1}, then the Markov transition $\mathbf{M_n} = \mathbf{M_{n,\mu_{n-1}}}$ defined in (4.27) as well as the potential function $\mathbf{G_{n-1}} = \mathbf{G_{n-1,\mu_{n-1}}}$ in the evolution Equation (4.28) will also depend on the current distribution μ_{n-1}. The resulting nonlinear semigroup is now given by the following equation

$$\mu_{n+1} = \mathbf{\Phi}_{n+1}(\mu_n) := \Psi_{\mathbf{G_{n,\mu_n}}}(\mu_n)\mathbf{M_{n+1,\mu_n}} \tag{4.29}$$

Our first remark suggests that the mean field IPS model (1.53), associated with a McKean interpretation (1.16) of the nonlinear evolution Equation (4.29), is consistent. Nevertheless, this observation does not provide any clue to analyze the performance of these particle models. Under some appropriate weak regularity conditions on the mappings $\mu_n \mapsto (\mathbf{G_{n,\mu_n}}, \mathbf{M_{n+1,\mu_n}})$, we can analyze the convergence of these models, using the perturbation analysis presented in Section 2.1 and further developed in Chapter 10.

4.3   Interacting Markov chain Monte Carlo algorithms

The third interpretation of Feynman-Kac IPS models relies on adaptive and interacting Markov chain Monte Carlo methodologies (*abbreviated i-MCMC*). These sampling techniques are one of the most important tools in Bayesian inference and numerical physics. As in the SMC methodology, we are given a sequence of complex target distributions on some multidimensional state space. The central idea is to find a sequence of judicious Markov chain samplers, with prescribed limiting behaviors as the time horizon tends to infinity. These stochastic sampling methods require various parameters to be tuned appropriately (such as some cooling schedule), for the Markov chains to converge reasonably well to their stationary measures.

Typically, we start running in a reasonable time some Markov chain to a first stationary state. Then, we slightly modify the Markov chain transitions to bring the chain to a second more complex and slightly different stationary state. We iterate this process with a sequence of stationary states with increasing complexity up to the desired final state. An alternative strategy is to run in parallel a sequence of Markov chains with a given stationary state. This mechanism can be interpreted as a mutation transition of a sequence of individual. Then, we update these chains with selecting the random states that better fit to the next target distribution.

4.3.1   Markov chain Monte Carlo

Markov chain Monte Carlo methodologies refer to a class of stochastic simulation algorithms based on sampling a dedicated Markov chain model with a prescribed target distribution. We refer to Section 3.1.2 for a detailed discussion on the numerical analysis and the application domains of MCMC models, including a description of the traditional Metropolis-Hastings algorithm and the Gibbs samplers. Next we propose a more formal description on the use of the MCMC technology when we are given a nonincreasing sequence of target probability measures $(\mu_p)_{0 \le p \le n}$ on some measurable state space E

$$\mu_n \ll \mu_{n-1} \ll \ldots \ll \mu_2 \ll \mu_1 \ll \mu_0 = \lambda \qquad (4.30)$$

with some initial probability measure λ on E and Radon-Nikodym derivatives associated with some measurable functions h_n

$$\frac{d\mu_{n+1}}{d\mu_n} \propto h_n \quad \Longleftrightarrow \quad \mu_{n+1} = \Psi_{h_n}(\mu_n) \qquad (4.31)$$

It is not difficult to check that these measures μ_n are given by

$$\mu_n = \left(\Psi_{h_{n-1}} \circ \ldots \circ \Psi_{h_0} \right)(\lambda) = \Psi_{H_n}(\lambda) \quad \text{with} \quad H_n := \prod_{0 \le p < n} h_p \qquad (4.32)$$

We also use the convention $\prod_\emptyset = 1$, so that $\mu_0 = \lambda$. In this notation, we have

$$d\mu_n = \frac{1}{\mathcal{Z}_n} \, H_n \, d\lambda \quad \text{with} \quad \mathcal{Z}_n = \lambda(H_n)$$

The prototype models we have in mind are the Boltzmann-Gibbs measures and the probability measure restrictions:

- The Boltzmann-Gibbs measures associated with some energy function V and some increasing inverse cooling schedule β_n starting at $\beta_0 = 0$ are given by choosing

$$h_p = \exp\left(-(\beta_{p+1} - \beta_p)V\right) \Rightarrow d\mu_n = \frac{1}{\mathcal{Z}_n} \, \exp\left(-\beta_n V\right) \, d\lambda \qquad (4.33)$$

- The restriction probabilities associated with a nonincreasing sequence of measurable subsets $A_n \subset E$ and some reference probability measure $\lambda \in \mathcal{P}(E)$ s.t. $\lambda(A_n) > 0$ are given by choosing

$$h_p = 1_{A_{p+1}} \Rightarrow d\mu_n = \frac{1}{\mathcal{Z}_n} \, 1_{A_n} \, d\lambda \qquad (4.34)$$

These couple of examples are discussed in some more detail in Section 4.3.4 and in Section 4.3.5.

We further assume that we have a dedicated Markov chain Monte Carlo transition M_n with prescribed target measures $\mu_n = \mu_n M_n$, at any time step $n \geq 1$. We design a sequential MCMC algorithm as follows:

We start running a sequence of random states $(X_p)_{0 \leq p \leq n_1}$ with transitions M_1 and initial distrbution $\lambda_0 = \mu_0$. For a sufficiently large time horizon n_1, both the occupation measure and the law of the terminal state approximate μ_1; that is, we have that

$$\frac{1}{n_1} \sum_{1 \leq p \leq n_1} \delta_{X_p} \quad \text{and} \quad \text{Law}(X_{n_1}) = \mu_0 M_1^{n_1} \quad \simeq_{n_1 \uparrow \infty} \quad \mu_1$$

In the second step, starting from X_{n_1}, we run a sequence of random states $(X_{n_1+p})_{0 \leq p \leq n_2}$ with transitions M_2. For a sufficiently large time horizon n_2, the occupation measure and the law of the terminal state

$$\frac{1}{n_2} \sum_{1 \leq p \leq n_1} \delta_{X_{n_1+p}} \quad \text{and} \quad \text{Law}(X_{n_1+n_2}) = \text{Law}(X_{n_1}) M_2^{n_2} \quad \simeq_{n_2 \uparrow \infty} \quad \mu_2$$

Iterating the construction, we define the time inhomogeneous Markov chain described in the following synthetic diagrams

$$X_0 \xrightarrow{M_1^{n_1}} X_{n_1} \xrightarrow{M_2^{n_2}} X_{n_2} \xrightarrow{M_3^{n_3}} X_{n_3} \xrightarrow{M_4^{n_4}} \dots \qquad (4.35)$$

and at the level of the distributions

$$\mu_0 \xrightarrow{M_1^{n_1}} \text{Law}(X_{n_1}) \simeq \mu_1 \xrightarrow{M_2^{n_2}} \text{Law}(X_{n_1+n_2}) \simeq \mu_2 \xrightarrow{M_3^{n_3}} \dots$$

Notice that the simulated annealing model corresponds to the choice of Metropolis-Hasting moves M_n in the Boltzmann-Gibbs example given in (4.33). We also refer the reader to the series of articles [115, 116, 118, 119, 120, 162] related to applications of this type of annealed models to the convergence analysis of genetic type particle models on finite state spaces with fixed population sizes.

A more detailed discussion on simulated annealing models and their i-MCMC version is provided in Section 4.3.5.

4.3.2 A nonlinear perfect sampling Markov model

Using (4.31), we readily check that

$$\mu_{n+1} = \Psi_{h_n}(\mu_n) \Rightarrow \mu_{n+1} = \mu_{n+1} M_{n+1} = \Psi_{h_n}(\mu_n) M_{n+1} \qquad (4.36)$$

Using remark 2.5.1, we conclude that μ_n coincides with the n-th time marginal η_n of the Feynman-Kac measures $\mathbb{Q}_n$ associated with the potential functions and the Markov transitions $(G_n, M_n) = (h_n, M_n)$ and defined in (1.37). More precisely, for any $f \in \mathcal{B}_b(E)$ we have that

$$\eta_n(f) = \mathbb{E}\left(f(X_n) \prod_{0 \le p < n} G_p(X_p) \right) \Big/ \mathbb{E}\left(\prod_{0 \le p < n} G_p(X_p) \right) = \mu_n(f) \quad (4.37)$$

In the above display, X_n is the reference Markov chain given by

$$\mathbb{P}\left(X_n \in dx \mid X_{n-1} \right) = M_n(X_{n-1}, dx)$$

with the initial distribution $\eta_0 = \lambda$.

In addition, we also have the product formula

$$H_{n+1} = H_n \times h_n \implies \mathcal{Z}_{n+1} = \lambda(H_n) \times \frac{\lambda(H_n h_n)}{\lambda(H_n)} = \mathcal{Z}_n \times \eta_n(h_n)$$

from which we conclude that $\mathcal{Z}_{n+1} = \prod_{0 \le p \le n} \eta_p(G_p)$. We recall that McKean interpretation of the flow of measures (4.36) is defined in terms of a selection transition S_{n,η_n} that satisfies the compatibility condition

$$\eta_n S_{n,\eta_n} = \Psi_{G_n}(\eta_n)$$

For a detailed discussion on these McKean transitions, we refer the reader to Section 1.4.2.1 in the opening chapter, as well as in Section 4.4.5. In this situation, we have

$$\eta_{n+1} = \eta_n K_{n+1,\eta_n} \quad \text{with} \quad K_{n+1,\eta_n} = S_{n,\eta_n} M_{n+1} \quad (4.38)$$

In this interpretation, $\eta_{n+1} = \mathrm{Law}(X_{n+1})$ can be interpreted as the distributions of the random states X_{n+1} of a Markov chain with transitions $X_n \rightsquigarrow X_{n+1}$ that depends on the distribution $\eta_n = \mathrm{Law}(X_n)$ of the random state X_n. By (4.38), the evolution of this Markov model is decomposed into two steps

$$X_0 \xrightarrow{S_{0,\eta_0}} \widehat{X}_0 \xrightarrow{M_1} X_1 \xrightarrow{S_{1,\eta_1}} \widehat{X}_1 \xrightarrow{M_2} X_2 \xrightarrow{S_{2,\eta_2}} \widehat{X}_2 \xrightarrow{M_3} X_3 \ldots$$

$$(4.39)$$

In contrast with the MCMC model discussed in Section 4.3.1, the Markov chain (4.39) can be interpreted as a perfect sampler, in the sense that at every time step the law of the random states coincides with the desired target measures; that is, for any $n \ge 1$, we have that

$$\mathrm{Law}(\widehat{X}_{n-1}) = \eta_n = \mathrm{Law}(X_n) = \mu_n.$$

4.3.3   A particle algorithm with recycling

In this context, the genetic approximation algorithm presented in Section 4.1.1 consists in evolving an N particle with a mutation transition M_n

and the selection fitness function h_n. In this case, the population of particles after the mutation transition is defined by

$$\xi_n = \left(\xi_n^1, \ldots, \xi_n^N\right) \in E^N \quad \text{with} \quad \mu_n^N := \frac{1}{N} \sum_{1 \le i \le N} \delta_{\xi_n^i} \simeq_{N \uparrow \infty} \mu_n$$

The pseudocode of this interacting MCMC *abbreviated i-MCMC* algorithm coincides with the genetic algorithm pseudocode described on page 114 when we consider the mutation transition M_n and the selection fitness function $G_n = h_n$.

Algorithm 4.3: i-MCMC pseudocode

<u>Initialization</u>
Fix some population size parameter $N \ge 1$.
Sample N **particles** $\xi_0 := (\xi_0^i)_{1 \le i \le N}$ **with law** $\mu_0 = \lambda$.

for $k = 1$ **to** n **do**
  {For each time step k}

  <u>Selection - Recycling</u>
  for $i = 1$ **to** N **do**
    {For each particle i}

$$\widehat{\xi}_{k-1}^i := \begin{cases} \xi_{k-1}^i, & \text{with probability } \epsilon_{k-1} h_{k-1}(\xi_{k-1}^i) \quad (1) \\[2mm] \xi_{k-1}^i, & \text{a r.v. with law } \sum_{i=1}^N \dfrac{h_{k-1}(\xi_{k-1}^i)}{\sum_{j=1}^N h_{k-1}(\xi_{k-1}^j)} \, \delta_{\xi_{k-1}^i}, \quad \text{otherwise } (2). \end{cases}$$

  end for

  <u>Mutation - MCMC moves</u>
  for $i = 1$ **to** N **do**
    {For each particle}
    ξ_k^i **a r.v. with law** $M_k(\widehat{\xi}_{k-1}^i, dx)$
  end for
end for

Here again, the parameter ϵ_{k-1} is a tuning parameter. For instance, we can choose $\epsilon_{k-1} = 1/\max_{1 \le j \le N} h_{k-1}(\xi_{k-1}^j)$. If we set $\epsilon_{k-1} = 0$, then we remove the line (1), so that the selection transition coincides with the classical proportional or roulette selection. For $[0,1]$-valued fitness functions h_{k-1}, we can choose $\epsilon_{k-1} = 1$ so that $\widehat{\xi}_{k-1}^i = \xi_{k-1}^i$ with probability $h_{k-1}(\xi_{k-1}^i)$.

A more synthetic picture of the i-MCMC algorithm is provided below.

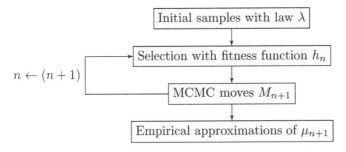

We end this section with an assorted collection of enriching comments on the i-MCMC methodology discussed above.

Conventional MCMC methods with time varying target measures μ_n can be seen as a particle model with a single particle, evolving with only mutation explorations according to the Markov transitions $M_n = K_n^{m_n}$, where

$K_n^{m_n}$ stands for the iteration of an MCMC transition K_n s.t. $\mu_n = \mu_n K_n$. In this situation, we choose a judicious increasing sequence m_n so that the non-homogeneous Markov chain is sufficiently stable, even if the target measures become more and more complex to sample. When the target measure is fixed, say of the form μ_T for some large T, the MCMC sampler again uses a single particle to behave as a Markov chain with time homogeneous transitions M_T. The obvious drawback with these two conventional MCMC samplers is that the user does not know how many steps are really needed to be close to the equilibrium target measure. A wrong choice will return samples with a distribution far from the desired target measure.

The i-MCMC methods run a population of MCMC samplers that interact with each other through a recycling-updating mechanism so that the occupation measure of the current measure converges to the target measure, when we increase the population sizes. In contrast with conventional MCMC methods, there are no burn-in time questions, nor any quantitative analysis to estimate the convergence to equilibrium of the MCMC chain.

In the next three sections, we illustrate these rather abstract models with three applications related, respectively, to probability restriction models, simulated annealing type models, and particle NCMC algorithms. For a more thorough discussion on these interacting MCMC models, and related sequential Monte Carlo methods, we refer the reader to [163, 165, 279, 280].

4.3.4 Probability restrictions samplers

If we choose Markov chain Monte Carlo type local moves

$$\mu_n = \mu_n M_n$$

with some prescribed target Boltzmann-Gibbs measures

$$\mu_n(dx) \propto 1_{A_n}(x)\ \lambda(dx)$$

associated with a sequence of decreasing subsets $A_n \downarrow$, and some reference measure λ, then we find that $\mu_n = \eta_n$ and $\mathcal{Z}_n = \lambda(A_n)$, as soon as the potential functions in (4.32) and in (4.37) are chosen so that

$$G_n = h_n = 1_{A_{n+1}}$$

This stochastic model arises in several application domains. In computer science literature, the corresponding particle approximation models are sometimes called subset methods, sequential sampling plans, randomized algorithms, or level splitting algorithms. They are used to solve complex NP-hard combinatorial counting problems [121], extreme quantile probabilities [123, 308], and uncertainty propagations in numerical codes [100].

4.3.5 Stochastic optimization algorithms

If we choose Markov chain Monte Carlo type local moves $\mu_n = \mu_n M_n$ with some prescribed target Boltzmann-Gibbs measures

$$\mu_n(dx) \propto e^{-\beta_n V(x)} \lambda(dx)$$

associated with a sequence of increasing inverse temperature parameters $\beta_n \uparrow$, and some reference measure λ, then we find that $\mu_n = \eta_n$ and $\mathcal{Z}_n = \lambda(e^{-\beta_n V})$ as soon as the potential functions in (4.32) and (4.37) are chosen so that

$$G_n = h_n = e^{-(\beta_{n+1} - \beta_n)V}$$

For instance, we can assume that the Markov transition $M_n = \mathcal{M}_{n,\beta_n}^{m_n}$ is the m_n-iterate of the following Metropolis-Hasting transitions

$$
\begin{aligned}
\mathcal{M}_{n,\beta_n}(x, dy) \quad = \quad &= K_n(x, dy) \ \min\left(1, e^{-\beta_n(V(y) - V(x))}\right) \\
&+ \left(1 - \int_z K_n(x, dz) \ \min\left(1, e^{-\beta_n(V(z) - V(x))}\right)\right) \delta_x(dy)
\end{aligned}
$$

$$(4.40)$$

4.3.6 Particle Markov chain Monte Carlo methods

We consider a collection of Markov transition and positive potential functions $(M_{\theta,n}, G_{\theta,n})$ that depend on some random variable $\Theta = \theta$, with distribution ν on some state space S. We let $\eta_{\theta,n}$ be the n-time marginal of the Feynman-Kac measures defined as in (1.37), by replacing (M_n, G_n) by $(M_{\theta,n}, G_{\theta,n})$. In other words, $\eta_{\theta,n}$ are defined as the Feynman-Kac measures η_n, by replacing the function G_n by $G_{\theta,n}$, and the reference Markov chain X_n by some Markov chain $X_{\theta,n}$ with elementary transitions $M_{\theta,n}$. We also denote by

$$\mathcal{Z}_n(\theta) = \mathbb{E}\left(\prod_{0 \le p < n} G_{\theta,p}(X_{\theta,p})\right) = \prod_{0 \le p < n} \eta_{\theta,p}(G_{\theta,p})$$

the corresponding normalizing constants.

Our next objective is to find a judicious sampling algorithm of the sequence of target measures π_n on S defined by

$$\pi_n(d\theta) := \frac{1}{\mathcal{Z}_n} \mathcal{Z}_n(\theta) \ \nu(d\theta)$$

where $\mathcal{Z}_n$ stands for some normalizing constant. These complex distributions arise in a variety of application domains. For instance, in the context of hidden Markov chain problems, π_{n+1} represents the posterior distribution of a random kinetic type parameter Θ w.r.t. to a sequence of observations Y_p, with $p \le n$

$$\pi_{n+1}(d\theta) = \mathbb{P}\left(\Theta \in d\theta \mid Y_0, \ldots, Y_n\right)$$

A more detailed description of this statistical inference problem is provided in Section 3.3.5. In the context of particle models evolving in a random absorbing medium that depends on some random parameter Θ, π_{n+1} represents the conditional distribution of a random kinetic type parameter Θ of a nonabsorbed particle

$$\pi_{n+1}(d\theta) = \mathbb{P}\left(\Theta \in d\theta \mid \mid T > n\right)$$

A more detailed description of this problem is provided in Section 3.3.3.

To get one step further, we consider the probability distribution $P(\theta, d\xi)$ of the N-particle genetic model

$$\xi := (\xi_{\theta,0}, \xi_{\theta,1}, \ldots, \xi_{\theta,T})$$

on the interval $[0, T]$, with mutation transitions $M_{\theta,n}$, and potential selection functions $G_{\theta,n}$, with $n \leq T$. By construction, the random variable

$$X := (\Theta, \xi) \in E := S \times \left(\prod_{0 \leq p \leq T} E_p^N\right)$$

is distributed on E with the probability measure

$$\lambda(dx) := \mathbb{P}\left(X \in dx\right) = \mathbb{P}\left((\Theta, \xi) \in d\left(\theta, \zeta\right)\right) = \nu(d\theta) \; P(\theta, d\zeta)$$

In the above display, $dx = d\left(\theta, \zeta\right)$ stands for an infinitesimal neighborhood of the point $x = (\theta, \zeta) \in E$, with $\zeta := (\zeta_0, \zeta_1, \ldots, \zeta_T) \in \left(\prod_{0 \leq p \leq T} E_p^N\right)$.

We consider the sequence of measures $(\mu_n)_{n \leq T}$ defined in (4.32) with the reference measure λ and the potential functions $(h_n)_{n \leq T}$ defined for any $x = (\theta, \zeta) \in E$ by the empirical mean potential

$$h_n(x) = h_n(\theta, \zeta) = \frac{1}{N} \sum_{1 \leq i \leq N} G_{\theta,n}(\zeta_n^i)$$

We observe that these target measures can be sampled using the Interacting Markov chain Monte Carlo methodology presented in Section 4.3.3. For instance, suppose we have a dedicated Markov transition $K(\theta, d\theta')$ that is reversible w.r.t. the measure ν on S. In this case, we can design easily a Metropolis-Hasting transition with target invariant measure μ_n. Firstly, we consider the following Markov transition

$$\mathbf{K_n}(x, dx') = K(\theta, d\theta') \times P(\theta', d\zeta')$$

as a proposition of the move $x \rightsquigarrow x'$, where dx' stands for an infinitesimal neighborhood of $x' = (\theta', \zeta')$ and $x = (\theta, \zeta)$. The Metropolis-Hasting acceptance rate of the transition is now given by the formula

$$\min\left(1, \frac{d\left(\mu_n \otimes \mathbf{K_n}\right)_1}{d\left(\mu_n \otimes \mathbf{K_n}\right)_0}(x, x')\right) = \min\left(1, \frac{\prod_{0 \leq p < n} h_p(\theta', \zeta')}{\prod_{0 \leq p < n} h_p(\theta, \zeta)}\right)$$

with the pair of measures on E^2 defined by

$$(\mu_n \otimes \mathbf{K_n})_1 (d(x, x')) := (\mu_n \otimes \mathbf{K_n})_0 (d(x', x))$$

with

$$(\mu_n \otimes \mathbf{K_n})_0 (d(x, x'))$$

$$= \mu_n(dx) \, \mathbf{K_n}(x, dx')$$

$$= \frac{1}{\mathcal{Z}_n} \left\{ \prod_{0 \leq p < n} h_p(\theta, \zeta) \right\} \times [\nu(d\theta) \, P(\theta, d\zeta)] \, [K(\theta, d\theta') \times P(\theta', d\zeta')]$$

Using the unbiased property of the particle free energy models presented in (4.9), we clearly have

$$\int P(\theta, d\xi) \left\{ \prod_{0 \leq p < n} h_p(\xi, \theta) \right\} = \mathbb{E} \left(\prod_{0 \leq p < n} \eta_{\theta,p}^N (G_{\theta,p}) \right)$$

$$= \prod_{0 \leq p < n} \eta_{\theta,p}(G_{\theta,p}) = \mathcal{Z}_n(\theta)$$

where $\eta_{\theta,n}^N = \frac{1}{N} \sum_{1 \leq i \leq N} \delta_{\xi_{\theta,n}^i}$ stands for the N-particle genetic model with mutation transition $M_{\theta,n}$ and selection fitness function $G_{\theta,n}$.

This implies that the Θ-marginal of μ_n coincides with the measure π_n. Notice that the acceptance Metroplis-Hasting ratio can be expressed in terms of the N-particle approximations

$$\min \left(1, \frac{d \, (\mu_n \otimes \mathbf{K_n})_1}{d \, (\mu_n \otimes \mathbf{K_n})_0} (x, x') \right) = \min \left(1, \mathcal{Z}_n^N(\theta') / \mathcal{Z}_n^N(\theta) \right)$$

for any $x' = (\theta', \zeta')$ and $x = (\theta, \zeta)$, with the particle normalizing constants

$$\mathcal{Z}_n^N(\theta) := \prod_{0 \leq p < n} \eta_{\theta,p}^N(G_{\theta,p})$$

4.4 Mean field interacting particle models

4.4.1 Introduction

The fourth interpretation of Feynman-Kac IPS models relies on abstract mean field particle models on general state spaces. In the opening chapter, Chapter 1, we have seen that Feynman-Kac genetic IPS models are particular instances of mean field particle interpretations of nonlinear evolution equations in distribution spaces. This class of particle models is, of course, much more general than the first three particle interpretation models developed in earlier sections. They can be used to design and to analyze a variety of IPS models, including McKean-Vlasov diffusion equations and McKean models of gases, as well as Feynman-Kac particle models.

The abstract theoretical framework presented in this section is the base of a unifying mean field theory that can be used without further work to analyze the performance of stochastic particle samplers, often presented as natural heuristic-like algorithms in applied scientific literature.

It is important to underline that the interpretations of the nonlinear evolution equations and their mean field IPS schemes depend on the problem at hand. In some instances, such as in fluid mechanics and condensed matter modeling, the mean field particle model is dictated by the physical evolution model under study. In much the same way, in the population dynamic modeling discussed in Section 4.1.3, the evolution of the individuals is dictated by the biological model at hand. In this situation, the limiting nonlinear evolution equation is interpreted as the infinite population model.

In other instances, such as in filtering problems arising in advanced signal processing, the mathematical structure of the particle scheme evolution is always decomposed into a prediction and an updating step. The choice of the prediction-mutation of the particle filters is rather flexible, but the limiting evolution equation always coincides with the optimal filter equations. Further details on these filtering models are provided in Chapter 8. In the SMC methodology presented in Section 4.2, the choice of the particle exploration depends on the choice of some importance sampling twisted transitions. In this context, the updating-resampling step is dictated with the corresponding Radon-Nikodym derivative weight functions, so that to preserve the structure of the limiting target measures (4.17). In the interacting MCMC models discussed in Section 4.3, the free exploration of the particles depends on the choice of some MCMC transition, with a prescribed target invariant measure. The updating-resampling stage depends on the nonincreasing sequence (4.30) of target measures we have chosen.

A more detailed discussion on general and abstract mean field model and their application areas is provided in the opening chapter, as well as in Chapter 8 and in Chapter 10.

In the present section, we merely content ourselves with presenting some mathematical foundations of these discrete generation mean field particle models. We also show how these models fit with the modeling of the genetic type models discussed in earlier sections.

The rest of this section is organized as follows:

Section 4.4.2 is dedicated to Markov-McKean interpretations of nonlinear evolution equations in distribution spaces. We illustrate these abstract models with the infinite population Feynman-Kac flows discussed in Section 4.1.3. Section 4.4.3 is dedicated to the definition of the mean field IPS model associated with a given Markov-McKean interpretation. In Section 4.4.4 we provide a discussion of the local sampling errors associated with these stochastic IPS algorithms. In the last section, Section 4.4.5, we discuss a series of Markov-McKean interpretations of Feynman-Kac evolution equations and the corresponding genetic type IPS algorithms.

More general Markov-McKean models associated with evolution equations in the space of positive measures are discussed in Section 2.3 in the opening chapter, as well as in Section 14.7, and Section 14.8.

4.4.2 Markov-McKean models

We consider a sequence of probability measures η_n on some state spaces E_n satisfying a nonlinear evolution equation of the following form

$$\eta_{n+1} = \Phi_{n+1}(\eta_n) \tag{4.41}$$

for some mapping Φ_{n+1}, the set $\mathcal{P}(E_n)$ of all probability measures on E_n, into $\mathcal{P}(E_{n+1})$. Several illustrations of these rather abstract evolution models related to Feynman-Kac semigroups, infinite genetic type population models, and interacting perfect sampling MCMC models are provided in Section 3.2.2, Section 4.1.3, and in Section 4.3.2.

For instance, in Section 3.2.2 we have proved that the flow of the n-th time marginals η_n of the measures $\mathbb{Q}_n$ introduced in (1.37) satisfies the evolution equation

$$\eta_{n+1} = \Phi_{n+1}(\eta_n) := \Psi_{G_n}(\eta_n) M_{n+1} \tag{4.42}$$

Discrete generation mean field particle interpretation models are based on the fact that the flow of probability measures η_n always satisfy a nonlinear evolution equation of the following form

$$\eta_{n+1} = \eta_n K_{n+1,\eta_n} \tag{4.43}$$

for some collection of Markov transitions $K_{n+1,\eta}$, indexed by the time parameter $n \geq 0$ and the set of probability measures $\mathcal{P}(E_n)$. Several examples of transitions $K_{n+1,\eta}$ are discussed in the opening chapter, as well as in the further development of Section 4.4.5 dedicated to Feynman-Kac models.

Note that we can define sequentially a Markov chain sequence $(\overline{X}_n)_{n\geq 0}$ such that

$$\mathbb{P}\left(\overline{X}_{n+1} \in dx \mid \overline{X}_n\right) = K_{n+1,\eta_n}\left(\overline{X}_n, dx\right) \quad \text{with} \quad \text{Law}(\overline{X}_n) = \eta_n \tag{4.44}$$

with the initial distribution $\text{Law}(\overline{X}_0) = \eta_0$.

From the practical point of view, this Markov chain can be seen as a perfect sampler of the flow of the distributions (4.43) of the random states $\overline{X}_n$, in the sense that the laws of the random states X_n coincide with the solution η_n of the nonlinear evolution Equation (4.41), at any time step $n \geq 0$.

Definition 4.4.1 *The Markov chain $\overline{X}_n$ defined in (4.44) is called a Mc-Kean, or a Markov-McKean, interpretation of the sequence of measures (4.41) associated with the collection of Markov transitions K_{n+1,η_n}. The distribution of the random trajectories $(\overline{X}_p)_{0 \leq p \leq n}$*

$$\mathbb{P}\left((\overline{X}_0, \ldots, \overline{X}_n) \in d(x_0, \ldots, x_n)\right)$$

$$= \eta_0(dx_0) \times K_{1,\eta_0}(x_0, dx_1) \times \cdots \times K_{n,\eta_{n-1}}(x_{n-1}, dx_n)$$

is called the McKean measure associated with the Markov chain $(\overline{X}_p)_{0 \leq p \leq n}$.

For a more thorough discussion on these nonlinear Markov chain models, we refer the reader to Section 2.5 in the book [163].

4.4.3   Mean field particle models

The mean field particle interpretation associated with the nonlinear measure valued model (4.43) is the E_n^N-valued Markov chain

$$\xi_n = \left(\xi_n^1, \xi_n^2, \ldots, \xi_n^N\right) \in E_n^N$$

with elementary transitions defined as

$$\mathbb{P}\left(\xi_{n+1} \in dx \mid \xi_n\right) = \prod_{i=1}^{N} K_{n+1,\eta_n^N}(\xi_n^i, dx^i) \tag{4.45}$$

with the occupation measure of the system $\eta_n^N := \frac{1}{N}\sum_{j=1}^{N} \delta_{\xi_n^j}$. In the above displayed formula, dx stands for an infinitesimal neighborhood of the point $x = (x^1, \ldots, x^N) \in E_{n+1}^N$. The initial system ξ_0 consists of N i.i.d. r.v. with common law η_0.

For regular McKean models, a basic convergence estimate is the following

$$\frac{1}{N}\sum_{i=1}^{N} \delta_{(\xi_0^i, \xi_1^i, \ldots, \xi_n^i)}(d(x_0, \ldots, x_n))$$

$$\longrightarrow_{N \to \infty} \eta_0(dx_0) \times K_{1,\eta_0}(x_0, dx_1) \times \cdots \times K_{n,\eta_{n-1}}(x_{n-1}, dx_n) \tag{4.46}$$

Definition 4.4.2 *The sequence of n-time marginals η_n^N of the occupation measure presented in (4.46) is called the flow of particle density profiles.*

Taking the marginal w.r.t. the terminal time horizon, we conclude that

$$\eta_n^N \longrightarrow_{N \to \infty} \eta_n$$

4.4.4 Local sampling errors

The local sampling errors induced by the mean field particle model (4.45) are expressed in terms of the empirical random field sequence V_n^N defined by

$$V_{n+1}^N = \sqrt{N} \left[\eta_{n+1}^N - \Phi_{n+1}\left(\eta_n^N\right) \right]$$

Notice that V_{n+1}^N is alternatively defined by the following stochastic perturbation formulae

$$\eta_{n+1}^N = \Phi_{n+1}\left(\eta_n^N\right) + \frac{1}{\sqrt{N}} V_{n+1}^N \qquad (4.47)$$

For $n = 0$, we also set

$$V_0^N = \sqrt{N} \left[\eta_0^N - \eta_0 \right] \Leftrightarrow \eta_0^N = \eta_0 + \frac{1}{\sqrt{N}} V_0^N$$

In this interpretation, the N-particle model can also be interpreted as a stochastic perturbation of the limiting system

$$\eta_{n+1} = \Phi_{n+1}\left(\eta_n\right)$$

It is rather elementary to check that

$$
\begin{aligned}
\mathbb{E}\left(V_{n+1}^N(f) \mid \xi_n\right) &= 0 \\
\mathbb{E}\left(V_{n+1}^N(f)^2 \mid \xi_n\right) &= \int \eta_n^N(dx)\, K_{n+1,\eta_n^N}\left(f - K_{n+1,\eta_n^N}(f)(x)\right)(x)^2 \\
&\leq \mathrm{osc}(f)^2
\end{aligned}
$$

4.4.5 Feynman-Kac particle models

The choice of the McKean transitions K_{n+1,η_n} associated with the Feynman-Kac evolution Equations (4.42) is far from being unique. Several choices of models can be underlined depending on the Markov transport formulation (0.3) of the Boltzmann-Gibbs transformations Ψ_{G_n}.

To describe with some conciseness these models, we recall that $\Psi_G : \mathcal{P}(E) \to \mathcal{P}(E)$ represents the Boltzmann-Gibbs transformation associated with some bounded positive potential function G defined for any $\eta \in \mathcal{P}(E)$ by the following formula

$$\Psi_G(\eta)(dx) = \frac{1}{\eta(G)} G(x)\, \eta(dx)$$

These transformations can be interpreted as a nonlinear Markov transport transformation of the following form

$$\Psi_G(\mu) = \mu S_{\mu,G} \qquad (4.48)$$

for some Markov transitions $S_{\mu,G}$ from E into itself. Several choices of models are discussed in (0.3). For $[0,1]$-valued potential functions we can choose

$$S_{\mu,G}(x, dy) := G(x)\, \delta_x(dy) + (1 - G(x))\ \Psi_G(\mu)(dy)$$

For $]1, \infty[$-valued potential functions we can choose

$$S_{\mu,G}(x, dy) := \epsilon_\mu G(x)\, \delta_x(dy) + (1 - \epsilon_\mu G(x))\ \Psi_G(\mu)(dy)$$

For more general models, we can always choose the transitions

$$S_{\mu,G}(x, dy) := (1 - a_{G,\mu}(x))\ \delta_x(dy) + a_{G,\mu}(x)\ \Psi_{(G-G(x))_+}(\mu)(dy) \quad (4.49)$$

with

$$
\begin{aligned}
a_{G,\mu}(x) \quad &:= \quad \mu\left((G - G(x))_+\right)/\mu\left(G\right) \\
&= \quad 1 - \mu\left(G \wedge G(x)\right)/\mu(G) \in [0,1] \quad (4.50)
\end{aligned}
$$

In all the situations discussed above, we have

$$\eta_{n+1} = \Psi_{G_n}(\eta_n)M_{n+1} = \eta_n K_{n+1,\eta_n} \quad \text{with} \quad K_{n+1,\eta_n} := S_{\eta_n,G_n}M_{n+1}$$

and the elementary transitions $\xi_n \rightsquigarrow \xi_{n+1}$ of the N-particle model defined in (4.45) is decomposed into two steps

$$\xi_n \xrightarrow{\ S_{\eta_n^N,G_n}\ } \widehat{\xi}_n := \left(\widehat{\xi}_n^i\right)_{1\le i\le N} \in E_n^N \xrightarrow{\ M_{n+1}\ } \xi_{n+1}$$

More formally, the selection transition is given by

$$\mathbb{P}\left(\widehat{\xi}_n \in dx \mid \xi_n\right) = \prod_{i=1}^N S_{\eta_n^N,G_n}(\xi_n^i, dx^i)$$

and the mutation transition of the selected particles $\widehat{\xi}_n$ is defined by

$$\mathbb{P}\left(\xi_{n+1} \in dx \mid \widehat{\xi}_n\right) = \prod_{i=1}^N M_{n+1}(\widehat{\xi}_n^i, dx^i)$$

Chapter 5

Continuous time Feynman-Kac models

5.1 Some operator aspects of Markov processes

5.1.1 Infinitesimal generators

We consider a time inhomogeneous Markov process X_t taking values in some Polish space E, with infinitesimal generators L_t defined on some common domain $D(L) \subset \mathcal{B}_b(E)$. We also assume that the domain $D(L)$ is an algebra, and we let $\mathcal{F} = (\mathcal{F}_t)_{t \geq 0}$, with $\mathcal{F}_t = \sigma(X_\tau, \ \tau \leq t)$, be the filtration of σ-algebras generated by the Markov process.

This abstract framework allows combining semigroup techniques with the principal theorems of stochastic differential calculus, e.g., Ito's formulae, and "carré du champ" operators (a.k.a. square fields), characterizing the predictable quadratic variations of the martingales that appear in Ito's formulae. We mention that a more general extended setup is developed in [375], as well as in [161, 178] in the context of Feynman-Kac models.

Several illustrations of these abstract continuous time processes are provided in the opening chapter, in Section 1.1.2 and Section 1.2.2, including pure jump processes with bounded jump rates with $D(L) = \mathcal{B}_b(E)$, or Euclidean diffusions on $E = \mathbb{R}^d$, with regular and Lipschitz coefficients. In this situation, we can take $D(L)$ as the set of $\mathcal{C}^\infty$-functions, with derivatives decreasing at infinity faster than any polynomial function.

By construction, for any $f \in \mathcal{C}^1([0, \infty[, D(L))$, we have the second order Taylor's type expansion

$$df_t(X_t) = \left(\frac{\partial}{\partial t} + L_t \right)(f_t)(X_t) \, dt + dM_t(f) \tag{5.1}$$

with a $\mathcal{F}$-martingale term $M_t(f)$, with predictable angle bracket given by

$$d\langle M(f)\rangle_t = \Gamma_{L_t}(f_t, f_t)(X_t) \, dt$$

In the above display formula Γ_{L_t} stands for the "carré du champ" associated to the infinitesimal generator L_t, defined for any $f \in D(L)$ by

$$\Gamma_{L_t}(f, f)(x) := L_t \left[(f - f(x))^2 \right](x) = L_t(f^2)(x) - 2f(x)L_t(f)(x)$$

We also recall that the predictable angle bracket $\langle M(f) \rangle_t$ of the $\mathcal{F}$-martingale is the $\mathcal{F}$-predictable process s.t. $M_t(f)^2 - \langle M(f) \rangle_t$ is again an $\mathcal{F}$-martingale. This formula has to be understood in the integral sense; that is, for any $s \leq t$, we have that

$$f_t(X_t) = f_s(X_s) + \int_s^t \left(\frac{\partial}{\partial \tau} + L_\tau \right)(f_\tau)(X_\tau) \, d\tau + M_t(f) - M_s(f) \qquad (5.2)$$

In Probability literature, these second order Taylor's type formulae (5.1) are also called the Ito's formulae, or sometimes the stochastic chain rule formulae.

5.1.2   A local characterization

In this section, we present an alternative, and more intuitive, local characterization of the infinitesimal generator, and its "carré du champ" operator in terms of limiting predictable averages of the increments of $f_t(X_t)$ and the ones of the martingale $M_t(f)$.

Firstly, using (5.2), we readily show that

$$\frac{1}{t-s} \, \mathbb{E}\left(f_t(X_t) - f_s(X_s) \mid \mathcal{F}_s\right)$$

$$= \frac{1}{t-s} \, \mathbb{E}\left(\int_s^t \left(\frac{\partial}{\partial \tau} + L_\tau \right)(f_\tau)(X_\tau) \, d\tau \mid \mathcal{F}_s \right)$$

This implies that

$$\frac{1}{t-s} \, \mathbb{E}\left(\, [f_t(X_t) - f_s(X_s)] \mid X_s = x \right) \longrightarrow_{t \to s} \left(\frac{\partial}{\partial s} + L_s \right)(f_s)(x)$$

On the other hand, using the fact that $M_t(f)$ and $M_t(f)^2 - \langle M(f) \rangle_t$ are $\mathcal{F}$-martingales, we prove that

$$\mathbb{E}\left([M_t(f) - M_s(f)]^2 \mid \mathcal{F}_s \right)$$

$$= \mathbb{E}\left(M_t(f)^2 \mid \mathcal{F}_s \right) - M_s(f)^2$$

$$= \underbrace{\mathbb{E}\left(M_t(f)^2 - \langle M(f) \rangle_t \mid \mathcal{F}_s \right) - \left(M_s(f)^2 - \langle M(f) \rangle_s \right)}_{=0}$$

$$\qquad\qquad\qquad\qquad\qquad + \mathbb{E}\left(\langle M(f) \rangle_t - \langle M(f) \rangle_s \mid \mathcal{F}_s \right)$$

$$= \mathbb{E}\left(\langle M(f) \rangle_t - \langle M(f) \rangle_s \mid \mathcal{F}_s \right)$$

from which we conclude that

$$\mathbb{E}\left([M_t(f) - M_s(f)]^2 \mid \mathcal{F}_s \right) = \mathbb{E}\left(\int_s^t \Gamma_{L_\tau}(f_\tau, f_\tau)(X_\tau) \, d\tau \mid X_s \right)$$

Arguing as above, we prove that

$$\frac{1}{t-s} \, \mathbb{E}\left([M_t(f) - M_s(f)]^2 \mid \mathcal{F}_s \right) \longrightarrow_{t\to s} \Gamma_{L_s}(f_s, f_s)(X_s)$$

or alternatively

$$\frac{1}{t-s} \, \mathbb{E}\left(\left[f_t(X_t) - f_s(X_s) - \int_s^t \left(\frac{\partial}{\partial \tau} + L_\tau \right)(f_\tau)(X_\tau)\, d\tau \right]^2 \mid X_s = x \right)$$

$$\longrightarrow_{t\to s} \Gamma_{L_s}(f_s, f_s)(x)$$

5.2 Feynman-Kac models

Feynman-Kac models and their weighted Monte Carlo estimates were originally presented by Mark Kac in 1949 [350]. Nowadays, they are of current use in molecular chemistry and computational physics to calculate the ground state energy of some Hamiltonian operators associated with some potential function V describing the energy of a molecular configuration (see for instance [91, 163, 175, 176, 246, 498], and references therein).

In the further development of this section, we briefly introduce the reader to the stochastic modeling and the numerical analysis of this class of continuous time models.

5.2.1 Path integral measures

We consider a Markov process $(\mathcal{X}_s)_{0 \le s \le t}$, and *bounded* potential function $\mathcal{V}_t$, on some some Polish state space E. We let $E_t := D([0, t], E)$ be the set of càdlàg trajectories from $[0, t]$ into E and $\mathbb{P}_t$ be for the distribution of the random trajectories

$$\mathcal{X}_t := (\mathcal{X}_s)_{0 \le s \le t} \in E_t = D([0, t], E)$$

The Feynman-Kac measures associated with $(\mathcal{X}_t, \mathcal{V}_t)$ are given by the following formula

$$d\mathbb{Q}_t = \frac{1}{\mathcal{Z}_t} \, \exp\left(\int_0^t \mathcal{V}_s(\mathcal{X}_s)\, ds \right) d\mathbb{P}_t \tag{5.3}$$

In other words, for any $f_t \in \mathcal{B}_b(E_t)$, the measure $\mathbb{Q}_t$ is defined by the following formulae

$$\mathbb{Q}_t(f_t) := \frac{\Lambda_t(f_t)}{\Lambda_t(1)} \quad \text{with} \quad \Lambda_t(f_t) = \mathbb{E}\left(f_t(\mathcal{X}_t) \, \exp\left(\int_0^t \mathcal{V}_s(\mathcal{X}_s) ds \right) \right) \tag{5.4}$$

We let (ν_t, μ_t) be the t-marginal of the measures $(\Lambda_t, \mathbb{Q}_t)$, and we further assume that $\mathcal{X}_t$ has infinitesimal generators L_t, defined on some common domain of functions $D(L) \subset \mathcal{B}_b(E)$.

In this situation, it is well known that the weak evolution equation of flow of measures ν_t is given, for sufficiently regular function $f \in D(L)$, by the following partial differential equation

$$\frac{d}{dt} \nu_t(f) := \nu_t(L_t^V(f)) \quad \text{with} \quad L_t^V = L_t + \mathcal{V}_t$$

In physics, this equation is called the imaginary time Schrödinger equation. In scientific computing literature, as well as in some branches of pure and applied mathematics, this equation is also called the "heat equation." We also mention that these evolution equations are closely related to the Zakai equations and the normalized Kushner-Stratonovitch models, arising in the description of nonlinear filtering models. A more thorough discussion on these equations can be found in the series of articles [145, 153, 161], as well as in the textbook of A. Bain and D. Crisan [30].

Using some elementary stochastic calculus manipulations, we readily prove that

$$\frac{d}{dt} \log \nu_t(1) = \frac{1}{\nu_t(1)} \mathbb{E}\left(\mathcal{V}_t(\mathcal{X}_t) \exp\left(\int_0^t \mathcal{V}_s(\mathcal{X}_s) \, ds\right)\right) = \mu_t(\mathcal{V}_t)$$

and therefore

$$\nu_t(1) = \exp\left(\int_0^t \mu_s(\mathcal{V}_s) \, ds\right) = \Lambda_t(1)$$

This implies that

$$\mu_t(f) = \mathbb{E}\left(f(\mathcal{X}_t) \exp\left(\int_0^t (\mathcal{V}_s(\mathcal{X}_s) - \mu_s(\mathcal{V}_s)) \, ds\right)\right)$$

We conclude that the normalized distribution μ_t satisfies the same evolution equation as ν_t, by replacing $\mathcal{V}_t$ by their centered versions $(\mathcal{V}_t - \mu_t(\mathcal{V}_t))$; that is, we have that

$$\frac{d}{dt} \mu_t(f) = \mu_t(L_t(f)) + \mu_t(\mathcal{V}_t f) - \mu_t(\mathcal{V}_t)\mu_t(f) \tag{5.5}$$

Further details on the derivation of these evolution equations can be found in the articles [161, 176, 179].

5.2.2   Time discretization models

We consider the mesh sequence $t_k = k/m$, $k \geq 0$, with time step $h = t_n - t_{n-1} = 1/m$ associated with some integer $m \geq 1$, and we let $\mathbb{Q}_{t_n}^{(m)}$ and

$\Lambda_{t_n}^{(m)}$ be the Feynman-Kac measures on E_{t_n} defined for any $f_n \in \mathcal{B}_b(E_{t_n})$, by the following formulae

$$\mathbb{Q}_{t_n}^{(m)}(f_n) := \Lambda_{t_n}^{(m)}(f_n)/\Lambda_{t_n}^{(m)}(1) \qquad (5.6)$$

with the unnormalized Feynman-Kac measures

$$\Lambda_{t_n}^{(m)}(f_n) = \mathbb{E}\left(f_n(\mathcal{X}_{t_n}) \prod_{0 \le p < n} e^{V_{t_p}(X_{t_p})/m}\right) \quad \text{and} \quad \mathcal{X}_{t_n} := (X_s)_{0 \le s \le t_n}$$

We also denote $(\nu_{t_n}^{(m)}, \mu_{t_n}^{(m)})$ the t_n-marginal of $(\Lambda_{t_n}^{(m)}, \mathbb{Q}_{t_n}^{(m)})$, and we consider the integral operators

$$Q_{t_n, t_{n+1}}^{(m)}(x, dy) := G_{t_n}^{(m)}(x) \, M_{t_n, t_{n+1}}^{(m)}(x, dy)$$

with the potential functions, and the Markov transitions

$$G_{t_n}^{(m)} := e^{V_{t_n}/m} \quad \text{and} \quad M_{t_n, t_{n+1}}^{(m)}(x, dy) := \mathbb{P}\left(\mathcal{X}_{t_{n+1}} \in dy \mid \mathcal{X}_n = x\right)$$

By construction, the evolution equation of these distributions is given by

$$
\begin{aligned}
\nu_{t_{n+1}}^{(m)} &= \nu_{t_n}^{(m)} Q_{t_n, t_{n+1}}^{(m)} \\
\mu_{t_{n+1}}^{(m)} &= \Phi_{t_n, t_{n+1}}^{(m)}\left(\mu_{t_n}^{(m)}\right) := \Psi_{G_{t_n}^{(m)}}\left(\mu_{t_n}^{(m)}\right) M_{t_n, t_{n+1}}^{(m)}
\end{aligned}
$$

with the Boltzmann-Gibbs transformation defined in (0.2). A detailed proof of these evolution equations is provided in Section 3.2.2.

Under appropriate regularity conditions, we have the following first order decompositions

$$\Lambda_{t_n}^{(m)} = \Lambda_{t_n} + \frac{1}{m} \, r_{m, t_n} \quad \text{and} \quad \mathbb{Q}_{t_n}^{(m)} = \mathbb{Q}_{t_n} + \frac{1}{m} \, \overline{r}_{m, t_n}$$

with some remainder signed measures $r_{m, t_n}, \overline{r}_{m, t_n}$ such that

$$\sup_{m \ge 1} \left[\|\overline{r}_{m, t_n}\|_{tv} \vee \|r_{m, t_n}\|_{tv}\right] < \infty$$

A detailed proof of these estimates can be found in [185].

The l.h.s. measures in the m-approximation model (5.6) of the measures (5.3) are defined on some time mesh sequence that can be thought of as a time discretization of the exponential path integrals, in the continuous time model (5.3). Nevertheless, the elementary Markov transitions of the Markov chain $(\mathcal{X}_{t_n})_{n \ge 0}$ are generally unknown, so that the mean field Feynman-Kac models cannot be sampled without another level of approximation.

To get some feasible Monte Carlo approximation scheme, we need to have a dedicated technique to sample the transitions of this chain. One natural strategy is to replace in (5.6) the reference Markov chain $(\mathcal{X}_{t_n})_{n \ge 0}$ by the Markov chain $(\mathcal{X}_{t_n}^{(m)})_{n \ge 0}$ associated with some Euler type discretization model with time step $\Delta t = 1/m$. The stochastic analysis of these models is discussed in some detail in a series of articles [154, 155, 173], including first order expansions, in terms of the size of the time mesh sequence.

5.2.3   Time dilation models

Let us assume that the potential function $\mathcal{V}_t$ and Markov process $\mathcal{X}_t$ in (5.3) and (5.6) is given by

$$\mathcal{X}_t = X_{\lfloor t \rfloor} \quad \text{and} \quad \mathcal{V}_t = \log G_{\lfloor t \rfloor} \tag{5.7}$$

where X_n, $n \in \mathbb{N}$ is an E-valued Markov chain, and G_n are Borel positive functions s.t. $\log G_n$ is bounded. In this situation, we also have that

$$\nu_n^{(m)} = \nu_n = \gamma_n \quad \text{and} \quad \mu_n^{(m)} = \mu_n = \eta_n$$

with the Feynman-Kac measures γ_n and η_n defined in Section 3.2.2.

In contrast to standard discrete generation particle models associated with the flow of measures η_n, the particle interpretation of the flow of measures $\mu_{t_n}^{(m)}$ is defined on a refined time mesh sequence between integers. This time mesh sequence can be interpreted as a time dilation. Between two integers, the particle evolution passes through an additional series of intermediate time evolutions steps. A dedicated Bernoulli acceptance-rejection trial, coupled with a recycling scheme, is performed at every one of these time steps.

When the time step decreases to 0, the geometric interacting jump Markov chain converges to an exponential interacting jump Markov process. A more thorough discussion on these models is provided in Section 5.4.4.

5.3   Continuous time McKean models

5.3.1   Integro-differential equations

We consider the continuous time Feynman-Kac models presented in Section 5.2. The mean field approximation of the flow of Feynman-Kac measures μ_t introduced in (5.3) depends on the interpretation of the correlation term in the r.h.s. of the evolution Equation (5.5), in terms of interacting jump type infinitesimal generators.

As in the discrete time case, the choice of these generators is far from being unique. In the further development of this section, we illustrate these models with three important situations. These three illustrations are the continuous time versions of the three cases discussed in Section 4.4.5. These models are also closely related to the nonlinear Markov transport interpretations of the Boltzmann-Gibbs transformations presented in (0.3).

Firstly, we assume that $\mathcal{V}_t = -\mathcal{U}_t$, for some nonnegative function μ_t. In this situation, we have the formula

$$\mu_t(\mathcal{V}_t f) - \mu_t(\mathcal{V}_t)\mu_t(f) = \mu_t(\mathcal{U}_t [\mu_t(f) - f]) = \mu_t\left(\widehat{L}_{t,\mu_t}(f)\right)$$

with the interacting jump generator

$$\widehat{L}_{t,\mu_t}(f)(x) = \mathcal{U}_t(x) \int [f(y) - f(x)] \, \mu_t(dy) \tag{5.8}$$

When $\mathcal{V}_t$ is a positive function, we also have the formula

$$\mu_t(\mathcal{V}_t f) - \mu_t(\mathcal{V}_t)\mu_t(f) = \mu_t\left(\widehat{L}_{t,\mu_t}(f)\right)$$

with the interacting jump generator

$$\begin{aligned}
\widehat{L}_{t,\mu_t}(f)(x) &= \int [f(y) - f(x)] \, \mathcal{V}_t(y) \, \mu_t(dy) \\
&= \mu_t(\mathcal{V}_t) \int [f(y) - f(x)] \, \Psi_{\mathcal{V}_t}(\mu_t)(dy) \tag{5.9}
\end{aligned}$$

Finally, for any bounded potential function $\mathcal{V}_t$ we have

$$\begin{aligned}
\mu_t(\mathcal{V}_t f) - \mu_t(\mathcal{V}_t)\mu_t(f) &= \int (f(y) - f(x)) \, (\mathcal{V}_t(y) - \mathcal{V}_t(x))_+ \, \mu_t(dx) \, \mu_t(dy) \\
&= \mu_t\left(\widehat{L}_{t,\mu_t}(f)\right)
\end{aligned}$$

with $a_+ = a \vee 0$, and with the interacting jump generator

$$\widehat{L}_{t,\mu_t}(f)(x) = \int [f(y) - f(x)] \, (\mathcal{V}_t(y) - \mathcal{V}_t(x))_+ \, \mu_t(dy) \tag{5.10}$$

In the three cases discussed above, for any test functions $f \in D(L)$, we have the evolution equation

$$\frac{d}{dt}\mu_t(f) = \mu_t(L_{t,\mu_t}(f)) \quad \text{with} \quad L_{t,\mu_t} := L_t + \widehat{L}_{t,\mu_t} \tag{5.11}$$

where L_t is the infinitesimal generator of the process $\mathcal{X}_t$ introduced in Section 5.2.1.

5.3.2 Nonlinear Markov jump processes

The integro-differential equations discussed in Section 5.3.1 can be interpreted as the evolution of the laws $\mathrm{Law}(\overline{\mathcal{X}}_t) = \mu_t$ of a time inhomogeneous Markov process $\overline{\mathcal{X}}_t$, with an infinitesimal generator $\overline{L}_t := L_{t,\mu_t}$. In this situation, it is essential to observe that L_{t,μ_t} depends on the distribution μ_t of the random state $\overline{\mathcal{X}}_t$.

The stochastic model $\overline{\mathcal{X}}_t$, or equivalently the collection of infinitesimal generator $L_{t,\mu}$, indexed by $\mu \in \mathcal{P}(E)$, is called a McKean interpretation of the nonlinear evolution equation in distribution space defined in (5.11).

In this framework, using Ito's formula (5.1), for any test function $f \in \mathcal{C}^1([0, \infty[, D(L))$, we readily check that

$$df_t(\overline{\mathcal{X}}_t) = \left(\frac{\partial}{\partial t} + L_{t,\mu_t}\right)(f_t)(\overline{\mathcal{X}}_t)dt + d\overline{M}_t(f) \qquad (5.12)$$

with a remainder, and second order $\mathcal{F}$-martingale term $\overline{M}_t(f)$ with predictable angle bracket given by

$$d\langle \overline{M}(f)\rangle_t/dt = \Gamma_{L_{t,\mu_t}}(f_t, f_t)(\overline{\mathcal{X}}_t)$$

Using the r.h.s. description of L_{t,μ_t} in (5.11), for any $f \in D(L)$ we notice that

$$\Gamma_{L_{t,\mu_t}}(f, f) = \Gamma_{L_t}(f, f) + \Gamma_{\widehat{L}_{t,\mu_t}}(f, f)$$

In this section, we provide a description of this Markov process in the three cases discussed in Section 5.3.1.

In case (5.8), between the jump times, the process $\overline{\mathcal{X}}_t$ evolves as a copy of the reference process $\mathcal{X}_t$, with generator L_t. The jump rate is defined by the function $\mathcal{U}_t$. In other words, the jump times $(T_n)_{n \geq 0}$ are given by the following recursive formulae

$$T_{n+1} = \inf \left\{ t \geq T_n : \int_{T_n}^t \mathcal{U}_s(\overline{\mathcal{X}}_s)\, ds \geq e_n \right\}$$

where $T_0 = 0$, and $(e_n)_{n \geq 0}$ stands for a sequence of i.i.d. exponential random variables with unit parameter. At the jump time T_n the process $\overline{\mathcal{X}}_{T_n-} = x$ jumps to new site $\overline{\mathcal{X}}_{T_n} = y$ randomly chosen with the distribution $\mu_{T_n-}(dy)$. For any $f \in D(L)$, it is also easily checked that

$$\Gamma_{\widehat{L}_{t,\mu_t}}(f, f)(x) = \widehat{L}_{t,\mu_t}\left((f - f(x))^2\right)(x) = \mathcal{U}_t(x) \int [f(y) - f(x)]^2 \, \mu_t(dy)$$

In this situation, an explicit expression of the time inhomogeneous semigroup $\overline{\mathcal{P}}_{s,t,\mu_s}$, $s \leq t$, of the process $\overline{\mathcal{X}}_t$ is provided by the following formula

$$\begin{aligned}
\overline{\mathcal{P}}_{s,t,\mu_s}(f)(x) &= \mathbb{E}\left(f(\overline{X}_t) \mid \overline{X}_s = x\right) \\
&= Q_{s,t}(1)(x)\, \Phi_{s,t}(\delta_x)(f) + (1 - Q_{s,t}(1)(x))\, \Phi_{s,t}(\mu_s)(f) \\
&= Q_{s,t}(f)(x) + (1 - Q_{s,t}(1)(x))\, \Phi_{s,t}(\mu_s)(f) \qquad (5.13)
\end{aligned}$$

The discrete time version of this formula is discussed in Section 12.3.2, dedicated to McKean interpretations of discrete time Feynman-Kac models.

We let $\overline{\mathcal{P}}^{(m)}_{t_n,t_{n+1},\mu_{t_n}}$, and $\Phi^{(m)}_{t_n,t_{n+1}}$, be the Markov transition and the transformation of probability measures defined as $\overline{\mathcal{P}}_{s,t,\mu_s}$, and $\Phi_{t_n,t_{n+1}}$, replacing $Q_{t_n,t_{n+1}}$ by the integral operator

$$Q^{(m)}_{t_n,t_{n+1}}(f)(x) = e^{-\mathcal{U}_{t_n}(x)/m}\, \mathbb{E}\left(f(\mathcal{X}_{t_{n+1}}) \mid \mathcal{X}_{t_n} = x\right)$$

Using (5.13), after some tedious but elementary calculations, we prove that

$$\overline{\mathcal{P}}^{(m)}_{t_n,t_{n+1},\mu_{t_n}} = \overline{\mathcal{P}}_{t_n,t_{n+1},\mu_{t_n}} + \frac{1}{m}\,\mathcal{R}^{(m)}_{t_n,t_{n+1},\mu_{t_n}} \tag{5.14}$$

with some remainder signed measures $\mathcal{R}^{(m)}_{t_n,t_{n+1},\mu_{t_n}}$ such that

$$\sup_{m\geq 1}\left\|\mathcal{R}^{(m)}_{t_n,t_{n+1},\mu_{t_n}}\right\|_{tv} \leq c_{t_n}$$

for some finite constant whose values only depend on the potential function $\mathcal{U}_t$. A detailed proof of these estimates can be found in [185].

In case (5.9), between the jump times, the process $\overline{\mathcal{X}}_t$ evolves as a copy of the reference process $\mathcal{X}_t$, with generator L_t. The rate of the jumps is now given by the parameter $\mu_t(\mathcal{V}_t)$. In other words, the jump times $(T_n)_{n\geq 0}$ are defined by the following recursive formulae

$$T_{n+1} = \inf\left\{t\geq T_n \; : \; \int_{T_n}^t \mu_s(\mathcal{V}_s)\,ds \geq e_n\right\}$$

where $T_0 = 0$, and $(e_n)_{n\geq 0}$ stands for a sequence of i.i.d. exponential random variables with unit parameter. At the jump time T_n the process $\overline{\mathcal{X}}_{T_n-} = x$ jumps to new site $\overline{\mathcal{X}}_{T_n} = y$ randomly chosen with the distribution $\Psi_{\mathcal{V}_{T_n-}}(\mu_{T_n-})(dy)$. In addition, for any $f\in D(L)$, we also have that

$$\Gamma_{\widehat{L}_{t,\mu_t}}(f,f)(x) = \widehat{L}_{t,\mu_t}\left((f-f(x))^2\right)(x) = \int [f(y)-f(x)]^2\,\mathcal{V}_t(y)\,\mu_t(dy)$$

In case (5.10), between the jump times the process $\overline{\mathcal{X}}_t$ evolves as a copy of the reference process $\mathcal{X}_t$, with generator L_t. The rate of the jumps is now given by the function

$$
\begin{aligned}
\mathcal{W}_{t,\mu_t}(x) \;\; &:= \;\; \mu_t((\mathcal{V}_t - \mathcal{V}_t(x))_+)\\
&= \;\; \mu_t(\mathcal{V}_t\,1_{\mathcal{V}_t\geq\mathcal{V}_t(x)}) - \mu_t(\mathcal{V}_t\geq\mathcal{V}_t(x))\,\mathcal{V}_t(x)\\
&= \;\; \mu_t(\mathcal{V}_t\geq\mathcal{V}_t(x))\,\Psi_{\mathcal{V}_t\geq\mathcal{V}_t(x)}(\mu_t)\,[\mathcal{V}_t-\mathcal{V}_t(x)]
\end{aligned}
$$

More formally, the jump times $(T_n)_{n\geq 0}$ are given by the following recursive formulae

$$T_{n+1} = \inf\left\{t\geq T_n \; : \; \int_{T_n}^t \mathcal{W}_{t,\mu_t}(\overline{\mathcal{X}}_s)\,ds \geq e_n\right\}$$

where $T_0 = 0$, and $(e_n)_{n\geq 0}$ stands for a sequence of i.i.d. exponential random variables with unit parameter. At the jump time T_n the process $\overline{\mathcal{X}}_{T_n-} = x$ jumps to new site $\overline{\mathcal{X}}_{T_n} = y$ randomly chosen with the distribution $\Psi_{(\mathcal{V}_{T_n}-\mathcal{V}_{T_n}(x))_+}(\mu_{T_n-})$.

In this situation, for any $f\in D(L)$, we also have that

$$
\begin{aligned}
\Gamma_{\widehat{L}_{t,\mu_t}}(f,f)(x) \;\; &= \;\; \widehat{L}_{t,\mu_t}\left((f-f(x))^2\right)(x)\\
&= \;\; \int [f(y)-f(x)]^2\,(\mathcal{V}_t(y)-\mathcal{V}_t(x))_+\,\mu_t(dy)
\end{aligned}
$$

and therefore

$$\mu_t \left[\Gamma_{\widehat{L}_{t,\mu_t}} (f, f) \right] = \int [f(y) - f(x)]^2 \, (\mathcal{V}_t(y) - \mathcal{V}_t(x))_+ \, \mu_t(dx)\mu_t(dy) \quad (5.15)$$

5.4   Mean field particle models

5.4.1   An interacting particle system model

This section is mainly concerned with the design of the mean field N-particle model $\xi_t := (\xi_t^i)_{1 \leq i \leq N}$ associated with a given collection of generators L_{t,μ_t} satisfying the weak Equation (5.11). This N-particle model is a Markov process in E^N, with infinitesimal generator defined, for sufficiently regular functions F on E^N, by the following formulae

$$\mathcal{L}_t(F)(x^1, \ldots, x^N) := \sum_{1 \leq i \leq N} L_{t,m(x)}^{(i)}(F)(x^1, \ldots, x^i, \ldots, x^N) \quad (5.16)$$

In the above display, $m(x) := \frac{1}{N} \sum_{1 \leq i \leq N} \delta_{x^i}$ stands for the occupation measure of the population $x = (x^i)_{1 \leq i \leq N} \in E^N$; and $L_{t,m(x)}^{(i)}$ stands for the operator $L_{t,m(x)}$, acting on the function $x^i \mapsto F(x^1, \ldots, x^i, \ldots, x^N)$.

Before entering into the detailed description of the interacting particle system model associated with the three cases presented in Section 5.3, we provide a brief discussion on the convergence analysis of these mean field stochastic models.

Firstly, using the Ito's formula (5.1), we have that

$$dF(\xi_t) = \mathcal{L}_t(F)(\xi_t) \, dt + d\mathcal{M}_t(F)$$

for some martingale $\mathcal{M}_t(F)$ with predictable increasing process defined by

$$\langle \mathcal{M}(F) \rangle_t := \int_0^t \Gamma_{\mathcal{L}_s} (F, F) (\xi_s) \, ds$$

We recall that the "carré du champ" operator $\Gamma_{\mathcal{L}_s}$ associated to $\mathcal{L}_s$ is defined by

$$\Gamma_{\mathcal{L}_s} (F, F) (x) := \mathcal{L}_s \left[(F - F(x))^2 \right] (x) = \mathcal{L}_s(F^2)(x) - 2F(x)\mathcal{L}_s(F)(x)$$

For empirical test functions of the following form

$$F(x) = m(x)(f) = \frac{1}{N} \sum_{i=1}^N f(x^i)$$

with $f \in D(L)$, we find that

$$
\begin{aligned}
\mathcal{L}_s(F)(x) &= m(x)(L_{s,m(x)}(f)) \\
\Gamma_{\mathcal{L}_s}(F,F)(x) &= \frac{1}{N} m(x)\left(\Gamma_{L_{s,m(x)}}(f,f)\right)
\end{aligned}
\tag{5.17}
$$

From this discussion, it should be clear that

$$
\mu_t^N := \frac{1}{N} \sum_{1 \le i \le N} \delta_{\xi_t^i} \implies d\mu_t^N(f) = \mu_t^N(L_{t,\mu_t^N}(f))\, dt + \frac{1}{\sqrt{N}}\, dM_t^N(f)
\tag{5.18}
$$

with the martingale

$$
M_t^N(f) = \sqrt{N}\, \mathcal{M}_t(F)
\tag{5.19}
$$

The predictable angle bracket is given by

$$
\begin{aligned}
\langle M^N(f) \rangle_t &:= \int_0^t \mu_s^N\left(\Gamma_{L_{s,\mu_s^N}}(f,f)\right)\, ds \\
&= \int_0^t \mu_s^N\left(\Gamma_{L_s}(f,f)\right)\, ds + \int_0^t \mu_s^N\left(\Gamma_{\widehat{L}_{s,\mu_s^N}}(f,f)\right)\, ds
\end{aligned}
$$

A more explicit description of the r.h.s. terms in the above display can be given in the three cases discussed in Section 5.3. For instance, in the third case, using formula (5.15) we find that

$$
\begin{aligned}
\mu_s^N\left(\Gamma_{\widehat{L}_{s,\mu_s^N}}(f,f)\right) &= \int [f(y) - f(x)]^2 \left(\mathcal{V}_s(y) - \mathcal{V}_s(x)\right)_+ \mu_s^N(dx)\mu_s^N(dy) \\
&\le \operatorname{osc}(f)^2 \operatorname{osc}(\mathcal{V}_s)
\end{aligned}
$$

From the r.h.s. perturbation formulae (5.18), we conclude that μ_t^N "almost solve," as $N \uparrow \infty$, the nonlinear evolution Equation (5.11). For a more thorough discussion on these continuous time models, we refer the reader to the review article [173], and the references therein.

5.4.2 Mutation-selection infinitesimal generators

By construction, the generator $\mathcal{L}_t$ associated with the nonlinear model (5.11) is divided into a mutation generator $\mathcal{L}_t^{mut}$, and an interacting jump generator $\mathcal{L}_t^{jump}$

$$
\mathcal{L}_t = \mathcal{L}_t^{mut} + \mathcal{L}_t^{jump}
$$

with $\mathcal{L}_t^{mut}$ and $\mathcal{L}_t^{jump}$ defined by

$$
\begin{aligned}
\mathcal{L}_t^{mut}(F)(x) &= \sum_{1 \le i \le N} L_t^{(i)}(F)(x^1, \ldots, x^i, \ldots, x^N) \\
\mathcal{L}_t^{jump}(F)(x) &= \sum_{1 \le i \le N} \widehat{L}_{t,m(x)}^{(i)}(F)(x^1, \ldots, x^i, \ldots, x^N)
\end{aligned}
$$

The mutation generator $\mathcal{L}_t^{mut}$ describes the evolution of the particles between the jumps. Between the jumps, the particles evolve independently with L_t-motions in the sense that they explore the state space as independent copies of the process $\mathcal{X}_t$ with generator L_t. The jump transition depends on the form of the generator $\widehat{L}_{t,\mu_t}$.

Arguing as in (5.17), for empirical test functions of the following form

$$F(x) = m(x)(f) = \frac{1}{N} \sum_{i=1}^{N} f(x^i)$$

with $f \in D(L)$, we find that

$$\mathcal{L}_t^{mut}(F)(x) = m(x)(L_t(f)) \quad \text{and} \quad \mathcal{L}_t^{jump}(F)(x) = m(x)(\widehat{L}_{t,m(x)}(f))$$

In the same vein, we have

$$\Gamma_{\mathcal{L}_t^{mut}}(F, F)(x) = \frac{1}{N} m(x) (\Gamma_{L_t}(f, f))$$

$$\Gamma_{\mathcal{L}_t^{jump}}(F, F)(x) = \frac{1}{N} m(x) \left(\Gamma_{\widehat{L}_{t,m(x)}}(f, f)\right)$$

This shows that the martingale defined in (5.19) is the composition of a couple of martingales

$$M_t^N(f) = M_t^{N,mut}(f) + M_t^{N,jump}(f)$$

with predictable angle brackets

$$\langle M^N(f) \rangle_t = \langle M^{N,mut}(f) \rangle_t + \langle M^{N,jump}(f) \rangle_t$$

given by

$$d\langle M^{N,mut}(f) \rangle_t = \mu_t^N (\Gamma_{L_t}(f, f)) \, dt$$

$$d\langle M^{N,jump}(f) \rangle_t = \mu_t^N \left(\Gamma_{\widehat{L}_{t,\mu_t^N}}(f, f)\right) \, dt$$

5.4.3 Some motivating illustrations

In this section, we illustrate the abstract models developed in Section 5.4.2 in three cases discussed in Section 5.3.

In case (5.8), the jump generator is given by

$$\mathcal{L}_t^{jump}(F)(x) = \sum_{1 \leq i \leq N} \mathcal{U}_t(x^i) \int [F(\theta_u^i(x)) - F(x)] \, m(x)(du)$$

with the population mappings θ_u^i defined below

$$\theta_u^i : x \in E^N \mapsto \theta_u^i(x) = (x^1, \ldots, x^{i-1}, \underbrace{u}_{\text{i-th}}, x^{i+1}, \ldots, x^N) \in E^N$$

The quantity $\mathcal{U}_t(\xi_t^i)$ represents the jump rate of the i-th particle ξ_t^i. More precisely, if we denote by T_n^i the n-th jump time of ξ_t^i, then we have

$$T_{n+1}^i = \inf\left\{ t \geq T_n^i \ : \ \int_{T_n^i}^t \mathcal{U}_s(\xi_s^i) \, ds \geq e_n^i \right\} \tag{5.20}$$

where $(e_n^i)_{1 \leq i \leq N, n \in \mathbb{N}}$ stands for a sequence of i.i.d. exponential random variables with unit parameter. At the jump time T_n^i the process $\xi_{T_n^i-}^i = x^i$ jumps to new site $\xi_{T_n^i}^i = y$ randomly chosen with the distribution $m(\xi_{T_n^i}-)(dy)$. In other words, at the jump times the particles jump randomly to a new state randomly chosen in the current population.

The probabilistic interpretation of the jump generator is far from being unique. For instance, it is easily checked that $\mathcal{L}_t^{jump}$ can be rewritten in the following form

$$\mathcal{L}_t^{jump}(F)(x) \ = \ \lambda_t(x) \int [F(y) - F(x)] \, \mathcal{P}_t(x, dy)$$

with the population jump rate $\lambda_t(x)$ and the Markov transition $\mathcal{P}_t(x, dy)$ on E^N given below

$$\lambda_t(x) := N m(x) \, (\mathcal{U}_t)$$

and

$$\mathcal{P}_t(x, dy) = \sum_{1 \leq i \leq N} \frac{\mathcal{U}_t(x^i)}{\sum_{1 \leq i' \leq N} \mathcal{U}_t(x^{i'})} \, \frac{1}{N} \sum_{1 \leq j \leq N} \delta_{\theta^i_{x_j}(x)}(dy)$$

In this interpretation, the individual jumps are replaced by population jumps at rate $\lambda_t(\xi_t)$. More precisely, the jump times T_n of the whole population are defined by

$$T_{n+1} = \inf\left\{ t \geq T_n \ : \ \int_{T_n}^t \left[\sum_{1 \leq i \leq N} \mathcal{U}_s(\xi_s^i) \right] ds \geq e_n \right\}$$

where $(e_n)_{n \in \mathbb{N}}$ stands for a sequence of i.i.d. exponential random variables with unit parameter. At the jump time T_n the population $\xi_{T_n-} = x$ jumps to new population $\xi_{T_n} = y$ randomly chosen with the distribution $\mathcal{P}_{T_n-}(\xi_{T_n-}, dy)$. In other words, at the jump times T_n, we select randomly a state $\xi_{T_n-}^i$ with a probability proportional to $\mathcal{U}_t(\xi_{T_n-}^i)$, and we replace this state by a randomly chosen state $\xi_{T_n-}^j$ in the population, with $1 \leq j \leq N$.

We end this description, with an alternative interpretation when $\|\mathcal{U}_t\| \leq C$ for some finite constant $C < \infty$. In this situation, we clearly have $\|\lambda_t\| \leq NC$ and

$$\mathcal{L}_t^{jump}(F)(x) = \lambda' \int [F(y) - F(x)] \, \mathcal{P}_t'(x, dy)$$

with the jump rate λ' and the Markov jump transitions $\mathcal{P}_t'$ defined below

$$\lambda' = NC$$

and
$$\mathcal{P}'_t(x, dy) := \frac{\lambda_t(x)}{NC} \, \mathcal{P}_t(x, dy) + \left(1 - \frac{\lambda_t(x)}{NC}\right) \, \delta_x(dy)$$

In this interpretation, the population jump times T_n arrive at the higher rate $\lambda' = NC$. At the jump time T_n the population $\xi_{T_n-} = x$ jumps to new population $\xi_{T_n} = y$ randomly chosen with the distribution $\mathcal{P}'_{T_n-}(\xi_{T_n-}, dy)$.

As usual, in the couple of models described above between the jump times T_n of the population, every particle evolves independently with L_t-motions.

In case (5.9), the jump generator is given by

$$\mathcal{L}^{jump}_t(F)(x) \quad = \quad \sum_{1 \leq i \leq N} m(x)(\mathcal{V}_t) \int [F(\theta^i_u(x)) - F(x)] \, \Psi_{\mathcal{V}_t}(m(x)) \, (du)$$

The particles have a common jump rate that is given by the empirical average $m(\xi_t)(\mathcal{V}_t)$. In other words, the jump times T^i_n of a particle ξ^i_t are given by the following recursive formulae

$$T^i_{n+1} = \inf \left\{ t \geq T^i_n \ : \ \int_{T^i_n}^t m(\xi_s)(\mathcal{V}_s) \, ds \geq e^i_n \right\}$$

where $(e^i_n)_{1 \leq i \leq N, n \geq 0}$ stands for a sequence of i.i.d. exponential random variables with unit parameter. At the jump time T^i_n, the process $\xi^i_{T_n-} = x^i$ jumps to new site $\xi^i_{T_n} = y$, randomly chosen with the weighted distribution $\Psi_{\mathcal{V}_{T_n-}}(m(\xi_{T_n-}))(dy)$.

As we mentioned in the first case, the probabilistic interpretation of the jump generator is not unique. In this situation, it is easily checked that $\mathcal{L}^{jump}_t$ can be rewritten in the following form

$$\mathcal{L}^{jump}_t(F)(x) \quad = \quad \lambda_t(x) \int [F(y) - F(x)] \, \mathcal{P}_t(x, dy)$$

with the population jump rate λ_t and the Markov transition $\mathcal{P}_t(x, dy)$ on E^N defined below
$$\lambda_t(x) := Nm(x)(\mathcal{V}_t)$$

and
$$\mathcal{P}_t(x, dy) = \frac{1}{N} \sum_{1 \leq i \leq N} \sum_{1 \leq j \leq N} \frac{\mathcal{V}_t(x^j)}{\sum_{1 \leq j' \leq N} \mathcal{V}_t(x^{j'})} \delta_{\theta^i_{x^j}(x)}(dy)$$

The description of the evolution of the population model follows the same lines as the ones given in case 2.

Finally, in case (5.10), the jump generator is given by

$$\mathcal{L}_t^{jump}(F)(x)$$

$$= \sum_{1\leq i\leq N} \int [F(\theta_u^i(x)) - F(x)]\ (\mathcal{V}_t(u) - \mathcal{V}_t(x^i))_+\ m(x)(du)$$

$$= \sum_{1\leq i\leq N} m(x)((\mathcal{V}_t - \mathcal{V}_t(x^i))_+)\ \int [F(\theta_u^i(x)) - F(x)]\ \Psi_{(\mathcal{V}_t-\mathcal{V}_t(x^i))_+}(m(x))(du)$$

In this interpretation, the jump rate of the i-th particle is given by the average potential variation of the j-th particles with higher values; that is, we have that

$$m(x)((\mathcal{V}_t - \mathcal{V}_t(x^i))_+) = \frac{1}{N} \sum_{1\leq j\leq N} 1_{\{\mathcal{V}_t(x^j)>\mathcal{V}_t(x^i)\}}\ (\mathcal{V}_t(x^j) - \mathcal{V}_t(x^i))$$

More precisely, if we denote by T_n^i the n-th jump time of ξ_t^i, we have

$$T_{n+1}^i = \inf\left\{ t \geq T_n^i\ :\ \int_{T_n^i}^t m(\xi_s)((\mathcal{V}_s - \mathcal{V}_s(\xi_s^i))_+)\ ds\ \geq\ e_n^i \right\}$$

where $(e_n^i)_{1\leq i\leq N, n\in\mathbb{N}}$ stands for a sequence of of i.i.d. exponential random variables with unit parameter. At the jump time T_n^i, the particle $\xi_{T_n^i-}^i = x^i$ jumps to new site $\xi_{T_n^i}^i$ randomly chosen with the distribution

$$\Psi_{(\mathcal{V}_{T_n^i-}-\mathcal{V}_{T_n^i-}(x^i))_+}(m(\xi_{T_n^i-}))$$

$$\propto \sum_{1\leq j\leq N} 1_{\left\{\mathcal{V}_{T_n^i-}(\xi_{T_n^i-}^j)>\mathcal{V}_{T_n^i-}(x^i)\right\}}\ \left(\mathcal{V}_{T_n^i-}(\xi_{T_n^i-}^j) - \mathcal{V}_{T_n^i-}(x^i)\right)\ \delta_{\xi_{T_n^i-}^j}$$

In other words, it chooses randomly a new site $\xi_{T_n^i}^i = \xi_{T_n^i-}^j$, among the ones with higher potential value, with a probability proportional to the difference of potential $\left(\mathcal{V}_{T_n^i-}(\xi_{T_n^i-}^j) - \mathcal{V}_{T_n^i-}(\xi_{T_n^i-}^i)\right)$.

As in the first two cases discussed above, we can interpret this jump generator at the level of the population. In this interpretation, we have

$$\mathcal{L}_t^{jump}(F)(x) = \lambda_t(x) \int [F(y) - F(x)]\ \mathcal{P}_t(x, dy)$$

with the population jump rate

$$\lambda_t(x) = N \int m(x)(du)\ m(x)(dv)\ (\mathcal{V}_t(u) - \mathcal{V}_t(v))_+$$

and the population jump transition

$$\mathcal{P}_t(x, dy) = \sum_{1 \leq i,j \leq N} \frac{(\mathcal{V}_t(x^j) - \mathcal{V}_t(x^i))_+}{\sum_{1 \leq i',j' \leq N} (\mathcal{V}_t(x^{j'}) - \mathcal{V}_t(x^{i'}))_+} \, \delta_{\theta^i_{x^j}(x)}(dy)$$

We end this section with a discussion on case (5.7). In this situation, the reference Markov process $\mathcal{X}_t = X_{\lfloor t \rfloor}$ has deterministic and fixed time jumps on every integer time so that the infinitesimal generator approach developed above does not apply directly. Nevertheless their probabilistic interpretation is defined in the same way: Between the jumps, the process $\overline{X}_t$ evolves as $\mathcal{X}_t$, and the N particles explore the state space as independent copies of the process $\mathcal{X}_t$. The rate of the jumps and their random spatial location are defined using the same interpretations as the ones given above.

The stochastic modeling, and the analysis of these continuous time models, and their particle interpretations, can be developed using the semigroup techniques provided in the articles [161, 178].

5.4.4   Geometric jump models

Next, we provide a brief description of the geometric type interacting jump particle models associated with the m-approximation Feynman-Kac model defined in (5.6). Firstly, if we set

$$\mathcal{M}_{t_n, t_{n+1}}(x, dy) = \mathbb{P}\left(\mathcal{X}_{t_{n+1}} \in dy \mid \mathcal{X}_{t_n} = x\right) \quad \text{and} \quad \mathcal{G}_{t_n} = \exp\left(\mathcal{V}_{t_n}/m\right)$$

then we recall that the flow of measures $\mu_{t_n}^{(m)}$ satisfies the following evolution equation

$$\mu_{t_{n+1}}^{(m)} = \Psi_{\mathcal{G}_{t_n}}(\mu_{t_n}^{(m)}) \mathcal{M}_{t_n, t_{n+1}} \tag{5.21}$$

with the Boltzmann-Gibbs transformation defined in (0.2).

In Section 4.4.5, we have seen that the mean field simulation of this model is dictated by the interpretation of the Botzmann-Gibbs transformation $\Psi_{\mathcal{G}_{t_n}}$ in terms of a Markov transport equation of the form

$$\Psi_{\mathcal{G}_{t_n}}(\mu) = \mu \mathcal{S}_{t_n, \mu} \tag{5.22}$$

In the above display, $\mathcal{S}_{t_n, \mu}$ stands for some (nonunique) collection of Markov transitions indexed by the time parameter t_n and the measures $\mu/in/Pa(E)$.

In this situation, we have

$$\mu_{t_{n+1}}^{(m)} = \Psi_{\mathcal{G}_{t_n}}(\mu_{t_n}^{(m)}) \mathcal{M}_{t_n, t_{n+1}} = \mu_{t_n}^{(m)} \mathcal{K}_{t_n, t_{n+1}, \mu_{t_n}^{(m)}} \tag{5.23}$$

with the Markov transitions $\mathcal{K}_{t_n, t_{n+1}, \mu_{t_n}^{(m)}} := \mathcal{S}_{t_n, \mu_{t_n}^{(m)}} \mathcal{M}_{t_n, t_{n+1}}$.

As in the continuous time case, these discrete time evolution equations can be interpreted as the evolution of the laws $\text{Law}\left(\overline{X}_{t_n}\right) = \mu_{t_n}^{(m)}$ of a time inhomogeneous Markov process $\overline{X}_{t_n}$ with Markov transitions generators $\mathcal{K}_{t_n, t_{n+1}, \mu_{t_n}^{(m)}}$

that depend on the distribution of the random states. These McKean interpretations of the evolution Equation (5.23) are discussed in some detail in Section 1.2.1 and in Section 4.4.2.

In this context, we recall that the elementary transitions of the Markov chain $\overline{X}_{t_n} \rightsquigarrow \overline{X}_{t_{n+1}}$ are decomposed into two separate transitions

$$\overline{X}_{t_n} \rightsquigarrow \widehat{X}_{t_n} \rightsquigarrow \overline{X}_{t_{n+1}}$$

Firstly, the state $\overline{X}_{t_n} = x$ jumps to a new location $\widehat{X}_{t_n} = y$ randomly chosen with the Markov transition $\mathcal{S}_{t_n,\mu_{t_n}^{(m)}}(x, dy)$. Then, the selected state $\widehat{X}_{t_n} = y$ evolves to a new site $\overline{X}_{t_{n+1}} = z$ randomly chosen with the Markov transition $\mathcal{M}_{t_n,t_{n+1}}(y, dz)$.

The mean field N-particle model $\xi_{t_n} := (\xi_{t_n}^i)_{1 \leq i \leq N}$ associated with the evolution Equation (5.23) is a Markov process in E^N with elementary transitions given by

$$\mathbb{P}\left(\xi_{t_{n+1}} \in dx \mid \xi_{t_n}\right) = \prod_{1 \leq i \leq N} \mathcal{K}_{t_n,t_{n+1},\mu_{t_n}^N}(\xi_{t_n}^i, dx^i) \quad \text{with} \quad \mu_{t_n}^N = \frac{1}{N} \sum_{1 \leq i \leq N} \delta_{\xi_{t_n}^i}$$
(5.24)

In the above display, $dx = dx^1 \times \ldots \times dx^N$ stands for an infinitesimal neighborhood of the point $x = (x^i)_{1 \leq i \leq N} \in E^N$.

We illustrate these rather abstract models with the following three classes of models corresponding to the models discussed in Section 4.4.5, and the three nonlinear Markov transport models presented in (0.3).

Firstly, if we have $\mathcal{V}_t = -\mathcal{U}_t$, for some nonnegative and bounded function $\mathcal{U}_t$, then formula (5.22) is met with the Markov transition

$$\mathcal{S}_{t_n,\mu}(x, dy) := e^{-\mathcal{U}_{t_n}(x)/m} \, \delta_x(dy) + \left(1 - e^{-\mathcal{U}_{t_n}(x)/m}\right) \, \Psi_{e^{-\mathcal{U}_{t_n}/m}}(\mu)(dy)$$

When the function $\mathcal{V}_t$ is nonnegative, the transport Equation (5.22) is also met with the Markov transition

$$\mathcal{S}_{t_n,\mu}(x, dy) := \frac{1}{\mu\left(e^{\mathcal{V}_{t_n}/m}\right)} \, \delta_x(dy) + \left(1 - \frac{1}{\mu\left(e^{\mathcal{V}_{t_n}/m}\right)}\right) \, \Psi_{\left(e^{\mathcal{V}_{t_n}/m}-1\right)}(\mu)(dy)$$

Finally, the Markov transport Equation (5.22) is always met with the transitions

$$\mathcal{S}_{t_n,\mu}(x, dy) := (1 - a_{t_n,\mu}(x)) \, \delta_x(dy) + a_{t_n,\mu}(x) \, \Psi_{\left(e^{\mathcal{V}_{t_n}/m}-e^{\mathcal{V}_{t_n}(x)/m}\right)_+}(\mu)(dy)$$

$$a_{t_n,\mu}(x) := \mu\left(\left(e^{\mathcal{V}_{t_n}/m} - e^{\mathcal{V}_{t_n}(x)/m}\right)_+\right)/\mu\left(e^{\mathcal{V}_{t_n}/m}\right) \in [0, 1]$$

In these three cases we have the following first order expansion

$$\mathcal{S}_{t_n,\mu} = Id + \frac{1}{m} \, \widehat{L}_{t_n,\mu} + \frac{1}{m^2} \, \widehat{R}_{t_n,\mu}$$
(5.25)

with some jump type generator $\widehat{L}_{t_n,\mu}$ and some integral operator $\widehat{R}_{t_n,\mu_{t_n}^{(m)}}$ s.t.

$$\sup \left\| \widehat{R}_{t_n,\mu} \right\|_{tv} < \infty$$

In the above display, the supremum is taken over all $m \geq 1$ and $\mu \in \mathcal{P}(E)$.

The jump generators $\widehat{L}_{t_n,\mu}$ corresponding to the three cases presented above are described, respectively, in (5.8), (5.9), and (5.10). The proofs of these expansions are provided in Section 9.6.2.

We further assume that condition (5.25) is satisfied with some jump type generator $\widehat{L}_{t_n,\mu}$. In this situation, for the continuous time Feynman-Kac models presented in Chapter 5, for any function $f \in D(L)$, and any $N \geq m \geq 1$, we have the nonasymptotic bias and variance estimates

$$\left| \mathbb{E} \left(\mu_{t_n}^N(f) \right) - \mu_{t_n}(f) \right| \leq c_{t_n}(f) \left[\frac{1}{N} + \frac{1}{m} \right]$$

and

$$\mathbb{E} \left(\left[\mu_{t_n}^N(f) - \mu_{t_n}(f) \right]^2 \right) \leq c_{t_n}(f) \left[\frac{1}{N} + \frac{1}{m^2} \right]$$

for some finite constant $c_{t_n}(f) < \infty$ that only depends on t_n and on f.

For the continuous time dilation model (5.7), for any $f \in \mathcal{B}_b(E)$, and for any $N \geq m \geq 1$ we have the nonasymptotic bias estimates

$$\left| \mathbb{E} \left(\mu_n^N(f) \right) - \eta_n(f) \right| \vee \mathbb{E} \left(\left[\mu_n^N(f) - \eta_n(f) \right]^2 \right) \leq c_n(f)/N$$

for some finite constant $c_n(f) < \infty$ that only depends on n and on f. A proof of these nonasymptotic variance estimates can be found in [185].

Chapter 6

Nonlinear evolutions of intensity measures

6.1 Intensity of spatial branching processes

The Feynman-Kac models presented in Section 1.4.2, and further developed in Chapter 3, were defined in terms of Markov chain distributions, weighted by some potential functions. This description is particularly useful to model conditional distributions of Markov chains w.r.t. a collection of conditioning events. In this section, we present a natural and alternative interpretation of these models in terms of spatial branching processes. We also extend the Feynman-Kac methodology to branching models, equipped with spontaneous birth rates. These extended Feynman-Kac distribution flows are defined as in (1.41), by adding at every time step a possibly different positive measure.

For a more detailed discussion on spatial branching processes, and their connections with Feynman-Kac models, we also refer the reader to [19, 319, 341]. Section 6.3 and the more recent studies [103, 163] also provide applications of these models to multiple object nonlinear filtering problems.

6.1.1 Spatial branching processes

Assume that, at a given time n, there are N_n individuals $(X_n^i)_{1 \leq i \leq N_n}$, taking values in some measurable state space E_n, enlarged with an auxiliary cemetery point c. As usual, we extend the measures γ_n on E_n and the bounded measurable functions f_n on E_n by setting $\gamma_n(\{c\}) = 0$ and $f_n(c) = 0$.

We emphasize that the state space E_n depends on the problem at hand. It may vary with the time parameter, and it can include all the characteristics of an individual, such as its type, its kinetic parameters as well as its complete path from the origin.

Each individual X_n^i has a survival probability, say $e_n(X_n^i) \in [0,1]$. When it dies, it goes instantly to the cemetery point c. We also use the convention $e_n(c) = 0$, so that a dead particle can only stay in the cemetery state.

Survival particles give birth to a random strictly positive number of individuals $h_n^i(X_n^i)$ where $\left(h_n^i(x_n)\right)_{1 \leq i \leq N_n}$ stands for a collection of independent random variables such that

$$\mathbb{E}\left(h_n^i(x_n)\right) = H_n(x_n)$$

for any $x_n \in E_n$, where H_n is a given collection of bounded nonnegative functions.

Notice that $H_n(x_n) \geq 1$ for any $x_n \in E_n$, since $h_n^i(x_n) \geq 1$. This branching transition is sometimes called spawning in signal processing and multi-target tracking literature.

After this branching transition, the system consists of a random number $\widehat{N}_n$ of individuals $(\widehat{X}_n^i)_{1 \leq i \leq \widehat{N}_n}$. Each of them evolves randomly, and independently, from state space to the next

$$\widehat{X}_n^i = x_n \in E_n \rightsquigarrow X_{n+1}^i = x_{n+1} \in E_{n+1}$$

according to a Markov transition $M_{n+1}(x_n, dx_{n+1})$ from E_n into E_{n+1}. Here again, we use the convention $M_{n+1}(c, \{c\}) = 1$, so that any dead particle remains in the cemetery state.

At the same time, an independent collection of new individuals is added to the current configuration.

We further assume that this additional spatial point process is modeled by a spatial Poisson process, with a prescribed intensity measure μ_{n+1} on E_{n+1}. It is used to model new particles entering the state space.

At the end of this transition, we obtain

$$N_{n+1} = \widehat{N}_n + N_{n+1}'$$

individuals $(X_{n+1}^i)_{1 \leq i \leq N_{n+1}}$, where N_{n+1}' is a Poisson random variable with parameters given by the total mass $\mu_{n+1}(1)$ of the positive measure μ_{n+1}, and $(X_{n+1}^{\widehat{N}_n + i})_{1 \leq i \leq N_{n+1}'}$ are independent and identically distributed random variables with common distribution

$$\overline{\mu}_{n+1} = \mu_{n+1}/\mu_{n+1}(1) \quad \text{where} \quad \mu_{n+1}(1) := \int_{E_{n+1}} \mu_{n+1}(dx) = \mathbb{E}\left(N_{n+1}'\right)$$

(6.1)

For a more thorough discussion on spatial Poisson point processes, we refer the reader to Section 6.3.3.

We end this section with some definitions, some conventions, and a few regularity conditions that will implicitly be assumed in the further development of this chapter.

Firstly, to simplify the presentation, we shall further assume that the initial configuration of the spatial branching process $(X_0^i)_{1 \leq i \leq N_0}$ is given by a spatial Poisson process, with a prescribed intensity measure μ_0 on E_0.

We denote by G_n the potential functions defined by

$$x_n \in E_n \mapsto G_n(x_n) = e_n(x_n)H_n(x_n)$$

To avoid unnecessary technical details, we further assume that the potential functions G_n are chosen so that for any $x \in E_n$

$$0 < g_{n,-} \leq G_n(x) \leq g_{n,+} < \infty \tag{6.2}$$

for any time parameter $n \geq 0$.

Note that this assumption is satisfied in most realistic spatial branching scenarios. Indeed, as $H_n(x) \geq 1$, the condition $g_{n,-} \leq G_n(x)$ essentially states that there exists $e_{n,-} > 0$ such that $e_n(x) \geq e_{n,-}$ for any $x \in E_n$. Loosely speaking, this condition ensures that every particle always has a small chance to survive.

On the other hand, the condition $G_n(x) \leq g_{n,+}$ states that there exists $H_{n,+} < \infty$ such that $H_n(x) \leq H_{n,+}$ for any $x \in E_n$. Loosely speaking, this condition allows controlling the total size of the branching process by some rather crude, but bounded finite constants.

In the unlikely scenario where (6.2) is not satisfied, the forthcoming analysis can be extended to more general models with general nonnegative potential functions, using the techniques developed in Section 4.4 in the monograph [163].

6.1.2 Intensity distribution flows

In this section we discuss the evolution equation of the intensity measures associated with the spatial branching model presented in Section 6.1.1.

Definition 6.1.1 *We denote by $\mathcal{X}_n$ the occupation measure of the branching particle model*

$$\mathcal{X}_n := \sum_{i=1}^{N_n} \delta_{X_n^i}$$

The intensity measure γ_n associated with this point process is given for any bounded measurable function f_n on $E_{n,c} = E_n \cup \{c\}$ by the following formula:

$$\gamma_n(f_n) := \mathbb{E}(\mathcal{X}_n(f)) \quad \text{with} \quad \mathcal{X}_n(f_n) := \int \mathcal{X}_n(dx_n) \, f_n(x_n)$$

To simplify the presentation, we suppose that the initial configuration of the particles is a spatial Poisson process with intensity measure μ_0 on the state space E_0.

Given the construction defined in Section 6.1.1, it follows almost straightforwardly that the intensity measures γ_n on E_n satisfy the following recursive equation.

Lemma 6.1.2 *For any $n \geq 0$, we have*

$$\gamma_{n+1} = \gamma_n Q_{n+1} + \mu_{n+1} \tag{6.3}$$

with the initial condition $\gamma_0 = \mu_0$. In the above displayed formulae, μ_{n+1} is the intensity measure of the spatial point process associated to the birth of new individuals at time $n+1$, while Q_{n+1} is the bounded and positive integral operator from E_n into E_{n+1}

$$Q_{n+1}(x_n, dx_{n+1}) := G_n(x_n)\, M_{n+1}(x_n, dx_{n+1}) \tag{6.4}$$

Proof:

For any bounded measurable function f on $E_{n+1} \cup \{c\}$, we have

$$\gamma_{n+1}(f) = \mathbb{E}\left(\sum_{i=1}^{\widehat{N}_n} f\left(X_{n+1}^i\right)\right) + \mathbb{E}\left(\sum_{i=\widehat{N}_n+1}^{\widehat{N}_n+N'_{n+1}} f\left(X_{n+1}^i\right)\right)$$

Thanks to the Poisson assumption, we have

$$\mathbb{E}\left(\sum_{i=\widehat{N}_n+1}^{\widehat{N}_n+N'_{n+1}} f\left(X_{n+1}^i\right)\right) = \mu_{n+1}(1)\,\overline{\mu}_{n+1}(f) = \mu_{n+1}(f)$$

with the normalized measures $\overline{\mu}_{n+1}$ defined in (6.1).

We let $\widehat{\mathcal{F}}_n$ be the σ-field generated by $(\widehat{X}_n^i)_{1\le i\le \widehat{N}_n}$ and $\mathcal{F}_n$ the σ-field generated by $(X_n^i)_{1\le i\le N_n}$. In this notation, we have that

$$\mathbb{E}\left(\sum_{i=1}^{\widehat{N}_n} f\left(X_{n+1}^i\right)\right) = \mathbb{E}\left(\mathbb{E}\left[\sum_{i=1}^{\widehat{N}_n} f\left(X_{n+1}^i\right)\,\bigg|\,\widehat{\mathcal{F}}_n\right]\right)$$

$$= \mathbb{E}\left(\mathbb{E}\left[\sum_{i=1}^{\widehat{N}_n} M_{n+1}(f)\left(\widehat{X}_n^i\right)\,\bigg|\,\mathcal{F}_n\right]\right)$$

$$= \mathbb{E}\left(\sum_{i=1}^{N_n} e_n\left(X_n^i\right) H_n(X_n^i) M_{n+1}(f)\left(X_n^i\right)\right)$$

from which we conclude that

$$\mathbb{E}\left(\sum_{i=1}^{\widehat{N}_n} f\left(X_{n+1}^i\right)\right) = \gamma_n\left(e_n H_n M_{n+1}(f)\right) = \gamma_n\left(G_n M_{n+1}(f)\right)$$

and therefore

$$\gamma_{n+1}(f) = \gamma_n\left(Q_{n+1}(f)\right) + \mu_{n+1}(f)$$

This ends the proof of the lemma. ∎

The flow of intensity measures γ_n is clearly more complex that the Feynman-Kac distribution flows (1.41) discussed in Section 1.4.2.1. Thus, we typically do not expect to find any closed-form expression to solve these equations. A natural way to approximate them numerically is to use a mean field particle interpretation of the associated sequence of normalized probability distributions.

Definition 6.1.3 *The normalized probability distributions associated with the intensity distributions γ_n are the probability measures $\eta_n \in \mathcal{P}(E_n)$ defined for any $f_n \in \mathcal{B}_b(E_n)$ by*

$$\eta_n(f_n) := \gamma_n(f_n)/\gamma_n(1)$$

We end this section, with a couple of remarks. When $\mu_n = 0$, the distributions (γ_n, η_n) coincide with the Feynman-Kac models (1.39) discussed in Section 1.4.2.1 and further developed in Chapter 3 and Chapter 12. On the other hand, we notice that

$$\mathbb{E}\,(N_n) = \gamma_n(1) \quad \text{and} \quad \eta_n(f_n) = \frac{1}{\mathbb{E}\,(N_n)}\,\mathbb{E}\left(\sum_{i=1}^{N_n} f_n(X_n^i)\right)$$

6.1.3 Nonlinear evolution equations

In this section, we discuss the evolution equations of the normalized probability measures η_n, introduced in Definition 6.1.3. The stability properties and the Lipschitz type regularity properties of these equations are discussed in some detail in Chapter 13.

In contrast with conventional Feynman-Kac Equations (1.40), these extended models are expressed in terms of updating-prediction transitions that depend on the total mass $\gamma_n(1)$ of the intensity measures γ_n.

To describe with some conciseness these models, we need another round of notation. In subsequent pages of this section, we identify the measures γ_n with a couple of parameters $(\gamma_n(1), \eta_n)$. The first component $\gamma_n(1)$ represents the total mass of γ_n, and η_n the normalized probability measure.

Definition 6.1.4 *We consider the collection of Markov transitions $M_{n+1,(m,\eta)}$ indexed by the parameters $m \in \mathbb{R}_+$ and the probability measures $\eta \in \mathcal{P}(E_n)$ defined by*

$$M_{n+1,(m,\eta)}(x, dy) := \alpha_n\,(m, \eta)\,M_{n+1}(x, dy) + (1 - \alpha_n\,(m, \eta))\,\overline{\mu}_{n+1}(dy) \quad (6.5)$$

with the collection of $[0,1]$-valued functions

$$\alpha_n \; : \; (m, \eta) \in (\mathbb{R}_+ \times \mathcal{P}(E_n)) \; \mapsto \; \alpha_n\,(m, \eta) = \frac{m\eta(G_n)}{m\eta(G_n) + \mu_{n+1}(1)}$$

Definition 6.1.5 *We let Λ_{n+1} be the mapping from $\mathbb{R}_+ \times \mathcal{P}(E_n)$ into $\mathbb{R}_+ \times \mathcal{P}(E_{n+1})$ given by*

$$\Lambda_{n+1}(m, \eta) = \left(\Lambda_{n+1}^1(m, \eta), \Lambda_{n+1}^2(m, \eta)\right) \quad (6.6)$$

with the pair of transformations:

$$\Lambda_{n+1}^{1}(m, \eta) = m\,\eta(G_n) + \mu_{n+1}(1)$$
$$\Lambda_{n+1}^{2}(m, \eta) = \Psi_{G_n}(\eta)M_{n+1,(m,\eta)}$$

In the above display, Ψ_{G_n} stands for the Boltzmann-Gibbs transformation associated with a potential function G_n, defined in (0.2).

The semigroup of the flow γ_n, or equivalently $(\gamma_n(1), \eta_n)$, is now expressed in terms of the mathematical objects defined above.

Proposition 6.1.6 *For any $n \geq 0$, we have the evolution equations*

$$(\gamma_n(1), \eta_n) = \Lambda_n(\gamma_{n-1}(1), \eta_{n-1}) \tag{6.7}$$

Proof:
Observe that for any function $f \in \mathcal{B}(E_{n+1})$, we have that

$$\eta_{n+1}(f) = \frac{\gamma_n(G_n M_{n+1}(f)) + \mu_{n+1}(f)}{\gamma_n(G_n) + \mu_{n+1}(1)} = \frac{\gamma_n(1)\,\eta_n(G_n M_{n+1}(f)) + \mu_{n+1}(f)}{\gamma_n(1)\,\eta_n(G_n) + \mu_{n+1}(1)}$$

from which we find that

$$\eta_{n+1} = \alpha_n\,(\gamma_n(1), \eta_n)\,\Psi_{G_n}(\eta_n)M_{n+1} + (1 - \alpha_n\,(\gamma_n(1), \eta_n))\,\overline{\mu}_{n+1}$$

From these observations, we prove (6.7). This ends the proof of the proposition. ■

6.1.4   McKean interpretations

In this section, we design a McKean interpretation of the measure valued process $(\gamma_n(1), \eta_n) \in (\mathbb{R}_+ \times \mathcal{P}(E_n))$ introduced in Section 6.1.3.

In Proposition 6.1.6, we have shown that the evolution Equation (6.7) of the sequence of probability measures $\eta_n \rightsquigarrow \eta_{n+1}$ is a combination of an updating type transition $\eta_n \rightsquigarrow \Psi_{G_n}(\eta_n)$, and an integral transformation w.r.t. a Markov transition $M_{n+1,(\gamma_n(1),\eta_n)}$ that depends on the current total mass $\gamma_n(1)$, as well as on the current probability distribution η_n.

The integral operator $M_{n+1,(\gamma_n(1),\eta_n)}$ defined in (6.5) is a mixture of the Markov transition M_{n+1} and the spontaneous birth normalized measure $\overline{\mu}_{n+1}$. Notice that for null spontaneous birth measures, this Markov transition reduces to the one of the free exploration of the particles; that is, we have that

$$\mu_{n+1} = 0 \implies M_{n+1,(\gamma_n(1),\eta_n)} = M_{n+1}$$

We let S_{n,η_n} be any Markov transition from E_n into itself satisfying the following compatibility condition

$$\Psi_{G_n}(\eta_n) = \eta_n S_{n,\eta_n}$$

Several examples of transitions S_{n,η_n} are discussed in (0.3) (cf. also Section 4.4.5).

By construction, we have the recursive formula

$$\eta_{n+1} = \eta_n K_{n+1,(\gamma_n(1),\eta_n)} \tag{6.8}$$

with the Markov transitions

$$K_{n+1,(\gamma_n(1),\eta_n)} = S_{n,\eta_n} M_{n+1,(\gamma_n(1),\eta_n)}$$

Using the same arguments as the ones we used in Section 4.4.2, the sequence of probability distributions η_n can be interpreted as the distributions of the random states $\overline{X}_n$ of a process defined, conditional upon $(\gamma_n(1), \eta_n)$, by the elementary transitions

$$\mathbb{P}\left(\overline{X}_{n+1} \in dx \mid \overline{X}_n\right) = K_{n,(\gamma_n(1),\eta_n)}\left(\overline{X}_n, dx\right) \quad \text{with} \quad \eta_n = \mathrm{Law}(\overline{X}_n)$$

and the auxiliary total mass evolution equation

$$\gamma_{n+1}(1) = \gamma_n(1)\, \eta_n(G_n) + \mu_{n+1}(1) \tag{6.9}$$

The transport formula presented in (6.8) provides a natural interpretation of the probability distributions η_n as the laws of a process $\overline{X}_n$ whose elementary transitions $\overline{X}_n \rightsquigarrow \overline{X}_{n+1}$ depend on the distribution $\eta_n = \mathrm{Law}(\overline{X}_n)$ as well as on the current mass $\gamma_n(1)$.

In contrast to the more traditional McKean type nonlinear Markov chains discussed in Section 1.2 and in Section 4.4.2, the dependency on the mass process induces a dependency on the whole sequence of measures η_p, from the origin $p = 0$ up to the current time $p = n$.

6.1.5 Mean field particle interpretation

In this section, we design a mean field interpretation of the McKean models developed in Section 6.1.4. The performance analysis of these mean field models is discussed in some detail in Section 14.7.

From now on, we will always assume that the mappings

$$\left(m, \left(x^i\right)_{1 \leq i \leq N}\right) \in \left(\mathbb{R}_+ \times E_n^N\right) \mapsto K_{n+1,\left(m, \frac{1}{N}\sum_{j=1}^N \delta_{x^j}\right)}\left(x^i, A_{n+1}\right)$$

are measurable w.r.t. the product σ-fields on $(\mathbb{R}_+ \times E_n^N)$, for any $n \geq 0$, $N \geq 1$, and $1 \leq i \leq N$, and any measurable subset $A_{n+1} \subset E_{n+1}$.

In this situation, the mean field particle interpretation of (6.8) and (6.9) is the Markov chain

$$\left(\gamma_n^N(1), \xi_n\right) \in \left(\mathbb{R}_+ \times E_n^N\right) \quad \text{with} \quad \xi_n = \left(\xi_n^i\right)_{1 \leq i \leq N} \in E_n^N$$

and with elementary transitions

$$
\begin{cases}
\mathbb{P}\left(\xi_{n+1} \in dx \mid (\gamma_n^N(1), \xi_n)\right) = \prod_{i=1}^{N} K_{n+1,(\gamma_n^N(1),\eta_n^N)}(\xi_n^i, dx^i) \\[2mm]
\gamma_{n+1}^N(1) = \gamma_n^N(1)\,\eta_n^N(G_n) + \mu_{n+1}(1)
\end{cases}
$$

$$(6.10)$$

with the infinitesimal neighborhood $dx = dx^1 \times \ldots \times dx^N$ of a point $x = (x^1, \ldots, x^N) \in E_{n+1}^N$. In the above displayed formula, (γ_n^N, η_n^N) stand for the couple of occupation measures defined for any $f_n \in \mathcal{B}_b(E_n)$ by

$$
\eta_n^N := \frac{1}{N} \sum_{i=1}^{N} \delta_{\xi_n^i} \quad \text{and} \quad \gamma_n^N(f_n) := \gamma_n^N(1)\,\eta_n^N(f_n)
$$

The initial system ξ_0 consists of N independent and identically distributed random variables with common law η_0, and we assume that the initial mass $\gamma_0^N(1) = \gamma_0(1) = \mu_0(1)$ is explicitly known. In this connection, we mention that the particle total mass model is also given by the following formula

$$
\gamma_n^N(1) = \sum_{p=0}^{n} \mu_p(1) \prod_{p \leq q < n} \eta_q^N(G_q)
$$

By definition of the two step McKean transitions (6.8), the mean particle evolution described by (6.10) is a "simple" combination of a selection and a mutation genetic type transition

$$
\xi_n \in E_n^N \quad \leadsto \quad \widehat{\xi}_n = (\widehat{\xi}_n^i)_{1 \leq i \leq N} \in E_n^N \quad \leadsto \quad \xi_{n+1} \in E_{n+1}^N
$$

During the selection transitions $\xi_n \leadsto \widehat{\xi}_n$, each particle $\xi_n^i \leadsto \widehat{\xi}_n^i$ evolves according to the selection type transition $S_{n,\eta_n^N}(\xi_n^i, dx)$. During the mutation stage, each of the selected particles $\widehat{\xi}_n^i \leadsto \xi_{n+1}^i$ evolves according to the transition

$$
M_{n+1,(\gamma_n^N(1),\eta_n^N)}(x, dy)
$$

$$
:= \alpha_n\left(\gamma_n^N(1), \eta_n^N\right) M_{n+1}(x, dy) + \left(1 - \alpha_n\left(\gamma_n^N(1), \eta_n^N\right)\right) \overline{\mu}_{n+1}(dy)
$$

6.2   Nonlinear equations of positive measures

This section is devoted to the applications of mean field simulation theory to the analysis of the nonlinear evolution equations presented in Section 1.18. This general class of nonlinear equations in the space of positive measures encapsulates all the discrete generation measure valued models we have discussed in earlier chapters.

In the first section, Section 6.2.1, we present a general, and abstract, model involving the time evolution of total mass process and the normalized distribution flows. We mention that these evolution models are defined as the Feynman-Kac flows (3.28), by replacing the integral operator Q_n, by some collection of integral operators Q_{n,γ_n} that may depend on the current intensity measure γ_n. The mean field IPS interpretation of these evolution equations is discussed in Section 6.2.2.

The applications of these abstract models to multiple objects filtering problems will be discussed in Section 6.3.

6.2.1 Measure valued evolution equations

We consider a general class of measure-valued processes $\gamma_n \in \mathcal{M}_+(E_n)$ defined by the following nonlinear equations

$$\gamma_n = \Xi_n(\gamma_{n-1}) := \gamma_{n-1} Q_{n,\gamma_{n-1}} \tag{6.11}$$

with initial measure $\gamma_0 \in \mathcal{M}_+(E_0)$. In the above display, $Q_{n,\gamma}$ stands for a collection of positive and bounded integral operators from E_{n-1} into E_n, indexed by the time parameter $n \geq 1$, and the set of measures $\gamma \in \mathcal{M}_+(E_{n-1})$.

One natural way to solve the nonlinear integral Equation (6.11) is to use a judicious probabilistic interpretation of the normalized distributions flow given for any $f_n \in \mathcal{B}_b(E_n)$ by

$$\eta_n(f_n) := \gamma_n(f_n)/\gamma_n(1)$$

To describe with some conciseness these stochastic models, it is important to observe that the pair process $(\gamma_n(1), \eta_n) \in (\mathbb{R}_+ \times \mathcal{P}(E_n))$ satisfies an evolution equation of the following form

$$(\gamma_n(1), \eta_n) = \Lambda_n(\gamma_{n-1}(1), \eta_{n-1}) \tag{6.12}$$

for some mapping

$$\Lambda_n \; : \; (m, \eta) \in (\mathbb{R}_+ \times \mathcal{P}(E_{n-1})) \mapsto \Lambda_n(m, \eta) \in (\mathbb{R}_+ \times \mathcal{P}(E_n))$$

We also denote by $(\Lambda_n^1, \Lambda_n^2)$, the first, and second, component mappings of the one step transformation Λ_n given by

$$\Lambda_n^1 \; : \; (\mathbb{R}_+ \times \mathcal{P}(E_n)) \; \to \; \mathbb{R}_+ \quad \text{and} \quad \Lambda_n^2 \; : \; (\mathbb{R}_+ \times \mathcal{P}(E_n)) \; \to \; \mathcal{P}(E_n)$$

By construction, we notice that the total mass process can be computed using the recursive formula

$$\gamma_{n+1}(1) = \gamma_n(G_{n,\gamma_n}) = \eta_n(G_{n,\gamma_n})\,\gamma_n(1) \quad \text{with} \quad G_{n,\gamma_n} := Q_{n+1,\gamma_n}(1) \tag{6.13}$$

On the other hand, for any $f \in \mathcal{B}_b(E_{n+1})$ we have that

$$\eta_{n+1} = \Psi_{G_{n,\gamma_n}}(\eta_n)\,M_{n+1,\gamma_n} \quad \text{with} \quad M_{n+1,\gamma_n}(f) := \frac{Q_{n+1,\gamma_n}(f)}{Q_{n+1,\gamma_n}(1)}$$

and the Boltzmann-Gibbs transformation Ψ_G associated with the potential function $G = G_{n,\gamma_n}$ and defined in (0.2). This implies that

$$\Lambda_{n+1}^1(m,\eta) = m\ \eta(G_{n,m\eta}) \quad \text{and} \quad \Lambda_{n+1}^2(m,\eta) = \Psi_{G_{n,m\eta}}(\eta_n)\ M_{n+1,m\eta} \tag{6.14}$$

We end this section with some comments on the applications and the stability analysis of these rather abstract models. We also provide some reference pointers to the sections of the book discussing in more detail these questions.

Illustrations in the context of multiple target tracking problems are presented in Section 6.3.2.

The stability properties and the Lipschitz type regularity of these models are developed in Section 13.2. In Section 13.3.2, we also present some regularity conditions, under which the nonlinear semigroups forget exponentially facts of the initial conditions. Illustrations of these stability results in the context of PHD equations are provided in Section 13.4. We also use this property in Section 14.8.3 to prove uniform estimates of mean field approximation schemes w.r.t. the time parameter.

6.2.2   Mean field particle models

The mean field particle model associated with the Equation (6.12) relies on the fact that the one step mappings Λ_{n+1}^2 given in (6.14) can be rewritten in terms of the nonlinear Markov transport equations

$$\Psi_{G_{n,\gamma}}(\eta)\ M_{n,\gamma} = \eta K_{n+1,\gamma} \quad \text{with} \quad K_{n+1,\gamma} := S_{n,\gamma} M_{n+1,\gamma} \tag{6.15}$$

for any $\gamma = m\eta \in \mathcal{M}_+(E_n)$. In the above displayed formula, $S_{n,\gamma}$ stands for any collection of Markov transitions, from E_n into itself, and index by $\gamma \in \mathcal{M}_+(E_n)$, satisfying the following compatibility condition

$$\Psi_{G_{n,\gamma}}(\eta) = \eta S_{n,\gamma}$$

Several examples of transitions $S_{n,m\eta}$ are discussed in (0.3).

These models provide a natural interpretation of the distribution laws η_n as the laws of a nonlinear Markov chain $\overline{X}_n$ whose elementary transitions $\overline{X}_n \rightsquigarrow \overline{X}_{n+1}$ depend on the distribution $\eta_n = \text{Law}(\overline{X}_n)$, as well as on the current mass process $\gamma_n(1)$. In contrast to the traditional McKean model, the dependency on the mass process induces a dependency of all the flow of measures η_p, for $0 \leq p \leq n$.

The mean field particle interpretation of this nonlinear measure-valued model is the Markov chain

$$(\gamma_n^N(1), \xi_n) \in (\mathbb{R}_+ \times E_n^N) \quad \text{with} \quad \xi_n = (\xi_n^i)_{1 \leq i \leq N} \in E_n^N$$

with elementary transitions

$$\mathbb{P}\left(\xi_{n+1} \in dx \mid (\gamma_n^N(1), \xi_n)\right) = \prod_{i=1}^{N} K_{n+1,\gamma_n^N}(\xi_n^i, dx^i) \qquad (6.16)$$

$$\gamma_{n+1}^N(1) = \gamma_n^N(1) \, \eta_n^N(G_{n,\gamma_n^N}) \qquad (6.17)$$

with the infinitesimal neighborhood $dx = dx^1 \times \ldots \times dx^N$ of a point $x = (x^1, \ldots, x^N) \in E_{n+1}^N$. In the above display, (γ_n^N, η_n^N) stands for the pair of measures defined for any $f_n \in \mathcal{B}_b(E_n)$ by

$$\eta_n^N := \frac{1}{N} \sum_{j=1}^{N} \delta_{\xi_n^j} \quad \text{and} \quad \gamma_n^N(f_n) := \gamma_n^N(1) \times \eta_n^N(f_n)$$

The initial system ξ_0 consists of N independent and identically distributed random variables with common law η_0. We also assume that $\gamma_0^N(1) = \gamma_0(1) = \mu_0(1)$ is explicitly known.

The performance analysis of these mean field models is discussed in Section 14.8.

6.3   Multiple-object nonlinear filtering equations

In this section we discuss in more detail the multiple-objects nonlinear filtering problems presented in Section 2.4.2. These nonlinear evolution models in distribution spaces are particular examples of the measure valued evolution equations discussed in Section 6.2.1.

6.3.1   A partially observed branching model

6.3.1.1   A signal branching model

Suppose that at a given time n there are N_n^X targets $(X_n^i)_{1 \leq i \leq N_n^X}$, each taking values in some measurable state space E_n.

A target X_n^i, at time n, survives to the next time step with probability $s_n(X_n^i) \in [0,1]$, and it evolves to a new random state according to a given elementary Markov transition M_{n+1}', from E_n into E_{n+1}.

In addition, any target X_n^i can spawn new targets at the next time, usually modeled by a spatial Poisson process with a given intensity measure $B_{n+1}(X_n^i, \cdot)$, on the state space E_{n+1}. At the same time, an independent collection of new targets is added to the scene. This additional and spontaneous branching process is often modeled by a spatial Poisson process with a prescribed intensity measure μ_{n+1} on E_{n+1}.

For any $n \geq 0$, and any $x_n \in E_n$, we set

$$\overline{\mu}_n(dx_n) \;=\; \mu_n(dx_n)/\mu_n(1)$$

$$b_n(x_n) \;=\; B_{n+1}(1)(x_n) \quad \text{and} \quad \overline{B}_{n+1}(x_n, dx_{n+1}) = \frac{B_{n+1}(x_n, dx_{n+1})}{B_{n+1}(1)(x_n)}$$

The signal process can be interpreted as a branching process of the same form as the one discussed in Section 6.1.1. The occupation measures of the branching process are given by

$$\mathcal{X}_n := \sum_{1 \leq i \leq N_n^X} \delta_{X_n^i}$$

Using the same arguments as in the proof of Lemma 6.1.2, for any $f_n \in \mathcal{B}_b(E_n)$ we prove that

$$\mathbb{E}\left(\mathcal{X}_n(f_n) \mid \mathcal{X}_{n-1}\right) = \mathcal{X}_{n-1} Q_n(f_n) + \mu_n(f_n)$$

with the nonnegative integral operator

$$Q_n(f_n) = s_{n-1}\, M_n'(f_n) + b_{n-1}\, \overline{B}_n(f_n)$$

We notice that the operators Q_{n+1} can be rewritten in terms of the Feynman-Kac integral operator defined in (3.27). More precisely, we have

$$Q_{n+1}(x_n, dx_{n+1}) = G_n(x_n)\, M_{n+1}(x_n, dx_{n+1}) \qquad (6.18)$$

with the potential function

$$G_n(x_n) = s_n(x_n) + b_n(x_n)$$

and the Markov transition

$$M_{n+1}(x_n, dx_{n+1})$$

$$:= \frac{s_n(x_n)}{s_n(x_n) + b_n(x_n)}\, M_{n+1}'(x_n, dx_{n+1}) + \frac{b_n(x_n)}{s_n(x_n) + b_n(x_n)}\, \overline{B}_{n+1}(x_n, dx_{n+1})$$

This shows that we can use the mean field particle techniques developed in Section 6.1 to compute the intensity distribution flows associated with the signal branching process.

6.3.1.2   A partial observation model

Given a realization of the branching process $\mathcal{X}_n$ defined in Section 6.3.1.1, with a probability $d_n(x)$ every random target $X_n^i = x$ generates an observation Y_n^i, on some possibly different state space E_n^Y, with distribution $L_n(x, dy)$, where $L_n(x, dy)$ stands for some Markov transition from E_n into E_n^Y. Otherwise, with a probability $(1 - d_n(x))$, the target disappears from the scene, and goes into an auxiliary cemetery or coffin state c. The $[0,1]$-valued function d_n is called the detection probability of the targets.

More formally, a given state x generates a random observation in the augmented state space $E_{n,c}^Y := E_n^Y \cup \{c\}$, with distribution

$$L_{n,c}(x, dy) := d_n(x) \, L_n(x, dy) + (1 - d_n(x)) \, \delta_c(dy) \qquad (6.19)$$

The resulting observation point process is the random measure

$$\mathcal{Y}_n = \sum_{1 \le i \le N_{n,c}^Y} \delta_{Y_n^i}$$

with $N_{n,c}^Y = N_n^X$, on the augmented state space $E_{n,c}^Y$.

In addition to this partial observation process, we also observe an additional, and independent of $(\mathcal{X}_p)_{p \le n}$, Poisson point process

$$\mathcal{Y}_n' := \sum_{1 \le i \le N_n'} \delta_{Y_n'^i}$$

with intensity measure ν_n on E_n^Y.

We further assume that $\nu_n \ll \lambda_n$ and $L_n(x, .) \ll \lambda_n$, for any $x \in E_n$, for some reference measure $\lambda_n \in \mathcal{M}(E_n^Y)$. We also assume that the Radon-Nikodym derivatives given by

$$g_n(x, y) = \frac{dL_n(x, .)}{d\lambda_n}(y) \quad \text{and} \quad h_n(y) := \frac{d\nu_n}{d\lambda_n}(y) \qquad (6.20)$$

are such that

$$h_n(y) + d_n(x) g_n(x, y) > 0$$

for any $(x, y) \in (E_n \times E_n^Y)$.

The full observation process on $E_{n,c}^Y$ is now given by the random measure

$$\mathcal{Y}_n'' = \mathcal{Y}_n + \mathcal{Y}_n'$$

The coffin state c being unobservable, the "real world" observation point process $\mathcal{Y}_n^o$ is the random measure on the state space E_n^Y, given by the trace $(\mathcal{Y}_n'')_{|E_n^Y}$ of the measure $\mathcal{Y}_n''$ on the set E_n^Y. More precisely, the observed random measure is given by the following formula

$$\mathcal{Y}_n^o := (\mathcal{Y}_n'')_{|E_n^Y} = (\mathcal{Y}_n)_{|E_n^Y} + \mathcal{Y}_n' \quad \text{with} \quad (\mathcal{Y}_n)_{|E_n^Y} := \sum_{1 \le i \le N_{n,c}^Y} 1_{E_n^Y}(Y_n^i) \, \delta_{Y_n^i}$$

$$(6.21)$$

Multi-target tracking problems are concerned with the sequential estimation of the random measures

$$\mathcal{X}_n = \sum_{1 \le i \le N_n^X} \delta_{X_n^i}$$

given the noisy and partial observation occupation measures

$$\mathcal{Y}_p^o = \sum_{1 \le i \le N_p^Y} \delta_{Y_p^i} \quad \text{with} \quad 0 \le p \le n$$

6.3.2 Probability hypothesis density equations

From the pure probabilistic viewpoint, multi-target tracking problems consist in estimating the conditional distributions of the occupation measures of spatial branching processes, given some noisy and partial observation random fields. This filtering problem is clearly much more complex than the single target filtering model discussed in Section 2.5.1.

Besides the fact that the underlying signal is a well defined and an easy to sample Markov chain model $\mathcal{X}_n$, the computation of the likelihood of the observation process $\mathcal{Y}_n^o$, given the random state $\mathcal{X}_n$, involves intractable combinatorial calculations. In this section, we present a Poisson approximation model that simplifies drastically the analysis.

The equations associated with these approximated filters are expressed in terms of nonlinear evolution equations of intensity measures, of the same type as the one discussed in Section 6.2. For a single target filtering problem, these equations reduce to the traditional single target optimal filter equations presented in Section 2.5.1.

We emphasize that the multiple-object filtering equations developed in this section are not optimal, in the sense that they only represent an "approximation" of the conditional distributions. The connections between these Poisson approximation models and the optimal filter are still an important, but difficult open research question.

The stability properties of the PHD nonlinear semigroups are developed in Section 13.4.

6.3.2.1 Poisson approximation models

Further on in this section, we assume that the initial random measure $\mathcal{X}_0$ is a Poisson point process, with intensity measure $\gamma_0 = \mu_0 \in \mathcal{M}_+(E_0)$, on the initial state space E_0. We also consider the sequence of integral operators Q_n defined in (6.18).

Given a realization of $\mathcal{X}_0$, the corresponding observation process $\mathcal{Y}_0^o$ on E_0^Y is defined as in (6.21) with some detection functions d_0 on E_0, some clutter intensity measures ν_0, and some Markov transitions $(L_{c,0}, L_0)$ defined as in (6.19). We also assume that the regularity condition (6.20) is satisfied, for some reference measures λ_0 and some clutter intensity functions h_0.

For any function $f \in \mathcal{B}(E_0)$, we have the

$$
\begin{aligned}
\widehat{\gamma}_0(f) &:= \mathbb{E}\left(\mathcal{X}_0(f) \mid \mathcal{Y}_0^o\right) \\
&= \gamma_0((1-\alpha_0)f) + \int \mathcal{Y}_0^o(dy)\,(1 - \beta_{0,\gamma_0}(y))\ \Psi_{\alpha_0 g_0(y,.)}(\gamma_0)(f)
\end{aligned}
$$

with the $[0,1]$-valued function β_γ on E_2 defined by

$$
\beta_{0,\gamma_0}(y) := h_0(y)/\left[h_0(y) + \gamma_0(d_0 g_0(.,y))\right]
$$

A detailed proof of this conditional formula relies on Poisson point process conditioning principles, and it is provided in Section 6.3.3.3 (cf. Corollary 6.3.12).

Suppose we have defined the measure valued process $(\widehat{\gamma}_p, \gamma_p)$ and the random signal-observation process $(\mathcal{X}_p, \mathcal{Y}_p^o)$, from the origin $p = 0$, up to a given time horizon $p = n$.

Given these values, we define the pair of random measures $(\mathcal{X}_{n+1}, \mathcal{Y}_{n+1}^o)$ as follows:

- Firstly, we let $\mathcal{X}_{n+1}$ be a spatial Poisson point process with intensity measure γ_{n+1} defined by the following recursions

$$\gamma_{n+1} \ := \ \widehat{\gamma}_n Q_n + \mu_n \tag{6.22}$$

$$\widehat{\gamma}_n(f) \ := \ \gamma_n((1 - d_n)f) + \int \mathcal{Y}_n^o(dy)\,(1 - \beta_{\gamma_n}(y))\ \Psi_{d_n g_n(y,.)}(\gamma_n)(f)$$

for any function $f \in \mathcal{B}(E_n)$, with the $[0, 1]$-valued parameters

$$\beta_{n,\gamma_n}(y) := \frac{h_n(y)}{[h_n(y) + \gamma_n(d_n g_n(\cdot, y))]}$$

- Given a realization of $\mathcal{X}_{n+1}$, the corresponding observation process $\mathcal{Y}_{n+1}^o$ is defined as in (6.21), for some detection $[0, 1]$-valued functions d_{n+1} on E_{n+1}, some clutter intensity measures ν_{n+1}, and some Markov transitions $(L_{c,(n+1)}, L_{n+1})$ defined as in (6.19); and satisfying (6.20), for some reference measures λ_{n+1} and some functions h_{n+1}.

We let $\mathcal{F}_n^Y = \sigma\left(\mathcal{Y}_p^o, 0 \leq p \leq n\right)$ be the filtration generated by the observation point processes $\mathcal{Y}_p^o$, from the origin $p = 0$, up to the current time $p = n$. By construction, for any function $f \in \mathcal{B}(E_{n+1})$, we clearly have that

$$\mathbb{E}\left(\mathcal{X}_{n+1}(f) \mid \mathcal{F}_n^Y\right) = \gamma_{n+1}(f)$$

In addition, using the same arguments as the ones we used at the initial time $n = 0$, we have the updating formulae

$$\widehat{\gamma}_{n+1}(f)$$

$$:= \mathbb{E}\left(\mathcal{X}_{n+1}(f) \mid \mathcal{F}_{n+1}^Y\right)$$

$$= \gamma_{n+1}((1 - d_{n+1})f) + \int \mathcal{Y}_{n+1}^o(dy)\,(1 - \beta_{\gamma_{n+1}}(y))\ \Psi_{d_{n+1} g_{n+1}(y,.)}(\gamma_{n+1})(f)$$

In summary, we have proved that the solution of the (PHD) Equations (6.22) coincides at any time step, with the desired conditional distributions

$$\widehat{\gamma}_n(f) = \mathbb{E}\left(\mathcal{X}_n(f) \mid \mathcal{F}_n^Y\right) \quad \text{and} \quad \mathbb{E}\left(\mathcal{X}_n(f) \mid \mathcal{F}_{n-1}^Y\right) = \gamma_n(f)$$

We end this section with a more synthetic description of the PHD equations. More precisely, using the decomposition (6.18), we can rewrite the PHD Equation (6.22) in terms of a nonlinear model of the form (6.11), combining in a single step the updating $\gamma_n \rightsquigarrow \widehat{\gamma}_n$ and the prediction transition $\widehat{\gamma}_n \rightsquigarrow \gamma_{n+1}$.

Proposition 6.3.1 *The PHD filter satisfies the integral Equation (6.11), with the integral operator given by*

$$Q_{n+1,\gamma_n}(x_n, dx_{n+1})$$

$$= g_{n,\gamma_n}(x_n) M_{n+1}(x_n, dx_{n+1}) + \gamma_n(1)^{-1} \mu_{n+1}(dx_{n+1}) \tag{6.23}$$

The likelihood function g_{n,γ_n} is given by

$$g_{n,\gamma_n} := r_n \times \widehat{g}_{n,\gamma_n} \quad with \quad r_n := (s_n + b_n) \tag{6.24}$$

and

$$\widehat{g}_{n,\gamma_n}(x_n)$$

$$:= (1 - d_n(x_n)) + d_n(x_n) \int \mathcal{Y}_n^o(dy) \; \frac{g_n(x_n, y_n)}{h_n(y_n) + \gamma_n(d_n g_n(\cdot, y_n))}$$

6.3.2.2　An updating-prediction formulation

In this section, we present a more traditional updating-prediction formulation of the PHD equations presented in (6.22). These updating-prediction models are often used in the literature of multiple target tracking.

We extend the observation state space E_n^Y by adding a virtual but cemetery type state $\{c\}$, and we consider the following likelihood functions on $E_{n,c}^Y = E_n^Y \cup \{c\}$

$$\widehat{g}_{n,\gamma}^c(\cdot, y) = \begin{cases} (1 - d_n) & \text{if} \quad y = c \\ \dfrac{d_n g_n(\cdot, y_n)}{h_n(y) + \gamma(d_n g_n(\cdot, y))} & \text{if} \quad y \neq c \end{cases}$$

In this interpretation, the state $y = c$ is considered as a virtual observable state, with a likelihood function $\widehat{g}_{n,\gamma}^c(x, c) = (1 - d_n(x))$ that measures the undetectability properties of the site x. The likelihood function is high in regions with low detectability conditions.

In this notation, we have

$$\gamma_{n+1} = \widehat{\gamma}_n Q_{n+1} + \mu_{n+1} \quad with \quad Q_{n+1}(f) := r_n \; M_{n+1}(f)$$

with the updated measures defined below

$$\widehat{\gamma}_n(f) := \gamma_n(\widehat{g}_{n,\gamma_n}^c f) \quad with \quad \widetilde{g}_{n,\gamma_n}^c = \int \mathcal{Y}_n^c(dy) \; \widehat{g}_{n,\gamma_n}^c(\cdot, y)$$

and $\mathcal{Y}_n^c = \mathcal{Y}_n^o + \delta_c$. Notice that

$$\widehat{\gamma}_n(1) = \gamma_n(\widetilde{g}_{n,\gamma_n}^c) \quad and \quad \widehat{\eta}_n(dx) := \widehat{\gamma}_n(dx)/\widehat{\gamma}_n(1) = \Psi_{\widetilde{g}_{\gamma_n,n}^c}(\eta_n)(dx)$$

from which we find the recursive formulae

$$\begin{pmatrix} \gamma_n(1) \\ \eta_n \end{pmatrix} \xrightarrow{\text{updating}} \begin{pmatrix} \widehat{\gamma}_n(1) \\ \widehat{\eta}_n \end{pmatrix} \xrightarrow{\text{prediction}} \begin{pmatrix} \gamma_{n+1}(1) \\ \eta_{n+1} \end{pmatrix}$$

The updating transition is defined by

$$\widehat{\gamma}_n(1) = \gamma_n(1) \; \eta_n(\widetilde{g}^c_{n,\gamma_n}) \quad \text{and} \quad \widehat{\eta}_n = \Psi_{\widetilde{g}^c_{\gamma_n,n}}(\eta_n)$$

It is instructive to observe that for fully detectable target models without clutter, we have

$$(d_n, h_n) = (1,0) \Rightarrow \widehat{g}^c_{n,\gamma_n}(x,y) = 1_{E^Y_n}(y) \; \frac{g_n(x,y_n)}{\gamma_n(g_n(\cdot,y))} \Rightarrow \widehat{\gamma}_n(1) = \gamma_n(1)$$

The prediction transitions are given by

$$\gamma_{n+1}(1) = \widehat{\gamma}_n(r_n) + \mu_{n+1}(1) \quad \text{and} \quad \eta_{n+1} = \Psi_{r_n}(\widehat{\eta}_n) \; M'_{n+1,\widehat{\gamma}_n}$$

In the above displayed formula, $M'_{n+1,\gamma}$ is the collection of Markov transitions defined by

$$M'_{n+1,\gamma}(x,\cdot) = \alpha'_n(\gamma) \; M_{n+1}(x,\cdot) + (1 - \alpha'_n(\gamma)) \; \overline{\mu}_{n+1}$$

with the collection of $[0,1]$-valued parameters

$$\alpha'_n(\gamma) = \gamma(r_n)/(\gamma(r_n) + \mu_{n+1}(1))$$

It is important to mention that the updating, as well as the prediction transitions, can be rewritten in terms of a measure valued equation of the same form as the one presented in Section 6.2.1. For instance, we can decompose the two step updating-prediction discussed above with intermediate time steps between integers

$$\gamma_n \quad \longrightarrow \quad \gamma_{n+1/2} := \widehat{\gamma}_n \quad \longrightarrow \quad \gamma_{n+1}$$

In this notation, we have a couple of one step transformations

$$\gamma_{n+1/2} = \Xi_{n+1/2}(\gamma_n) \quad \text{and} \quad \gamma_{n+1} = \Xi_{n+1}(\gamma_{n+1/2})$$

with the one step transformations, defined for any $\gamma \in \mathcal{M}_+(E_n)$ by

$$\begin{aligned}
\Xi_{n+1/2}(\gamma) &= \gamma(\widetilde{g}^c_{n,\gamma}) \; \Psi_{\widetilde{g}^c_{n,\gamma}}(\gamma) \\
\Xi_{n+1}(\gamma) &= [\gamma(r_n) + \mu_{n+1}(1)] \; \Psi_{r_n}(\gamma) \; M'_{n+1,\gamma}
\end{aligned}$$

Now, it should be clear that the updating and the prediction transitions can be approximated using mean field particle models presented in Section 6.2.2. Therefore, the performances of these updating-prediction mean field algorithms are direct consequences of the convergence analysis of the general particle model discussed in Section 6.2.2.

6.3.3 Spatial Poisson processes

In this short section, we recall some more or less well known results on spatial Poisson point processes, including restriction techniques and conditioning principles for partially observed models.

6.3.3.1   Preliminary results

We consider a measure $\gamma \in \mathcal{M}(E)$ on some measurable state space E s.t. $\gamma(1) > 0$. We let N be a integer valued Poisson random variable with parameter $\gamma(1)$. We also denote by $X = (X^i)_{i \geq 1}$ a sequence of independent and identically distributed random variables with common distribution $\eta(dx) := \gamma(dx)/\gamma(1)$. We assume that N and X are independent.

Definition 6.3.2 *The Poisson point process $\mathcal{X}$ with intensity measure γ is the random measure defined below*

$$\mathcal{X} := m_N(X) = \sum_{1 \leq i \leq N} \delta_{X^i} \in \mathcal{P}(E)$$

One of the main simplifications of Poisson point processes comes from the fact that their expectation measure coincides with their intensity measures:

$$\mathbb{E}\left(\mathcal{X}(f)\right) = \mathbb{E}\left(\mathbb{E}\left(\mathcal{X}(f) \mid N\right)\right) = \mathbb{E}\left(N\eta(f)\right) = \gamma(1)\eta(f) = \gamma(f)$$

Definition 6.3.3 *For every sequence of point $x = (x^i)_{i \geq 1}$ in E, any $A \in \mathcal{E}$, and every $p \geq 0$, we denote by $m_{p,A}(x)$ the restriction of the occupation measure $m_p(x)$ to the set A.*

$$m_{p,A}(x)(dy) = m_p(x)(dy)1_A(y) = \sum_{1 \leq i \leq p} 1_A(x^i)\delta_{x^i}(dy)$$

Notice that

$$m_p(x)(A) = \sum_{1 \leq i \leq p} 1_{A_i}(x^i) > 0 \implies m_{p,A}(x) = m_p(x)(A) \ \Psi_{1_A}(m(x))$$

with the Boltzmann-Gibbs transformation Ψ_G associated with the indicator function $G = 1_A$, defined in (3.8).

Lemma 6.3.4 *Let $(\mathcal{X}_j)_{j \geq 1}$ be a sequence of independent Poisson point processes with intensity measure $(\gamma_i)_{i \geq 1}$ on some common measurable state space E. For any $d \geq 1$, $\mathcal{X}$ is a Poisson point process with intensity measure $\sum_{1 \leq i \leq d} \gamma_i$ if, and only if, $\mathcal{X}$ is equal in law to the Poisson point process $\sum_{1 \leq i \leq d} \mathcal{X}_i$.*

Proof:
By symmetry arguments, we have for any $F \in \mathcal{B}(\mathcal{M}(E))$ and any $d \geq 1$

$$\mathbb{E}\left(F\left(\sum_{1 \leq i \leq d} \mathcal{X}_i\right)\right) =$$

$$e^{-\sum_{1 \leq i \leq d} \gamma_i(1)} \sum_{p_1, \ldots, p_d \geq 0} \frac{1}{p_1! \ldots p_d!} \int F\left(\sum_{1 \leq i \leq d} m_{p_i}(x_i)\right) \prod_{1 \leq i \leq d} \gamma_i^{\otimes p_i}(dx_i)$$

This implies that

$$
\mathbb{E}\left(F\left(\sum_{1\leq i\leq d} \mathcal{X}_i\right)\right) =
$$
$$
= e^{-\sum_{1\leq i\leq d}\gamma_i(1)} \sum_{s\geq 0} \frac{1}{s!}
$$

$$
\sum_{p_1+\ldots+p_d=s} \frac{s!}{p_1!\ldots p_d!} \int F\left(m_s(x)\right)\ \left(\gamma_1^{\otimes p_1}\otimes\ldots\otimes\gamma_d^{\otimes p_d}\right)(dx)
$$

In the above displayed integral $dx = dx^1\times\ldots\times dx^s$ stands for an infinitesimal neighborhood of the point $x = (x^i)_{1\leq i\leq s}$. This implies that

$$
\mathbb{E}\left(F\left(\sum_{1\leq i\leq d} \mathcal{X}_i\right)\right) = e^{-\sum_{1\leq i\leq d}\gamma_i(1)} \sum_{s\geq 0} \frac{1}{s!}\int F\left(m_s(x)\right)\ \left(\sum_{i=1}^d \gamma_i\right)^{\otimes s}(dx)
$$
$$
(6.25)
$$

This shows that $\sum_{1\leq i\leq d}\mathcal{X}_i$ is a Poisson point process with intensity measure $\sum_{1\leq i\leq d}\gamma_i$. In addition, by (6.25), any Poisson point process with such an intensity measure has the same law as $\sum_{1\leq i\leq d}\mathcal{X}_i$. ∎

The next result is a direct consequence of Lemma 6.3.4.

Lemma 6.3.5 *Let $\mathcal{X} := \sum_{1\leq i\leq N}\delta_{X^i}$ be a Poisson point process with intensity measure γ that is the random measure on E. We consider a measurable subset $A\subset E$, such that $\gamma(A) > 0$. Then, the restriction, or the trace, $\mathcal{X}_A = m_{N,A}(\mathcal{X})$ of $\mathcal{X}$ on the set A is again a Poisson point process with intensity measure $\gamma_A(dx) := 1_A(x)\gamma(dx)$.*

In addition, the conditional distribution of $\mathcal{X}$ given $\mathcal{X}_A$ is given for any $F\in\mathcal{B}(\mathcal{M}(E))$ by the following formula

$$
\mathbb{E}\left(F(\mathcal{X})\,|\mathcal{X}_A\right) = e^{-\gamma(A^c)} \sum_{p\geq 0} \frac{1}{p!}\int F\left(\mathcal{X}_A + m_p(x)\right)\ \gamma_{A^c}^{\otimes p}(dx)
$$

In the above, displayed integral $dx = dx^1\times\ldots\times dx^p$ stands for an infinitesimal neighborhood of the point $x = (x^i)_{1\leq i\leq p}$.

Proof:
Using the decomposition

$$
\gamma(dx) = 1_A(x)\gamma(dx) + 1_{A^c}(x)\gamma(dx)\ \Rightarrow\ \gamma = \gamma_A + \gamma_{A^c}
$$

for any $F\in\mathcal{B}(\mathcal{M}(E))$ we find that

$$
\mathbb{E}\left(F\left(\mathcal{X}_A\right)\right) = e^{-\gamma(1)} \sum_{s\geq 0} \frac{1}{s!}\int F\left(m_{s,A}(x)\right)\ \left(\gamma_A + \gamma_{A^c}\right)^{\otimes s}(dx)
$$

By symmetry arguments, this implies that

$$\mathbb{E}\left(F\left(\mathcal{X}_A\right)\right) \;=\; e^{-\gamma(1)} \sum_{s\geq 0} \frac{1}{s!} \sum_{p+q=s} \frac{s!}{p!q!} \int F\left(m_{s,A}(x)\right) \left[\gamma_A^{\otimes p} \otimes \gamma_{A^c}^{\otimes(s-p)}\right](dx)$$

from which we find that

$$\mathbb{E}\left(F\left(\mathcal{X}_A\right)\right) \;=\; e^{-\gamma(1)} \sum_{p\geq 0} \frac{1}{p!} \left(\sum_{s\geq p} \frac{\gamma(A^c)^{s-p}}{(s-p)!}\right) \int F\left(m_p(x)\right) \gamma_A^{\otimes p}(dx)$$

$$=\; e^{-(\gamma(E)-\gamma(A^c))} \sum_{p\geq 0} \frac{1}{p!} \int F\left(m_p(x)\right) \gamma_A^{\otimes p}(dx)$$

The last assertion is a direct consequence of Lemma 6.3.4, applied to $d=2$, replacing $(\mathcal{X}_1, \mathcal{X}_2)$ by $(\mathcal{X}_A, \mathcal{X}_{A^c})$. This ends the proof of the lemma. ∎

6.3.3.2 Some conditioning principles

We consider a measure $\gamma \in \mathcal{M}(E_1)$ on some measurable space $(E_1, \mathcal{E}_1)$ and a bounded positive integral operator Q from $(E_1, \mathcal{E}_1)$ into an auxiliary measurable space $(E_2, \mathcal{E}_2)$.

We further assume that $Q(1) > 0$, γ-a.e., and we let

$$\mathcal{X} := m_N(X_1, X_2) = \sum_{1\leq i\leq N} \delta_{(X_1^i, X_2^i)} \tag{6.26}$$

be the Poisson point process on some product space $(E_1 \times E_2, \mathcal{E}_1 \otimes \mathcal{E}_2)$ with intensity measure Γ of the following form

$$\Gamma(d(x_1, x_2)) := \gamma(dx_1)\, Q(x_1, dx_2)$$

It is immediate to check that the marginal random measures $\mathcal{X}_j := m_N(X_j)$ are Poisson point processes on E_j, $j = 1, 2$, with intensity measures

$$\gamma_1(dx) := Q(1)(x)\, \gamma(dx) \quad \text{and} \quad \gamma_2 := \gamma Q$$

Our next objective is to describe the conditional distributions of the random measures $\mathcal{X}_i$ w.r.t. $\mathcal{X}_j$, with $i \neq j$.

Lemma 6.3.6 *For any $f \in \mathcal{B}(E_1)$ we set $\gamma_f(dx) := f(x)\, \gamma(dx)$. The integral operators*

$$\overline{Q}(x_1, dx_2) = \frac{Q(x_1, dx_2)}{Q(x_1, E_2)} \quad \text{and} \quad f \in \mathcal{B}(E_1) \mapsto Q_\gamma(f) := \frac{d\gamma_f Q}{d\gamma Q}$$

are well defined Markov transitions from E_1 into E_2, resp. from E_2 into E_1.

In addition, if we have $Q(x_1, \cdot) \ll \gamma Q$, for every $x_1 \in E_1$, then we have the following explicit formula

$$Q_\gamma(x_2, dx_1) := \gamma(dx_1) \, \frac{dQ(x_1, \cdot)}{d\gamma Q}(x_2)$$

Proof:

The fact that $\overline{Q}$ is a well defined Markov transition is immediate. To check the second assertion, we use the fact that

$$\gamma_f(dx) := f(x) \, \gamma(dx) \ll \gamma(dx) \Longrightarrow \gamma_f Q \ll \gamma Q$$

for any $f \in \mathcal{B}(E_1)$. The r.h.s. assertion comes from the following series of implications

$$\gamma Q(A) = 0 \quad \Rightarrow \quad Q(1_A) = 0 \quad \gamma - \text{almost everywhere}$$
$$\Rightarrow \quad Q(1_A) = 0 \quad \gamma_f - \text{almost everywhere} \Rightarrow \gamma_f Q(A) = 0$$

Using this property, we define the following operator from $\mathcal{B}(E_1)$ into $\mathcal{B}(E_2)$:

$$\forall f \in \mathcal{B}(E_1) \quad \forall x \in E_2 \qquad Q_\gamma(f)(x_2) := \frac{d\gamma_f Q}{d\gamma Q}(x_2)$$

Notice that $Q_\gamma(1)(x_2) = 1$, and for any $(f, g) \in \mathcal{B}(E_1)^2$, we have

$$\gamma_{f+g} = \gamma_f + \gamma_g \Rightarrow Q_\gamma(f + g) = Q_\gamma(f) + Q_\gamma(g)$$

Using the fact that $\lim_{n \to \infty} \gamma_{1_{A_n}} = 0$ for every decreasing sequence of subsets $A_n \in E_1$ s.t. $\lim_{n \to \infty} A_n = \emptyset$, we prove that $\lim_{n \to \infty} Q_\gamma(1_{A_n})(x_2) = 0$, γQ-a.e. This implies that $A \in \mathcal{E}_1 \mapsto Q_\gamma(1_A)(x_2)$ is a well defined probability measure $Q_\gamma(x_2, dx_1)$ on the set $(E_1, \mathcal{E}_1)$, and we have the following γQ-a.e. Lebesgue integral representation

$$Q_\gamma(f)(x_2) = \int Q_\gamma(x_2, dx_1) f(x_1)$$

This ends the proof of the lemma. ∎

By construction, we have the equivalent time reversal formulae

$$\gamma(dx_1) \, Q(x_1, dx_2) = (\gamma Q) \, (dx_2) \, Q_\gamma(x_2, dx_1) \tag{6.27}$$

Notice that

$$\gamma(dx_1) \, Q(x_1, dx_2) = \gamma(G) \, \Psi_G(\eta)(dx_1) \, \overline{Q}(x_1, dx_2)$$

with

$$\eta(dx_1) = \gamma(dx_1)/\gamma(1) \quad \text{and} \quad G := Q(1)$$

In the same vein, we also have that

$$(\gamma Q)\,(dx_2)\,Q_\gamma(x_2, dx_1) = \gamma(G)\,(\Psi_G(\eta)\overline{Q})(dx_2)\,Q_\gamma(x_2, dx_1)$$

This implies that

$$\Psi_G(\eta)(dx_1)\,\overline{Q}(x_1, dx_2) = (\Psi_G(\eta)\overline{Q})(dx_2)\,Q_\gamma(x_2, dx_1)$$

from which we conclude that $Q_\gamma = \overline{Q}_{\Psi_G(\eta)}$, where $\overline{Q}_\gamma$ is defined as Q_γ by replacing Q by $\overline{Q}$.

Using rather elementary manipulations, we prove the following lemma.

Lemma 6.3.7 *For any functions $F_j \in \mathcal{B}(\mathcal{M}(E_j))$, with $j = 1, 2$, we have the almost sure formulae:*

$$\mathbb{E}\left(F_1(\mathcal{X}_1) \mid \mathcal{X}_2\right) = \int F_1\left(m_N(x_1)\right) \prod_{1 \leq i \leq N} Q_\gamma(X_2^i, dx_1^i)$$

and

$$\mathbb{E}\left(F_2(\mathcal{X}_2) \mid \mathcal{X}_1\right) = \int F_2\left(m_N(x_2)\right) \prod_{1 \leq i \leq N} \overline{Q}(X_1^i, dx_2^i)$$

In the above displayed formula $\overline{Q}$ and Q_γ stand for the pair of Markov transitions introduced in Lemma 6.3.6.

Proof:
To prove the second assertion, we recall that $\mathcal{X}_1 := m_N(X_1) = \sum_{1 \leq i \leq N} \delta_{X_1^i}$ is a Poisson point process on E_1, with intensity measure $\gamma_1(dx) := Q(1)(x)\,\gamma(dx)$. From this result, we find that

$$\mathbb{E}\left(F_1(\mathcal{X}_1) \left\{\int F_2\left(m_N(x_2)\right) \prod_{1 \leq i \leq N} \overline{Q}(X_1^i, dx_2^i)\right\}\right)$$

$$= e^{-\gamma Q(1)} \sum_{p \geq 0} \frac{1}{p!}$$

$$\times \int F_1\left(m_p(x_1)\right)\,F_2\left(m_p(x_2)\right) \prod_{1 \leq i \leq p} \left[\overline{Q}(x_1^i, dx_2^i)\,Q(1)(x_1^i)\,\gamma(dx_1^i)\right]$$

$$= e^{-\gamma Q(1)} \sum_{p \geq 0} \frac{1}{p!} \int F_1\left(m_p(x_1)\right)\,F_2\left(m_p(x_2)\right) \prod_{1 \leq i \leq p} \left[\gamma(dx_1^i)Q(x_1^i, dx_2^i)\right]$$

$$= \mathbb{E}\left(F_1(\mathcal{X}_1)F_2(\mathcal{X}_2)\right)$$

This ends the proof of the second assertion. Using the time reversal decomposition formula (6.27), and recalling that Q_γ is a Markov transition, the first assertion is a direct consequence of the second one. This ends the proof of the lemma. ∎

6.3.3.3 Partially observed models

We consider a spatial branching signal model defined by a Poisson point process $\mathcal{X} := \sum_{1 \leq i \leq N} \delta_{X^i}$, with intensity measure γ on some measurable state space $(E_1, \mathcal{E}_1)$, and we set $\eta(f) := \gamma(f)/\gamma(1)$, for any $f \in \mathcal{B}(E_1)$.

The random variable $\mathcal{X}$ is partially observed, on some possibly different measurable state space E_2. The observation is defined by a spatial point process. It consists in a collection of random observation variables, directly generated by some random points in the support of $\mathcal{X}$, plus some random observations unrelated to $\mathcal{X}$, sometimes called the clutter. We use the partial observation model presented in Section 6.3.1.2.

For the convenience of the reader, we briefly recall the description of this model in this static framework. Given a realization of $\mathcal{X}$, every random state $X^i = x$ generates an observation Y^i on $E_2 \cup \{c\}$ with distribution

$$L_c(x, dy) := d(x) \ L(x, dy) + (1 - d(x)) \ \delta_c(dy)$$

The function d represents the detectability degree of the states, and $L(x, .)$ stands for the distribution of the random observations on E_2 generated by the point x in E_1. The resulting observation process is the random measure $\mathcal{Y} = \sum_{1 \leq i \leq N} \delta_{Y^i}$ on the augmented state space $E_2 \cup \{c\}$.

In addition to this partial observation model, we also observe an additional, and independent of $\mathcal{X}$, clutter Poisson point process $\mathcal{Y}' := \sum_{1 \leq i \leq N'} \delta_{Y'^i}$, with intensity measure ν on E_2. As in (6.19), we further assume that ν and $L(x, .) \ll \lambda$, for some $\lambda \in \mathcal{M}(E_2)$, and we set

$$g(x, y) = \frac{dL(x, .)}{d\lambda}(y) \quad \text{and} \quad h := d\nu/d\lambda \tag{6.28}$$

We also suppose that $h(y) + \gamma(dg(., y)) > 0$, for any $y \in E_2$.

The full observation process on $E_2 \cup \{c\}$ is given by the random measure $\mathcal{Y}'' = \mathcal{Y} + \mathcal{Y}'$, while the "real world" observation $\mathcal{Y}^o$ is given the random measure

$$\mathcal{Y}^o := \mathcal{Y}''_{|E_2} = \mathcal{Y}_{|E_2} + \mathcal{Y}' \implies \mathcal{Y} + \mathcal{Y}' = \mathcal{Y}^o + N_c \ \delta_c \quad \text{with} \quad N_c := \mathcal{Y}(\{c\}) \tag{6.29}$$

The following proposition results from the construction of the observation process $\mathcal{Y}''$ on $E_2 \cup \{c\}$.

Proposition 6.3.8 *A version of the conditional distribution of the random measure $\mathcal{Y}''$ given $\mathcal{X} := \sum_{1 \leq i \leq N} \delta_{X^i}$ is given for any function $F \in \mathcal{B}(\mathcal{M}(E_2 \cup \{c\}))$ by*

$$\mathbb{E}\left(F\left(\mathcal{Y}''\right) | \mathcal{X}\right)$$

$$= e^{-\nu(1)} \sum_{p \geq 0} \frac{1}{p!} \int F\left(m_p(y') + m_N(y)\right) \ \nu^{\otimes p}(dy') \ \prod_{1 \leq i \leq N} L_c(X^i, dy^i)$$

Our next objective is to compute the conditional distribution of $\mathcal{X}$, given the observation process $\mathcal{Y}^o$.

Definition 6.3.9 *We let $\mathcal{Z}$ be the spatial point process defined by*

$$\mathcal{Z} := \sum_{1 \leq i \leq N} \delta_{(X^i, Y^i)} + \sum_{1 \leq i \leq N'} \delta_{(c, Y'^i)} := \sum_{1 \leq i \leq N''} \delta_{(Z_1^i, Z_2^i)} \tag{6.30}$$

For any $i \in \{1, 2\}$, we denote by $\mathcal{B}_b(E_i \cup \{c\})$ the set of functions $f \in \mathcal{B}_b(E_i)$, extended to $E_i \cup \{c\}$, by setting $f(c) = 0$.

We observe that

$$\mathcal{Z}_1 = \sum_{1 \leq i \leq N''} \delta_{Z_1^i} = \mathcal{X} + N' \delta_c$$

$$\mathcal{Z}_2 = \sum_{1 \leq i \leq N''} \delta_{Z_2^i} = (\mathcal{Z}_2)_{|E_2} + (\mathcal{Z}_2)_{|\{c\}} = \mathcal{Y}^o + N_c \delta_c$$

By construction, the random measure $\mathcal{Z}$ is a Poisson point process taking values in the state space

$$E_c = [(E_1 \cup \{c\}) \times (E_2 \cup \{c\})]$$

with intensity distribution given by the factorization formulae

$$\gamma(dx) L_c(x, dy) + \delta_c(dx)\nu(dy)$$

$$= \left[\underbrace{\gamma(dx)\, 1_{E_1}(x) + \nu(1)\, \delta_c(dx)}_{=\gamma_c(dx)}\right] \left[\underbrace{1_{E_1}(x)\, L_c(x, dy) + 1_c(x)\frac{\nu(dy)}{\nu(1)}}_{M_c(x, dy)}\right]$$

The marginal of the above distribution w.r.t. the second component is given by

$$\gamma_c M_c(dy) = (\gamma L_c + \nu)(dy)$$
$$= [\gamma(dg(.,y)) + h(y)]\, \lambda(dy) + \gamma(1 - d)\, \delta_c(dy)$$

On the other hand we have the decomposition

$$\gamma_c(dx)\, M_c(x, dy)$$

$$= [\gamma(dx)\, d(x)\, g(x, y) + \delta_c(dx)h(y)]\, \lambda(dy) + \gamma(dx)(1 - d(x))\, \delta_c(dy)$$

This yields the Bayes' type formula

$$\gamma_c(dx)\, M_c(x, dy) = \gamma_c M_c(dy)\, M_{c, \gamma_c}(y, dx)$$

with the Markov transition

$$M_{c, \gamma_c}(y, dx)$$

$$:= 1_{E_2}(y)\, \frac{[\gamma(dx)\, d(x)\, g(x, y) + \delta_c(dx)h(y)]}{[\gamma(dg(.,y)) + h(y)]} + 1_c(dy)\, \frac{\gamma(dx)(1 - d(x))}{\gamma(1 - d)}$$

This implies that

$$M_{c,\gamma_c}(y, dx) = 1_{E_2}(y) \, Q_{\gamma_c}(y, dx) + 1_c(dy) \, \Psi_{1-d}(\eta)(dx) \tag{6.31}$$

with the $[0,1]$-valued function β_γ on E_2 defined by

$$\beta_\gamma(y) := h(y)/\left[h(y) + \gamma(dg(\cdot, y))\right] \tag{6.32}$$

and the Markov transition Q_{γ_c} from E_2 into $E_1 \cup \{c\}$ defined by the following formula

$$Q_{\gamma_c}(y, dx) = (1 - \beta_\gamma(y)) \, \Psi_{dg(y,\cdot)}(\eta)(dx) + \beta_\gamma(y) \, \delta_c(dx) \tag{6.33}$$

We summarize the above discussion with the following proposition.

Proposition 6.3.10 *A version of the conditional distribution of the random measure $\mathcal{Z}_1$ given $\mathcal{Z}_2$ is given for any function $F \in \mathcal{B}(\mathcal{M}(E_1 \cup \{c\}))$ by*

$$\mathbb{E}\left(F(\mathcal{Z}_1) \mid \mathcal{Z}_2\right) = \int F\left(m_{\mathcal{Z}_2(1)}(x)\right) \prod_{1 \leq i \leq \mathcal{Z}_2(1)} M_{c,\gamma_c}(\mathcal{Z}_2^i, dx^i)$$

with the Markov M_{c,γ_c} transition, from $E_2 \cup \{c\}$ into $E_1 \cup \{c\}$, defined in (6.31).

Using the fact that

$$\mathcal{Z}_2 = (\mathcal{Z}_2)_{|E_2} + (\mathcal{Z}_2)_{|\{c\}} = \mathcal{Y}^o + N_c \delta_c$$

for any function $F \in \mathcal{B}(\mathcal{M}(E_2 \cup \{c\}))$, we also prove the following equation

$$\mathbb{E}\left(F(\mathcal{Z}_2) \mid \mathcal{Y}^o\right) = e^{-\gamma(1-d)} \sum_{p \geq 0} \frac{\gamma(1-d)^p}{p!} \, F\left(p\delta_c + \mathcal{Y}^o\right)$$

If we set

$$\mathcal{Y}^o := \sum_{1 \leq i \leq N_1} \delta_{Y_1^i} = \sum_{1 \leq i \leq N} 1_{E_2}(Y^i) \, \delta_{Y^i} + \mathcal{Y}'$$

then we find that

$$\mathbb{E}\left(F(\mathcal{Z}_1) \mid (\mathcal{Z}_2)_{|E_2}\right)$$

$$= \mathbb{E}\left(\mathbb{E}\left(F(\mathcal{Z}_1) \mid \mathcal{Z}_2\right) \mid (\mathcal{Z}_2)_{|E_2}\right)$$

$$= e^{-\gamma(1-d)} \sum_{p \geq 0} \frac{\gamma(1-d)^p}{p!}$$

$$\times \int F\left(m_p(x') + m_{\mathcal{Y}^o(1)}(x)\right) \Psi_{(1-d)}(\eta)^{\otimes p}(dx') \prod_{i=1}^{\mathcal{Y}^o(1)} Q_{\gamma_c}(Y_1^i, dx^i)$$

We summarize the above discussion with the following theorem.

Theorem 6.3.11 *A version of the conditional distribution of*

$$\mathcal{X}_c := \mathcal{X} + N'\delta_c$$

given the observation point process $\mathcal{Y}^o = \sum_{1 \le i \le N_1} \delta_{Y_1^i}$ *is given for any function* $F \in \mathcal{B}(\mathcal{M}(E_1 \cup \{c\}))$ *by*

$$\mathbb{E}\left(F\left(\mathcal{X}_c\right) | \mathcal{Y}^o\right)$$

$$= e^{-\gamma(1-d)} \sum_{p \ge 0} \frac{\gamma(1-d)^p}{p!}$$

$$\times \int F\left(m_p(x') + m_{N_1}(x)\right) \, \Psi_{(1-d)}(\eta)^{\otimes p}\,(dx') \prod_{i=1}^{N_1} Q_{\gamma_c}\left(Y_1^i, dx^i\right)$$

with the Markov Q_{γ_c} *transition, from* E_2 *into* $E_1 \cup \{c\}$, *defined in (6.33).*

The conditional expectation measures of the random point processes $\mathcal{Y}_c$ and $\mathcal{Y}^o$ resp. $\mathcal{X}_c$ and $\mathcal{X}$, given the point process $\mathcal{X}$, resp. $\mathcal{Y}^o$, are now easily computed.

Corollary 6.3.12 *For any function* $f \in \mathcal{B}(E_2 \cup \{c\})$, *we have the almost sure integral representation formula*

$$\mathbb{E}\left(\mathcal{Y}_c(f) \mid \mathcal{X}\right) \;=\; \mathbb{E}\left(\mathcal{Y}^o(f) \mid \mathcal{X}\right) = \mathcal{X}(dL(f)) + \nu(f)$$

and for any function $f \in \mathcal{B}(E_1 \cup \{c\})$ *we have the almost sure integral representation formula*

$$\mathbb{E}\left(\mathcal{X}_c(f) \mid \mathcal{Y}^o\right) \;=\; \mathbb{E}\left(\mathcal{X}(f) \mid \mathcal{Y}^o\right)$$

$$= \; \gamma((1-d)f) + \int \mathcal{Y}^o(dy)\,(1 - \beta_\gamma(y))\,\Psi_{dg(y,\cdot)}(\eta)(f)$$

with the $[0,1]$-*valued function* β_γ *defined in (6.32).*

In particular, the conditional mean value of the number of states N given the spatial point observation is given below:

$$\mathbb{E}\left(N | \mathcal{Y}^o\right) = \mathbb{E}(N)\,\eta((1-d)) + \mathcal{Y}^o\,(1 - \beta_\gamma) \tag{6.34}$$

Notice that the first term in the r.h.s. of (6.34) represents the mean value of N times the nondetection probability. Roughly speaking, the second term represents the $\mathcal{Y}^o$-probability that observations do not cause the clutter. In this connection, models with no clutter and fully detectable states are described below.

Corollary 6.3.13 *In the situation where* $d = 1$ *and* $\nu = 0$ *we have*

$$\mathcal{X}_c = \mathcal{X} \quad \text{and} \quad \mathcal{Y}^o = \mathcal{Y} = \sum_{1 \le i \le N} \delta_{Y^i}$$

In addition, for any function $F \in \mathcal{B}(\mathcal{M}(E_1))$, we have the following almost sure integral representation formula

$$\mathbb{E}\left(F\left(\mathcal{X}\right)|\mathcal{Y}\right) = \int F\left(m_N(x)\right) \prod_{1 \leq i \leq N} \Psi_{g(Y^i,.)}(\eta)(dx^i)$$

For any function $f \in \mathcal{B}(E_1 \cup \{c\})$ we also have the almost sure integral representation formula

$$\mathbb{E}\left(\mathcal{X}_c(f) \mid \mathcal{Y}\right) = \mathbb{E}\left(\mathcal{X}(f) \mid \mathcal{Y}\right) = \int \mathcal{Y}(dy) \, \Psi_{g(y,.)}(\eta)(f)$$

6.4 Association tree based measures

This section is concerned with association tree based measures and their mean field approximations. The central idea behind these filtering association models is to solve the data association problem; that is, to find the right sequences of observations delivered by every target track.

In the first section, Section 6.4.1, we design a new class of evolution equations in the set of measures on finite association trees. This class of models has the same form as the ones discussed in Section 6.2.1. They can also be interpreted as an extended version of the Boltzmann-Gibbs models discussed in Section 2.5.2 to association positive measures.

Then, we examine two situations:

Firstly, for a given data association trajectory, we assume that the optimal single target tracking problem can be solved using Kalman filters, or by some auxiliary particle filter. In this context, the central problem is to find a judicious way of reducing the set of all possible associations to a reasonable finite number, with high likelihood value. In Section 6.4.2, we design a mean field solution to this problem, in the spirit of the i-MCMC methodologies presented in Section 4.3.

In more general situations, even given the exact sequence of observations of a given target trajectory, the optimal filtering problem associated with this data cannot be computed explicitly. In Section 6.4.3, we couple the data association mean field model discussed in Section 6.4.2 with mean field type particle filters.

To the best of our knowledge, these mean field particle approximations of association tree based measures are one of the most performant algorithms, for solving multiple target tracking problems. We refer to the series of articles [137, 459, 460, 461] for numerical experiments and comparisons between these mean field models.

6.4.1 Nonlinear evolution equations

We let $(\mathcal{A}_n)_{n\geq 0}$ be a sequence of finite sets equipped with some finite positive measures $(\nu_n)_{n\geq 0}$. We let $\eta^{(a)}$, $Q^{(a)}$, and $f^{(a)}$ be some collection of measures, integral operators, and measurable functions, on some state spaces, indexed by the parameter a in some finite set $\mathcal{A}$.

To clarify the presentation, for any measure ν on $\mathcal{A}$, we set

$$\eta^{(\nu)} := \int \nu(da)\, \eta^{(a)} \tag{6.35}$$

$$Q^{(\nu)} := \int \nu(da)\, Q^{(a)} \quad \text{and} \quad f^{(\nu)} := \int \nu(da)\, f^{(a)} \tag{6.36}$$

We return to the general measure valued model (γ_n, η_n) defined in Section 6.2.1. We further assume that the initial distribution γ_0 and the integral operators Q_{n+1,γ_n} in (6.11) have the following form

$$\gamma_0 = \eta_0^{(\nu_0)} \quad \text{and} \quad Q_{n+1,\gamma_n} = Q_{n+1,\gamma_n}^{(\nu_{n+1})}$$

for some collection of probability measures $\eta_0^{(a)}$, and positive and bounded integral operators $Q_{n+1,\gamma_n}^{(a)}$, indexed by $a \in \mathcal{A}_{n+1}$. In this situation, we have

$$\gamma_0(1) = \nu_0(1) \quad \text{and} \quad \eta_0 = \eta_0^{(A_0)} \quad \text{with} \quad A_0(da) := \nu_0(da)/\nu_0(1)$$

We also assume that the following property is met

$$G_{n,\gamma}^{(a)} := Q_{n+1,\gamma}^{(a)}(1) \propto G_n^{(a)} \quad \text{and} \quad Q_{n+1,\gamma}^{(a)}(f)/Q_{n+1,\gamma}^{(a)}(1) := M_{n+1}^{(a)}(f) \tag{6.37}$$

for some function $G_n^{(a)}$ on E_n, and some Markov transitions $M_{n+1}^{(a)}$ from E_n into E_{n+1} whose values do not depend on the measures γ.

Example 6.4.1 *We illustrate these rather abstract conditions in the context of the multiple target tracking equation presented in (6.23).*

In this situation, it is convenient to add a pair of virtual observation states c, c' to E_n^Y. Using this notation, the above conditions are satisfied with the finite sets $\mathcal{A}_{n+1}$ and their counting measures ν_{n+1} defined below

$$\mathcal{A}_{n+1} = \{Y_n^i, 1 \leq i \leq N_n^Y\} \cup \{c, c'\}$$

and

$$\nu_{n+1} = \mathcal{Y}_n^o + \delta_c + \delta_{c'} \in \mathcal{M}(\mathcal{A}_{n+1})$$

Using (6.23) and (6.24), we check that (6.37) is met with a couple of potential functions and Markov transitions defined by

$$(G_n^{(a)}, M_{n+1}^{(a)}) = \begin{cases} (r_n d_n g_n(\cdot, a)\,,\; M_{n+1}) & \text{for} \quad a \notin \{c, c'\} \\ (r_n(1 - d_n)\,,\; M_{n+1}) & \text{for} \quad a = c \\ (1\,,\; \overline{\mu}_{n+1}) & \text{for} \quad a = c' \end{cases}$$

In this case, we observe that

$$Q_{n+1,\gamma_n}^{(a)}(x_n, \cdot) = G_{n,\gamma_n}^{(a)}(x_n) \, M_{n+1}^{(a)}(x_n, \cdot)$$

with the potential function $G_{n,\gamma_n}^{(a)}$ *defined below*

$$G_{n,\gamma_n}^{(a)}/G_n^{(a)} = \begin{cases} [h_n(a) + \gamma_n(d_n g_n(\cdot, a))]^{-1} & \text{for} \quad a \notin \{c, c'\} \\ 1 & \text{for} \quad a = c \\ \mu_{n+1}(1)/\gamma_n(1) & \text{for} \quad a = c' \end{cases} \qquad (6.38)$$

Definition 6.4.2 *We consider the collection of probability measures* $\eta_n^{(\mathbf{a}_n)} \in \mathcal{P}(E_n)$, *indexed by sequences of parameters*

$$\mathbf{a}_n = (a_0, \ldots, a_n) \in \mathbf{A_n} := (\mathcal{A}_0 \times \ldots \times \mathcal{A}_n)$$

and defined by the following equations

$$\eta_n^{(\mathbf{a}_n)} = \left(\Phi_n^{(a_n)} \circ \ldots \circ \Phi_1^{(a_1)} \right) \left(\eta_0^{(a_0)} \right) \qquad (6.39)$$

with the mappings $\Phi_n^{(a)} : \mathcal{P}(E_{n-1}) \to \mathcal{P}(E_n)$, *indexed by* $a \in \mathcal{A}_n$, *and defined by the updating-prediction Feynman-Kac transformation*

$$\Phi_n^{(a)}(\eta) = \Psi_{G_{n-1}^{(a)}}(\eta) \, M_n^{(a)}$$

Given some $\mathbf{a} = (a_0, \ldots, a_n) \in \mathbf{A_n}$, *we set* $|\mathbf{a}| = n$. *In this situation, we simplify the notation, and write*

$$\eta^{(\mathbf{a})} := \eta_{|\mathbf{a}|}^{(\mathbf{a}_{|\mathbf{a}|})} = \eta_n^{(\mathbf{a}_n)} \qquad (6.40)$$

the measure defined in (6.39).

Definition 6.4.3 *We let* Ω_{n+1} *be the mapping*

$$\Omega_{n+1} : (m, A) \in (]0, \infty[\times \mathcal{P}(\mathbf{A_n})) \mapsto \Omega_{n+1}(m, A) \in \mathcal{P}(\mathbf{A}_{n+1})$$

defined by the following formula

$$\Omega_{n+1}(m, A) = \Psi_{\mathcal{G}_{m,A}} (A \otimes \nu_{n+1}) \qquad (6.41)$$

with the Boltzmann-Gibbs transformations $\Psi_{\mathcal{G}_{m,A}}$ *from* $\mathcal{P}(\mathbf{A_n} \times \mathbb{R}_+)$, *into itself defined in (0.2), associated with the potential function*

$$\mathcal{G}_{m,A}(a, b) = \eta_n^{(a)} \left(G_{n,m\eta^{(A)}}^{(b)} \right) \qquad \text{with} \quad \eta^{(A)} \text{ given by (6.36) and (6.40).}$$

Definition 6.4.4 *We consider the collection of integral operators* $\mathcal{Q}_{n+1,B}$ *from* $\mathbf{A}_n$ *into* $\mathbf{A}_{n+1}$, *indexed by* $B \in \mathcal{M}_+(\mathbf{A}_n)$, *and defined, for any* $a \in \mathbf{A}_n$ *by*

$$\mathcal{Q}_{n+1,B}(a, d(a', b)) := [\delta_a \otimes \nu_{n+1}] \, (d(a', b)) \; \mathcal{G}_{B(1),\overline{B}}(a', b) \qquad (6.42)$$

with the normalized distribution

$$\overline{B} = B/B(1) \in \mathcal{P}(\mathbf{A}_n)$$

In the above display formulae, $d(a', b) = da' \times db$ *stands for an infinitesimal neighborhood of the point* $(a', b) \in \mathbf{A}_{n+1} := (\mathbf{A}_n \times \mathbf{A}_{n+1})$.

Proposition 6.4.5 *The solution of (6.2.1) has the form*

$$\eta_n = \eta_n^{(A_n)} = \eta^{(A_n)}$$

with the sequence of association measures $A_n \in \mathcal{P}(\mathbf{A}_n)$ *defined, for any* $F \in \mathcal{B}_b(\mathbf{A}_{n+1})$, *by the evolution equations*

$$A_{n+1}(F) = \Omega_{n+1}(\gamma_n(1), A_n)(F) := \frac{A_n \mathcal{Q}_{n+1,\gamma_n(1)A_n}(F)}{A_n \mathcal{Q}_{n+1,\gamma_n(1)A_n}(1)}$$

In addition, the flow of unnormalized measures $(B_n)_{n\geq 0} \in \mathcal{M}_+(\mathbf{A}_n)$ *defined by* $B_n := \gamma_n(1) \times A_n$ *satisfies the same type of equation as in (6.11); that is, we have that*

$$B_{n+1} = B_n \mathcal{Q}_{n+1,B_n} \qquad (6.43)$$

Proof:
We check the first assertion using an inductive proof on the time parameter.

For $n = 0$, we have set $\eta_0 = \eta^{(A_0)}$, so that the assertion is met at the time $n = 0$. We further assume that $\eta_n = \eta_n^{(A_n)}$, for some $A_n \in \mathcal{P}(\mathbf{A}_n)$.

Using (6.11), we find that it is simply based on the fact that

$$\eta_{n+1} \quad \propto \quad \eta^{(A_n)} Q_{n+1,\gamma_n(1)\eta^{(A_n)}}^{(\nu_{n+1})}$$

$$= \quad \int [A_n \otimes \nu_{n+1}](d(a, b)) \, \eta_n^{(a)} Q_{n+1,\gamma_n(1)\eta^{(A_n)}}^{(b)}$$

Using (6.37), we have that

$$Q_{n+1,\gamma_n(1)\eta^{(A_n)}}^{(b)}(1) = G_{n,\gamma_n(1)\eta^{(A_n)}}^{(b)} \propto G_n^{(b)}$$

and

$$Q_{n+1,\gamma_n(1)\eta^{(A_n)}}^{(b)}(f)/Q_{n+1,\gamma_n(1)\eta^{(A_n)}}^{(b)}(1) = M_{n+1}^{(b)}(f)$$

for any $f \in \mathcal{B}_b(E_{n+1})$. This implies that

$$\eta_{n+1}^{(a,b)}(f) = \Phi_{n+1}^{(b)}\left(\eta_n^{(a)}\right)(f) = \frac{\eta_n^{(a)}\left(G_n^{(b)} M_{n+1}^{(b)}(f)\right)}{\eta_n^{(a)}\left(G_n^{(b)}\right)} = \frac{\eta_n^{(a)} Q_{n+1,\gamma_n(1)\eta^{(A_n)}}^{(b)}(f)}{\eta_n^{(a)} Q_{n+1,\gamma_n(1)\eta^{(A_n)}}^{(b)}(1)}$$

from which we conclude that

$$\eta_{n+1} \propto \int \underbrace{[A_n \otimes \nu_{n+1}](d(a,b))\ \eta_n^{(a)} \left(G^{(b)}_{n,\gamma_n(1)\eta^{(A_n)}}\right)}_{=\Omega_{n+1}(\gamma_n(1),A_n)(d(a,b))} \eta_{n+1}^{(a,b)}$$

In terms of the second coordinate mapping Λ_{n+1}^2 defined in (6.14), we have proved that

$$\eta_{n+1} = \Lambda_{n+1}^2 \left(\gamma_n(1), \eta_n^{(A_n)}\right) = \eta_{n+1}^{(\Omega_{n+1}(\gamma_n(1),A_n))} = \eta_{n+1}^{(A_{n+1})}$$

This ends the proof of the first assertion.
On the other hand, we have that

$$A_n \mathcal{Q}_{n+1,\gamma_n(1)A_n}(F)(a) = \int A_n(da)\ \nu_{n+1}(db)\ \mathcal{G}_{\gamma_n(1),A_n}(a,b)\ F(a,b)$$

$$\propto \int A_{n+1}(d(a,b))\ F(a,b)$$

This ends the proof of the proposition. ∎

Using Proposition 6.4.5, we readily prove the following theorem.

Theorem 6.4.6 *The solution of (6.2.1) has the form*

$$\gamma_n = \gamma_n(1) \times \eta^{(A_n)}$$

with the process $(\gamma_n(1), A_n) \in (\mathbb{R}_+ \times \mathcal{P}(\boldsymbol{A}_n))$ defined by

$$\begin{cases} A_{n+1} &= \Psi_{\mathcal{G}_{\gamma_n(1),A_n}}(A_n \otimes \nu_{n+1}) \\ \\ \gamma_{n+1}(1) &= \gamma_n(1) \times [A_n \otimes \nu_{n+1}](\mathcal{G}_{\gamma_n(1),A_n}) \end{cases} \qquad (6.44)$$

In the above display, $\mathcal{G}_{\gamma_n(1),A_n}$ stands for the potential function presented in Definition 6.4.3.

We end this section with some comments on the evolution Equation (6.44). The first equation is only defined in terms of the Boltzmann-Gibbs transformations $\Psi_{\mathcal{G}_{\gamma_n(1),A_n}}$. As a result, we cannot expect the measure valued equation to have some nice stability properties.

These only updating type models have the same form as the Boltzmann-Gibbs measures discussed in Section 2.5.2.

Loosely speaking, we can stabilize these equations adding some MCMC steps between the updating Boltzmann-Gibbs transformations. These stabilizing techniques have been developed in the *i*-MCMC methodologies presented in Section 4.3.

More formally, we can add, at every time step, an MCMC transition $\boldsymbol{M}_{n+1}$

on the set $\mathcal{A}_{n+1}$ with invariant measure $\Psi_{\mathcal{G}_{\gamma_n(1),A_n}}$. The resulting equation, is now given by

$$A_{n+1} = \Psi_{\mathcal{G}_{\gamma_n(1),A_n}} \left(A_n \otimes \nu_{n+1} \right) M_{n+1}$$

Further details on the weak regularity properties of the flow of measures A_n are discussed in the end of Section 13.2.1.

6.4.2 Mean field particle model

We further assume that $\eta_n^{(a)} \left(G_{n,\gamma_n(1)\eta^{(A_n)}}^{(b)} \right)$ are explicitly known for any sequence of parameters $((a, A_n), b) \in \left(\mathcal{A}_n^2 \times \mathcal{A}_{n+1} \right)$.

This rather strong condition is satisfied for the multiple target tracking model discussed above as long as the quantities

$$\eta_n^{(a_0, y_0, \dots, y_{n-1})}(r_n d_n g_n(., y_n)) \qquad \eta_n^{(a_0, y_0, \dots, y_{n-1})} \left(r_n (1 - d_n) \right)$$

and

$$\eta_n^{(a_0, y_0, \dots, y_{n-1})}(d_n g_n(., y_n))$$

are explicitly known. This condition is clearly met for linear Gaussian target evolution and observation sensors, as long as the survival and detection probabilities, s_n and d_n, are state independent, and the spontaneous birth $\bar{\mu}_n$ as well as the spawned targets branching rates b_n are Gaussian mixtures.

In this situation, the collections of measures $\eta_n^{(a_0, y_0, \dots, y_{n-1})}$ are Gaussian distributions and the Equation (6.39) coincides with the traditional updating-prediction transitions of the discrete generation Kalman-Bucy filter.

We let $A_0^N = \frac{1}{N} \sum_{i=1}^N \delta_{a_0^i}$ be the empirical measure associated with N independent and identically distributed random variables $(a_0^i)_{1 \leq i \leq N}$ with common distribution A_0. By construction, we have

$$\eta_0^N := \int A_0^N(da) \, \eta_0^{(a)} \simeq_{N \uparrow \infty} \eta_0$$

We further assume that $\gamma_0(1)$ is known and we set $\gamma_0^N = \gamma_0(1) \, \eta_0^N$,

$$\gamma_1^N(1) = \gamma_0^N(1) \, \eta_0^N(G_{0,\gamma_0^N}), \quad \text{and} \quad \eta_1^N := \int A_1^N(da) \, \eta_1^{(a)}$$

with the occupation measure $A_1^N = \frac{1}{N} \sum_{i=1}^N \delta_{a_1^i}$ associated with N conditionally independent and identically distributed random variables $a_1^i := (a_{0,1}^i, a_{1,1}^i)$ with common law $\Omega_1 \left(\gamma_0^N(1), A_0^N \right)$. By construction, we also have

$$\eta_1^N \simeq_{N \uparrow \infty} \int \Omega_1 \left(\gamma_0^N(1), A_0^N \right)(da) \, \eta_1^{(a)} = \Lambda_1^2 \left(\gamma_0^N(1), \eta_0^N \right)$$

Iterating this procedure, we define by induction a sequence of N-particle approximation measures

$$\gamma_n^N(1) = \gamma_{n-1}^N(1) \, \eta_{n-1}^N(G_{n-1,\gamma_{n-1}^N}) \quad \text{and} \quad \eta_n^N := \int A_n^N(da) \, \eta_n^{(a)}$$

with the occupation measure $A_n^N = \frac{1}{N} \sum_{i=1}^N \delta_{a_n^i}$ associated with N condition-ally independent and identically distributed random variables

$$a_n^i := (a_{0,n}^i, a_{1,n}^i, \ldots, a_{n,n}^i)$$

with common law $\Omega_n \left(\gamma_{n-1}^N(1), A_{n-1}^N \right)$. Arguing as above, we find that

$$\eta_n^N \simeq_{N\uparrow\infty} \int \Omega_n \left(\gamma_{n-1}^N(1), A_{n-1}^N \right)(da) \, \eta_n^{(a)} = \Lambda_n^2 \left(\gamma_{n-1}^N(1), \eta_{n-1}^N \right)$$

6.4.3 Mixed particle association models

We consider the association mapping

$$\boldsymbol{\Omega_{n+1}} : (m, A, \eta) \in \left( ]0, \infty[\times \boldsymbol{A_n} \times \mathcal{P}(E_n)^{\boldsymbol{A_n}} \right) \mapsto \Omega_{n+1}(m, A, \eta) \in \mathcal{P}(\boldsymbol{A_{n+1}})$$

defined for any $(m, A) \in \left( ]0, \infty[\times \boldsymbol{A_n} \right)$, and any mapping

$$\eta : a \in \mathrm{Supp}(A) \mapsto \eta^{(a)} \in Pa(E_n)$$

by the following equation

$$\boldsymbol{\Omega_{n+1}}\left(m, A, \eta\right)(d(a,b)) \propto A(da) \, \nu_{n+1}(db) \, \eta^{(a)} \left(G_{n,m \int A(da) \, \eta^{(a)}}^{(b)} \right)$$

By construction, for any discrete measure $A \in \mathcal{P}(\boldsymbol{A_{n-1}})$, and any mapping $a \in \mathrm{Supp}(A) \mapsto \eta^{(a)} \in \mathcal{P}(E_{n-1})$, we have the formula

$$\Lambda_n^2 \left(m, \int A(da) \, \eta^{(a)} \right) = \int \Omega_n \left(m, A, \eta^{(\cdot)} \right)(d(a,b)) \; \Phi_n^{(b)} \left(\eta^{(a)} \right)$$

We also mention that the updating-prediction transformation defined in (6.39) can be rewritten in terms of nonlinear transport equations

$$\Phi_n^{(a)}(\eta) = \Psi_{G_{n-1}}^{(a)}(\eta) \, M_n^{(a)} = \eta K_{n,\eta}^{(a)} \quad \text{with} \quad K_{n,\eta}^{(a)} = \mathcal{S}_{n-1,\eta}^{(a)} M_n^{(a)} \qquad (6.45)$$

In the above displayed formula, $\mathcal{S}_{n-1,\eta}^{(a)}$ stands for some updating Markov transition, from E_{n-1} into itself, satisfying the compatibility condition

$$\eta \mathcal{S}_{n-1,\eta}^{(a)} = \Psi_{G_{n-1}}^{(a)}(\eta)$$

We let $A_0^N = \frac{1}{N} \sum_{i=1}^N \delta_{a_0^i}$ be the empirical measure associated with N independent and identically distributed random variables $(a_0^i)_{1 \le i \le N}$ with common distribution A_0. We set

$$\eta_0^N := \int A_0^N(da) \, \eta_0^{(a,N')}$$

with the empirical measure $\eta_0^{(a,N')} = \frac{1}{N'} \sum_{i=1}^{N'} \delta_{\xi_0^{[a,j]}}$ associated with N' random variables $\xi_0^{[a]} = \left(\xi_0^{[a,j]}\right)_{1 \le j \le N'}$, with common law $\eta_0^{(a)}$. We further assume that $\gamma_0(1)$ is known, and we set

$$\gamma_0^N := \gamma_0(1)\, \eta_0^N \quad \text{and} \quad \gamma_1^N(1) := \gamma_0^N(1)\, \eta_0^N(G_{0,\gamma_0^N})$$

By construction, we have

$$\int A_0^N(da)\, \eta_0^{(a,N')} \simeq_{N'\uparrow\infty} \int A_0^N(da)\, \eta_0^{(a)} \Rightarrow \eta_0^N \simeq_{N,N'\uparrow\infty} \eta_0$$

Using (6.45), for any $a_1 = (a_0, a_1) \in \mathbf{\mathcal{A}_1}$ we find that

$$\Phi_1^{(a_1)}\left(\eta_0^{(a_0,N')}\right) = \eta_0^{(a_0,N')} K_{n,\eta_0^{(a_0,N')}}^{(a_1)}$$

We let $A_1^N = \frac{1}{N} \sum_{i=1}^N \delta_{a_1^i}$ be the occupation measure associated with N conditionally i.i.d. r.v. $a_1^i := (a_{0,1}^i, a_{1,1}^i)$ with common law

$$\mathbf{\Omega_1}\left(\gamma_0^N(1), A_0^N, \eta_0^{(\cdot,N')}\right)$$

In the above displayed formula $\eta_0^{(\cdot,N')}$ stands for the mapping

$$a_0 \in \mathcal{A}_0 \mapsto \eta_0^{(a_0,N')} \in \mathcal{P}(E_0)$$

We consider a sequence of conditionally independent random variables $\xi_1^{[a_0,a_1,j]}$ with distribution $K_{n,\eta_0^{(a_0,N')}}^{(a_1)}\left(\xi_0^{[a_0,j]},.\right)$, with $1 \le j \le N'$, and we set

$$\eta_1^{((a_0,a_1),N')} = \frac{1}{N'} \sum_{i=1}^{N'} \delta_{\xi_1^{[(a_0,a_1),j]}} \quad \text{and} \quad \eta_1^N := \int A_1^N(da)\, \eta_1^{(a,N')}$$

Arguing as before, we find that

$$\begin{aligned} \eta_1^N &\simeq_{N\uparrow\infty} \int \mathbf{\Omega_1}\left(\gamma_0^N(1), A_0^N, \eta_0^{(\cdot,N')}\right)(d(a_0,a_1))\, \Phi_1^{(a_1)}\left(\eta_0^{(a_0,N')}\right) \\ &= \Lambda_1^2\left(\gamma_0^N(1), \eta_0^N\right) \end{aligned}$$

Iterating this procedure, we define by induction a sequence of N-particle approximation measures

$$\gamma_n^N(1) = \gamma_{n-1}^N(1)\, \eta_{n-1}^N(G_{n-1,\gamma_{n-1}^N}) \quad \text{and} \quad \eta_n^N := \int A_n^N(da)\, \eta_n^{(a,N')}$$

with the occupation measure $A_n^N = \frac{1}{N} \sum_{i=1}^N \delta_{a_n^i}$ associated with N conditionally independent and identically distributed random variables

$$a_n^i := (a_{0,n}^i, a_{1,n}^i, \ldots, a_{n,n}^i)$$

with common law

$$\Omega_n\left(\gamma_{n-1}^N(1), A_{n-1}^N, \eta_{n-1}^{(\cdot,N')}\right)$$

Arguing as above, we find that

$$
\eta_n^N \simeq_{N\uparrow\infty} \int \Omega_n\left(\gamma_{n-1}^N(1), A_{n-1}^N, \eta_{n-1}^{(\cdot,N')}\right)(d(a,b))\,\Phi_n^{(b)}\left(\eta_{n-1}^{(a,N')}\right)
$$

$$
= \Lambda_n^2\left(\gamma_{n-1}^N(1), \eta_{n-1}^N\right)
$$

As before, the N-particle occupation measures A_n^N converge as N tends to ∞ to the association probability measures A_n.

Part III

Application domains

Chapter 7

Particle absorption models

7.1 Particle motions in absorbing medium

In probability theory, particle absorption models are represented by Markov chains evolving in a deterministic, or in a random, environment associated with some absorption rate functions.

The interpretation of the absorption event clearly depends on their application models. In optical ray propagation problems, the event of interest is related to photon absorptions [482]. In particle physics or in chemistry, the absorption rate is dictated by the energy of an electronic or macro-molecular configuration. For a more detailed discussion on these models, and their applications to the computation of Schrödinger ground state energies, we refer the reader to the series of articles [91, 161, 175, 176, 498]. In natural evolution theory, as well as in population model analysis, the absorption event is often related to an extinction probability. Further details on these applications can be found in the series of articles [420, 432, 561, 562, 563].

Absorption and critical type events can also be thought of as network overflows in complex queueing systems [240] and production systems [478]. Absorbed Markov chain also used in Web engineering [479] and bio-chemistry [397], as well as in environmental analysis [576], and in many others scientific disciplines.

This rather extraordinary variety of application domains is not really surprising, since all of these absorption models can be represented by a Feynman-Kac model of the same form as in the one (1.37) discussed in Section 3.2 (cf. also Section 1.4.2, in the opening chapter).

Inversely, as we already mentioned on page 25, we emphasize that *any* Feynman-Kac model (1.37) can be interpreted as the distribution of the random trajectories of a Markov chain evolving in an absorbing environment.

7.1.1 Killed Markov chain models

We return to the particle absorption model (3.30) discussed in the opening chapter (see also Section 1.4.2.2, on page 28). We consider a collection of measurable state spaces E_n and an auxiliary coffin, or cemetery, state c. We set $E_{n,c} = E_n \cup \{c\}$. We also denote by G_n some $[0,1]$-valued potential functions on E_n, and M_{n+1} some Markov transitions from E_n, into E_{n+1}. We define an $E_{n,c}$-valued Markov chain X_n^c with two separate killing/exploration transitions:

$$X_n^c \xrightarrow{\text{killing}} \widehat{X}_n^c \xrightarrow{\text{exploration}} X_{n+1}^c \tag{7.1}$$

These killing/exploration mechanisms are defined as follows:

- **Killing:** If $X_n^c = c$, we set $\widehat{X}_n^c = c$. Otherwise the particle X_n^c is still alive. In this case, with a probability $G_n(X_n^c)$, it remains in the same site, so that $\widehat{X}_n^c = X_n^c$; and with a probability $1 - G_n(X_n^c)$, it is killed, and we set $\widehat{X}_n^c = c$.

- **Exploration:** Once a particle has been killed, it cannot be brought back to life; so if $\widehat{X}_n^c = c$, then we set $\widehat{X}_p^c = X_p = c$, for any $p > n$. Otherwise, the particle $\widehat{X}_n^c \in E_n$ evolves to a new location X_{n+1}^c in E_{n+1}, randomly chosen according to the distribution $M_{n+1}(X_n^c, dx_{n+1})$.

Definition 7.1.1 *The Markov chain X_n^c defined above is called a Markov chain with the absorption rates $(1 - G_n)$, and the free exploration transitions M_n, on the state spaces E_n.*

Notice that the Markov chain X_n^c on the augmented state spaces $E_{n,c}$ can be interpreted as a conventional Markov chain, with a single absorbing state $\{c\}$, as soon as $M_n(x_n, \{x_n\}) \neq 1$ for any $x_n \in E_n$. Inversely, any Markov chain with a single absorbing state can be represented in this form.

In branching processes and population dynamics literature, the model X_n^c often represents the number of individuals of a given species [255, 291, 531]. Each individual can die or reproduce. The state $0 \in E_n = \mathbb{N}$ is interpreted as a trap, or as a hard obstacle, in the sense that the species disappears as soon as X_n^c hits 0. For a more thorough discussion on particle motions in an absorbing medium with hard and soft obstacles and their application domains, we refer the reader to the series of articles [176, 175, 498], the monograph [163], and the more recent lecture notes [186].

7.1.2   Feynman-Kac models

The aim of this section is to provide an equivalent Feynman-Kac formulation of particle absorption models. We denote by X_n the Markov chain on E_n, with elementary transitions M_n. In this notation, the Feynman-Kac measures $\mathbb{Q}_n$ associated with the parameters (G_n, M_n), and defined in (1.37), represent the conditional distributions of the random paths of a nonabsorbed Markov particle. To see this claim, we let T be the killing time; that is, the first time at which the particle enters in the cemetery state

$$T = \inf \{n \geq 0 \; ; \; \widehat{X}_n^c = c\}$$

By construction, we have

$$
\begin{aligned}
\mathbb{P}(T \geq n) &= \mathbb{P}(\widehat{X}_0^c \in E_0, \ldots, \widehat{X}_{n-1}^c \in E_{n-1}) \\
&= \int_{E_0 \times \ldots \times E_{n-1}} \eta_0(dx_0) \, G_0(x_0) \prod_{1 \leq p < n} (M_p(x_{p-1}, dx_p) G_p(x_p))
\end{aligned}
$$

This shows that the normalizing constants $\mathcal{Z}_n$ of the Feynman-Kac measures $\mathbb{Q}_n$ represent the probability for the particle to be alive at time $n-1$; that is, we have that

$$
\mathcal{Z}_n = \mathbb{P}(T \geq n) = \mathbb{E}\left(\prod_{0 \leq p < n} G_p(X_p) \right)
$$

In the above display, X_n stands for a Markov chain on E_n, with initial distribution η_0 and elementary Markov transitions M_n.

In the same vein, in terms of the n-th time marginal Feynman-Kac models (3.28) we have

$$
\mathbb{E}(f(X_n^c) \, 1_{T \geq n}) = \gamma_n(f_n) := \mathbb{E}\left[f_n(X_n) \left\{ \prod_{0 \leq p < n} G(X_p) \right\} \right] \quad (7.2)
$$

$$
\mathbb{E}(f(X_n^c) \mid T \geq n) = \eta_n(f_n) := \gamma_n(f_n)/\gamma_n(1) \quad (7.3)
$$

Using these formulae, we find that

$$
\mathcal{Z}_n = \mathbb{P}(T \geq n) = \gamma_n(1) = \mathbb{E}\left(\prod_{0 \leq p < n} G_p(X_p) \right) = \prod_{0 \leq p < n} \eta_p(G_p)
$$

We let η_n^N be the particle density profiles associated with some McKean particle interpretation of the Feynman-Kac measures $\mathbb{Q}_n$, and defined in Section 4.4.5. Using (4.8), a particle approximation of the nonabsorption probabilities is given by

$$
\mathcal{Z}_n^N = \prod_{0 \leq p < n} \eta_p^N(G_p)
$$

More generally, similar arguments yield that is the distribution of a particle conditional upon being alive at time $n-1$ that is defined by the Feynman-Kac model introduced in (1.37); that is, we have that

$$
\mathbb{Q}_n(d(x_0, \ldots, x_n)) = \mathbb{P}\left((X_0^c, \ldots, X_n^c) \in d(x_0, \ldots, x_n) \mid T \geq n \right)
$$

Inversely, any Feynman-Kac model of the form (1.37) associated with some bounded potential functions G_n can be interpreted in terms of a particle absorption model. To prove this claim, we further assume that $\|G_n\| \leq c_n$ for some finite constant $c_n < \infty$. We let X_n^c be the Markov chain on $E_{n,c}$ defined in (7.1) with absorption rate $(1 - G_n(x_n)/c_n)$. By construction, we readily check that

$$\mathbb{Q}_n := \text{Law}\left((X_0^c, \ldots, X_n^c) \mid T \geq n\right)$$

In the context of interacting MCMC models (4.32) discussed in Section 4.3.1, the particle absorption interpretation of the Feynman-Kac model (4.37) can be seen as a novel sequential acceptance-refection algorithm. More precisely, if we consider the killing rate $G_n = h_n$ (see for instance (4.33) and (4.34)) and the MCMC transitions M_n s.t. $\mu_n = \mu_n M_n$ then we have

$$\mu_n = \text{Law}(X_n^c \mid T \geq n)$$

In contrast to conventional MCMC algorithms (4.35), nonabsorbed particles are prefect samples w.r.t. the desired target measures.

7.2 Mean field particle models

This section is dedicated to mean field simulation of nonabsorbed particle motions, in terms of genealogical tree based occupation measures. In Section 7.2.2, and Section 7.2.3, we provide particle estimates of absorption time distributions and occupation measures of nonabsorbed particles. In section 7.2.4, we design a backward particle integration model of nonabsorbed trajectories. Some illustrations of these models are presented in Section 7.3, including Markov chains evolving in a given tube, polymers models, and self-avoiding walks in some lattice.

7.2.1 Genealogical tree based measures

The law of a nonabsorbed Markov chain model can be estimated using the genealogical tree occupation measures η_n^N associated with some McKean particle interpretation of the Feynman-Kac measures $\mathbb{Q}_n$. A precise description of these models is provided in Section 2.2.3 and in section 4.1.2. More formally, for any $\mathbf{f_n} \in \mathcal{B}_b(E_0 \times \ldots \times E_n)$, we have the almost sure convergence result

$$\lim_{N \to \infty} \eta_n^N(\mathbf{f_n}) = \mathbb{E}\left(\mathbf{f_n}(X_0^c, \ldots, X_n^c) \mid T \geq n\right)$$

We also notice that

$$\text{Law}((X_0^c, \ldots, X_n^c) \mid T > n) = \widehat{\mathbb{Q}}_n$$

with the updated Feynman-Kac measures $\widehat{\mathbb{Q}}_n$ defined by

$$d\widehat{\mathbb{Q}}_n := \frac{1}{\widehat{Z}_n} \left\{ \prod_{0 \leq p \leq n} G_p(X_p) \right\} d\mathbb{P}_n$$

In the above display, $\widehat{\mathcal{Z}}_n$ is a normalizing constant and $\mathbb{P}_n$ is the distribution of the random paths

$$\mathbf{X_n} = (X_0, \dots, X_n) \in \mathbf{E_n} := (E_0 \times \dots \times E_n)$$

of the Markov process X_p from the origin $p = 0$ with initial distribution η_0, up to the current time $p = n$. When the potential functions G_n take values in $[0, 1]$, the set $E_n^0 = G_n^{-1}(\{0\})$ can be interpreted as a hard obstacle, in the sense that a particle with null potential is instantly killed. We set

$$\widehat{E}_n = E_n - E_n^0 = \{x_n \in E_n \ : \ G_n(x_n) > 0\} \quad \text{and} \quad \widehat{E}_n := (\widehat{E}_0 \times \dots \times \widehat{E}_n)$$

By construction, we have the following identity of measures on $\widehat{E}_n$

$$\eta_0(dx_0) G_0(x_0) \left\{ \prod_{1 \leq p \leq n} (M_p(x_{p-1}, dx_p) G_p(x_p)) \right\}$$

$$= \eta_0(G_0) \, \widehat{\eta}_0(dx_0) \left\{ \prod_{1 \leq p \leq n} \left(\widehat{G}_{p-1}(x_{p-1}) \widehat{M}_p(x_{p-1}, dx_p) \right) \right\}$$

with the updated measure $\widehat{\eta}_0 = \Psi_{G_0}(\eta_0)$, the potential functions $\widehat{G}_n$, and Markov transitions $\widehat{M}_n$ defined by

$$\widehat{G}_{n-1} := M_n(G_n) \quad \text{and} \quad \widehat{M}_n(x_{n-1}, dx_n) \propto M_n(x_{n-1}, dx_n) G_n(x_n)$$

This implies that

$$d\widehat{\mathbb{Q}}_n := \frac{1}{\widehat{\mathcal{Z}}_n} \left\{ \prod_{0 \leq p < n} \widehat{G}_p(\widehat{X}_p) \right\} d\widehat{\mathbb{P}}_n \tag{7.4}$$

where $\widehat{\mathbb{P}}_n$ stands for the distribution of the random paths $(\widehat{X}_0, \dots, \widehat{X}_n) \in \widehat{E}_n$ of the Markov process $\widehat{X}_p$, with initial distribution $\widehat{\eta}_0$, and elementary Markov transitions $\widehat{M}_n$.

This representation shows that the measures $\widehat{\mathbb{Q}}_n$ can be computed using the mean field genealogical model associated with the Feynman-Kac measures (1.37), with the parameters $(\widehat{G}_n, \widehat{M}_n)$. More precisely, if we denote by $\widehat{\eta}_n^N$ the occupation measures of this genealogical tree model, then for any $\mathbf{f_n} \in \mathcal{B}_b(E_n)$, we have the almost sure convergence result

$$\lim_{N \to \infty} \widehat{\eta}_n^N(\mathbf{f_n}) = \mathbb{E}\left(\mathbf{f_n}(X_0^c, \dots, X_n^c) \mid T > n\right)$$

In practice, the IPS genetic type model associated with the fitness potential functions and the mutation Markov transitions $(\widehat{G}_n, \widehat{M}_n)$ is more efficient than the one associated with the couple of parameters (G_n, M_n). For instance, for indicator potential functions $G_n = 1_A$, with some measurable subset $A \subset E = E_n$, the N-IPS model associated with (G_n, M_n) may be stopped when all the particles after the mutation step fail to enter into the desired set A. The analysis of this stopping time is developed in the monograph [163] (cf. Section 7.2.2, and Section 7.4.1). In this situation, the potential function

$$\widehat{G}_n(x_n) = M_{n+1}(x_n, A)$$

evaluates the chances to stay in the set A after a mutation step from x_n. The mutation transition

$$\widehat{M}_n(x_{n-1}, dx_n) = \mathbb{P}\left(X_n \in dx_n \mid X_{n-1} = x_{n-1}, \ X_n \in A\right)$$

is the local conditional transition w.r.t. the event that the transition remains in the desired subset A. In other words, the Markov chain $\widehat{X}_n$ starting in A, with elementary transitions $\widehat{M}_n$ is reflected at the boundary of the set A.

In the above discussion, it is implicitly assumed that we can evaluate the potential function $\widehat{G}_n$, and we can sample transitions according to the Markov transitions $\widehat{M}_n$. In general, these assumptions are not satisfied and we need to resort to additional levels of approximations. We shall return to this question in Section 7.5.2. Further details on these models and their particle interpretations can also be found in [152, 161, 163]. We mention that the Markov transitions $\widehat{M}_n$ only depend on a single potential function. Using a natural state space enlargement, we can also extend these models to Markov transitions that may depend on a finite collection of potential functions [152].

7.2.2 Absorption time distributions

Various statistical properties of the absorption process can be expressed in terms of the normalized distributions η_n. For instance, we have

$$\begin{aligned}
\mathbb{P}\left(T = n\right) &= \mathbb{P}(T \geq n) - \mathbb{P}\left(T \geq n+1\right) \\
&= \gamma_n(1) - \gamma_{n+1}(1) = \gamma_n(1) \ (1 - \eta_n(G_n))
\end{aligned}$$

from which we prove that

$$\mathbb{P}\left(T = n\right) = \left\{ \prod_{0 \leq p < n} \eta_p(G_p) \right\} \ (1 - \eta_n(G_n))$$

In other words, T can be thought as a nonhomogeneous geometric type absorption time, with parameters $(1 - \eta_n(G_n))$. In much the same way, the mean absorption time is given by the formula

$$\begin{aligned}
\mathbb{E}(T) &= \sum_{n \geq 1} n \ \mathbb{P}(T = n) \\
&= \sum_{n \geq 1} \mathbb{P}(T \geq n) = \sum_{n \geq 1} \gamma_n(1) = \sum_{n \geq 1} \left\{ \prod_{0 \leq p < n} \eta_p(G_p) \right\}
\end{aligned}$$

We let η_n^N be the particle density profiles associated with some McKean particle interpretation of the Feynman-Kac measures $\mathbb{Q}_n$ (cf. Section 4.4.5). In this situation, we have the almost sure convergence results

$$\lim_{N\to\infty} \sum_{n\geq 1} \left\{ \prod_{0\leq p<n} \eta_p^N(G_p) \right\} = \mathbb{E}(T)$$

$$\lim_{N\to\infty} \left\{ \prod_{0\leq p<n} \eta_p^N(G_p) \right\} \left(1 - \eta_n^N(G_n)\right) = \mathbb{P}\left(T=n\right)$$

Furthermore, using (4.9) the above particle estimates satisfy the unbiasedness property.

7.2.3 Occupation measures

The occupation measures of the nonabsorbed Markov chain evolution are given for any collection of functions $f_n \in \mathcal{B}_b(E_n)$ by the following formulae

$$\mathbb{E}\left(\sum_{0\leq p\leq n} f_p(X_p^c) 1_{E_p}(X_p^c) \right) = \sum_{0\leq p\leq n} \mathbb{E}\left(f_p(X_p^c)\, 1_{T\geq p}\right)$$

$$= \sum_{0\leq p\leq n} \gamma_p(f_pG_p) = \sum_{0\leq p\leq n} \eta_0 Q_{0,p}(f_pG_p)$$

with the Feynman-Kac semigroup

$$Q_{p,n}(f_n)(x_p) := \mathbb{E}\left(f_n(X_n) \prod_{p\leq q<n} G_q(X_q) \mid X_p = x_p \right)$$

We can also rewrite these formulae as follows:

$$\mathbb{E}\left(\sum_{0\leq p\leq n} f_p(X_p^c) 1_{E_p}(X_p^c) \right) = \sum_{0\leq p\leq n} \gamma_p(1)\, \eta_p(f_pG_p)$$

$$= \sum_{0\leq p\leq n} \gamma_{p+1}(1)\, \Psi_{G_p}(\eta_p)(f_p)$$

We let η_n^N be the particle density profiles associated with some McKean particle interpretation of the Feynman-Kac measures $\mathbb{Q}_n$ (cf. Section 4.4.5). In this situation, we have the almost sure convergence result

$$\lim_{N\to\infty} \sum_{0\leq p\leq n} \left\{ \prod_{0\leq q<p} \eta_q^N(G_q) \right\} \eta_p^N(f_pG_p) = \mathbb{E}\left(\sum_{0\leq p\leq n} f_p(X_p^c) 1_{E_p}(X_p^c) \right)$$

Using (4.9), the above particle estimates satisfy the unbiasedness property.

7.2.4　Backward Markov chain models

We further assume that the Markov transitions M_n are absolutely continuous with respect to some measures λ_n on E_n, and for any $(x, y) \in (E_{n-1} \times E_n)$ we have

$$H_n(x, y) := \frac{dM_n(x, \cdot)}{d\lambda_n}(y) > 0 \tag{7.5}$$

For any $n \geq 1$, we set

$$H_{n-1,n}(x_{n-1}, x_n) := G_{n-1}(x_{n-1}) \, H_n(x, y)$$

Using the backward representation formulae (1.50) presented in Section 1.4.3.4, we find that

$$\mathbb{P}\left((X_0^c, \ldots, X_n^c) \in d(x_0, \ldots, x_n) \mid T \geq n\right)$$

$$= \eta_n(dx_n) \, \mathbb{M}_{n,\eta_{n-1}}(x_n, dx_{n-1}) \cdots \mathbb{M}_{1,\eta_0}(x_1, dx_0) := \mathbb{Q}_n(d(x_0, \ldots, x_n))$$

with the time reversal Markov transitions $\mathbb{M}_{n+1,\eta_n}$ defined by

$$\mathbb{M}_{n+1,\eta_n}(x_{n+1}, dx_n) \propto \eta_n(dx_n) H_{n,n+1}(x_n, dx_{n+1})$$

For any collection of functions $f_n \in \mathcal{B}_b(E_n)$, the occupation measure of the particle model before the absorption time is now given by the formula

$$\mathbb{E}\left(\frac{1}{n+1} \sum_{0 \leq p \leq n} f_p(X_p^c) \mid T \geq n\right) = \frac{1}{n+1} \sum_{0 \leq p \leq n} \eta_n \mathbb{M}_{n,\eta_{n-1}} \cdots \mathbb{M}_{p+1,\eta_p}(f_p)$$

We recall from (2.13) that the N-particle approximation measure $\mathbb{Q}_n^N$ associated with these backward models is given by

$$\mathbb{Q}_n^N(d(x_0, \ldots, x_n)) = \eta_n^N(dx_n) \prod_{q=1}^n \mathbb{M}_{q,\eta_{q-1}^N}(x_q, dx_{q-1})$$

In the above display, η_n^N stands for the particle density profiles associated with some McKean particle interpretation of the Feynman-Kac measures $\mathbb{Q}_n$ (cf. Section 4.4.5). In this situation, we have the almost sure convergence result

$$\lim_{N \to \infty} \frac{1}{n+1} \sum_{0 \leq p \leq n} \eta_n^N \mathbb{M}_{n,\eta_{n-1}^N} \cdots \mathbb{M}_{p+1,\eta_p^N}(f_p)$$

$$= \mathbb{E}\left(\frac{1}{n+1} \sum_{0 \leq p \leq n} f_p(X_p^c) \mid T \geq n\right)$$

For a detailed discussion on these backward particle models, we refer the reader to Section 2.2.6, Section 4.1.6, and Chapter 17, dedicated to the convergence properties of these models.

7.3 Some illustrations

7.3.1 Restriction of a Markov chain in a tube

One of the simplest examples of Feynman-Kac model (1.37) is given by choosing indicator functions $G_n = 1_{A_n}$ of measurable subsets $A_n \subset E_n$ s.t. $\mathbb{P} (\forall 0 \leq p < n \ X_p \in A_p) > 0$. In this situation, we have

$$
\begin{aligned}
\mathbb{Q}_n &= \text{Law}((X_0, \ldots, X_n) \mid \forall 0 \leq p < n \ X_p \in A_p) \qquad (7.6) \\
\mathcal{Z}_n &= \mathbb{P} (\forall 0 \leq p < n \ X_p \in A_p)
\end{aligned}
$$

This Markov chain restriction model fits into the particle absorption model (3.30) presented in Section 7.1. We illustrate the rather abstract model with an elementary symmetric random walk X_n on the integers $\mathbb{Z}$ starting at the origin $X_0 = 0$. More formally, we take independent random variables U_n, where $\mathbb{P} (U_n = 1) = \mathbb{P} (U_n = -1) = 1/2$. In this situation, we can take $X_n = X_0 + \sum_{1 \leq p \leq n} U_p$, and $A = [-a, a] \cap \mathbb{Z}$, with some $a \in \mathbb{N}$.

7.3.2 Polymers and self-avoiding walks

Conformation models of polymers in a chemical solvent can be seen as the realization of a Feynman-Kac distribution of a free Markov chain, weighted by some Boltzmann-Gibbs exponential weight function. These potential functions reflect the attraction, or the repulsion, forces between the monomers.

Self-avoiding walks are defined in terms of the historical process

$$
\mathbf{X_n} = (X_0, \ldots, X_n) \in \boldsymbol{E_n} = (E_0 \times \ldots \times E_n)
$$

of some random walk X_n evolving in some lattice $E_n = \mathbb{Z}^d$; and the repulsive potential functions

$$
\mathbf{G_n}(\mathbf{X_n}) = 1_{\notin \{X_p, \ p < n\}}(X_n)
$$

In this situation, the Feynman-Kac measures $\mathbb{Q}_n$, and their n-th time marginal defined in (1.37), and resp. in (1.39), have been given by

$$
\begin{aligned}
\mathbb{Q}_n &= \text{Law}(\mathbf{X_n} \mid \forall 0 \leq p < n \quad \mathbf{X_p} \in A_p) \\
\eta_n &= \text{Law}((X_0, \ldots, X_n) \mid \forall 0 \leq p < q < n \quad X_p \neq X_q) \\
\mathcal{Z}_n &= \mathbb{P}(\forall 0 \leq p < q < n \quad X_p \neq X_q)
\end{aligned}
$$

with the set $A_n = \mathbf{G_n}^{-1}(\{1\}) \subset \boldsymbol{E_n}$. We can also show that this self-avoiding walk model fits into the Markov restriction model discussed in Section 7.3.1.

7.4   Absorption models in random environments

This section is dedicated to particle absorption models in random environments. In section 7.4.1, we introduce quenched and annealed Feynman-Kac models. The random environment evolution is represented by a Markov chain in the set of environments. Quenched models correspond to nonabsorbed conditional distributions w.r.t. a given environment sequence. The annealed models are derived by integrating the environment sequences.

In some instances, such as for linear-Gaussian models and quadratic energy functions, these quenched distribution flows can be computed using Kalman type recursions. In more general situations, the quenched models associated with a fixed value of the environment can be also computed using the mean field simulation techniques discussed in section 7.2.

The computation of annealed models is much more involved, mainly because it requires the integration of the environment realizations. In section 7.4.2 we design a mean field strategy to sample the annealed Feynman-Kac models, in terms of interacting and mean field quenched Feynman-Kac flows.

7.4.1   Quenched and annealed models

Markov chains evolving in an absorbing random environment are defined in terms of the quenched and annealed models discussed in the opening chapter and Section 2.3.2. These models are defined in terms of a Markov chain model

$$\boldsymbol{X_n} := (\Theta_n, X_n)$$

taking values in some product state spaces $\boldsymbol{E_n} := (\Omega_n \times E_n)$. We let $\boldsymbol{M_n}$ be the elementary transitions of the chain $\boldsymbol{X_n}$ from $\boldsymbol{E_{n-1}}$ into $\boldsymbol{E_n}$. We also consider the sequence of $[0,1]$-valued potential functions

$$\boldsymbol{G_n} \; : \; \boldsymbol{x_n} \in \boldsymbol{E_n} \; \mapsto \; \boldsymbol{G_n(x_n)} \in [0,1]$$

The first component Θ_n of the chain $\boldsymbol{X_n}$ represents the random environment evolution, and it is assumed to be a Markov chain taking values in the set of environments Ω_n. Given a sequence of random environments $\Theta := (\Theta_n)_{n \geq 0}$, we also assume that the second component X_n is a Markov chain with elementary transitions

$$\mathbb{P}\left(X_n \in dx_n \mid X_{n-1}, \; \Theta\right) = M_{n,\Theta_n}(X_{n-1}, dx_n)$$

The Markov chain $\boldsymbol{X_n} := (\Theta_n, X_n)$ is sometimes called the *bichain* (cf. [70]). We let $\boldsymbol{X_n^c}$ be the particle absorption model on $\boldsymbol{E_{n,c}} = \boldsymbol{E_n} \cup \{c\}$ with absorption rate $(1 - \boldsymbol{G_n})$ and absorption time T. We also have by $(\mathbb{Q}_n, \boldsymbol{\gamma_n}, \boldsymbol{\eta_n})$

the annealed Feynman-Kac models associated with the potential functions and the Markov transitions (G_n, M_n).

By construction, we have

$$\mathbb{Q}_n(d(x_0, \ldots, x_n)) = \mathbb{P}\left((X_0^c, \ldots, X_n^c) \in d(x_0, \ldots, x_n) \mid T \geq n\right)$$

as well as the formula

$$\mathbb{P}(T \geq n) = \mathbb{E}\left(\prod_{0 \leq p < n} G_p(X_p)\right) = \gamma_n(1) = \prod_{0 \leq p < n} \eta_p(G_p)$$

Definition 7.4.1 *Given a realization of the environment*

$$(\Theta_n)_{n \geq 0} = \Theta = \theta = (\theta_n)_n \in \prod_{n \geq 0} \Omega_n$$

we let $X_{\theta,n}^c$ be the particle $E_{c,n}$-valued Markov chain with absorption probabilities $(1 - G_{n,\theta_n}(x_n))$, with the potential function

$$G_{n,\theta}(x_n) := G_n(\theta_n, x_n)$$

We also have by $(\mathbb{Q}_{n,\theta}, \gamma_{n,\theta}, \eta_{n,\theta})$ the quenched Feynman-Kac models associated with the potential functions and the Markov transitions $(G_{n,\theta_n}, M_{n,\theta_n})$.

By construction, we have

$$\mathbb{Q}_{n,\theta}(d(x_0, \ldots, x_n)) = \mathbb{P}\left((X_0^c, \ldots, X_n^c) \in d(x_0, \ldots, x_n) \mid T \geq n, \ \Theta = \theta\right)$$

as well as the formula

$$\mathbb{P}(T \geq n \mid \Theta = \theta) = \mathbb{E}\left(\prod_{0 \leq p < n} G_{p,\Theta}(X_p) \mid \Theta = \theta\right) = \prod_{0 \leq p < n} \eta_{p,\theta}(G_{p,\theta}) \tag{7.7}$$

Given a realization of the environment $\Theta = \theta$, we recall by (3.28) that $\eta_{n,\theta}$ satisfies the deterministic measure valued evolution equation

$$\eta_{n+1,\theta} = \Psi_{G_{n,\theta_n}}(\eta_{n,\theta}) M_{n+1,\theta_{n+1}} \tag{7.8}$$

Remark 7.4.2 *In some situations, the environment is associated with a static random parameter Θ whose values do not change w.r.t. the time parameter. This situation is discussed in Section 3.3.2.*

The IPS simulation of the corresponding conditional distributions is presented in Section 4.3.6.

We also mention that the sensitivity measures associated with these quenched absorption models are discussed in Section 3.5.

7.4.2   Feynman-Kac models in distribution spaces

The quenched, and the annealed, particle absorption models discussed in Section 7.4.1 can be computed using the mean field IPS models presented in Chapter 4. In this Section, we present an alternative description of the annealed Feynman-Kac models in terms of the Feynman-Kac models defined below.

Definition 7.4.3 *We let $\mathcal{Q}_n$ be the Feynman-Kac measures associated with the reference Markov chain*

$$\mathcal{X}_n := \left(\mathcal{X}_n^1, \mathcal{X}_n^2\right) = (\Theta_n, \eta_{n,\Theta}) \in (\Omega_n \times \mathcal{P}(E_n)) \tag{7.9}$$

and the potential functions

$$\mathcal{G}_n(\mathcal{X}_n) = \eta_{n,\Theta}(G_{n,\Theta_n}) \tag{7.10}$$

We denote by $\mathcal{M}_n$ the elementary Markov transitions of the chain $\mathcal{X}_n$.

Theorem 7.4.4 *For any $n \geq 0$ and any function $f_n \in \mathcal{B}_b(\prod_{0 \leq p \leq n} \Omega_p)$ we have the Feynman-Kac formulae*

$$\mathbb{E}\left(f_n(\Theta_0, \ldots, \Theta_n) \prod_{0 \leq p < n} \boldsymbol{G_p}(\Theta_p, X_p)\right)$$

$$= \mathbb{E}\left(f_n(\mathcal{X}_0^1, \ldots, \mathcal{X}_n^1) \prod_{0 \leq p < n} \mathcal{G}_p(\mathcal{X}_p)\right)$$

In addition, for any function $f_n \in \mathcal{B}_b(\prod_{0 \leq p \leq n} E_p)$ we have the Feynman-Kac formulae

$$\mathbb{E}\left(f_n(X_0, \ldots, X_n) \prod_{0 \leq p < n} \boldsymbol{G_p}(\Theta_p, X_p)\right)$$

$$= \mathbb{E}\left(\mathbb{Q}_{n,\mathcal{X}^1}(f_n) \prod_{0 \leq p < n} \mathcal{G}_p(\mathcal{X}_p)\right)$$

Proof:
We use (7.7) to prove that

$$\mathbb{E}\left(f_n(\Theta_0, \ldots, \Theta_n) \prod_{0 \leq p < n} G_p(\Theta_p, X_p)\right)$$

$$= \mathbb{E}\left(f_n(\Theta_0, \ldots, \Theta_n) \mathbb{E}\left(\prod_{0 \leq p < n} G_p(\Theta_p, X_p) | \Theta\right)\right)$$

$$= \mathbb{E}\left(f_n(\Theta_0, \ldots, \Theta_n) \prod_{0 \leq p < n} \eta_{p,\Theta}(G_{p,\Theta})\right)$$

This ends the proof of the first assertion. We prove the second claim using the fact that

$$\mathbb{E}\left(f_n(X_0,\ldots,X_n)\ \prod_{0\le p<n} G_p(\Theta_p, X_p)\right)$$

$$= \mathbb{E}\left(\mathbb{E}\left(\prod_{0\le p<n} G_p(\Theta_p, X_p)\,|\Theta\ \right)\ \frac{\mathbb{E}\left(f_n(X_0,\ldots,X_n)\ \prod_{0<p<n} G_p(\Theta_p,X_p)|\Theta\ \right)}{\mathbb{E}\left(\prod_{0\le p<n} G_p(\Theta_p,X_p)|\Theta\ \right)}\right)$$

$$= \mathbb{E}\left(\mathbb{Q}_{n,\Theta}(f_n)\ \prod_{0\le p<n} \eta_{p,\Theta}(G_{p,\Theta_p})\ \right)$$

This ends the proof of the theorem. ∎

Definition 7.4.5 *We denote by* $\overline{\mathcal{Q}}_n$ *the probability measure on the path space* $\overline{E}_n := (E_0 \times \ldots \times E_n)$ *defined for any function* $\mathbf{f_n} \in \mathcal{B}_b(\overline{E}_n)$ *by*

$$\overline{\mathcal{Q}}_n(\mathbf{f_n}) \propto \mathbb{E}\left(\mathbb{Q}_{n,\Theta}(\mathbf{f_n})\ \prod_{0\le p<n} \eta_{p,\Theta}(G_{p,\Theta_p})\right)$$

Corollary 7.4.6 *We let* $\mathbb{Q}_n$ *be the Feynman-Kac measures associated with the potential functions and the Markov transitions* $(\mathbf{G_n}, \mathbf{M_n})$ *defined in Section 7.4.1. We also denote by* $\mathcal{Q}_n$ *the Feynman-Kac measures associated with the potential functions and the Markov transitions* $(\mathcal{G}_n, \mathcal{M}_n)$.
In this situation, the following assertions hold:

- *The* $(\Theta_0,\ldots,\Theta_n)$-*marginals of the Feynman-Kac measures* $\mathbb{Q}_n$ *and* $\mathcal{Q}_n$ *coincide with the conditional distributions of the random environment sequence* $(\Theta_0,\ldots,\Theta_n)$ *w.r.t. the nonabsorption event* $\{T \ge n\}$; *that is, we have that*

$$\mathbb{Q}_n \circ (\Theta_0,\ldots,\Theta_n)^{-1} = \mathcal{Q}_n \circ (\Theta_0,\ldots,\Theta_n)^{-1}$$
$$= \text{Law}\left((\Theta_0,\ldots,\Theta_n)\ |\ T \ge n\ \right)$$

- *The* $(X_0,\ldots,X_n)$-*marginals of the Feynman-Kac measures* $\mathbb{Q}_n$ *coincide with the conditional distributions of the random states* $(X_0,\ldots,X_n)$ *w.r.t. the nonabsorption event* $\{T \ge n\}$; *that is, we have that*

$$\mathbb{Q}_n \circ (X_0,\ldots,X_n)^{-1} = \overline{\mathcal{Q}}_n = \text{Law}\left((X_0,\ldots,X_n)\ |\ T \ge n\ \right)$$

The sampling of Markov chain X_n defined in (7.9) requires sampling random environments transitions $\Theta_{n-1} \rightsquigarrow \Theta_n$ and solving the measure valued Equation (7.8). In some instances, such as for linear-Gaussian models and Gaussian type potential functions, the solving of (7.8) can be done explicitly using the traditional Kalman updating and prediction steps (c.f. Section 8.3

and Section 8.4). In this situation, the N-particle model associated with the Feynman-Kac models defined in Proposition 7.4.4 is a genetic type mean field particle model, with fitness potential function $\mathcal{G}_n$ and mutation transitions given by the Markov chain $\mathcal{X}_n$. For a detailed description of these IPS models, we refer the reader to Section 4.1.1 and to Section 4.4.5. An illustration of these models in the context of interacting Kalman filters arising in advanced signed processing is provided in Section 8.4.

7.5   Particle Feynman-Kac models

7.5.1   Particle quenched models

We return to the Feynman-Kac models discussed in Section 7.4.2. As we mentioned in the end of this Section, the sampling of the Markov chain $\mathcal{X}_n$ defined in (7.9) requires solving a Feynman-Kac evolution equation in distribution spaces. In the present Section, we present a particle Feynman-Kac model that combines the evolution of the environment parameter with the mean field sampling schemes described in Section 4.4.5.

Definition 7.5.1 *Given a realization of the environment*

$$(\Theta_n)_{n\geq 0} = \Theta = \theta = (\theta_n)_n \in \prod_{n\geq 0} \Omega_n := \overline{\Omega}_n.$$

we let $\eta_{n,\theta}^N$ be the N-particle density profiles of the IPS approximation of the Feynman-Kac model associated with the potential functions and the Markov transitions $(G_n, M_n) := (G_{n,\theta_n}, M_{n,\theta_n})$.
        We also denote by $\boldsymbol{\eta}_{\boldsymbol{\theta},n}^{\boldsymbol{N}}$, and resp. $\mathbb{Q}_{\theta,n}^N$, the genealogical tree based measures, and resp. the backward particle model, on $\overline{\boldsymbol{E}}_{\boldsymbol{n}} = (E_0 \times \ldots \times E_n)$.

More precise descriptions of these genealogical, and complete ancestral lines, measures are provided in (2.10) and (2.13). We also refer to Section 4.1.2, and Section 4.1.6. A more detailed description of the mean field schemes is provided in Section 4.1.1 and in Section 4.4.5.

In this section, we present an alternative description of the annealed Feynman-Kac models in terms of the particle measures defined below.

Definition 7.5.2 *We consider Feynman-Kac measures $\boldsymbol{\mathcal{Q}}_{\boldsymbol{n}}^{\boldsymbol{N}}$ associated with the reference Markov chain*

$$\mathcal{X}_n^N := \left(\Theta_n, \eta_{n,\Theta}^N\right) \in \left(\Omega_n \times \mathcal{P}(E_n)\right) \tag{7.11}$$

and the potential functions

$$\mathcal{G}_n(\mathcal{X}_n^N) = \eta_{n,\Theta}^N(G_{n,\Theta_n})$$

We denote by $\mathcal{M}_n^N$ the elementary Markov transitions of the chain $\mathcal{X}_n^N$.

Using the unbiasedness properties presented in (2.12) and (2.14), and invoking Theorem 7.4.4, we prove the following theorem.

Theorem 7.5.3 *For any $n \geq 0$ and any function $f_n \in \mathcal{B}_b(\overline{\Omega}_n)$ we have the Feynman-Kac formulae*

$$\mathbb{E}\left(f_n(\Theta_0,\ldots,\Theta_n) \prod_{0 \leq p < n} G_p(\Theta_p, X_p)\right)$$

$$= \mathbb{E}\left(\varphi_n(\mathcal{X}_0^N,\ldots,\mathcal{X}_n^N) \prod_{0 \leq p < n} \mathcal{G}_p(\mathcal{X}_p^N)\right)$$

with the function

$$\varphi_n(\mathcal{X}_0^N,\ldots,\mathcal{X}_n^N) = f_n(\Theta_0,\ldots,\Theta_n)$$

In addition, for any function $f_n \in \mathcal{B}_b(\overline{E}_n)$ we have the Feynman-Kac formulae

$$\mathbb{E}\left(f_n(X_0,\ldots,X_n) \prod_{0 \leq p < n} G_p(\Theta_p, X_p)\right)$$

$$= \mathbb{E}\left(\psi_n(\mathcal{X}_0^N,\ldots,\mathcal{X}_n^N) \prod_{0 \leq p < n} \mathcal{G}_p(\mathcal{X}_p^N)\right)$$

with the function

$$\psi_n(\mathcal{X}_0^N,\ldots,\mathcal{X}_n^N) = \mathbb{Q}_{n,\Theta}^N(f_n)$$

Definition 7.5.4 *We denote by $\overline{\mathcal{Q}}_n^N$ the probability measure on $\overline{E}_n$ defined for any function $\mathbf{f_n} \in \mathcal{B}_b(\overline{E}_n)$ by*

$$\overline{\mathcal{Q}}_n^N(\mathbf{f_n}) \propto \mathbb{E}\left(\mathbb{Q}_{n,\Theta}^N(\mathbf{f_n}) \prod_{0 \leq p < n} \eta_{p,\Theta}^N(G_{p,\Theta_p})\right)$$

The particle version of Corollary 7.4.6 now takes the following form.

Corollary 7.5.5 *We let $\mathbb{Q}_n$ be the Feynman-Kac measures associated with the potential functions and the Markov transitions (G_n, M_n) defined in Section 7.4.1. We also denote by $\mathcal{Q}_n^N$ the Feynman-Kac measures associated with the potential functions and the Markov transitions $(\mathcal{G}_n, \mathcal{M}_n^N)$*
In this situation, the following assertions hold:

- *The $(\Theta_0,\ldots,\Theta_n)$-marginals of the Feynman-Kac measures $\mathbb{Q}_n$ and $\mathcal{Q}_n^N$ coincide with the conditional distributions of the random environment sequence $(\Theta_0,\ldots,\Theta_n)$ w.r.t. the nonabsorption event $\{T \geq n\}$; that is, we have that*

$$\mathbb{Q}_n \circ (\Theta_0,\ldots,\Theta_n)^{-1} = \mathcal{Q}_n^N \circ (\Theta_0,\ldots,\Theta_n)^{-1}$$
$$= \text{Law}\left((\Theta_0,\ldots,\Theta_n) \mid T \geq n\right)$$

- *The $(X_0, \ldots, X_n)$-marginals of the Feynman-Kac measures $\mathbb{Q}_n$ coincide with the conditional distributions of the random states $(X_0, \ldots, X_n)$ w.r.t. the nonabsorption event $\{T \geq n\}$; that is, we have that*

$$\mathbb{Q}_n \circ (X_0, \ldots, X_n)^{-1} \;=\; \overline{\mathcal{Q}}_n^N = \mathrm{Law}\left((X_0, \ldots, X_n) \mid T \geq n\right)$$

The sampling of Markov chain $\mathcal{X}_n^N$ defined in (7.11) now requires sampling random environments transitions $\Theta_{n-1} \rightsquigarrow \Theta_n$ and running the mean field N-IPS transitions $\eta_{n-1,\Theta}^N \rightsquigarrow \eta_{n,\Theta}^N$.

In this situation, the N-particle model associated with the Feynman-Kac models defined in Proposition 7.5.3 is a genetic type particle model with fitness potential function $\mathcal{G}_n$ and mutation transitions given by the Markov chain $\mathcal{X}_n^N$. For a detailed description of these IPS models, we refer the reader to Section 4.1.1 and to Section 4.4.5. An illustration of these models in the context of course grained nonlinear filtering models is provided in Section 8.5.

7.5.2   Particle updated measures

We return to the updated Feynman-Kac measures $\widehat{\mathbb{Q}}_n$ presented in (7.4). To simplify the presentation, we further assume that $G_0 = 1$.

Definition 7.5.6 *We consider the Markov chain*

$$X_n = (\mathcal{X}_n^1, \mathcal{X}_n^2) := \left(X_n^N, m(Y_{n+1})\right) \in (E_n \times \mathcal{P}(E_{n+1})) \tag{7.12}$$

defined as follows:

- *For $n = 0$, we set $X_0^N = X_0$.*

- *For any $n \geq 1$, $m(Y_n) = \frac{1}{N}\sum_{i=1}^{N} \delta_{Y_n^i}$ stands for the empirical measure associated with N independent random samples $Y_n := (Y_n^i)_{1 \leq i \leq N}$ with common distribution $M_n(X_{n-1}^N, \cdot)$.*

- *For any $n \geq 1$, X_n^N is a random variable with distribution $m(Y_n)$.*

We also consider the potential functions $\mathcal{G}_n$ on $(E_n \times \mathcal{P}(E_{n+1}))$ defined for any $n \geq 0$ by

$$\mathcal{G}_n(\mathcal{X}_n) = G_n(X_n^N)$$

The following lemma states an important unbiasedness property.

Lemma 7.5.7 *For any $n \geq 0$ and any function $\mathbf{f_n} \in \mathcal{B}_b(E_n)$ we have*

$$\widehat{\mathbb{Q}}_n(\mathbf{f_n}) \propto \mathbb{E}\left(\mathbf{f_n}(\mathcal{X}_0^1, \ldots, \mathcal{X}_n^1) \prod_{0 \leq p \leq n} \mathcal{G}_p(\mathcal{X}_p)\right)$$

Proof:

The lemma is a direct consequence of the following formula

$$\mathbb{E}\left(\mathbf{f_n}(X_0^N,\ldots,X_n^N)\ \prod_{0\leq p\leq n} G_p(X_p^N)\right)$$

$$= \int \mathbf{f_n}(x_0,\ldots,x_n)\ \left\{\prod_{0\leq p\leq n} G_p(x_p)\right\}$$

$$\times \eta_0(dx_0)\ \prod_{1\leq p\leq n}\left(M_p^{\otimes N}(x_{p-1},dy_p)m(y_p)(dx_p)\right)$$

In the above display $m(y_p)$ stands for the empirical measures $m(y_p) = \frac{1}{N}\sum_{i=1}^{N}\delta_{y_p^i}$. We end the proof of desired assertion integrating the y_p-variables. This ends the proof of the lemma. ∎

Definition 7.5.8 *We consider the Markov chain*

$$\widehat{\mathcal{X}}_n = \left(\widehat{\mathcal{X}}_n^1,\widehat{\mathcal{X}}_n^2\right) := \left(\widehat{X}_n^N, m(Y_{n+1})\right) \in (E_n \times \mathcal{P}(E_{n+1})) \tag{7.13}$$

defined as follows:

- *For $n = 0$, $\widehat{X}_0^N = X_0$ is a r.v. with distribution $\Psi_{G_0}(\eta_0) = \eta_0$.*

- *For $n \geq 1$, $m(Y_n) = \frac{1}{N}\sum_{i=1}^{N}\delta_{Y_n^i}$ is the empirical measure associated with N independent r.v. $Y_n := (Y_n^i)_{1\leq i\leq N}$ with common distribution $M_n(\widehat{X}_{n-1}^N, \cdot)$.*

- *For any $n \geq 1$, $\widehat{X}_n^N$ is a r.v. with distribution $\Psi_{G_n}(m(Y_n))$.*

We also consider the potential functions $\mathcal{G}_n$ on $(E_n \times \mathcal{P}(E_{n+1}))$ defined for any $n \geq 1$ by

$$\widehat{\mathcal{G}}_{n-1}(\widehat{\mathcal{X}}_{n-1}) = m(Y_n)(G_n)$$

Notice that for $n = 1$, we have

$$\widehat{\mathcal{G}}_0(\widehat{\mathcal{X}}_0) = m(Y_1)(G_1)$$

We also mention that the sampling of Markov chain

$$\widehat{\mathcal{X}}_{n-1} = \left(\widehat{X}_{n-1}^N, m(Y_n)\right) \rightsquigarrow \widehat{\mathcal{X}}_n = \left(\widehat{X}_n^N,, m(Y_{n+1})\right)$$

defined in (7.13) now requires sampling at every time step a r.v. $\widehat{X}_n^N$ with distribution $\Psi_{G_n}(m(Y_n))$ and N random states $(Y_{n+1}^i)_{1\leq i\leq N}$ with common distribution $M_{n+1}(\widehat{X}_n^N, dx_{n+1})$.

Definition 7.5.9 *We denote by* $\widehat{\mathcal{Q}}_n$ *the Feynman-Kac measures on* $\prod_{0\leq p\leq n}(E_p \times \mathcal{P}(E_{p+1}))$ *associated with the reference Markov chain* $\widehat{\mathcal{X}}_n$ *and the potential functions* $\widehat{\mathcal{G}}_n$

$$d\widehat{\mathcal{Q}}_n = \frac{1}{\widehat{\mathcal{Z}}_n}\left\{\prod_{0\leq p<n}\widehat{\mathcal{G}}_p(\widehat{\mathcal{X}}_p)\right\}d\widehat{\mathcal{P}}_n$$

where $\widehat{\mathcal{P}}_n$ *stands for the distribution of the random paths* $\left(\widehat{\mathcal{X}}_0,\ldots,\widehat{\mathcal{X}}_n\right)$*, and* $\widehat{\mathcal{Z}}_n$ *stands for some normalizing constant.*

The N-particle model associated with the Feynman-Kac models $\widehat{\mathcal{Q}}_n$ is a genetic type particle model with fitness potential function $\widehat{\mathcal{G}}_n$ and mutation transitions given by the Markov chain $\widehat{\mathcal{X}}_n$. For a detailed description of these IPS models, we refer the reader to Section 4.1.1 and Section 4.4.5.

Lemma 7.5.10 *For any* $n \geq 0$*, and any bounded measurable function* φ_n *on the product space* $\prod_{0\leq p\leq n}(E_p \times \mathcal{P}(E_{p+1}))$ *we have*

$$\widehat{\mathcal{Q}}_n(\varphi_n) \propto \mathbb{E}\left(\varphi_n(\mathcal{X}_0,\ldots,\mathcal{X}_n)\prod_{0\leq p\leq n}\mathcal{G}_p(\mathcal{X}_p)\right)$$

Proof:
By construction, we readily check that

$$\prod_{1\leq p\leq n}\left\{m(y_p)(G_p)\Psi_{G_p}(m(y_p))(dx_p)\right\} = \prod_{1\leq p\leq n}(m(y_p)(dx_p)G_p(x_p))$$

This implies that

$$\eta_0(dx_0)\prod_{1\leq p\leq n}\left(M_p^{\otimes N}(x_{p-1},dy_p)m(y_p)(dx_p)G_p(x_p)\right)$$

$$= \left\{\prod_{0\leq p<n}m(y_{p+1})(G_{p+1})\right\}$$

$$\times\eta_0(dx_0)\prod_{1\leq p\leq n}\left(M_p^{\otimes N}(x_{p-1},dy_p)\ \Psi_{G_p}(m(y_p))(dx_p)\right)$$

$$= \left\{\prod_{0\leq p<n}\widehat{\mathcal{G}}_p(x_p,m(y_{p+1}))\right\}$$

$$\times\eta_0(dx_0)\prod_{1\leq p\leq n}\left(M_p^{\otimes N}(x_{p-1},dy_p)\ \Psi_{G_p}(m(y_p))(dx_p)\right)$$

This ends the proof of the lemma. ∎

Using the above lemmas, we readily check the following proposition.

Proposition 7.5.11 *The* $(\widehat{X}_0^N, \ldots, \widehat{X}_n^N)$*-marginals of the Feynman-Kac measures* $\widehat{\mathcal{Q}}_n$ *coincide with the conditional distributions of the random states* $(X_0, \ldots, X_n)$ *w.r.t. the nonabsorption event* $\{T > n\}$*; that is, we have that*

$$\widehat{\mathcal{Q}}_n \circ (\widehat{\mathcal{X}}_0^1, \ldots, \widehat{\mathcal{X}}_n^1)^{-1} \;=\; \text{Law} \left((X_0, \ldots, X_n) \mid T > n \right) = \widehat{\mathbb{Q}}_n$$

7.5.3 Some state space enlargements

In the further development of this section, $r \geq 1$ stands for a given fixed parameter and X'_n some Markov chain taking values in some state spaces E'_n, with elementary transitions M'_n and initial distribution η'_0. We also denote by $\mathbb{P}'_n$ the distributions of the random paths $(X'_0, \ldots, X'_n)$.

In Section 7.5.2 we have designed a particle methodology to approximate a given updated measure using a genetic type IPS scheme equipped with a mutation transition that depends on a single potential function. In this section, we present a natural state space enlargement that allows using a collection of r potential functions.

To describe with some conciseness these models, we need to introduce some notations.

Definition 7.5.12 *We consider the Markov chain defined by*

$$X_n := \left(X'_p \right)_{nr \leq p \leq (n+1)r} \in E_n := \prod_{nr \leq p \leq (n+1)r} E'_p$$

and we let M_n *be the elementary transitions of* X_n *and* $\eta_0 := \text{Law}(X_0)$ *its initial distribution. We also let* G_n *be the potential functions on* E_n *defined by*

$$G_n(X_n) := \prod_{nr \leq p < (n+1)r} G'_p \left(X'_p \right)$$

with some collection of potential functions G'_n *on* E'_n*.*

Definition 7.5.13 *For any* $n \geq 1$ *we denote by* $\widehat{X}_n$ *the* E_n*-valued Markov chain with elementary transitions defined for any* $x_{n-1} \in E_{n-1}$ *by*

$$\widehat{M}_n(x_{n-1}, dy_n) := \frac{M_n(x_{n-1}, dy_n) G_n(y_n)}{M_n(G_n)(x_{n-1})}$$

We also consider the potential functions on E_{n-1} *defined by*

$$\widehat{G}_{n-1}(x_{n-1}) = M_n(G_n)(x_n)(x_{n-1})$$

In this notation, we readily prove the following proposition.

Proposition 7.5.14 *For any $n \geq 0$, we have the following equivalent formulations*

$$d\widehat{\mathbb{Q}}_n \quad := \quad \frac{1}{\mathcal{Z}_n} \left\{ \prod_{0 \leq p \leq n} G_p(X_p) \right\} d\mathbb{P}_n = \frac{1}{\widehat{\mathcal{Z}}_n} \left\{ \prod_{0 \leq p < n} \widehat{G}_p(\widehat{X}_p) \right\} d\widehat{\mathbb{P}}_n$$

$$= \frac{1}{\mathcal{Z}'_{(n+1)r}} \left\{ \prod_{0 \leq p < (n+1)r} G'_p(X'_p) \right\} d\mathbb{P}'_{(n+1)r}$$

In the above display, $\mathcal{Z}_n$, $\widehat{\mathcal{Z}}_n$, and $\mathcal{Z}'_{(n+1)r}$ stand for some normalizing constants. The measures $\mathbb{P}_n$, and resp. $\widehat{\mathbb{P}}_n$, are the distributions of the chain $(X_0, \ldots, X_n)$ with the initial law η_0 and the Markov transitions M_n, and resp. the distribution of the random paths $(\widehat{X}_0, \ldots, \widehat{X}_n)$ with the initial law $\widehat{\eta}_0 := \Psi_{G_0}(\eta_0)$ and the Markov transitions $\widehat{M}_n$.

This proposition allows applying without further work the updated IPS simulation models developed in Section 7.5.2.

We end this section with an alternative representation. We denote by $\mathbb{P}'_{nr,(n+1)r}$ the conditional distributions

$$\mathbb{P}'_{nr,(n+1)r}(x'_{nr}, dy_n) = \mathbb{P}\left(\left(X'_{nr}, \ldots, X'_{(n+1)r} \right) \in dy_n \mid X'_{nr} = x'_{nr} \right)$$

where dy_n stands for an infinitesimal neighborhood of the point $y_n \in E_n$.

Definition 7.5.15 *We also let $\mathbb{Q}'_{nr,(n+1)r}(x'_{nr}, \cdot)$ be the Feynman-Kac measures on E_n, defined as the measure $\mathbb{Q}_r$ given in (1.37), by replacing*

$$(\eta_0, (X_p)_{0 \leq p \leq r}; (G_p)_{0 \leq p < r}) \quad by \quad (\delta_{x'_{nr}}; (X'_{nr+p})_{0 \leq p \leq r}; (G'_{nr+p})_{0 \leq p < r})$$

More formally, for any $n \geq 0$, and any $x'_{nr} \in E'_{nr}$, we have

$$\mathbb{Q}'_{nr,(n+1)r}(x'_{nr}, dy_n) \propto \left\{ \prod_{nr \leq p < (n+1)r} G'_p(y'_p) \right\} \mathbb{P}'_{nr,(n+1)r}(x'_{nr}, dy_n)$$

$$(7.14)$$

up to some normalizing constant

$$\mathcal{Z}'_{nr,(n+1)r}(x'_{nr}) = \mathbb{E}\left(\prod_{nr \leq p < (n+1)r} G'_p(X'_p) \, \| X'_{nr} = x'_{nr} \right)$$

In the above display, dy_n stands for an infinitesimal neighborhood of $y_n = (y'_{nr}, \ldots, y'_{(n+1)r}) \in E_n$.

The main advantage of this construction comes from the following easily checked technical lemma.

Lemma 7.5.16 *For any* $x_{n-1} = (x'_{(n-1)r}, \ldots, x'_{nr}) \in E_{n-1}$, *we have that*

$$\widehat{M}_n(x_{n-1}, \cdot) = Q'_{nr,(n+1)r}(x'_{nr}, \cdot) \quad and \quad \widehat{G}_{n-1}(x_{n-1}) = Z_{nr,(n+1)r}(x'_{nr})$$

This lemma indicates that we can alternatively approximate the potential functions and the Markov transitions $\left(\widehat{G}_n, \widehat{M}_n\right)$ by considering the mean field IPS approximation of these quantities.

More precisely, given some initial state x'_{nr} we denote by

$$0 \leq p \leq r \mapsto \eta_{nr+p}^{(x'_{nr},N)} \in \mathcal{P}(E'_{nr+p})$$

the particle density profiles associated with some mean field interpretation of the Feynman-Kac measures $Q'_{nr,(n+1)r}(x'_{nr}, \cdot)$. We also denote by

$$\eta^N_{x'_{nr},(n+1)r} \in \mathcal{P}(E_n)$$

the genealogical tree occupation measure at the terminal time.

In this notation, we can use the N-approximations

$$\widehat{G}_{n-1}(x_{n-1}) \quad \simeq_{N \uparrow \infty} \quad \widehat{G}^N_{n-1}(x_{n-1}) := \prod_{nr \leq p < (n+1)r} \eta^N_{x'_{nr},nr+p}(G'_p)$$

$$\widehat{M}_n(x_{n-1}, \cdot) \quad \simeq_{N \uparrow \infty} \quad \widehat{M}^N_n(x_{n-1}, \cdot) := \eta^N_{x'_{nr},(n+1)r}$$

Using the unbiasedness property of the unnormalized Feynman-Kac measures, for any $f_n \in \mathcal{B}_b(E_n)$ we prove that the

$$\mathbb{E}\left(\widehat{G}^N_{n-1}(x_{n-1}) \, \widehat{M}^N_n(f_n)(x_{n-1})\right) = \widehat{G}_{n-1}(x_{n-1}) \, \widehat{M}_n(f_n)(x_{n-1})$$

Using the same arguments as the ones we used in Section 7.5.2, we can use this important observation to design an unbiased mean field particle simulation of the updated Feynman-Kac.

7.6 Doob h-processes

This section is dedicated to time homogeneous absorption models and quasi-invariant measures. These mathematical objects are intimately related to the computation of the top of the spectrum of Schrödinger operator and related ground state energies.

For a more thorough discussion on these limiting measures and their IPS samplers we refer to the series of articles [91, 176, 175, 498].

7.6.1 Time homogeneous Feynman-Kac models

We consider the time homogeneous Feynman-Kac model $(\Gamma_n, \mathbb{Q}_n)$, associated with the parameters $(E_n, G_n, M_n) = (E, G, M)$ on some measurable state space E, defined in (1.37). We also set

$$Q(x, dy) := G(x)M(x, dy)$$

We also assume that G is uniformly bounded above and below by some positive constant, and the Markov transition M is reversible w.r.t. some probability measure μ on E, with $M(x, .) \simeq \mu$ and $dM(x, .)/d\mu \in \mathbb{L}_2(\mu)$. We denote by λ the largest eigenvalue of the integral operator Q on $\mathbb{L}_2(\mu)$, and by $h(x)$ a positive eigenvector

$$Q(h) = \lambda h$$

Under some regularity conditions on (G, M), there exists some constant $\rho \geq 1$ such that

$$1/\rho \leq h(x)/h(y) \leq \rho \qquad (7.15)$$

for any $x, y \in E$. For instance, when the regularity condition $\mathbf{H_m(G, M)}$, stated on page 372, is met for some parameters $\chi_m, g < \infty$, and some integer $m \geq 1$, we have

$$Q^m(h)(x)/Q^m(h)(y) = h(x)/h(y) \leq \rho \quad \text{with} \quad \rho \leq g^m \chi_m$$

Definition 7.6.1 *The Doob h-process, corresponding to the ground state eigenfunction h defined above, is a Markov chain X_n^h, with initial distribution $\eta_0^h = \Psi_h(\eta_0)$, and the Markov transition*

$$M^h(x, dy) := \frac{1}{\lambda} \times h^{-1}(x)Q(x, dy)h(y) = \frac{M(x, dy)h(y)}{M(h)(x)}$$

We also denote by η_n^h the distribution of the random state X_n^h starting with initial distribution η_0^h; that is, we have that

$$\text{Law}(X_n^h) = \eta_n^h = \eta_0^h (M^h)^n$$

Our next objective is to connect the distribution of the paths of the h-process

$$\mathbb{P}\left((X_0^h, \ldots, X_n^h) \in d(x_0, \ldots, x_n)\right) = \eta_0^h(dx_0)M^h(x_0, dx_1)\ldots M^h(x_{n-1}, dx_n)$$

with the Feynman-Kac measures Γ_n and $\mathbb{Q}_n$. Firstly, by construction we have

$$G = \lambda \times h/M(h)$$

and therefore

$$
\begin{aligned}
\Gamma_n(d(x_0, \ldots, x_n)) &= \eta_0(dx_0) \left\{ \prod_{0 \leq p < n} G(x_p) \right\} \times \left\{ \prod_{1 \leq p \leq n} M(x_{p-1}, dx_p) \right\} \\
&= \lambda^n \eta_0(dx_0) \, h(x_0) \left\{ \prod_{1 \leq p \leq n} \frac{M(x_{p-1}, dx_p)h(x_p)}{M(h)(x_{p-1})} \right\} \frac{1}{h(x_n)}
\end{aligned}
$$

We conclude that

$$\Gamma_n(d(x_0, \ldots, x_n)) = \lambda^n \ \eta_0(h) \ \mathbb{P}_n^h(d(x_0, \ldots, x_n)) \ \frac{1}{h(x_n)}$$

where $\mathbb{P}_n^h$ stands for the law of the historical process

$$\mathbf{X_n^h} = (X_0^h, \ldots, X_n^h)$$

This clearly implies that

$$d\mathbb{Q}_n = \frac{1}{\mathbb{E}(h^{-1}(X_n^h))} \ h^{-1}(X_n^h) \ d\mathbb{P}_n^h$$

with the normalizing constants

$$\mathcal{Z}_n = \lambda^n \ \eta_0(h) \ \mathbb{E}(h^{-1}(X_n^h))$$

Under condition (7.15), using the multiplicative formula (1.47) (cf. also (3.29), Section 3.2.2), we also have that

$$\frac{1}{n} \log \mathcal{Z}_n = \frac{1}{n} \sum_{0 \le p < n} \log \eta_p(G) = \log \lambda + \frac{1}{n} \log \left(\eta_0(h) \ \mathbb{E}(h^{-1}(X_n^h)) \right)$$

and therefore

$$\log \lambda - \frac{1}{n} \log \rho \le \frac{1}{n} \log \mathcal{Z}_n = \frac{1}{n} \sum_{0 \le p < n} \log \eta_p(G) \le \log \lambda + \frac{1}{n} \log \rho \quad (7.16)$$

from which we conclude that

$$\lim_{n \to \infty} \frac{1}{n} \sum_{0 \le p < n} \log \eta_p(G) = \log \lambda$$

In terms of the h-process, the n-th time marginal γ_n of the Feynman-Kac measures Γ_n takes the following form

$$\gamma_n(f) = \lambda^n \ \eta_0(h) \ \eta_0^h (M^h)^n (f/h) = \lambda^n \ \eta_0(h) \ \eta_n^h(f/h)$$

Using these observations, we readily prove the following h-transformation formulae between Feynman-Kac type and h-process distribution flows. For any $n \ge 0$ we have

$$\eta_n = \Psi_{1/h}(\eta_n^h) \quad \text{and} \quad \eta_n^h = \Psi_h(\eta_n)$$

7.6.2 Quasi-invariant measures

We return to the time homogeneous Feynman-Kac models introduced in Section 7.6. Using the particle absorption interpretation (3.30) we have

$$\text{Law}((X_0^c, \ldots, X_n^c) \mid T^c \ge n) = \frac{1}{\mathbb{E}(h^{-1}(X_n^h))} \ h^{-1}(X_n^h) \ d\mathbb{P}_n^h$$

and

$$\mathcal{Z}_n = \mathbb{P}\left(T^c \geq n\right) = \lambda^n \, \eta_0(h) \, \mathbb{E}(h^{-1}(X_n^h)) \longrightarrow_{n\uparrow\infty} 0 \qquad (7.17)$$

Whenever it exists, the Yaglom limit of the measure η_0 is defined as the limiting of measure

$$\eta_n \longrightarrow_{n\uparrow\infty} \eta_\infty = \Psi_G(\eta_\infty)M \qquad (7.18)$$

of the Feynman-Kac flow η_n, when n tends to infinity. We also say that η_0 is a quasi-invariant measure as we have $\eta_0 = \eta_n$, for any time step. When the Feynman-Kac flow η_n is asymptotically stable, in the sense that it forgets its initial conditions, we also say that the quasi-invariant measure η_∞ is the Yaglom measure.

For instance, when the regularity condition $\mathbf{H_m(G, M)}$ stated on page 372 is met for some parameters $\chi_m, g < \infty$, and some integer $m \geq 1$, we shall prove in Section 12.2.2 that the semigroup $\Phi(\eta) := \Psi_G(\eta)M$ is exponentially stable, in the sense that

$$\|\Phi^n(\eta) - \Phi^n(\mu)\|_{tv} \leq a \, \exp\left(-b\, n\right)$$

for some positive constants (a, b) that depend on the parameters (m, χ_m, g). If we take $\mu = \eta_\infty$, then we find that

$$\|\eta_n - \eta_\infty\|_{tv} \leq a \, \exp\left(-b\, n\right) \qquad (7.19)$$

If we set $\eta_0 = \eta_\infty$ in (7.16), we find that

$$\forall n \geq 1 \quad |\log \eta_\infty(G) - \log \lambda| \leq \frac{1}{n} \log \rho \Longrightarrow \log \eta_\infty(G) = \log \lambda$$

In much the same way, when (7.19) is met we have

$$\left|\frac{1}{n} \log \mathcal{Z}_n - \log \lambda\right| \leq \frac{1}{n} \sum_{0 \leq p < n} |\log \eta_p(G) - \log \eta_\infty(G)| \leq c/n$$

for some finite constant $c < \infty$ that depends on the parameters (m, χ_m, g).

Whenever it exists, we let η_∞^h be the invariant measure of the h-process X_n^h. Under our assumptions, it is now a simple exercise to check that

$$\eta_\infty = \Psi_{M(h)}(\mu) \qquad \text{and} \qquad \eta_\infty^h := \Psi_h(\eta_\infty) = \Psi_{hM(h)}(\mu)$$

Quantitative convergence estimates of the limiting formulae (7.17) and (7.18) can be derived using the stability properties of the Feynman-Kac models developed in Section 12.2. For a more thorough discussion on these particle absorption models, we refer the reader to the series of articles of the author with A. Guionnet [169, 170], L. Miclo [161, 176] and A. Doucet [175], as well as the monograph [163].

Chapter 8

Signal processing and control systems

8.1 Nonlinear filtering problems

Signal processing is the art and science of analyzing time varying signals, extracting information from partial and noisy observations delivered by some sensors, such as, for example, radar, sonar, digital electronic cameras, acoustic or laser sensors. The exact evolution equation of the signal and the various disturbance sources are usually unknown, but it is reasonable to assume that we know their structure and the statistical nature of the perturbations.

From the pure probability point of view, the signal is represented by some Markov chain model. The signal noise reflects the mathematical model uncertainties, as well as any unknown quantities including initial conditions and kinetic parameters, including unknown control sequences in tracking problems. The filtering problem consists in computing *the conditional distributions* of the signal given the sequence of observations. These estimation problems arise in various scientific disciplines, including speech recognition, target tracking, seismology, econometrics, and more recently in mathematical finance, microbiology, and biomedical engineering. It is clearly out of the scope of this chapter to discuss all of these application domains. For a more detailed discussion, with a list of references, we refer the reader to the opening chapter.

Each applied scientific discipline tends to use a different language to express and analyze the same filtering problem. In the present chapter we have chosen to adopt standard notation from probability theory. When dealing with complex multivariate distributions, to avoid unnecessary technicalities and clarify the presentation, sometimes we use slightly abusive but natural Bayesian notation.

8.1.1 Description of the models

In this section, we present the probabilistic description of filtering problems in terms of signal-observation Markov chain in general state spaces. We also provide a Feynman-Kac description of the conditional distributions of the signal w.r.t. the observations, and we give a brief discussion on their interpretations in terms of mean field IPS models.

Suppose that at every time step the state of the Markov chain X_n taking values in some state spaces E_n^X is partially observed according to the following schematic picture

$$
\begin{array}{ccccccc}
X_0 & \longrightarrow & X_1 & \longrightarrow & X_2 & \longrightarrow & \ldots \\
\downarrow & & \downarrow & & \downarrow & & \\
Y_0 & & Y_1 & & Y_2 & & \ldots
\end{array}
$$

The typical model is given by a reference Markov chain model X_n and some partial and noisy observation Y_n. We denote by M_n the elementary Markov transitions of the Markov chain X_n. The pair process (X_n, Y_n) usually forms a Markov chain on some product measurable state space $\left(E_n^X \times E_n^Y\right)$ with elementary transitions given

$$\mathbb{P}\left((X_n, Y_n) \in d(x, y) \mid (X_{n-1}, Y_{n-1})\right) = M_n(X_{n-1}, dx)\, g_n(x, y)\, \lambda_n(dy) \tag{8.1}$$

for some positive likelihood function g_n, and some reference probability measure λ_n on E_n^Y. In the further development of this section, we fix the observation sequence $Y_n = y_n$, for $n \geq 0$. As traditional in nonlinear filtering literature, when there is no possible confusion, we slightly abuse the notation and we suppress as much as we can the dependence on the observation sequence. For any $n \geq 0$, we set

$$G_n(x_n) = g_n(x_n, y_n) \tag{8.2}$$

With a slight abuse of notation, sometimes we denote by $p(y_0, \ldots, y_n)$ the density of the observations $(Y_0, \ldots, Y_n)$ with respect to the product measure $\otimes_{0 \leq p \leq n} \lambda_p$; that is, we have that

$$\mathbb{P}\left((Y_0, \ldots, Y_n) \in d(y_0, \ldots, y_n)\right) = p(y_0, \ldots, y_n)\, \lambda_0(dy_0) \ldots \lambda_n(dy_n)$$

By construction, the Feynman-Kac measures (1.37) associated with the pair (G_n, M_n) are given by

$$\mathbb{Q}_n = \mathrm{Law}((X_0, \ldots, X_n) \mid \forall 0 \leq p < n \quad Y_p = y_p) \tag{8.3}$$

In addition, the normalizing constants defined in (1.37) take the following form $\mathcal{Z}_{n+1} = p(y_0, \ldots, y_n)$. To underline the dependence on the observation sequence, sometimes we write $\mathcal{Z}_{n+1}(y)$, the normalizing constants associated with a given sequence of observations $y = (y_p)_{0 \leq p \leq n}$.

In this context, the optimal one step predictor η_n and the optimal filter $\widehat{\eta}_n$ are given by the n-th time marginal distribution defined by

$$
\begin{aligned}
\eta_n &= & \mathrm{Law}\left(X_n \mid \forall 0 \leq p < n \quad Y_p = y_p\right) & \tag{8.4} \\
\widehat{\eta}_n = \Psi_{G_n}(\eta_n) &= & \mathrm{Law}\left(X_n \mid \forall 0 \leq p \leq n \quad Y_p = y_p\right) & \tag{8.5}
\end{aligned}
$$

We notice that we can combine these filtering models with the Markov restriction models (7.6) discussed in Section 7.3. For instance, if we replace the potential likelihood function G_n defined in (8.2) by the function

$$G_n(x_n) = g_n(x_n, y_n) \, 1_{A_n}(x_n)$$

for some measurable subset $A_n \in E_n^X$, then we find that

$$\mathbb{Q}_n = \mathrm{Law}((X_0, \ldots, X_n) \mid \forall 0 \le p < n \quad Y_p = y_p, \ X_p \in A_p)$$

The mean field IPS simulations of the conditional distributions η_n are often referred to as particle filters in signal processing and statistics (see for instance [151, 152, 161, 163, 222], and references therein). These IPS models can also be used to approximate the log-likelihood functions using the multiplicative formula (3.29). More precisely, using (4.8) the log-likelihood functions $L_n(y) := \log \mathcal{Z}_n(y)$ are approximated using the particle approximations

$$L_n^N(y) := \log \mathcal{Z}_n^N(y) = \sum_{0 \le p < n} \log \eta_p^N(G_p)$$

where η_n^N stands for the N-particle density profiles discussed in Chapter 4.

Smoothing problems consist of estimating some values of the signal X_p at some time p, given s series of observations $Y_q = y_q$, with $0 \le q \le n$, and $p \le n$. The conditional distributions defined in (8.3) can be estimated using the genealogical tree based models, or the complete ancestral tree models, presented in Section 4.1.2, and in Section 4.1.6.

8.1.2 Signal-noise estimation

This section is concerned with the estimation of the conditional distributions of the signal-noise given the observations in terms of a genealogical tree based model. We also design a mean field IPS approximation of the conditional maximum likelihood of signal noise sequences. These mathematical objects are closely related to optimal regulation problems. In this situation, we have a dedicated controlled complex system and a given reference trajectory. The problem is to compute the optimal sequence of controls that minimizes some energy cost function and drives the system as close as possible to some reference trajectory.

These regulation problems can be interpreted in terms of the maximum likelihood of a dual filtering problem. Inversely, the logarithm of the conditional distributions of a filtering model in path space can be interpreted in terms of the cost associated with a given optimal regulation model. In this context the mean field IPS genealogical tree model can be interpreted as a genealogical tree based decision tree algorithm.

For a detailed discussion on these models, we refer the reader to the appendix of the book [373]. Several real-world application of these decision tree models to optimal thermal processing, w.r.t. specified temperature trajectories, and robotic optimal control problems, can be found in the series of articles [337, 338, 447].

We further assume the signal process given by recursive equations on some spaces E_n of the following form

$$X_n := F_n(X_{n-1}, U_n) \tag{8.6}$$

In the above display, U_n stands for a sequence of independent, and independent of X_0, random variables with distribution ν_n on some state spaces $\mathcal{U}_n$. We also assume that F_n is a measurable function from $(E_{n-1} \times \mathcal{U}_n)$ into E_n. For $n = 0$, we set $X_0 = U_0 \in E_0 = \mathcal{U}_0$. We denote by $X_n := \psi_n(U_0, \ldots, U_n)$ the stochastic semigroup associated with the random system (8.6).

We let $\mathbb{Q}_n$ be the Feynman-Kac model (1.37) associated with the reference Markov chain $\mathcal{X}_n$ and the potential functions $\mathcal{G}_n$

$$\mathcal{X}_n := (U_0, \ldots, U_n) \quad \text{and} \quad \mathcal{G}_n := G_n \circ \psi_n$$

By construction, we have

$$\mathbb{Q}_n = \mathrm{Law}\left((U_0, (U_0, U_1), \ldots, (U_0, \ldots, U_n))\right) \mid \forall 0 \le p < n \quad Y_p = y_p)$$

Thus, the n-th time marginal measures η_n are given by

$$\eta_n = \mathrm{Law}\left((U_0, \ldots, U_n)\right) \mid \forall 0 \le p < n \quad Y_p = y_p)$$

These conditional distributions can be estimated using the genealogical tree based IPS measures

$$\eta_n^N := \frac{1}{N} \sum_{i=1}^N \delta_{\xi_n^i} \quad \text{with} \quad \xi_n^i := (\xi_{0,n}^i, \xi_{1,n}^i, \ldots, \xi_{n,n}^i)$$

discussed in Section 4.1.2. The occupation measures of these ancestral trees are illustrated below for $(N, n) = (4, 5)$, for any $i_0 \in \{1, 2, 3, 4\}$, $i_2 \in \{2, 3, 4\}$, $i_3 \in \{2, 3\}$, and any $i_4 \in \{2, 3\}$:

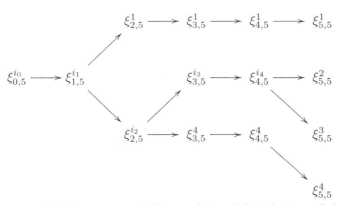

The ancestral lines represent the conditional distribution of the sequence $(U_0, U_1, U_2, U_3, U_4, U_5)$ w.r.t. the sequence of observations $(Y_0, Y_1, Y_2, Y_3, Y_4)$, in terms of a likely initial condition $\xi_{0,5}^{i_0}$, and 4 likely signal-noise sequences $(\xi_{1,5}^i, \xi_{2,5}^i, \xi_{3,5}^i, \xi_{4,5}^i, \xi_{5,5}^i)_{i=1,2,3,4}$.

8.1.3   Maximum likelihood trajectories

This section is concerned with conditional maximum likelihood of the signal noise filtering problem discussed in Section 8.1.2. We further assume the random variables U_n have a density $h_n(u)$ w.r.t. to some reference distribution μ_n on $\mathcal{U}_n$, and we set $\boldsymbol{\mu_n} = \otimes_{0 \leq p \leq n} \mu_p$. By construction, we have $\eta_n \ll \boldsymbol{\mu_n}$ and

$$
\frac{d\eta_n}{d\boldsymbol{\mu_n}}(u_0, \ldots, u_n) \propto \left\{ \prod_{0 \leq p \leq n} h_p(u_p) \right\} \times \left\{ \prod_{0 \leq p < n} g_p(\psi_p(u_0, \ldots, u_p), y_p) \right\}
$$

The log-likelihood function

$$
V_n(u_0, \ldots, u_n) := -\log\left(\frac{d\eta_n}{d\boldsymbol{\mu_n}}(u_0, \ldots, u_n) \right)
$$

can be rewritten as follows

$$
V_n(u_0, \ldots, u_n) = \sum_{0 \leq p \leq n} c_p(u_p) + \sum_{0 \leq p < n} C_p\left(\psi_p(u_0, \ldots, u_p), y_p \right)
$$

with the local logarithmic cost functions $(c_n, C_n) = (-\log h_n, -\log g_n)$. The function $-c_n$ can be interpreted as an energy type function on the control sequence u_n, and the energy function $-C_n$ measures the difference between the controlled semigroup of the system $\psi_n(u_0, \ldots, u_n)$ and the reference observation state y_n.

The optimal sequence of control that maximizes the log-likelihood function can be computed using the genealogical tree occupation measure. The estimate is defined by

$$
\eta_n^N - \text{ess inf } V_n := \min_{1 \leq i \leq N} V_n(\xi_{0,n}^i, \xi_{1,n}^i, \ldots, \xi_{n,n}^i)
$$

In Chapter 9, dedicated to the modeling and the analysis of mean field IPS models, we shall prove the convergence in probability

$$
\eta_n^N - \text{ess inf } V_n \longrightarrow_{N \to \infty} \eta_n - \text{ess inf } V_n
$$

The details of the proof of the above result can be found in Theorem 9.5.7.

8.1.4   Approximate filtering models

This section is concerned with approximate filtering models. These filtering models are closely related to Approximate Bayesian computation (*abbreviated ABC*). These Bayesian inference methods are currently used to evaluate posterior distributions without having to calculate likelihoods.

For instance, in biology applications and more particularly in predictive bacteriology and food risk analysis, the observations of a kinetic biological complex system are given by counting bacteria individuals after successive dilutions of a food sample coming from an in vitro culture [244, 245, 267, 268]. Of course, this experimental observation process is often modeled by a series of Poisson type dependent random variables but the computation of the likelihood function often requires successive summations over the set of all the integers. In this situation likelihood functions are computationally intractable, or too costly to estimate in a reasonable time.

One of the central ideas of ABC methods is to replace the evaluation of the likelihood function by a simulation-based procedure of the observation process coupled with a numerical comparison between the observed and simulated data. This strategy is rather well known in particle filtering literature; see for instance [154, 155, 156]. In the same vein, these additional levels of simulation-based approximations can also be extended to compute the posterior distribution of fixed parameters in hidden Markov chain models. In signal processing literature, these ABC type mean field IPS models are sometimes called convolution particle filters; see for instance [89, 90, 497, 558].

More formally, in some instance the likelihood functions $x_n \mapsto g_n(x_n, y_n)$ in (8.2) are computationally intractable, or too expensive to evaluate in a reasonable computational time. To solve this problem, a natural solution is to sample pseudo-observations. The central idea is to sample the signal-observation Markov chain $\boldsymbol{X_n} = (X_n, Y_n) \in \boldsymbol{E_n^X} = (E_n^X \times E_n^Y)$, and compare the values of the sampled observations with the real observations sequence.

To describe with some conciseness these approximate filtering models, we notice that the transitions of $\boldsymbol{X_n}$ are given by

$$\boldsymbol{M_n}(\boldsymbol{X_{n-1}}, d(x_n, y_n)) = M_n(X_{n-1}, dx_n) \, g_n(x_n, y_n) \, \lambda_n(dy_n)$$

To simplify the presentation, we further assume that $E_n^Y = \mathbb{R}^d$, for some $d \geq 1$, and we let g be a Borel bounded nonnegative function on $\mathbb{R}^d$ such that

$$\int g(u) \, du = 1 \quad \int u \, g(u) \, du = 0 \quad \text{and} \quad \int |u|^3 \, g(u) \, du < \infty$$

Then, we set for any $\epsilon > 0$, any $\mathbf{x} = (x, y) \in \boldsymbol{E_n^X} = (E_n^X \times E_n^Y)$, and $\mathbf{z} \in \mathbb{R}^d$

$$g_{\epsilon,n}(\mathbf{x}, \mathbf{z}) = g_{\epsilon,n}((x, y), z) = \epsilon^{-d} \, g\left((y - z)/\epsilon\right)$$

Finally, we consider a Markov chain $(\boldsymbol{X_n}, \boldsymbol{Y_n^\epsilon})$ on the augmented state space $\left(\boldsymbol{E_n^X} \times E_n^Y\right) = \left((E_n^X \times \mathbb{R}^d) \times \mathbb{R}^d\right)$ with transitions given

$$\mathbb{P}\left((\boldsymbol{X_n}, \boldsymbol{Y_n^\epsilon}) \in d(\boldsymbol{x_n}, \boldsymbol{y_n}) \mid (\boldsymbol{X_{n-1}}, \boldsymbol{Y_{n-1}^\epsilon})\right)$$

$$= \boldsymbol{M_n}(\boldsymbol{X_{n-1}}, d\boldsymbol{x_n}) \, g_{\epsilon,n}(\boldsymbol{x_n}, \boldsymbol{y_n}) \, d\boldsymbol{y_n}$$

(8.7)

This approximated filtering problem has exactly the same form as the one introduced in (3.31). Here, the particle approximation model is defined in terms of signal-observation valued particles, and the selection potential function is given by the pseudo-likelihood functions $g_{\epsilon,n}(., y_n)$, where y_n stands for the value of the observation sequence at time n.

8.2 Linear-Gaussian models

This section is dedicated to linear-Gaussian filtering models and the derivation of the traditional forward-backward Kalman filters. The origins of these optimal filters starts with the seminal article by R. E. Kalman [353] and the earlier pioneering works by R. L. Startonovich [533, 534, 535, 536].

The first historical application of the Kalman filter was developed by S. F. Schmidt [508], in the Apollo program of NASA Ames Research Center, to solve nonlinear navigation equations of the manned lunar mission. The Kalman filter has been applied in the design of a variety of defense navigation and guidance systems, including ballistic submarines and the U.S. Navy's Tomahawk as well as in the U.S. Air Force's air launched cruise missiles. It is also used in the NASA Space Shuttle, as well as in the attitude control systems of the International Space Station.

Originally developed for spacecraft navigation systems, the Kalman filter and its extended version are one of the most commonly and routinely used tools to remove noise from partially observed sequences of random variables. Its range of application has been extended to almost every scientific discipline, including in financial mathematics, econometrics, and computational biology, as well as in Bayesian statistics and in various branches of engineering sciences.

Nevertheless, for nonlinear and/or non-Gaussian state space models, Kalman type observers are no more optimal, in the sense that the conditional distributions of the signal w.r.t. the observations are generally not Gaussian, and they cannot be encoded by a couple of mean-variance style evolution equation. One natural way to turn extended Kalman type observers into *optimal* filters for nonlinear state space models, is to use the importance sampling SMC methodologies developed in Section 4.2. More precisely, the Gaussian elementary transitions associated with the Kalman type observers can be used as importance twisted transitions in (4.21) to design a sequence of *mean field interaction of Kalman type SMC filters*.

8.2.1 Forward Kalman filters

We consider an $\mathbb{R}^{p+q}$-valued Markov chain (X_n, Y_n) defined by the recursive relations

$$\begin{cases} X_n &= A_n X_{n-1} + B_n W_n, & n \geq 1 \\ Y_n &= C_n X_n + D_n V_n, & n \geq 0 \end{cases} \tag{8.8}$$

for some $\mathbb{R}^{d_w}$ and $\mathbb{R}^{d_v}$-valued independent random sequences W_n and V_n, independent of X_0, and some matrices A_n, B_n, C_n, D_n with appropriate dimensions. We further assume that W_n and V_n are centered Gaussian random sequences with covariance matrices R_n^v, R_n^w, and X_0 is a Gaussian random variable in $\mathbb{R}^p$, with a mean $\widehat{X}_0^-$, and covariance matrix $\widehat{P}_0^-$.

Definition 8.2.1 *We denote by $\mathcal{N}(m, R)$ a Gaussian distribution a d-dimensional space $\mathbb{R}^d$ with mean vector $m \in \mathbb{R}^d$ and covariance matrix $R \in \mathbb{R}^{d \times d}$*

$$\mathcal{N}(m, R)(dx) = \frac{1}{(2\pi)^{d/2}\sqrt{|R|}} \exp\left[-2^{-1}(x - m)R^{-1}(x - m)'\right] dx$$

With a slight abuse of notation, sometimes we denote by

$$\mathcal{N}[m, R](x) := \frac{1}{(2\pi)^{d/2}\sqrt{|R|}} \exp\left[-2^{-1}(x - m)R^{-1}(x - m)'\right]$$

the density of a Gaussian distribution w.r.t. the Lebesgue measure dx, so that

$$\mathcal{N}(m, R)(dx) = \mathcal{N}[m, R](x)dx$$

In this notation, we have

$$
\begin{aligned}
\eta_n &:= \operatorname{Law}(X_n \mid Y_0, \ldots, Y_{n-1}) = \mathcal{N}(\widehat{X}_n^-, P_n^-) \\
\widehat{\eta}_n &:= \operatorname{Law}(X_n \mid Y_0, \ldots, Y_{n-1}, Y_n) = \mathcal{N}(\widehat{X}_n, P_n)
\end{aligned}
$$

With a slight abuse of notation, we denote by $p(x_n|Y_0, \ldots, Y_{n-1})$ and $p(x_n|Y_0, \ldots, Y_n)$, the densities (w.r.t. the Lebesgue measure dx_n on $\mathbb{R}^p$) of the conditional distributions η_n and $\widehat{\eta}_n$. We also denote by $p(y_n \mid x_n)$, resp. $p(x_{n+1} \mid x_n)$, the density (w.r.t. the Lebesgue measure dy_n on $\mathbb{R}^q$, resp. dx_{n+1} on $\mathbb{R}^p$) of the conditional distribution of the random state Y_n, resp. X_{n+1}, given $X_n = x_n$. More formally, we have that

$$
\begin{aligned}
p(y_n \mid x_n) &:= \mathcal{N}\left[C_n x_n; D_n R_n^v D_n'\right](y_n) \\
p(x_{n+1} \mid x_n) &:= \mathcal{N}\left[A_n x_n; B_n R_n^w B_n'\right](x_{n+1})
\end{aligned}
$$

and for any bounded function f on $\mathbb{R}^d$ we have

$$
\begin{aligned}
\mathbb{E}(f(X_n) \mid Y_0, \ldots, Y_{n-1}) &= \int f(x_n) \, p(x_n|Y_0, \ldots, Y_{n-1}) \, dx_n \\
\mathbb{E}(f(X_n) \mid Y_0, \ldots, Y_{n-1}, Y_n) &= \int f(x_n) \, p(x_n|Y_0, \ldots, Y_{n-1}, Y_n) \, dx_n
\end{aligned}
$$

By construction, we have

$$
\begin{aligned}
p(x_n|Y_0, \ldots, Y_{n-1}) &= \mathcal{N}\left[\widehat{X}_n^-, P_n^-\right](x_n) \\
p(x_n|Y_0, \ldots, Y_{n-1}, Y_n) &= \mathcal{N}\left[\widehat{X}_n, P_n\right](x_n)
\end{aligned}
$$

The traditional updating-prediction filtering evolution equation of the conditional distributions η_n and $\widehat{\eta}_n$ are given by the schematic picture

$$\eta_n \xrightarrow{\text{updating}} \widehat{\eta}_n := \Psi_{G_n}(\eta_n) \xrightarrow{\text{prediction}} \eta_{n+1} \quad \text{with} \quad G_n(x_n) := p_n(Y_n \mid x_n) \tag{8.9}$$

The prediction transition is defined by

$$\eta_{n+1}(dx_{n+1}) = \int \hat{\eta}_n(dx_n) \, M_{n+1}(x_n, dx_{n+1})$$

with the elementary transitions of the signal $X_n \rightsquigarrow X_{n+1}$ given by

$$M_{n+1}(x_n, dx_{n+1}) := p(x_{n+1} \mid x_n) \, dx_{n+1}$$

For linear-Gaussian models, the synthesis of the conditional mean and covariance matrices is carried out using the traditional Kalman-Bucy recursive updating-prediction equations

$$\left(\widehat{X}_n^-, P_n^- \right) \xrightarrow{\text{updating}} \left(\widehat{X}_n, P_n \right) \xrightarrow{\text{prediction}} \left(\widehat{X}_{n+1}^-, P_{n+1}^- \right) \qquad (8.10)$$

Theorem 8.2.2 *The updating step is defined by the couple of equations*

$$\widehat{X}_n = \widehat{X}_n^- + \mathbf{G}_n \, (Y_n - C_n \widehat{X}_n^-) \quad \text{and} \quad P_n = (Id - \mathbf{G}_n C_n) P_n^-$$

with the Kalman gain matrix

$$\mathbf{G}_n = P_n^- C_n' (C_n P_n^- C_n' + D_n R_n^v D_n')^{-1}$$

The prediction step is defined by the couple of equations

$$\widehat{X}_{n+1}^- = A_{n+1} \widehat{X}_n \quad \text{and} \quad P_{n+1}^- = A_{n+1} \, P_n \, A_{n+1}' + B_{n+1} \, R_{n+1}^w \, B_{n+1}'$$

Before getting into the proof of the theorem, we observe that the updating step of the Kalman filter can be interpreted as a Boltzmann-Gibbs transformation.

Lemma 8.2.3 *Given an observation state $y \in \mathbb{R}^q$, some matrix $C \in \mathbb{R}^{q \times p}$, and some covariance matrix $R \in \mathbb{R}^{q \times p}$, we have*

$$G(x) := \mathcal{N} \left[Cx; R_0 \right](y) \implies \Psi_G \left(\mathcal{N}(m_1, R_1) \right) = \mathcal{N}(m_2, R_2) \qquad (8.11)$$

with

$$m_2 = m_1 + \mathbf{G} \, (y - Cm_1) \quad \text{and} \quad R_2 = (Id - \mathbf{G}C) R_1$$

with the gain matrix

$$\mathbf{G} = R_1 C' (R_0 + C R_1 C')^{-1}$$

Now, we come to the proof of the theorem.

Proof of Theorem 8.2.2:

The proof of the updating recursion equation is based on the fact that

$$\widehat{X}_n := \widehat{X}_n^- + \mathbf{G}_n \, (Y_n - \widehat{Y}_n^-) \quad \text{with} \quad \widehat{Y}_n^- = \mathbb{E}(Y_n \mid Y_0, \ldots, Y_{n-1}) = C_n \widehat{X}_n^-$$

Since $\mathbb{E}((X_n - \widehat{X}_n)(Y_n - \widehat{Y}_n^-)') = 0$, we find

$$\mathbb{E}((X_n - \widehat{X}_n^-)(Y_n - \widehat{Y}_n^-)') = \mathbf{G}_n \, \mathbb{E}((Y_n - \widehat{Y}_n^-)(Y_n - \widehat{Y}_n^-)')$$

from which we find the gain matrix. Finally using the decomposition

$$X_n - \widehat{X}_n = (X_n - \widehat{X}_n^-) + (\widehat{X}_n^- - \widehat{X}_n)$$

and by symmetry argument we conclude that

$$\begin{aligned} P_n &= P_n^- - \mathbb{E}((\widehat{X}_n^- - \widehat{X}_n)(\widehat{X}_n^- - \widehat{X}_n)') \\ &= P_n^- - \mathbf{G}_n \mathbb{E}((Y_n - \widehat{Y}_n^-)(Y_n - \widehat{Y}_n^-)')\mathbf{G}_n' = P_n^- - \mathbf{G}_n C_n P_n^- \end{aligned}$$

The proof of the prediction recursion is rather elementary. The first assertion is clear. The second one comes from the fact that

$$P_{n+1}^-$$

$$= \mathbb{E}((A_{n+1}(X_n - \widehat{X}_n) + B_{n+1}W_{n+1})(A_{n+1}(X_n - \widehat{X}_n) + B_{n+1}W_{n+1})')$$

$$= A_{n+1} \, P_n \, A_{n+1}' + B_{n+1} \, R_{n+1}^w \, B_{n+1}'$$

$$\blacksquare$$

It is also useful to observe that

$$\mathrm{Law}(Y_n \mid Y_0, \ldots, Y_{n-1}) = \mathcal{N}\left(C_n \widehat{X}_n^-, \Sigma_n(P_n^-)\right)$$

with the covariance matrix

$$\Sigma_n(P_n^-) := C_n \, P_n^- \, C_n' + D_n \, R_n^v \, D_n'$$

We prove this claim using the fact that, given $(Y_0, \ldots, Y_{n-1})$, the current observation takes the form

$$Y_n = C_n \tilde{X}_n + D_n V_n$$

with some variable $\tilde{X}_n$ such that

$$\mathrm{Law}\left(\tilde{X}_n \mid Y_0, \ldots, Y_{n-1}\right) := \mathcal{N}(\widehat{X}_n^-, P_n^-)$$

The density $p(y_0, \ldots, y_n)$ of the sequence of observation $(Y_0, \ldots, Y_n)$ evaluated at the random observation path $(Y_0, \ldots, Y_n)$ is given by

$$p(Y_0, \ldots, Y_n) = \prod_{k=0}^{n} \mathcal{N}\left[C_k \widehat{X}_k^-, \Sigma_k(P_k^-)\right](Y_k)$$

In Bayesian inference literature, this formula is sometimes written in the following form

$$p(Y_0, \ldots, Y_n) \quad = \quad p(Y_n \mid Y_0, \ldots, Y_{n-1}) \times p(Y_0, \ldots, Y_{n-1})$$

$$= \quad \prod_{k=0}^{n} p(Y_k \mid Y_0, \ldots, Y_{k-1})$$

where $p(y_n \mid y_0, \ldots, y_{n-1})$ stands for the density (w.r.t. the Lebesgue measure dy_n on $\mathbb{R}^q$) of the conditional distribution of the random variable Y_n given the observations $Y_p = y_p$, for $0 \le p < n$.

8.2.2   Backward Kalman smoother

With a slight abuse of the notation, we denote by

$$p((x_0, \ldots, x_n) \mid (y_0, \ldots, y_{n-1}))$$

the density (w.r.t. the Lebesgue measure $dx_0 \times \ldots \times dx_n$ on $(\mathbb{R}^p)^{n+1}$) of the conditional distribution of the random variable $(X_0, \ldots, X_n)$ given the observations $Y_p = y_p$, for $0 \le p < n$. We also denote by

$$p(x_k \mid x_{k+1}, (y_0, \ldots, y_k))$$

the density (w.r.t. the Lebesgue measure dx_k on $\mathbb{R}^p$) of the conditional distribution of the random variable X_k given the observations $Y_p = y_p$, for $0 \le p \le k$, and the random state $X_{k+1} = x_{k+1}$.

In these Bayesian notation, we have the conditional density formulae

$$p((x_0, \ldots, x_n) \mid (y_0, \ldots, y_{n-1}))$$

$$= p(x_n \mid (y_0, \ldots, y_{n-1})) \, p(x_{n-1} \mid x_n, (y_0, \ldots, y_{n-1}))$$

$$\times p(x_{n-2} \mid x_{n-1}, (y_0, \ldots, y_{n-2})) \ldots p(x_1 \mid x_2, (y_0, y_1)) \, p(x_0 \mid x_1, y_0)$$
$$(8.12)$$

These backward density formulae coincide with the backward Markov chain models presented in Section 1.4.3.4. Before getting into the proof of this assertion, we provide an analytic expression of the conditional transitions.

Running backward in time, given the sequence of observations $(y_0, \ldots, y_{n-1})$ the random state X_{n-1} has a conditional density

$$p(x_{n-1} \mid (y_0, \ldots, y_{n-1})) = \mathcal{N}\left[\widehat{X}_{n-1}, P_{n-1}\right](x_{n-1})$$

and the density of the random state

$$X_n = A_n x_{n-1} + B_n W_n$$

is given by the Gaussian density

$$p(x_n \mid x_{n-1}) := \mathcal{N}\left[A_n x_{n-1}, B_n R_n^w B_n'\right](x_n)$$

Using the updating transformation (8.11) stated in Theorem 8.2.2, we find that

$$p(x_{n-1} \mid (y_0, \ldots, y_{n-1}), x_n) = \mathcal{N}\left[m_{n-1}(x_n), \Sigma_{n-1}\right](x_{n-1}) \qquad (8.13)$$

The linear mean function is given by

$$m_{n-1}(x_n) = \widehat{X}_{n-1} + \overline{\mathbf{G}}_n \left(x_n - A_n \widehat{X}_{n-1}\right) = \overline{\mathbf{G}}_n\, x_n + \left(I - \overline{\mathbf{G}}_n A_n\right)\widehat{X}_{n-1}$$

The covariance matrix

$$\Sigma_{n-1} = \left(I - \overline{\mathbf{G}}_n A_n\right) P_{n-1}$$

is defined in terms of the gain matrix

$$\overline{\mathbf{G}}_n = P_{n-1} A_n' (A_n P_{n-1} A_n' + B_n R_n^w B_n')^{-1}$$

Iterating backward in time the updating formula (8.11), we obtain the backward evolution equation of the Kalman filter.

Now, we check that (8.12) coincides with the backward Markov chain models presented in Section 1.4.3.4.

Firstly, we observe that

$$\eta_n(dx_n) = p(x_n \mid (y_0, \ldots, y_{n-1}))\, dx_n$$

and for any $0 \le k < n$ we have

$$
\begin{aligned}
p(x_k, x_{k+1}, (y_0, \ldots, y_k)) &= p(x_{k+1} \mid x_k) \times p(x_k, (y_0, \ldots, y_k))\\
&\propto p(x_{k+1} \mid x_k)\, p(y_k \mid x_k)\, p(x_k \mid (y_0, \ldots, y_{k-1}))
\end{aligned}
$$

This implies that

$$p(x_k \mid x_{k+1}, (y_0, \ldots, y_k)) = \frac{p(x_{k+1} \mid x_k)\, p(y_k \mid x_k)\, \eta_k(dx_k)}{\displaystyle\int p(x_{k+1} \mid x_k')\, p(y_k \mid x_k')\, \eta_k(dx_k')}$$

In other words, recalling that $G_k(x_k) = p(y_k \mid x_k)$, if we set

$$H_{k+1}(x_k, x_{k+1}) = p(x_{k+1} \mid x_k)$$

then we prove that

$$
\begin{aligned}
\mathbb{M}_{k+1, \eta_k}(x_{k+1}, dx_k) &:= p(x_k \mid x_{k+1}, (y_0, \ldots, y_k))\, dx_k \qquad (8.14)\\
&= \frac{\eta_k(dx_k)\, G_k(x_k) H_{k+1}(x_k, x_{k+1})}{\displaystyle\int \eta_k(dx_k')\, G_k(x_k')\, H_{k+1}(x_k', x_{k+1})} \qquad (8.15)
\end{aligned}
$$

This shows that (8.12) is equivalent to the backward Markov models (1.50) presented in Section 1.4.3.4. This ends the proof of the desired result.

8.2.3   Ensemble Kalman filters

Given some probability measure η on $\mathbb{R}^d$, whenever they exist, we denote by m_η and P_η the mean value and the covariance matrix defined by

$$m_\eta = \int \varphi(x)\ \eta(dx) \quad \text{and} \quad P_\eta := \eta\left(\left[\varphi - \eta(\varphi)\right]\ \left[\varphi - \eta(\varphi)\right]'\right)$$

with the column identity vector $\varphi(x) = x \in \mathbb{R}^p$. Using Lemma 8.2.3, the Kalman recursion (8.9) can alternatively be written as follows

$$\eta_n \xrightarrow{\text{e-updating}} \widetilde{\eta}_n := \widetilde{\Psi}(\eta_n) \xrightarrow{\text{prediction}} \eta_{n+1} = \widetilde{\eta}_n M_{n+1} \qquad (8.16)$$

where $\widetilde{\Psi}(\eta_n) = \widetilde{\eta}_n$ stands for the distribution of the random variable

$$\widetilde{X}_n := \overline{X}_n + \mathbf{G}_{n,\eta_n}(y_n - C_n\overline{X}_n - D_n V_n)$$

with $\mathbf{G}_{n,\eta_n} = P_{\eta_n} C_n' (C_n P_{\eta_n} C_n' + D_n R_n^v D_n')^{-1}$ and $\mathrm{Law}(\overline{X}_n) = \eta_n$. In the above display formulae, V_n stands for a sequence of independent centered Gaussian random sequences with covariance matrices R_n^v. In other words, we have that

$$\widetilde{\Psi}(\eta_n)(f) = \int f\left[x + \mathbf{G}_{n,\eta_n}(y_n - C_n x - D_n v)\right]\ \eta_n(dx)\ \mathcal{N}(0, R_n^v)(dv)$$

For linear-Gaussian models, the evolution Equations (8.16) and (8.9) are equivalent. Indeed, by Lemma 8.2.3, if we set $G_n(x) := \mathcal{N}\left[C_n x; D_n R_n^v D_n'\right](y_n)$, then we have

$$\eta_n = \mathcal{N}(m_{\eta_n}, P_{\eta_n}) \Longrightarrow \Psi_{G_n}(\eta_n) = \mathcal{N}(m_{\widetilde{\eta}_n}, P_{\widetilde{\eta}_n}) = \widetilde{\Psi}(\eta_n)$$

with $m_{\widetilde{\eta}_n} = m_{\eta_n} + \mathbf{G}_{n,\eta_n}(y_n - C_n m_{\eta_n})$ and $P_{\widetilde{\eta}_n} = (Id - \mathbf{G}_{n,\eta_n} C_n) P_{\eta_n}$.

Of course for non Gaussian models, the e-updating transition is not equivalent to the Bayes' rule. We consider the two step Markov chain model on $\mathbb{R}^p$ defined by the following synthetic diagram

$$\overline{X}_n \to \widetilde{X}_n = \overline{X}_n + \mathbf{G}_{n,\eta_n}(y_n - C_n\overline{X}_n - D_n V_n) \to \overline{X}_{n+1} = A_{n+1}\widetilde{X}_n + B_{n+1}W_{n+1}$$

with the initial condition $\overline{X}_0 = X_0$, and the same Gaussian r.v. (X_0, V_n, W_n) as the ones defined in (8.8). The elementary transition $\overline{X}_n \to \widetilde{X}_n$ is given by

$$\widetilde{S}_{n,\eta_n}(x_n, d\widetilde{x}_n) = \mathbb{P}\left(\widetilde{X}_n \in d\widetilde{x}_n \mid \overline{X}_n = x_n\right)$$

so that

$$\widetilde{\eta}_n = \widetilde{\Psi}(\eta_n) = \eta_n \widetilde{S}_{n,\eta_n}$$

This yields the McKean interpretation of the Kalman filter

$$\eta_{n+1} = \eta_n K_{n+1,\eta_n} \quad \text{with} \quad K_{n+1,\eta_n} := \widetilde{S}_{n,\eta_n} M_{n+1}$$

These discrete generation McKean models are discussed in Section 1.2.1 and in Section 4.4. The mean field particle model (1.53) associated with this McKean model is defined by evolving an $(\mathbb{R}^p)^N$-valued and two step Markov chain model

$$\xi_n := (\xi_n^i)_{1 \leq i \leq N} \xrightarrow{\widetilde{S}_{n,\eta_n^N}} \widetilde{\xi}_n := \left(\widetilde{\xi}_n^i\right)_{1 \leq i \leq N} \xrightarrow{M_{n+1}} \xi_{n+1} := (\xi_{n+1}^i)_{1 \leq i \leq N}$$

with $\eta_n^N := \frac{1}{N} \sum_{1 \leq i \leq N} \delta_{\xi_n^i}$. More formally, the elementary transitions (1.53) are decomposed into the following steps

$$\xi_n^i \to \widetilde{\xi}_n^i = \xi_n^i + \mathbf{G}_{n,\eta_n^N}(y_n - C_n\xi_n^i - D_n V_n^i) \to \xi_{n+1}^i = A_{n+1}\widetilde{\xi}_n^i + B_{n+1}W_{n+1}^i$$

for any $i \in \{1, \ldots, N\}$, with i.i.d. copies $(V_n^i, W_n^i)_{1 \leq i \leq N}$ of (V_n, W_n), and the empirical approximation of the covariance function

$$P_{\eta_n^N} := \eta_n^N \left([\varphi - \eta_n^N(\varphi)] \ [\varphi - \eta_n^N(\varphi)]' \right)$$

The main advantage of this mean field formulation of the Kalman filter comes from the fact that for large dimension signals intractable covariance prediction error matrices P_n^- are now computed using the sampled mean field particle empirical matrices $P_{\eta_n^N}$.

We emphasize, that for nonlinear and/or non-Gaussian models, we can turn these ensemble Kalman filters into *optimal* filters, using the importance sampling SMC methodologies developed in Section 4.2 and the adaptive importance sampling models presented in Section 4.2.5.3. The Gaussian elementary mean field transitions can be used as importance twisted transitions in (4.21) to design a sequence of *mean field interacting of Kalman type SMC filters*.

These mean field approximations are widely used in meteorological forecasting problems, where the signal process comes from a grid type approximation of Navier-Stokes' partial different equations arising in fluid mechanics.

Further details of these mean field particle filters, and their performance analysis, can be found in the recent articles by F. Le Gland, V. Monbet, and V. D. Tran [388, 389], as well as in the Ph.D. thesis of V. D. Tran [551] in 2009, and the one by Ch. Baehr [27] in 2008. We also refer the reader to the pioneering work by G. Evensen on ensemble Kalman filters [247, 248, 249], and a series of articles [10, 88, 325] on the numerical performance of these models in forecasting data assimilation problems.

We end this section with a simple and natural technique to turn the Ensemble Kalman Filters into an Ensemble Kalman smoother. The key idea is to replace in the backward Markov models (1.50) the sequence of optimal prediction measures η_n by the N-approximation measures $\eta_n^N := \frac{1}{N} \sum_{1 \leq i \leq N} \delta_{\xi_n^i}$ defined above. Using the same notation as in (8.15), for linear Gaussian models we readily check that the resulting particle backward measures

$$\mathbb{Q}_n^N(d(x_0, \ldots, x_n))$$

$$:= \eta_n^N(dx_n)\mathbb{M}_{n,\eta_{n-1}^N}(x_n, dx_{n-1}) \cdots \mathbb{M}_{2,\eta_1^N}(x_2, dx_1)\mathbb{M}_{1,\eta_0^N}(x_1, dx_0)$$

converge as $N \uparrow \infty$ to the distribution of the random paths of the signal $(X_0, \ldots, X_n)$, given the observations $Y_p = y_p$, with $p < n$. Notice that we can also replace in (8.15) the updated measures $\eta_k(dx_k) \, G_k(x_k)$ by the updated Ensemble Kalman filters $\widetilde{\eta}_k = \frac{1}{N}\sum_{i=1}^N \delta_{\widetilde{\xi}_k^i}$. We can alternatively use the $\widetilde{\Psi}$-version of the backward updating transition (8.13).

8.3 Interacting Kalman filters

In this section, we present a mean field interacting Kalman filter. These particle approximations are often referred to as particle methods in path space, or as Rao-Blackwellized particle filters in signal processing literature, and Bayesian inference (see for instance [163, 177, 222], and references therein).

8.3.1 Partial linear-Gaussian models

We consider a Markov chain Θ_n taking values in some measurable state space E_n, and a collection of matrices

$$A_n(\theta), B_n(\theta), C_n(\theta), D_n(\theta) \tag{8.17}$$

indexed by $\theta \in E_n$, of the same dimension as the matrices (A_n, B_n, C_n, D_n) introduced in (8.8).

We let (Θ_n, X_n, Y_n) be the $(E \times \mathbb{R}^{p+q})$-valued Markov chain defined by the same recursive relations as in (8.8):

$$\begin{cases} X_n = A_n(\Theta_n)X_{n-1} + B_n(\Theta_n)W_n, & n \geq 1 \\ Y_n = C_n(\Theta_n)X_n + D_n(\Theta_n)V_n, & n \geq 0 \end{cases} \tag{8.18}$$

Arguing as above, given a realization of the chain $\Theta = (\Theta_n)_{n \geq 0}$ we have

$$\begin{aligned} \text{Law}(X_n \mid \Theta, \, Y_0, \ldots, Y_{n-1}) &= \mathcal{N}(\widehat{X}_n^{\Theta,-}, P_n^{\Theta,-}) \\ \text{Law}(X_n \mid \Theta, \, Y_0, \ldots, Y_{n-1}, Y_n) &= \mathcal{N}(\widehat{X}_n^{\Theta}, P_n^{\Theta}) \end{aligned}$$

with some parameters $(\widehat{X}_n^{\Theta,-}, P_n^{\Theta,-})$ and $(\widehat{X}_n^{\Theta}, P_n^{\Theta})$ that can be computed using the same Kalman recursions given above by replacing the matrices (A_n, B_n, C_n, D_n) by the matrices (8.17).

With a slight abuse of notation, we shall use the following Bayesian formulations:

The density (w.r.t. the Lebesgue measure dx_n on $\mathbb{R}^p$) of the conditional distribution of the random variable X_n given the random sequences $(Y_0, \ldots, Y_{n-1})$ and $(\Theta_0, \ldots, \Theta_{n-1}))$ is denoted by

$$p(x_n | (Y_0, \ldots, Y_{n-1}), (\Theta_0, \ldots, \Theta_n)) = \mathcal{N}\left[\widehat{X}_n^{\Theta,-}, P_n^{\Theta,-}\right](x_n)$$

The density (w.r.t. the Lebesgue measure dy_n on $\mathbb{R}^q$) of the conditional distribution of the random variable Y_n, given the random variable Θ_n and $X_n = x_n$ is denoted by

$$p(y_n | x_n, \Theta_n) := \mathcal{N}\left[C_n(\Theta_n)x_n; D_n(\Theta_n)R_n^v D_n'(\Theta_n)\right](y_n) \qquad (8.19)$$

The density (w.r.t. the Lebesgue measure dy_n on $\mathbb{R}^q$) of the conditional distribution of the random variable Y_n, given the sequence of observation $(Y_0, \ldots, Y_{n-1})$ and a realization of the chain $(\Theta_0, \ldots, \Theta_n)$, is denoted by

$$p(y_n | (Y_0, \ldots, Y_{n-1}), (\Theta_0, \ldots, \Theta_n))$$

Finally, the density (w.r.t. the Lebesgue measure $dy_0 \times \ldots \times dy_n$ on $(\mathbb{R}^q)^{n+1}$) of the conditional distribution of the sequence of observation $(Y_0, \ldots, Y_n)$ given a realization of the parameters $(\Theta_0, \ldots, \Theta_n) = (\theta_0, \ldots, \theta_n)$ is denoted by

$$p((y_0, \ldots, y_n) \mid (\theta_0, \ldots, \theta_n)) \qquad (8.20)$$

8.3.2 Normalizing constants

Using the Bayesian notation presented in Section 8.3.1, we readily check the following formulae:

$$p(Y_n | (Y_0, \ldots, Y_{n-1}), (\Theta_0, \ldots, \Theta_n))$$

$$= \int p(Y_n | x_n, \Theta_n) \, p(x_n | (Y_0, \ldots, Y_{n-1}), (\Theta_0, \ldots, \Theta_n)) \, dx_n \qquad (8.21)$$

$$= \mathcal{N}\left[C_n(\Theta_n)\widehat{X}_n^{\Theta,-}, \Sigma_n(\Theta_n, P_n^{\Theta,-})\right](Y_n)$$

with the covariance matrices

$$\Sigma_n(\Theta_n, P_n^{\Theta,-}) := C_n(\Theta_n) \, P_n^{\Theta,-} \, C_n'(\Theta_n) + D_n(\Theta_n)R_n^v D_n(\Theta_n)'$$

We observe that $(\widehat{X}_n^{\Theta,-}, P_n^{\Theta,-})$ only depends on the random sequence $(\Theta_0, \ldots, \Theta_n)$. This shows that the following random potential functions

$$G_{n,Y_n}(\Theta_0, \ldots, \Theta_n) := \mathcal{N}\left[C_n(\Theta_n)\widehat{X}_n^{\Theta,-}, \Sigma_n(\Theta_n, P_n^{\Theta,-})\right](Y_n) \qquad (8.22)$$

are well defined functions on the product state space $\prod_{0 \leq p \leq n} E_p$.

Using Bayesian notations, we also easily check the following multiplicative formula

$$
\begin{aligned}
p((Y_0, \ldots, Y_n) \mid (\theta_0, \ldots, \theta_n)) &= \prod_{k=0}^{n} p(Y_k \mid (Y_0, \ldots, Y_{k-1}), (\theta_0, \ldots, \theta_k)) \\
&= \prod_{k=0}^{n} G_{k,Y_k}(\theta_0, \ldots, \theta_n) \quad (8.23)
\end{aligned}
$$

8.3.3 Path space measures

If we denote by $\mathbb{P}_n(d(\theta_0, \ldots, \theta_n))$ the probability measure of the sequence of random parameters $(\Theta_0, \ldots, \Theta_n)$, then using Bayes' rule we find that the probability measure

$$
\mathbb{Q}_n(d(\theta_0, \ldots, \theta_n)) := \frac{1}{\mathcal{Z}_{n,Y}} \left\{ \prod_{0 \leq k < n} G_{k,Y_k}(\theta_0, \ldots, \theta_n) \right\} \mathbb{P}_n(d(\theta_0, \ldots, \theta_n))
$$

(8.24)

(with some normalizing constant $\mathcal{Z}_{n,Y}$) coincides with the conditional distribution of the random sequence $(\Theta_0, \ldots, \Theta_n)$ given the observations $(Y_0, \ldots, Y_{n-1})$; that is, we have that

$$
\mathbb{Q}_n = \mathrm{Law}\left((\Theta_0, \ldots, \Theta_n) \mid (Y_0, \ldots, Y_{n-1})\right)
$$

As the reader may have noticed, if we only want to estimate the distribution of the random states Θ_n given the sequence of observations, it is not really needed to store the full ancestral lines of the historical process $(\Theta_0, \ldots, \Theta_n)$. Indeed, we have

$$
\mathrm{Law}\left(X_n \mid (\Theta_0, \ldots, \Theta_n), \ (Y_0, \ldots, Y_{n-1})\right) = \mathcal{N}\left(\widehat{X}_n^{\Theta,-}, P_n^{\Theta,-}\right)
$$

with some pair of parameters $\left(\widehat{X}_n^{\Theta}, P_n^{\Theta}\right)$ whose values only depend on the sequence $(\Theta_0, \ldots, \Theta_n)$ and they can be computed using the Kalman recursion up to the n-th prediction transition. From this observation, we readily check that

$$
\mathcal{X}_n := \left(\Theta_n, \left(\widehat{X}_n^{\Theta,-}, P_n^{\Theta,-}\right)\right)
$$

forms a Markov chain. If we set

$$
\mathcal{G}_{n,Y_n}(\mathcal{X}_n) = \mathcal{N}\left[C_n(\Theta_n)\widehat{X}_n^{\Theta,-}, \Sigma_n(\Theta_n, P_n^{\Theta,-})\right](Y_n) \quad (8.25)
$$

and $F(\mathcal{X}_n) = f(\Theta_n)$ then we have the Feynman-Kac representation

$$
\mathbb{E}\left(f(\Theta_n) \mid \forall p < n \ \ Y_p = y_p\right) \propto \mathbb{E}\left(F(\mathcal{X}_n) \left\{ \prod_{0 \leq k < n} \mathcal{G}_{p,y_p}(\mathcal{X}_k) \right\}\right)
$$

Furthemore, for test functions of the following form

$$F(\theta, m, P) = \int \mathcal{N}(m, P)(dx) f(x)$$

we can also prove that

$$\mathbb{E}\left(f(X_n) \mid \forall p < n \quad Y_p = y_p\right) \propto \mathbb{E}\left(F(\mathcal{X}_n) \left\{ \prod_{0 \le k < n} \mathcal{G}_{p, y_p}(\mathcal{X}_k) \right\}\right) \qquad (8.26)$$

Path space conditional distributions can also be expressed in terms of the backward Kalman filter described in Section 8.2.2. More precisely, we have the formulae

$$\mathbb{E}\left(f_n(X_0, \ldots, X_n) \mid (Y_0, \ldots, Y_{n-1}) = (y_0, \ldots, y_{n-1})\right)$$

$$= \mathbb{E}\left[\;\mathbb{E}\left(f_n(X_0, \ldots, X_n) \mid \Theta, \;(Y_0, \ldots, Y_{n-1})\right) \\ \mid (Y_0, \ldots, Y_{n-1}) = (y_0, \ldots, y_{n-1})\right] \qquad (8.27)$$

$$= \mathbb{E}\left(\mathbb{Q}_{\Theta, n}(f_n) \left\{ \prod_{0 \le k < n} \mathcal{G}_{p, y_p}(\mathcal{X}_k) \right\}\right)$$

with the backward Kalman smoother given by

$$\mathbb{Q}_{\Theta, n}(d(x_0, \ldots, x_n)) = \eta_{n, \Theta}(dx_n) \prod_{1 \le p \le n} \mathbb{M}_{p, \Theta, \eta_{p-1, \Theta}}(x_p, dx_{p-1})$$

and the Markov transitions $\mathbb{M}_{p, \Theta, \eta_{p-1, \Theta}}$ given for any realization of the chain $\Theta = \theta$ by

$$\mathbb{M}_{n, \theta, \eta_{n-1, \theta}}(x_n, dx_{n-1})$$

$$:= p(x_{n-1} \mid x_n, (y_0, \ldots, y_{n-1}) \, \theta) \, dx_{n-1}$$

$$\propto \eta_{n-1, \theta}(dx_{n-1}) \, p(y_{n-1}|x_{n-1}, \theta_{n-1}) \, p(x_n \mid x_{n-1}, \theta_n)$$

This shows that

$$\mathbb{Q}_{\Theta, n}(f_n) = \mathcal{F}_n(\mathcal{X}_0, \ldots, \mathcal{X}_n)$$

with the function

$$\mathcal{F}_n(\mathcal{X}_0, \ldots, \mathcal{X}_n)$$

$$= \int \eta_{n, \Theta}(dx_n) \left\{ \prod_{1 \le p \le n} \mathbb{M}_{p, \Theta, \eta_{p-1, \Theta}}(x_p, dx_{p-1}) \right\} f_n(x_0, \ldots, x_n)$$

The conditional distributions defined in (8.26) and (8.27) can be estimated using the genealogical tree based models presented in Section 4.1.2. The resulting mean field IPS model is a genetic type algorithm with mutation transitions associated with the Markov chain transitions $\mathcal{X}_{n-1} \leadsto \mathcal{X}_n$ and the selection fitness functions $\mathcal{G}_{n, Y_n}(\mathcal{X}_n)$. By construction, this IPS algorithm can be interpreted as a sequence of interacting Kalman filters.

8.3.4 Maximum likelihood trajectories

This section is concerned with conditional maximum likelihood of the distributions (8.24) presented in Section 8.3.3. We further assume the random variables $(\Theta_0, \ldots, \Theta_n)$ have a density w.r.t. to some reference distribution μ_n on $\prod_{0 \leq k \leq n} E_k$ of the following form

$$\mathbb{P}\left((\Theta_0, \ldots, \Theta_n) \in d(\theta_0, \ldots, \theta_n)\right) = \exp\left(\sum_{0 \leq k \leq n} c_k(\theta_{k-1}, \theta_k)\right) \mu_n(d(\theta_0, \ldots, \theta_n))$$

for some likelihood functions c_k on $E_{k-1} \times E_k$. For $k = 0$, we use the convention $c_0(\theta_{-1}, \theta_0) = c_0(\theta_0)$, with some function c_0 on E_0.

We let

$$\psi_n(\Theta) = \left(\psi_n^1(\Theta), \psi_n^2(\Theta)\right) = \left(\widehat{X}_n^{\Theta, -}, P_n^{\Theta, -}\right)$$

According to (8.22), we have the formula

$$G_{n, y_n}(\Theta_0, \ldots, \Theta_n) = \frac{1}{(2\pi)^{q/2}\sqrt{|\Sigma_n(\Theta_n, \psi_n^2(\Theta))|}}$$

$$\exp\left[-2^{-1}\left(Y_n - C_n(\Theta_n)\psi_n^1(\Theta)\right)\Sigma_n(\Theta_n, \psi_n^2(\Theta))^{-1}\left(y - C_n(\Theta_n)\psi_n^1(\Theta)\right)'\right]$$

If we set

$$G_{n, y_n}(\Theta_0, \ldots, \Theta_n) := \exp\left(C_n(y_n, \psi_n(\Theta))\right)$$

and

$$V_n(\theta_0, \ldots, \theta_k) \quad = \quad \sum_{0 \leq k \leq n} c_k(\theta_{k-1}, \theta_k) + \sum_{0 \leq k < n} C_k(y_k, \psi_k(\Theta))$$

then we find that $\mathbb{Q}_n \ll \mu_n$, with the Radon-Nikodym derivative

$$\frac{d\mathbb{Q}_n}{d\mu_n} := \frac{1}{\mathcal{Z}_{n,Y}} \exp V_n \quad \Longleftrightarrow \quad \mathbb{Q}_n = \Psi_{e^{V_n}}(\mu_n)$$

The function $-c_n$ can be interpreted as an energy type function on the control sequence θ_n, and the energy functions $-C_n$ measure the difference between the controlled semigroup of the Kalman predictor $\psi_n(u_0, \ldots, u_n)$ and the reference observation state y_n.

The path space version of the conditional distributions defined in (8.26) can be estimated using the genealogical tree based models presented in Section 4.1.2. The resulting mean field IPS model is a genetic type algorithm with mutation transitions associated with the Markov chain transitions $\mathcal{X}_{n-1} \rightsquigarrow \mathcal{X}_n$ and the selection fitness functions $\mathcal{G}_{n, Y_n}(\mathcal{X}_n)$. In this context, we recall that every ancestral line

$$\xi_n^i := \left(\xi_{0,n}^i, \xi_{1,n}^i, \ldots, \xi_{n,n}^i\right)$$

of the IPS genealogical tree model has three components

$$\xi_{p,n}^i := \left(\Theta_{p,n}^i, \left(\widehat{X}_{p,n}^{\Theta^i, -}, P_{p,n}^{\Theta^i, -}\right)\right) \in \left(E_n \times \mathbb{R}^p \times \mathbb{R}^{p \times p}\right)$$

Using the same lines of arguments as the ones we used in Section 8.1.3, we prove the convergence in probability

$$\max_{1 \le i \le N} V_n(\Theta_{0,n}^i, \ldots, \Theta_{n,n}^i) \longrightarrow_{N \to \infty} \mathbb{Q}_n - \text{ess sup } V_n$$

These mean field IPS maximum estimation techniques provide a way of solving optimization problems w.r.t. average type criteria. To be more precise, let us assume that $D_n(\theta_n) = Id$. In this situation, using (8.19) we can rewrite the likelihood function V_n in terms of the average likelihood function of the observation sequence w.r.t. the partial linear-Gaussian system defined in (8.18); that is, we have that

$$V_n(\theta_0, \ldots, \theta_k) + \sum_{0 \le k < n} \log \left[(2\pi)^{q/2} \sqrt{|R_k^v|} \right]$$

$$= \sum_{0 \le k \le n} c_k(\theta_{k-1}, \theta_k) + \log \mathbb{E} \left(\exp \left[-\frac{1}{2} \sum_{0 \le k < n} \|y_k - C_k(\Theta_k) X_k\|_k^2 \right] \mid \Theta = \theta \right)$$

with the norm on $\mathbb{R}^q$ defined by $\|y\|_k^2 := y \, (R_k^v)^{-1} y'$.

8.4 Quenched and annealed filtering models

Suppose that at every time step the state of a Markov chain with two coordinates (Θ_n, X_n) is partially observed according to the following schematic picture

$$
\begin{array}{ccccccc}
\Theta_0 & \longrightarrow & \Theta_1 & \longrightarrow & \Theta_2 & \longrightarrow & \cdots \\
\downarrow & & \downarrow & & \downarrow & & \\
X_0 & \longrightarrow & X_1 & \longrightarrow & X_2 & \longrightarrow & \cdots \\
\downarrow & & \downarrow & & \downarrow & & \\
Y_0 & & Y_1 & & Y_2 & & \cdots
\end{array}
$$

We assume that Θ_n is a Markov chain on some state spaces E_n^Θ with initial distribution ν_0 and Markov transitions

$$K_n(\theta_{n-1}, d\theta_n) := \mathbb{P}(\Theta_n \in d\theta_n \mid \Theta_{n-1} = \theta_{n-1})$$

Given a realization of the Markov chain $\Theta_n = \theta_n$, the second coordinate of the signal X_n is a Markov chain taking values in some state spaces E_n^X, with initial distribution $\mu_{(0,\theta_0)}$, and elementary Markov transitions

$$M_{n,\theta_n}(x_{n-1}, dx_n) := \mathbb{P}(X_n \in dx_n \mid X_{n-1} = x_{n-1}, \Theta_n = \theta_n)$$

We denote by $\mathbb{P}_{n,\theta}$ the law of the random trajectories of the chain $(X_0, \ldots, X_n)$ given the parameter $\Theta = \theta$.

Given a realization of the Markov chain $\Theta_n = \theta_n$, and the random state $X_n = x_n$, the observation Y_n is a random variable taking values on some finite state space E_n^Y, with distribution

$$\mathbb{P}\left(Y_n \in dy_n \mid X_n = x_n, \Theta_n = \theta_n\right) := g_{n,\theta_n}(x_n, y_n)\, \lambda_n(dy_n) \qquad (8.28)$$

for some reference positive measure λ_n on E_n^Y.

We fix a realization $\theta = (\theta_n)_{n \geq 0}$ of the chain $\Theta = (\Theta_n)_{n \geq 0}$, and we set

$$G_{n,\theta_n} = g_{n,\theta_n}(x_n, y_n)$$

In this notation, the conditional distributions

$$\mathbb{Q}_{n,\theta} := \mathrm{Law}((X_0, \ldots, X_n) \mid \Theta = \theta,\ \forall p < n\ \ Y_p = y_p)$$

can be rewritten in terms of the following Feynman-Kac measures

$$d\mathbb{Q}_{n,\theta} = \frac{1}{\mathcal{Z}_{n,\theta}}\, d\Gamma_{n,\theta} \quad \text{with} \quad d\Gamma_{n,\theta} \left\{ \prod_{0 \leq p < n} G_{p,\theta_p}(X_p) \right\} d\mathbb{P}_{n,\theta}$$

In the above display, $\mathcal{Z}_{n,\theta}$ stands for some normalizing constant. This shows that the n-th time marginal measures

$$\eta_{n,\theta} := \mathrm{Law}(X_n \mid \Theta = \theta,\ \forall p < n\ \ Y_p = y_p)$$

are also defined by

$$\eta_{n,\theta}(f_n) = \gamma_{n,\theta}(f_n)/\gamma_{n,\theta}(1)$$

with the unnormalized distributions defined by

$$\gamma_{n,\theta}(f) := \mathbb{E}\left(f_n(X_n) \left\{ \prod_{0 \leq p < n} G_{p,\theta_p}(X_p) \right\} \mid \Theta = \theta \right)$$

We recall that these distributions satisfy the nonlinear filtering equation

$$\eta_{n+1,\theta} = \Psi_{G_{n,\theta_n}}(\eta_{n,\theta})\, M_{n+1,\theta_{n+1}} \qquad (8.29)$$

On the other hand, we also have the following multiplicative formula

$$\gamma_{n,\theta}(1) = \prod_{0 \leq p < n} \eta_{p,\theta}(G_{p,\theta_p}) \qquad (8.30)$$

from which we prove that

$$\gamma_{n,\theta}(f_n) = \mathbb{E}\left(f_n(X_n) \left\{ \prod_{0 \leq p < n} G_{p,\theta_p}(X_p) \right\} \mid \Theta = \theta \right) = \eta_{n,\theta}(1)\, \gamma_{n,\theta}(1)$$

Using these observations, we readily prove that the quenched conditional distributions

$$\eta_n(f_n) := \mathbb{E}(f_n(X_n) \mid \forall p < n \ \ Y_p = y_p) = \gamma_n(f_n)/\gamma_n(1)$$

have the following form

$$\gamma_n(f_n) \ := \ \mathbb{E}\left(f_n(X_n)\left\{\prod_{0 \leq p < n} G_{p,\Theta_p}(X_p)\right\}\right) = \mathbb{E}\left(\eta_{\Theta,n}(f_n)\,\gamma_{\theta,n}(1)\right)$$

We conclude that the quenched measures also have the following Feynman-Kac formulation

$$\gamma_n(f_n) = \mathbb{E}\left(\eta_{\Theta,n}(f_n)\prod_{0 \leq p < n}\eta_{\Theta,p}(G_{p,\Theta_p})\right)$$

Using the same arguments, the annealed conditional distributions

$$
\begin{aligned}
\mathbb{Q}_n &:= \ \mathrm{Law}((X_0,\ldots,X_n) \mid \forall p < n \ \ Y_p = y_p)\\
\overline{\mathbb{Q}}_n &:= \ \mathrm{Law}((\Theta_0,\ldots,\Theta_n) \mid \forall p < n \ \ Y_p = y_p)
\end{aligned}
$$

can be expressed in terms of the quenched models. More precisely, for any $\mathbf{f_n} \in \mathcal{B}_b(\prod_{0 \leq p \leq n} E_p^X)$ we have

$$\mathbb{Q}_n(\mathbf{f_n}) \propto \mathbb{E}\left(\mathbb{Q}_{n,\Theta}(\mathbf{f_n})\prod_{0 \leq p < n}\eta_{\Theta,p}(G_{p,\Theta_p})\right)$$

and for any $\mathbf{f_n} \in \mathcal{B}_b(\prod_{0 \leq p \leq n} E_p^\Theta)$

$$\overline{\mathbb{Q}}_n(\mathbf{f_n}) \propto \mathbb{E}\left(f_n(\Theta_0,\ldots,\Theta_n)\prod_{0 \leq p < n}\eta_{\Theta,p}(G_{p,\Theta_p})\right)$$

If we consider the Markov chain model

$$\mathcal{X}_n := (\Theta_n, \eta_{\Theta,n}) \in \left(E_n^\Theta \times \mathcal{P}(E_n^X)\right)$$

then we find that

$$\mathbb{Q}_n(\mathbf{f_n}) \propto \mathbb{E}\left(\mathcal{F}_n(\mathcal{X}_0,\ldots,\mathcal{X}_n)\prod_{0 \leq p < n}\mathcal{G}_p(\mathcal{X}_p)\right) \tag{8.31}$$

for any $\mathbf{f_n} \in \mathcal{B}_b(\prod_{0 \leq p \leq n} E_p^X)$, with

$$\mathcal{F}_n(\mathcal{X}_0,\ldots,\mathcal{X}_n) := \mathbb{Q}_{n,\Theta}(\mathbf{f_n}) \quad \text{and} \quad \mathcal{G}_n(\mathcal{X}_n) := \eta_{\Theta,n}(G_{n,\Theta_n})$$

In the above display $\mathbb{Q}_{n,\Theta}$ stands for the backward formulation of the Feynman-Kac measure on path space in terms of the flow of measures $\eta_{\Theta,p}$, with $p \leq n$.

In much the same way, we prove that

$$\overline{\mathbb{Q}}_n(\boldsymbol{f}_n) \propto \mathbb{E}\left(\mathcal{F}_n \left(\mathcal{X}_0, \ldots, \mathcal{X}_n \right) : \prod_{0 \leq p < n} \mathcal{G}_p(\mathcal{X}_p) \right) \tag{8.32}$$

for any $\boldsymbol{f}_n \in \mathcal{B}_b(\prod_{0 \leq p \leq n} E_p^\Theta)$ with

$$\mathcal{F}_n \left(\mathcal{X}_0, \ldots, \mathcal{X}_n \right) := f_n(\Theta_0, \ldots, \Theta_n)$$

The conditional distributions defined in (8.31) and (8.32) can be estimated using the genealogical tree based models presented in Section 4.1.2, as soon as we have a dedicated simulation scheme of the Markov chain $\mathcal{X}_{n-1} \rightsquigarrow \mathcal{X}_n$, and when the potential functions $\mathcal{G}_{n,Y_n}(\mathcal{X}_n)$ are explicitly known for any state variable $\mathcal{X}_n$. In this situation, the resulting mean field IPS model is a genetic type algorithm with mutation transitions associated with the Markov chain transitions $\mathcal{X}_{n-1} \rightsquigarrow \mathcal{X}_n$ and the selection fitness functions $\mathcal{G}_{n,Y_n}(\mathcal{X}_n)$. In general the chain $\mathcal{X}_n$ and the potential functions are unknown and we need to resort to another approximation scheme. In the next section, Section 8.5, we design a particle quenched model that allows applying these mean field IPS techniques.

8.5 Particle quenched and annealed models

We return to the quenched and annealed models discussed in Section 8.4. We denote by $\left(\Gamma_{n,\Theta}^N, \mathbb{Q}_{n,\Theta}^N, \gamma_{n,\Theta}^N, \eta_{n,\Theta}^N\right)$ a mean field IPS approximation of the Feynman-Kac measures $(\Gamma_{n,\Theta}, \mathbb{Q}_{n,\Theta}, \gamma_{n,\Theta}, \eta_{n,\Theta})$. We also denote by $\boldsymbol{\eta}_{n,\Theta}^N$ the genealogical tree based model associated with the genetic type IPS sampler.

We consider the Markov chain model

$$\mathcal{X}_n^N := \left(\Theta_n, \eta_{\Theta,n}^N \right) \in \left(E_n^\Theta \times \mathcal{P}(E_n^X) \right)$$

Using the unbiasedness properties (2.11) and (2.14) of the unnormalized distributions $\left(\gamma_{n,\Theta}^N, \Gamma_{n,\Theta}^N \right)$, we prove that

$$\mathbb{Q}_n(\boldsymbol{f}_n) \propto \mathbb{E}\left(\mathcal{F}_n \left(\mathcal{X}_0^N, \ldots, \mathcal{X}_n^N \right) \prod_{0 \leq p < n} \mathcal{G}_p(\mathcal{X}_p^N) \right)$$

for any $\boldsymbol{f}_n \in \mathcal{B}_b(\prod_{0 \leq p \leq n} E_p^X)$, with

$$\mathcal{F}_n \left(\mathcal{X}_0^N, \ldots, \mathcal{X}_n^N \right) := \mathbb{Q}_{n,\Theta}^N(\boldsymbol{f}_n) \quad \text{or} \quad \mathcal{F}_n \left(\mathcal{X}_0^N, \ldots, \mathcal{X}_n^N \right) := \boldsymbol{\eta}_{n,\Theta}^N(\boldsymbol{f}_n)$$

and the potential functions $\mathcal{G}_n(\mathcal{X}_n^N) := \eta_{\Theta,n}^N(G_{n,\Theta_n})$.

In much the same way, we prove that

$$\overline{\mathbb{Q}}_n(\boldsymbol{f_n}) \propto \mathbb{E}\left(\mathcal{F}_n\left(\mathcal{X}_0^N, \ldots, \mathcal{X}_n^N\right) \prod_{0 \leq p < n} \mathcal{G}_p(\mathcal{X}_p^N)\right) \qquad (8.33)$$

for any $\boldsymbol{f_n} \in \mathcal{B}_b(\prod_{0 \leq p \leq n} E_p^\Theta)$ with $\mathcal{F}_n\left(\mathcal{X}_0^N, \ldots, \mathcal{X}_n^N\right) := f_n(\Theta_0, \ldots, \Theta_n)$.

The above formulae allow computing the annealed Feynman-Kac models $\overline{\mathbb{Q}}_n$ in terms of genetic type mean field IPS models with mutation transitions dictated by the reference processes $\mathcal{X}_n^N$ and a selection fitness function $\mathcal{G}_n$.

8.6 Parameter estimation in hidden Markov models

This section is dedicated to fixed parameter estimation in hidden Markov chain models and their mean field IPS interpretations. Section 8.6.1 is concerned with partial linear-Gaussian models and interacting MCMC techniques. More general situations are developed in Section 8.6.2, dedicated to general mean field particle models. Section 8.6.3 and Section 8.6.4 are concerned with gradient type methods, including expected maximization techniques and related stochastic gradient models.

8.6.1 Partial linear-Gaussian models

We return to the partial linear-Gaussian models (8.18) discussed in Section 8.3.1. We further assume that $\Theta_n = \Theta$, for any $n \geq 0$, where Θ stands for a r.v. with distribution λ on some measurable state space E. With a slight abuse of notation we set

$$\Theta_n = \Theta \implies p((Y_0, \ldots, Y_n) | (\Theta_0, \ldots, \Theta_n)) = p((Y_0, \ldots, Y_n) | \Theta)$$

and

$$h_n(\Theta) = p(Y_n | \Theta, Y_0, \ldots, Y_{n-1})$$

By (8.22), the function h_k is explicitly given by the following formula

$$h_n(\Theta) := \mathcal{N}\left[C_n(\Theta)\widehat{X}_n^{\Theta,-}, \Sigma_n(\Theta, P_n^{\Theta,-})\right](Y_n)$$

Furthermore, using (8.24), we readily check that the conditional distribution μ_n of the r.v. Θ given the observations $(Y_0, \ldots, Y_n)$ is given by

$$\mu_n(d\theta) := \frac{1}{\mathcal{Z}_n}\left\{\prod_{0 \leq k \leq n} h_k(\theta)\right\}\lambda(d\theta) \qquad (8.34)$$

with some normalizing constant $\mathcal{Z}_n$. These models have exactly the same form as the Boltzmann-Gibbs measures (2.25) discussed in Section 2.5.2 in the opening chapter. The mean field IPS models $\xi_n = \left(\xi_n^i\right)_{1 \leq i \leq N}$ associated with these models are defined in terms of a series of MCMC transitions M_n with fixed point measures $\mu_n = \mu_n M_n$ and with the potential functions $G_n = h_{n+1}$.

Using the same line of arguments as the ones presented in Section 8.1.3 and Section 8.3.4, we prove that

$$\max_{1 \leq i \leq n} V_n(\xi_n^i) \longrightarrow_{N \to \infty} \mu_n - \text{ess sup } V_n$$

with the likelihood function $V_n := \sum_{0 \leq k \leq n} \log h_k$.

8.6.2 Mean field particle models

We return to the quenched and annealed models (8.18) discussed in Section 8.4 and in Section 8.5 (see also Section 2.5.2.3).

We further assume that $\Theta_n = \Theta$, for any $n \geq 0$, where Θ stands for a r.v. with distribution λ on some measurable state space E. We denote by $\xi := (\xi_n)_{n \geq 0}$ a mean field N-IPS model associated with the Feynman-Kac model $\mathbb{Q}_{n,\Theta}$, and we denote by Υ the probability distribution of the random variable

$$\mathcal{X} := (\Theta, \xi) \in \mathcal{E} := \left(E \times \prod_{n \geq 0} (E_n^X)^N \right)$$

Notice that Υ can be disintegrated as follows

$$\Upsilon(d(\theta, \xi)) = \lambda(d\theta) \, \mathcal{P}(\theta, d\xi)$$

where $\mathcal{P}(\theta, d\xi)$ stands for the conditional distribution of the mean field IPS $\xi := (\xi_n)_{n \geq 0}$, given a realization of the parameter $\Theta = \theta$.

We consider the Boltzmann-Gibbs measures μ_n on $\mathcal{E}$ defined by

$$\mu_n(d(\theta, \xi)) = \frac{1}{\mathcal{Z}_n} \times \left\{ \prod_{0 \leq p < n} h_p(\theta, \xi) \right\} \Upsilon(d(\theta, \xi))$$

with some normalizing constant $\mathcal{Z}_n$ and with the potential functions

$$h_n(\theta, \xi) = \frac{1}{N} \sum_{i=1}^{N} G_{n,\theta}(\xi_n^i)$$

By (8.33), for any bounded measurable function f on E, we have

$$\text{Law} \left(\Theta \mid \forall p < n, \ Y_p = y_p \right) = \mu_n \circ \Theta^{-1}$$

These hidden Markov chain models and their interacting particle interpretations are developed in Section 4.3.6. We also refer the reader to the seminal article by C. Andrieu, A. Doucet, and R. Holenstein [11], presenting an MCMC methodology in a statistical framework to sample target distributions using particle auxiliary variables.

8.6.3   Expected maximization models

We consider some parameter $\theta \in \mathbb{R}^d$ a Markov chain X_n, with elementary transitions $M_{n,\theta}$ on some measurable state spaces E_n with initial distribution $\eta_{0,\theta}$. We also consider a sequence of positive and bounded potential functions $G_{n,\theta}$ on the set E_n. We denote by $\mathbb{Q}_{n,\theta}$ the Feynman-Kac path measures (1.37) associated with the pairs $(M_{n,\theta}, G_{n,\theta})$.

We further assume that $\eta_{0,\theta} \ll \lambda_0$, and $M_{n,\theta}(x,.) \ll \lambda_n$, for some $\lambda_n \in \mathcal{M}_+(E_n)$ and we have

$$H_{0,\theta} = d\eta_{0,\theta}/d\lambda_0 > 0 \quad \text{and} \quad H_{n,\theta}(x,.) := dM_{n,\theta}(x,.)/d\lambda_n > 0$$

By construction, we have $\mathbb{Q}_{n,\theta} \ll \boldsymbol{\lambda_n} := \otimes_{0 \leq p \leq n} \lambda_p$ and the Radon-Nikodym derivates are given by the multiplicative formula

$$\frac{d\mathbb{Q}_{n,\theta}}{d\boldsymbol{\lambda_n}}(x_0,\ldots,x_n) := \mathbb{H}_{n,\theta}(x_0,\ldots,x_n) = \frac{1}{\mathcal{Z}_{n,\theta}} \prod_{0 \leq p \leq n} \boldsymbol{H}_{p,\theta}(x_{p-1},x_p)$$

for some normalizing constant $\mathcal{Z}_{n,\theta}$.
In the above display, $\boldsymbol{H}_{n,\theta}$ stands for the functions defined for any by

$$\boldsymbol{H}_{n,\theta}(x,y) = G_{n-1,\theta}(x)H_{n,\theta}(x,y)$$

with the convention $\boldsymbol{H}_{0,\theta}(x_{-1},x_0) = H_{0,\theta}(x_0)$, for $n = 0$.

These models arise in various scientific disciplines. The prototype of model we have in mind is the parameter estimation in Hidden Markov chain problem discussed in Section 8.6. In this situation, we are given a pair of signal-observation processes that depend on some random parameter Θ. The distributions $\mathbb{Q}_{n,\theta}$ represent the conditional distribution of the random states, given a realization of the parameter $\Theta = \theta$, and their normalizing constants $\mathcal{Z}_{n,\theta}$ coincide with the distribution of the observations given $\Theta = \theta$. In this context, we are given a series of observation data related to some unknown θ and we want to maximize the mapping $\theta \mapsto \mathcal{Z}_{n,\theta}$ so that to find the parameter θ that "explains" these data.

One way to solve this problem is to use the celebrated expected maximization model. This statistical search model is a recursive gradient type algorithm that improves sequentially its solution computing the parameter that maximizes the expected log-likelihood function.

We briefly recall the principle of this gradient based approch. For any pair of parameters (θ, θ') we find that

$$\text{Ent}\left(\mathbb{Q}_{n,\theta} | \mathbb{Q}_{n,\theta'}\right) \geq 0 \Rightarrow \mathbb{Q}_{n,\theta}\left(\log \mathbb{H}_{n,\theta}\right) \geq \mathbb{Q}_{n,\theta}\left(\log \mathbb{H}_{n,\theta'}\right)$$

Also observe that
$$\log \mathbb{H}_{n,\theta} = -\log \mathcal{Z}_{n,\theta} + \mathbb{L}_{n,\theta}$$

with the additive functional

$$\mathbb{L}_{n,\theta}(x_0,\ldots,x_n) := \sum_{p=0}^{n} \log \boldsymbol{H}_{p,\theta}(x_{p-1},x_p)$$

One concludes that for any pair of parameters (θ,θ')

$$\mathbb{Q}_{n,\theta}\left(\log \mathbb{H}_{n,\theta'}\right) = -\log \mathscr{Z}_{n,\theta'} + \mathbb{Q}_{n,\theta}\left(\mathbb{L}_{n,\theta'}\right) \leq -\log \mathscr{Z}_{n,\theta} + \mathbb{Q}_{n,\theta}\left(\mathbb{L}_{n,\theta}\right)$$

and therefore

$$\mathbb{Q}_{n,\theta}\left(\mathbb{L}_{n,\theta}\right) - \mathbb{Q}_{n,\theta}\left(\mathbb{L}_{n,\theta'}\right) \geq \log \left(\mathscr{Z}_{n,\theta}/\mathscr{Z}_{n,\theta'}\right)$$

In other words, we have

$$\mathbb{Q}_{n,\theta}\left(\mathbb{L}_{n,\theta}\right) \leq \mathbb{Q}_{n,\theta}\left(\mathbb{L}_{n,\theta'}\right) \Rightarrow \mathscr{Z}_{n,\theta} \leq \mathscr{Z}_{n,\theta'}$$

We denote by $(\theta_k)_{k\geq 0}$ a sequence of parameters starting at some state θ_0 and defined by the following recursion

$$\theta_k := \max_{\theta'} \mathbb{Q}_{n,\theta_{k-1}}\left(\mathbb{L}_{n,\theta'}\right) \implies \mathscr{Z}_{n,\theta_k} \geq \mathscr{Z}_{n,\theta_{k-1}}$$

If $\mathbb{Q}_{n,\theta}$ is in the exponential family, then the maximization step is usually straightforward. More precisely, there exists a collection of functions $(f_{n,\theta}^{(i)})_{i\in\mathcal{I}}$, indexed by some finite set $\mathcal{I}$, on $\boldsymbol{E}_n$ and some $F_{n,\theta} : \mathbb{R}^{\mathcal{I}} \mapsto \mathbb{R}^d$ such that

$$\theta_k = F_{n,\theta_{k-1}}\left(\left[\mathbb{Q}_{n,\theta_{k-1}}\left(f_{n,\theta_{k-1}}^{(i)}\right)\right]_{i\in\mathcal{I}}\right) \tag{8.35}$$

The set of functions $(f_n^{(i,\theta)})_{i\in\mathcal{I}}$ is sometimes referred to as sufficient statistics in the literature.

As shown in [194], one way to approximate the recursive Equation (8.35) consists of replacing the measures $\mathbb{Q}_{n,\theta_{n-1}}$ by the backward particle measures $\mathbb{Q}_{n,\theta_{n-1}}^N$ defined in (2.13) or by the genealogical tree occupation measures $\eta_{n,\theta_{n-1}}^N$ defined in (2.10). The corresponding mean field IPS approximation models are defined by

$$\theta_k^N = F_{n,\theta_{k-1}^N}\left(\left[\mathbb{Q}_{n,\theta_{k-1}}^N\left(f_{n,\theta_{k-1}^N}^{(i)}\right)\right]_{i\in\mathcal{I}}\right)$$

or by

$$\theta_k^N = F_{n,\theta_{k-1}^N}\left(\left[\boldsymbol{\eta}_{n,\theta_{n-1}}^N\left(f_{n,\theta_{k-1}^N}^{(i)}\right)\right]_{i\in\mathcal{I}}\right)$$

For a more thorough discussion on these stochastic algorithms and their convergence analysis, we refer the reader to the article [194].

8.6.4   Stochastic gradient algorithms

We come back to parametrized models presented in Section 8.6.3. One alternative way of computing the maximum value of the mapping $\theta \mapsto \mathcal{Z}_{n,\theta}$ is to introduce a more conventional gradient type steepest descent model

$$\theta_k = \theta_{k-1} + \tau_k \, \nabla \log \mathcal{Z}_{n,\theta_{k-1}}$$

with a positive real sequence of parameters τ_n such that $\sum_k \tau_k = \infty$ and $\sum_k \tau_k^2 < \infty$. Using (3.56), we have

$$\nabla \log \mathcal{Z}_{n,\theta} = \mathbb{Q}_{n,\theta}(\Lambda_{n,\theta})$$

with the additive functional

$$\Lambda_{n,\theta}(x_0, \ldots, x_n) := \sum_{p=0}^{n} \nabla \log \boldsymbol{H}_{p,\theta}(x_{p-1}, x_p)$$

Using the particle derivation model discussed in the end of Section 4.1.6, we can approximate these equations using the following equations

$$\theta_k^N = \theta_{k-1}^N + \tau_k \, \nabla_N \log \mathcal{Z}_{n,\theta_{k-1}} = \theta_{k-1}^N + \tau_k \, \mathbb{Q}_{n,\theta_{k-1}}^N(\Lambda_{n,\theta_{k-1}^N})$$

with the particle derivatives associated with the backward particle measures $\mathbb{Q}_{n,\theta_{n-1}}^N$ defined in (2.13). We can alternatively use the recursion

$$\theta_k^N = \theta_{k-1}^N + \tau_k \, \nabla_N \log \mathcal{Z}_{n,\theta_{k-1}} = \theta_{k-1}^N + \tau_k \, \boldsymbol{\eta}_{n,\theta_{k-1}}^N(\Lambda_{n,\theta_{k-1}^N})$$

with the genealogical tree occupation measures $\boldsymbol{\eta}_{n,\theta_{n-1}}^N$ defined in (2.10). For a more thorough discussion on these particle steepest descent algorithms and their connections with filter derivative models, we refer the reader to the article [195]. The convergence analysis of these models can be developed using the stochastic analysis techniques presented in the textbook [37].

8.7   Optimal stopping problems

As their name indicates, optimal stopping problems consist in choosing a judicious stopping time to optimize some expected reward or some cost energy type function. These estimation problems arise in a variety of application domains including biology [15], physics [470] and Bayesian statistics [72, 81, 356, 357, 517, 524, 527, 528, 529, 530, 570, 571], applied probability and reliability theory [234, 235], as well as in economics and mathematical finance [3, 97, 354, 453, 196, 197, 496]. For a more detailed historical account on this subject and the applications of optimal stopping, we refer the reader to the book [136].

In discrete time framework, the computation of optimal stopping time amounts to solving Snell envelope equations. These nonlinear and backward evolution equations in the set of nonnegative functions, with a terminal condition dictated by the reward function, are discussed in Section 8.7.1. In Section 8.7.2, we shall also provide some change of measures formulae expressing the original optimal stopping time in terms of an importance sampling strategy, with twisted reward functions [196].

In the last two decades, several numerical methodologies of the Snell envelope equations have been proposed, mostly in mathematical finance literature, including Longstaff-Schwartz's functional regression style methods [105, 139, 407, 553], refined singular values decomposition strategies [64], Monte Carlo simulation style methods [80, 191, 196, 197, 404], and the quantization grid technology [462, 463, 464, 465, 466].

8.7.1 Snell envelope equations

In a discrete time setting, we are given real valued stochastic process $(Z_k)_{0\leq k\leq n}$, adapted to some Markov chain sequence $(X_k)_{0\leq k\leq n}$ taking values in some measurable state space $(E_k)_{0\leq k\leq n}$, representing the available information at any time $0 \leq k \leq n$. In other words, we assume that $Z_k = F_k(X_0, \ldots, X_k)$, for some reward function F_k. In the further development of this section n stands for a given fixed final time horizon.

For any $k \in \{0, \ldots, n\}$, we let $\mathcal{T}_k$ be the set of all stopping times τ taking values in $\{k, \ldots, n\}$. The Snell envelope of $(Z_k)_{0\leq k\leq n}$ is the stochastic process $(U_k)_{0\leq k\leq n}$ defined for any $0 \leq k < n$ by the following backward equation

$$U_k = Z_k \vee \mathbb{E}(U_{k+1}|(X_0, \ldots, X_k))$$

with the terminal condition $U_n = Z_n$. The main property of this stochastic process is that

$$U_k = \sup_{\tau \in \mathcal{T}_k} \mathbb{E}\left(Z_\tau \mid (X_0, \ldots, X_k)\right) = \mathbb{E}\left(Z_{\tau_k^*}|(X_0, \ldots, X_k)\right)$$

with $\qquad \tau_k^* = \min\{k \leq l \leq n : U_l = Z_l\} \in \mathcal{T}_k \qquad (8.36)$

Notice that $U_k \geq Z_k$, for any $0 \leq k \leq n$ and τ_k^*, is given by the following backward formula

$$\tau_k^* = k\ \mathbf{1}_{Z_k \geq U_k} + \tau_{k+1}^*\ \mathbf{1}_{Z_k < U_k} \quad \text{with} \quad \tau_n^* = n$$

For a more precise derivation of these formulae, we refer the reader to the articles [196, 197].

To get one step closer, we let $\eta_0 = \text{Law}(X_0)$ be the initial distribution on E_0, and we denote by M_k the elementary transition of the Markov chain X_k

from E_{k-1} into E_k. We also assume that

$$\forall\,,0 \le k \le n \qquad Z_k = F_k(X_0,\dots,X_k) := f_k(X_k) \prod_{0 \le l < k} G_l(X_l) \qquad (8.37)$$

for some collection of nonnegative functions f_n and G_n on E_n. In this situation, the expected reward function is given by the Feynman-Kac formulae

$$\mathbb{E}\left(Z_\tau \mid (X_0,\dots,X_k)\right) = \left\{ \prod_{0 \le l < k} G_l(X_l) \right\} \mathbb{E}\left(f_\tau(X_\tau) \prod_{k \le l < \tau} G_l(X_l) \mid X_k \right)$$

Definition 8.7.1 *We consider the Feynman-Kac integral operators*

$$Q_{k+1}(x_k, dx_{k+1}) = G_k(x_k)\, M_{k+1}(x_k, dx_{k+1}) \qquad (8.38)$$

We denote by $Q_{k,l}$ the composition operator such that $Q_{k,l} := Q_{k+1}Q_{k+2}\cdots Q_l$, for any $k \le l$, with the convention $Q_{k,k} = Id$, the identity operator.

In this notation, the computation of the Snell envelope amounts to solving the following backward and nonlinear functional equation

$$u_k = \mathcal{H}_{k+1}(u_{k+1}) \qquad (8.39)$$

for any $0 \le k < n$, with the terminal value $u_n = f_n$, and the functional transformations

$$\mathcal{H}_{k+1}(u_{k+1}) := f_k \vee Q_{k+1}(u_{k+1})$$

We also mention that the Snell envelope U_k defined in (8.36) is given by

$$U_k = u_k(X_k) \prod_{0 \le p < k} G_p(X_p) \Rightarrow \tau_k^* = \min\{k \le l \le n \;:\; u_l(X_l) = f_l(X_l)\}$$

With this definition, one can check that a necessary and sufficient condition for the existence of the Snell envelope $(u_k)_{0 \le k \le n}$ is that

$$\forall k \le l \le n \quad \forall x_k \in E_k \qquad Q_{k,l}(f_l)(x_k) < \infty \qquad (8.40)$$

To check this claim, we simply notice that

$$f_k \le u_k \le f_k + Q_{k+1}(u_{k+1}) \;\Longrightarrow\; f_k \le u_k \le \sum_{k \le l \le n} Q_{k,l}(f_l)$$

8.7.2   Some changes of measures

For potential functions $G_l(X_l, X_{l+1})$ depending on the local transitions (X_l, X_{l+1}) of the reference process, the sequence of functions $(u_k)_{0 \le k \le n}$ satisfies the backward recursion:

$$u_k = f_k \vee \int M_{k+1}(.,dx_{k+1})\, G_k(.,x_{k+1})\, u_{k+1}(x_{k+1})$$

for any $0 \leq k < n$, with the terminal value $u_n = f_n$. This equation has exactly the same form as (8.39), by replacing in (8.38) the function $G_k(x_k)$ by the function $G_k(x_k, x_{k+1})$. We illustrate these properties in two situations.

The first one concerns the design of more general change of reference measure. For instance, let us suppose we are given a judicious Markov transition $M'_k(x_{k-1}, \cdot) \ll M_k(x_{k-1}, \cdot)$. We let X'_k be a Markov chain with initial condition $\eta'_0 = \eta_0$) and elementary transitions M'_k. In this notation, we have the Feynman-Kac formula

$$\mathbb{E}\left(f_n(X_n) \prod_{0 \leq p < n} G_p(X_p)\right) = \mathbb{E}\left(f_n(X'_n) \prod_{0 \leq p < n} G'_p(X'_p, X'_{p+1})\right)$$

with the potential functions $G'_p(x_p, x_{p+1}) = G_p(x_p) \times \frac{dM_{p+1}(x_p, \cdot)}{dM'_{p+1}(x_p, \cdot)}(x_{p+1})$.

Our second example concerns the design of an importance sampling strategy. Suppose we are given a sequence of positive payoff functions $(f_k)_{0 \leq k \leq n}$, with $f_0 \equiv 1$. In this situation, we have the Feynman-Kac formula

$$\mathbb{E}(f_n(X_n)) = \mathbb{E}\left(\prod_{0 \leq p < n} G_p(X_p, X_{p+1})\right)$$

with the potential function $G_p(x_p, x_{p+1}) = f_{p+1}(x_{p+1})/f_p(x_p)$.

From the pure mathematical point of view, using a natural state space enlargement, these classes of models are equivalent to the one discussed above (8.39). To check this claim, it suffices to replace in (8.37) the reference Markov chain X_k by the Markov chain $\mathcal{X}_k = (X_k, X_{k+1})$, on the state space $\mathcal{E}_k = (E_k \times E_{k+1})$.

8.7.3   A perturbation analysis

In this section, we present an original and powerful perturbation analysis of nonlinear semigroups. Section 8.7.3.1 is dedicated to the proof of a key robustness lemma.

This perturbation analysis is a natural as well as a fundamental tool for the analysis of the Snell envelope approximations. It can be used sequentially to obtain nonasymptotic estimates for models combining several levels of approximations. It is also useful to reduce the analysis of Snell approximation models on compact state spaces or even on finite but possibly large quantization trees or Monte Carlo type grids.

This perturbation analysis is clearly not new; it has been used with success in [152, 170, 387, 555] in the context of nonlinear filtering semigroups and particle approximation models. In the context of optimal stopping problems and numerical quantization schemes, these techniques were also used for instance in the papers of Egloff [241] and Gobet, Lemor, and Warin [285] or

Pagès [465]. To the best of our knowledge, the general and abstract formulation given above was first developed in the recent article [197].

Section 8.7.3.2 is dedicated to cutoff models because of their importance in practice. When computing the Snell envelope, it is often useful to approximate the state space by a compact set. For instance, it is shown in [283] that for standard unbounded models, as the Black-Scholes model, Monte Carlo estimation schemes require samples of exponential size in the number of variables of the value function, and the bounded state space assumption enables us to estimate the Snell envelope using samples of polynomial size w.r.t. the number of variables.

In other contexts, the article [242] presents a new algorithm that first requires a cutoff step which consists of replacing the price process by another process killed at first exit from a given bounded set. However, no bound is provided for the error induced by this cutoff approximation.

8.7.3.1   A robustness lemma

This section is concerned with the regularity properties of the Snell envelope w.r.t. the choice of the parameters (E_n, G_n, M_n).

Definition 8.7.2 *We let $\mathcal{H}_{k,l} = \mathcal{H}_{k+1} \circ \mathcal{H}_{k+1,l}$, with $k \leq l \leq n$, be the nonlinear semigroups associated with the backward Equations (8.39). We use the convention $\mathcal{H}_{k,k} = Id$, the identity operator.*

Using the elementary inequality

$$|(a \vee a') - (b \vee b')| \leq |a - b| + |a' - b'| \tag{8.41}$$

which is valid for any $a, a', b, b' \in \mathbb{R}$, we also easily check the following Lipschitz properties

$$|\mathcal{H}_k(u) - \mathcal{H}_k(v)| \leq Q_{k+1}(|u - v|) \implies |\mathcal{H}_{k,l}(u) - \mathcal{H}_{k,l}(v)| \leq Q_{k,l}(|u - v|) \tag{8.42}$$

for any functions u, v on E_l, and any $k \leq l \leq n$.

Even if it looks innocent, the numerical solving of the recursion (8.39) often requires extensive calculations. The central problem is to compute the conditional expectations $M_{k+1}(u_{k+1})$ on the whole state space E, at every time step $0 \leq k < n$.

For Markov chain models taking values in some finite state spaces with a reasonably large cardinality, the above expectations can be easily computed by a simple backward inspection of the whole realization tree that lists all possible outcomes and every transition of the chain. In more general situations, we need to resort to some approximation strategy.

Most of the numerical approximation schemes amount to replacing the pair $(f_k, Q_k)_{0 \leq k \leq n}$ by some possibly random approximation model $(\widehat{f}_k, \widehat{Q}_k)_{0 \leq k \leq n}$ on some possibly reduced finite and possibly random subsets $\widehat{E}_k \subset E_k$.

Definition 8.7.3 *We let $\widehat{u}_k$ be the Snell envelope on $\widehat{E}_k$ associated with the functions $\widehat{f}_k$ and the sequence of integral operators $\widehat{M}_k$ from $\widehat{E}_{k-1}$ into $\widehat{E}_k$; that is, we have that*

$$\widehat{u}_k = \widehat{\mathcal{H}}_{k+1}(\widehat{u}_{k+1}) := \widehat{f}_k \vee \widehat{Q}_{k+1}(\widehat{u}_{k+1}) \tag{8.43}$$

We let $\widehat{\mathcal{H}}_{k,l} = \widehat{\mathcal{H}}_{k+1} \circ \widehat{\mathcal{H}}_{k+1,l}$, with $k \leq l < n$, be the nonlinear semigroups associated with the backward Equations (8.43), so that $\widehat{u}_k = \widehat{\mathcal{H}}_{k,l}(\widehat{u}_l)$, for any $k \leq l \leq n$.

For any $0 \leq k < n$ and for any functions u on E_{k+1}, using (8.41) one readily gets the local approximation inequality

$$\left| \mathcal{H}_{k+1}(u) - \widehat{\mathcal{H}}_{k+1}(u) \right| \leq |f_k - \widehat{f}_k| + |(Q_{k+1} - \widehat{Q}_{k+1})(u)| \tag{8.44}$$

To transfer these local estimates to the semigroups $\mathcal{H}_{k,l}$ and $\widehat{\mathcal{H}}_{k,l}$ we use a perturbation analysis. The difference between the approximate and the exact Snell envelope can be written as a telescoping sum

$$u_k - \widehat{u}_k = \sum_{l=k}^{n} \left[\widehat{\mathcal{H}}_{k,l}(\mathcal{H}_{l+1}(u_{l+1})) - \widehat{\mathcal{H}}_{k,l}(\widehat{\mathcal{H}}_{l+1}(u_{l+1})) \right]$$

setting for simplicity $\mathcal{H}_{n+1}(u_{n+1}) = u_n$ and $\widehat{\mathcal{H}}_{n+1}(u_{n+1}) = \widehat{u}_n$, for $l = n$. The Lipschitz property (8.42) of the semigroup $\widehat{\mathcal{H}}_{k,l}$, in conjunction with the local estimate (8.44), leads to the following estimates.

Lemma 8.7.4 *For any $k \leq n$ we have the robustness inequalities*

$$|u_k - \widehat{u}_k| \leq \sum_{l=k}^{n} \widehat{Q}_{k,l} \left| f_l - \widehat{f}_l \right| + \sum_{l=k}^{n-1} \widehat{Q}_{k,l} \left[\left| (Q_{l+1} - \widehat{Q}_{l+1})(u_{l+1}) \right| \right]$$

8.7.3.2 Cutoff models

We suppose that E_n are topological spaces with σ-fields $\mathcal{E}_n$ that contain the Borel σ-field on E_n. Our next objective is to find conditions under which we can reduce the backward functional Equation (8.39) to a sequence of compact sets $\widehat{E}_n$.

To this end, we further assume that the initial measure η_0 and the Markov transition M_n of the chain X_n satisfy the following tightness property: For every sequence of positive numbers $\epsilon_n \in [0, 1[$, there exists a collection of compact subsets $\widehat{E}_n \subset E_n$ s.t. for any $n \geq 0$ we have

$$(\mathcal{T}) \qquad \eta_0(\widehat{E}_0^c) \leq \epsilon_0 \quad \text{and} \quad \sup_{x_n \in \widehat{E}_n} M_{n+1}(x_n, \widehat{E}_{n+1}^c) \leq \epsilon_{n+1}$$

For instance, this condition is clearly met for regular Gaussian type transitions on the Euclidean space, for some collection of increasing compact balls.

In this situation, a natural cutoff consists in considering the original integral operator M_k by their restrictions $\widehat{M}_k$ to the compact sets $\widehat{E}_k$, defined for any $x \in \widehat{E}_{k-1}$ by

$$\widehat{M}_k(x, dy) := M_k(x, dy) \, 1_{\widehat{E}_k}(y)/M_k(1_{\widehat{E}_k})(x)$$

We set

$$\widehat{Q}_k(x, dy) := G_{k-1}(x) \, \widehat{M}_k(x, dy)$$

Notice that these transitions are well defined as soon as $M_k(x, \widehat{E}_k) > 0$, for any $x \in \widehat{E}_{k-1}$.

Theorem 8.7.5 *We assume that the tightness condition $(\mathcal{T})$ is met, for every sequence of positive numbers $\epsilon_n \in [0, 1[$, and for some collection of compact subsets $\widehat{E}_n \subset E_n$. In this situation, for any $0 \le k \le n$, we have that*

$$\|u_k - \widehat{u}_k\|_{\widehat{E}_k} \le \sum_{l=k+1}^{n} \|G_{l-1}\|_{\widehat{E}_{l-1}} \left[\frac{\epsilon_l}{1 - \epsilon_l} \, \|M_l(u_l)\|_{\widehat{E}_{l-1}} + \epsilon_l^{1/2} \, \|M_l(u_l^2)\|_{\widehat{E}_{l-1}}^{1/2} \right]$$

Proof:
Lemma 8.7.4, combined with the decompositions

$$[\widehat{Q}_k - Q_k](u_k) = G_{k-1} \, [\widehat{M}_k - M_k](u_k)$$

and

$$
\begin{aligned}
{[\widehat{M}_k - M_k](u_k)} &= \widehat{M}_k(u_k) - M_k(1_{\widehat{E}_k} u_k) - M_k(1_{\widehat{E}_k^c} u_k) \\
&= \left(\frac{1}{M_k(1_{\widehat{E}_k})} - 1 \right) M_k(u_k 1_{\widehat{E}_k}) - M_k(1_{\widehat{E}_k^c} u_k) \\
&= \frac{M_k(1_{\widehat{E}_k^c})}{M_k(1_{\widehat{E}_k})} \times M_k(u_k 1_{\widehat{E}_k}) - M_k(1_{\widehat{E}_k^c} u_k) \; .
\end{aligned}
$$

leads to the following estimates

$$\|u_k - \widehat{u}_k\|_{\widehat{E}_k}$$

$$:= \sup_{x \in \widehat{E}_k} |u_k(x) - \widehat{u}_k(x)|$$

$$\le \sum_{l=k+1}^{n} \|G_{l-1}\|_{\widehat{E}_{l-1}} \left[\left\| \frac{M_l(1_{\widehat{E}_l^c})}{M_l(1_{\widehat{E}_l})} \right\|_{\widehat{E}_{l-1}} \|M_l(u_l 1_{\widehat{E}_l})\|_{\widehat{E}_{l-1}} + \|M_l(u_l 1_{\widehat{E}_l^c})\|_{\widehat{E}_{l-1}} \right]$$

The end of the proof is now clear. This ends the proof of the theorem.   ∎

Remark 8.7.6 *Using (8.40) we readily prove the following rather crude upper bounds $u_k \leq c(n) \sum_{k \leq l \leq n} M_{k,l}(f_l)$, with $c(n) \leq \left(\prod_{0 \leq l < n} (\|G_l\| \vee 1) \right)$. This implies that*

$$\left\| M_k(u_k^2) \right\|_{\widehat{E}_{k-1}} \leq c'(n) \sum_{l=k}^{n} \left\| M_{k-1,l}(f_l)^2 \right\|_{\widehat{E}_{k-1}} \quad \text{with} \quad c'(n) \leq (n+1) \, c(n)^2$$

Consequently, using the estimates presented in Theorem 8.7.5, one can find sets $(\widehat{E}_l)_{k<l\leq n}$ so that $\|u_k - \widehat{u}_k\|_{\widehat{E}_k}$ is as small as one wants as soon as $\|M_{k,l}(f_l)^2\|_{\widehat{E}_k} < \infty$, for any $0 \leq k < l \leq n$. A similar cutoff approach was introduced and analyzed in Bouchard and Touzi [66], but the cutoff was operated on some regression functions and not on the transition kernels.

8.7.4 Mean field particle approximations

This section is dedicated to mean field IPS solving of Snell envelope evolution equations. In Section 8.7.4.1 we present a formulation of the Snell envelope equations in terms of a natural importance sampling type Feynman-Kac model. The mean field IPS is defined by replacing the Feynman-Kac distribution flow by particle density profiles associated with a mean field interpretation of the Feynman-Kac model. In Section 8.7.4.2, we present some nonasymptotic estimates.

8.7.4.1 Description of the models

The nature of the evolution Equation (8.39), with the integral operators Q_k defined in (8.38), indicates that it is not relevant to compute precisely the conditional expectation $M_{k+1}(u_{k+1})$ for null or too small values of the potential functions G_k or the reward function f_k. The Feynman-Kac measures $\mathbb{Q}_n$ defined in (1.37) are natural importance sampling distributions if we want to design a mean field IPS approximation scheme. We further assume that the Markov transitions $M_k(x_{k-1}, .) \ll \lambda_k$ for some reference probability measures λ_k on E_k, and we have

$$\frac{dM_k(x_{k-1}, .)}{d\lambda_k}(x_k) := H_k(x_{k-1}, x_k) > 0 \qquad (8.45)$$

We also assume that

$$\sup_{x,y \in E_{k-1}} \frac{H_k(x,z)}{H_k(y,z)} := h_k(z) \quad \text{with} \quad \|h_k\| < \infty \quad \text{and} \quad \sup_{x,y \in E_k} \frac{G_k(x)}{G_k(y)} = g_k < \infty \qquad (8.46)$$

In this situation, we have $Q_k(x_{k-1}, .) \ll \lambda_k$ with the Radon-Nikodym derivatives

$$\frac{dQ_k(x_{k-1}, .)}{d\lambda_k}(x_k) = G_{k-1}(x_{k-1})H_k(x_{k-1}, x_k) := H_{k-1,k}(x_{k-1}, x_k)$$

The following lemma expresses the Snell envelope equations in terms of the parameters $(\eta_k, G_k, H_{k,k+1})$.

Lemma 8.7.7 *The Snell envelope Equation (8.39) can be rewritten in the following form*

$$u_k = f_k \vee \left[\eta_k(G_k) \int \eta_{k+1}(dx_{k+1}) \times \frac{H_{k,k+1}(x_k, x_{k+1})}{\eta_k\left(H_{k,k+1}(\cdot, x_{k+1})\right)} \, u_{k+1}(x_{k+1}) \right]$$

(8.47)

for any $0 \leq k < n$, with the terminal value $u_n = f_n$.

Proof:
Under our assumption, we have that $\eta_k \ll \lambda_k$, for any $k \leq n$ with

$$
\begin{aligned}
\eta_{k+1}(dx_{k+1}) &= \Phi_{k+1}(\eta_k)(dx_{k+1}) = \Psi_k(\eta_k) M_{k+1}(dx_{k+1}) \\
&= \frac{1}{\eta_k(G_k)} \, \eta_k(H_{k,k+1}(\cdot, x_{k+1})) \, \lambda_{k+1}(dx_{k+1})
\end{aligned}
$$

from which we conclude that

$$Q_{k+1}(x_k, dx_{k+1}) = \eta_k(G_k) \times \eta_{k+1}(dx_{k+1}) \times \frac{H_{k,k+1}(x_k, x_{k+1})}{\eta_k\left(H_{k,k+1}(\cdot, x_{k+1})\right)}$$

The end of the proof is now a direct consequence of (8.39). This ends the proof of the lemma. ∎

Definition 8.7.8 *We let $(\eta_k^N)_{0 \leq k \leq n}$ be the particle density profiles associated with some McKean particle interpretation of the Feynman-Kac measures $(\eta_k)_{0 \leq k \leq n}$ defined in Section 4.4.5.*

The mean field IPS approximation of the backward Snell envelope model (8.47) is defined by the backward equation

$$u_k^N = f_k \vee \left[\eta_k^N(G_k) \int \eta_{k+1}^N(dx_{k+1}) \times \frac{H_{k,k+1}(\cdot, x_{k+1})}{\eta_k^N\left(H_{k,k+1}(\cdot, x_{k+1})\right)} \, u_{k+1}^N(x_{k+1}) \right]$$

for any $0 \leq k < n$, with the terminal value $u_n^N = f_n (= u_n)$.

8.7.4.2   Some convergence estimates

The aim of this section is to express u_k^N in terms of the local sampling random fields

$$V_k^N := \sqrt{N} \, \left(\eta_k^N - \Phi_k(\eta_k^N) \right)$$

and the random integral operators defined below.

Definition 8.7.9 *We let Q_{k+1}^N be the random integral operator defined by*

$$Q_{k+1}^N = Q_{k+1} + \frac{1}{\sqrt{N}} \, R_{k+1}^N$$

with the random integral operator R_{k+1}^N defined by

$$R_{k+1}^N(v) := \int V_{k+1}^N(dx_{k+1}) \frac{H_{k,k+1}(\cdot, x_{k+1})}{\eta_k^N(H_{k,k+1}(\cdot, x_{k+1}))} v(x_{k+1})$$

for any measurable function v on E_{k+1}. We let

$$Q_{k,l}^N := Q_{k+1}^N Q_{k+2}^N \cdots Q_l^N$$

with $k \leq l$, the semigroup associated with the random integral operators Q_k^N. We use the convention $Q_{k,k}^N = Id$, the identity operator.

In the next lemma we express the Snell envelope equation in terms of the integral operators Q_k^N.

Lemma 8.7.10 *For any $k \leq l < n$, and any measurable function v on E_l, we have*

$$\mathbb{E}\left(Q_{k,l}^N(v) \mid \eta_{k-1}^N\right) = Q_{k,l}(v)$$

In addition, the particle Snell envelope $(u_k^N)_{0 \leq k \leq n}$ satisfies for any $0 \leq k < n$ the backward recursion:

$$u_k^N = f_k \vee Q_{k+1}^N\left(u_{k+1}^N\right)$$

Proof:
The first assertion is immediate. The second one is a direct consequence of the fact that

$$\eta_{k+1}^N = \Phi_{k+1}(\eta_k^N)(f_{k+1}) + \frac{1}{\sqrt{N}} V_{k+1}^N$$

and

$$\eta_k^N(G_k) \int \Phi_{k+1}(\eta_k^N)(dx_{k+1}) \times \frac{H_{k,k+1}(\cdot, x_{k+1})}{\eta_k^N(H_{k,k+1}(\cdot, x_{k+1}))} u_{k+1}^N(x_{k+1})$$

$$= \int \lambda_{k+1}(dx_{k+1}) \times H_{k,k+1}(\cdot, x_{k+1}) u_{k+1}^N(x_{k+1}) = Q_{k+1}(u_{k+1}^N)$$

This ends the proof of the lemma. ∎

Remark 8.7.11 *Under condition (8.40), the first assertion of the lemma ensures that the particle Snell envelope u_k^N is well defined, and we have $\mathbb{E}\left(u_k^N(x)\right) < \infty$, for any $k \leq n$. For a more thorough discussion on these particle models, including general state space models, their convergence analysis, and a variety of related approximation grid type models, we refer the reader to the pair of articles [196, 197].*

The next result shows that the mean field IPS model overestimates the Snell envelope.

Proposition 8.7.12 *For any $0 \leq k \leq n$ and any $x_k \in E_k$*

$$\mathbb{E}\left(u_k^N(x_k)\right) \geq u_k(x_k) \tag{8.48}$$

Proof:
We can easily prove this inequality with a simple backward induction. The result is immediate at the terminal n. Assuming the inequality at time $k+1$, then the Jensen's inequality implies

$$\begin{aligned}
\mathbb{E}\left(u_k^N(x_k)\right) &\geq f_k(x_k) \vee \mathbb{E}\left(Q_{k+1}^N(u_{k+1}^N)(x_k)\right) \\
&= f_k(x_k) \vee \mathbb{E}\left(Q_{k+1}(u_{k+1}^N)(x_k)\right)
\end{aligned}$$

By the induction assumption at time $k+1$, we have

$$\mathbb{E}\left(Q_{k+1}(u_{k+1}^N)(x_k)\right) \geq Q_{k+1}(u_{k+1})(x_k)$$

from which we conclude that

$$\mathbb{E}\left(u_k^N(x_k)\right) \geq f_k(x_k) \vee Q_{k+1}(u_{k+1})(x_k) = u_k(x_k)$$

This ends the proof of the proposition.   ∎

We are now in position to state and to prove the main result of this section.

Theorem 8.7.13 *For any $k \leq n$ and any $N \geq 1$, we have*

$$\sqrt{N}\mathbb{E}\left[\left|u_k(x_k) - u_k^N(x_k)\right|\right] \leq 2 \sum_{k \leq l < n} g_l\, Q_{k,l}(h_l)(x_k)$$

In addition, for any $m \geq 2$, we have

$$\sqrt{N}\mathbb{E}\left[\left|u_k(x_k) - u_k^N(x_k)\right|^m\right]^{1/m} \leq b(m)\, a_{k,n}(m)$$

In the above display, $a_{k,n}(m)$ stands for some finite constants such that

$$a_{k,n}(m) \leq 2 \sum_{k \leq l < n} g_l\left(c_{k,l} \wedge \left(c_{k,l}^{1-\frac{1}{2m}}\, Q_{k,l}(h_l^{2m})(x_k)^{1/(2m)}\right)\right)$$

with the parameters $c_{k,l} := \prod_{k < p \leq l}(g_{p-1}\|h_p\|)$.

Proof:
We let $\mathcal{F}_k^N$ be the σ-field generated by the particle density profiles η_l^N, with $l \leq k$. The proof of the theorem is based on the following decomposition

$$\sqrt{N}\left|u_k - u_k^N\right| \leq \sum_{k \leq l < n} Q_{k,l}^N\left(\left|R_{l+1}^N(u_{l+1})\right|\right)$$

In Section 4.4.4 (cf. Proposition 9.5.5), we shall prove that for any test function $f_k \in \mathrm{Osc}(E_k)$, and any $m \geq 1$, we have

$$\mathbb{E}\left(\left|V_k^N(f_k)\right|^m \left|\mathcal{F}_{k-1}^N\right.\right)^{\frac{1}{m}} \leq b(m)$$

with the constant $b(m)$ defined in (0.6). Under our regularity conditions (8.46), using these estimates we readily check that

$$\mathbb{E}\left(\left|R_{k+1}^N(u_{k+1})(x_k)\right|^m \left|\mathcal{F}_k^N\right.\right)^{1/m} \leq 2\, b(m)\, g_k\, h_k(x_k)$$

These $\mathbb{L}_m$-mean error estimates combined with the generalized Minkowski integral inequality yield

$$\mathbb{E}\left(\left[Q_{k,l}^N\left(\left|R_{l+1}^N(u_{l+1})\right|\right)(x_k)\right]^m \left|\mathcal{F}_l^N\right.\right)^{1/m}$$

$$\leq \int Q_{k,l}^N(x_k, dx_l)\, \mathbb{E}\left(\left|R_{l+1}^N(u_{l+1})(x_l)\right|^m \left|\mathcal{F}_l^N\right.\right)^{1/m} \leq 2\, b(m)\, g_l\, Q_{k,l}^N(h_l)(x_k)$$

Using Lemma 8.7.10, we conclude that

$$\mathbb{E}\left(\left[Q_{k,l}^N\left(\left|R_{l+1}^N(u_{l+1})\right|\right)(x_k)\right]^m\right)^{1/m} \leq 2\, b(m)\, g_l\, \mathbb{E}\left(\left(Q_{k,l}^N(h_l)(x_k)\right)^m\right)^{1/m}$$

Using Lemma 8.7.10, the proof of the first assertion is immediate.
In addition, we have

$$\mathbb{E}\left(\left[Q_{k,l}^N\left(\left|R_{l+1}^N(u_{l+1})\right|\right)(x_k)\right]^m \left|\mathcal{F}_l^N\right.\right)^{1/m} \leq 2\, b(m)\, g_l\, c_{k,l}$$

This implies that

$$\sqrt{N}\mathbb{E}\left[\left|u_k(x_k) - u_k^N(x_k)\right|^m\right]^{1/m} \leq 2\, b(m) \sum_{k \leq l < n} g_l\, c_{k,l}$$

Now we come to the proof of the $\mathbb{L}_m$-mean error estimates. We use the decomposition

$$\left(Q_{k,l}^N(h_l)(x_k)\right)^m = Q_{k,l}^N(1)(x_k)^m \left[\left(\frac{Q_{k,l}^N(h_l)(x_k)}{Q_{k,l}^N(1)(x_k)}\right)^{2m}\right]^{\frac{1}{2}}$$

$$\leq Q_{k,l}^N(1)(x_k)^{m-\frac{1}{2}} \left[Q_{k,l}^N(h_l^{2m})(x_k)\right]^{\frac{1}{2}}$$

More generally, we have

$$\mathbb{E}\left(\left[Q_{k,l}^N\left(\left|R_{l+1}^N(u_{l+1})\right|\right)(x_k)\right]^m\right)^{1/m}$$

$$\leq 2\, b(m)\, g_l\, \left(\mathbb{E}\left(Q_{k,l}^N(1)(x_k)^{2m-1}\right)\right)^{1/(2m)} Q_{k,l}(h_l^{2m})(x_k)^{1/(2m)}$$

$$\leq 2\, b(m)\, g_l\, c_{k,l}^{1-\frac{1}{2m}}\, Q_{k,l}(h_l^{2m})(x_k)^{1/(2m)}$$

The end of the proof of the $\mathbb{L}_m$-mean error bounds is now clear. This ends the proof of the theorem. ∎

The following Orlicz norm estimates are direct consequences of Theorem 8.7.13 combined with Lemma 11.3.1 and Lemma 11.3.2.

Corollary 8.7.14 *For any $k \leq n$ and any $N \geq 1$, we have the Orlicz norm estimates*

$$\sqrt{N} \; \pi_\psi \left(\left| u_k(x_k) - u_k^N(x_k) \right| \right) \leq a_{k,n}$$

and

$$\sqrt{N} \; \pi_\psi \left(\left\| u_k - u_k^N \right\|_{\mathbb{L}_1(\lambda_k)} \right) \leq a_{k,n} \quad with \quad a_{k,n} \leq 2 \sum_{k \leq l < n} g_l \, c_{k,l}$$

In addition, for any $y \geq 0$, $k \leq n$, $N \geq 1$, and any $x_k \in E_k$, the probability of the following events is greater than $1 - e^{-y}$

$$\left| u_k(x_k) - u_k^N(x_k) \right| \leq a_{k,n} \sqrt{(y + \log 2)/N}$$
$$\left\| u_k - u_k^N \right\|_{\mathbb{L}_1(\lambda_k)} \leq a_{k,n} \sqrt{(y + \log 2)/N}$$

with the parameters $a_{k,n}$ defined in Theorem 8.7.13.

8.7.5 Partial observation models

8.7.5.1 Snell envelope model

We consider the partially observed Markov chain model discussed in (3.31). The Snell envelope associated with an optimal stopping problem with finite horizon, payoff style function $f_n(X_n, Y_n)$, and noisy observations Y_n as some Markov process, is given by

$$U_k := \sup_{\tau \in \mathcal{T}_k^Y} \mathbb{E}(f_\tau(X_\tau, Y_\tau) | (Y_0, \ldots, Y_k))$$

where $\mathcal{T}_k^Y$ stands for the set of all stopping times τ taking values in $\{k, \ldots, n\}$, whose values are measurable w.r.t. the sigma field generated by the observation sequence Y_p, from $p = 0$ up to the current time k. We denote by $\eta_n^{[y_0, \ldots, y_{n-1}]}$ and $\widehat{\eta}_n^{[y_0, \ldots, y_n]}$ the conditional distributions defined in (8.4) and (8.5). In this notation, for any $0 \leq k \leq n$ we have that

$$\mathbb{E}(f_\tau(X_\tau, Y_\tau) | (Y_0, \ldots, Y_k)) = \mathbb{E}\left(F_\tau \left(Y_\tau, \widehat{\eta}_\tau^{[Y_0, \ldots, Y_\tau]} \right) \, \middle| \, (Y_0, \ldots, Y_k) \right) \quad (8.49)$$

with the conditional payoff function

$$F_p \left(Y_p, \widehat{\eta}_p^{[Y_0, \ldots, Y_p]} \right) = \int \widehat{\eta}_p^{[Y_0, \ldots, Y_p]}(dx) \, f_p(X_p, Y_p)$$

It is rather well known that $\mathcal{X}_p := \left(Y_p, \widehat{\eta}_p^{[Y_0,\ldots,Y_p]}\right)$ is a Markov chain with elementary transitions defined by

$$\mathbb{E}\left[F_p\left(Y_p, \widehat{\eta}_p^{[Y_0,\ldots,Y_p]}\right) \;\Big|\; \left(Y_{p-1}, \widehat{\eta}_{p-1}^{[Y_0,\ldots,Y_{p-1}]}\right) = (y, \mu)\right]$$

$$= \int \lambda_p(dy_p) \; \mu M_p\left(g_p(\cdot, y_p)\right) \; F_p\left(y_p, \Psi_{g_p(\cdot, y_p)}\left(\mu M_p\right)\right)$$

A detailed proof of this assertion can be found in any textbook on advanced stochastic filtering theory. For instance, the book of W. Runggaldier and L. Stettner [507] provides a detailed treatment on discrete time nonlinear filtering and related partially observed control models.

Roughly speaking, using some abusive Bayesian notation, we have

$$\begin{aligned}
\eta_p^{[y_0,\ldots,y_{p-1}]}(dx_p) &= dp(x_p \mid (y_0,\ldots,y_{p-1})) \\
&= \int dp(x_p \mid x_{p-1}) \times dp(x_{p-1} \mid (y_0,\ldots,y_{p-1})) \\
&= \widehat{\eta}_{p-1}^{[y_0,\ldots,y_{p-1}]} M_p(dx_p)
\end{aligned}$$

and

$$\Psi_{g_p(\cdot, y_p)}\left(\widehat{\eta}_{p-1}^{[y_0,\ldots,y_{p-1}]} M_p\right)(dx_p)$$

$$= \frac{p(y_p \mid x_p)}{\int p(y_p \mid x_p') \, dp(x_p' \mid (y_0,\ldots,y_{p-1}))} \, dp(x_p \mid (y_0,\ldots,y_{p-1}))$$

$$= dp(x_p \mid (y_0,\ldots,y_{p-1}, y_p))$$

from which we prove that

$$\begin{aligned}
\mu M_p(g_p(\cdot, y_p)) &= \int p(y_p \mid x_p) \, dp(x_p \mid (y_0,\ldots,y_{p-1})) \\
&= p(y_p \mid (y_0,\ldots,y_{p-1}))
\end{aligned}$$

and

$$\Psi_{g_p(\cdot, y_p)}\left(\mu M_p\right) = \widehat{\eta}_p^{[y_0,\ldots,y_p]}$$

as soon as $\mu = \widehat{\eta}_{p-1}^{[y_0,\ldots,y_{p-1}]}$ $\left(\Rightarrow \mu M_p = \eta_p^{[y_0,\ldots,y_{p-1}]}\right)$

8.7.5.2 Separation principle

From the above discussion, we can rewrite (8.49) as the Snell envelope of a fully observed augmented Markov chain sequence

$$\mathbb{E}(f_\tau(X_\tau, Y_\tau) \mid (Y_0,\ldots,Y_k)) = \mathbb{E}\left(F_\tau\left(\mathcal{X}_\tau\right) \mid (\mathcal{X}_0,\ldots,\mathcal{X}_k)\right)$$

The Markov chain $\mathcal{X}_n$ takes values in an infinite dimensional state space, and it can rarely be sampled without additional levels of approximations. Using the N-particle approximation models, we can replace the chain $\mathcal{X}_n$ by the N-particle approximation model defined by

$$\mathcal{X}_n^N := \left(Y_p, \widehat{\eta}_p^{([Y_0,\ldots,Y_p],N)} \right)$$

where

$$\widehat{\eta}_p^{([Y_0,\ldots,Y_p],N)} := \Psi_{g_p(.,Y_p)} \left(\eta_{p-1}^{([Y_0,\ldots,Y_{p-1},N)]} \right)$$

stands for the updated measure associated with the likelihood selection functions $g_p(.,Y_p)$. The N-particle approximation of the Snell envelope is now given by

$$\mathbb{E}(f_\tau(X_\tau,Y_\tau)|(Y_0,\ldots,Y_k)) \simeq_{N\uparrow\infty} \mathbb{E}\left(F_\tau\left(\mathcal{X}_\tau^N\right) \mid (\mathcal{X}_0^N,\ldots,\mathcal{X}_k^N) \right)$$

In this interpretation, the N-approximated optimal stopping problem amounts to compute the quantities

$$U_k^N := \sup_{\tau \in \mathcal{T}_k^N} \mathbb{E}\left(F_\tau\left(\mathcal{X}_\tau^N\right) \mid (\mathcal{X}_0^N,\ldots,\mathcal{X}_k^N) \right)$$

where $\mathcal{T}_k^N$ stands for the set of all stopping times τ taking values in $\{k,\ldots,n\}$, whose values are measurable w.r.t. the sigma field generated by the Markov chain sequence $\mathcal{X}_k^N$, from $p=0$ up to time k.

Part IV

Theoretical aspects

Chapter 9

Mean field Feynman-Kac models

9.1 Feynman-Kac models

In this opening section, we review some basic definitions on Feynman-Kac models, their normalized and unnormalized distributions, and their evolution equations on general state space models. For a more detailed description, with a more refined analysis, we refer to Chapter 12, dedicated to the regularity and the stability properties of Feynman-Kac semigroups.

We consider a collection of Markov transitions M_n and nonnegative potential functions G_n on some measurable state spaces $(E_n, \mathcal{E}_n)$, with $n \geq 0$, and we let η_0, some probability measure on E_0.

We let $(\Gamma_n, \mathbb{Q}_n)$ be the Feynman-Kac path measures associated with the pairs (G_n, M_n) introduced in (1.37); and we let (γ_n, η_n) be their n-th time marginals.

We recall that the marginal measures are defined for any $n \geq 0$ and any function $f_n \in \mathcal{B}_b(E_n)$ by the following formula

$$\eta_n(f_n) := \gamma_n(f_n)/\gamma_n(1) \quad \text{with} \quad \gamma_n(f_n) = \mathbb{E}\left(f_n(X_n) \prod_{0 \leq p < n} G_p(X_p) \right) \tag{9.1}$$

In addition, as we have seen in (1.42), the semigroup of the unnormalized distributions

$$\gamma_n = \gamma_p Q_{p,n} \tag{9.2}$$

is defined for any $p \leq n$, $x_p \in E_p$, and any $f_n \in \mathcal{B}_b(E_n)$ by the Feynman-Kac formulae

$$Q_{p,n}(f_n)(x_p) := \mathbb{E}\left(f_n(X_n) \prod_{p \leq q < n} G_q(X_q) \mid X_p = x_p \right)$$

We also recall from (3.29) that the normalizing constants $\gamma_n(1)$ are expressed in terms of the flow of measures $(\eta_p)_{0 \leq p < n}$ with the following multiplicative formula

$$\gamma_n(1) = \mathbb{E}\left(\prod_{0 \le p < n} G_p(X_p)\right) = \prod_{0 \le p < n} \eta_p(G_p)$$

This also implies that the unnormalized Feynman-Kac measures can be rewritten in the following form:

$$\gamma_n(f) = \eta_n(f)\,\gamma_n(1) \quad \text{with} \quad \gamma_n(f) = \eta_n(f) \prod_{0 \le p < n} \eta_p(G_p) \qquad (9.3)$$

In Section 4.1.3 and Section 4.4 we have shown that these flows of measures satisfy the following measure valued equations

$$\eta_{n+1} = \Phi_{n+1}(\eta_n) := \Psi_{G_n}(\eta_n) M_{n+1} \qquad (9.4)$$

with the Boltzmann-Gibbs transformations Ψ_{G_n} associated with the potential functions defined in (0.2).

9.2   McKean-Markov chain models

In this section, we provide a short reminder of some notation, concepts, and definitions related to the general class of McKean-Markov chain models presented in Section 1.2.1 and Section 4.4.

Definition 9.2.1 *We let Φ_{n+1} be a sequence of mappings from $\mathcal{P}(E_n)$ into $\mathcal{P}(E_{n+1})$. We consider a collection of Markov transitions $K_{n+1,\eta}$ from E_n into E_{n+1}, indexed by $\eta \in \mathcal{P}(E_n)$, and satisfying the compatibility condition*

$$\Phi_{n+1}(\eta) = \eta K_{n+1,\eta} \qquad (9.5)$$

for any $\eta \in \mathcal{P}(E_n)$, and any $n \ge 0$, are called McKean transitions associated with the nonlinear evolution equation defined by

$$\eta_{n+1} = \Phi_{n+1}(\eta_n) \qquad (9.6)$$

By construction, it is immediate to check that the flow of probability measures η_n in (9.6) satisfies the nonlinear evolution equation

$$\eta_{n+1} = \eta_n K_{n+1,\eta_n} \qquad (9.7)$$

for any choice of the McKean transitions associated with the Equation (9.6).

We also recall that choice of the McKean transitions K_{n+1,η_n} is not unique. Several choices of McKean transitions are provided in (0.3). For instance, in the context of Feynman-Kac models, we can choose

$$K_{n+1,\eta_n}(x, dy) = \Phi_{n+1}(\eta_n)(dy)$$

as well as

$$K_{n+1,\eta_n}(x, dy)$$

$$= \epsilon(\eta_n)G_n(x) \ M_{n+1}(x, dy) + (1 - \epsilon(\eta_n)G_n(x)) \ \Phi_{n+1}(\eta_n)(dy)$$

for any $\epsilon(\eta_n)$ s.t. $\epsilon(\eta_n)G_n(x) \in [0, 1]$. We leave the reader to check that these Markov transitions satisfy the compatibility condition (9.5).

More generally, for any bounded positive function G_n, we can take

$$K_{n+1,\eta_n} = S_{n,\eta_n} M_{n+1} \tag{9.8}$$

for some selection Markov transition S_{n,η_n} from E_n into itself satisfying the compatibility condition

$$\eta_n S_{n,\eta_n} = \Psi_{G_n}(\eta_n) \tag{9.9}$$

For instance, we can take the selection transition

$$S_{n,\eta_n}(x, dy) := (1 - a_{n,\eta_n}(x)) \ \delta_x(dy) + a_{n,\eta_n}(x) \ \Psi_{(G_n - G_n(x))_+}(\eta_n)(dy) \tag{9.10}$$

with the rejection rate potential function

$$a_{n,\eta_n}(x) \quad := \quad \eta_n \left((G_n - G_n(x))_+ \right)/\eta_n (G_n) \in [0, 1]$$

A more detailed discussion on the choice of these McKean transitions is provided in (0.3).

Definition 9.2.2 *Given a collection of McKean transitions $K_{n+1,\eta}$ associated with the Equation (9.6), we define (sequentially) the E_n-valued Markov chain $\overline{X}_n$ with elementary transitions given by*

$$\mathbb{P}\left(\overline{X}_{n+1} \in dx \mid \overline{X}_n\right) = K_{n+1,\eta_n}\left(\overline{X}_n, dx\right) \quad \text{with} \quad \text{Law}(\overline{X}_n) = \eta_n \tag{9.11}$$

This Markov chain is called a McKean interpretation of (9.7) or sometimes the McKean-Markov model associated with the sequence of measures defined by the evolution Equation (9.7).

The law of the random trajectories of the chain

$$\mathbb{P}\left(\overline{X}_0, \ldots, \overline{X}_n\right) \in d(x_0, \ldots, x_n))$$

$$= \eta_0(dx_0) \times K_{1,\eta_0}(x_0, dx_1) \times \cdots \times K_{n,\eta_{n-1}}(x_{n-1}, dx_n)$$

is called the McKean measures on the path spaces $\boldsymbol{E_n} = \prod_{0 \leq p \leq n} E_p$.

We prove that the McKean-Markov chain model is well defined using the following easily checked formula

$$\left[\eta_0 K_{1,\eta_0} \ldots K_{n-1,\eta_{n-2}}\right] K_{n,\eta_{n-1}} = \eta_{n-1} K_{n,\eta_{n-1}} = \eta_n$$

for any $n \geq 1$. For a more thorough discussion on these nonlinear Markov chain models, we refer the reader to Section 2.5 in the book [163], as well as Section 1.2 in the opening chapter, Section 4.4.5, and the Boltzmann-Gibbs transport formulae (0.3) in the beginning of the book.

The following easily checked technical lemma provides a description of the semigroup of the time inhomogeneous Markov chain model $\overline{X}_n$.

Lemma 9.2.3 *The Markov transition of the chain $\overline{X}_p \rightsquigarrow \overline{X}_n$, starting at time p with the distribution $\eta_p := \eta$, is given by the composition formulae*

$$K_{p,n}^\eta := K_{p+1,\eta} K_{p+2,\Phi_{p,p+1}(\eta)} \cdots K_{n,\Phi_{p,n-1}(\eta)} \qquad (9.12)$$

with the convention $K_{n,n}^\eta = Id$, for $p = n$.

Notice that for $p = (n-1)$, we have that $K_{n-1,n}^\eta = K_{n,\eta}$, for any $n \geq 1$. In addition, the semigroup $\Phi_{p,n}(\eta_p) = \eta_n$ associated with the McKean-Markov chain model $\overline{X}_n$ is given by the formula

$$\Phi_{p,n}(\mu) = \mu K_{p,n}^{\eta_p}$$

A more thorough discussion on these semigroups is provided in Section 12.3.

9.3   Perfect sampling models

In this section we present a perfect sampling interpretation of the McKean models defined in (9.11). Section 9.3.1 discusses a Monte Carlo model based on independent copies of the McKean model. This perfect Monte Carlo algorithm cannot be used in practice, since its elementary transitions depend on the unknown distributions of the random states of chains. In Section 9.4.1, we shall use this interpretation to design a mean field approximation scheme defined as the perfect sampling model, by replacing in the elementary transitions the exact distributions of the random states by their finite particle approximations.

In the last section, Section 9.3.2, we analyze the fluctuations of the perfect sampling model in terms of the local sampling errors, using the backward decomposition formula presented in (2.5).

9.3.1 Independent McKean models

From the "practical point of view," the Markov chain defined in (9.11) can be seen as a perfect sampler of the flow of the distributions (9.7) of the random states $\overline{X}_n$. Indeed at a first level of analysis, let us suppose that the elementary transitions $\overline{X}_{n-1} \rightsquigarrow \overline{X}_n$ are easy to sample.

In this situation, we can approximate the flow of measures $\eta_n = \mathrm{Law}(\overline{X}_n)$ by sampling a series of N independent copies $(\overline{X}_n^i)_{1 \le i \le N}$ of the chain $\overline{X}_n$.

A synthetic description of the parallel simulation of this perfect sampling model is given in the following diagram:

$$
\begin{array}{ccc}
\overline{X}_n^1 & \xrightarrow{\quad K_{n+1,\eta_n} \quad} & \overline{X}_{n+1}^1 \\
\vdots & & \vdots \\
\overline{X}_n^i & \xrightarrow{\quad K_{n+1,\eta_n} \quad} & \overline{X}_{n+1}^i \\
\vdots & & \vdots \\
\overline{X}_n^N & \xrightarrow{\quad K_{n+1,\eta_n} \quad} & \overline{X}_{n+1}^N
\end{array}
\tag{9.13}
$$

By the well-known law of large numbers, in some sense we have for any time horizon n and any function f

$$
\eta_n^N(f) := \frac{1}{N} \sum_{i=1}^{N} f(\overline{X}_n^i) \xrightarrow{\;N\uparrow\infty\;} \eta_n(f) = \mathbb{E}\left(f(\overline{X}_n)\right)
\tag{9.14}
$$

More generally, for any function f_n on the path space $E_n = \prod_{0 \le p \le n} E_p$, we have

$$
\mathbb{P}_n^N(f_n) := \frac{1}{N} \sum_{i=1}^{N} f_n(\overline{X}_0^i, \dots, \overline{X}_n^i) \xrightarrow{\;N\uparrow\infty\;} \mathbb{E}\left(f_n(\overline{X}_0, \dots, \overline{X}_n)\right) := \mathbb{P}_n(f_n)
$$

This class of time inhomogeneous Markov chain is of the same type as the abstract Markov chain models discussed in Section 1.1.1, in the opening chapter. In this situation, the concentration inequality (1.3) is satisfied, and for any functions $f_n \in \mathcal{B}_b(E_n)$, we have the unbiasedness property $\mathbb{E}\left(\mathbb{P}_n^N(f_n)\right) = \mathbb{P}_n(f_n)$.

9.3.2 A local perturbation analysis

By construction, the measures η_n^N defined in (9.14) are the occupation measures associated with N independent copies $(\overline{X}_n^i)_{1 \le i \le N}$ of the McKean-Markov chain $\overline{X}_n$. Therefore, invoking the central limit theorem, for any $n \ge 0$ the sequence of empirical random fields

$$
W_n^{\eta,N} := \sqrt{N} \left[\eta_n^N - \eta_n\right]
$$

converges in law, as $N \to \infty$, to a centered Gaussian fields W_n^η with a covariance functional given for any functions $f_n \in \mathcal{B}_b(E_n)$ by the formula

$$\mathbb{E}\left(W_n^\eta(f_n)^2\right) = \eta_n\left((f_n - \eta_n(f_n))^2\right)$$

In addition, in this particular situation we also have the well known nonasymptotic variance results

$$\mathbb{E}\left(W_n^{\eta,N}(f_n)\right) = 0$$

and

$$\mathbb{E}\left(W_n^{\eta,N}(f_n)^2\right) = \mathbb{E}\left(W_n^\eta(f_n)^2\right)$$

It is instructive to express the limiting Gaussian fields W_n^η in terms of the local sampling random fields V_n^N defined for any $n \geq 1$ by

$$V_n^N = \sqrt{N}\ [\eta_n^N - \eta_{n-1}^N K_{n,\eta_{n-1}}] \iff \eta_n^N = \eta_{n-1}^N K_{n,\eta_{n-1}} + \frac{1}{\sqrt{N}}\ V_n^N$$

and, for $n = 0$, we set $V_0^N = \sqrt{N}\ [\eta_0^N - \eta_0]$. We let $\mathcal{F}_n^N$ be the σ-field generated by the random variables $(\overline{X}_p^i)_{1 \leq i \leq N}$, from the origin $p = 0$, up to the final time $p = n$. By construction, we have $\mathbb{E}\left(V_{n+1}^N(f_{n+1}) \,\middle|\, \mathcal{F}_n^N\right) = 0$, and

$$\mathbb{E}\left(V_{n+1}^N(f_{n+1})^2 \,\middle|\, \mathcal{F}_n^N\right) \;=\; \eta_n^N\ K_{n+1,\eta_n}\left[(f_{n+1} - K_{n+1,\eta_n}(f_{n+1}))^2\right]$$

On the other hand, for any $p < q$, and any $(f_p, f_q) \in (\mathcal{B}_b(E_p) \times \mathcal{B}_b(E_q))$ we have

$$\mathbb{E}\left(V_p^N(f_p)V_q^N(f_q) \,\middle|\, \mathcal{F}_p^N\right) = V_p^N(f_p)$$

and

$$\mathbb{E}\left(V_q^N(f_q) \,\middle|\, \mathcal{F}_p^N\right) = 0$$

as well as $\mathbb{E}\left(V_{n+1}^N(f_{n+1})\right) = 0$, and

$$\mathbb{E}\left(V_{n+1}^N(f_{n+1})^2\right) \;=\; \eta_n\ K_{n+1,\eta_n}\left[(f_{n+1} - K_{n+1,\eta_n}(f_{n+1}))^2\right]$$

From these observations, we easily prove the following proposition.

Proposition 9.3.1 *The random fields* $(V_n^N)_{n \geq 0}$ *converge in distribution, as N tends to ∞, to a sequence of independent and centered Gaussian fields* $(V_n)_{n \geq 0}$ *with covariance function given for any $f_n \in \mathcal{B}_b(E_n)$ and $n \geq 1$ by*

$$\mathbb{E}\left(V_n(f_n)^2\right) = \eta_{n-1}\ K_{n,\eta_{n-1}}\left[(f_n - K_{n,\eta_{n-1}}(f_n))^2\right] = \mathbb{E}\left(V_n^N(f_n)^2\right)$$

For $n = 0$, V_0 is a centered Gaussian field with covariance function given for any $f_0 \in \mathcal{B}_b(E_0)$ by $\mathbb{E}\left(V_0(f_0)^2\right) = \eta_0\left[(f_0 - \eta_0(f_0))^2\right]$.

A proof of this result for general mean field IPS models can be found in Section 10.6.

To take the final step, we observe that

$$\Phi_{p,n}(\mu) = \mu K_{p,n}^{\eta_p} \implies \Phi_{p,n}(\mu_1) - \Phi_{p,n}(\mu_2) = (\mu_1 - \mu_2) d_{\mu_2} \Phi_{p,n}$$

with the first order integral operator $d_{\mu_2} \Phi_{p,n}$ from $\mathcal{B}_b(E_n)$ into $\mathcal{B}_b(E_p)$ defined by the equation

$$d_{\mu_2} \Phi_{p,n} = [Id - \mu_2] K_{p,n}^{\eta_p}$$

Notice that

$$d_{\Phi_p(\mu)} \Phi_{p,n} = [Id - \mu K_{p,\eta_{p-1}}] K_{p,n}^{\eta_p} = K_{p,n}^{\eta_p} - K_{p-1,n}^{\eta_{p-1}}$$

In this context, the backward semigroup decomposition formula presented in (2.5) takes the following form

$$
\begin{aligned}
W^{\eta,N} &= \sum_{q=0}^{n} \sqrt{N} \, [\eta_q^N - \eta_{q-1}^N K_{q,\eta_{q-1}}][Id - \eta_{q-1}^N K_{q,\eta_{q-1}}] K_{q,n}^{\eta_q} \\
&= \sum_{q=0}^{n} V_q^N K_{q,n}^{\eta_q}
\end{aligned}
$$

This clearly implies the following fluctuation theorem.

Theorem 9.3.2 *The sequence of random fields $W_n^{\eta,N}$ converge in distribution, as N tends to ∞, to the centered Gaussian fields W_n^{η} defined by*

$$W_n^{\eta} := \sum_{q=0}^{n} V_q K_{q,n}^{\eta_q}$$

The following decompositions show how the local fluctuation variances combine with the semigroup of the McKean-Markov model:

$$
\begin{aligned}
\mathbb{E}\left(W_n^{\eta}(f_n)^2\right) &= \mathbb{E}\left(W_n^{\eta,N}(f_n)^2\right) = \sum_{q=0}^{n} \mathbb{E}\left(V_q\left(K_{q,n}^{\eta_q}(f_n)\right)^2\right) \\
&= \eta_0 \left[\left(K_{1,n}^{\eta_1}(f_n) - \eta_0 K_{0,n}^{\eta_0}(f_n)\right)^2\right] \\
&\quad + \sum_{q=1}^{n} \eta_{q-1} K_{q,\eta_{q-1}}\left[\left(K_{q,n}^{\eta_q}(f_n) - K_{q-1,n}^{\eta_{q-1}}(f_n)\right)^2\right] \\
&= \eta_0\left(K_{1,n}^{\eta_1}(f_n)^2\right) - \left(\eta_0 K_{0,n}^{\eta_0}(f_n)\right)^2 \\
&\quad + \sum_{q=1}^{n}\left(\eta_q\left[K_{q,n}^{\eta_q}(f_n)^2\right] - \eta_{q-1}\left[K_{q-1,n}^{\eta_{q-1}}(f_n)^2\right]\right) \\
&= \eta_n\left(f_n^2\right) - \eta_n\left(f_n\right)^2
\end{aligned}
$$

9.4 Interacting particle systems

This section revolveds around particle approximations of the measure valued process discussed in (9.6). In Section 9.4.1, we start with a short reminder on the mean field IPS models introduced in (1.53) in the opening chapter and further developed in Section 4.4.

In Section 9.4.2 we illustrate these models with a genetic type particle approximation model. A more advanced analysis is developed in Chapter 14, as well as in Chapter 16, including Orlicz's norm estimates, functional fluctuations theorems, and uniform concentration inequalities w.r.t. the time horizon.

Genealogical tree based models are presented in Section 9.4.3. A more thorough discussion on these path space models is developed in Chapter 15, as well as in Chapter 17, including Burkholder-Davis-Gundy type mean error inequalities, functional fluctuations theorems, and exponential concentration inequalities.

The last section, Section 9.4.4, is dedicated to particle backward Markov chain models.

9.4.1 Mean field particle models

Definition 9.4.1 *We let $K_{n,\eta}$ be a collection of McKean transitions associated with the nonlinear evolution Equation (9.6). The mean field particle model associated with these transitions is the valued model in the E_n^N-valued Markov chain $\xi_n = \left(\xi_n^1, \xi_n^2, \ldots, \xi_n^N\right) \in E_n^N$, with elementary transitions defined as*

$$\mathbb{P}\left(\xi_{n+1} \in dx \mid \xi_n\right) = \prod_{i=1}^{N} K_{n+1,\eta_n^N}(\xi_n^i, dx^i) \quad with \quad \eta_n^N := \frac{1}{N} \sum_{j=1}^{N} \delta_{\xi_n^j} \quad (9.15)$$

In the above displayed formula, dx stands for an infinitesimal neighborhood of the point $x = (x^1, \ldots, x^N) \in E_{n+1}^N$. The initial system ξ_0 consists of N i.i.d. r.v. with common law η_0.

Roughly speaking, under some appropriate regularity conditions on the Markov transitions K_{n+1,η_n}, we have the following approximations

$$\eta_n^N \simeq_{N\uparrow\infty} \eta_n \implies K_{n+1,\eta_n^N} \simeq_{N\uparrow\infty} K_{n+1,\eta_n}$$

A precise discussion on the regularity conditions and a detailed description of these estimates are provided in Section 10.2 in Chapter 10 dedicated to general classes of mean field models. In Section 10.4, we also illustrate this framework with a couple of examples related to McKean model of gases and McKean-Vlasov type models. Applications to Feynman-Kac models are discussed in the end of Section 10.2.

From the previous discussion, the central and natural idea in the design of the mean field model (9.15) is to replace the limiting measures η_n, in the perfect sampling model of $\overline{X}_n$, by the empirical occupation measures η_n^N.

The elementary transitions of the resulting particle Markov model are now defined as in the diagram (9.13) by replacing the measures η_n by their finite approximation occupation measures η_n^N; that is, we have that

$$\xi_n^i \xrightarrow{\quad K_{n+1,\eta_n^N} \quad} \xi_{n+1}^i$$

Given some McKean model, the occupation measures of the N-particle system from the origin up to a given time horizon converge, as $N \to \infty$, to the distribution of the random paths of the McKean-Markov chain model defined in (9.11); that is, we have that

$$\frac{1}{N} \sum_{i=1}^{N} \delta_{(\xi_0^i, \xi_1^i, \dots, \xi_n^i)} \longrightarrow_{N \to \infty} \mathrm{Law}\left(\overline{X}_0, \dots, \overline{X}_n\right) \tag{9.16}$$

Further details on the convergence of these measures can be found in [163]. In the further development of this book, we are mainly concerned with the convergence analysis of the n-time marginal measures

$$\eta_n^N := \frac{1}{N} \sum_{i=1}^{N} \delta_{\xi_n^i} \longrightarrow_{N \to \infty} \eta_n = \mathrm{Law}\left(\overline{X}_n\right)$$

Remark 9.4.2 *Besides the fact that the limiting distribution McKean particle measures defined in (9.16) depend on the choice of the McKean interpretation model, we observe that the limiting distribution η_n of the particle density profiles η_n^N does not depend on the choice of the McKean interpretation.*

Mimicking the multiplicative formula (9.3), we design a particle approximation defined as in (9.3) by replacing the flow of measures $(\eta_p)_{0 \le p \le n}$ by their finite approximations $(\eta_p^N)_{0 \le p \le n}$.

Definition 9.4.3 *The N-particle approximations γ_n^N of the unnormalized Feynman-Kac measures γ_n defined in (9.1) are given by the unbiased particle measures given for any $f_n \in \mathcal{B}_b(E_n)$ by*

$$\gamma_n^N(f_n) := \eta_n^N(f_n) \times \prod_{0 \le p < n} \eta_p^N(G_p) \tag{9.17}$$

The proof of the unbiasedness property of the measures γ_n^N is provided in Section 14.7.3. We also refer to Chapter 16 for a detailed discussion on the convergence of these unnormalized particle measures.

9.4.2   Genetic type particle models

In this section we provide a more detailed description of the mean field IPS model associated with the McKean transitions K_{n+1,η_n} given by the Equations (9.8) and (9.10). In this context, the flow of Feynman-Kac measures η_n evolves according to a two-step updating/prediction transition

$$\eta_n \xrightarrow{\ S_{n,\eta_n}\ } \widehat{\eta}_n = \eta_n S_{n,\eta_n} = \Psi_{G_n}(\eta_n) \xrightarrow{\ M_{n+1}\ } \eta_{n+1} = \widehat{\eta}_n M_{n+1} \quad (9.18)$$

The mean field particle sampling approximation of these couple of transitions is defined in terms of a two-step selection-mutation genetic type model transition in product state spaces

$$\xi_n \in E^N \xrightarrow{\ selection\ } \widehat{\xi}_n \in E^N \xrightarrow{\ mutation\ } \xi_{n+1} \in E^N \quad (9.19)$$

The genetic type evolution of the system is summarized by the following synthetic picture:

$$
\begin{bmatrix} \xi_n^1 \\ \vdots \\ \xi_n^i \\ \vdots \\ \xi_n^N \end{bmatrix}
\xrightarrow{\ S_{n,\eta_n^N}\ }
\begin{bmatrix} \widehat{\xi}_n^1 \\ \vdots \\ \widehat{\xi}_n^i \\ \vdots \\ \widehat{\xi}_n^N \end{bmatrix}
\xrightarrow{M_{n+1}}
\begin{bmatrix} \xi_{n+1}^1 \\ \vdots \\ \xi_{n+1}^i \\ \vdots \\ \xi_{n+1}^N \end{bmatrix}
$$

The selection of the particles amounts to sampling N independent r.v. $\widehat{\xi}_n^i$ with distribution

$$S_{n,\eta_n^N}(\xi_n^i, dx) := \left(1 - a_{n,\eta_n^N}(\xi_n^i)\right) \delta_{\xi_n^i}(dx) + a_{n,\eta_n^N}(\xi_n^i)\, \Psi_{(G_n - G_n(\xi_n^i))_+}(\eta_n^N)(dx)$$

In the above display, a_{n,η_n^N} stands for the rejection rate function defined by

$$a_{n,\eta_n^N}(\xi_n^i) := \eta_n^N\left(\left(G_n - G_n(\xi_n^i)\right)_+\right)/\eta_n^N(G_n) = 1 - \frac{\eta_n^N\left(\left(G_n \wedge G_n(\xi_n^i)\right)\right)}{\eta_n^N(G_n)}$$

For $]0,1]$-valued potential functions G_n, we recall that we can also take the selection transition

$$S_{n,\eta_n}(x, dy) = G_n(x)\, \delta_x(dy) + (1 - G_n(x))\, \Phi_{n+1}(\eta_n)(dy)$$

In this situation, the selection transitions $\xi_n^i \to \widehat{\xi}_n^i$ can alternatively be described by the following acceptation/rejection mechanism:

$$\widehat{\xi}_n^i = g_n^i\, \xi_n^i + (1 - g_n^i)\, \widetilde{\xi}_n^i \quad (9.20)$$

where g_n^i is a collection of conditionally independent Bernoulli random variables with respective distributions

$$\mathbb{P}(g_n^i = 1 \mid \xi_n) = 1 - \mathbb{P}(g_n^i = 0 \mid \xi_n) = G_n(\xi_n^i)$$

If we let τ^i be the first time we have $g_n^i = 0$ then it is readily checked that for any $n \geq 0$

$$\mathbb{P}(\tau^i \geq n) = \mathbb{E} \left(\prod_{p=0}^{n-1} G_p(X_p) \right)$$

and therefore

$$\mathrm{Law}\left((\xi_0^i, \ldots, \xi_n^i) \mid \tau^i \geq n \right) = \mathbb{Q}_n$$

More generally on the intersection of events $\cap_{i=1}^{q} (\tau^i \geq n)$ the path particles $(\xi_0^i, \ldots, \xi_n^i)_{1 \leq i \leq q}$ are independent and identically distributed with common law $\mathbb{Q}_n$. Loosely speaking, up to the first interaction time, the interacting particle model produces independent samples according to the desired distribution on path space.

9.4.3 Genealogical tree models

In this section we provide a discussion on the occupation measures of the genealogical tree model associated with the genetic type branching scheme associated with these particle models. We consider the historical process

$$\mathbf{X_n} := (X_0, \ldots, X_n) \in \mathbf{E_n} := (E_0 \times \ldots \times E_n)$$

associated with the reference E_n-valued Markov chain X_n, and we let $\mathbf{G_n}$ be the sequence of positive and bounded potential functions on $\mathbf{E_n}$ defined by

$$\mathbf{G_n} \; : \; \mathbf{x_n} = (x_0, \ldots, x_n) \in \mathbf{E_n} \mapsto \mathbf{G_n}(\mathbf{x_n}) = G_n(x_n) \qquad (9.21)$$

Using the equivalence principles (3.45) presented in Section 3.4, we can check that the genealogical tree model discussed above coincides with the mean field N-particle interpretation of the Feynman-Kac measures η_n associated with the pair $(\mathbf{G_n}, \mathbf{M_n})$ on the path spaces $\mathbf{E_n}$.

To be more precise, we consider the Feynman-Kac measures $(\mathbf{\Gamma_n}, \mathbf{Q_n})$ introduced in (1.37) associated with the pairs $(\mathbf{G_n}, \mathbf{M_n})$, on the path spaces $(\mathbf{E_0} \times \ldots \times \mathbf{E_n})$. We also denote by (γ_n, η_n) the n-th time marginal of $(\mathbf{\Gamma_n}, \mathbf{Q_n})$. By construction, we have the evolution equation

$$\eta_{n+1} = \Psi_{G_n}(\eta_n) M_{n+1} \qquad (9.22)$$

Under our assumption on the potential functions (9.21), we also recall that

$$(\gamma_n, \eta_n) = (\Gamma_n, \mathbb{Q}_n)$$

In this situation, the mean field particle model $\xi_n = (\xi_n^i)_{1 \leq i \leq N}$ associated with the evolution Equations (9.22) is given by path-valued particles

$$\xi_n^i := (\xi_{0,n}^i, \xi_{1,n}^i, \ldots, \xi_{n,n}^i) \in \mathbf{E_n} := (E_0 \times \ldots \times E_n)$$

It is not difficult to check that these path-valued particles coincide with the ancestral lines of the mean field genetic type particle model associated with the evolution equations of the n-th time marginals η_n presented in (9.4). Thus, the corresponding N-particle occupation measures are now given by

$$\eta_n^N := \frac{1}{N} \sum_{i=1}^{N} \delta_{\xi_n^i} := \frac{1}{N} \sum_{i=1}^{N} \delta_{(\varsigma_{0,n}^i, \varsigma_{1,n}^i, \dots, \varsigma_{n,n}^i)} \in \mathcal{P}(\mathbf{E_n}) = \mathcal{P}(E_0 \times \dots \times E_n)$$
(9.23)

In much the same way, the particle unnormalized Feynman-Kac measures on path space take the following form

$$
\begin{aligned}
\gamma_n^N &:= \left\{ \prod_{0 \leq p < n} \eta_p^N(G_p) \right\} \times \eta_n^N \\
&= \left\{ \prod_{0 \leq p < n} \eta_p^N(G_p) \right\} \times \eta_n^N \in \mathcal{M}(\mathbf{E_n})
\end{aligned}
$$

For a more thorough discussion on the regularity properties and the convergence analysis of these genealogical tree based measures, we refer the reader to Chapter 15.

9.4.4 Backward Markov chain models

We further assume that the Markov transitions M_n are absolutely continuous with respect to some measures λ_n on E_n, and for any $(x, y) \in (E_{n-1} \times E_n)$ we have

$$H_n(x, y) := \frac{dM_n(x, \cdot)}{d\lambda_n}(y) > 0 \qquad (9.24)$$

In this situation, the Feynman-Kac measures $\mathbb{Q}_n$ on path space can also be approximated using the backward particle models presented in Section 1.4.3.4 and in Section 4.1.6. To more precise, when the regularity condition (9.24) is met, we proved in Section 3.4.1 the following backward representation formula

$$\mathbb{Q}_n(d(x_0, \dots, x_n)) = \eta_n(dx_n) \prod_{q=1}^{n} \mathbb{M}_{q, \eta_{q-1}}(x_q, dx_{q-1}) \qquad (9.25)$$

with the collection of Markov transitions defined by

$$\mathbb{M}_{n+1, \eta_n}(x, dy) \propto G_n(y) \, H_{n+1}(y, x) \, \eta_n(dy) \qquad (9.26)$$

Mimicking the backward formula (9.25), we consider the following particle measures

$$\mathbb{Q}_n^N(d(x_0, \dots, x_n)) = \eta_n^N(dx_n) \prod_{q=1}^{n} \mathbb{M}_{q, \eta_{q-1}^N}(x_q, dx_{q-1}) \qquad (9.27)$$

In this situation, we have the following convergence result

$$\lim_{N \to \infty} \mathbb{Q}_n^N = \mathbb{Q}_n \qquad (9.28)$$

For a more thorough discussion on the regularity properties and the convergence analysis of these backward particle models, we refer the reader to Chapter 17.

9.5 Some convergence estimates

In this section, we introduce some of the simplest convergence estimates that can be obtained using the local perturbation theory presented in Section 2.1. The first section, Section 9.5.1, is dedicated to the variance analysis of the local sampling random field models (2.1) associated with a given mean field simulation scheme. We also present $\mathbb{L}_m$-mean error estimates, as well as a functional central limit theorem describing the limiting fluctuations in terms of a sequence of independent Gaussian random fields with covariance functions that depend on the choice of the McKean interpretation model.

In Section 9.5.2, we present some rather crude nonasymptotic $\mathbb{L}_m$-mean error estimates that can be derived using an elementary inductive proof w.r.t. the time parameter. We end this section with a discussion of the particle estimation of the essential supremum of some function w.r.t. the limiting Feynman-Kac measures. This asymptotic analysis is used in Section 8.1.2 to design maximum likelihood estimates of Feynman-Kac models on path spaces. It is also used in Section 14.4 to analyze the fluctuations of mean field IPS models associated with selection transitions (0.4) that depend on the $\eta_n^N - ess\sup G_n$.

In Section 9.5.3, we present a first order perturbation analysis in terms of the backward semigroup techniques presented in (2.5). We compare the limiting Gaussian random field models with the ones obtained in the fluctuation analysis of the perfect sampling algorithm discussed in Section 9.3.2.

9.5.1 Local sampling errors

This section is concerned with the analysis of the local sampling errors associated with the elementary mean field particle transitions defined in (9.15). In the further development of the section, we let $\mathcal{F}_n^N := \sigma\left(\xi_0, \ldots, \xi_n\right)$ be the natural filtration associated with the N-particle approximation model defined in (9.15).

Definition 9.5.1 *The local sampling random field models associated with the mean field particle model (9.15) are defined by the empirical random fields V_n^N*

given by

$$V_n^N = \sqrt{N} \left[\eta_n^N - \Phi_n \left(\eta_{n-1}^N \right) \right] \tag{9.29}$$

with the convention $\Phi_0 \left(\eta_{-1}^N \right) = \eta_0$, *for* $n = 0$.

Notice that V_n^N can also be defined by the stochastic perturbation formulae

$$\eta_n^N = \Phi_n \left(\eta_{n-1}^N \right) + \frac{1}{\sqrt{N}} V_n^N \tag{9.30}$$

In this interpretation, the N-particle model is interpreted as a stochastic perturbation of the limiting system

$$\eta_n = \Phi_n \left(\eta_{n-1} \right)$$

The following technical lemma gives some information on the conditional variance of the local sampling random field models

Lemma 9.5.2 *For any* $n \geq 0$, $f \in \mathcal{B}_b(E_n)$, *and any* $N \geq 1$ *we have the formulae*

$$\mathbb{E} \left(V_{n+1}^N(f) \,\big|\, \mathcal{F}_n^N \right) = 0$$
$$\mathbb{E} \left(V_{n+1}^N(f)^2 \,\big|\, \mathcal{F}_n^N \right) = \eta_n^N \left[K_{n+1,\eta_n^N} \left(f - K_{n+1,\eta_n^N}(f) \right)^2 \right]$$

Proof:
By construction, for any $n \geq 0$ and any $f \in \mathcal{B}_b(E_{n+1})$ we have that

$$\mathbb{E} \left(\eta_{n+1}^N(f) \,\big|\, \mathcal{F}_n^N \right) = \eta_n^N K_{n+1,\eta_n^N}(f) = \Phi_{n+1}(\eta_n^N)(f)$$

and

$$\mathbb{E} \left(\left[\eta_{n+1}^N(f) - \Phi_{n+1}(\eta_n^N)(f) \right]^2 \,\big|\, \mathcal{F}_n^N \right) = \frac{1}{N} \eta_n^N \left[K_{n+1,\eta_n^N} \left(f - K_{n+1,\eta_n^N}(f) \right)^2 \right]$$

We prove the last claim using the decomposition

$$\eta_{n+1}^N(f) - \Phi_{n+1}(\eta_n^N)(f) = \frac{1}{N} \sum_{i=1}^N \left(f(\xi_{n+1}^i) - K_{n+1,\eta_n^N}(f)(\xi_n^i) \right)$$

and recalling that the sequence of random variables

$$\left(f(\xi_{n+1}^i) - K_{n+1,\eta_n^N}(f)(\xi_n^i) \right)$$

is $\mathcal{F}_n^N$-conditionally independent with

$$\mathbb{E} \left(\left[f(\xi_{n+1}^i) - K_{n+1,\eta_n^N}(f)(\xi_n^i) \right]^2 \,\big|\, \mathcal{F}_n^N \right) = K_{n+1,\eta_n^N} \left(f - K_{n+1,\eta_n^N}(f) \right)^2 (\xi_n^i)$$

This ends the proof of the lemma.   ∎

The next theorem shows that the local random perturbations behave asymptotically as Gaussian random perturbations. The details of the proof of this functional central limit theorem can be found in [163]. We also refer the reader to Section 10.6 for a proof of this theorem for general mean field particle models.

Theorem 9.5.3 *For any fixed time horizon $n \geq 0$, the sequence of random fields V_n^N converges in law, in the sense of convergence of finite dimensional distributions, as the number of particles N tends to infinity, to a sequence of independent, Gaussian, and centered random fields V_n, with, for any f, and $n \geq 0$,*

$$\mathbb{E}(V_n(f)^2) = \eta_{n-1} K_{n, \eta_{n-1}}([f - K_{n, \eta_{n-1}}(f)]^2)$$

Definition 9.5.4 *We denote by σ_n^2 the uniform local variance parameter given by*

$$\sigma_n^2 := \sup \mu \left(K_{n, \mu} [f_n - K_{n, \mu}(f_n)]^2 \right) (\leq 1) \tag{9.31}$$

In the above displayed formula the supremum is taken over all functions $f_n \in \text{Osc}(E_n)$, and all probability measures μ on E_n, with $n \geq 1$. For $n = 0$, we set

$$\sigma_0^2 = \sup_{f_0 \in \text{Osc}(E_0)} \eta_0 \left([f_0 - \eta_0(f_0)]^2 \right) (\leq 1)$$

Working a little harder, we obtain the following $\mathbb{L}_m$-mean error estimates, which are valid for any mean field particle model of the form (9.15).

Proposition 9.5.5 *For any $N \geq 1$, $m \geq 1$, $n \geq 0$ and any test function $f_n \in \text{Osc}(E_n)$ we have the almost sure estimate*

$$\mathbb{E} \left(|V_n^N(f_n)|^m \, \big| \, \mathcal{F}_{n-1}^N \right)^{\frac{1}{m}} \leq b(m) \, (1 \wedge c(m)) \tag{9.32}$$

with the constant $b(m)$ defined in (0.6) and

$$c(m) = 6b(m) \, \max \left(\sqrt{2} \sigma_n, \left[\frac{2\sigma_n^2}{N^{\frac{m'}{2}-1}} \right]^{1/m'} \right) \tag{9.33}$$

In the above display, m' stands for the smallest even integer $m' \geq m$. In addition, for $m \geq 2$, we have

$$\mathbb{E} \left(|V_n^N(f)|^m \, \big| \, \mathcal{F}_{n-1}^N \right)^{\frac{1}{m}} \leq 2 \, b(m) \, \Phi_n \left(\eta_{n-1}^N \right) \left(f^{m'} \right)^{\frac{1}{m'}} \tag{9.34}$$

The above proposition is a direct consequence of the Theorem 11.2.1 on empirical random fields stated in Chapter 11 devoted to the analysis of empirical processes.

9.5.2 $\mathbb{L}_m$-mean error estimates

In this section, we provide a direct derivation of some interesting $\mathbb{L}_m$-mean error estimates using an elementary proof by induction w.r.t. the time parameter n. We end the section with a convergence analysis of particle essential supremums.

To simplify the presentation, and to avoid some unnecessary technical conditions, we further assume that $g_n := \sup_{x,y} (G_n(x)/G_n(y)) < \infty$. This condition can be relaxed by assuming that G_n is a bounded potential function with $\eta_n(G_n) > 0$.

Proposition 9.5.6 *For any $N \geq 1$, $n \geq 0$, and any function $f_n \in \mathrm{Osc}(E_n)$ we have*

$$\left| \mathbb{E}\left(\eta_n^N(f_n)\right) - \eta_n(f_n) \right| \leq c(n)/N \tag{9.35}$$

for some finite constant $c(n) < \infty$. In addition, for any integer $m \geq 1$, we have the variance estimate

$$\sqrt{N}\, \mathbb{E}\left(\left| \eta_n^N(f_n) - \eta_n(f_n) \right|^m \right)^{\frac{1}{m}} \leq b(m)\, c(n) \tag{9.36}$$

for some $c(n) < \infty$ and the collection of constants $b(m)$ defined in (0.6).

Proof:
The proof is based on the regularity properties of Feynman-Kac transformations presented in Section 3.1.4.

We let G be a positive and bounded potential function on E and M some Markov transition from E into some possibly different measurable state space $(F, \mathcal{F})$. We denote by Φ the one step Feynman-Kac mapping from $\mathcal{P}(E)$ into $\mathcal{P}(F)$ defined for any $\mu \in \mathcal{P}(E)$ by

$$\Phi(\mu) = \Psi_G(\mu)M$$

We recall from Lemma 3.1.10 that

$$\left| [\Phi(\mu) - \Phi(\nu)]\,(f) \right| \leq g\, \left| (\mu - \nu)(d'_\nu \Phi(f)) \right| \quad \text{with} \quad g := \sup_{x,y} (G(x)/G(y))$$

In the above display $d'_\nu \Phi$ is the integral operator from $\mathcal{B}_b(F)$ into $\mathcal{B}_b(E)$ defined by

$$d'_\nu \Phi(f) := G'\,(M(f) - \Phi(\nu)(f)) \quad \text{and} \quad G' := G/\|G\|$$

and we have the estimate

$$\|d'_\nu \Phi(f)\| \leq \beta(M)\, \mathrm{osc}(f)$$

We also quote the following first order decomposition

$$[\Phi(\mu) - \Phi(\nu)]\,(f)$$

$$= (\mu - \nu)(d_\nu \Phi(f)) - \tfrac{1}{\mu(G_\nu)}\,(\mu - \nu)^{\otimes 2}\,[G_\nu \otimes d_\nu \Phi(f)] \quad \text{with } G_\nu = \tfrac{G}{\nu(G)}.$$

If we take

$$(G, M, \mu, \nu) = (G_n, M_{n+1}, \eta_n^N, \eta_n) \Rightarrow \Phi = \Phi_{n+1}$$

then we find that

$$\left| \Phi_{n+1}\left(\eta_n^N\right)(f) - \Phi_{n+1}(\eta_n)(f) \right| \leq g_n \left| (\eta_n^N - \eta_n)(d'_{\eta_n}\Phi_{n+1}(f)) \right|$$

Using the decomposition

$$\eta_{n+1}^N(f) - \eta_{n+1}(f)$$

$$= \left[\eta_{n+1}^N(f) - \Phi_{n+1}\left(\eta_n^N\right)(f) \right] + \left[\Phi_{n+1}\left(\eta_n^N\right)(f) - \Phi_{n+1}(\eta_n)(f) \right]$$

we conclude that

$$\left| \eta_{n+1}^N(f) - \eta_{n+1}(f) \right| \leq \frac{1}{\sqrt{N}} \left| V_{n+1}^N(f) \right| + g_n \left| (\eta_n^N - \eta_n)(d'_{\eta_n}\Phi_{n+1}(f)) \right|$$

and

$$\| d'_{\eta_n}\Phi_{n+1}(f) \| \leq \beta(M_{n+1})\ \mathrm{osc}(f)$$

The proof of (9.36) follows by induction using Kintchine's type inequalities (11.6) at rank $n = 0$.

We prove the bias estimates (9.35) using first order decomposition

$$\left[\Phi_{n+1}(\eta_n^N) - \Phi_{n+1}(\eta_n) \right](f)$$

$$= (\eta_n^N - \eta_n)(d_{\eta_n}\Phi_{n+1}(f)) - \frac{1}{\eta_n^N(G_{n,\eta_n})} (\eta_n^N - \eta_n)^{\otimes 2} \left[G_{n,\eta_n} \otimes d_{\eta_n}\Phi_{n+1}(f) \right]$$

with $G_{n,\eta_n} = \frac{G_n}{\eta_n(G_n)}$ This yields

$$\mathbb{E}\left(\left[\eta_{n+1}^N - \eta_{n+1} \right](f) \right) = \mathbb{E}\left\{ \left[\eta_n^N - \eta_n \right](d_{\eta_n}\Phi_{n+1}(f)) \right\}$$

$$- \mathbb{E}\left(\frac{1}{\eta_n^N(G_{n,\eta_n})} (\eta_n^N - \eta_n)^{\otimes 2} \left[G_{n,\eta_n} \otimes d_{\eta_n}\Phi_{n+1}(f) \right] \right)$$

from which we conclude that

$$\left| \mathbb{E}\left\{ \left[\eta_{n+1}^N - \eta_{n+1} \right](f) \right\} \right|$$

$$\leq \left| \mathbb{E}\left\{ \left[\eta_n^N - \eta_n \right](d_{\eta_n}\Phi_{n+1}(f)) \right\} \right|$$

$$+ g_n\ \mathbb{E}\left(\left[(\eta_n^N - \eta_n)(G_{\eta_n}) \right]^2 \right)^{1/2} \mathbb{E}\left(\left[(\eta_n^N - \eta_n)(d_{\eta_n}\Phi_{n+1}(f)) \right]^2 \right)^{1/2}$$

The end of the proof of the biased estimates is now clear. This ends the proof of the proposition. ∎

The next theorem ensures that the particle η_n^N-essential supremum of functions converges to the corresponding η_n-essential supremum.

Theorem 9.5.7 *For any $n \geq 0$, and any measurable function V_n on E_n s.t. $\eta_n - ess \ sup \ V_n < \infty$, we have the convergence in probability*

$$\eta_n^N - ess \ sup \ V_n = \max_{1 \leq i \leq N} V_n(\xi_n^i) \longrightarrow_{N \to \infty} \eta_n - ess \ sup \ V_n$$

The proof of the theorem is partly based on the following technical lemma.

Lemma 9.5.8 *For any nonnegative function $f_n \in \mathcal{B}_b(E_n)$, we have*

$$c_1(n) \ \eta_n(f_n) \leq \mathbb{E}\left(\eta_n^N(f_n)\right) \leq c_2(n) \ \eta_n(f_n) \qquad (9.37)$$

for some finite constants $c_i(n)$, with $i = 1, 2$.

Proof:
The r.h.s. upper bound is clearly met for $n = 0$. We further assume that it is satisfied at rank n. Firstly, we observe that

$$\mathbb{E}\left(\eta_{n+1}^N(f_{n+1})\right) = \mathbb{E}\left(\Psi_{G_n}(\eta_n^N)M_{n+1}(f_{n+1})\right) \leq g_n \ \mathbb{E}\left(\eta_n^N M_{n+1}(f_{n+1})\right)$$

Using the induction hypothesis, we prove that

$$\mathbb{E}\left(\eta_{n+1}^N(f_{n+1})\right) \leq c_2(n)g_n \ \mathbb{E}\left(\eta_n M_{n+1}(f_{n+1})\right)$$

Arguing as above, we also have that

$$g_n^{-1} \ \mathbb{E}\left(\eta_n M_{n+1}(f_{n+1})\right) \leq \mathbb{E}\left(\Psi_{G_n}(\eta_n) M_{n+1}(f_{n+1})\right)$$

from which we conclude that

$$\mathbb{E}\left(\eta_{n+1}^N(f_{n+1})\right) \leq c_2(n+1) \ \eta_{n+1}(f_{n+1}) \quad \text{with} \quad c_2(n+1) \leq c_2(n) \ g_n^2$$

The proof of the lower bounds follows the same arguments; thus it is omitted. This ends the proof of the lemma. ∎

We are now in position to prove the theorem.
Proof of Theorem 9.5.7: We set $a = \eta_n - ess \ sup \ V_n$. Using Lemma 9.5.8, for any $\epsilon > 0$ we have

$$
\begin{aligned}
\mathbb{P}\left(\eta_n^N - ess \ sup V_n \geq a + \epsilon\right) &= \mathbb{P}\left(\exists 1 \leq i \leq N \ : \ V_n(\xi_n^i) \geq a + \epsilon\right) \\
&= \sum_{i=1}^{N} \mathbb{E}\left(1_{V_n^{-1}([a+\epsilon,\infty[)}(\xi_n^i)\right) \\
&\leq N \ c_2(n) \ \eta_n(V_n^{-1}([a+\epsilon,\infty[)) = 0
\end{aligned}
$$

On the other hand, for any function $f_n \in \mathcal{B}_b(E_n)$ we have

$$\mathbb{E}\left(\prod_{1 \leq i \leq q} f_n(\xi_n^i)\right) = \mathbb{E}\left((\eta_n^N)^{\odot q}(f_n^{\otimes q})\right)$$

with the symmetric statistic type particle empirical measures

$$\left(\eta_n^N\right)^{\odot q} := \frac{1}{(N)_q} \sum_{a \in \langle q, N \rangle} \delta_{\left(\xi_n^{\alpha(1)}, \ldots, \xi_n^{\alpha(q)}\right)}$$

where $\langle q, N \rangle$ stands for the set of all $(N)_q = N!/(N-q)!$ one to one mappings from $\langle q \rangle = \{1, \ldots, q\}$ into $\langle N \rangle = \{1, \ldots, N\}$.

On the other hand, it is more or less well known that

$$\left\| \left(\eta_n^N\right)^{\odot q} - \left(\eta_n^N\right)^{\otimes q} \right\|_{tv} \leq (q-1)^2/N$$

For a proof of this result, we refer the reader to Proposition 8.6.1 in the monograph [163]. This implies that

$$\left| \mathbb{E}\left(\prod_{1 \leq i \leq q} f_n(\xi_n^i) \right) - \mathbb{E}\left(\eta_n^N(f_n)^q \right) \right| \leq (q-1)^2/N$$

for any $f_n \in \mathcal{B}_1(E_n)$, we consider the decomposition

$$\begin{aligned}
\mathbb{E}\left(\eta_n^N(f_n)^q\right) &= \mathbb{E}\left(\left[\eta_n(f_n) + \frac{1}{\sqrt{N}} W_n^{\eta,N}(f_n) \right]^q \right) \\
&= \eta_n(f_n)^q + \sum_{1 \leq p \leq q} \binom{p}{q} \frac{1}{N^{p/2}} \mathbb{E}\left(W_n^{\eta,N}(f_n)^p \right) \eta_n(f_n)^{q-p}
\end{aligned}$$

with

$$W_n^{\eta,N} := \sqrt{N} \left[\eta_n^N - \eta_n \right]$$

Using Proposition 9.5.6, for any $f_n \in \mathcal{B}_b(E_n)$, with $\|f_n\| \leq 1$, we also prove that

$$\left| \mathbb{E}\left(\eta_n^N(f_n)^q \right) - \eta_n(f_n)^q \right| \leq c(n) \sum_{1 \leq p \leq q} \binom{p}{q} \frac{1}{N^{p/2}} b(p)^p$$

for some $c(n) < \infty$, and the collection of constants $b(m)$ defined in (0.6). This implies that

$$\lim_{N \to \infty} \mathbb{E}\left(\prod_{1 \leq i \leq q} f_n(\xi_n^i) \right) = \eta_n(f_n)^q \tag{9.38}$$

On the other hand, we have

$$\mathbb{P}\left(\eta_n^N - \text{ess sup} V_n \leq a - \epsilon \right) \leq \mathbb{P}\left(\forall 1 \leq i \leq q \ \ V_n(\xi_n^i) \leq a - \epsilon \right)$$

and by (9.38) we have that

$$\lim_{N \to \infty} \mathbb{P}\left(\forall 1 \leq i \leq q \ \ V_n(\xi_n^i) \leq a - \epsilon \right) = \eta_n\left(\{V_n \leq a - \epsilon\} \right)^q$$

We conclude that

$$\limsup_{N \uparrow \infty} \mathbb{P}\left(\eta_n^N - \text{ess sup} V_n \le a - \epsilon\right) \le \eta_n\left(\{V_n \le a - \epsilon\}\right)^q$$

for any $q \ge 1$. Since $\eta_n\left(\{V_n \le a - \epsilon\}\right) < 1$ (by definition of a), letting $q \uparrow \infty$ we conclude that

$$\lim_{N \to \infty} \mathbb{P}\left(\eta_n^N - \text{ess sup} V_n \le a - \epsilon\right) = 0$$

This ends the proof of the theorem.   ∎

9.5.3   A local perturbation analysis

It is instructive to compare the local fluctuation decompositions of the perfect sampling model presented in Section 9.3.2 with the ones associated with the backward Feynman-Kac semigroup decompositions presented in (2.5).

In the further development of this section, we assume that the potential functions G_n take values in $]0, 1]$, and the McKean transitions $K_{n+1,\mu}$, with $\mu \in \mathcal{P}(E_n)$, are defined by (9.8), with the collection of selection transition $S_{n,\eta}$ defined by

$$S_{n,\mu}(x, dy) = G_n(x)\, \delta_x(dy) + (1 - G_n(x))\, \Psi_{G_n}(\mu)(dy)$$

We also denote by

$$\forall 0 \le p \le n \qquad \Phi_{p,n} := \Phi_n \circ \Phi_{n-1} \circ \ldots \circ \Phi_{p+1}$$

the Feynman-Kac semigroup $\Phi_{p,n}(\eta_p) = \eta_n$ associated with the flow of measures η_n defined in (9.4). We also use the convention $\Phi_{n,n} = Id$, the identity mapping, for $p = n$.

In terms of the Markov transitions $K_{p,n}^\mu$ of the McKean model $\overline{X}_n$ defined in (9.12), we have that

$$\Phi_{p,n}(\mu) = \mu K_{p,n}^\mu \tag{9.39}$$

In Section 12.3.2, we also prove that

$$K_{p,n}^\mu(x_p, dx_n) \quad = \quad G_{p,n}(x_p)\, P_{p,n}(x_p, dx_n) + (1 - G_{p,n}(x_p))\, \Phi_{p,n}(\mu)(dx_n)$$

On the other hand, using (9.2), the semigroup $\Phi_{p,n}$ is alternatively defined by the formula

$$\Phi_{p,n}(\mu)(f_n) = \mu Q_{p,n}(f_n)/\mu Q_{p,n}(1) = \Psi_{G_{p,n}}(\mu) P_{p,n}(f_n) \tag{9.40}$$

for any $p \le n$, and any $f_n \in \mathcal{B}_b(E_n)$, with the potential functions $G_{p,n}$ and the Markov transitions $P_{p,n}$ defined by

$$G_{p,n} = Q_{p,n}(1) \quad \text{and} \quad P_{p,n}(f_n) = Q_{p,n}(f_n)/Q_{p,n}(1)$$

Using Lemma 3.1.10, we also have the first order expansion

$$\Phi_{p,n}(\mu_1) - \Phi_{p,n}(\mu_2) \simeq (\mu_1 - \mu_2)d_{\mu_2}\Phi_{p,n} \tag{9.41}$$

with the integral operator $d_{\mu_2}\Phi_{p,n}$ from $\mathcal{B}_b(E_n)$ into $\mathcal{B}_b(E_p)$ defined by

$$
\begin{aligned}
d_{\mu_2}\Phi_{p,n}(f_n) \quad &:= \quad \frac{G_{p,n}}{\mu_2(G_{p,n})} \, P_{p,n}\left(f_n - \Phi_{p,n}(\mu_2)(f_n)\right) \\
&= \quad \frac{1}{\mu_2(G_{p,n})} \left[K_{p,n}^{\mu_2}(f_n) - \Phi_{p,n}(\mu_2)(f_n) \right] \tag{9.42}
\end{aligned}
$$

This clearly implies that

$$d_{\mu}\Phi_{p,n}(f_n) = \left[K_{p,n}^{\mu}(f_n) - \Phi_{p,n}(\mu)(f_n) \right] + (1 - \mu(G_{p,n})) \, d_{\mu}\Phi_{p,n}(f_n) \tag{9.43}$$

In summary, we have proved that

$$d_{\mu}\Phi_{p,n} = [Id - \mu] \, K_{p,n}^{\mu} + (1 - \mu(G_{p,n})) \, d_{\mu}\Phi_{p,n}$$

We are now in position to state and to prove the main result of this section. The proof of the theorem is only sketched using stochastic perturbation arguments. A more detailed proof, which applies for any McKean interpretation models, is provided in Section 10.6.3, dedicated to a more general class of mean field IPS models. We also mention that more precise and nonasymptotic second order expansions are discussed in Section 14.2.

Theorem 9.5.9 *For any $n \geq 0$ the sequence of empirical random fields*

$$W_n^{\eta,N} := \sqrt{N} \, \left[\eta_n^N - \eta_n \right]$$

converges in law, as $N \to \infty$, to a centered Gaussian field W_n^{η} defined by

$$W_n^{\eta} \quad = \quad \sum_{p=0}^{n} V_p d_{\eta_p} \Phi_{p,n} = \sum_{p=0}^{n} V_p K_{p,n}^{\eta_p} + \sum_{p=0}^{n} (1 - \eta_p(G_{p,n})) \, V_p d_{\eta_p} \Phi_{p,n}$$

with the integral operators $d_{\eta_p}\Phi_{p,n}$ defined in (9.42). In addition, for any $n \geq 0$, and any $f_n \in \mathcal{B}_b(E_n)$, we have the asymptotic variance formula

$$\mathbb{E}\left(W_n^{\eta}(f_n)^2 \right)$$

$$= \sum_{p=0}^{n} \mathbb{E}\left(V_p \left[d_{\eta_p}\Phi_{p,n}(f_n) \right]^2 \right)$$

$$= \eta_n \left[(f_n - \eta_n(f_n))^2 \right] + \sum_{p=0}^{n} (1 - \eta_p(G_{p,n})) \, \mathbb{E}\left(V_p \left[d_{\eta_p}\Phi_{p,n}(f_n) \right]^2 \right)$$

Sketch of the proof:

Using the backward semigroup decomposition formula presented in (2.5), we find the first order expansion

$$W_n^{\eta,N} \simeq \sum_{p=0}^{n} V_p^N d_{\Phi_p(\eta_{p-1}^N)} \Phi_{p,n}$$

$$= \sum_{p=0}^{n} V_p^N \left[Id - \eta_{p-1}^N K_{p,\eta_{p-1}^N} \right] K_{p,n}^{\Phi_p(\eta_{p-1}^N)}$$

$$+ \sum_{p=0}^{n} (1 - \Phi_p(\eta_{p-1}^N)(G_{p,n})) V_p^N d_{\Phi_p(\eta_{q-p-1}^N)} \Phi_{p,n}$$

from which we conclude that

$$W_n^{\eta,N} \simeq \sum_{p=0}^{n} V_p^N K_{p,n}^{\Phi_p(\eta_{p-1}^N)} + \sum_{p=0}^{n} (1 - \Phi_p(\eta_{p-1}^N)(G_{p,n})) V_p^N d_{\Phi_p(\eta_{p-1}^N)} \Phi_{p,n}$$

Letting $N \uparrow \infty$, we have the almost sure convergence $\Phi_p(\eta_{p-1}^N)(f_p) \to \eta_p(f_p)$, for any function $f_p \in \mathcal{B}_b(E_p)$. We end the proof of the theorem combining Theorem 9.5.3 with some perturbation arguments so that to replace $\left(K_{p,n}^{\Phi_p(\eta_{p-1}^N)}, d_{\Phi_p(\eta_{p-1}^N)} \Phi_{p,n} \right)$ by their limiting values $(K_{p,n}^{\eta_p}, d_{\eta_p} \Phi_{p,n})$. This ends the proof of the first assertion.

To prove the variance formula, we use Theorem 9.3.2 to check that

$$\mathbb{E}\left(W_n^{\eta}(f_n)^2 \right) = \eta_n \left[(f_n - \eta_n(f_n))^2 \right]$$

$$+ \sum_{p=0}^{n} (1 - \eta_p(G_{p,n}))^2 \, \mathbb{E}\left(V_p \left[d_{\eta_p} \Phi_{p,n}(f_n) \right]^2 \right)$$

$$+ \sum_{p=0}^{n} (1 - \eta_p(G_{p,n})) \, \mathbb{E}\left(V_p \left[K_{p,n}^{\eta_p}(f_n) \right] \, V_p \left[d_{\eta_p} \Phi_{p,n}(f_n) \right] \right)$$

Using (9.43) we have that

$$(1 - \eta_p(G_{p,n})) \, V_p \left[d_{\eta_p} \Phi_{p,n}(f_n) \right] + V_p \left[K_{p,n}^{\eta_p}(f_n) \right] = V_p \left[d_{\eta_p} \Phi_{p,n}(f_n) \right]$$

The end of the proof of the variance formula is now clear. This ends the proof of the theorem. ∎

9.6 Continuous time models

In this section, we illustrate some results with a brief discussion on the continuous time interacting jump type models presented in Chapter 5. We also refer the reader to Section 1.2.2, for a synthetic description of a more general class of continuous mean field IPS models.

9.6.1 Time discretization models

In this section, we provide a brief discussion on the uniform local variance parameters σ_n^2 introduced in (9.31) in the context of continuous time discretization models. In this situation, we show that σ_n^2 is proportional to the time step of the time discretization mesh.

We start with a discussion on a discrete time Markov transition $K_{n,\mu}$ coming from a time discretization of a continuous time jump model with time step $\Delta t = h(\leq 1)$. In other words, we assume that

$$K_{n,\mu} = Id + hL_{n,\mu} + \mathrm{O}(h^2) \qquad (9.44)$$

for some infinitesimal bounded generator $L_{n,\mu}$. In terms of generators, we have

$$L_{K_{n,\mu}} := K_{n,\mu} - Id = hL_{n,\mu} + \mathrm{O}(h^2)$$

Our next objective is to express the uniform local variance parameters σ_n^2 in terms of the parameter h. We start with a brief reminder on the carré du champ operator

$$\Gamma_{L_K}(f,f)(x) := L_K((f - f(x))^2)(x) = L_K(f^2)(x) - 2f(x)L_K(f)(x)$$

associated with a Markov generator $L_K = K - Id$, for some Markov transition K. By construction, we notice that

$$
\begin{aligned}
K([f - K(f)]^2) &= K(f^2) - K(f)^2 \\
&= L_K(f^2) - 2fL_K(f) - (L_K(f))^2 \\
&= \Gamma_{L_K}(f,f) - (L_K(f))^2
\end{aligned}
$$

When $K = K_{n,\mu}$ and $L_{K_{n,\mu}} = hL_{n,\mu} + \mathrm{O}(h^2)$, this yields the following formula

$$
\begin{aligned}
\mu\left[K_{n,\mu}[f_n - K_{n,\mu}(f_n)]^2\right] &= \mu\Gamma_{L_{K_{n,\mu}}}(f,f)\,h - h^2\,\mu(L_{K_{n,\mu}}(f)^2) \\
&= h\,\mu\left(\Gamma_{L_{n,\mu}}(f,f)\right) + \mathrm{O}(h^2)
\end{aligned}
$$

from which we conclude that σ_n^2 is proportional to the parameter h. In the next section, we refine the above analysis in the case of diffusion discretization style generators.

9.6.2   Geometric jump models

We further assume that the Markov transitions M_{n+1} are associated with the elementary transitions of a Markov process $\mathcal{X}_{t_n}$ on some time mesh sequence t_n with time step $t_{n+1} - t_n = h$, with $n \geq 0$. More formally, we have

$$M_{n+1}(x, dy) = \mathbb{P}\left(\mathcal{X}_{t_n+h} \in dy \mid \mathcal{X}_{t_n} = x\right) := \mathcal{M}_{t_n, t_{n+1}}(x, dy)$$

We also assume that for any function f in some sub-algebra $D(L) \subset \mathcal{B}_b(E)$ of functions we have

$$L_{t_n}^{(h)}(f) := \left[\mathcal{M}_{t_n, t_{n+1}} - Id\right](f) = L_{t_n}(f) \, h + R_{t_n}^h(f) \, h^2 \qquad (9.45)$$

for some infinitesimal generators $L_t : D(L) \to D(L)$, and some remainder operator R_{t_n} such that

$$\|R_{t_n}^h(f)\| \leq c_{t_n} \, \|f\|$$

with some norm of $\|f\| \, (\geq \|f\| \vee \|L_{t_n}(f)\|, \, \forall n)$ on $D(L)$, and for some finite constant $c_{t_n} < \infty$, whose values do not depend on the parameter h.

These regularity conditions hold for pure jump processes with bounded jump rates with $D(L) = \mathcal{B}_b(E)$, for Euclidian diffusions on $E = \mathbb{R}^d$ with regular and Lipschitz coefficients by taking $D(L)$ as the set of $\mathcal{C}^\infty$-functions with derivatives decreasing at infinity faster than any polynomial function. These regularity conditions allow using the principal theorems of stochastic differential calculus. For instance, they allow considering the "carré du champ" operator that characterizes the predictable quadratic variations of the martingales that appear in Ito's formulae.

We consider a potential function of the following form $G_n = e^{V_{t_n} h}$, for some bounded potential function. We further assume that the selection transition given in (9.9) is given in terms of a Markov transition $S_{n,\mu} := S_{t_n, \mu}$ that satisfies the following first order expansion

$$\widehat{L}_{t_n, \mu}^{(h)} := S_{t_n, \mu} - Id = h \, \widehat{L}_{t_n, \mu} + h^2 \, \widehat{R}_{t_n, \mu}^h \qquad (9.46)$$

for some jump type bounded generator $\widehat{L}_{t_n, \mu}$, and some remainder integral operator $\widehat{R}_{t_n, \mu}^h$ s.t.

$$\sup \left\|\widehat{R}_{t_n, \mu}^h(f)\right\| < \infty$$

where the supremum is taken over all $h > 0$, $\mu \in \mathcal{P}(E)$, and all $f \in \mathrm{Osc}(E)$.

Finally, we denote by $L_{t_n, \mu}^{(h)}$ the generator associated with the McKean transition defined in (9.8) with $(M_{n+1}, S_{n,\mu}) := (\mathcal{M}_{t_n, t_{n+1}}, S_{t_n, \mu})$; that is, we have that

$$L_{t_n, \mu}^{(h)} = \mathcal{K}_{t_n, t_{n+1}, \mu} - Id \quad \text{with} \quad \mathcal{K}_{t_n, t_{n+1}, \mu} = S_{t_n, \mu} \mathcal{M}_{t_n, t_{n+1}}$$

We illustrate the regularity condition (9.46) with the three cases presented in (0.3).

In the first case, we assume that $\mathcal{V}_t = -\mathcal{U}_t$, for some nonnegative and bounded function $\mathcal{U}_t$ and we set

$$\mathcal{S}_{t_n,\mu}(x, dy) := e^{-\mathcal{U}_{t_n}(x)h}\, \delta_x(dy) + \left(1 - e^{-\mathcal{U}_{t_n}(x)h}\right)\, \Psi_{e^{-\mathcal{U}_{t_n}h}}(\mu)(dy)$$

In this situation, we find that

$$h^{-1}\,[\mathcal{S}_{t_n,\mu}(f)(x) - f(x)]$$

$$= h^{-1}\,\left(1 - e^{-\mathcal{U}_{t_n}(x)h}\right)\,\left[\Psi_{e^{-\mathcal{U}_{t_n}h}}(\mu)(f) - f(x)\right] = \widehat{L}_{t_n,\mu}(f) + O(1)$$

with the bounded operator

$$\widehat{L}_{t_n,\mu}(f)(x) := \mathcal{U}_{t_n}(x)\int\,(f(y) - f(x))\,\mu(dy)$$

Much more is true, if we set

$$\widehat{R}^h_{t_n,\mu}(f)$$

$$:= h^{-1}\left[h^{-1}\,[\mathcal{S}_{t_n,\mu}(f) - f] - \widehat{L}_{t_n,\mu}(f)\right]$$

$$= h^{-1}\left[h^{-1}\,\left(1 - e^{-\mathcal{U}_{t_n}h}\right)\,\left[\Psi_{e^{-\mathcal{U}_{t_n}h}}(\mu)(f) - f\right] - \mathcal{U}_{t_n}\,[\mu(f) - f]\right]$$

then we find that

$$\left|\widehat{R}^h_{t_n,\mu}(f)\right| \leq h^{-1}\left|h^{-1}\,\left(1 - e^{-\mathcal{U}_{t_n}h}\right) - \mathcal{U}_{t_n}\right| \times \left|\Psi_{e^{-\mathcal{U}_{t_n}h}}(\mu)(f) - f\right|$$

$$+ \mathcal{U}_{t_n}\,h^{-1}\left|[\Psi_{e^{-\mathcal{U}_{t_n}h}}(\mu)(f) - \mu(f)]\right|$$

Using the fact that

$$h^{-1}\,\left[\Psi_{e^{-\mathcal{U}_{t_n}h}}(\mu) - \mu\right](f) = \frac{1}{\mu\left(e^{-\mathcal{U}_{t_n}h}\right)}\,\mu\left(h^{-1}\,\left[e^{-\mathcal{U}_{t_n}h} - 1\right]\,[f - \mu(f)]\right)$$

we prove the following first order expansion

$$\mathcal{S}_{t_n,\mu}(f) - f = \widehat{L}_{t_n,\mu}(f)\,h + \widehat{R}^h_{t_n,\mu}(f)\,h^2 \qquad (9.47)$$

with some integral operator $\widehat{R}_{t_n,\mu}$ such that

$$\sup_{h>0}\left\|\widehat{R}^h_{t_n,\mu}(f)\right\| \leq c\,\|\mathcal{U}_{t_n}\|^2\,\mathrm{osc}(f)$$

for some finite constant $c < \infty$.

In the second case, we assume that $\mathcal{V}_t$ is a positive and bounded function and we set

$$\mathcal{S}_{t_n,\mu}(x, dy) := \frac{1}{\mu\left(e^{\mathcal{V}_{t_n}h}\right)}\, \delta_x(dy) + \left(1 - \frac{1}{\mu\left(e^{\mathcal{V}_{t_n}h}\right)}\right)\, \Psi_{\left(e^{\mathcal{V}_{t_n}h} - 1\right)}(\mu)(dy)$$

In this situation, we notice that

$$h^{-1} \left[\mathcal{S}_{t_n,\mu}(f)(x) - f(x) \right]$$

$$= h^{-1} \left(1 - \frac{1}{\mu(e^{\mathcal{V}_{t_n} h})} \right) \left[\Psi_{(e^{\mathcal{V}_{t_n} h} - 1)}(\mu)(f) - f(x) \right]$$

$$= \widehat{L}_{t_n,\mu}(f)(x) + O(1)$$

with the bounded operator

$$\widehat{L}_{t_n,\mu}(f)(x) := \int \left(f(y) - f(x) \right) \, \mathcal{V}_{t_n}(y) \, \mu(dy)$$

Using some elementary calculations, we also prove a first order expansion of the same form as in (9.47).

Finally, in the third case we consider the transitions

$$\mathcal{S}_{t_n,\mu}(x,dy) := \left(1 - \frac{\mu\left(\left(e^{\mathcal{V}_{t_n} h} - e^{\mathcal{V}_{t_n}(x)h} \right)_+ \right)}{\mu(e^{\mathcal{V}_{t_n} h})} \right) \, \delta_x(dy)$$

$$+ \frac{\mu\left(\left(e^{\mathcal{V}_{t_n} h} - e^{\mathcal{V}_{t_n}(x)h} \right)_+ \right)}{\mu\left(e^{\mathcal{V}_{t_n} h} \right)} \, \Psi_{\left(e^{\mathcal{V}_{t_n} h} - e^{\mathcal{V}_{t_n}(x)h} \right)_+}(\mu)(dy)$$

For bounded potential functions we observe that

$$h^{-1} \left[\mathcal{S}_{t_n,\mu}(f)(x) - f(x) \right]$$

$$= h^{-1} \frac{\mu\left(\left(e^{\mathcal{V}_{t_n} h} - e^{\mathcal{V}_{t_n}(x)h} \right)_+ \right)}{\mu\left(e^{\mathcal{V}_{t_n} h} \right)} \left[\Psi_{\left(e^{\mathcal{V}_{t_n} h} - e^{\mathcal{V}_{t_n}(x)h} \right)_+}(\mu)(f) - f(x) \right]$$

$$= \widehat{L}_{t_n,\mu}(f)(x) + O(1)$$

with the bounded operator

$$\widehat{L}_{t_n,\mu}(f)(x) := \int \left[f(y) - f(x) \right] \left(\mathcal{V}_{t_n}(y) - \mathcal{V}_{t_n}(x) \right)_+ \, \mu(dy)$$

The proof of the following technical lemma is elementary; thus it is skipped.

Lemma 9.6.1 *We have the decomposition*

$$L_{t_n,\mu}^{(h)} \;=\; L_{t_n}^{(h)} + \widehat{L}_{t_n,\mu}^{(h)} + \widehat{L}_{t_n,\mu}^{(h)} L_{t_n}^{(h)}$$

In addition, for any $f \in \mathcal{B}_b(E)$, $\mu \in \mathcal{P}(E)$, and any $x \in E$, we have

$$K_{t_n,t_{n+1},\mu}\left(\left[f - K_{t_n,t_{n+1},\mu}(f)(x) \right]^2 \right)(x) = \Gamma_{L_{t_n,\mu}^{(h)}}(f,f)(x) - \left(L_{t_n,\mu}^{(h)}(f)(x) \right)^2$$

We are now in position to state and to prove the main result of this section.

Proposition 9.6.2 *For any $n \geq 0$, we have the first order expansion*

$$L_{t_n,\mu}^{(h)}(f) = h \, L_{t_n,\mu}(f) + h^2 \, R_{t_n,\mu}^h(f) \tag{9.48}$$

for any function $f \in D(L)$, with some remainder operator $R_{t_n,\mu}^h(f)$ such that

$$\sup_{\mu \in \mathcal{P}(E)} \left\| R_{t_n,\mu}^h(f) \right\| \leq c_{t_n} \, \|f\|$$

with some norm of $\|.\|$ on $D(L)$, and for some finite constant $c_{t_n} < \infty$, whose values do not depend on the parameter h.

In addition, we have

$$\mathcal{K}_{t_n,t_{n+1},\mu}\left(\left[f - \mathcal{K}_{t_n,t_{n+1},\mu}(f)\right]^2\right) = \Gamma_{L_{t_n,\mu}}(f,f) \; h + \mathcal{R}_{t_n,\mu}^h(f,f) \; h^2 \tag{9.49}$$

with some remainder operator s.t.

$$\sup_{\mu \in \mathcal{P}(E)} \left\| \mathcal{R}_{t_n,\mu}^h(f,f) \right\| \leq c_{t_n} \, \left\| f^2 \right\| \qquad \text{for some finite constant } c_{t_n} < \infty$$

Before entering into the proof of the proposition, it is useful to make a couple of remarks. When the parameter h tends to 0, the generators and their carré du champ operators converge to some limiting operators that are only defined on the domain of functions $D(L)$ equipped with the norm $\|.\|$. In this situation, we define the uniform local variance parameters σ_n^2 as follows

$$\sigma_n^2 := \sup_{f \in D(L), \|f^2\| \leq 1} \mu\left[\mathcal{K}_{t_n,t_{n+1},\mu}\left(\left[f - \mathcal{K}_{t_n,t_{n+1},\mu}(f)\right]^2\right)\right]$$

$$\leq h \left[\sup_{f \in D(L), \|f^2\| \leq 1} \mu\left[\Gamma_{L_{t_n,\mu}}(f,f)\right] + h \, c_{t_n}\right]$$

for some finite constant $c_{t_n} < \infty$.

Now we come to the proof of the proposition.

Proof of Proposition 9.6.2:

Using Lemma 9.6.1 and the first order decompositions (9.45) and (9.47) we have

$$L_{t_n,\mu}^{(h)} = L_{t_n,\mu} \, h + R_{t_n,\mu}^h \, h^2$$

with

$$R_{t_n,\mu}^h = R_{t_n}^h + \widehat{R}_{t_n,\mu}^h + h^{-2}\left[\widehat{L}_{t_n,\mu} \, h + \widehat{R}_{t_n,\mu}^h \, h^2\right]\left[L_{t_n} \, h + R_{t_n}^h \, h^2\right]$$

On the other hand, we have the decomposition

$$h^{-2}\left[\widehat{L}_{t_n,\mu} \, h + \widehat{R}_{t_n,\mu}^h \, h^2\right]\left[L_{t_n} \, h + R_{t_n}^h \, h^2\right] = \widehat{L}_{t_n,\mu}\left[L_{t_n} + R_{t_n}^h \, h\right] + \widehat{R}_{t_n,\mu}^h L_{t_n}^{(h)}$$

with

$$\left\| \widehat{R}^h_{t_n,\mu} L^{(h)}_{t_n}(f) \right\| \leq c^1_{t_n} \|f\|$$

and

$$\left\| \widehat{L}_{t_n,\mu} \left[L_{t_n} + R^h_{t_n} h \right](f) \right\| \quad \leq \quad c^2_{t_n} \left\| L_{t_n}(f) + R^h_{t_n}(f) h \right\|$$
$$\leq \quad c^3_{t_n} \left(\|L_{t_n}(f)\| + h \|f\| \right)$$

for some finite constant $c^i_{t_n} < \infty$ with $i = 1, 2, 3$. This implies that

$$\|R^h_{t_n,\mu}(f)\| \leq c_{t_n} \left(\|f\| + \|f\| + \|L_{t_n}(f)\| \right)$$

This ends the proof of (9.48).

The proof of the last assertion is based on the decomposition

$$\mathcal{K}_{t_n,t_{n+1},\mu} \left(\left[f - \mathcal{K}_{t_n,t_{n+1},\mu}(f)(x) \right]^2 \right)(x)$$

$$= \Gamma_{L^{(h)}_{t_n,\mu}}(f, f)(x) - \left(L^{(h)}_{t_n,\mu}(f)(x) \right)^2$$

and the fact that

$$\Gamma_{L^{(h)}_{t_n,\mu}}(f, f)(x) = \Gamma_{L_{t_n,\mu}}(f, f)(x) h + R^h_{t_n,\mu} \left((f - f(x))^2 \right)(x) h^2$$

This ends the proof of the proposition.

∎

Chapter 10

A general class of mean field models

10.1 Description of the models

Let $(E_n)_{n\geq 0}$ be a sequence of measurable spaces. We consider a collection of transformations $\Phi_{n+1} : \mathcal{P}(E_n) \to \mathcal{P}(E_{n+1})$, $n \geq 0$, and we denote by $(\eta_n)_{n\geq 0}$ a sequence of probability measures on E_n satisfying a nonlinear equation of the following form

$$\eta_{n+1} = \Phi_{n+1}(\eta_n) \qquad (10.1)$$

The mean field type interacting particle system associated with the Equation (10.1) relies on the fact that the one step mappings can be rewritten in terms of a Markov transport equation

$$\Phi_{n+1}(\eta_n) = \eta_n K_{n+1,\eta_n} \qquad (10.2)$$

for some collection of Markov transitions $K_{n+1,\mu}$ indexed by the time parameter $n \geq 0$, and the set of measures μ on the space E_n.

We recall that the choice of the Markov transitions $K_{n,\eta}$ is not unique. Several examples of McKean transitions are presented in Section 9.2 in the context of Feynman-Kac semigroups. The McKean-Markov model associated with these transitions is the Markov chain $\overline{X}_n$ defined in (9.11).

We further assume that we have a dedicated Monte Carlo simulation tool to sample according to the elementary transitions $K_{n,m(y)}(x_{n-1}, dx_n)$, for any occupation measure $m(y) := \frac{1}{N}\sum_{1\leq i\leq N}\delta_{y^i}$, associated with N given states $(y^i)_{1\leq i\leq N} \in E_{n-1}^N$. In this situation, the mean field particle interpretation of this nonlinear measure valued model is the E_n^N-valued Markov chain $\xi_n = (\xi_n^i)_{1\leq i\leq N}$, with elementary transitions defined as in (9.15).

We recall that the local sampling errors associated with the corresponding mean field particle model are expressed in terms of the centered random fields

$$V_n^N = \sqrt{N}\left[\eta_n^N - \Phi_n\left(\eta_{n-1}^N\right)\right]$$

given by (9.29). We emphasize that the centered random fields V_n^N have the same regularity properties as the local sampling Feynman-Kac models discussed in Section 4.4.4. In particular, the $\mathbb{L}_m$-mean error estimates stated in Proposition 9.5.5 remain valid for this general class of models.

10.2   Some weak regularity properties

To analyze the propagation properties of these local sampling errors, *up to a second order remainder measure*, we introduce the following weak regularity condition.

Definition 10.2.1 *We let* $\Upsilon(E, F)$ *be the set of mappings* Φ *from* $\mathcal{P}(E)$, *into* $\mathcal{P}(F)$, *satisfying the first order decomposition*

$$\Phi(\mu) - \Phi(\eta) = (\mu - \eta)d_\eta\Phi + \mathcal{R}^\Phi(\mu, \eta) \tag{10.3}$$

In the above displayed formula, the first order operator $(d_\eta\Phi)_{\eta \in \mathcal{P}(E)}$ *is some collection of bounded integral operators from* $\mathcal{B}_b(F)$, *into* $\mathcal{B}_b(E)$, *such that*

$$(d_\eta\Phi)(1)(x) = 0 \quad and \quad \beta\,(d\Phi) := \sup\nolimits_{\eta \in \mathcal{P}(E)} \beta\,(d_\eta\Phi) < \infty\,. \tag{10.4}$$

for any $(x, \eta) \in (E \times \mathcal{P}(E))$. *The collection of second order remainder signed measures* $(\mathcal{R}^\Phi(\mu, \eta))_{(\mu,\eta) \in \mathcal{P}(E^2)}$ *on* F *is such that*

$$\left|\mathcal{R}^\Phi(\mu, \eta)(f)\right| \leq \int \left|(\mu - \eta)^{\otimes 2}(g)\right|\, R_\eta^\Phi(f, dg) \tag{10.5}$$

for some integral operators R_η^Φ *from* $\mathcal{B}_b(F)$ *into* $(\mathrm{Osc}(E) \otimes \mathrm{Osc}(E))$ *s.t.*

$$\sup_{\eta \in \mathcal{P}(E)} R_\eta^\Phi(1)(f) \leq \mathrm{osc}(f)\,\delta\left(R^\Phi\right) \quad for\ some \quad \delta\left(R^\Phi\right) < \infty \tag{10.6}$$

An alternative description of these regularity conditions in terms of Fréchet and Gâteaux differentials of semigroups on measure spaces is provided in section 10.3.

These rather weak first order regularity properties are satisfied for a large class of one step transformation Φ_n associated with a nonlinear measure valued process (10.1). Illustrative examples are provided in Section 10.4. For instance, using the first order regularity properties of the Feynman-Kac transformations Φ_n developed in Section 3.1.4 we readily check that $\Phi_n \in \Upsilon(E_{n-1}, E_n)$ with the first order operator

$$d_\mu\Phi_n(f) := \frac{1}{\mu Q_n(1)}\, Q_n\,(f - \Phi_n(\mu)(f))$$

In the further development of this section, we assume that the one step mappings

$$\Phi_n \,:\, \mu \in \mathcal{P}(E_{n-1}) \longrightarrow \Phi_n(\mu) := \mu K_{n,\mu} \in \mathcal{P}(E_n)$$

governing the Equation (10.1) are chosen s.t. $\Phi_n \in \Upsilon(E_{n-1}, E_n)$.

The assumption that R_η^Φ maps $\mathcal{B}_b(F)$ into the subset of tensor product functions $(\mathrm{Osc}(E) \otimes \mathrm{Osc}(E)) \subset \mathcal{B}_b(E^2)$ is only made to simplify the concentration analysis of the general class of mean field models developed in section 14.6.

More precisely, we shall apply these first order expansions to empirical measures

$$\mu = \eta^N := \frac{1}{N} \sum_{1 \leq i \leq N} \delta_{X^i}$$

based on independent samples $(X^i)_{1 \leq i \leq N}$, and their limiting measure η, as $N \uparrow \infty$. As we mentioned in section 2.1.1, the fluctuations of the normalized measures

$$V^N := \sqrt{N} \, (\eta^N - \eta)$$

behave as gaussian random fields. In this interpretation, the term

$$(\mu - \eta)d_\eta \Phi = \frac{1}{\sqrt{N}} \, V^N d_\eta \Phi$$

represents the first order local fluctuations of the backward semigroup expansions (2.5), and the second order remainder term $\mathcal{R}^\Phi(\eta^N, \eta)$ represent the bias second order term in the expansion (2.5). When R_η^Φ maps $\mathcal{B}_b(F)$ into $\mathrm{Osc}(E) \otimes \mathrm{Osc}(E)$, the stochastic analysis of the bias terms reduces to the estimation of second order terms of the form

$$(V^N)^{\otimes 2}(g_1 \otimes g_2) = V^N(g_1) V^N(g_2) \quad \text{with} \quad g = (g_1 \otimes g_2) \in (\mathrm{Osc}(E) \otimes \mathrm{Osc}(E))$$

The concentration analysis of these quadratic terms can be reduced to the one of a product of two Gaussian random variables. For more general functions $f \in \mathcal{B}_b(E^2)$, the second order term $(V^N)^{\otimes 2}(f)$ is a U-symmetric statistic of order 2. The concentration properties, and the $\mathbb{L}_m$-moment analysis of these quantities for general functions f is rather well known in statistical literature, but it requires much more technical tools than the one for tensor product functions $f = (g_1 \otimes g_2)$. The analysis of nonlinear semigroups satisfying condition (10.6) for some second order measure R_η^Φ mapping $\mathcal{B}_b(F)$ into $\mathcal{B}(E^2)$ can be developed combining the backward semigroup decompositions (2.5) with the statistical tools on the concentration analysis of degenerate U-statistics developed in [75, 278, 318, 328], and the series of articles by P. Major [414, 415, 416, 417].

We also observe that under the condition (10.3), we clearly have the Lipschitz property

$$(\Phi) \qquad |[\Phi(\mu) - \Phi(\eta)](f)| \leq \int |(\mu - \eta)(h)| \; T_\eta^\Phi(f, dh)$$

for some integral operators T_η^Φ from $\mathcal{B}(F)$ into the set $\mathrm{Osc}(E)$ such that

$$\int \mathrm{osc}(h) \, T_\eta^\Phi(f, dh) \leq T_\eta^\Phi(1)(f) \leq \mathrm{osc}(f) \, \delta\left(T^\Phi\right) \quad \text{for some} \quad \delta\left(T^\Phi\right) < \infty$$

$$(10.7)$$

Lemma 10.2.2 *For any pair of mappings* $\Phi_1 \in \Upsilon(E_0, E_1)$ *and* $\Phi_2 \in \Upsilon(E_1, E_2)$ *the composition mapping* $(\Phi_2 \circ \Phi_1) \in \Upsilon(E_0, E_2)$ *and we have the first order derivation type formula*

$$d_\eta (\Phi_2 \circ \Phi_1) = d_\eta \Phi_1 \, d_{\Phi_1(\eta)} \Phi_2 \tag{10.8}$$

In this case, condition (Φ) *is also met with* $\delta \left(T^{\Phi_2 \circ \Phi_1} \right) \leq \delta \left(T^{\Phi_2} \right) \times \delta \left(T^{\Phi_1} \right)$

Proof:
Using (Φ), we easily check that (10.8) is met with

$$\beta \left(d \left(\Phi_2 \circ \Phi_1 \right) \right) \leq \beta \left(d\Phi_2 \right) \, \beta \left(d\Phi_1 \right)$$

and $\delta \left(R^{\Phi_2 \circ \Phi_1} \right) \leq \delta \left(T^{\Phi_1} \right) + \delta \left(T^{\Phi_1} \right)^2 \delta \left(R^{\Phi_2} \right)$. This ends the proof of the lemma. ∎

We also mention that for any pair of mappings $\Phi_1 : \eta \in \mathcal{P}(E_0) \mapsto \Phi_1 \in \mathcal{P}(E_1)$ and $\Phi_2 : \eta \in \mathcal{P}(E_1) \mapsto \Phi_1 \in \mathcal{P}(E_2)$, the composition mapping $\Phi = \Phi_2 \circ \Phi_1$ satisfies condition (Φ) as soon as this condition is met for each mapping.

The main advantage of the regularity condition (10.3) comes from the fact that $\Phi_{p,n} \in \Upsilon(E_p, E_n)$ with the first order decomposition type formula

$$\Phi_{p,n}(\eta) - \Phi_{p,n}(\mu) = [\eta - \mu] d_\mu \Phi_{p,n} + \mathcal{R}^{\Phi_{p,n}}(\eta, \mu) \tag{10.9}$$

for some collection of bounded integral operators $d_\mu \Phi_{p,n}$ from E_p into E_n and some second-order remainder signed mesures $\mathcal{R}^{\Phi_{p,n}}(\eta, \mu)$. For further use, we let r_n be the second order stochastic perturbation term related to the quadratic remainder measures $R^{\Phi_{p,n}}$ and defined by $r_n := \sum_{p=0}^{n} \delta(R^{\Phi_{p,n}})$.

Definition 10.2.3 *We say that a collection of Markov transitions* K_η *from a measurable space* $(E, \mathcal{E})$ *into another* $(F, \mathcal{F})$ *satisfies condition* (K) *as soon as the following Lipschitz type inequality is met for every* $f \in \mathrm{Osc}(F)$:

$$(K) \qquad \| [K_\mu - K_\eta] (f) \| \leq \int |(\mu - \eta)(h)| \, T_\eta^K(f, dh) \tag{10.10}$$

In the above display, T_η^K *stands for some collection of bounded integral operators from* $\mathcal{B}_b(F)$ *into* $\mathcal{B}_b(E)$ *such that*

$$\sup_{\eta \in \mathcal{P}(E)} \int \mathrm{osc}(h) \, T_\eta^K(f, dh) \leq \mathrm{osc}(f) \, \delta \left(T^K \right) \tag{10.11}$$

for some finite constant $\delta \left(T^\Phi \right) < \infty$. *In the special case where* $K_\eta(x, dy) = \Phi(\eta)(dy)$, *for some mapping* $\Phi : \eta \in \mathcal{P}(E) \mapsto \Phi(\eta) \in \mathcal{P}(F)$, *condition* (10.10) *coincides with* (Φ).

Suppose we are given a mapping Φ defined in terms of a nonlinear transport formula $\Phi(\eta) = \eta K_\eta$, with a collection of Markov transitions K_η from a measurable space $(E, \mathcal{E})$ into another $(F, \mathcal{F})$ satisfying condition (K). Using the decomposition

$$\Phi(\mu) - \Phi(\eta) = [\eta - \mu] K_\eta + \mu [K_\mu - K_\eta]$$

we readily check that

$$(K) \implies (\Phi) \quad \text{with} \quad T_\eta^\Phi(f, dh) = \delta_{K_\eta(f)}(dh) + T_\eta^K(f, dh)$$

We further assume that we are given a collection of McKean transitions $K_{n,\eta}$ satisfying the weak Lipschitz type condition stated in (10.10). In this situation, we notice that the corresponding one step mappings $\Phi_n(\eta) = \eta K_{n,\eta}$ and the corresponding semigroup $\Phi_{p,n}$ satisfy condition $(\Phi_{p,n})$ for some collection of bounded integral operators $T_\eta^{\Phi_{p,n}}$.

In the context of Feynman-Kac type models, it is not difficult to check that condition (Φ_n) is equivalent to the fact that the McKean transitions $K_{n,\eta}$ satisfy the Lipschitz condition (10.10). The latter is also met for the McKean type model of gases discussed in Section 10.4.3, as well as for the Gaussian transitions discussed in Section 10.4.4, and the nonlinear diffusion models presented in section 10.4.5, as soon as the drift and the diffusion function are sufficiently regular.

10.3 A differential calculus on measure spaces

In this section, we develop a differential calculus of semigroups on measure spaces equipped with the total variation norm. This framework provides an alternative description of the regularity properties presented in Section 10.2 in terms of Fréchet and Gâteaux differentials.

Firstly, we observe that any mapping $\Phi \in \Upsilon(E, F)$ is Fréchet differentiable in the sense that

$$\Phi(\mu) - \Phi(\eta) - (\mu - \eta) d_\eta \Phi = o(\mu - \eta)$$

where $\lim_{\mu \to \eta} \|(\mu - \eta)\|_{\mathrm{tv}}^{-1} \|o(\mu - \eta)\|_{\mathrm{tv}}$. This implies that Φ is also Gâteaux differentiable at any $\mu \in \mathcal{P}(E)$ in any direction $\nu = (\eta - \mu) \in \mathcal{M}_0(E)$, with $\eta \in \mathcal{P}(E)$, in the sense that

$$\lim_{\epsilon \downarrow 0} \left\| \frac{1}{\epsilon} [\Phi(\mu + \epsilon\nu) - \Phi(\mu)] - \nu d_\mu \Phi \right\|_{\mathrm{tv}} = 0$$

One practical way to compute the integral operator $d_\mu \Phi$ is to check that

$$\frac{d}{d\epsilon} \Phi(\mu + \epsilon\nu)(f)_{|\epsilon=0} = \nu d_\mu \Phi(f)$$

for any $f \in \mathcal{B}_b(F)$. In this interpretation, Formula (10.8) in Lemma 10.2.2 is equivalent to the chain rule property

$$\nu \, d_\mu(\Phi_2 \circ \Phi_1) = (\nu \, d_\mu\Phi_1) \, d_{\Phi_1(\mu)}\Phi_2 = \nu \, \left(d_\mu\Phi_1 d_{\Phi_1(\mu)}\Phi_2\right) := \nu \, d_\mu\Phi_1 d_{\Phi_1(\mu)}\Phi_2$$

In this interpretation the Gâteaux derivative of the semigroup $\Phi_{p,n}$ in (10.9) satisfies the semigroup properties

$$d_\mu\Phi_{p,n} = d_\mu\Phi_{p,p+1} d_{\Phi_{p,p+1}(\mu)}\Phi_{p+1,n} = d_\mu\Phi_{p,n-1} d_{\Phi_{p,n-1}(\mu)}\Phi_{n-1,n}$$

Inversely, let Φ be a Gâteaux differentiable mapping at any $\mu \in \mathcal{P}(E)$ in any direction $\nu = (\eta - \mu) \in \mathcal{M}_0(E)$, with $\eta \in \mathcal{P}(E)$, such that

$$\lim_{\epsilon \downarrow 0} \left\| \frac{1}{\epsilon} [\nu d_{\mu+\epsilon\nu}\Phi - \nu d_\mu\Phi] - \nu^{\otimes 2} d_\mu^2 \Phi \right\|_{tv} = 0$$

for some bounded integral operator $d_\mu^2\Phi$ from $\mathcal{B}_b(F)$ into $\mathcal{B}_b(E \times E)$. Here again, one way to compute $d_\mu\Phi$ is to check that

$$\frac{d^2}{d\epsilon^2}\Phi(\mu + \epsilon\nu)(f)_{|\epsilon=0} = \frac{d}{d\epsilon}\nu d_{\mu+\epsilon\nu}\Phi(f)_{|\epsilon=0} = \nu^{\otimes 2} d_\mu^2\Phi(f)$$

We further assume that the mappings $(\nu, \mu) \mapsto \nu^{\otimes 2} d_\mu^2\Phi$ and $(\nu, \mu) \mapsto \nu d_\mu\Phi$ are continuous. In this situation, Φ is C^2-Gâteaux differentiable mapping at any $\mu \in \mathcal{P}(E))$ in any direction $\nu = (\eta - \mu) \in \mathcal{M}_0(E)$, with $\eta \in \mathcal{P}(E)$, and we have the Taylor's theorem with integral remainder

$$\Phi(\mu + \nu) = \Phi(\mu) + \nu d_\mu\Phi + \int_0^1 (1-t) \, \nu^{\otimes 2} d_{\mu+t\nu}^2\Phi \; dt$$

The r.h.s. integral is the Gelfand-Pettis weak sense integral [78]. This yields the first order decomposition

$$\Phi(\eta) = \Phi(\mu) + (\eta - \mu)d_\mu\Phi + \mathcal{R}^\Phi(\eta, \mu)$$

with the second order remainder measure

$$\mathcal{R}^\Phi(\eta, \mu) = \int_0^1 (1-t) \, (\eta - \mu)^{\otimes 2} d_{\mu+t(\eta-\mu)}^2 \Phi \; dt$$

We can extend Taylor's expansions at any order. For instance, whenever Φ is a C^3-Gâteaux differentiable mapping at any $\mu \in \mathcal{P}(E))$ in any direction $\nu = (\eta - \mu) \in \mathcal{M}_0(E)$, with $\eta \in \mathcal{P}(E)$, we have

$$\lim_{\epsilon \downarrow 0} \left\| \frac{1}{\epsilon} [\nu^{\otimes 2} d_{\mu+\epsilon\nu}^2\Phi - \nu^{\otimes 2} d_\mu^2\Phi] - \nu^{\otimes 3} d_\mu^3\Phi \right\|_{tv} = 0$$

for some bounded integral operator $d_\mu^3\Phi$ from $\mathcal{B}_b(F)$ into $\mathcal{B}_b(E \times E)$, such that $(\nu, \mu) \mapsto \nu^{\otimes 3} d_\mu^3\Phi$ is continuous. This yields the Taylor's theorem

$$\Phi(\mu + \nu) = \Phi(\mu) + \nu d_\mu\Phi + \frac{1}{2} \, \nu^{\otimes 2} d_\mu^2\Phi + \frac{1}{2}\int_0^1 (1-t)^2 \, \nu^{\otimes 3} d_{\mu+t\nu}^3\Phi \; dt \quad (10.12)$$

10.4   Some illustrative examples

10.4.1   Feynman-Kac semigroups

We consider the Feynman-Kac flows (9.4) discussed in Section 9.1. We recall from (9.40) that the semigroup $\Phi_{p,n}$ is given for any $f_n \in \mathcal{B}_b(E_n)$ by the following formula

$$\eta_n(f_n) = \frac{\eta_p(Q_{p,n}(f_n))}{\eta_p(Q_{p,n}(1))} \quad \text{with} \quad Q_{p,n} = Q_{p+1} \ldots Q_{n-1} Q_n$$

For $p = n$, we use the convention $Q_{n,n} = Id$, the identity operator.

The aim of this section is to show that $\Phi_{p,n} \in \Upsilon(E_p, E_n)$.

Using Lemma 3.1.10, we prove that for any couple of measures $(\mu, \eta) \in \mathcal{P}(E_p)$ we have

$$[\Phi_{p,n}(\mu) - \Phi_{p,n}(\eta)](f_n) = \frac{1}{\mu(G_{p,n,\eta})} \ (\mu - \eta) d_\eta \Phi_{p,n}(f_n)$$

with the collection of first order integral operators $d_\eta \Phi_{p,n}$ from $\mathcal{B}_b(E_n)$ into $\mathcal{B}_b(E_p)$ defined by

$$d_\eta \Phi_{p,n}(f_n) := G_{p,n,\eta} \ P_{p,n} \ (f_n - \Phi_{p,n}(\eta)(f_n))$$

In the above display $G_{p,n,\eta}$ and $P_{p,n}$ stand for the potential function and the Markov operator given by

$$G_{p,n,\eta} := Q_{p,n}(1)/\eta(Q_{p,n}(1)) \quad \text{and} \quad P_{p,n}(f) = Q_{p,n}(f)/Q_{p,n}(1)$$

It is now easy to check that

$$\mathcal{R}^{\Phi_{p,n}}(\mu, \eta)(f_n) := -\frac{1}{\mu(G_{p,n,\eta})} \ [\mu - \eta]^{\otimes 2}(G_{p,n,\eta} \otimes d_\eta \Phi_{p,n}(f_n))$$

Using the fact that

$$d_\eta \Phi_{p,n}(f_n)(x) = G_{p,n,\eta}(x) \int \ [P_{p,n}(f)(x) - P_{p,n}(f)(y)] \ G_{p,n,\eta}(y) \ \eta(dy)$$

we find that

$$\forall f_n \in \mathrm{Osc}(E_n) \qquad \|d_\eta \Phi_{p,n}(f_n)\| \leq g_{p,n} \ \beta(P_{p,n})$$

with $g_{p,n} = \sup_{x,y} Q_{p,n}(1)(x)/Q_{p,n}(1)(y)$. This implies that $\beta(d\Phi_{p,n}) := \sup_{\eta \in \mathcal{P}(E_p)} \beta(d_\eta \Phi_{p,n}) \leq 2 \ g_{p,n} \ \beta(P_{p,n})$. Finally, we observe that

$$\left| \mathcal{R}^{\Phi_{p,n}}(\mu, \eta)(f_n) \right| \leq (2 \ q_{p,n} \beta(d\Phi_{p,n})) \ \left| [\mu - \eta]^{\otimes 2} \left(\frac{G_{p,n,}}{2\|G_{p,n}\|} \otimes \frac{d_\eta \Phi_{p,n}(f_n)}{\beta(d\Phi_{p,n})} \right) \right|$$

This shows that (10.5) and (10.6) are satisfied with

$$\delta(R^{\Phi_{p,n}}) \leq 2\, g_{p,n}\, \beta(d\Phi_{p,n}) \leq 4\, g_{p,n}^2\, \beta(P_{p,n})$$

We end this section, we an equivalent description of these regularity conditions in terms of Gâteaux differentials. For any $\mu \in \mathcal{P}(E_p)$, any direction $\nu = (\eta - \mu) \in \mathcal{M}_0(E_p)$, with $\eta \in \mathcal{P}(E_p)$, and any test function $f_n \in \mathcal{B}_b(E_n)$, we have

$$\Phi_{p,n}(\mu + \epsilon\nu)(f_n) = \frac{\mu Q_{p,n}(f_n) + \epsilon\nu Q_{p,n}(f_n)}{\mu Q_{p,n}(1) + \epsilon\nu Q_{p,n}(1)}$$

from which we find that

$$\frac{d}{d\epsilon}\Phi_{p,n}(\mu + \epsilon\nu)(f_n)$$

$$= \frac{1}{(\mu + \epsilon\nu)Q_{p,n}(1)}\left[\nu Q_{p,n}(f_n) - \Phi_{p,n}(\mu + \epsilon\nu)(f_n)\, \nu Q_{p,n}(1)\right]$$

$$= \nu\left(\frac{Q_{p,n}}{(\mu+\epsilon\nu)Q_{p,n}(1)}\,(f_n - \Phi_{p,n}(\mu + \epsilon\nu)(f_n))\right) = \nu d_{\mu+\epsilon\nu}\Phi_{p,n}(f_n)$$

In much the same way, we find that

$$\frac{d^2}{d\epsilon^2}\Phi_{p,n}(\mu + \epsilon\nu)(f_n)$$

$$= -2\,\nu^{\otimes 2}\left(\frac{Q_{p,n}(1)}{(\mu+\epsilon\nu)Q_{p,n}(1)} \otimes \frac{Q_{p,n}}{(\mu+\epsilon\nu)Q_{p,n}(1)}\,(f_n - \Phi_{p,n}(\mu + \epsilon\nu)(f_n))\right)$$

and

$$\frac{d^3}{d\epsilon^3}\Phi_{p,n}(\mu + \epsilon\nu)(f_n)$$

$$= 6\,\nu^{\otimes 3}\left(\left(\frac{Q_{p,n}(1)}{(\mu + \epsilon\nu)Q_{p,n}(1)}\right)^{\otimes 2} \otimes \frac{Q_{p,n}}{(\mu + \epsilon\nu)Q_{p,n}(1)}\,(f_n - \Phi_{p,n}(\mu + \epsilon\nu)(f_n))\right)$$

We conclude that

$$\frac{d^2}{d\epsilon^2}\Phi_{p,n}(\mu + \epsilon\nu)(f_n) = \nu^{\otimes 2}d^2_{\mu+\epsilon\nu}\Phi_{p,n}(f_n)$$

and

$$\frac{d^3}{d\epsilon^3}\Phi_{p,n}(\mu + \epsilon\nu)(f_n) = \nu^{\otimes 3}d^3_{\mu+\epsilon\nu}\Phi_{p,n}(f_n)$$

with the integral operators

$$d^2_{\mu+\epsilon\nu}\Phi_{p,n}(f_n) = -2!\,\frac{Q_{p,n}(1)}{(\mu + \epsilon\nu)Q_{p,n}(1)} \otimes d_{\mu+\epsilon\nu}\Phi_{p,n}(f_n)$$

and

$$d_{\mu+\epsilon\nu}^3 \Phi_{p,n}(f_n) = 3! \left(\frac{Q_{p,n}(1)}{(\mu+\epsilon\nu)Q_{p,n}(1)} \right)^{\otimes 2} \otimes d_{\mu+\epsilon\nu}\Phi_{p,n}(f_n)$$

Combining the estimate

$$\left| \nu^{\otimes 3} d_{\mu+\epsilon\nu}^3 \Phi_{p,n}(f_n) \right| \leq \|d_{\mu+\epsilon\nu}\Phi_{p,n}(f_n)\| \times \left(\nu \left(\frac{Q_{p,n}(1)}{(\mu+\epsilon\nu)Q_{p,n}(1)} \right) \right)^2$$

$$\leq g_{p,n}^3 \, \beta(P_{p,n}) \left(\nu \left(\frac{Q_{p,n}(1)}{\mu Q_{p,n}(1)} \right) \right)^2$$

with the Taylor expansion (10.12), we find that

$$\left\| \Phi_{p,n}(\eta) - \Phi_{p,n}(\mu) - (\eta-\mu)d_\mu\Phi_{p,n} - \tfrac{1}{2}\,(\eta-\mu)^{\otimes 2}d_\mu^2\Phi_{p,n} \right\|_{\mathrm{tv}}$$

$$\leq g_{p,n}^3 \, \beta(P_{p,n}) \left(\nu \left(\tfrac{Q_{p,n}(1)}{\mu Q_{p,n}(1)} \right) \right)^2$$

On the other hand, we have

$$\frac{1}{2} \left| (\eta-\mu)^{\otimes 2}d_\mu^2\Phi_{p,n} \right| \leq \left| (\eta-\mu)^{\otimes 2} \left(\frac{Q_{p,n}(1)}{\mu Q_{p,n}(1)} \otimes d_\mu\Phi_{p,n}(f_n) \right) \right|$$

from which we conclude that $\Phi_{p,n} \in \Upsilon(E_p, E_n)$. More precisely, we have the first order decomposition

$$\Phi_{p,n}(\eta) - \Phi(\mu) = (\eta-\mu)d_\mu\Phi_{p,n} + \mathcal{R}^{\Phi_{p,n}}(\eta,\mu)$$

with some remainder measure satisfying the rather crude upper bound

$$\left| \mathcal{R}^{\Phi_{p,n}}(\eta,\mu)(f_n) \right|$$

$$\leq g_{p,n}^3 \, \beta(P_{p,n}) \, (\eta-\mu)^{\otimes 2} \left(\left(\tfrac{Q_{p,n}(1)}{\mu Q_{p,n}(1)} \right)^{\otimes 2} \right)$$

$$+ \left| (\eta-\mu)^{\otimes 2} \left(\tfrac{Q_{p,n}(1)}{\mu Q_{p,n}(1)} \otimes d_\mu\Phi_{p,n}(f_n) \right) \right|$$

10.4.2 Interacting jump models

We let (G_n, H_n) be a couple of $]0,1]$-valued potential functions on some measurable state spaces E_n. We also consider a sequence of Markov transitions M_{n+1} from E_n into E_{n+1}. We associate with these objects the collection of McKean transitions

$$K_{n+1,\mu}(x,dy) := G_n(x)\,\delta_x(dy) + (1-G_n(x))\,\Phi_{n+1,H_n}(\mu)(dy) \qquad (10.13)$$

with the measures

$$\Phi_{n+1,H_n}(\mu) := \Psi_{H_n}(\mu)M_{n+1}$$

The McKean model $\overline{X}_n$ associated with the collection of Markov transition $K_{n+1,\mu}$ is an interacting jump model with jump rate G_n. At geometric random times, the process jumps to a new location randomly chosen with the distribution $\Psi_{H_n}(\eta_n)$, where $\eta_n = \text{Law}(\overline{X}_n)$. Then, it evolves from this site to a new random state in E_{n+1} according to the Markov transition M_{n+1}.

We denote by Φ_{n+1} the one step mappings associated with the McKean transitions $K_{n+1,\mu}$; that is, we have that

$$\Phi_{n+1}(\mu) := \mu K_{n+1,\mu}$$

By construction, we observe that

$$[K_{n+1,\mu_1} - K_{n+1,\mu_2}] (f_{n+1})$$

$$= (1 - G_n) \, [\Phi_{n+1,H_n}(\mu_1) - \Phi_{n+1,H_n}(\mu_2)] (f_{n+1})$$

for any couple of measures $(\mu_1, \mu_2) \in \mathcal{P}(E_n)$. On the other hand, using Lemma 3.1.5, for any $f_n \in \mathcal{B}_b(E_n)$ we prove the decompositions

$$[\Psi_{H_n}(\mu_1) - \Psi_{H_n}(\mu_2)] (f_n)$$

$$= (\mu_1 - \mu_2)(d_{\mu_2} \Psi_{H_n}(f_n)) - \frac{1}{\mu_1(H_{n,\mu_2})} \, (\mu_1 - \mu_2)^{\otimes 2} [H_{n,\mu_2} \otimes d_{\mu_2} \Psi_{H_n}(f_n)]$$

with the first order integral operator

$$d_{\mu_2} \Psi_{H_n}(f_n) := H_{n,\mu_2} \, (Id - \Psi_{H_n}(\mu_2))(f_n) \quad \text{and} \quad H_{n,\mu_2} := H_n/\mu_2(H_n)$$

This clearly implies that

$$[K_{n+1,\mu_1} - K_{n+1,\mu_2}] (f_{n+1})$$

$$= (1 - G_n) \, (\mu_1 - \mu_2) d_{\mu_2} \Phi_{n+1,H_n}(f_{n+1})$$

$$-(1 - G_n) \, \frac{1}{\mu_1(H_{n,\mu_2})} \, (\mu_1 - \mu_2)^{\otimes 2} [H_{n,\mu_2} \otimes d_{\mu_2} \Phi_{n+1,H_n}(f_{n+1})]$$

with the integral operator

$$d_{\mu_2} \Phi_{n+1,H_n} := d_{\mu_2} \Psi_{H_n} M_{n+1}$$

Using the decomposition

$$\begin{aligned} \Phi_{n+1}(\mu_1) - \Phi_{n+1}(\mu_2) \ = \ & (\mu_1 - \mu_2) K_{n+1,\mu_2} \\ & + \mu_2 \, (K_{n+1,\mu_1} - K_{n+1,\mu_2}) \\ & + [\mu_1 - \mu_2] \, (K_{n+1,\mu_1} - K_{n+1,\mu_2}) (10.14) \end{aligned}$$

after some elementary but tedious calculations, we conclude that $\Phi_{n+1} \in \Upsilon(E_n, E_{n+1})$ with the first order integral operators

$$d_\mu \Phi_{n+1}(f) := K_{n+1,\mu}(f - \Phi_{n+1}(\mu)(f)) + (1 - \mu(G_n)) \, d_\mu \Phi_{n+1,H_n} \quad (10.15)$$

10.4.3 A McKean model of gases

We consider a measurable state space $(S_n, \mathcal{S}_n)$ with a countably generated σ-field and an $(\mathcal{S}_n \otimes \mathcal{E}_n)$-measurable mapping a_n be from $(S_n \times E_n)$ into $\mathbb{R}_+$ such that $\int \nu_n(ds) a_n(s, x) = 1$, for any $x \in E_n$, and some bounded positive measure $\nu_n \in \mathcal{M}(S_n)$. To illustrate this abstract model, we can take a partition of the state $E_n = \cup_{s \in S_n} A_s$ associated with a countable set S_n equipped with the counting measure $\nu_n(s) = 1$, and set $a_n(s, x) = 1_{A_s}(x)$. We let $K_{n+1, \eta}$ be the McKean transition defined by

$$K_{n+1, \eta}(x, dy) = \int \nu_n(ds) \, \eta(du) \, a_n(s, u) \, M_{n+1}((s, x), dy) \qquad (10.16)$$

In the above displayed formula, M_n stands for some Markov transition from $(S_n \times E_n)$ into E_{n+1}. The discrete time version of the McKean's 2-velocities model for Maxwellian gases corresponds to the time homogenous model on $E_n = S_n = \{-1, +1\}$ associated with the counting measure ν_n and the pair of parameters

$$a_n(s, x) = 1_s(x) \quad \text{and} \quad M_{n+1}((s, x), dy) = \delta_{sx}(dy)$$

In this situation, the measure valued Equation (10.1) takes the following quadratic form:

$$\eta_{n+1}(+1) = \eta_n(+1)^2 + (1 - \eta_n(+1))^2$$

Our next objective is to check that the first order decomposition (10.3) is met with the first order operator defined by

$$
\begin{aligned}
d_\mu \Phi_{n+1}(f)(x) \;=\;& [K_{n+1, \mu}(f)(x) - \Phi_{n+1}(\mu)(f)] \\
& + \int \nu_n(ds) \, [a(s, x) - \mu(a(s, \cdot))] \, \mu(M_{n+1}(f)(s, \cdot))
\end{aligned}
$$

To simplify the presentation, we consider time homogeneous models and we surpress the time index. In this notation, we find that

$$[K_\eta - K_\mu](f)(x) = \int \nu(ds) \, [\eta - \mu] (a(s, \cdot)) \, M(f)(s, x)$$

Observe that

$$[\eta - \mu](K_\eta - K_\mu)(f)(x) = \int \nu(ds) \, [\eta - \mu] (a(s, \cdot)) \, [\eta - \mu] (M(f)(s, \cdot))$$

Using the decomposition (10.14), we readily check that $\Phi \in \Upsilon(E, E)$ with the first order operator

$$
\begin{aligned}
d_\mu \Phi(f)(x) \;=\;& [K_\mu(f)(x) - \Phi(\mu)(f)] \\
& + \int \nu(ds) \, [a(s, x) - \mu(a(s, \cdot))] \, \mu \, (M(f)(s, \cdot))
\end{aligned}
$$

and the second order remainder measure

$$\mathcal{R}^{\Phi}(\mu, \eta)(f) = \int [\eta - \mu]^{\otimes 2} (g_s) \ \nu(ds) \quad \text{with} \quad g_s = a(s, .) \otimes M(f)(s, .).$$

In this situation, we notice that

$$\beta(d\Phi) \le \left[\beta(M) + \int \nu(ds) \ \mathrm{osc}(a(s, .))\right]$$

and

$$\delta(R^{\Phi}) \le \beta(M) \int \nu(ds) \ \mathrm{osc}(a(s, .))$$

10.4.4   Gaussian mean field models

We consider the McKean models associated with a collection of multivariate Gaussian type Markov transitions on $E_n = \mathbb{R}^d$, defined by

$$K_{n,\eta}(x, dy)$$

$$= \frac{1}{\sqrt{(2\pi)^d \det(Q_n)}} \ \exp\left\{-\tfrac{1}{2} (y - d_n(x, \eta))' \ Q_n^{-1} (y - d_n(x, \eta))\right\} dy$$

(10.17)

with a nonsingular, positive, and semi-definite covariance matrix Q_n and some sufficiently regular drift mapping $d_n : (x, \eta) \in \mathbb{R}^d \times \mathcal{P}(\mathbb{R}^d) \mapsto d(x, \eta) \in \mathbb{R}^d$.

Our next objective is to check that these transitions satisfy the first order regularity condition (10.3).

To simplify the presentation, we only discuss time homogenous and one dimensional models. We consider the one dimensional Gaussian transitions on $E = \mathbb{R}$ defined below

$$K_\eta(x, dy) = \frac{1}{\sqrt{2\pi}} \ \exp\left\{-\frac{1}{2} (y - d(x, \eta))^2\right\} dy$$

with some linear drift function d_n of the form

$$d(x, \eta) = a(x) + \eta(b)c(x)$$

with some measurable (and nonnecessarily bounded) function a, and some pair of functions b and $c \in \mathcal{B}_b(\mathbb{R})$. In this situation the first order decomposition (10.3) is met with the first order operator defined by

$$d_\mu \Phi_n(f)(x) \quad := \quad [K_\mu(f)(x) - \Phi_n(\mu)(f)]$$

$$+ b(x) \int \mu(dy) \ c(y) \ K_\mu(y, dz) \ f(z) \ (z - d(y, \mu))$$

We prove this claim using the decomposition

$$[K_\eta - K_\mu] (f)(x) \quad = \quad \int K_\mu(x, dy) \ \Theta (\Delta_{\mu,\eta}(x, y)) \ f(y)$$

$$+ \int K_\mu(x, dy) \ \Delta_{\mu,\eta}(x, y) \ f(y)$$

with $\Theta(u) = e^u - 1 - u$ and the function $\Delta_{\mu,\eta}(x,y)$ defined by

$$\Delta_{\mu,\eta}(x,y) = \log \frac{dK_\eta(x,\cdot)}{dK_\mu(x,\cdot)}(y) = \Delta^{(1)}_{\mu,\eta}(x,y) + \Delta^{(2)}_{\mu,\eta}(x,y)$$

with

$$\begin{aligned}
\Delta^{(1)}_{\mu,\eta}(x,y) &:= [d(x,\eta) - d(x,\mu)] \; [y - d(x,\mu)] \\
&= c(x) \; (\eta - \mu)(b) \; [y - d(x,\mu)] \\
\Delta^{(2)}_{\mu,\eta}(x,y) &:= -\frac{1}{2}[d(x,\eta) - d(x,\mu)]^2 = -\frac{1}{2} c(x)^2 \; [(\eta - \mu)(b)]^2
\end{aligned}$$

Under our assumptions on the drift function d, we have

$$\left|\Delta^{(1)}_{\mu,\eta}(x,y)\right| \leq \|c\| \; \mathrm{osc}(b) \; |y - d(x,\mu)| \quad \text{and} \quad \left|\Delta^{(2)}_{\mu,\eta}(x,y)\right| \leq \|c\|^2 \; \mathrm{osc}(b)^2/2$$

Using the fact that $|\Theta(u)| \leq e^{|u|}u^2/2$, after some elementary manipulations we prove that

$$\sup_{x \in \mathbb{R}} \left|[K_\eta - K_\mu](f)(x) - \int K_\mu(x,dy) \, \Delta^{(1)}_{\mu,\eta}(x,y) \, f(y)\right| \leq C \; [(\eta - \mu)(b)]^2 \; \|f\|$$

with some finite constant $C < \infty$ whose values only depend on $\|c\|$ and $\mathrm{osc}(b)$. On the other hand, we have

$$\int (\eta - \mu)(dx) \int K_\mu(x,dy) \, \Delta^{(1)}_{\mu,\eta}(x,y) \, f(y) = (\eta - \mu)^{\otimes 2} \left(b \otimes (K'_\mu(f))\right)$$

and

$$\int \mu(dx) \int K_\mu(x,dy) \, \Delta^{(1)}_{\mu,\eta}(x,y) \, f(y) = (\eta - \mu)(b) \; \mu \left(K'_\mu(f)\right)$$

with the bounded integral operator K'_μ defined by

$$K'_\mu(f)(x) = c(x) \int K_\mu(x,dy) \; [y - d(x,\mu)] \; f(y)$$

Using the decomposition (10.14) we prove that

$$\Phi(\eta)(f - \Phi(\mu)(f)) = (\eta - \mu)d_\mu\Phi(f) + \mathcal{R}^\Phi(\eta,\mu)(f)$$

with the first order operator

$$d_\mu\Phi(f) = K_\mu \left(f - \Phi(\mu)(f)\right) + b \; \mu \left(K'_\mu(f)\right) \tag{10.18}$$

and a second order remainder term such that

$$\left|\mathcal{R}^\Phi(\eta,\mu)(f)\right| \leq C' \; \left[\left|(\eta - \mu)^{\otimes 2} \left(b \otimes (K'_\mu(f))\right)\right| + [(\eta - \mu)(b)]^2 \; \mathrm{osc}(f)\right]$$

with some finite constant $C' < \infty$ whose values only depend on $\|c\|$ and $osc(b)$. Using the fact that

$$K'_\mu(1) = 0 \quad \text{and} \quad \|K'_\mu(f)\| = \|K'_\mu(f - \Phi(\mu)(f))\| \le \|c\| \ osc(f)$$

we conclude that (10.5) and (10.6) are met with

$$\delta\left(R^\Phi\right) \le C' osc(b)(2\|c\| + osc(b))$$

and condition (10.4) is satisfied with $\beta(d\Phi) \le 1 + \|c\| \ osc(b)$.

10.4.5   McKean-Vlasov diffusion models

We consider the nonlinear diffusion models presented in Section 1.4.1.1. In this context, the nonlinear semigroup of the law of the stochastic process defined in (1.34) is given by the one step transformation

$$\Phi_n(\mu)(f) = \int \mu(dx) \ K_{n,\mu}(f)(x) \tag{10.19}$$

with the collection of Markov transitions $K_{n,\mu}$ defined for any $f \in \mathcal{B}_b(\mathbb{R})$ by

$$K_{n,\mu}(f)(x) := \mathbb{E}\left(f\left(x + \mathbf{a_n}(x,\mu) + \boldsymbol{\sigma_n}(x,\mu) \ W\right)\right)$$

where W is a centered Gaussian random variables with unit variance. We further assume that the functions a_n and σ_n defined in (1.35) are bounded, and $\sigma_n(x,y) \ge \sigma_{n,\star}$, for some positive constant $\sigma_{n,\star}$.

In this context, for any $\mu \in \mathcal{P}(\mathbb{R})$, any direction $\nu = (\eta - \mu) \in \mathcal{M}_0(\mathbb{R})$, with $\eta \in \mathcal{P}(\mathbb{R})$, and any $\epsilon \in [0,1]$, we have

$$X^x_{\mu+\epsilon\nu} := \mathbf{a_n}(x, \mu+\epsilon\nu) + \boldsymbol{\sigma_n}(x, \mu+\epsilon\nu) \ W = X^x_\mu + \epsilon \ X^x_\nu$$

with

$$X^x_\mu = \mathbf{a_n}(x,\mu) + \boldsymbol{\sigma_n}(x,\mu) \ W \quad \text{and} \quad X^x_\nu = \mathbf{a_n}(x,\nu) + \boldsymbol{\sigma_n}(x,\nu) \ W$$

In this notation, we find that

$$\Phi_n(\mu)(f) = \int \mu(dx) \ \mathbb{E}\left(f\left(x + X^x_\mu + \epsilon \ X^x_\nu\right)\right)$$

We also notice that for constant diffusion functions $\sigma_n(x,y) = \sigma_n > 0$, we have $\boldsymbol{\sigma_n}(x,\nu) = \sigma_n\nu(1) = 0$, and therefore $X^x_\nu = \mathbf{a_n}(x,\nu)$.

Using elementary computations, for any smooth test function f, and for any $k \ge 1$ we have that

$$\frac{d^k}{d\epsilon^k}\Phi_n(\mu + \epsilon\nu)(f) = k \ \nu\left(\frac{d^{k-1}}{d\epsilon^{k-1}}K_{n,\mu+\epsilon\nu}(f)\right) + (\mu + \epsilon\nu)\left[\frac{d^k}{d\epsilon^k}K_{n,\mu+\epsilon\nu}(f)\right]$$

$$\tag{10.20}$$

and

$$\frac{d^k}{d\epsilon^k} K_{n,\mu+\epsilon\nu}(f)(x) := \mathbb{E}\left(f^{(k)}\left(x + X^x_{\mu+\epsilon\nu}\right)(X^x_\nu)^k\right)$$

Our next objective to extend these differential operators to non necessarily smooth functions. To this end, we use the following technical lemma.

Lemma 10.4.1 *We let D be the differential operator defined by*

$$DG(w) = wG(w) - G'(w)$$

For any $k \geq 1$, we let D^k be k-th iterate of D defined by $D^k = DD^{k-1} = D^{k-1}D$. For any $k \geq 1$, and any couple of smooth functions (F, G), such that $\lim_{x \to +/-\infty} F^{(l)}(w)D^{k-l-1}G(w) \, e^{-w^2/2} = 0$, for any $0 \leq l < k$, we have the integration by part formula

$$\mathbb{E}\left(F^{(k)}(W)G(W)\right) = \mathbb{E}\left(F(W)D^kG(W)\right) \tag{10.21}$$

where W is a centered Gaussian random variables with unit variance.

Proof:
For any smooth functions (F, G) s.t. $\lim_{x \to +/-\infty} F(w)G(w) \, e^{-w^2/2} = 0$, we have the integration by part formula

$$\mathbb{E}\left(F'(W)G(W)\right) = \mathbb{E}\left(F(W)DG(W)\right)$$

We prove this formula using the fact that

$$\int F'(w) \, G(w) \, \frac{1}{\sqrt{2\pi}} \, e^{-w^2/2}dw \;\; = \;\; -\int F(w) \, \frac{d}{dw}\left[G(w) \, \frac{1}{\sqrt{2\pi}} \, e^{-w^2/2}\right]dw$$

$$= \;\; \int F(w) \, D(G)(w) \, \frac{1}{\sqrt{2\pi}} \, e^{-w^2/2}dw$$

The proof of (10.21) follows a simple induction on the parameter k. This ends the proof of the lemma. ∎

For instance, D^2 and D^3 are given by

$$D^2G(w) \;\; = \;\; wD(G)(w) - D(G)'(w) = (w^2 - 1)G(w) - 2wG'(w) + G''(w)$$
$$D^3G(w) \;\; = \;\; wD^2(G)(w) - (D^2(G))'(w)$$
$$= \;\; (w^3 - 3w)G(w) + 3(1 - w^2)G'(w) + 3wG''(w) - G'(w)$$

The following result is a direct consequence of Lemma 10.4.1.

Proposition 10.4.2 *For any function $f \in \mathcal{B}_b(\mathbb{R})$, and any $k \geq 1$, we have*

$$\frac{d^k}{d\epsilon^k} K_{n,\mu+\epsilon\nu}(f)(x)$$

$$= \sum_{0 \leq l \leq k} \binom{k}{l} \nu^{\otimes k}\left(\mathbf{a_n}(x, .)^{\otimes(k-l)} \otimes \boldsymbol{\sigma}_n(x, .)^{\otimes l}\right) U^{(k,l)}_{n,\mu+\epsilon\nu}(f)(x)$$

with the integral operators $U_\eta^{(k,l)}$ indexed by $\eta \in \mathcal{P}(\mathbb{R})$, from $\mathcal{B}_b(\mathbb{R})$ into itself, and defined by

$$U_{n,\eta}^{(k,l)}(f)(x) := \frac{1}{\boldsymbol{\sigma}_n(x,\eta)^k} \ \mathbb{E}\left(f\left(x + X_\eta^x\right) D^k W^l\right)$$

Under our assumptions, for any $\eta \in \mathcal{P}(\mathbb{R})$ we have that $U_{n,\eta}^{(k,l)}(1) = 0$, and

$$\left\| U_{n,\eta}^{(k,l)} \right\| := \sup_{\|f\| \le 1} \left\| U_{n,\eta}^{(k,l)}(f) \right\| \le \sigma_{n,\star}^{-k} \ \mathbb{E}\left(\left|D^k W^l\right|\right)$$

Definition 10.4.3 *For any $k \ge 0$, and $\eta \in \mathcal{P}(E)$ we denote by $J_{n,\eta}^{(k+1)}$ the integral operator from $\mathcal{B}_b(\mathbb{R})$ into $\mathcal{B}_b(\mathbb{R}^{k+1})$ defined for any $f \in \mathcal{B}_b(\mathbb{R})$ and any $(x,y) \in (\mathbb{R} \times \mathbb{R}^k)$ by*

$$J_{n,\eta}^{(k+1)}(f)(x,y)$$

$$= \sum_{0 \le l \le k} \binom{k}{l} \left[\prod_{1 \le p \le k-l} a_n(x,y_p) \right] \times \left[\prod_{k-l < q \le k} \sigma_n(x,y_q) \right] U_{n,\eta}^{(k,l)}(f)(x)$$

with the convention $J_{n,\eta}^1(f)(x) = U_{n,\eta}^{(0,0)}(f)(x) = K_{n,\eta}(f)(x)$, for $k = 0$.

In terms of the Gâteaux derivative of the mapping $\mu \mapsto \delta_x K_{n,\mu}$, we find that

$$\frac{d^k}{d\epsilon^k} K_{n,\mu+\epsilon\nu}(f)(x) = \frac{d^k}{d\epsilon^k} \left(\delta_x K_{n,\mu+\epsilon\nu}\right)(f) = \nu^{\otimes k} d_{\mu+\epsilon\nu}^k \left(\delta_x K_{n,.}\right)(f)$$

with the integral operators $d_\eta^k \left(\delta_x K_{n,.}\right)$ from $\mathcal{B}_b(\mathbb{R})$ into $\mathcal{B}_b(\mathbb{R}^k)$ defined for any $y = (y_1, \ldots, y_k) \in \mathbb{R}^k$ by

$$d_\eta^k \left(\delta_x K_{n,.}\right)(f)(y) = J_{n,\eta}^{(k+1)}(f)(x,y)$$

The following proposition provides a description of Gâteaux differentials (10.20) for any bounded functions $f \in \mathcal{B}_b(\mathbb{R})$.

Proposition 10.4.4 *For any $k \ge 1$, and any function $f \in \mathcal{B}_b(\mathbb{R})$ we have*

$$\frac{d^k}{d\epsilon^k} \Phi_n(\mu + \epsilon\nu)(f) = \nu^{\otimes k} d_{\mu+\epsilon\nu}^k \Phi_n(f)$$

with the collection of integral operator $d_\eta^k \Phi_n$, from $\mathcal{B}_b(\mathbb{R})$ into $\mathcal{B}_b(\mathbb{R}^k)$, indexed by $\eta \in \mathcal{P}(E)$ and defined for any $y \in \mathbb{R}^k$ by the formula

$$d_\eta^k \Phi_n(f)(y) = k \ J_{n,\eta}^k(f)(y) + \int \eta(dx) \ J_{n,\eta}^{(k+1)}(f)(x,y) \tag{10.22}$$

Replacing in the above display f by the centered function $(f - \Phi_n(\mu)(f))$, we define a operator $d_\mu \Phi_n(f)$ satisfying the property $d_\mu \Phi_n(1) = 0$. For instance, we have

$$J_{n,\eta}^2(f)(x,y) = a_n(x,y) \ U_{n,\eta}^{(1,0)}(f)(x) + \sigma_n(x,y) \ U_{n,\eta}^{(1,1)}(f)(x)$$

so that

$$d_\eta \Phi_n(f)(y)$$

$$= K_{n,\eta}(f)(y) + \int \eta(dx) \ \left[a_n(x,y) \ U_{n,\eta}^{(1,0)}(f)(x) + \sigma_n(x,y) \ U_{n,\eta}^{(1,1)}(f)(x) \right]$$
(10.23)

In the same way, we have

$$J_{n,\eta}^3(f)(x,(y_1,y_2))$$

$$= a_n(x,y_1) \ a_n(x,y_2) \ U_{n,\eta}^{(2,0)}(f)(x) \ + \ 2 \ a_n(x,y_1) \ \sigma_n(x,y_2) \ U_{n,\eta}^{(2,1)}(f)(x)$$

$$+ \sigma_n(x,y_1) \ \sigma_n(x,y_2) \ U_{n,\eta}^{(2,2)}(f)(x)$$

so that

$$d_\eta^2 \Phi_n(f)(y_1,y_2)$$

$$= 2 \ \left[a_n(y_1,y_2) \ U_{n,\eta}^{(1,0)}(f)(y_1) + \sigma_n(y_1,y_2) \ U_{n,\eta}^{(1,1)}(f)(y_1) \right]$$

$$+ \int \eta(dx) \ a_n(x,y_1) \ a_n(x,y_2) \ U_{n,\eta}^{(2,0)}(f)(x)$$

$$+ 2 \int \eta(dx) \ a_n(x,y_1) \ \sigma_n(x,y_2) \ U_{n,\eta}^{(2,1)}(f)(x)$$

$$+ \int \eta(dx) \ \sigma_n(x,y_1) \ \sigma_n(x,y_2) \ U_{n,\eta}^{(2,2)}(f)(x)$$

Proposition 10.4.5 *Assume that the functions (a_n, σ_n) have the following form*

$$a_n = \int_{\mathcal{V}_{a,n}} \left(a_{\theta,n}^{(1)} \otimes a_{\theta,n}^{(2)} \right) \Xi_{a,n}(\theta) \quad and \quad \sigma_n = \int_{\mathcal{V}_{\sigma,n}} \left(\sigma_{\theta,n}^{(1)} \otimes \sigma_{\theta,n}^{(2)} \right) \Xi_{\sigma,n}(\theta)$$

for some probability measures $(\Xi_{a,n}, \Xi_{\sigma,n})$ on some measurable spaces $(\mathcal{V}_{a,n}, \mathcal{V}_{\sigma,n})$, and some bounded functions $\left(a_{n,\theta}^{(i)}, \sigma_{n,\theta}^{(i)} \right)$ on $\mathbb{R}$, with $i = 1, 2$.

In this situation, the semigroup Φ_n defined in (10.19) is Gâteaux differentiable at any order, at any $\mu \in \mathcal{P}(E)$, in any direction $\nu = (\eta - \mu) \in \mathcal{M}_0(E)$, with $\eta \in \mathcal{P}(E)$. In addition, for any $k \geq 1$ we have the Taylor's theorem

$$\Phi_n(\eta) = \Phi_n(\mu) + \sum_{0 \leq l < k} \frac{1}{l!} \ (\eta - \mu)^{\otimes l} \ d_\mu^l \Phi_n + \mathcal{R}_k^\Phi \ (\eta, \mu)$$

with the integral operators $d_\mu^k \Phi_n$ defined as in (10.22) by replacing f by $(f - \Phi_n(\mu)(f))$, and with a remainder measure $\mathcal{R}_k^\Phi(\eta, \mu) \in \mathcal{P}(\mathbb{R})$ such that

$$\left| \mathcal{R}_k^\Phi(\eta, \mu)(f) \right| \leq \int \left| (\eta - \mu)^{\otimes k}(g) \right| \ R_{k,\mu}^\Phi(f, dg)$$

for some collection of integral operators $R_{k,\mu}^\Phi$ from $\mathcal{B}_b(\mathbb{R})$ into $\mathrm{Osc}(\mathbb{R})^k$ s.t.

$$\sup_{\eta \in \mathcal{P}(\mathbb{R})} R_{k,\mu}^\Phi(1)(f) \leq \mathrm{osc}(f) \ \delta\left(R_k^\Phi\right) \quad \text{for some} \quad \delta\left(R_k^\Phi\right) < \infty$$

In particular, we have $\Phi_n \in \Upsilon(\mathbb{R}, \mathbb{R})$ with the first order integral operators $d_\eta \Phi_n$ defined in (10.23).

Proof:

For any $1 \leq l, m \leq k$, we set

$$g_{\theta, m, n}^{(k,l)} := 1_{1 \leq m \leq k-l} \ a_{\theta, n}^{(1)} + 1_{k-l < m \leq n} \ \sigma_{\theta, n}^{(1)}$$

$$h_{\theta, m, n}^{(k,l)} := 1_{1 \leq m \leq k-l} \ a_{\theta, n}^{(2)} + 1_{k-l < m \leq n} \ \sigma_{\theta, n}^{(2)}$$

In this notation, we have

$$J_{n,\eta}^{(k+1)}(f)(x, y)$$

$$= \sum_{0 \leq l \leq k} \binom{k}{l} \int_{\overline{\mathcal{V}}_{k,l}} \left[\prod_{1 \leq m \leq k} h_{\theta_m, m, n}^{(k,l)}(y_m) \right] \overline{U}_{n,\theta,\eta}^{(k,l)}(f)(x) \ \overline{\Xi}_{k,l}(d\theta)$$

with the probability measure $\overline{\Xi}_{k,l}$ on $\overline{\mathcal{V}}_{k,l} := \left(\mathcal{V}_{a,n}^{(k-l)} \times \mathcal{V}_{\sigma,n}^l \right)$ given by

$$\overline{\Xi}_{k,l}(d\theta_1, \ldots, \theta_k) = \prod_{1 \leq m \leq k-l} \Xi_{a,n}(d\theta_m) \prod_{k-l < m \leq k} \Xi_{\sigma,n}(d\theta_m)$$

and the collection of functions

$$\overline{U}_{n,\theta,\eta}^{(k,l)}(f)(x) = \left[\prod_{1 \leq m \leq k} g_{\theta_m, m, n}(x) \right] U_{n,\eta}^{(k,l)}(f)(x)$$

This implies that

$$\left(\eta \otimes \nu^{\otimes k} \right) \left(J_{n,\eta}^{(k+1)}(f) \right)$$

$$= \sum_{0 \leq l \leq k} \binom{k}{l} \int_{\overline{\mathcal{V}}_{k,l}} \left[\prod_{1 \leq p \leq k} \nu\left(h_{\theta_p, p, n}^{(k,l)} \right) \right] \eta\left(\overline{U}_{n,\theta,\eta}^{(k,l)}(f) \right) \overline{\Xi}_{k,l}(d\theta)$$

from which we find the upper bound

$$\left\| \left(\eta \otimes \nu^{\otimes k} \right) J_{n,\eta}^{(k+1)} \right\|_{\mathrm{tv}} \leq \sum_{0 \leq l \leq k} c_{k,l} \int_{\mathcal{V}_{k,l}} \prod_{1 \leq p \leq k} \left| \nu \left(h_{\theta_p, p, n}^{(k,l)} \right) \right| \; \Xi_{k,l}(d\theta)$$

with some finite constants

$$c_{k,l} \leq \binom{k}{l} \sigma_{n,\star}^{-k} \left[\prod_{1 \leq p \leq k-l} \sup_{\theta \in \mathcal{V}_{a,n}} \left\| a_{\theta,n}^{(1)} \right\| \right]$$

$$\left[\prod_{k-l < p \leq k} \sup_{\theta \in \mathcal{V}_{\sigma,n}} \left\| \sigma_{\theta,n}^{(1)} \right\| \right] \; \mathbb{E}(|D^k W^l|)$$

The end of the proof of the proposition is now a direct consequence of (10.22). This ends the proof of the proposition. ∎

10.4.6   McKean-Vlasov models with jumps

The Gaussian mean field models with interacting jumps defined in this section are a simple combination of the McKean models on $E_n = \mathbb{R}^d$ defined in Section 10.4.2 and in Section 10.4.4. More precisely, we let $K_{n+1,\eta}^{(1)}$, and resp. $K_{n+1,\eta}^{(2)}$, be the collection of Markov transitions defined in (10.17), and respectively in (10.13). We denote by $\Phi_{n+1}^{(1)}$, and resp. $\Phi_{n+1}^{(2)}$, the one step mappings associated with the McKean transitions $K_{n+1,\eta}^{(1)}$, and resp. $K_{n+1,\eta}^{(2)}$; that is, for any $i \in \{1,2\}$ we have

$$\Phi_{n+1}^{(i)}(\mu) := \mu K_{n+1,\mu}^{(i)}$$

We consider the flow of measures η_n defined by the following synthetic diagram:

$$\eta_0 \xrightarrow{K_{1,\eta_0}^{(1)}} \eta_1 \xrightarrow{K_{2,\eta_1}^{(2)}} \eta_2 \xrightarrow{K_{3,\eta_0}^{(1)}} \eta_3 \xrightarrow{K_{4,\eta_1}^{(2)}} \eta_4 \xrightarrow{K_{5,\eta_1}^{(1)}} \cdots$$

More formally, for any $n \geq 0$ we have

$$\eta_{2n+1} := \eta_{2n} K_{2n+1,\eta_{2n}}^{(1)} = \Phi_{n+1}^{(1)}(\eta_{2n})$$

$$\eta_{2(n+1)} := \eta_{2n+1} K_{2(n+1),\eta_{2n+1}}^{(2)} = \Phi_{2(n+1)}^{(2)}(\eta_{2n+1})$$

The McKean model $\overline{X}_n$ associated with the collection of Markov transition $K_{n+1,\mu}^{(i)}$, with $i \in \{1,2\}$, is decomposed into two steps. The odd transitions $\overline{X}_{2n} \rightsquigarrow \overline{X}_{2n+1}$ are sampled according to the Gaussian mean field transitions $K_{2n+1,\eta_{2n}}^{(1)}$; the even ones $\overline{X}_{2n+1} \rightsquigarrow \overline{X}_{2(n+1)}$ are sampled according to the interacting jump transitions $K_{2(n+1),\eta_{2n+1}}^{(2)}$.

By construction, we have

$$\eta_{n+1} = \Phi_{n+1}(\eta_n) := \eta_n K_{n+1,\eta_n}$$

with the collection of Markov transitions

$$K_{2n+1,\mu} = K_{2n+1,\mu}^{(1)} \quad \text{and} \quad K_{2(n+1),\mu} = K_{2(n+1),\mu}^{(2)}$$

By Lemma 10.2.2 using the analysis developed in Section 10.4.2 and Section 10.4.4, we conclude that $\Phi_{n+1} \in \Upsilon(\mathbb{R}^d, \mathbb{R}^d)$. For $d = 1$, the first order operators are defined using the first order decompositions given in (10.18) and (10.15).

10.5   A stochastic coupling technique

In this section, we provide an alternative stochastic coupling technique to analyze the deviations of the mean field IPS model presented in (1.36) (see also Section 1.5.3). We consider a class of McKean-Markov chain models on $E_n = \mathbb{R}^d$, for some $d \geq 1$, given by the recursive formulae

$$X_n = F_n(X_{n-1}, \eta_{n-1}, W_n) \quad \text{with} \quad \eta_{n-1} := \mathrm{Law}(X_{n-1}) \qquad (10.24)$$

In the above display, W_n stands for some collection of independent, and independent of $(X_p)_{0 \leq p < n}$, random variables taking values in some state space $\mathcal{W}_n$, and F_n is a collection of some measurable mapping from $(\mathbb{R}^d \times \mathcal{P}(\mathbb{R}^d) \times \mathcal{W}_n)$ into $\mathbb{R}^d$. We let $(\xi_n^{N,i})_{1 \leq i \leq N}$ be the mean field particle model defined in (1.56).

Theorem 10.5.1 *We assume that the collection of functions F_n satisfy the following Lipschitz regularity conditions*

$$\|F_n(x, \eta, w) - F_n(y, \eta, w)\| \leq c_n \|x - y\|$$

$$\|F_n(x, \eta, w) - F_n(x, \mu, w)\| \leq c_n \int |\eta(\varphi) - \mu(\varphi)| \, \Lambda_n((x, w); d\varphi)$$

for some finite constants $c_n < \infty$, and some Markov transition Λ_n from $(\mathbb{R}^d \times \mathcal{W}_n)$ into the set of Lipschitz functions φ on $\mathbb{R}^d$, with unit Lipschitz constant, and such that $\|\varphi\| \leq 1$. In this situation, for any $m \geq 1$ and any $n \geq 0$ we have

$$\mathbb{E}\left(\left\|\xi_n^{(N,1)} - X_n\right\|^m\right)^{1/m} \leq b(m) \, c(n)/\sqrt{N} \qquad (10.25)$$

for some finite constant $c(n)$, and with the parameters $b(m)$ that are given in (0.6).

Proof:
We let $\bar{\eta}_n^N = \frac{1}{N} \sum_{i=1}^N \delta_{X_n^i}$ be the occupation measure associated with N *independent* Markov chains X_n^i defined as in (10.24) associated with the random perturbations W_n^i and starting at the locations $X_0^i = \xi_0^i$. For a given fixed parameter $m \geq 1$, we also set

$$I_n := \mathbb{E}\left(\left\|\xi_n^{(N,1)} - X_n\right\|^m\right)^{1/m}$$

In this notation, we have the decomposition

$$\begin{aligned}
\xi_n^{(N,1)} - X_n^1 &= \left[F_n\left(\xi_{n-1}^{(N,1)}, \eta_{n-1}^N, W_n^1\right) - F_n\left(X_{n-1}^1, \eta_{n-1}^N, W_n^1\right)\right] \\
&\quad + \left[F_n\left(X_{n-1}^1, \eta_{n-1}^N, W_n^1\right) - F_n\left(X_{n-1}^1, \bar{\eta}_{n-1}^N, W_n^1\right)\right] \\
&\quad + \left[F_n\left(X_{n-1}^1, \bar{\eta}_{n-1}^N, W_n^1\right) - F_n\left(X_{n-1}^1, \bar{\eta}_{n-1}, W_n^1\right)\right]
\end{aligned}$$

from which we prove that

$$\begin{aligned}
\left\|\xi_n^{(N,1)} - X_n^1\right\| &\leq c_n \left\|\xi_{n-1}^{(N,1)} - X_{n-1}^1\right\| \\
&\quad + c_n \int \left|\eta_{n-1}^N(\varphi) - \bar{\eta}_{n-1}^N(\varphi)\right| \Lambda_n\left((X_{n-1}^1 W_n^1); d\varphi\right) \\
&\quad + c_n \int \left|\bar{\eta}_{n-1}^N(\varphi) - \bar{\eta}_{n-1}(\varphi)\right| \Lambda_n\left((X_{n-1}^1 W_n^1); d\varphi\right)
\end{aligned}$$

To take the final step, we observe that

$$\left|\eta_{n-1}^N(\varphi) - \bar{\eta}_{n-1}^N(\varphi)\right| \leq \frac{1}{N} \sum_{1 \leq i \leq N} \left\|\xi_{n-1}^{(N,i)} - X_{n-1}^i\right\|$$

and

$$\begin{aligned}
\left|\bar{\eta}_{n-1}^N(\varphi) - \bar{\eta}_{n-1}(\varphi)\right| &\leq \frac{1}{N} \varphi(X_{n-1}^1) \\
&\quad + \left(1 - \frac{1}{N}\right)\left(\frac{1}{N-1}\sum_{i=2}^N \varphi(X_{n-1}^i) - \bar{\eta}_{n-1}(\varphi)\right)
\end{aligned}$$

Using the generalized Minkowski inequality (cf. for instance [192]), and conventional $\mathbb{L}_m$-mean error estimates for empirical measures associated with i.i.d. sequences , we prove that

$$I_n \leq 2c_n I_{n-1} + \frac{c_n}{N} + \left(1 - \frac{1}{N}\right)\frac{2c_n b(m)}{\sqrt{N}} \leq 2c_n\, I_{n-1} + \frac{3c_n b(m)}{\sqrt{N}}$$

The end of the proof is now a direct application of Gronwald's lemma. This ends the proof of the theorem. ∎

10.6   Fluctuation analysis

10.6.1   Some $\mathbb{L}_m$-mean error estimates

Combining the Lipschitz property $(\Phi_{p,n})$ of the semigroup $\Phi_{p,n}$ with the decomposition

$$\left[\eta_n^N - \eta_n\right] = \sum_{p=0}^{n} \left[\Phi_{p,n}(\eta_p^N) - \Phi_{p,n}\left(\Phi_p(\eta_{p-1}^N)\right)\right]$$

we find that

$$\sqrt{N}\ \left|\left[\eta_n^N - \eta_n\right](f_n)\right| = \sum_{p=0}^{n} \int\ |V_p^N(h)|\ T_{\Phi_p(\eta_{p-1}^N)}^{\Phi_{p,n}}(f, dh)$$

In the above displayed formulae, we have used the convention $\Phi_0(\eta_{-1}^N) = \eta_0$, for $p = 0$. From the conditional $\mathbb{L}_{2m}$-mean error estimates presented in Proposition 9.5.5, (see also Theorem 11.2.1), we readily conclude that

$$\sup_{N \geq 1} \sqrt{N}\ \mathbb{E}\left(\left|\left[\eta_n^N - \eta_n\right](f_n)\right|^{2m}\right)^{\frac{1}{2m}} \leq b(2m)\ (1 \wedge c(2m)) \sum_{p=0}^{n} \delta(T^{\Phi_{p,n}})$$

$$\tag{10.26}$$

with the constants $b(m)$ and $c(m)$ defined in (0.6) and (9.33).

10.6.2   Local sampling random fields

This section is concerned with a detailed proof of the central limit theorem, Theorem 9.5.3, for this abstract class of local sampling models.

Theorem 10.6.1 *For any fixed time horizon $n \geq 0$, the sequence of random fields V_n^N converges in law, in the sense of convergence of finite dimensional distributions, as the number of particles N tends to infinity, to a sequence of independent, Gaussian, and centered random fields V_n, with, for any $f_n \in \mathcal{B}_b(E_n)$, and $n \geq 0$,*

$$\mathbb{E}(V_n(f_n)^2) = \eta_{n-1}K_{n,\eta_{n-1}}([f_n - K_{n,\eta_{n-1}}(f_n)]^2)$$

Proof:
Let $\mathcal{F}^N = \{\mathcal{F}_n^N\ ;\ n \geq 0\}$ be the natural filtration associated with the N-particle system ξ_n. The first class of martingales that arises naturally in our context is the $\mathbb{R}^d$-valued and $\mathcal{G}^N$-martingale $M_n^N(f)$ defined by

$$M_n^N(f) = \sum_{p=0}^{n} \left[\eta_p^N(f_p) - \Phi_p(\eta_{p-1}^N)(f_p)\right] \tag{10.27}$$

where $f_p : x_p \in E_p \mapsto f_p(x_p) = (f_p^u(x_p))_{u=1,\ldots,d} \in \mathbb{R}^d$ is a d-dimensional and bounded measurable function. By direct inspection, we see that the vth component of the martingale $M_n^N(f) = (M_n^N(f^u))_{u=1,\ldots,d}$ is the d-dimensional and $\mathcal{F}^N$-martingale defined for any $u = 1,\ldots,d$ by the formula

$$M_n^N(f^u) = \sum_{p=0}^{n} \left[\eta_p^N(f_p^u) - \Phi_p(\eta_{p-1}^N)(f_p^u) \right] = \sum_{p=0}^{n} \left[\eta_p^N(f_p^u) - \eta_{p-1}^N K_{p,\eta_{p-1}^N}(f_p^u) \right]$$

with the usual convention $K_{0,\eta_{-1}^N} = \eta_0 = \Phi_0(\eta_{-1}^N)$ for $p = 0$. The idea of the proof consists in using the CLT for triangular arrays of $\mathbb{R}^d$-valued random variables (Theorem 3.33, p. 437 in [340]). We first rewrite the martingale $\sqrt{N}\, M_n^N(f)$ in the following form:

$$\sqrt{N}\, M_n^N(f) = \sum_{i=1}^{N} \sum_{p=0}^{n} \frac{1}{\sqrt{N}} \left(f_p(\xi_p^i) - K_{p,\eta_{p-1}^N}(f_p)(\xi_{p-1}^i) \right)$$

This readily yields $\sqrt{N}\, M_n^N(f) = \sum_{k=1}^{(n+1)N} U_k^N(f)$ where for any $1 \le k \le (n+1)N$ with $k = pN + i$ for some $i = 1,\ldots,N$ and $p = 0,\ldots,n$

$$U_k^N(f) = \frac{1}{\sqrt{N}} \left(f_p(\xi_p^i) - K_{p,\eta_{p-1}^N}(f_p)(\xi_{p-1}^i) \right)$$

We further denote by $\mathcal{H}_k^N$ the σ-algebra generated by the random variables ξ_p^j for any pair index (j,p) such that $pN + j \le k$. It can be checked that, for any $1 \le u < v \le d$ and for any $1 \le k \le (n+1)N$ with $k = pN + i$ for some $i = 1,\ldots,N$ and $p = 0,\ldots,n$, we have $\mathbb{E}(U_k^N(f^u) \mid \mathcal{H}_{k-1}^N) = 0$ and

$$\mathbb{E}(U_k^N(f^u) U_k^N(f^v) \mid \mathcal{H}_{k-1}^N)$$

$$= \tfrac{1}{N} K_{p,\eta_{p-1}^N}[(f_p^u - K_{p,\eta_{p-1}^N} f_p^u) \ (f_p^v - K_{p,\eta_{p-1}^N} f_p^v)](X_{p-1}^{(N,i)})$$

This also yields that

$$\sum_{k=pN+1}^{pN+N} \mathbb{E}(U_k^N(f^u) U_k^N(f^v) \mid \mathcal{H}_{k-1}^N)$$

$$= \eta_{p-1}^N [K_{p,\eta_{p-1}^N}[(f_p^u - K_{p,\eta_{p-1}^N} f_p^u) \ (f_p^v - K_{p,\eta_{p-1}^N} f_p^v)]]$$

Our aim is now to describe the limiting behavior of the martingale $\sqrt{N}\, M_n^N(f)$ in terms of the process $X_t^N(f) \overset{\text{def.}}{=} \sum_{k=1}^{[Nt]+N} U_k^N(f)$. By the definition of the particle model associated with a given mapping Φ_n and using the fact that $\left[\frac{[Nt]}{N}\right] = [t]$, one gets that for any $1 \leq u, v \leq d$

$$\sum_{k=1}^{[Nt]+N} E\left(U_k^N(f^u)U_k^N(f^v)\,\big|\,\mathcal{H}_{k-1}^N\right)$$

$$= C_{[t]}^N(f^u, f^v) + \frac{[Nt]-N[t]}{N}\left(C_{[t]+1}^N(f^u, f^v) - C_{[t]}^N(f^u, f^v)\right)$$

where, for any $n \geq 0$ and $1 \leq u, v \leq d$,

$$C_n^N(f^u, f^v) = \sum_{p=0}^{n} \eta_{p-1}^N\left[K_{p,\eta_{p-1}^N}\left(\left(f_p^u - K_{p,\eta_{p-1}^N}f_p^u\right)\left(f_p^v - K_{p,\eta_{p-1}^N}f_p^v\right)\right)\right]$$

Under our regularity conditions on the McKean transitions, this implies that for any $1 \leq i, j \leq d$,

$$\sum_{k=1}^{[Nt]+N} E\left(U_k^N(f^u)U_k^N(f^v)\,\big|\,\mathcal{H}_{k-1}^N\right) \xrightarrow[N \to \infty]{P} C_t(f^u, f^v)$$

with

$$C_n(f^u, f^v) = \sum_{p=0}^{n} \eta_{p-1}[K_{p,\eta_{p-1}}\left(\left(f_p^u - K_{p,\eta_{p-1}}f_p^u\right)\left(f_p^v - K_{p,\eta_{p-1}}f_p^v\right)\right)]$$

and, for any $t \in \mathbb{R}_+$,

$$C_t(f^u, f^v) = C_{[t]}(f^u, f^v) + \{t\}\left(C_{[t]+1}(f^u, f^v) - C_{[t]}(f^u, f^v)\right)$$

Since $\left\|U_k^N(f)\right\| \leq \frac{2}{\sqrt{N}}(\vee_{p \leq n}\|f_p\|)$, for any $1 \leq k \leq [Nt] + N$, the conditional Lindeberg condition is clearly satisfied and therefore one concludes that the $\mathbb{R}^d$-valued martingale $\{X_t^N(f)\ ;\ t \in \mathbb{R}_+\}$ converges in law to a continuous Gaussian martingale $\{X_t(f)\ ;\ t \in \mathbb{R}_+\}$ such that, for any $1 \leq u, v \leq d$ and $t \in \mathbb{R}_+$

$$\langle X(f^u), X(f^v)\rangle_t = C_t(f^u, f^v)$$

Recalling that

$$X_{[t]}^N(f) = \sqrt{N}\,M_{[t]}^N(f)$$

we conclude that the $\mathbb{R}^d$-valued and $\mathcal{F}^N$-martingale $\sqrt{N}\,M_n^N(f)$ converges in law to an $\mathbb{R}^d$-valued and Gaussian martingale $M_n(f) = (M_n(f^u))_{u=1,\ldots,d}$ such that for any $n \geq 0$ and $1 \leq u, v \leq d$

$$\langle M(f^u), M(f^v)\rangle_n = \sum_{p=0}^{n} \eta_{p-1}[K_{p,\eta_{p-1}}\left(\left(f_p^u - K_{p,\eta_{p-1}}f_p^u\right)\left(f_p^v - K_{p,\eta_{p-1}}f_p^v\right)\right)]$$

with the convention $K_{0,\eta_{-1}} = \eta_0$ for $p = 0$.

To take the final step, we let $(\varphi_n)_{n \geq 0}$ be a sequence of bounded measurable functions respectively in $\mathcal{B}(E_n)^{d_n}$. We associate with $\varphi = (\varphi_n)_n$ the sequence of functions $f = (f_p)_{0 \leq p \leq n}$ defined for any $0 \leq p \leq n$ by the following formula

$$
\begin{aligned}
f_p &= (f_p^u)_{u=0,\ldots,n} \\
&= (0,\ldots,0,\varphi_p,0,\ldots,0) \in \mathcal{B}(E_p)^{d_0 + \ldots + d_p + \ldots + d_n}
\end{aligned}
$$

In the above display, 0 stands for the null function in $\mathcal{B}(E_p)^{d_q}$ (for $q \neq p$). By construction, we have $f_u^u = \varphi_u$ and for any $0 \leq u \leq n$, we have that

$$
f^u = (f_p^u)_{0 \leq p \leq n} = (0,\ldots,0,\varphi_u,0,\ldots,0) \in \mathcal{B}(E_0)^{d_0} \times \ldots \times \mathcal{B}(E_u)^{d_u} \times \ldots \mathcal{B}(E_n)^{d_n}
$$

so that

$$
\sqrt{N} \, M_n^N(f^u) = \sqrt{N} \, [\eta_u^N(\varphi_u) - \eta_{u-1}^N K_{u,\eta_{u-1}^N}(\varphi_u)] = V_u^N(\varphi_u)
$$

and therefore

$$
\sqrt{N} \, M_n^N(f) := (\sqrt{N} \, M_n^N(f^u))_{0 \leq u \leq n} = (V_u^N(\varphi_u))_{0 \leq u \leq n} := V_n^N(\varphi)
$$

We conclude that $V_n^N(\varphi)$ converges in law to an $(n+1)$-dimensional and centered Gaussian random field $V_n(\varphi) = (V_u(\varphi_u))_{0 \leq u \leq n}$ with, for any $0 \leq u, v \leq n$,

$$
\mathbb{E}(V_u(\varphi_u^1) V_v(\varphi_v^2))
$$

$$
= 1_u(v) \, \eta_{u-1}[K_{u,\eta_{u-1}} (\varphi_u^1 - K_{u,\eta_{u-1}} \varphi_u^1) K_{u,\eta_{u-1}} (\varphi_u^2 - K_{u,\eta_{u-1}} \varphi_u^2)]
$$

This ends the proof of Theorem 10.6.1. ∎

10.6.3 A functional central limit theorem

This section is concerned with proving a functional central limit theorem for the random fields

$$
W_n^{\eta,N} := \sqrt{N} \, [\eta_n^N - \eta_n] \tag{10.28}
$$

This fluctuation theorem takes basically the following form.

Theorem 10.6.2 *For any fixed time horizon $n \geq 0$, the sequence of random fields $W_n^{\eta,N}$ converges in law, as the number of particles N tends to infinity, to Gaussian and centered random field*

$$
W_n^{\eta} = \sum_{p=0}^{n} V_p d_{p,n}
$$

In the above display, $d_{p,n}$ stands for the semigroup associated with the integral operator $d_n = d_{\eta_{n-1}} \Phi_n$.

Proof:

Using the decomposition

$$W_n^{\eta,N} \;=\; V_n^N + W_{n-1}^{\eta,N} d_n + \sqrt{N}\; R^{\Phi_n}\left(\eta_{n-1}^N, \eta_{n-1}\right)$$

we readily prove that

$$W_n^{\eta,N} = \sum_{p=0}^{n} V_p^N d_{p,n} + \frac{1}{\sqrt{N}}\; \mathcal{R}_n^N \tag{10.29}$$

with the remainder second order measure

$$\mathcal{R}_n^N := N \sum_{p=0}^{n-1} R_{p+1}^{\Phi_{p+1}}\left(\eta_p^N, \eta_p\right) d_{p+1,n}$$

In the above display, $d_{p,n} = d_{p+1}\ldots d_n$ stands for the semigroup associated with the integral operators $d_n := d_{\eta_{n-1}}\Phi_n$, with the usual convention $d_{n,n} = Id$, for $p = n$. Using a first order derivation formula for the semigroup $\Phi_{p,n}$ (cf. for instance Lemma 10.2.2 on page 302), it is readily checked that

$$d_{\eta_p}\Phi_{p,n} = (d_{\eta_p}\Phi_{p+1})(d_{\eta_{p+1}}\Phi_{p+1,n}) = d_{p+1}(d_{\eta_p}\Phi_{p,n}) = d_{p,n}$$

Using the fact that

$$\left|\mathcal{R}_n^N(f_n)\right| \le \sum_{p=0}^{n-1} \int \left|\left(W_p^{\eta,N}\right)^{\otimes 2}(g)\right| R_{\eta_p}^{\Phi_{p+1}}(d_{p+1,n}(f), dg)$$

we conclude that, for any $m \ge 1$, we have

$$\mathbb{E}\left(\left|\mathcal{R}_n^N(f_n)\right|^m\right)^{1/m} \le b(2m)^2 \sum_{p=0}^{n-1} \beta(d_{p+1,n}) \left(\sum_{q=0}^{p} \delta(T^{\Phi_{q,p}})\right)^2 \delta\left(R^{\Phi_{p+1}}\right)$$

We prove the r.h.s estimate combining the backward semigroup decomposition (2.5) with the weak Lipschitz property $(\Phi_{p,n})$ of the semigroup $\Phi_{p,n}$ stated on page 450, and applying the $\mathbb{L}_m$-estimates presented in Proposition 9.5.5 (see also Theorem 11.2.1).

This clearly implies that $\frac{1}{\sqrt{N}}\mathcal{R}_n^N$ converge in law to the null measure, in the sense that $\frac{1}{\sqrt{N}}\mathcal{R}_n^N(f_n)$ converge in law to zero, for any bounded test function f_n on E_n. Using the fact that V_n^N converges in law to the sequence of n independent, random fields V_n, the proposition is now a direct consequence of the decomposition formula (10.29). This ends the proof of Theorem 10.6.2. ∎

10.6.4 A note on genetic type particle models

In this section we present some direct applications of these general fluctuation theorems to Feynman-Kac mean field and genetic type IPS models (9.15) discussed in Chapter 9. In this context, the particle model associated with the McKean transitions $K_{n+1,\eta_n} = S_{n,\eta_n} M_{n+1}$ introduced in (9.8) is decomposed in two steps (9.19). More formally, the particle profiles evolve according to the following synthetic diagram

$$\eta_n^N \xrightarrow{\;S_{n,\eta_n^N}\;} \widehat{\eta}_n^N \xrightarrow{\;M_{n+1}\;} \eta_{n+1}^N \tag{10.30}$$

The corresponding limiting evolution is given by

$$\eta_n \xrightarrow{\;S_{n,\eta_n}\;} \widehat{\eta}_n = \eta_n S_{n,\eta_n} \xrightarrow{\;M_{n+1}\;} \eta_{n+1} = \widehat{\eta}_n M_{n+1}$$

The local sampling random fields models associated with these two steps IPS simulation are defined by

$$V_{n+1}^{N,mut} := \sqrt{N}\ \left[\eta_{n+1}^N - \widehat{\eta}_n^N M_{n+1}\right] \quad \text{and} \quad V_n^{N,select} := \sqrt{N}\ \left[\widehat{\eta}_n^N - \Psi_{G_n}(\eta_n^N)\right]$$

By Theorem 10.6.1, for any fixed time horizon $n \geq 0$, the sequence of random fields $(V_n^{N,select}, V_{n+1}^{N,mut})$ converges in law, in the sense of convergence of finite dimensional distributions, as the number of particles N tends to infinity, to a sequence of independent, Gaussian, and centered random fields $(V_n^{select}, V_{n+1}^{mut})$, with, for any $n \geq 0$, and any $(f_n, f_{n+1}) \in (\mathcal{B}_b(E_n) \times \mathcal{B}_b(E_{n+1}))$

$$\begin{aligned}
\mathbb{E}(V_n^{select}(f_n)^2) &= \eta_n S_{n,\eta_n}([f_n - S_{n,\eta_n}(f_n)]^2) \\
\mathbb{E}(V_{n+1}^{mut}(f_{n+1})^2) &= \eta_n M_{n+1}([f_{n+1} - M_{n+1}(f_{n+1})]^2)
\end{aligned}$$

In this notation, the random fields defined in (9.29) are given by

$$V_{n+1}^N = V_{n+1}^{N,mut} + V_n^{N,select} M_{n+1}$$

We conclude that the sequence of random fields V_n^N converges in law, in the sense of convergence of finite dimensional distributions, as the number of particles N tends to infinity, to a sequence of independent, Gaussian, and centered random fields $V_n = V_n^{mut} + V_{n-1}^{select} M_n$, with the convention $V_0 = V_0^{mut}$, for $n = 0$. By Theorem 10.6.2, we also find that the sequence of random fields

$$W_n^{\eta,N} = \sqrt{N}\ \left[\eta_n^N - \eta_n\right]$$

converges in law, as the number of particles N tends to infinity, to Gaussian and centered random fields

$$W_n^\eta = \sum_{p=0}^n V_p d_{p,n} = \sum_{0 \leq p \leq n} V_p^{mut} d_{p,n} + \sum_{0 \leq p < n} V_p^{select} M_{p+1} d_{p+1,n}$$

In the above display, $d_{p,n}$ stands for the semigroup associated with the operator $d_n = d_{\eta_{n-1}} \Phi_n$

Chapter 11

Empirical processes

11.1 Description of the models

We let $(\mu^i)_{i \geq 1}$ be a sequence of probability measures on a given measurable state space $(E, \mathcal{E})$. During the further development of this chapter, we fix an integer $N \geq 1$. To clarify the presentation, with a slight abuse of notation denoted respectively, by

$$m(X) = \frac{1}{N} \sum_{i=1}^{N} \delta_{X^i} \quad \text{and} \quad \mu = \frac{1}{N} \sum_{i=1}^{N} \mu^i$$

the N-empirical measure is associated with a collection of independent random variables $X = (X^i)_{i \geq 1}$, with respective distributions $(\mu^i)_{i \geq 1}$, and the N-averaged measure is associated with the sequence of measures $(\mu^i)_{i \geq 1}$.

Definition 11.1.1 *We consider the empirical random field sequences*

$$V(X) = \sqrt{N} \ (m(X) - \mu)$$

on $\mathcal{B}_b(E)$, and the co-variance functional σ^2 given by

$$\sigma(f)^2 := \mathbb{E}\left(V(X)(f)^2\right) = \frac{1}{N} \sum_{i=1}^{N} \mu^i([f - \mu^i(f)]^2) \qquad (11.1)$$

When N is large, the empirical random measures $m(X)$ are very close to the average measure μ, in the sense that

$$\lim_{N \to \infty} [m(X) - \mu](f) = 0 \qquad \mathbb{P} - a.e. \qquad (11.2)$$

for any $f \in \mathrm{Osc}(E)$. One classical way to prove this result is to combine the Borel-Cantelli lemma with the fourth moment estimate

$$\mathbb{E}\left(V(X)(f)^4\right) \ \leq \ \sigma(f)^2 \left(\frac{1}{N} \ \mathrm{osc}(f)^2 + 6 \ \sigma(f)^2\right)$$

The rather crude estimate given above is a direct consequence of the decomposition

$$\mathbb{E}\left(V(X)(f)^4\right) \;=\; \frac{1}{N}\,\frac{1}{N}\sum_{i=1}^{N}\mu_i\left((f-\mu_i(f))^4\right)$$

$$+6\,\frac{1}{N^2}\sum_{i\neq j}\mu_i\left((f-\mu_i(f))^2\right)\mu_j\left((f-\mu_j(f))^2\right)$$

The r.h.s. term is upper bounded by $6\sigma(f)^4$. We end the proof using the fact that $(f-\mu_i(f))^4 \le (f-\mu_i(f))^2$, as soon as $\mathrm{osc}(f)\le 1$.

From the above discussion, the random field $V(X)$ represents the fluctuations of $m(X)$ around the "limiting" average measure μ. The following formula illustrates this assertion

$$m(X) = \mu + \frac{1}{\sqrt{N}}\,V(X)$$

Before entering into the description of more mathematical objects, we illustrate these abstract empirical processes with the local sampling random fields models associated with a mean field particle model discussed in Section 4.4.4.

To be more precise, given the information on the N-particle model at time $(n-1)$, the sequence of random variables ξ_n^i is independent random sequences with a distribution that depends on the current state ξ_{n-1}^i. That is, at any given fixed time horizon n and given $\mathcal{G}_{n-1}^N$, we have

$$X^i = \xi_n^i \in E = E_n \quad \text{and} \quad \mu^i(dx) := K_{n,\eta_{n-1}^N}(\xi_{n-1}^i, dx) \tag{11.3}$$

In this case, we find that $m(X) = \eta_n^N$ and $V(X) = V_n^N$, with the variance parameter

$$\sigma(f)^2 \;=\; \mathbb{E}\left(V_{n+1}^N(f)^2 \mid \mathcal{G}_n^N\right) = \eta_n^N\left[K_{n+1,\eta_n^N}\left(f - K_{n+1,\eta_n^N}(f)\right)^2\right]$$

We end this section with some notation used further on in this chapter.

Definition 11.1.2 *Let $\mathcal{F}$ be a given collection of measurable functions $f :$ $E \to \mathbb{R}$ such that $\|f\| \le 1$. We associate with $\mathcal{F}$ the Zolotarev seminorm on $\mathcal{P}(E)$ defined by*

$$\|\mu - \nu\|_{\mathcal{F}} = \sup\{|\mu(f) - \nu(f)|; \; f \in \mathcal{F}\},$$

A detailed discussion on these seminorms is provided for instance in [481]).

No generality is lost and much convenience is gained by supposing that the unit and the null functions $f = 1$ and $f = 0 \in \mathcal{F}$. Furthermore, to avoid some unnecessary technical measurability questions, we shall also suppose that $\mathcal{F}$ is separable in the sense that it contains a countable and dense subset.

We measure the size of a given class $\mathcal{F}$ in terms of the covering numbers $N(\epsilon, \mathcal{F}, \mathbb{L}_2(\mu))$ defined as the minimal number of $\mathbb{L}_2(\mu)$-balls of radius $\epsilon > 0$ needed to cover $\mathcal{F}$. We shall also use the following uniform covering numbers and entropies.

Definition 11.1.3 *By* $\mathcal{N}(\epsilon, \mathcal{F})$, $\epsilon > 0$, *and by* $I(\mathcal{F})$ *we denote the uniform covering numbers and entropy integral given by*

$$\mathcal{N}(\epsilon, \mathcal{F}) = \sup\{N(\epsilon, \mathcal{F}, \mathbb{L}_2(\eta)); \eta \in \mathcal{P}(E)\}$$

$$I(\mathcal{F}) = \int_0^2 \sqrt{\log(1 + \mathcal{N}(\epsilon, \mathcal{F}))}\, d\epsilon$$

Definition 11.1.4 *We let* $\pi_\psi[Y]$ *be the Orlicz norm of an* $\mathbb{R}$-*valued random variable* Y *associated with the convex function* $\psi(u) = e^{u^2} - 1$, *and defined by*

$$\pi_\psi(Y) = \inf\{a \in (0, \infty) : \mathbb{E}(\psi(|Y|/a)) \le 1\} \tag{11.4}$$

with the convention $\inf_\emptyset = \infty$. *Notice that* $\pi_\psi(Y) \le c \iff \mathbb{E}(\psi(Y/c)) \le 1$.

For instance, the Orlicz norm of a Gaussian and centered random variable U, s.t. $E(U^2) = 1$, is given by

$$\pi_\psi(U) = \sqrt{8/3} \tag{11.5}$$

The concentration inequalities presented in this chapter are expressed in terms of the inverse of a couple of functions defined below.

Definition 11.1.5 *We let* (ϵ_0, ϵ_1) *be the functions on* $\mathbb{R}_+$ *defined by*

$$\epsilon_0(\lambda) = \frac{1}{2}(\lambda - \log(1 + \lambda)) \quad and \quad \epsilon_1(\lambda) = (1 + \lambda)\log(1 + \lambda) - \lambda$$

Rather crude estimates can be derived using the following upper bounds

$$\epsilon_0^{-1}(x) \le 2(x + \sqrt{x}) \quad and \quad \epsilon_1^{-1}(x) \le \frac{x}{3} + \sqrt{2x}$$

A proof of these elementary inequalities and refined estimates can be found in the article [166].

11.2 Nonasymptotic theorems

In this opening section, we state and comment on the main theorems developed in this chapter. Section 11.2.1 is concerned with finite marginal models, and Section 11.2.2 provides extensions at the level of the empirical process. These sections also contain some perturbation theorems that apply to nonlinear functional or empirical processes. We already mentioned that these inequalities are more crude, with greater constants than the ones for marginal models.

11.2.1 Finite marginal models

The main result of this section is a quantitative concentration inequality for the finite marginal models $f \mapsto V(X)(f)$. In the next theorem, we provide Kintchine's type mean error bound, and related Orlicz norm estimates. The proofs of these results are housed in Section 11.4. The last quantitative concentration inequality is a direct consequence of (11.9), and it is proved in Remark 11.6.7.

Theorem 11.2.1 *For any integer $m \geq 1$ and any measurable function f we have the $\mathbb{L}_m$-mean error estimates*

$$\mathbb{E}(|V(X)(f)|^m)^{1/m} \leq b(m)\ c_m(f) \tag{11.6}$$

with the constant

$$c_m(f) := \operatorname{osc}(f) \wedge \left[2\ \mu(|f|^{m'})^{1/m'}\right] \wedge \left(6b(m)\ \max\left(\sqrt{2}\sigma(f), \left[\frac{2\sigma(f)^2}{N^{\frac{m'}{2}-1}}\right]^{1/m'}\right)\right) \tag{11.7}$$

In the above display, m' stands for the smallest even integer $m' \geq m$, and $b(m)$ the collection of constants defined in (0.6).

In particular, for any $f \in \operatorname{Osc}(E)$, we have

$$\pi_\psi(V(X)(f)) \leq \sqrt{8/3} \tag{11.8}$$

and for any N s.t. $2\sigma^2(f)N \geq 1$ we have

$$\mathbb{E}\left(|V(X)(f)|^m\right)^{\frac{1}{m}} \leq 6\sqrt{2}\ b(m)^2\sigma(f) \tag{11.9}$$

In addition, the probability of the event $|V(X)(f)| \leq 6\sqrt{2}\ \sigma(f)\left[1 + \epsilon_0^{-1}(x)\right)$ is greater than $1 - e^{-x}$, for any $x \geq 0$.

In Section 11.3.2 dedicated to concentration properties of random variables Y with finite Orlicz norms $\pi_\psi(Y) < \infty$, we shall prove that the probability of the event

$$Y \leq \pi_\psi(Y)\ \sqrt{y + \log 2}$$

is greater than $1 - e^{-y}$, for any $y \geq 0$ (cf. Lemma 11.3.2). This implies that the probability of the events

$$|V(X)(f)| \leq 2\sqrt{2(x + \log 2)/3}$$

is greater than $1 - e^{-x}$, for any $x \geq 0$. Our next objective is to derive concentration inequalities for nonlinear functionals of the empirical random field $V(X)$. To introduce with some conciseness these objects, we need some definitions.

Definition 11.2.2 *For any measure ν, and any sequence of measurable functions $f = (f_1, \ldots, f_d)$, we write $\nu(f) := [\nu(f_1), \ldots, \nu(f_d)]$. We associate with* the second order smooth function F on $\mathbb{R}^d$, for some $d \geq 1$, the random functionals defined by

$$f = (f_i)_{1 \leq i \leq d} \in \mathrm{Osc}(E)^d$$

$$\mapsto F(m(X)(f)) = F(m(X)(f_1), \ldots, m(X)(f_d)) \in \mathbb{R}$$

(11.10)

Definition 11.2.3 *Given a probability measure ν, and a collection of functions $(f_i)_{1 \leq i \leq d} \in \mathrm{Osc}(E)^d$, we set*

$$D_\nu(F)(f) = \nabla F(\nu(f)) \ f^\top$$

(11.11)

We also introduce the following constants

$$\|\nabla F(\nu(f))\|_1 \quad := \quad \sum_{i=1}^{d} \left| \frac{\partial F}{\partial u^i}(\nu(f)) \right|$$

$$\|\nabla^2 F_f\|_1 \quad := \quad \sum_{i,j=1}^{d} \sup \left| \frac{\partial^2 F}{\partial u^i \partial u^j}(\nu(f)) \right|$$

(11.12)

In the r.h.s. display, the supremum is taken over all probability measures $\nu \in \mathcal{P}(E)$.

Notice that

$$\mathrm{osc}\,(D_\nu(F)(f)) \leq \|\nabla F(\nu(f))\|_1$$

(11.13)

In the next proposition we present some bias and $\mathbb{L}_m$-mean error estimates. The proof of this proposition is provided on page 348.

Proposition 11.2.4 *For any smooth functional $F(m(X)(f))$ of the form (11.10) we have*

$$|\mathbb{E}\,(F(m(X)(f))) - F(\mu(f))| \leq \frac{1}{2N}\ \sigma^2(f)\ \|\nabla^2 F_f\|_1$$

with the uniform variance parameter

$$\sigma^2(f) := \sup_{1 \leq i \leq d} \sigma^2(f_i)$$

(11.14)

In addition, for $m \geq 1$, we have the $\mathbb{L}_m$-mean error bound

$$\mathbb{E}\,(|[F(m(X)(f)) - F(\mu(f))]|^m)^{1/m}$$

$$\leq \frac{1}{\sqrt{N}}\ b(m)\ \|\nabla F(\mu(f))\|_1 + \frac{1}{2N}\ b(2m)^2\ \|\nabla^2 F_f\|_1$$

with the collection of constants $b(m)$ defined in (0.6).

The next theorem extends the exponential inequalities stated in Theorem 11.2.1 to this class of nonlinear functionals. It also provides more precise concentration properties in terms of the variance functional σ defined in (11.1). The detailed proof is provided in Section 11.7.

Theorem 11.2.5 *Let F be a second order smooth function on $\mathbb{R}^d$, for some $d \geq 1$. For any collection of functions $(f_i)_{1 \leq i \leq d} \in \mathrm{Osc}(E)^d$, and any $N \geq 1$, the probability of the events*

$$[F(m(X)(f)) - F(\mu(f))]$$

$$\leq \frac{1}{2N} \left\| \nabla^2 F_f \right\|_1 \left[3/2 + \epsilon_0^{-1}(x) \right]$$

$$+ \left\| \nabla F(\mu(f)) \right\|_1^{-1} \sigma^2(D_\mu(F)(f)) \; \epsilon_1^{-1} \left(\frac{x \left\| \nabla F(\mu(f)) \right\|_1^2}{N \sigma^2(D_\mu(F)(f))} \right)$$

is greater than $1 - e^{-x}$, for any $x \geq 0$. In the above display, $D_\mu(F)(f)$ stands for the first order function defined in (11.11).

11.2.2   Empirical processes

The objective of this section is to extend the quantitative concentration theorems, Theorem 11.2.1 and Theorem 11.2.5, at the level of the empirical process associated with a class of function $\mathcal{F}$. These processes are given by the mapping $f \in \mathcal{F} \mapsto V(X)(f)$. Our main result in this direction is the following theorem, whose proof is postponed to Section 11.5.

Theorem 11.2.6 *For any class of functions $\mathcal{F}$, with $I(\mathcal{F}) < \infty$, we have*

$$\pi_\psi \left(\left\| V(X) \right\|_{\mathcal{F}} \right) \leq 12^2 \int_0^2 \sqrt{\log \left(8 + \mathcal{N}(\mathcal{F}, \epsilon)^2 \right)} \; d\epsilon$$

Remark 11.2.7 *Using the fact that $\log \left(8 + x^2 \right) \leq 4 \log x$, for any $x \geq 2$, we obtain the rather crude estimate*

$$\int_0^2 \sqrt{\log \left(8 + \mathcal{N}(\mathcal{F}, \epsilon)^2 \right)} \; d\epsilon \leq 2 \int_0^2 \sqrt{\log \mathcal{N}(\mathcal{F}, \epsilon)} \; d\epsilon$$

Various examples of classes of functions with finite covering and entropy integral are given in the book of Van der Vaart and Wellner [559] (see, for instance p. 86, p. 129, p. 135, and exercise 4 on p. 150 and p. 155). The estimation of the quantities introduced above often depends on several deep results on combinatorics that are not discussed here.

To illustrate these mathematical objects, we mention that, for the set of indicator functions

$$\mathcal{F} = \left\{ 1_{\prod_{i=1}^d (-\infty, x_i]} \; ; \; (x_i)_{1 \leq i \leq d} \in \mathbb{R}^d \right\} \tag{11.15}$$

of cells in $E = \mathbb{R}^d$, we have $\mathcal{N}(\epsilon, \mathcal{F}) \leq c\,(d+1)(4e)^{d+1}\,\epsilon^{-2d}$, for some universal constant $c < \infty$. This implies that

$$\sqrt{\log \mathcal{N}(\epsilon, \mathcal{F})} \leq \sqrt{\log\left[c(d+1)(4e)^{d+1}\right]} + \sqrt{(2d)}\,\sqrt{\log\left(1/\epsilon\right)}$$

An elementary calculation gives

$$\int_0^2 \sqrt{\log\left(1/\epsilon\right)} \leq 2 \int_0^\infty x^2 e^{-x^2}\, dx = \sqrt{\pi/4} \leq 1$$

from which we conclude that

$$\int_0^2 \sqrt{\log \mathcal{N}(\mathcal{F}, \epsilon)}\, d\epsilon \leq 2\sqrt{\log\left[c(d+1)(4e)^{d+1}\right]} + \sqrt{(2d)} \leq c'\sqrt{d} \qquad (11.16)$$

for some universal constant $c < \infty$. For $d = 1$, we also have that $\mathcal{N}(\epsilon, \mathcal{F}) \leq 2/\epsilon^2$ (cf. p. 129 in [559]) and therefore

$$\int_0^2 \sqrt{\log \mathcal{N}(\mathcal{F}, \epsilon)}\, d\epsilon \leq 3\sqrt{2}$$

Remark 11.2.8 *In this chapter, we have assumed that the class of functions $\mathcal{F}$ is such that $\sup_{f \in \mathcal{F}} \|f\| \leq 1$. When $\sup_{f \in \mathcal{F}} \|f\| \leq c_{\mathcal{F}}$, for some finite constant $c_{\mathcal{F}}$, using Theorem 11.2.6, it is also readily checked that*

$$\pi_\psi\left(\|V(X)\|_{\mathcal{F}}\right) \leq 12^2 \int_0^{2c_{\mathcal{F}}} \sqrt{\log\left(8 + \mathcal{N}(\mathcal{F}, \epsilon)^2\right)}\, d\epsilon \qquad (11.17)$$

We mention that the uniform entropy condition $I(\mathcal{F}) < \infty$ is required in Glivenko-Cantelli and Donsker theorems for empirical processes associated with nonnecessarily independent random sequences [160].

Arguing as above, we prove that the probability of the events

$$\|V(X)\|_{\mathcal{F}} \leq I_1(\mathcal{F})\,\sqrt{x + \log 2}$$

is greater than $1 - e^{-x}$, for any $x \geq 0$, with some constant

$$I_1(\mathcal{F}) \leq 12^2 \int_0^2 \sqrt{\log\left(8 + \mathcal{N}(\mathcal{F}, \epsilon)^2\right)}\, d\epsilon$$

As for marginal models 11.10, our next objective is to extend Theorem 11.2.5 to empirical processes associated with some classes of functions.

Definition 11.2.9 *We consider the empirical processes $f \in \mathcal{F}_i \mapsto m(X)(f)$, associated with d classes of functions $\mathcal{F}_i$, $1 \leq i \leq d$, defined in Section 11.1. We further assume that $\|f_i\| \vee \mathrm{osc}(f_i) \leq 1$, for any $f_i \in \mathcal{F}_i$, and we set*

$$\mathcal{F} := \prod_{1 \leq i \leq d} \mathcal{F}_i \quad \text{and} \quad \pi_\psi(\|V(X)\|_{\mathcal{F}}) := \sup_{1 \leq i \leq d} \pi_\psi(\|V(X)\|_{\mathcal{F}_i})$$

Using Theorem 11.2.6, we mention that

$$\pi_\psi(\|V(X)\|_{\mathcal{F}}) \leq 12^2 \int_0^2 \sqrt{\log\left(8 + \mathcal{N}(\mathcal{F}, \epsilon)^2\right)} \, d\epsilon$$

with $\mathcal{N}(\mathcal{F}, \epsilon) := \sup_{1 \leq i \leq d} \mathcal{N}(\mathcal{F}_i, \epsilon)$.

Definition 11.2.10 *For any second order smooth function F on $\mathbb{R}^d$, for some $d \geq 1$, we set*

$$\|\nabla F_\mu\|_\infty := \sup \left| \frac{\partial F}{\partial u^i}(\mu(f)) \right| \quad \text{and} \quad \|\nabla^2 F\|_\infty = \sup \left| \frac{\partial^2 F}{\partial u^i \partial u^j}(\nu(f)) \right|$$

The supremum in the l.h.s. is taken over all $1 \leq i \leq d$ and all $f \in \mathcal{F}$; and the supremum in the r.h.s. is taken over all $1 \leq i, j \leq d$, $\nu \in \mathcal{P}(E)$, and all $f \in \mathcal{F}$.

We are now in position to state the final main result of this section. The proof of the next theorem is housed in the end of Section 11.7.

Theorem 11.2.11 *Let F be a second order smooth function on $\mathbb{R}^d$, for some $d \geq 1$. For any classes of functions $\mathcal{F}_i$, $1 \leq i \leq d$, and for any $x \geq 0$, the probability of the following event is greater than $1 - e^{-x}$*

$$\sup_{f \in \mathcal{F}} |F(m(X)(f)) - F(\mu(f))|$$

$$\leq \frac{d}{\sqrt{N}} \, \pi_\psi\left(\|V(X)\|_{\mathcal{F}}\right) \, \|\nabla F_\mu\|_\infty \left(1 + 2\sqrt{x}\right)$$

$$+ \frac{1}{2N} \|\nabla^2 F\|_\infty \left(d \, \pi_\psi(\|V(X)\|_{\mathcal{F}})\right)^2 \left(1 + \epsilon_0^{-1}\left(\frac{x}{2}\right)\right)$$

11.3   A reminder on Orlicz's norms

In this section, we have collected some important properties of Orlicz's norms. The first section, Section 11.3.1, is concerned with rather elementary comparison properties. In Section 11.3.2, we present a natural way to obtain Laplace estimates, and related concentration inequalities, using simple Orlicz's norm upper bounds.

11.3.1   Comparison properties

This short section is mainly concerned with the proof of the following three comparison properties.

Lemma 11.3.1 *For any nonnegative variables (Y_1, Y_2) we have*

$$Y_1 \leq Y_2 \implies \pi_\psi(Y_1) \leq \pi_\psi(Y_2)$$

as well as $\left(\forall m \geq 0 \quad \mathbb{E}\left(Y_1^{2m}\right) \leq \mathbb{E}\left(Y_2^{2m}\right)\right) \Rightarrow \pi_\psi(Y_1) \leq \pi_\psi(Y_2)$ (11.18)

In addition, for any pair of independent random variables (X, Y) on some measurable state space, and any measurable function f, we have

$$(\pi_\psi(f(x, Y)) \leq c \quad \text{for } \mathbb{P}\text{-a.e. x}) \Longrightarrow \pi_\psi(f(X, Y)) \leq c \qquad (11.19)$$

Proof:
The first assertion is immediate, and the second assertion comes from the fact that

$$\mathbb{E}\left(\exp\left(\frac{Y_1}{\pi_\psi(Y_2)}\right)^2 - 1\right) \leq \sum_{m \geq 1} \frac{1}{m!} \frac{\mathbb{E}(Y_2^{2m})}{\pi_\psi(Y_2)^{2m}} = \mathbb{E}\left(\psi\left(\frac{Y_2}{\pi_\psi(Y_2)}\right)\right) \leq 1$$

The last assertion comes from the fact that

$$\mathbb{E}\left(\mathbb{E}\left(\psi(f(X, Y)/c) \,|\, X\right)\right) \leq 1 \Rightarrow \pi_\psi(f(X, Y)) \leq c$$

This ends the proof of the lemma. ∎

11.3.2 Concentration properties

The following lemma provides a simple way to transfer a control on Orlicz's norm into moment or Laplace estimates, which in turn can be used to derive quantitative concentration inequalities

Lemma 11.3.2 *For any nonnegative random variable Y, and any integer $m \geq 0$, we have*

$$\mathbb{E}\left(Y^{2m}\right) \leq m! \, \pi_\psi(Y)^{2m} \quad \text{and} \quad \mathbb{E}\left(Y^{2m+1}\right) \leq (m+1)! \, \pi_\psi(Y)^{2m+1} \quad (11.20)$$

In addition, for any $t \geq 0$ we have the Laplace estimates

$$\mathbb{E}\left(e^{tY}\right) \leq \min\left(2 \, e^{\frac{1}{4}(t\pi_\psi(Y))^2} \, , \, (1 + t\pi_\psi(Y)) \, e^{(t\pi_\psi(Y))^2}\right)$$

In particular, for any $x \geq 0$ the probability of the following event is greater than $1 - e^{-x}$

$$Y \leq \pi_\psi(Y) \, \sqrt{x + \log 2} \qquad (11.21)$$

Remark 11.3.3 *For a Gaussian and centered random variable Y, s.t. $E(Y^2) = 1$, we recall that $\pi_\psi(Y) = \sqrt{8/3}$. In this situation, letting $y = \sqrt{8(x + \log 2)/3}$ in (11.21), we find that $\mathbb{P}(|Y| \geq y) \leq 2 \, e^{-\frac{1}{2} \frac{3}{4} y^2}$. Working directly with the Laplace Gaussian function $\mathbb{E}\left(e^{tY}\right) = e^{t^2/2}$, we remove the factor $3/4$. In this sense, we lose a factor $3/4$ using the Orlicz's concentration* *property (11.21). In this situation, the l.h.s. moment estimate in (11.20) takes the form*

$$b(2m)^{2m} = \frac{(2m)!}{m!} \, 2^{-m} \le m! \, (8/3)^m \qquad (11.22)$$

while using Stirling's approximation of the factorials we obtain the estimate

$$\frac{(2m)!}{m!^2} \simeq \sqrt{2/m} \, 4^m \, (\le (8/3)^m)$$

Remark 11.3.4 *Given a sequence of independent Gaussian and centered random variables Y_i, s.t. $E(Y_i^2) = 1$, for $i \ge 1$, and any sequence of nonnegative numbers a_i, we have*

$$\pi_\psi \left(\sum_{i=1}^n a_i Y_i \right) = \sqrt{8/3} \, \sqrt{\sum_{i=1}^n a_i^2} := \sqrt{8/3} \, \|a\|_2$$

while

$$\sum_{i=1}^n a_i \pi_\psi (Y_i) = \sqrt{8/3} \sum_{i=1}^n a_i := \sqrt{8/3} \, \|a\|_1$$

Notice that $\|a\|_2 \le \|a\|_1 \le \sqrt{n} \, \|a\|_2$. When the coefficients a_i are almost equal, we can lose a factor $\sqrt{n}$ using the triangle inequality, instead of estimating directly with the Orlicz norm of the Gaussian mixture. In this sense, it is always preferable to avoid the use of the triangle inequality, and to estimate directly the Orlicz norms of linear combinations of "almost Gaussian" random variables.

Now, we come to the proof of the lemma.
Proof of Lemma 11.3.2:
For any $m \ge 1$, we have

$$x^{2m} \le m! \sum_{n \ge 1} \frac{x^{2n}}{n!} = m! \, \psi(x)$$

$$\Downarrow$$

$$\mathbb{E} \left(\left[\frac{Y}{\pi_\psi(Y)} \right]^{2m} \right) \le m! \, \mathbb{E} \left(\psi \left(\frac{Y}{\pi_\psi(Y)} \right) \right) \le m!$$

For odd integers, we simply use Cauchy-Schwartz's inequality to check that

$$\mathbb{E} \left(Y^{2m+1} \right)^2 \le \mathbb{E} \left(Y^{2m} \right) \mathbb{E} \left(Y^{2(m+1)} \right) \le (m+1)!^2 \, \pi_\psi(Y)^{2(2m+1)}$$

This ends the proof of the first assertion.

Recalling that $(2m)! \geq m!^2$, we find that

$$
\begin{aligned}
\mathbb{E}\left(e^{tY}\right) &= \sum_{m \geq 0} \frac{t^{2m}}{(2m)!} \mathbb{E}\left(Y^{2m}\right) + \sum_{m \geq 0} \frac{t^{2m+1}}{(2m+1)!} \mathbb{E}\left(Y^{2m+1}\right) \\
&\leq \sum_{m \geq 0} \frac{t^{2m}}{m!} \pi_\psi(Y)^{2m} + \sum_{m \geq 0} \frac{t^{2m+1}}{m!} \pi_\psi(Y)^{(2m+1)} \\
&= (1 + t\pi_\psi(Y)) \ \exp\left(t\pi_\psi(Y)\right)^2
\end{aligned}
$$

On the other hand, using the estimate

$$
tY = \left(\frac{t\pi_\psi(Y)}{\sqrt{2}}\right) \left(\frac{\sqrt{2}\,Y}{\pi_\psi(Y)}\right) \leq \frac{(t\pi_\psi(Y))^2}{4} + \left(\frac{Y}{\pi_\psi(Y)}\right)^2
$$

we prove that

$$
\mathbb{E}\left(e^{tY}\right) \leq 2 \ \exp\left(\frac{(t\pi_\psi(Y))^2}{4}\right)
$$

The end of the proof of the Laplace estimates is now completed. To prove the last assertion, we use the fact that for any $y \geq 0$

$$
\begin{aligned}
\mathbb{P}(Y \geq y) &\leq 2 \ \exp\left(-\sup_{t \geq 0}\left(ty - (t\pi_\psi(Y))^2/4\right)\right) \\
&= 2 \ \exp\left[-(y/\pi_\psi(Y))^2\right]
\end{aligned}
$$

This implies that

$$
\mathbb{P}\left(Y \geq \pi_\psi(Y)\sqrt{x + \log 2}\right) \leq 2\exp\left[-(x + \log 2)\right] = e^{-x}
$$

This ends the proof of the lemma. ∎

11.3.3 Maximal inequalities

Let us now put together the Orlicz's norm properties derived in Section 11.3 to establish a series of more or less well known maximal inequalities. More general results can be found in the books [480, 559] or in the lecture notes [277].

We emphasize that in the literature on empirical processes, maximal inequalities are often presented in terms of universal constant c without further information on their magnitude. In the present section, we shall try to estimate some of these universal constants explicitly. To begin with, we consider a couple of maximal inequalities over finite sets.

Lemma 11.3.5 *For any finite collection of nonnegative random variables* $(Y_i)_{i \in I}$, *and any collection of nonnegative numbers* $(a_i)_{i \in I}$, *we have*

$$\sup_{i \in I} \mathbb{E}(\psi(Y_i/a_i)) \leq 1 \Rightarrow \mathbb{E}\left(\max_{i \in I} Y_i\right) \leq \psi^{-1}(|I|) \times \max_{i \in I} a_i$$

Proof:

We check this claim using the following estimates

$$\psi\left(\frac{\mathbb{E}\left(\max_{i \in I} Y_i\right)}{\max_{i \in I} a_i}\right) \leq \psi\left(\mathbb{E}\left(\max_{i \in I}(Y_i/a_i)\right)\right)$$

$$\leq \mathbb{E}\left(\psi\left(\max_{i \in I}(Y_i/a_i)\right)\right)$$

$$\leq \mathbb{E}\left(\sum_{i \in I} \psi(Y_i/a_i)\right) \leq |I|$$

This ends the proof of the lemma. ∎

Working a little harder, we prove the following lemma.

Lemma 11.3.6 *For any finite collection of nonnegative random variables* $(Y_i)_{i \in I}$, *we have*

$$\pi_\psi\left(\max_{i \in I} Y_i\right) \leq \sqrt{6 \log(8 + |I|)} \ \max_{i \in I} \pi_\psi(Y_i)$$

Proof:

Without loss of generality, we assume that $\max_{i \in I} \pi_\psi(Y_i) \leq 1$, and $I = \{1, \ldots, |I|\}$. In this situation, it suffices to check that

$$\pi_\psi\left(\frac{\max_{1 \leq i \leq |I|} Y_i}{\sqrt{6 \log(8 + |I|)}}\right) \leq \pi_\psi\left(\max_{1 \leq i \leq |I|} \frac{Y_i}{\sqrt{6 \log(8 + i)}}\right) \leq 1$$

Firstly, we notice that for any $i \geq 1$ and $x \geq 3/2$ we have

$$\frac{1}{\log(8 + i)} + \frac{1}{\log x} \leq \frac{1}{\log 9} + \frac{1}{\log(3/2)} \leq 3$$

and therefore

$$3 \log(8 + i) \log(x) \geq \log(x(8 + i))$$

We can check the first estimate using the fact that

$$\log(3) \leq 5 \log(3/2) \Rightarrow \log(3) + \log(3/2) \leq 6 \log(3/2) \leq 3 \log(3/2) \log(9)$$

Using these observations, we have

$$\mathbb{P}\left(\max_{1\leq i\leq |I|}\left(\frac{Y_i}{\sqrt{6\log(8+i)}}\right)^2 > \log x\right)$$

$$= \mathbb{P}\left(\max_{1\leq i\leq |I|}\left(\frac{Y_i}{\sqrt{6\log(x)\log(8+i)}}\right)^2 > 1\right)$$

$$\leq \mathbb{P}\left(\max_{1\leq i\leq |I|}\frac{Y_i}{\sqrt{2\log(x(8+i))}} > 1\right)$$

$$\leq \sum_{i=1}^{|I|}\mathbb{P}\left(Y_i > \sqrt{2\log(x(8+i))}\right) \leq \sum_{i=1}^{|I|} e^{-2\log(x(8+i))}\,\mathbb{E}\left(e^{Y_i^2}\right)$$

This implies that

$$\mathbb{P}\left(\max_{1\leq i\leq |I|}\left(\frac{Y_i}{\sqrt{6\log(8+i)}}\right)^2 > \log x\right)$$

$$\leq \frac{2}{x^2}\sum_{i=1}^{|I|}\frac{1}{(8+i)^2} \leq \frac{2}{x^2}\int_8^\infty \frac{1}{u^2}\,du = \frac{1}{(2x)^2}$$

If we set

$$Z_I := \exp\left\{\left(\max_{1\leq i\leq |I|}\frac{Y_i}{\sqrt{6\log(8+i)}}\right)^2\right\}$$

then we have

$$\mathbb{E}(Z_I) = \int_0^\infty \mathbb{P}(Z_I > x)\,dx$$

$$\leq \frac{3}{2} + \int_{\frac{3}{2}}^\infty \frac{1}{(2x)^2}\,dx = \frac{3}{2}\left(1 + \frac{1}{4}\right) = \frac{15}{8} \leq 2$$

and therefore

$$\pi_\psi\left(\max_{1\leq i\leq |I|}\left(\frac{Y_i}{\sqrt{6\log(8+i)}}\right)\right) \leq 1$$

This ends the proof of the lemma. ∎

The following technical lemma is pivotal in the analysis of maximal inequalities for sequences of random variables indexed by infinite but separable subsets equipped with a pseudo-metric, under some Lipschitz regularity conditions w.r.t. the Orlicz's norm.

Lemma 11.3.7 *We assume that the index set (I, d) is a separable and totally bounded pseudo-metric space, with finite diameter*

$$d(I) := \sup_{(i,i) \in I^2} d(i, j) < \infty$$

We let $(Y_i)_{i \in I}$ be a separable and $\mathbb{R}$-valued stochastic process indexed by I and such that

$$\pi_\psi(Y_i - Y_j) \le c \, d(i, j)$$

for some finite constant $c < \infty$. We also assume that $Y_{i_0} = 0$, for some $i_0 \in I$. Then, we have

$$\pi_\psi \left(\sup_{i \in I} Y_i \right) \le 12 \, c \int_0^{d(I)} \sqrt{6 \log \left(8 + \mathcal{N}(I, d, \epsilon)^2 \right)} \, d\epsilon$$

Proof:
Replacing Y_i by $Y_i/d(I)$, and d by $d/d(I)$, there is no loss of generality to assume that $d(I) \le 1$. In the same way, replacing Y_i by Y_i/c, we can also assume that $c \le 1$. For a given finite subset $J \subset I$, with $i_0 \in J$, we let $J_k = \{i_1^k, \ldots, i_{n_k}^k\} \subset J$ be the centers of $n_k = \mathcal{N}(J, d, 2^{-k})$ balls of radius at most 2^{-k} covering J. For $k = 0$, we set $J_0 = \{i_0\}$. We also consider the mapping $\theta_k : i \in J \mapsto \theta_k(i) \in J_k$ s.t.

$$\sup_{i \in J} d(\theta_k(i), i) \le 2^{-k}$$

The set J being finite, there exist some sufficient integer $k_J^\star$ s.t. $d(\theta_k(i), i) = 0$, for any $k \ge k_J^\star$; and therefore $Y_i = Y_{\theta_k(i)}$, for any $i \in J$, and any $k \ge k_J^\star$. This implies that

$$Y_i = \sum_{k=1}^{k_J^\star} \left[Y_{\theta_k(i)} - Y_{\theta_{k-1}(i)} \right]$$

We also notice that

$$d(\theta_k(i), \theta_{k-1}(i)) \le d(\theta_k(i), i) + d(i, \theta_{k-1}(i)) \le 2^{-k} + 2^{-(k-1)} = 3 \times 2^{-k}$$

and

$$\sup_{(i,j) \in (J_k \times J_{k-1}) \, : \, d(i,j) \le 3 \times 2^{-k}} \pi_\psi (Y_i - Y_j) \le 3 \times 2^{-k}$$

Using Lemma 11.3.6 we prove that

$$\pi_\psi \left(\sup_{i \in J} Y_i \right) \le \sum_{k=1}^{k_J^\star} \pi_\psi \left(\sup_{i \in J} \left[Y_{\theta_k(i)} - Y_{\theta_{k-1}(i)} \right] \right)$$

$$\le 3 \sum_{k=1}^{k_J^\star} \sqrt{6 \log \left(8 + \mathcal{N}(J, d, 2^{-k})^2 \right)} \, 2^{-k}$$

On the other hand, we have $2 \left(2^{-k} - 2^{-(k+1)} \right) = 2^{-k}$, and

$$\sqrt{6 \log \left(8 + \mathcal{N}(J, d, 2^{-k})^2 \right)} \ 2^{-k} \leq 2 \int_{2^{-(k+1)}}^{2^{-k}} \sqrt{6 \log \left(8 + \mathcal{N}(J, d, \epsilon)^2 \right)} \ d\epsilon$$

from which we conclude that

$$\pi_\psi \left(\sup_{i \in J} Y_i \right) \leq 6 \int_0^{1/2} \sqrt{6 \log \left(8 + \mathcal{N}(J, d, \epsilon)^2 \right)} \ d\epsilon$$

Using the fact that the ϵ-balls with center in I and intersecting J are necessarily contained in a (2ϵ)-ball with center in J, we also have

$$\mathcal{N}(J, d, 2\epsilon) \leq \mathcal{N}(I, d, \epsilon)$$

This implies that

$$\pi_\psi \left(\sup_{i \in J} Y_i \right) \leq 12 \int_0^1 \sqrt{6 \log \left(8 + \mathcal{N}(I, d, \epsilon)^2 \right)} \ d\epsilon$$

The end of the proof is now a direct consequence of the monotone convergence theorem with increasing series of finite subsets exhausting I. This ends the proof of the lemma. ∎

11.4 Finite marginal inequalities

This section is mainly concerned with the proof of the Theorem 11.2.1. This result is a more or less direct consequence of the following technical lemma of separate interest.

Lemma 11.4.1 *Let $M_n := \sum_{0 \leq p \leq n} \Delta_p$ be a real valued martingale with symmetric and independent increments $(\Delta_n)_{n \geq 0}$. For any integer $m \geq 1$, and any $n \geq 0$, we have*

$$\mathbb{E} \left(|M_n|^m \right)^{\frac{1}{m}} \leq b(m) \ \mathbb{E} \left([M]_n^{m'/2} \right)^{\frac{1}{m'}} \qquad (11.23)$$

with the smallest even integer $m' \geq m$, the bracket process $[M]_n := \sum_{0 \leq p \leq n} \Delta_p^2$, and the collection of constants $b(m)$ defined in (0.6). In addition, for any $m \geq 2$, we have

$$\mathbb{E} \left(|M_n|^m \right)^{\frac{1}{m}} \leq b(m) \ \sqrt{(n+1)} \left(\frac{1}{n+1} \sum_{0 \leq p \leq n} \mathbb{E} \left(|\Delta_p|^{m'} \right) \right)^{\frac{1}{m'}} \qquad (11.24)$$

Proof of Theorem 11.2.1: We consider a collection of independent copies $X' = (X'^i)_{i \geq 1}$ of the random variables $X = (X^i)_{i \geq 1}$. We consider the martingale sequence $M = (M_i)_{1 \leq i \leq N}$ with symmetric and independent increments defined for any $1 \leq j \leq N$ by the following formula

$$M_j := \frac{1}{\sqrt{N}} \sum_{i=1}^{j} [f(X^i) - f(X'^i)]$$

By construction, we have

$$V(X)(f) = \frac{1}{\sqrt{N}} \sum_{i=1}^{N} (f(X^i) - \mu^i(f)) = \mathbb{E}(M_N \,|\, X)$$

Combining this conditioning property with the estimates provided in Lemma 11.4.1, the proof of the first assertion is now easily completed.

The Orlicz norm estimate (11.8) comes from the fact that for any $f \in \mathrm{Osc}(E)$, we have

$$\mathbb{E}(|V(X)(f)|^{2m}) \leq b(2m)^{2m} = \mathbb{E}(U^{2m})$$

for a Gaussian and centered random variable U, s.t. $\mathbb{E}(U^2) = 1$. Using the comparison lemma, Lemma 11.3.1, we find that

$$\pi_\psi(V(X)(f)) = \pi_\psi(|V(X)(f)|) \leq \pi_\psi(U) = \sqrt{8/3}$$

Applying Kintchine's inequalities (11.23), we prove that

$$\mathbb{E}(|V(X)(f)|^m)^{\frac{1}{m}} \leq b(m) \, \mathbb{E}\left(\left[\frac{1}{N} \sum_{i=1}^{N} [f(X^j) - f(X'^j)]^2\right]^{m'/2}\right)^{1/m'}$$

By construction, we notice that for any $f \in \mathrm{Osc}(E)$, and any $p \geq 2$, we have

$$\frac{1}{N} \sum_{j=1}^{N} \mathbb{E}\left([f(X^j) - f(X'^j)]^p\right) \leq 2\sigma(f)^2$$

By the Rosenthal type inequality stated in theorem 2.5 in [347], for any sequence of nonnegative, independent, and bounded random variables $(Y_i)_{i \geq 1}$, we have the rough estimate

$$\mathbb{E}\left[\sum_{i=1}^{N} Y_i^p\right]^{1/p} \leq 2p \max\left(\sum_{i=1}^{N} \mathbb{E}(Y_i), \left[\sum_{i=1}^{N} \mathbb{E}(Y_i^p)\right]^{1/p}\right)$$

for any $p \geq 1$. If we take $p = m'/2$, and $Y_i := \frac{1}{N} [f(X^i) - f(X'^i)]^2$, we prove

$$\mathbb{E}\left(\left[\frac{1}{N} \sum_{i=1}^{N} [f(X^i) - f(X'^i)]^2\right]^{m'/2}\right)^{2/m'}$$

$$\leq 4m \max\left(2\sigma(f)^2, \frac{1}{N^{1 - \frac{2}{m'}}} [2\sigma(f)^2]^{2/m'}\right)$$

for any $f \in \mathrm{osc}(E)$. Using Stirling's approximation of factorials

$$\sqrt{2\pi n} \ n^n \ e^{-n} \ \leq n! \leq e \ \sqrt{2\pi n} \ n^n \ e^{-n}$$

for any $p \geq 1$ we have $(2p)^p/b(2p)^{2p} = 2^{2p}p^p p!/(2p)! \leq e^{p+1} \leq 3^{2p}$, and

$$(2p+1)^{p+1/2}/b(2p+1)^{2p+1} = (2p+1)^{p+1}2^p p!/(2p+1)! \leq e^{p+2} \leq 3^{2p+1}$$

This implies that $m^{m/2}/b(m)^m \leq 3^m \Rightarrow \sqrt{m} \ b(m) \leq 3b(m)^2$, for any $m \geq 1$. This ends the proof of the theorem. ∎

Now, we come to the proof of the lemma.

Proof of Lemma 11.4.1:

We prove the lemma by induction on the parameter n. The result is clearly satisfied for $n = 0$. Suppose the estimate (11.23) is true at rank $(n-1)$. To prove the result at rank n, we use the binomial decomposition

$$(M_{n-1} + \Delta_n)^{2m} = \sum_{p=0}^{2m} \binom{2m}{p} M_{n-1}^{2m-p} (\Delta_n)^p$$

Using the symmetry condition, all the odd moments of Δ_n are null. Consequently, we find that

$$\mathbb{E}\left((M_{n-1} + \Delta_n)^{2m}\right) = \sum_{p=0}^{m} \binom{2m}{2p} \mathbb{E}\left(M_{n-1}^{2(m-p)}\right) \mathbb{E}(\Delta_n^{2p})$$

Using the induction hypothesis, we prove that the above expression is upper bounded by the quantity

$$\sum_{p=0}^{m} \binom{2m}{2p} 2^{-(m-p)} (2(m-p))_{(m-p)} \mathbb{E}\left([M]_{n-1}^{m-p}\right) \mathbb{E}(\Delta_n^{2p})$$

To take the final step, we use the fact that

$$\binom{2m}{2p} 2^{-(m-p)} (2(m-p))_{(m-p)} = \frac{2^{-m} (2m)_m}{2^{-p} (2p)_p} \binom{m}{p}$$

and $(2p)_p \geq 2^p$, to conclude that

$$\mathbb{E}\left((M_{n-1} + \Delta_n)^{2m}\right) \leq 2^{-m} (2m)_m \sum_{p=0}^{m} \binom{m}{p} \mathbb{E}\left([M]_{n-1}^{m-p}\right) \mathbb{E}(\Delta_n^{2p})$$

$$= 2^{-m} (2m)_m \mathbb{E}([M]_n^m)$$

For odd integers we use twice the Cauchy-Schwarz inequality to deduce that

$$\mathbb{E}(|M_n|^{2m+1})^2 \leq \mathbb{E}(M_n^{2m}) \ \mathbb{E}(M_n^{2(m+1)})$$

$$\leq 2^{-(2m+1)} (2m)_m (2(m+1))_{(m+1)} \mathbb{E}\left([M]_n^{m+1}\right)^{\frac{2m+1}{m+1}}$$

We conclude that

$$\mathbb{E}(|M_n|^{2m+1}) \leq 2^{-(m+1/2)} \frac{(2m+1)_{(m+1)}}{\sqrt{m+1/2}} \mathbb{E}\left([M]_n^{m+1}\right)^{1-\frac{1}{2(m+1)}}$$

The proof of (11.23) is now completed. Now, we come to the proof of (11.24). For any $m' \geq 2$ we have

$$\left[\frac{1}{n+1} \sum_{0 \leq p \leq n} \Delta_p^2\right]^{m'/2} \leq \frac{1}{n+1} \sum_{0 \leq p \leq n} \mathbb{E}\left(|\Delta_p|^{m'}\right)$$

and therefore

$$\mathbb{E}\left([M]_n^{m'/2}\right)^{\frac{1}{m'}} \leq (n+1)^{1/2} \left(\frac{1}{n+1} \sum_{0 \leq p \leq n} \mathbb{E}\left(|\Delta_p|^{m'}\right)\right)^{\frac{1}{m'}}$$

This ends the proof of the lemma. ∎

11.5   Maximal inequalities

The main goal of this section is to prove Theorem 11.2.6. We begin with the basic symmetrization technique. We consider a collection of independent copies $X' = (X'^i)_{i \geq 1}$ of the random variables $X = (X^i)_{i \geq 1}$. Let $\epsilon = (\epsilon_i)_{i \geq 1}$ constitute a sequence that is independent and identically distributed with

$$P(\epsilon_1 = +1) = P(\epsilon_1 = -1) = 1/2$$

We also consider the empirical random field sequences $V_\epsilon(X) := \sqrt{N} \, m_\epsilon(X)$. We also assume that (ϵ, X, X') are independent. We associate with the pairs (ϵ, X) and (ϵ, X') the random measures $m_\epsilon(X) = \frac{1}{N} \sum_{i=1}^N \epsilon_i \, \delta_{X^i}$ and $m_\epsilon(X') = \frac{1}{N} \sum_{i=1}^N \epsilon_i \, \delta_{X'^i}$. We notice that

$$\begin{aligned}
\|m(X) - \mu\|_{\mathcal{F}}^p &= \sup_{f \in \mathcal{F}} |m(X)(f) - \mathbb{E}(m(X')(f))|^p \\
&\leq \mathbb{E}(\|m(X) - m(X')\|_{\mathcal{F}}^p \, |X)
\end{aligned}$$

and in view of the symmetry of the random variables $(f(X^i) - f(X'^i))_{i \geq 1}$ we have

$$\mathbb{E}(\|m(X) - m(X')\|_{\mathcal{F}}^p) = \mathbb{E}(\|m_\epsilon(X) - m_\epsilon(X')\|_{\mathcal{F}}^p)$$

from which we conclude that

$$E\left(\|V(X)\|_{\mathcal{F}}^p\right) \le 2^p \ E\left(\|V_\epsilon(X)\|_{\mathcal{F}}^p\right) \tag{11.25}$$

By using the Chernov-Hoeffding inequality for any $x^1, \ldots, x^N \in E$, the empirical process

$$f \longrightarrow V_\epsilon(x)(f) := \sqrt{N} \ m_\epsilon(x)(f)$$

is sub-Gaussian for the norm $\|f\|_{L_2(m(x))} = m(x)(f^2)^{1/2}$. Namely, for any couple of functions f, g and any $\delta > 0$ we have

$$\mathbb{E}\left([V_\epsilon(x)(f) - V_\epsilon(x)(g)]^2\right) = \|f - g\|_{L_2(m(x))}^2$$

and by Hoeffding's inequality

$$\mathbb{P}\left(|V_\epsilon(x)(f) - V_\epsilon(x)(g)| \ge \delta\right) \le 2 \ e^{-\frac{1}{2}\delta^2/\|f-g\|_{L_2(m(x))}^2}$$

If we set $Z = \left(\dfrac{V_\epsilon(x)(f)}{\sqrt{6}\|f\|_{L_2(m(x))}}\right)^2$, then we find that

$$
\begin{aligned}
\mathbb{E}\left(e^Z\right) - 1 &= \int_0^\infty e^t \ \mathbb{P}\left(Z \ge t\right) \ dt \\
&= \int_0^\infty e^t \ \mathbb{P}\left(|V_\epsilon(x)(f)| \ge \sqrt{6t} \ \|f\|_{L_2(m(x))}\right) \ dt \\
&\le 2 \int_0^\infty e^t \ e^{-3t} \ dt = 1
\end{aligned}
$$

from which we conclude that

$$\pi_\psi \left(V_\epsilon(x)(f) - V_\epsilon(x)(g)\right) \le \sqrt{6}\|f - g\|_{L_2(m(x))}$$

Combining the maximal inequalities stated in Lemma 11.3.7 and the conditioning property (11.19) we find that

$$\pi_\psi \left(\|V_\epsilon(X)\|_{\mathcal{F}}\right) \le J(\mathcal{F})$$

with

$$J(\mathcal{F}) \le 2 \ 6^2 \int_0^2 \sqrt{\log\left(8 + \mathcal{N}(\mathcal{F}, \epsilon)^2\right)} \ d\epsilon \le c \ I(\mathcal{F}) < \infty$$

for some finite universal constant $c < \infty$. Combining (11.25) with (11.18), this implies that

$$\pi_\psi \left(\|V(X)\|_{\mathcal{F}}\right) \le 2 \ J(\mathcal{F})$$

This ends the proof of the theorem. ∎

11.6   Cramér-Chernov inequalities

11.6.1   Some preliminary convex analysis

In this section, we present some basic Cramér-Chernov tools to derive quantitative concentration inequalities. We begin by recalling some preliminary convex analysis on Legendre-Fenchel transforms. We associate with any convex function

$$L \ : \ t \in \mathrm{Dom}(\mathrm{L}) \mapsto L(t) \in \mathbb{R}_+$$

defined in some domain $\mathrm{Dom}(\mathrm{L}) \subset \mathbb{R}_+$, with $L(0) = 0$, the Legendre-Fenchel transform $L^\star$ defined by the variational formula

$$\forall \lambda \geq 0 \qquad L^\star(\lambda) := \sup_{t \in \mathrm{Dom}(\mathrm{L})} (\lambda t - L(t))$$

Note that $L^\star$ is a convex increasing function with $L^\star(0) = 0$ and its inverse $(L^\star)^{-1}$ is a concave increasing function.

We let L_A be the log-Laplace transform of a random variable A defined on some domain $\mathrm{Dom}(L_A) \subset \mathbb{R}_+$ by the formula

$$L_A(t) := \log \mathbb{E}(e^{tA})$$

Hölder's inequality implies that L_A is convex. Using the Cramér-Chernov-Chebychev inequality, we find that

$$\log \mathbb{P}\left(A \geq \lambda\right) \leq -L_A^\star(\lambda) \quad \text{and} \quad \mathbb{P}\left(A \geq (L_A^\star)^{-1}(x)\right) \leq e^{-x}$$

for any $\lambda \geq 0$ and any $x \geq 0$.

The next lemma provides some key properties of Legendre-Fenchel transforms that will be used in several places in the further development of the lecture notes.

Lemma 11.6.1   • *For any convex functions (L_1, L_2), such that*

$$\forall t \in \mathrm{Dom}(L_2) \quad L_1(t) \leq L_2(t) \quad \text{and} \quad \mathrm{Dom}(L_2) \subset \mathrm{Dom}(L_1)$$

we have
$$L_2^\star \leq L_1^\star \quad \text{and} \quad (L_1^\star)^{-1} \leq (L_2^\star)^{-1}$$

• *If we have for any $t \in v^{-1}\mathrm{Dom}(L_2) = \mathrm{Dom}(L_1)$, $L_1(t) = u \, L_2(v \, t)$, for some positive numbers $(u, v) \in \mathbb{R}_+^2$, then, for any $x, \lambda \geq 0$ we have*

$$L_1^\star(\lambda) = u \, L_2^\star\left(\frac{\lambda}{uv}\right) \quad \text{and} \quad (L_1^\star)^{-1}(x) = uv \, (L_2^\star)^{-1}\left(\frac{x}{u}\right)$$

- Let A be a random variable with a finite log-Laplace transform. For any $a \in \mathbb{R}$, we have $L_A(t) = -at + L_{A+a}(t)$, as well as

$$L_A^\star(\lambda) = L_{A+a}^\star(\lambda + a) \quad \text{and} \quad \left(L_A^\star\right)^{-1}(x) = -a + \left(L_{A+a}^\star\right)^{-1}(x)$$

We illustrate this technical lemma with the detailed analysis of three convex increasing functions of current use in the further development of these notes

- $L(t) = t^2/(1-t)$, $t \in [0, 1[$
- $L_0(t) := -t - \frac{1}{2} \log(1 - 2t)$, $t \in [0, 1/2[$.
- $L_1(t) := e^t - 1 - t$

In the first situation, we readily check that

$$L'(t) = \frac{1}{(1-t)^2} - 1 \quad \text{and} \quad L''(t) = \frac{2}{(1-t)^3}$$

An elementary manipulation yields that $L^\star(\lambda) = \left(\sqrt{\lambda+1} - 1\right)^2$, and

$$\left(L^\star\right)^{-1}(x) = \left(1 + \sqrt{x}\right)^2 - 1 = x + 2\sqrt{x}$$

In the second situation, we have $L_0'(t) = \frac{1}{1-2t} - 1$ and $L_0''(t) = \frac{2}{(1-2t)^2}$, from which we find that $L_0^\star(\lambda) = \frac{1}{2}(\lambda - \log(1 + \lambda))$. We also notice that

$$L_0(t) = t^2 \sum_{p \geq 0} \frac{2}{2+p}(2t)^p \leq \overline{L}_0(t) := \frac{t^2}{1 - 2t} = \frac{1}{4} L(2t)$$

for every $t \in [0, 1/2[$. Using Lemma 11.6.1, we prove that

$$\overline{L}_0^\star(\lambda) = \frac{1}{4} L^\star(2\lambda) \leq L_0^\star(\lambda)$$

$$\left(L_0^\star\right)^{-1}(x) \leq \left(\overline{L}_0^\star\right)^{-1}(x) = \frac{1}{2}\left(L^\star\right)^{-1}(4x) = 2(x + \sqrt{x}) \quad (11.26)$$

In the third situation, we have $L_1'(t) = e^t - 1$, and $L_1''(t) = e^t$ from which we conclude that

$$L_1^\star(\lambda) = (1 + \lambda) \log(1 + \lambda) - \lambda$$

On the other hand, using the fact that $2 \times 3^p \leq (p+2)!$, for any $p \geq 0$, we prove that we have

$$L_1(t) = \frac{t^2}{2} \sum_{p \geq 0} \frac{2 \times 3^p}{(p+2)!} \left(\frac{t}{3}\right)^p \leq \overline{L}_1(t) := \frac{t^2}{2(1 - t/3)} = \frac{9}{2} L\left(\frac{t}{3}\right)$$

for every $t \in [0, 1/3[$. This implies that $\overline{L}_1^{\star}(\lambda) = \frac{9}{2} L^{\star}\left(\frac{2\lambda}{3}\right) \leq L_1^{\star}(\lambda)$, and therefore

$$(L_1^{\star})^{-1}(x) \leq \left(\overline{L}_1^{\star}\right)^{-1}(x) = \frac{3}{2} (L^{\star})^{-1}\left(\frac{2x}{9}\right) = \left(\frac{x}{3} + \sqrt{2x}\right) \qquad (11.27)$$

Another crucial ingredient in the concentration analysis of the sum of two random variables is a deep technical lemma of J. Bretagnolle and E. Rio [492]. In the further development of this chapter, we use this argument to obtain a large family of concentration inequalities that are asymptotically "almost sharp" in a wide variety of situations.

Lemma 11.6.2 (J. Bretagnolle and E. Rio [492]) *For any pair of random variables A and B with finite log-Laplace transform in a neighborhood of 0, we have*

$$\forall x \geq 0 \quad (L_{A+B}^{\star})^{-1}(x) \leq (L_A^{\star})^{-1}(x) + (L_B^{\star})^{-1}(x) \qquad (11.28)$$

We also quote the following reverse type formulae that allows turning most of the concentration inequalities developed in these notes into Bernstein style exponential inequalities.

Lemma 11.6.3 *For any $(u, v) \in \mathbb{R}_+$, we have*

$$u \, (L_0^{\star})^{-1}(x) + v \, (L_1^{\star})^{-1}(x) \leq \left(L_{a(u,v),b(u,v)}^{\star}\right)^{-1}(x)$$

with the functions

$$a(u, v) := \left(2u + \frac{v}{3}\right) \quad and \quad b(u, v) := \left(\sqrt{2} \, u + v\right)^2$$

and the Laplace function

$$L_{a,b}(t) = \frac{b}{2a^2} L(at) \quad with \quad L_{a,b}^{\star}(\lambda) \geq \frac{\lambda^2}{2(b + \lambda a)}$$

Proof:
Using the estimates (11.26) and (11.27) we prove that

$$\begin{aligned} u \, (L_0^{\star})^{-1}(x) + v \, (L_1^{\star})^{-1}(x) &\leq 2 \, u \, (x + \sqrt{x}) + v \, \left(\frac{x}{3} + \sqrt{2x}\right) \\ &= a(u, v) \, x + \sqrt{2x \, b(u, v)} \end{aligned}$$

with

$$a(u, v) := \left(2u + \frac{v}{3}\right) \quad and \quad b(u, v) := \left(\sqrt{2} \, u + v\right)^2$$

Now, using Lemma 11.6.1, we observe that

$$a \, x + \sqrt{2xb} = (L_{a,b}^{\star})^{-1}(x) \quad with \quad L_{a,b}(t) = \frac{b}{2a^2} L(at) \qquad (11.29)$$

Finally, we have

$$L^\star(\lambda) = \left(\sqrt{\lambda+1} - 1\right)^2 \geq \frac{(\lambda/2)^2}{(1+\lambda/2)}$$

The r.h.s. inequality can be easily checked using the fact that

$$\left(\sqrt{1+2\lambda} - 1\right)^2 = 2\left(\frac{(1+\lambda)^2 - (1+2\lambda)}{(1+\lambda) + \sqrt{1+2\lambda}}\right)$$

$$\geq \frac{\lambda^2}{(1+\lambda)} \qquad \left(\Leftarrow \sqrt{1+2\lambda} \leq (1+\lambda)\right)$$

This implies that

$$L^\star_{a,b}(\lambda) = \frac{b}{2a^2} L^\star\left(\frac{2a}{b}\lambda\right) \geq \frac{\lambda^2}{2(b+\lambda a)}$$

This ends the proof of the lemma. ∎

11.6.2 Concentration inequalities

In this section, we investigate some elementary concentration inequalities for bounded and chi-square type random variables. We also apply these results to empirical processes associated with independent random variables.

Proposition 11.6.4 *Let A be a centered random variable such that $A \leq 1$. If we set $\sigma_A = \mathbb{E}(A^2)^{1/2}$, then for any $t \geq 0$, we have*

$$L_A(t) \leq \sigma_A^2 \, L_1(t) \tag{11.30}$$

In addition, the probability of the following events

$$A \leq \sigma_A^2 \, (L_1^\star)^{-1}\left(\frac{x}{\sigma_A^2}\right) \leq \frac{x}{3} + \sigma_A \sqrt{2x}$$

is greater than $1 - e^{-x}$, for any $x \geq 0$.

Proof:
To prove (11.30) we use the fact that decomposition

$$\mathbb{E}\left(e^{tA} - 1 - A\right) = \mathbb{E}\left(L_1(tA)1_{X<0}\right) + \mathbb{E}\left(L_1(tA)1_{X\in[0,1]}\right)$$

Since we have

$$\forall x \leq 0 \quad L_1(tx) \leq (tx)^2/2$$

and

$$\forall x \in [0,1] \quad L_1(tx) = x^2 \sum_{n\geq 2} x^{n-2}t^n/n! \leq x^2 L_1(t)$$

we conclude that

$$\mathbb{E}\left(e^{tA}\right) \leq 1 + \frac{t^2}{2}\,\mathbb{E}(A^2 1_{A<0}) + L_1(t)\mathbb{E}\left(A^2 1_{A\in[0,1]}\right)$$
$$\leq 1 + L_1(t)\sigma_A^2 \leq e^{L_1(t)\sigma_A^2}$$

Using Lemma 11.6.1, we readily prove that

$$(L_A^\star)^{-1}(x) \leq \sigma_A^2\,(L_1^\star)^{-1}\left(\frac{x}{\sigma_A^2}\right) \leq \frac{x}{3} + \sigma_A\,\sqrt{2x}$$

This ends the proof of the proposition.   ∎

Proposition 11.6.5 *For any measurable function f, with $0 < \mathrm{osc}(f) \leq a$, any $N \geq 1$, and any $t \geq 0$, we have*

$$L_{\sqrt{N}V(X)(f)}(t) \leq N\,\sigma^2(f/a)\,L_1(at) \tag{11.31}$$

In addition, the probability of the following events

$$V(X)(f) \leq a^{-1}\sigma^2(f)\sqrt{N}\,(L_1^\star)^{-1}\left(\frac{xa^2}{N\sigma^2(f)}\right) \tag{11.32}$$

$$\leq \frac{xa}{3\sqrt{N}} + \sqrt{2x\sigma(f)^2} \tag{11.33}$$

is greater than $1 - e^{-x}$, for any $x \geq 0$.

Proof:
Replacing f by f/a, there is no loss of generality to assume that $a = 1$. Using the same arguments as the ones we used in the proof of Proposition 11.6.4, we find that

$$\log\mathbb{E}\left(e^{t(f(X^i)-\mu^i(f))}\right) \leq \mu^i\left(\left[f - \mu^i(f)\right]^2\right)\,L_1(t)$$

from which we conclude that

$$L_N(t) := \log\mathbb{E}\left(e^{t\sqrt{N}\,V(X)(f)}\right)$$
$$= \sum_{i=1}^{N}\log\mathbb{E}\left(e^{t(f(X^i)-\mu^i(f))}\right) \leq \overline{L}_N(t) := N\,\sigma^2(f)\,L_1(t)$$

By Lemma 11.6.1, we have

$$(L_N^\star)^{-1}(x) \leq \left(\overline{L}_N^\star\right)^{-1}(x) = N\sigma^2(f)\,(L_1^\star)^{-1}\left(\frac{x}{N\sigma^2(f)}\right)$$

This ends the proof of the proposition.   ∎

Proposition 11.6.6 *For any random variable B such that*

$$\mathbb{E}\left(|B|^m\right)^{1/m} \le b(2m)^2 \, c \quad \text{with} \quad c < \infty$$

for any $m \ge 1$, with the finite constants $b(m)$ defined in (0.6), we have

$$L_B(t) \le ct + L_0(ct) \tag{11.34}$$

for any $0 \le ct < 1/2$. In addition, the probability of the following events

$$B \le c\left[1 + (L_0^\star)^{-1}(x)\right] \le c\left[1 + 2(x + \sqrt{x})\right]$$

is greater than $1 - e^{-x}$, for any $x \ge 0$.

Proof:
Replacing B by B/c, there is no loss of generality to assume that $c = 1$. We recall that $b(2m)^{2m} = \mathbb{E}(U^{2m})$ for every centered Gaussian random variable with $\mathbb{E}(U^2) = 1$ and

$$\forall t \in [0, 1/2) \quad \sum_{m \ge 0} \frac{t^m}{m!} b(2m)^{2m} = \frac{1}{\sqrt{1 - 2t}} = \mathbb{E}(\exp\{tU^2\})$$

This implies that

$$\mathbb{E}(\exp\{tB\}) \le \sum_{m \ge 0} \frac{t^m}{m!} b(2m)^{2m} = \frac{1}{\sqrt{1 - 2t}}$$

for any $0 \le t < 1/2$. In other words, we have

$$\begin{aligned}
L_{B-1}(t) &:= \log \mathbb{E}(\exp\{t(B - 1)\}) \le L_0(t) \\
L_B(t) &= t + L_{B-1}(t) \le t + L_0(t)
\end{aligned}$$

from which we conclude that

$$L_B^\star(\lambda) = L_{B-1}^\star(\lambda - 1) \Rightarrow (L_B^\star)^{-1}(x) = 1 + \left(L_{B-1}^\star\right)^{-1}(x) \le 1 + (L_0^\star)^{-1}(x)$$

This ends the proof of the proposition. ∎

Remark 11.6.7 *We end this section with some comments on the estimate (11.9). Using the fact that $b(m) \le b(2m)$ (see for instance (0.7)) we readily deduce from (11.9) that*

$$\mathbb{E}\left(|V(X)(f)|^m\right)^{\frac{1}{m}} \le 6\sqrt{2}\, b(2m)^2 \sigma(f)$$

for any $m \ge 1$, and for any N s.t. $2\sigma^2(f)N \ge 1$. Thus, if we set

$$B = |V(X)(f)| \quad \text{and} \quad c = 6\sqrt{2}\, \sigma(f)$$

in Proposition 11.6.6, we prove that for any N s.t. $2\sigma^2(f)N \geq 1$, and for any $0 \leq t < 1/(12\sqrt{2}\,\sigma(f))$

$$L_{|V(X)(f)|}(t) \leq 6\sqrt{2}\,\sigma(f)\,t + L_0(6\sqrt{2}\,\sigma(f)\,t)$$

In addition, the probability of the following events

$$|V(X)(f)| \quad \leq \quad 6\sqrt{2}\,\sigma(f)\left[1 + (L_0^\star)^{-1}(x)\right] \leq 6\sqrt{2}\,\sigma(f)\left[1 + 2(x + \sqrt{x})\right]$$

is greater than $1 - e^{-x}$, for any $x \geq 0$.

　　When N is chosen so that $2\sigma^2(f)N \geq 1$, using (11.33) we improve the above inequality. Indeed, using this concentration inequality implies that for any $f \in \mathrm{Osc}(E)$, the probability of the following events

$$V(X)(f) \quad \leq \quad \sqrt{2}\,\sigma(f)\left(\frac{x}{3} + \sqrt{x}\right) \tag{11.35}$$

is greater than $1 - e^{-x}$, for any $x \geq 0$.

11.7　Perturbation analysis

　　This section is mainly concerned with the proof of Theorem 11.2.5 and Theorem 11.2.11. We recall that for any second order smooth function F on $\mathbb{R}^d$, for some $d \geq 1$, $F(m(X)(f))$ stands for the random functionals

$$f = (f_i)_{1 \leq i \leq d} \in \mathrm{Osc}(E)^d$$

$$\mapsto F(m(X)(f)) = F(m(X)(f_1), \ldots, m(X)(f_d)) \in \mathbb{R}$$

　　Both results rely on the following second order decomposition of independent interest. The following proposition is a slight extension of Proposition 11.2.4.

Proposition 11.7.1 *For any $N \geq 1$, we have the decomposition*

$$\sqrt{N}\left[F(m(X)(f)) - F(\mu(f))\right] = V(X)\left[D_\mu(F)(f)\right] + \frac{1}{\sqrt{N}}\,R(X)(f)$$

with a first order functional $D_\mu(F)(f)$ defined in (11.11), and a second order term $R(X)(f)$ such that

$$\mathbb{E}\left(|R(X)(f)|^m\right)^{1/m} \leq \frac{1}{2}\,b(2m)^2\,\|\nabla^2 F_f\|_1$$

for any $m \geq 1$, with the parameter defined in (11.12). In addition, we have bias

$$|\mathbb{E}\left(F(m(X)(f))\right) - F(\mu(f))| \leq \frac{1}{2N}\,\sigma^2(f)\,\|\nabla^2 F_f\|_1$$

with the constant $\sigma^2(f)$ defined in (11.14), and the $\mathbb{L}_m$-mean error estimates

$$\mathbb{E}\left(|[F(m(X)(f)) - F(\mu(f))]|^m\right)^{1/m}$$

$$\leq \frac{1}{\sqrt{N}} \, b(m) \, \|\nabla F(\mu(f))\|_1 + \frac{1}{2N} \, b(2m)^2 \, \|\nabla^2 F_f\|_1$$

with the collection of constants $b(m)$ defined in (0.6).

Proof:
The last assertion is a direct consequence of the $\mathbb{L}_m$-mean error estimates (11.6) stated in Theorem 11.2.1, combined with the estimate (11.13). Using a Taylor's first order expansion, we have

$$\sqrt{N}\,[F(m(X)(f)) - F(\mu(f))] = \nabla F(\mu(f)) \, V(X)(f)^\top + \frac{1}{\sqrt{N}} \, R(X)(f)$$

with the second order remainder term

$$R(X)(f)$$

$$:= \int_0^1 (1-t)V(X)(f) \, \nabla^2 F\left(tm(X)(f) + (1-t)\mu(f)\right) \, V(X)(f)^\top \, dt$$

We notice that

$$\nabla F(\mu(f)) \, V(X)(f)^\top \;=\; V(X)\left[\nabla F(\mu(f)) \, f^\top\right]$$

and

$$\mathrm{osc}\left(\nabla F(\mu(f)) \, f^\top\right) \;\leq\; \sum_{i=1}^{d} \left|\frac{\partial F}{\partial u^i}(\mu(f))\right|$$

It is also easily checked that

$$\mathbb{E}\left(|R(X)(f)|^m\right)^{1/m}$$

$$\leq \tfrac{1}{2}\sum_{i,j=1}^{d} \, \sup_{\nu \in \mathcal{P}(E)} \left|\frac{\partial^2 F}{\partial u^i \partial u^j}(\nu(f))\right| \, \mathbb{E}\left(|V(X)(f_i)V(X)(f_j)|^m\right)^{1/m}$$

and for any $1 \leq i, j \leq d$, we have

$$\mathbb{E}\left(|V(X)(f_i)V(X)(f_j)|^m\right)^{1/m}$$

$$\leq \mathbb{E}\left(V(X)(f_j)^{2m}\right)^{1/(2m)} \mathbb{E}\left(V(X)(f_j)^{2m}\right)^{1/(2m)} \leq b(2m)^2$$

This clearly ends the proof of the proposition. ∎

We are now in position to prove Theorem 11.2.5.

Proof of Theorem 11.2.5:
We set

$$N\left[F(m(X)(f)) - F(\mu(f))\right] = A + B$$

with

$$A = \sqrt{N}\, V(X)\left[D_\mu(F)(f)\right] \quad \text{and} \quad B = R(X)(f)$$

Combining Proposition 11.6.5 with Proposition 11.6.6, if we set

$$g = D_\mu(F)(f) \quad a = \|\nabla F(\mu(f))\|_1 \quad \text{and} \quad c = \frac{1}{2}\|\nabla^2 F_f\|_1$$

then we have

$$\begin{aligned}
L_A(t) &\leq N\sigma^2(g/a)L_1(at) \\
L_B(t) &= ct + L_{B-c}(t) \quad \text{with} \quad L_{B-c}(t) \leq L_0(ct)
\end{aligned}$$

On the other hand, we have

$$(L_A^\star)^{-1}(x) \leq N\, a\, \sigma^2(g/a)\, (L_1^\star)^{-1}\left(\frac{x}{N\sigma^2(g/a)}\right)$$

and, using the fact that $L_B(t) = ct + L_{B-c}(t)$, we prove that

$$L_B^\star(\lambda) = L_{B-c}^\star(\lambda - c) \Rightarrow (L_B^\star)^{-1}(x) = c + (L_{B-c}^\star)^{-1}(x)$$
$$\leq c\left(1 + (L_0^\star)^{-1}(x)\right)$$

Using Bretagnolle-Rio's lemma, we find that

$$\begin{aligned}
(L_{A+B}^\star)^{-1}(x) &\leq (L_A^\star)^{-1}(x) + (L_B^\star)^{-1}(x) \\
&\leq N\, a^{-1}\, \sigma^2(g)\, (L_1^\star)^{-1}\left(\frac{xa^2}{N\sigma^2(g)}\right) + c\left(1 + (L_0^\star)^{-1}(x)\right)
\end{aligned}$$

This ends the proof of the theorem.   ∎

Now, we come to the proof of Theorem 11.2.11.
Proof of Theorem 11.2.11:
We consider the empirical processes $f \in \mathcal{F}_i \mapsto m(X)(f) \in \mathbb{R}$ associated with d classes of functions $\mathcal{F}_i$, $1 \leq i \leq d$, and defined in Section 11.1. We further assume that $\|f_i\| \vee \mathrm{osc}(f_i) \leq 1$, for any $f_i \in \mathcal{F}_i$, and we set

$$\pi_\psi(\|V(X)\|_{\mathcal{F}}) := \sup_{1 \leq i \leq d} \pi_\psi(\|V(X)\|_{\mathcal{F}_i})$$

Using Theorem 11.2.6, we have that

$$\pi_\psi(\|V(X)\|_{\mathcal{F}}) \leq 12^2 \int_0^2 \sqrt{\log\left(8 + \mathcal{N}(\mathcal{F}, \epsilon)^2\right)}\, d\epsilon$$

with $\mathcal{N}(\mathcal{F}, \epsilon) := \sup_{1 \leq i \leq d} \mathcal{N}(\mathcal{F}_i, \epsilon)$. Using Proposition 11.7.1, for any collection of functions $f = (f_i)_{1 \leq i \leq d} \in \mathcal{F} := \prod_{i=1}^d \mathcal{F}_i$, we have

$$\sqrt{N} \sup_{f \in \mathcal{F}} |F(m(X)(f)) - F(\mu(f))|$$

$$\leq \|\nabla F_\mu\|_\infty \sum_{i=1}^d \|V(X)\|_{\mathcal{F}_i} + \frac{d}{2\sqrt{N}} \|\nabla^2 F\|_\infty \sum_{i=1}^d \|V(X)\|_{\mathcal{F}_i}^2$$

If we set

$$A := \|\nabla F_\mu\|_\infty \sum_{i=1}^d \|V(X)\|_{\mathcal{F}_i}$$

then we find that

$$\pi_\psi(A) \leq \|\nabla F_\mu\|_\infty \sum_{i=1}^d \pi_\psi \left(\|V(X)\|_{\mathcal{F}_i} \right)$$

By Lemma 11.3.2, this implies that

$$\mathbb{E}\left(e^{tA}\right) \leq (1 + t\pi_\psi(A)) \, e^{(t\pi_\psi(A))^2} \leq e^{at + \frac{1}{2}t^2 b}$$

with $b = 2a^2$ and

$$a = \pi_\psi(A) \leq \|\nabla F_\mu\|_\infty \sum_{i=1}^d \pi_\psi \left(\|V(X)\|_{\mathcal{F}_i} \right)$$

Notice that

$$L_{A-a}(t) \leq L(t) = \frac{1}{2} t^2 b$$

Recalling that

$$L^\star(\lambda) = \frac{\lambda^2}{2b} \quad \text{and} \quad (L^\star)^{-1}(x) = \sqrt{2bx}$$

we conclude that

$$\begin{aligned} (L_A^\star)^{-1}(x) &= a + \left(L_{A-a}^\star\right)^{-1}(x) \\ &\leq a + \sqrt{2bx} = \pi_\psi(A)\left(1 + 2\sqrt{x}\right) \end{aligned}$$

Now, we come to the analysis of the second order term defined by

$$B = \frac{d}{2\sqrt{N}} \|\nabla^2 F\|_\infty \sum_{i=1}^d \|V(X)\|_{\mathcal{F}_i}^2$$

Using the inequality

$$\left(\sum_{i=1}^d a_i\right)^m \leq d^{m-1} \sum_{i=1}^d a_i^m$$

which is valid for any $d \geq$, any $m \geq 1$, and any sequence of real numbers $(a_i)_{1 \leq i \leq d} \in \mathbb{R}^d_+$, we prove that

$$\mathbb{E}(B^m) \leq \beta^m \ d^{m-1} \sum_{i=1}^{d} \mathbb{E}\left(\|V(X)\|^{2m}_{\mathcal{F}_i}\right) \quad \text{with} \quad \beta := \frac{d}{2\sqrt{N}} \left\|\nabla^2 F\right\|_\infty$$

Combining Lemma 11.3.2 with Theorem 11.2.6, we conclude that

$$\mathbb{E}(B^m) \leq m! \ \left(\beta \ d \ \pi_\psi(\|V(X)\|_{\mathcal{F}})^2\right)^m$$

If we set $b := \beta \ d \ \pi_\psi(\|V(X)\|_{\mathcal{F}})^2$, then we have that

$$\mathbb{E}\left(e^{tB}\right) \leq \sum_{m \geq 0} (bt)^m = \frac{1}{1 - bt} = e^{bt} \times e^{2L_0(bt/2)}$$

for any $0 \leq t < 1/b$ with the convex increasing function L_0 introduced on page 347, so that

$$2L_0(bt/2) = -bt - \log(1 - bt)$$

Using Lemma 11.6.1, we prove that $L_{B-b}(t) \leq 2L_0(bt/2)$, and

$$
\begin{aligned}
(L^\star_B)^{-1}(x) &= b + \left(L^\star_{B-b}\right)^{-1}(x) \\
&\leq b\left(1 + (L^\star_0)^{-1}\left(\frac{x}{2}\right)\right) \\
&= \frac{1}{2\sqrt{N}} \left\|\nabla^2 F\right\|_\infty (d \ \pi_\psi(\|V(X)\|_{\mathcal{F}}))^2 \left(1 + (L^\star_0)^{-1}\left(\frac{x}{2}\right)\right)
\end{aligned}
$$

Finally, using the Bretagnolle-Rio's lemma, we prove that

$$
\begin{aligned}
\left(L^\star_{A+B}\right)^{-1}(x) \leq \ & d\pi_\psi\left(\|V(X)\|_{\mathcal{F}}\right) \ \left[\|\nabla F_\mu\|_\infty (1 + 2\sqrt{x})\right. \\
& \left. + \frac{1}{2\sqrt{N}} \left\|\nabla^2 F\right\|_\infty (d \ \pi_\psi(\|V(X)\|_{\mathcal{F}})) \left(1 + (L^\star_0)^{-1}\left(\frac{x}{2}\right)\right)\right]
\end{aligned}
$$

This ends the proof of the Theorem 11.2.11. ∎

11.8   Interacting processes

11.8.1   Introduction

This section is devoted to the exponential concentration analysis of sequences of empirical processes associated with conditionally independent random variables.

In preparation for the concentration analysis of Feynman-Kac particle models, we consider a general class of interaction particle processes with non-necessarily mean field type dependency. The nonasymptotic theorems developed in the present section are used in Section 14.5, Section 16.6, and in Section 17.9, to deduce concentration inequalities for the particle density profiles (9.15), the occupation measures of genealogical tree models (9.23), the particle free energies, and the unnormalized particle models (9.17), as well as the backward particle models (9.27).

This section is organized as follows: Firstly, we analyze the concentration properties of integrals of local sampling error sequences, with general random but predictable measurable test functions. These results will be used to analyze the concentration properties of the first order fluctuation terms of the particle models.

We also present a stochastic perturbation technique to analyze the second order type decompositions. We consider finite marginal models and empirical processes. We close the chapter with the analysis of the covering numbers and the entropy parameters of linear transformation of classes of functions. We end this introductory section, with the precise description of the main mathematical objects we shall analyze in the further development of the chapter.

Definition 11.8.1 *We let $X_n^{(N)} = (X_n^{(N,i)})_{1 \leq i \leq N}$ be a Markov chain on some product state spaces E_n^N, for some $N \geq 1$. We also let $\mathcal{F}_n^N$ be the increasing σ-field generated by the random sequence $(X_p^{(N)})_{0 \leq p \leq n}$. We further assume that $(X_n^{(N,i)})_{1 \leq i \leq N}$ are conditionally independent, given $\mathcal{F}_{n-1}^N$.*

When there is no possible confusion, we simplify notation and suppress the index $(.)^{(N)}$ so that we write $(X_n, X_n^i, \mathcal{F}_n)$ instead of $(X_n^{(N)}, X_n^{(N,i)}, \mathcal{F}_n^N)$.

In this simplified notation, we also denote by μ_n^i the conditional distribution of the random state X_n^i given the $\mathcal{F}_{n-1}$; that is, we have that

$$\mu_n^i = \mathrm{Law}(X_n^i \mid \mathcal{F}_{n-1})$$

Notice that the conditional distributions

$$\mu_n := \frac{1}{N} \sum_{i=1}^N \mu_n^i$$

represent the local $\mathcal{F}_{n-1}$-conditional means of the occupation measures

$$m(X_n) := \frac{1}{N} \sum_{i=1}^N \delta_{X_n^i}$$

At this level of generality, we cannot obtain any kind of concentration properties for the deviations of the occupation measures $m(X_n)$ around some deterministic limiting value. In Section 14.5, dedicated to particle approximations of Feynman-Kac measures η_n, we shall deal with mean field type random measures μ_n^i, in the sense that the randomness only depends on the location of the random state X_{n-1}^i and on the current occupation measure $m(X_{n-1})$. In this situation, the fluctuation of $m(X_n)$ around the limiting deterministic measures η_n will be expressed in terms of second order Taylor's type expansions w.r.t. the local sampling errors

$$V(X_p) = \sqrt{N}\left(m(X_p) - \mu_p\right)$$

from the origin $p = 0$, up to the current time $p = n$.

The first order terms will be expressed in terms of integral formulae of predictable functions f_p w.r.t. the local sampling error measures $V(X_p)$. These stochastic first order expansions are defined below.

Definition 11.8.2 *For any sequence of $\mathcal{F}_{n-1}$-measurable random function $f_n \in \mathrm{Osc}(E_n)$, and any collection of numbers $a_n \in \mathbb{R}_+$, we set*

$$V_n(X)(f) = \sum_{p=0}^{n} a_p \, V(X_p)(f_p) \tag{11.36}$$

For any $\mathcal{F}_{n-1}$-measurable random function $f_n \in \mathrm{Osc}(E_n)$, we have

$$\mathbb{E}\left(V(X_n)(f_n) \,|\, \mathcal{F}_{n-1}\right) \;\; = \;\; 0$$

$$\mathbb{E}\left(V(X_n)(f_n)^2 \,|\, \mathcal{F}_{n-1}\right) \;\; = \;\; \sigma_n^N(f_n)^2 := \frac{1}{N}\sum_{i=1}^{N} \mu_n^i\left(\left[f_n - \mu_n^i(f_n)\right]^2\right)$$

Without further mention, we always assume that we have an almost sure estimate of the following form

$$\sup_{N \geq 1} \sigma_n^N(f_n)^2 \leq \sigma_n^2 \quad \text{for some positive constant } \sigma_n^2 \leq 1. \tag{11.37}$$

The concentration inequalities developed in this section are expressed in terms of the parameters defined below.

Definition 11.8.3 *Given some integer $n \geq 0$, and a sequence of nonnegative parameters $(\sigma_p)_{p \geq 0}$ and $(a_p)_{p \geq 0}$, we set*

$$\overline{\sigma}_n^2 := \sum_{0 \leq p \leq n} \sigma_p^2 \quad \text{and} \quad a_n^\star := \max_{0 \leq p \leq n} a_p \quad \text{and} \quad |a|_{n,1} := \sum_{p=0}^{n} a_p$$

Under this condition, using Theorem 11.2.1 we check that

$$\mathbb{E}(|V_n(X)(f)|^m)^{1/m} \leq b(m) \, |a|_{n,1}$$

for any integer $m \geq 1$. In much the same way, using Lemma 11.3.6 and (11.8), we readily prove the following proposition.

Proposition 11.8.4 *We let U_n be some $\mathcal{F}_n$-measurable random variable satisfying the following condition*

$$|U_n| \leq \sum_{p=0}^{n} a_p \ |V(X_p)(f_p)|$$

for any sequence of $\mathcal{F}_{n-1}$-measurable random function $f_n \in \mathrm{Osc}(E_n)$, and some collection of numbers $a_n \in \mathbb{R}_+$. In this case, for any integer $m \geq 1$ we have

$$\mathbb{E}(|U_n|^m)^{1/m} \leq b(m) \ |a|_{n,1} \quad \text{with} \quad |a|_{n,1} := \sum_{p=0}^{n} a_p$$

In addition, we have the Olicz norm estimates

$$\pi_\psi \left(\sup_{0 \leq p \leq n} |U_p| \right) \leq 4 \ \sqrt{\log{(9+n)}} \ |a|_{n,1}$$

11.8.2 Finite marginal models

We will now derive a quantitative contraction inequality for the general random fields models of the following form

$$W_n(X)(f) = V_n(X)(f) + \frac{1}{\sqrt{N}} \ R_n(X)(f) \tag{11.38}$$

with $V_n(X)(f)$ defined in (11.36), and a second order term such that

$$\mathbb{E} \left(|R_n(X)(f)|^m \right)^{1/m} \leq b(2m)^2 \ c_n$$

for any $m \geq 1$, for some finite constant $c_n < \infty$ whose values only depend on the parameter n.

For a null remainder term $R_n(X)(f) = 0$, these concentration properties are easily derived using Proposition 11.6.5.

Proposition 11.8.5 *We let $V_n(X)(f)$ be the random field sequence defined in (11.36). For any $t \geq 0$, we have that*

$$L_{\sqrt{N}V_n(X)(f)}(t) \leq N \ \bar{\sigma}_n^2 \ L_1(ta_n^\star)$$

with the parameters $\bar{\sigma}_n$ and $a_n^\star$ introduced in Definition 11.8.3. In addition, the probability of the following events

$$
\begin{aligned}
V_n(X)(f) \ &\leq \ \sqrt{N} \ a_n^\star \ \bar{\sigma}_n^2 \ (L_1^\star)^{-1} \left(\frac{x}{N\bar{\sigma}_n^2} \right) \\
&\leq \ a_n^\star \left(\frac{x}{3\sqrt{N}} + \sqrt{2\bar{\sigma}_n^2 \ x} \right)
\end{aligned}
$$

is greater than $1 - e^{-x}$, for any $x \geq 0$.

Proof:
By Proposition 11.6.5, we have

$$\mathbb{E}\left(e^{t\sqrt{N}V_n(X)(f)} \,|\, \mathcal{F}_{n-1}\right) = e^{t\sqrt{N}V_{n-1}(X)(f)} \; \mathbb{E}\left(e^{(ta_n)\sqrt{N}V(X_n)(f_n)} \,|\, \mathcal{F}_{n-1}\right)$$

with

$$\log \mathbb{E}\left(e^{(ta_n)\sqrt{N}V(X_n)(f_n)} \,|\, \mathcal{F}_{n-1}\right) \leq N \, \sigma_n^2 \, L_1(ta_n^\star)$$

This clearly implies that

$$L_{\sqrt{N}V_n(X)(f)}(t) \leq \overline{L}_1(t) := N \, \overline{\sigma}_n^2 \, L_1(ta_n^\star)$$

Using Lemma 11.6.1, we conclude that

$$\left(L_{\sqrt{N}V_n(X)(f)}^\star\right)^{-1}(x) \;\; \leq \;\; \left(\overline{L}_1^\star\right)^{-1}(x) = N \, a_n^\star \, \overline{\sigma}_n^2 \, (L_1^\star)^{-1}\left(\frac{x}{N\overline{\sigma}_n^2}\right)$$

The last assertion is a direct consequence of (11.27). This ends the proof of the proposition.   ∎

Theorem 11.8.6 *We let $W_n(X)(f)$ be the random field sequence defined in (11.38). In this situation, the probability of the events*

$$\sqrt{N} \, W_n(X)(f) \leq c_n \, \left(1 + (L_0^\star)^{-1}(x)\right) + N \, a_n^\star \, \overline{\sigma}_n^2 \, (L_1^\star)^{-1}\left(\frac{x}{N\overline{\sigma}_n^2}\right)$$

is greater than $1 - e^{-x}$, for any $x \geq 0$. In the above display, $\overline{\sigma}_n$ and $a_n^\star$ stand for the parameters introduced in Definition 11.8.3

Proof:
We set $\sqrt{N} \, W_n(X)(f) = A_n + B_n$, with

$$A_n = \sqrt{N}V_n(X)(f) \quad \text{and} \quad B_n = R_n(X)(f)$$

By Proposition 11.6.6 and Proposition 11.8.5, we have

$$\begin{aligned}
L_{A_n}(t) &\leq \overline{L}_{A_n}(t) := N \, \overline{\sigma}_n^2(f) \, L_1(ta_n^\star) \\
L_{B_n - c_n}(t) &\leq \overline{L}_{B_n - c_n}(t) := L_0(c_n t)
\end{aligned}$$

We recall that

$$L_{B_n}(t) = c_n t + L_{B_n - c_n}(t) \Rightarrow L_{B_n}^\star(\lambda) = L_{B_n - c_n}^\star(\lambda - c_n)$$

Using Lemma 11.6.1, we also have that

$$\begin{aligned}
\left(\overline{L}_{B_n}^\star\right)^{-1}(x) \;\; &= \;\; c_n + \left(L_{B_n - c_n}^\star\right)^{-1}(x) \\
&\leq \;\; c_n + \left(\overline{L}_{B_n - c_n}^\star\right)^{-1}(x) = c_n \left(1 + (L_0^\star)^{-1}(x)\right)
\end{aligned}$$

Arguing as in the end of the proof of Proposition 11.8.5, we also have

$$\left(L^{\star}_{\sqrt{N}V_n(X)(f)}\right)^{-1}(x) \ \leq \ N \, a_n^{\star} \, \overline{\sigma}_n^2 \, (L_1^{\star})^{-1}\left(\frac{x}{N\overline{\sigma}_n^2}\right)$$

The end of the proof is now a direct consequence of the Bretagnolle-Rio's lemma. This ends the proof of the theorem. ∎

11.8.3   Empirical processes

We let $V_n(X)$ be the random field sequence defined in (11.36), and we consider a sequence of classes of $\mathcal{F}_{n-1}$-measurable random functions $\mathcal{F}_n$, such that $\|f_n\| \vee \mathrm{osc}(f_n) \leq 1$, for any $f_n \in \mathcal{F}_n$.

Definition 11.8.7 *For any $f = (f_n)_{n\geq 0} \in \mathcal{F} := (\mathcal{F}_n)_{n\geq 0}$, and any sequence of numbers $a = (a_n)_{n\geq 0} \in \mathbb{R}_+^{\mathbb{N}}$, we set*

$$V_n(X)(f) = \sum_{p=0}^{n} a_p \, V(X_p)(f_p) \quad and \quad \|V_n(X)\|_{\mathcal{F}} = \sup_{f\in\mathcal{F}} |V_n(X)(f)|$$

We further assume that for any $n \geq 0$, and any $\epsilon > 0$, we have an almost sure estimate

$$\mathcal{N}\left(\mathcal{F}_n, \epsilon\right) \leq \mathcal{N}_n(\epsilon) \tag{11.39}$$

for some nonincreasing function $\mathcal{N}_n(\epsilon)$ such that

$$b_n := 12^2 \int_0^2 \sqrt{\log\left(8 + \mathcal{N}_n(\epsilon)^2\right)} \, d\epsilon < \infty$$

In this situation, we have

$$\pi_\psi\left(\|V_n(X)\|_{\mathcal{F}}\right) \leq \sum_{p=0}^{n} a_p \, \pi_\psi\left(\|V(X_p)\|_{\mathcal{F}_p}\right)$$

Using Theorem 11.2.6, given $\mathcal{F}_{n-1}$ we have the almost sure upper bound

$$\pi_\psi\left(\|V(X_p)\|_{\mathcal{F}_p}\right) \ \leq \ 12^2 \int_0^2 \sqrt{\log\left(8 + \mathcal{N}(\mathcal{F}_p, \epsilon)^2\right)} \, d\epsilon \leq b_p$$

Combining Lemma 11.3.1 with Lemma 11.3.2, we prove the following theorem.

Theorem 11.8.8 *For any classes of $\mathcal{F}_{n-1}$-measurable random functions $\mathcal{F}_n$ satisfying the entropy condition (11.39), we have*

$$\pi_\psi\left(\|V_n(X)\|_{\mathcal{F}}\right) \leq c_n := \sum_{p=0}^{n} a_p b_p$$

In particular, the probability of the events

$$\|V_n(X)\|_{\mathcal{F}} \le c_n \sqrt{x + \log 2}$$

is greater than $1 - e^{-x}$, for any $x \ge 0$.

Next, we consider classes of nonrandom functions $\mathcal{F} = (\mathcal{F}_n)_{n \ge 0}$. We further assume that $\|f_n\| \vee osc(f_n) \le 1$, for any $f_n \in \mathcal{F}_n$, and

$$I_1(\mathcal{F}) := 12^2 \int_0^2 \sqrt{\log\left(8 + \mathcal{N}(\mathcal{F}, \epsilon)^2\right)}\, d\epsilon < \infty$$

with $\mathcal{N}(\mathcal{F}, \epsilon) = \sup_{n \ge 0} \mathcal{N}(\mathcal{F}_n, \epsilon) < \infty$.

Theorem 11.8.9 *We let $W_n(X)(f)$, $f \in \mathcal{F}$, be the random field sequence defined by*

$$W_n(X)(f) = V_n(X)(f) + \frac{1}{\sqrt{N}} R_n(X)(f)$$

with a second order term such that

$$\mathbb{E}\left(\sup_{f \in \mathcal{F}} |R_n(X)(f)|^m\right) \le m!\, c_n^m$$

for any $m \ge 1$, for some finite constant $c_n < \infty$ whose values only depend on the parameter n. In this situation, the probability of the events

$$\|W_n(X)\|_{\mathcal{F}} \le \left[\sum_{p=0}^n a_p\right] I_1(\mathcal{F})\left(1 + 2\sqrt{x}\right) + \frac{c_n}{\sqrt{N}}\left(1 + (L_0^\star)^{-1}\left(\frac{x}{2}\right)\right)$$

is greater than $1 - e^{-x}$, for any $x \ge 0$.

Proof:
We set $\sqrt{N}\, \|W_n(X)\|_{\mathcal{F}} \le A_n + B_n$, with

$$A_n = \sqrt{N}\, \|V_n(X)\|_{\mathcal{F}} \quad \text{and} \quad B_n = \sup_{f \in \mathcal{F}} |R_n(X)(f)|$$

Using the fact that

$$\sup_{f \in \mathcal{F}} |V_n(X)(f)| \le \sum_{p=0}^n a_p\, \|V(X_p)\|_{\mathcal{F}_p}$$

by Lemma 11.3.1, we have

$$\pi_\psi\left(\|V_n(X)\|_{\mathcal{F}}\right) \le \sum_{p=0}^n a_p\, \pi_\psi\left(\|V(X_p)\|_{\mathcal{F}_p}\right)$$

Using Theorem 11.2.6, we also have that

$$\pi_\psi \left(\|V(X_p)\|_{\mathcal{F}_p} \right) \le I_1(\mathcal{F}) := 12^2 \int_0^2 \sqrt{\log \left(8 + \mathcal{N}(\mathcal{F}, \epsilon)^2\right)} \, d\epsilon$$

with $\mathcal{N}(\mathcal{F}, \epsilon) = \sup_{n \ge 0} \mathcal{N}(\mathcal{F}_n, \epsilon)$. This implies that

$$\pi_\psi (A_n) \le \bar{a}_n \sqrt{N} I_1(\mathcal{F}) \quad \text{with} \quad \bar{a}_n := \sum_{p=0}^n a_p$$

By Lemma 11.3.2, we have

$$\mathbb{E} \left(e^{tA_n}\right) \le (1 + t\pi_\psi(A_n)) \; e^{(t\pi_\psi(A_n))^2} \le e^{\alpha_n t + \frac{1}{2}t^2 \beta_n}$$

with $\beta_n = 2\alpha_n^2$, and $\alpha_n = \pi_\psi(A_n)$. Notice that

$$L_{A_n - \alpha_n}(t) \le L_n(t) := \frac{1}{2} t^2 \beta_n$$

Recalling that

$$L_n^\star(\lambda) = \frac{\lambda^2}{2\beta_n} \quad \text{and} \quad (L_n^\star)^{-1}(x) = \sqrt{2\beta_n x}$$

we conclude that

$$\begin{aligned}
\left(L_{A_n}^\star\right)^{-1}(x) &= \alpha_n + \left(L_{A_n - \alpha_n}^\star\right)^{-1}(x) \\
&\le \alpha_n + \sqrt{2\beta_n x} = \pi_\psi(A_n) \left(1 + 2\sqrt{x}\right)
\end{aligned}$$

On the other hand, under our assumption, we also have that

$$\mathbb{E} \left(e^{tB_n}\right) \le \sum_{m \ge 0} (c_n t)^m = \frac{1}{1 - c_n t} = e^{c_n t} \times e^{2L_0(c_n t/2)}$$

for any $0 \le t < 1/c_n$ with the convex increasing function L_0 introduced on page 347, so that

$$2L_0(c_n t/2) = -c_n t - \log \left(1 - c_n t\right)$$

Using Lemma 11.6.1, we conclude that

$$L_{B_n - c_n}(t) \le 2L_0(c_n t/2)$$

and

$$\begin{aligned}
\left(L_{B_n}^\star\right)^{-1}(x) &= c_n + \left(L_{B_n - c_n}^\star\right)^{-1}(x) \\
&\le c_n \left(1 + (L_0^\star)^{-1}\left(\frac{x}{2}\right)\right)
\end{aligned}$$

The end of the proof is now a direct consequence of the Bretagnolle-Rio's lemma. ∎

11.8.4 Covering numbers and entropy methods

In this final section, we derive some properties of covering numbers for some classes of functions. These two results are the key to derive uniform concentration inequalities w.r.t. the time parameter for Feynman-Kac particle models. This subject is investigated in Section 14.5, Section 16.6, and in Section 17.9 dedicated, respectively, to particle density profiles models, particle free energies, and backward particle interpretations.

We let $(E_n, \mathcal{E}_n)_{n=0,1}$ be a pair of measurable state spaces, and $\mathcal{F}$ be a separable collection of measurable functions $f : E_1 \to \mathbb{R}$ such that $\|f\| \leq 1$ and $\mathrm{osc}(f) \leq 1$.

We consider a Markov transition $M(x_0, dx_1)$ from E_0 into E_1, a probability measure μ on E_0, and a function G from E_0 into $[0, 1]$.

Definition 11.8.10 *We denote by $G \cdot M(\mathcal{F})$ and $G \cdot (M - \mu M)(\mathcal{F})$ the classes of functions*

$$G \cdot M(\mathcal{F}) = \{G\ M(f)\ :\ f \in \mathcal{F}\}$$

and

$$G \cdot (M - \mu M)(\mathcal{F}) = \{G\ [M(f) - \mu M(f)]\ :\ f \in \mathcal{F}\}$$

Lemma 11.8.11 *For any $\epsilon > 0$, we have*

$$\mathcal{N}\left[G \cdot M(\mathcal{F}), \epsilon\right] \leq \mathcal{N}(\mathcal{F}, \epsilon)$$

Proof:
For any probability measure η on E_0, we let $\{f_1, \ldots, f_{n_\epsilon}\}$ be the centers of $n_\epsilon = \mathcal{N}(\mathcal{F}, \mathbb{L}_2(\eta), \epsilon)$ $\mathbb{L}_2(\eta)$-balls of radius at most ϵ covering $\mathcal{F}$. For any $f \in \mathcal{F}$, there exists some $1 \leq i \leq n_\epsilon$ such that

$$\eta\left([G(f - f_i)]^2\right)^{1/2} \leq \eta\left([(f - f_i)]^2\right)^{1/2} \leq \epsilon$$

This implies that

$$\mathcal{N}\left(G \cdot \mathcal{F}, \mathbb{L}_2(\eta), \epsilon\right) \leq \mathcal{N}(\mathcal{F}, \mathbb{L}_2(\eta), \epsilon)$$

In much the same way, we let $\{f_1, \ldots, f_{n_\epsilon}\}$ be the $n_\epsilon = \mathcal{N}(\mathcal{F}, \mathbb{L}_2(\eta M), \epsilon)$ centers of $\mathbb{L}_2(\eta M)$-balls of radius at most ϵ covering $\mathcal{F}$. In this situation, for any $f \in \mathcal{F}$, there exists some $1 \leq i \leq n_\epsilon$ such that

$$\eta\left([(M(f) - M(f_i))]^2\right)^{1/2} \leq \eta M\left([(f - f_i)]^2\right)^{1/2} \leq \epsilon$$

This implies that

$$\mathcal{N}\left(M(\mathcal{F}), \mathbb{L}_2(\eta), \epsilon\right) \leq \mathcal{N}(\mathcal{F}, \mathbb{L}_2(\eta M), \epsilon)$$

This ends the proof of the lemma. ∎

One of the simplest ways to control the covering numbers of the second class of functions is to assume that M satisfies the following condition $M(x, dy) \geq \delta\nu(dy)$, for any $x \in E_0$, and for some measure ν, and some $\delta \in]0, 1[$. Indeed, in this situation we observe that

$$M_\delta(x, dy) = \frac{M(x, dy) - \delta\nu(dy)}{1 - \delta}$$

is a Markov transition and

$$(1 - \delta) \; [M_\delta(f)(x) - M_\delta(f)(y)] = [M(f)(x) - M(f)(y)]$$

This implies that

$$(1 - \delta) \; [M_\delta(f)(x) - \mu M_\delta(f)] = [M(f)(x) - \mu M(f)]$$

and

$$\eta \left[(M(f)(x) - \mu M(f))^2 \right] \leq 2(1 - \delta) \; \eta M_{\delta,\mu}(|f|^2)$$

with the Markov transition

$$M_{\delta,\mu}(x, dy) = \frac{1}{2} \; [M_\delta(x, dy) + \mu M_\delta(x, dy)]$$

We let $\{f_1, \ldots, f_{n_\epsilon}\}$ be the $n_\epsilon = \mathcal{N}(\mathcal{F}, \mathbb{L}_2 \, (\eta M_{\delta,\mu}) \,, \epsilon/2)$ centers of $\mathbb{L}_2 \, (\eta M_{\delta,\mu})$-balls of radius at most ϵ covering $\mathcal{F}$. If we set

$$\overline{f} = M(f) - \mu M(f) \quad \text{and} \quad \overline{f}_i = M(f_i) - \mu M(f_i)$$

then we find that

$$\overline{f} - \overline{f}_i = M(f - f_i) - \mu M(f - f_i)$$

from which we prove that

$$\eta \left[(\overline{f} - \overline{f}_i)^2 \right]^{1/2} \leq 2(1 - \delta) \; \left[\eta M_{\delta,\mu}(|f - f_i|^2) \right]^{1/2}$$

We conclude that

$$\mathcal{N} \left((M - \mu M)(\mathcal{F}), \mathbb{L}_2(\eta), 2\epsilon(1 - \delta) \right) \leq \mathcal{N}(\mathcal{F}, \mathbb{L}_2 \, (\eta M_{\delta,\mu}) \,, \epsilon)$$

and therefore

$$\mathcal{N} \left((M - \mu M)(\mathcal{F}), 2\epsilon(1 - \delta) \right) \leq \mathcal{N}(\mathcal{F}, \epsilon)$$

or equivalently

$$\mathcal{N} \left(\frac{1}{1 - \delta} \, (M - \mu M)(\mathcal{F}), \epsilon \right) \leq \mathcal{N}(\mathcal{F}, \epsilon/2)$$

In more general situations, we quote the following result.

Lemma 11.8.12 *For any $\epsilon > 0$, we have*

$$\mathcal{N}\left[G \cdot (M - \mu M)(\mathcal{F}), 2\epsilon\beta(M)\right] \leq \mathcal{N}(\mathcal{F}, \epsilon)$$

Proof:

We consider a Hahn-Jordan orthogonal decomposition

$$(M(x, dy) - \mu M(dy)) = M_\mu^+(x, dy) - M_\mu^-(x, dy)$$

with

$$M_\mu^+(x, dy) = (M(x, .) - \mu M)^+ \quad \text{and} \quad M_\mu^-(x, dy) = (M(x, .) - \mu M)^-$$

with

$$\|M(x, .) - \mu M\|_{\text{tv}} = M_\mu^+(x, E_1) = M_\mu^-(x, E_1) \leq \beta(M)$$

By construction, we have

$$M(f)(x) - \mu M(f) = M_\mu^+(x, E_1) \left(\overline{M}_\mu^+(f)(x) - \overline{M}_\mu^-(f)(x)\right)$$

with

$$\overline{M}_\mu^+(x, dy) := \frac{M_\mu^+(x, dy)}{M_\mu^+(x, E_1)} \quad \text{and} \quad \overline{M}_\mu^-(x, dy) := \frac{M_\mu^-(x, dy)}{M_\mu^-(x, E_1)}$$

This implies that

$$|M(f)(x) - \mu M(f)| \leq 2\beta(M) \, \overline{M}_\mu(|f|)(x)$$

with

$$\overline{M}_\mu(x, dy) = \frac{1}{2}\left(\overline{M}_\mu^+(x, dy) + \overline{M}_\mu^-(x, dy)\right)$$

One concludes that

$$\eta\left[(M(f)(x) - \mu M(f))^2\right]^{1/2} \leq 2\beta(M) \left[\eta\overline{M}_\mu(|f|^2)\right]^{1/2}$$

We let $\{f_1, \ldots, f_{n_\epsilon}\}$ be the $n_\epsilon = \mathcal{N}(\mathcal{F}, \mathbb{L}_2\left(\eta\overline{M}_\mu\right), \epsilon/2)$ centers of $\mathbb{L}_2\left(\eta\overline{M}_\mu\right)$-balls of radius at most ϵ covering $\mathcal{F}$.

If we set

$$\overline{f} = M(f) - \mu M(f) \quad \text{and} \quad \overline{f}_i = M(f_i) - \mu M(f_i)$$

then we find that

$$\overline{f} - \overline{f}_i = M(f - f_i) - \mu M(f - f_i)$$

from which we prove that

$$\eta\left[(\overline{f} - \overline{f}_i)^2\right]^{1/2} \leq 2\beta(M) \left[\eta\overline{M}_\mu(|f - f_i|^2)\right]^{1/2}$$

In this situation, for any $\overline{f} \in (M - \mu M)(\mathcal{F})$, there exists some $1 \leq i \leq n_\epsilon$ such that

$$\eta \left[(\overline{f} - \overline{f}_i)^2 \right]^{1/2} \leq \beta(M) \, \epsilon$$

We conclude that

$$\mathcal{N} \left((M - \mu M)(\mathcal{F}), \mathbb{L}_2(\eta), \epsilon\beta(M) \right) \leq \mathcal{N}(\mathcal{F}, \mathbb{L}_2 \left(\eta \overline{M}_\mu \right), \epsilon/2)$$

and therefore

$$\mathcal{N} \left(G \cdot (M - \mu M)(\mathcal{F}), \epsilon\beta(M) \right) \quad \leq \quad \mathcal{N} \left((M - \mu M)(\mathcal{F}), \epsilon\beta(M) \right) \leq \mathcal{N}(\mathcal{F}, \epsilon/2)$$

This ends the proof of the lemma. ∎

We consider the one step Feynman-Kac mapping Φ from $\mathcal{P}(E_0)$ into $\mathcal{P}(E_1)$ and the integral operator $d'_\nu \Phi$ from $\mathcal{B}_b(E_1)$ into $\mathcal{B}_b(E_0)$ defined in (3.1.7) and (3.1.8). We recall that these functional mappings are defined by

$$\Phi(\mu) = \Psi_G(\mu)M \quad \text{and} \quad d'_\nu \Phi(f) := G' \, (Id - \Psi_G(\nu))M(f)$$

Using Lemma 11.8.12, we readily prove the following lemma.

Lemma 11.8.13 *We let $\mathcal{F}$ be a separable collection of measurable functions $f \in \mathrm{Osc}(E_1)$ s.t. $\|f\| \leq 1$. For any Markov transition M from E_0 into E_1, we set*

$$d'_\nu \Phi(\mathcal{F}) := \{d'_\nu \Phi(f) \; : \; f \in \mathcal{F}\}$$

In this situation, we have the uniform estimate

$$\sup_{\nu \in \mathcal{P}(E)} \mathcal{N} \left[d'_\nu \Phi(\mathcal{F}), 2\epsilon\beta(M) \right] \leq \mathcal{N}(\mathcal{F}, \epsilon) \tag{11.40}$$

Chapter 12

Feynman-Kac semigroups

12.1 Description of the models

This section is concerned with the semigroup structure, and the weak regularity properties, of Feynman-Kac models (1.37) associated with a sequence of Markov transitions M_n, and some potential functions G_n, on some measurable state spaces E_n.

Definition 12.1.1 *We let $\overline{G}_n$ be the normalized potential functions*

$$\overline{G}_n := G_n/\eta_n(G_n) \tag{12.1}$$

Definition 12.1.2 *We denote by*

$$\Phi_{p,n}(\eta_p) = \eta_n \quad \text{and} \quad \gamma_p Q_{p,n} = \gamma_n$$

with $0 \leq p \leq n$, the linear, and the nonlinear, semigroup associated with the unnormalized and the normalized Feynman-Kac measures. For $p = n$, we use the convention $\Phi_{n,n} = Q_{n,n} = Id$, the identity mapping.

We recall from (1.42) that $Q_{p,n}$ has the following functional representation

$$Q_{p,n}(f_n)(x_p) := \mathbb{E}\left(f_n(X_n) \prod_{p \leq q < n} G_q(X_q) \mid X_p = x_p\right)$$

Definition 12.1.3 *We denote by $\overline{Q}_{p,n}$ the normalized Feynman-Kac semigroup defined by*

$$\overline{Q}_{p,n}(f) = Q_{p,n}(f)/\eta_p Q_{p,n}(1) \tag{12.2}$$

The semigroup property comes from the fact that, for any $0 \leq p \leq q \leq n$, we have

$$\eta_p Q_{p,n}(1) = \prod_{p \leq q < n} \eta_q(G_q) = \eta_p Q_{p,q}(1) \times \eta_q Q_{q,n}(1)$$

from which we conclude that

$$\overline{Q}_{p,n}(f) = \frac{1}{\eta_p Q_{p,q}(1)} Q_{p,q}\left(\frac{1}{\eta_q Q_{q,n}(1)} Q_{q,n}(f)\right) = \overline{Q}_{p,q}(\overline{Q}_{q,n}(f))$$

Definition 12.1.4 *We also let* $G_{p,n}$, *and* $P_{p,n}$, *be the potential functions, and the Markov transitions, defined by*

$$Q_{p,n}(1)(x) = G_{p,n}(x) \quad and \quad P_{p,n}(f) = Q_{p,n}(f)/Q_{p,n}(1) \tag{12.3}$$

and we set

$$g_{p,n} := \sup_{x,y} \frac{G_{p,n}(x)}{G_{p,n}(y)} \quad and \quad \beta(P_{p,n}) = \sup \operatorname{osc}(P_{p,n}(f))$$

The r.h.s. supremum is taken the set of functions $\operatorname{Osc}(E_p)$. *To simplify notation, for* $n = p + 1$ *we have also set*

$$G_{p,p+1} = Q_{p,p+1}(1) = G_p$$

and sometimes we write g_p *instead of* $g_{p,p+1}$.

The concentration inequalities developed in the further development of this book will be expressed in terms of the following parameters.

Definition 12.1.5 *For any* $k, l \geq 0$, *we also set*

$$\tau_{k,l}(n) := \sum_{0 \leq p \leq n} g_{p,n}^k \, \beta(P_{p,n})^l \quad and \quad \kappa(n) := \sup_{0 \leq p \leq n} (g_{p,n}\beta(P_{p,n})) \tag{12.4}$$

Whenever they exist, we consider the parameters

$$\overline{\tau}_{k,l} := \sup_{n \geq 0} \tau_{k,l}(n) \quad and \quad \overline{\kappa} := \sup_{n \geq 0} \kappa(n) \tag{12.5}$$

Using the fact that

$$
\begin{aligned}
\eta_n(f_n) \quad &:= \quad \frac{\eta_p Q_{p,n}(f_n)}{\eta_p Q_{p,n}(1)} \\
&= \quad \frac{1}{\eta_p(Q_{p,n}(1))} \times \eta_p \left(Q_{p,n}(1) \times \frac{Q_{p,n}(f_n)}{Q_{p,n}(1)} \right) \tag{12.6}
\end{aligned}
$$

we readily prove the following lemma.

Lemma 12.1.6 *For any* $0 \leq p \leq n$, *we have*

$$\Phi_{p,n}(\eta_p) = \Psi_{G_{p,n}}(\eta_p) P_{p,n}$$

As a direct consequence of (3.20) and (3.21), we quote the following weak regularity property of the Feynman-Kac semigroups.

Proposition 12.1.7 *For* $[0,1]$-*valued potential function* G_n, *any* $p \leq n$, *and any couple of measures* ν, μ *on the set* E_p *s.t.* $\mu(G_{p,n}) \wedge \nu(G_{p,n}) > 0$, *we have the decomposition*

$$\Phi_{p,n}(\mu) - \Phi_{p,n}(\nu) = \frac{1}{\nu(G_{p,n})} (\mu - \nu) S_{G_{p,n},\mu} P_{p,n}$$

with the collection of Markov Transitions $S_{G_{p,n},\mu}$ defined in (3.12). In addition, we have the following Lipschitz estimates

$$\|\Phi_{p,n}(\mu) - \Phi_{p,n}(\nu)\|_{\text{tv}} \leq \frac{\|G_{p,n}\|}{\mu(G_{p,n}) \vee \nu(G_{p,n})} \, \beta(P_{p,n}) \, \|\mu - \nu\|_{\text{tv}}$$

and

$$\sup_{\mu,\nu} \|\Phi_{p,n}(\mu) - \Phi_{p,n}(\nu)\|_{\text{tv}} = \beta(P_{p,n})$$

12.2 Stability properties

12.2.1 Regularity conditions

In this section we present one of the simplest quantitative contraction estimates we know for the normalized Feynman-Kac semigroups $\Phi_{p,n}$. There are several levels of assumptions used in this book. In the more general settings, we consider the Dobrushin coefficient of the Markov transition $P_{p,n}$ defined in (12.3). By Proposition 12.1.7, we know that this parameter reflects the uniform stability properties of Feynman-Kac semigroups w.r.t. the initial conditions; that is, we have that

$$\sup_{\mu,\nu} \|\Phi_{p,n}(\mu) - \Phi_{p,n}(\nu)\|_{\text{tv}} = \beta(P_{p,n})$$

Our first standing and minimal regularity condition only depends on the parameter $\beta(P_{p,n})$.

H(G, P) The parameters g_p and $\beta(P_{p,n})$ satisfy the following conditions

$$\upsilon_1 := \sup_{p \geq 0} \sum_{n \geq p} (g_n - 1) \, \beta(P_{p,n}) < \infty$$

and

$$\upsilon_2 := \sup_{n \geq 0} \left[\sum_{0 \leq p \leq n} \beta(P_{p,n}) \right] \vee \sup_{p \geq 0} \left[\sum_{n \geq p} \beta(P_{p,n}) \right] < \infty$$

Most of the stability results presented in this chapter rely on this rather weak condition. This regularity condition is also used in Chapter 14, Chapter 15, Chapter 16, and Chapter 17 dedicated, respectively, to the stochastic analysis of particle density profiles, genealogical tree models, particle free energy models, and backward particle Markov models. Under this condition, we derive uniform convergence estimates w.r.t. the time parameter, as well as linear complexity inequalities for particle tree based empirical measures.

Remark 12.2.1   • *It is easily checked that*

$$\mathbf{H}(\mathbf{G}, \mathbf{P}) \Longrightarrow \forall p \geq 0 \quad \beta(P_{p,n}) \to_{n \to \infty} 0$$

- *In addition, in Proposition 12.2.4 we shall prove that*

$$\mathbf{H}(\mathbf{G}, \mathbf{P}) \Longrightarrow \forall k \geq 0, \ \forall l \geq 1 \quad \overline{\tau}_{k,l} < \infty \quad and \quad \overline{\kappa} < \infty \qquad (12.7)$$

  with the parameters $\overline{\tau}_{k,l}$ and $\overline{\kappa}$ introduced in Definition 12.1.5.

- *For time homogenous models $(G_n, M_n) = (G, M)$, for any $n \geq p$, we have*

$$(g_n, \Phi_{p,n}, Q_{p,n}, P_{p,n}) = \left(g, \Phi_{0,(n-p)}, Q_{0,(n-p)}, P_{0,(n-p)}\right)$$

In this situation, we have

$$\upsilon_1 \leq (g-1)\upsilon_2 \quad with \quad \upsilon_2 := \sum_{n \geq 0} \beta(P_{0,n})$$

from which we conclude that

$$\mathbf{H}(\mathbf{G}, \mathbf{P}) \Longleftrightarrow \sum_{n \geq 0} \beta(P_{0,n}) < \infty$$

The condition $\mathbf{H}(\mathbf{G},\mathbf{P})$ is related to the stability properties of nonlinear Feynman-Kac semigroups. Roughly speaking, $\mathbf{H}(\mathbf{G},\mathbf{P})$ is met as soon as the Feynman-Kac semigroup forgets initial conditions sufficiently fast.

These stability properties can be analyzed using different stochastic techniques. Next, we provide a class of rather strong conditions related to the oscillations of the potential functions, and to the stability properties of the reference Markov chain model X_n with probability transition M_n. Both of them implies that the chain X_n tends to merge exponentially fast toward the random states starting from *any* two different locations.

$\mathbf{H_0}(\mathbf{G}, \mathbf{M})$ The following couple of regularity properties are satisfied

$$\rho := \sup_{n \geq 0} \left(g_n \beta(M_{n+1})\right) < 1 \quad and \quad g := \sup_n g_n < \infty \qquad (12.8)$$

$\mathbf{H_m}(\mathbf{G}, \mathbf{M})$ There exists some integer $m \geq 1$, such that for any $n \geq 0$, and any $((x, x'), A) \in \left(E_n^2 \times \mathcal{E}_n\right)$ and any $n \geq 0$, we have

$$M_{n,n+m}(x, A) \leq \chi_m \, M_{n,n+m}(x', A) \quad and \quad g := \sup_{n \geq 0} g_n < \infty$$

for some finite parameters $\chi_m, g < \infty$, and some integer $m \geq 1$.

In Section 12.2.2, we shall prove that

$$(\exists\ m \geq 0\ \text{s.t.}\ \mathbf{H_m(G, M)}\ )\ \Longrightarrow\ \mathbf{H(G, P)}$$

The above assertion is readily checked using the estimates of $g_{p,n}$ and $\beta(P_{p,n})$ provided in Theorem 12.2.5 and in Theorem 12.2.6.

One natural strategy to obtain some useful quantitative contraction estimates for the Markov transitions $P_{p,n}$ is to write this transition in terms of the composition of Markov transitions.

Lemma 12.2.2 *For any $0 \leq p \leq q \leq n$, we have*

$$P_{p,n} = R_{p,q}^{(n)} P_{q,n} \quad and \quad P_{p,n} = R_{p+1}^{(n)} R_{p+2}^{(n)} \dots R_{n-1}^{(n)} R_n^{(n)}$$

with the triangular array of Markov transitions $R_{p,q}^{(n)}$ and $(R_q^{(n)})_{1 \leq q \leq n}$ defined for any $f \in \mathcal{B}_b(E_q)$ by the integral operators

$$R_{p,q}^{(n)}(f) := \frac{Q_{p,q}(G_{q,n}f)}{Q_{p,q}(G_{q,n})} = \frac{P_{p,q}(G_{q,n}f)}{P_{p,q}(G_{q,n})}$$

and

$$R_p^{(n)}(f) = \frac{Q_p(G_{p,n}f)}{Q_p(G_{p,n})} = \frac{M_p(G_{p,n}f)}{M_p(G_{p,n})}$$

In addition, for any $0 \leq p \leq q \leq r \leq n$ we have the semigroup property

$$R_{p,r}^{(n)} = R_{p,q}^{(n)} R_{q,r}^{(n)}$$

and the regularity properties

$$\beta\left(R_{p,q}^{(n)}\right) \leq g_{q,n}\ \beta\left(P_{p,q}\right) \quad and \quad \log g_{p,n} \leq \sum_{p \leq q < n} (g_q - 1)\ \beta(P_{p,q}) \quad (12.9)$$

Proof:
Using the decomposition

$$Q_{p,n}(f) = Q_{p,q}(Q_{q,n}(f)) = Q_{p,q}(Q_{q,n}(1)\ P_{q,n}(f))$$

we easily check the first assertion. For any $p \leq q \leq r$, and any $f \in \mathcal{B}_b(E_r)$, we notice that

$$R_{p,q}^{(n)}\left(R_{q,r}^{(n)}(f)\right) = \frac{Q_{p,q}\left(Q_{q,n}(1) \frac{Q_{q,r}(G_{r,n}f)}{Q_{q,r}(Q_{r,n}(1))}\right)}{Q_{p,q}(Q_{q,r}(Q_{r,n}(1)))} = \frac{Q_{p,r}(G_{r,n}f)}{Q_{p,r}(G_{r,n})} = R_{p,r}^{(n)}(f)$$

The l.h.s. inequality in (12.9) is a direct consequence of (3.6). Using (3.29), the proof of the r.h.s. inequality in (12.9) is based on the fact that

$$\frac{G_{p,n}(x)}{G_{p,n}(y)} = \frac{\prod_{p \leq q < n} \Phi_{p,q}(\delta_x)(G_q)}{\prod_{p \leq q < n} \Phi_{p,q}(\delta_y)(G_q)}$$

$$= \exp\left\{\sum_{p \leq q < n} (\log \Phi_{p,q}(\delta_x)(G_q) - \log \Phi_{p,q}(\delta_y)(G_q))\right\}$$

Using the fact that

$$\log y - \log x = \int_0^1 \frac{(y-x)}{x + t(y-x)} \, dt$$

for any positive numbers x, y, we prove that

$$\frac{G_{p,n}(x)}{G_{p,n}(y)}$$

$$= \exp\left\{ \sum_{p \leq q < n} \int_0^1 \frac{(\Phi_{p,q}(\delta_x)(G_q) - \Phi_{p,q}(\delta_y)(G_q))}{\Phi_{p,q}(\delta_y)(G_q) + t(\Phi_{p,q}(\delta_x)(G_q) - \Phi_{p,q}(\delta_y)(G_q))} \, dt \right\}$$

$$\leq \exp\left\{ \sum_{p \leq q < n} \tilde{g}_q \times \left(\Phi_{p,q}(\delta_x)(\tilde{G}_q) - \Phi_{p,q}(\delta_y)(\tilde{G}_q) \right) \right\}$$

with

$$\tilde{G}_q := G_q / \mathrm{osc}(G_q) \quad \text{and} \quad \tilde{g}_q := \mathrm{osc}(G_q)/\inf G_q \leq g_q - 1$$

We end the proof of the desired estimates using (3.6) and Proposition 12.1.7. This completes the proof of the lemma. ∎

Combining Lemma 12.2.2 with Proposition 12.1.7 we prove the following regularity estimate.

Proposition 12.2.3 *For bounded and positive potential function G_n, and any measures ν, μ on the set E_p, we have the strong Lipschitz estimate*

$$\|\Phi_{p,n}(\mu) - \Phi_{p,n}(\nu)\|_{\mathrm{tv}} \leq \beta(P_{p,n}) \, \exp\left\{ \sum_{p \leq q < n} (g_q - 1)\, \beta(P_{p,q}) \right\} \|\mu - \nu\|_{\mathrm{tv}}$$

Proposition 12.2.3 shows that the semigroup $\Phi_{p,n}$ is exponentially stable as soon as $\beta(P_{p,n})$ decreases exponentially fast to 0, when $(n - p) \uparrow \infty$. We consider the following regularity condition.

$\mathbf{H_m(R)}$ There exists some $m \geq 1$, and some collection of positive measures $\nu_{p,p+m}^{(n)}$ on E_{p+m}, with $p + m \leq n$, such that for any $0 \leq p + m \leq n$ and any $x_p \in E_p$ we have

$$R_{p,p+m}^{(n)}(x_p, dx_{p+m}) \geq \epsilon_{p,p+m}^{(n)} \, \nu_{p,p+m}^{(n)}(dx_{p+m})$$

for some $\epsilon_{p,p+m}^{(n)}$ s.t. for any p we have

$$\epsilon_{p,\star} := \liminf_{n \to \infty} \frac{1}{n} \sum_{1 \leq k \leq n} \epsilon_{p+(k-1)m, p+km}^{(p+nm)} > 0$$

In this situation, using Lemma 12.2.2, for any $k, m \geq 1$ we have

$$P_{p,p+nm} = R_{p,p+m}^{(p+nm)} R_{p+m,p+2m}^{(p+nm)} \cdots R_{p+(n-1)m,p+nm}^{(p+nm)}$$

and therefore

$$
\begin{aligned}
\beta(P_{p,p+nm}) &\leq \prod_{k=1}^{n} \beta\left(R_{p+(k-1)m,p+km}^{(p+nm)}\right) \\
&\leq \prod_{k=1}^{n} \left(1 - \epsilon_{p+(k-1)m,p+km}^{(p+nm)}\right)
\end{aligned}
$$

This clearly implies that

$$\limsup_{n\to\infty} \frac{1}{n} \log \beta(P_{p,p+nm}) \leq -\epsilon_{p,\star}$$

For a more thorough discussion on these exponential stability properties, including for continuous time Feynman-Kac models, we refer the reader to [161, 163, 170, 180].

The next proposition provides some useful estimates of the parameters $\tau_{k,l}(n)$ and $\kappa(n)$ defined in (12.1.5) in terms of the regularity condition $\mathbf{H(G,P)}$ stated on page 371.

Proposition 12.2.4 *We assume that condition* $\mathbf{H(G,P)}$ *is satisfied for some finite constants* υ_1, υ_2. *In this situation, for any* $k \geq 0$ *and* $l \geq 1$, *we have the uniform estimates*

$$g_{p,n} \leq \exp \upsilon_1 \qquad \overline{\tau}_{k,l} \leq \upsilon_2 \exp\left(k\upsilon_1\right) \quad and \quad \overline{\kappa} \leq \exp \upsilon_1 \qquad (12.10)$$

Proof:
Using (12.9) we find that

$$g_{p,n} \leq \exp \sum_{p \leq q < n} (g_q - 1) \; \beta(P_{p,q}) \leq \exp \upsilon_1$$

This implies that

$$\tau_{k,l}(n) := \sum_{0 \leq p \leq n} g_{p,n}^k \; \beta(P_{p,n})^l \leq \exp\left(k\upsilon_1\right) \times \upsilon_2$$

and

$$\kappa(n) := \sup_{0 \leq p \leq n} \left(g_{p,n}\beta(P_{p,n})\right) \leq \sup_{0 \leq p \leq n} g_{p,n} \leq \exp \upsilon_1$$

This ends the proof of the proposition. ∎

12.2.2   Quantitative contraction theorems

This section is mainly concerned with the proof of two contraction theorems that can be derived under the couple of regularity conditions presented in Section 12.2.1.

Theorem 12.2.5 *We assume that condition* $\mathbf{H_m(G, M)}$ *is satisfied for some finite parameters* $\chi_m, g < \infty$, *and some integer* $m \geq 1$. *In this situation, we have the uniform estimates*

$$\sup_{0 \leq p \leq n} g_{p,n} \leq \iota(m) := (\chi_m g^m) \wedge \exp\left(\frac{(g-1)}{(1-\kappa_m)} \frac{1}{\left[1 - (1-\kappa_m)^{1/m}\right]}\right) \quad (12.11)$$

and

$$\sup_{p \geq 0} \beta(P_{p,p+km}) \leq (1-\kappa_m)^k \quad \text{with} \quad \kappa_m := g^{-(m-1)} \chi_m^{-2} \quad (12.12)$$

In addition, for any couple of measures $\nu, \mu \in \mathcal{P}(E_p)$, *and for any* $f \in \mathrm{Osc}(E_n)$ *we have the weak Lipschitz regularity estimate*

$$|[\Phi_{p,n}(\mu) - \Phi_{p,n}(\nu)](f)| \leq \rho_m (1-\kappa_m)^{(n-p)/m} \quad |(\mu - \nu)D_{p,n,\mu}(f)| \quad (12.13)$$

for some function $D_{p,n,\mu}(f) \in \mathrm{Osc}(E_p)$ *whose values only depend on the parameters* (p, n, μ) *and some parameters* $\rho_m < \infty$ *such that*

$$\rho_m \leq \iota(m)/(1-\kappa_m) \quad (12.14)$$

Proof:
For any nonnegative function f, we notice that

$$
\begin{aligned}
R^{(n)}_{p,p+m}(f)(x) &= \frac{Q_{p,p+m}(G_{p+m,n}f)(x)}{Q_{p,p+m}(G_{p+m,n})(x)} \\
&\geq g^{-(m-1)}\chi_m^{-2} \frac{M_{p,p+m}(G_{p+m,n}f)(x')}{M_{p,p+m}(G_{p+m,n})(x')}
\end{aligned}
$$

and for any $p + m \leq n$

$$\frac{G_{p,n}(x)}{G_{p,n}(x')} = \frac{Q_{p,p+m}(G_{p+m,n})(x)}{Q_{p,p+m}(G_{p+m,n})(x')} \leq g^m \frac{M_{p,p+m}(G_{p+m,n})(x)}{M_{p,p+m}(G_{p+m,n})(x')} \leq \chi_m g^m$$

For $p \leq n \leq p + m$, this upper bound remains valid. We conclude that

$$G_{p,n}(x) \leq \chi_m g^m \, G_{p,n}(x') \quad \text{and} \quad \beta\left(R^{(n)}_{p,p+m}\right) \leq 1 - g^{-(m-1)}\chi_m^{-2}$$

In the same way as above, we have

$$n = km \Rightarrow P_{p,p+km} = R^{(n)}_{p,p+m} R^{(n)}_{p+m,p+2m} \cdots R^{(n)}_{p+(k-1)m,p+km}$$

and

$$\beta(P_{p,p+km}) \leq \prod_{1 \leq l \leq k} \beta(R^{(n)}_{p+(l-1)m,p+lm}) \leq \left(1 - g^{-(m-1)}\chi_m^{-2}\right)^k$$

This ends the proof of (12.12). Combining (12.12) with the estimates (12.9) we find that

$$\log g_{p,n} \leq \frac{(g-1)}{(1-\kappa_m)} \sum_{p \leq q < n} (1-\kappa_m)^{(q-p)/m} \leq \frac{(g-1)}{(1-\kappa_m)} \frac{1}{1-(1-\kappa_m)^{1/m}}$$

This ends the proof of (12.11).

The proof of (12.13) is based on the decomposition

$$(\mu - \nu)S_{G_{p,n},\mu}P_{p,n}(f) = \beta(S_{G_{p,n},\mu}P_{p,n}) \times (\mu - \nu)D_{p,n,\mu}(f)$$

with

$$D_{p,n,\mu}(f) := S_{G_{p,n},\mu}P_{p,n}(f)/\beta(S_{G_{p,n},\mu}P_{p,n})$$

On the other hand, we have

$$S_{G_{p,n},\mu}(x,y) \geq (1 - \|G_{p,n}\|) \Rightarrow \beta(S_{G_{p,n},\mu}) \leq \|G_{p,n}\|$$

and

$$\beta(S_{G_{p,n},\mu})/\nu(G_{p,n}) \leq g_{p,n} \leq \iota(m)$$

Finally, we observe that

$$\beta(P_{p,n}) \leq \beta(P_{p,p+\lfloor (n-p)/m \rfloor})$$

from which we conclude that

$$\frac{1}{\nu(G_{p,n})} \beta(S_{G_{p,n},\mu}P_{p,n}) \leq \iota(m) \beta(P_{p,p+\lfloor (n-p)/m \rfloor})$$

The end of the proof is now a direct consequence of the contraction estimates (12.11) and (12.12). This ends the proof of the theorem. ∎

Theorem 12.2.6 *We assume that condition* $\mathbf{H}_0(\mathbf{G}, \mathbf{M})$ *is satisfied for some* $\rho < 1$. *In this situation, for any couple of measures* $\nu, \mu \in \mathcal{P}(E_p)$, *and for any* $f \in \mathrm{Osc}(E_n)$ *we have the estimate*

$$|[\Phi_{p,n}(\mu) - \Phi_{p,n}(\nu)](f)| \leq \rho^{(n-p)} \ |(\mu - \nu)D_{p,n,\mu}(f)|$$

for some function $D_{p,n,\mu}(f) \in \mathrm{Osc}(E_p)$, *whose values only depend on the parameters* (p, n, μ). *In addition, for any* $0 \leq p \leq n$, *we have the estimates*

$$\beta(P_{p,n}) \leq \rho^{n-p} \quad \text{and} \quad g_{p,n} \leq \exp\left((g-1)(1-\rho^{n-p})/(1-\rho)\right)$$

Proof:

Using Proposition 12.1.7, and recalling that $\beta(S_{G_{n-1},\mu}) \leq \|G_{n-1}\|$, we readily prove that

$$
\begin{aligned}
|[\Phi_n(\mu) - \Phi_n(\nu)](f)| &\leq g_{n-1}\beta(M_n) \; |(\mu - \nu)D_{n,\mu}(f)| \\
&\leq \rho \; |(\mu - \nu)D_{n,\mu}(f)|
\end{aligned}
$$

with the function

$$
D_{n,\mu}(f) = S_{G_{n-1},\mu}M_n(f)/\beta(S_{G_{n-1},\mu}M_n) \in \mathrm{Osc}(E_{n-1})
$$

Now, we can prove the theorem by induction on the parameter $n \geq p$. For $n = p$, the desired result follows from the above discussion. Suppose we have

$$
|[\Phi_{p,n-1}(\mu) - \Phi_{p,n-1}(\nu)](f)| \leq \rho^{(n-p-1)} \; |(\mu - \nu)D_{p,n-1,\mu}(f)|
$$

for any $f \in \mathrm{Osc}(E_{n-1})$, and some functions $D_{p,n-1,\mu}(f) \in \mathrm{Osc}(E_p)$. In this case, we have

$$
|[\Phi_n(\Phi_{p,n-1}(\mu)) - \Phi_n(\Phi_{p,n-1}(\nu))](f)|
$$

$$
\leq g_{n-1}\beta(M_n) \; |(\Phi_{p,n-1}(\mu) - \Phi_{p,n-1}(\nu))D_{n,\Phi_{p,n-1}(\mu)}(f)|
$$

for any $f \in \mathrm{Osc}(E_n)$, with $D_{n,\Phi_{p,n-1}(\mu)}(f) \in \mathrm{Osc}(E_{n-1})$.

Under our assumptions, we conclude that

$$
|[\Phi_n(\Phi_{p,n-1}(\mu)) - \Phi_n(\Phi_{p,n-1}(\nu))](f)| \leq \rho^{(n-p)} \; |(\mu - \nu)D_{p,n,\mu}(f)|
$$

with the function

$$
D_{p,n,\mu}(f) := D_{p,n-1,\mu}\left(D_{n,\Phi_{p,n-1}(\mu)}(f)\right) \in \mathrm{Osc}(E_p)
$$

The proof of the second assertion is a direct consequence of Proposition 12.1.7 and Lemma 12.2.2. This ends the proof of the theorem. ∎

Corollary 12.2.7 *Under any of the conditions* $\mathbf{H}_m(\mathbf{G}, \mathbf{M})$, *with* $m \geq 0$, *the functions* $\tau_{k,l}$ *and* κ *defined in (12.4) are uniformly bounded; that is, for any* $k, l \geq 1$ *we have that*

$$
\overline{\tau}_{k,l}(m) := \sup_{n \geq 0} \sup_{0 \leq p \leq n} \tau_{k,l}(p) < \infty \quad \text{and} \quad \overline{\kappa}(m) := \sup_{n \geq 0} \kappa(n) < \infty \quad (12.15)
$$

In addition, for any $m \geq 1$, *we have*

$$
\begin{aligned}
\overline{\kappa}(m) &\in [1, \iota(m)] \\
\overline{\tau}_{k,l}(m) &\leq m \, \iota(m)^k / \left(1 - (1 - \kappa_m)^l\right)
\end{aligned}
$$

with the parameters $\iota(m)$ *and* κ_m *defined in (12.11) and (12.12). In addition, for* $m = 0$, *we have the estimates*

$$
\begin{aligned}
\overline{\kappa}(0) &\leq \exp\left((g-1)/(1-\rho)\right) \\
\overline{\tau}_{k,l}(0) &\leq \left[\exp\left(k(g-1)/(1-\rho)\right)\right]/(1-\rho^l)
\end{aligned}
$$

12.2.3 Some illustrations

We illustrate the regularity conditions presented in Section 12.2.1 with three different types of Feynman-Kac models related, respectively, to time discretization techniques, simulated annealing type schemes, and path space models. Of course, a complete analysis of the regularity properties of the Feynman-Kac application models presented in Part III of this book would lead here to a too long discussion.

In some instances, the regularity conditions stated in Section 12.2.1 can be directly translated into regularity properties of the reference Markov chain model and the adaptation potential function. In other instances, the regularity properties of the Feynman-Kac semigroup depend on some important tuning parameters, including discretization time steps and cooling schedule in simulated annealing time models. In Section 12.2.3.1, we illustrate the regularity property $\mathbf{H_0}(\mathbf{G}, \mathbf{M})$ stated in (12.8) in the context of time discretization models with geometric style clocks. In Section 12.2.3.2, we present some tools to tune the cooling parameters of the annealing model discussed in (4.40), so that the resulting semigroups are exponentially stable.

For degenerate indicator style functions, we can use a one step integration technique to transform the model into a Feynman-Kac model on smaller state spaces with positive potential functions. In terms of particle absorption models, this technique allows turning a hard obstacle model into a soft obstacle particle model. Further details on this integration technique can be found in [163, 175].

Last, but not least, in some important applications, including Feynman-Kac models on path spaces, the limiting semigroups have very poor stability properties, in the sense that they do not forget their initial conditions. Nevertheless, in some situations it is still possible to control uniformly in time the quantities $g_{p,n}$.

12.2.3.1 Time discretization models

We consider the parameters (G_n, M_n) defined by

$$G_n(x) = e^{-V_n(x)h} \qquad M_n(x, dy) = (1 - \lambda_n h)\, \delta_x(dy) + \lambda_n h\, K_n(x, dy) \quad (12.16)$$

for some nonnegative and bounded function V_n, some positive parameter $\lambda_n \leq 1/h$ with $h > 0$, and some Markov transition K_n s.t.

$$\beta(K_n) \leq \kappa_n < 1, \qquad h \leq h_n = (1 - \kappa_n)/[v_{n-1} + \alpha] \quad \text{and} \quad \lambda_n \in {]}0, 1/h]$$

for some $\alpha > 0$. In the above display, $v_n = \mathrm{osc}(V_n)$ stands for the oscillations of V_n. We further assume that $v = \sup_n v_n < \infty$. In this situation, for any $\lambda_n \in \left[\frac{1}{h_n}, \frac{1}{h}\right]$, we have

$$
\begin{aligned}
g_{n-1}\beta(M_n) &\leq e^{v_{n-1}h}\left(1 - \lambda_n h\,(1 - \kappa_n)\right) \\
&\leq e^{-h(\lambda_n(1-\kappa_n) - v_{n-1})} \leq e^{-\alpha h}
\end{aligned}
$$

from which we conclude that $\mathbf{H_0(G, M)}$ is met with

$$g = \sup_n g_n \le e^{hv} \quad \text{and} \quad \rho \le e^{-\alpha h}$$

for some time step $h \in [0, 1]$ and some positive parameters (α, v). In this situation, using the fact that

$$e^{hv} - 1 \le hv \ e^{hv} \quad \text{and} \quad 1 - e^{-\alpha h} \ge \alpha h \ e^{-\alpha h}$$

we also prove the following upper bounds

$$(g - 1)/(1 - \rho) \le (v/\alpha) \ e^{(v+\alpha)h} \quad \text{and} \quad 1 - \rho^l \ge l\alpha h \ e^{-l\alpha h}$$

Using Theorem 12.2.6, and Corollary 12.2.7, we conclude that

$$g_{p,n} \vee \overline{\kappa}(0) \le \exp\left((v/\alpha) \ e^{(v+\alpha)h}\right) \quad \text{and} \quad \beta\left(P_{p,n}\right) \le e^{-\alpha(n-p)h}$$

and

$$\overline{\tau}_{k,l}(0) \le \frac{1}{l\alpha h} \ \exp\left(l\alpha h + k(v/\alpha) \ e^{(v+\alpha)h}\right) \tag{12.17}$$

12.2.3.2   Interacting simulated annealing model

We consider the Feynman-Kac annealing model discussed in (4.40). We further assume that $K_n^{k_n}(x, \cdot) \ge \epsilon_n \ \nu_n$, for some $k_n \ge 1$, some $\epsilon_n > 0$, and some measure ν_n. In this situation, we have

$$\mathcal{M}_{n,\beta_n}^{k_n}(x, \cdot) \ge K_n^{k_n}(x, \cdot) \ e^{-\beta_n k_n v} \ge \epsilon_n \ e^{-\beta_n k_n v} \ \nu_n$$

with $v := \mathrm{osc}(V)$. If we choose $m_n = k_n l_n$, this implies that

$$\beta(M_n) = \beta\left(\mathcal{M}_{n,\beta_n}^{m_n}\right) \le \beta\left(\mathcal{M}_{n,\beta_n}^{k_n}\right)^{l_n} \le \left(1 - \epsilon_n \ e^{-\beta_n k_n v}\right)^{l_n}$$

Therefore, for any given $\rho' \in \ ]0, 1[$ we can choose l_n such that

$$l_n \ge \frac{\log\left(1/\rho'\right) + v(\beta_n - \beta_{n-1})}{\log 1/(1 - \epsilon_n \ e^{-\beta_n k_n v})}$$

so that $g_{n-1}\beta(M_n) \le e^{v(\beta_n - \beta_{n-1})} \times \left(1 - \epsilon_n \ e^{-\beta_n k_n v}\right)^{l_n} \le \rho' \Rightarrow \rho \le \rho'$.

For any function $\beta \ : \ x \in [0, \infty[\mapsto \beta(x)$, with a decreasing derivative $\beta'(x)$ s.t. $\lim_{x \to \infty} \beta'(x) = 0$ and $\beta'(0) < \infty$, we also notice that

$$g = \sup_{n \ge 0} g_n \le \sup_{n \ge 0} e^{v\beta'(n)} \le e^{v\beta'(0)}$$

12.2.3.3 Historical processes

We return to the historical Feynman-Kac models introduced in Section 3.4. Using the equivalence principle (3.45), we have proved that the n-time marginal models associated with a Feynman-Kac model on path space coincide with the original Feynman-Kac measure defined in (1.37).

Definition 12.2.8 *We write* $\mathbf{Q}_{p,n}$ *and* $\mathbf{P}_{p,n}$ *the Feynman-Kac semigroups defined as* $Q_{p,n}$ *and* $P_{p,n}$, *by replacing* (G_n, M_n) *by* $(\mathbf{G}_n, \mathbf{M}_n)$.

By construction, for any $\mathbf{x}_p = (x_0, \ldots, x_p) \in E_p$, we have

$$\mathbf{Q}_{p,n}(1)(\mathbf{x}_p) = Q_{p,n}(1)(x_p)$$

and therefore

$$\mathbf{G}_{p,n}(\mathbf{x}_p) := \mathbf{Q}_{p,n}(1)(\mathbf{x}_p) \Longrightarrow \mathbf{g}_{p,n} := \sup_{\mathbf{x}_p, \mathbf{y}_p} \frac{\mathbf{G}_{p,n}(\mathbf{x}_p)}{\mathbf{G}_{p,n}(\mathbf{y}_p)} = \sup_{x_p, y_p} \frac{G_{p,n}(x_p)}{G_{p,n}(y_p)} = g_{p,n}$$

We emphasize that we cannot expect the Dobrushin's ergodic coefficient of the historical process semigroup to decrease, but we always have $\beta(\mathbf{P}_{p,n}) \leq 1$. Using Proposition 12.2.4, we readily prove the following proposition.

Proposition 12.2.9 *Assume that the reference Markov chain* X_n *satisfies the condition* $\mathbf{H}(\mathbf{G}, \mathbf{P})$, *stated on page 371, for some finite constants* v_1, v_2. *In this situation, we have*

$$\mathbf{g}_{p,n} \leq \exp v_1 \quad and \quad \beta(\mathbf{P}_{p,n}) \leq 1$$

as well as

$$\tau_{k,l}(n) \leq (n+1) \exp(k v_1) \quad and \quad \kappa(n) \leq \exp v_1 \qquad (12.18)$$

with the functions $\tau_{k,l}$ *and* κ *introduced, respectively, in (12.4) and (12.12).*

In addition, when the reference Markov chain X_n *satisfies the condition* $\mathbf{H}_\mathbf{m}(\mathbf{G}, \mathbf{M})$ *stated on page 372, for some* $m \geq 1$, *we have the estimates*

$$\mathbf{g}_{p,n} \leq \iota(m) \qquad \tau_{k,l}(n) \leq (n+1)\iota(m)^k \quad and \quad \kappa(n) \leq \iota(m) \qquad (12.19)$$

with the function $\iota(m)$ *introduced in (12.11).*

We end this section with some Markov chain Monte Carlo technique often used in practice to stabilize the genealogical tree based approximation model. To describe with some conciseness this stochastic method, we consider the Feynman-Kac measures $\eta_n \in \mathcal{P}(\mathbf{E}_n)$ associated with the potential function $\mathbf{G}_n$ and the Markov transitions $\mathbf{M}_n$ of the historical process defined, respectively, in (3.44) and in (3.1). In this notation, we notice that η_n satisfy the updating-prediction equation

$$\eta_n = \Psi_{\mathbf{G}_n}(\eta_n)\mathbf{M}_n$$

This equation on the set of measures on path spaces is unstable, in the sense that its initial condition is always kept in memory by the historical Markov transitions $\mathbf{M_n}$. One idea to stabilize this system is to incorporate an additional Markov chain Monte Carlo move at every time step. More formally, let us suppose that we have a dedicated Markov chain Monte Carlo transition $\mathbf{K_n}$ from the set $\mathbf{E_n}$ into itself, and such that

$$\eta_n = \eta_n \mathbf{K_n}$$

In this situation, we also have that

$$\eta_n = \Psi_{\mathbf{G}_n}(\eta_n) M'_n \quad \text{with} \quad M'_n := \mathbf{M_n K_n} \tag{12.20}$$

By construction, the mean field particle approximation of the Equation (12.20) is a genealogical tree type evolution model with path space particles on the state spaces $\mathbf{E_n}$. The updating-selection transitions are related to the potential function $\mathbf{G_n}$ on the state spaces $\mathbf{E_n}$, and the mutation-exploration mechanisms from $\mathbf{E_n}$ into $\mathbf{E_{n+1}}$ are dictated by the Markov transitions M'_{n+1}.

Notice that this mutation transition is decomposed into two different stages. Firstly, we extend the selected path-valued particles with an elementary move according to the Markov transition M_n. Then, from every of these extended paths, we perform a Markov chain Monte Carlo sample according to the Markov transition $\mathbf{K_n}$.

12.3   Semigroups of nonlinear Markov chain models

12.3.1   Description of the models

In Section 9.2, we presented a nonlinear type Markov chain interpretation of the flow of Feynman-Kac measures η_n, associated with a pair of potential functions and Markov transitions (G_n, M_n), on some measurable state spaces E_n. In this interpretation, for any $n \geq 0$, we recall that

$$\eta_n = \text{Law}(\overline{X}_n) \in \mathcal{P}(E_n)$$

where $(\overline{X}_n)_{n\geq 0}$ is a Markov chain on E_n, with an initial distribution $\eta_0 = \text{Law}(X_0) \in \mathcal{P}(E_0)$, and the elementary Markov transitions given by

$$\mathbb{P}\left(\overline{X}_{n+1} \in dx_{n+1} \mid \overline{X}_n = x_n\right) = K_{n+1,\eta_n}(x_n, dx_{n+1}) \quad \text{with} \quad \text{Law}(\overline{X}_n) = \eta_n$$

In the above display, K_{n+1,η_n} stands for a collection of Markov transitions satisfying the following compatibility conditions

$$\eta K_{n+1,\eta} = \Phi_{n+1}(\eta) = \Psi_{G_n}(\eta) M_{n+1}$$

for any $n \geq 0$ and any $\eta \in \mathcal{P}(E_n)$. Several examples of McKean models are discussed in (0.3). In the further development of this section, we assume that G_n are $]0,1]$-valued potential functions, and the transitions are given by

$$
K_{n+1,\eta_n}(x_n, dx_{n+1})
$$

$$
= G_n(x_n)\, M_{n+1}(x_n, dx_{n+1}) + (1 - G_n(x_n))\, \Psi_{G_n}(\eta_n) M_{n+1}(dx_{n+1})
$$
(12.21)

Definition 12.3.1 *For any $0 \leq p \leq n$ and any probability measure η on E_p we denote by $K^\eta_{p,n}$ the transition of the Markov chain $\overline{X}_p \rightsquigarrow \overline{X}_n$ starting at time p with the distribution η on E_p; that is we have that*

$$
\mathbb{P}\left(\overline{X}_n \in dx_n \mid \overline{X}_p = x_p\right) = K^\eta_{p,n}(x_p, dx_n) \quad \text{with} \quad \text{Law}(\overline{X}_p) = \eta
$$

For $p = n$, we use the convention $K^\eta_{n,n} = Id$, the identity operator.

An alternative description is given by the semigroup formulae

$$
\begin{aligned}
K^\eta_{p,n} &= K_{p+1,\eta} K^{\Phi_{p,p+1}(\eta)}_{p+1,n} \\
&= K_{p+1,\eta} K_{p+2,\Phi_{p,p+1}(\eta)} K_{p+3,\Phi_{p,p+2}(\eta)} \cdots K_{n,\Phi_{p,n-1}(\eta)}
\end{aligned}
$$

for any $0 \leq p \leq n$, with $K^\eta_{n-1,n} = K_{n,\eta}$.
By construction, we have the following semigroup properties.

Lemma 12.3.2 *For any $0 \leq p \leq q \leq n$ and any $\eta \in \mathcal{P}(E_p)$, we have the formulae*

$$
K^\eta_{p,n} = K^\eta_{p,q} K^{\Phi_{p,q}(\eta)}_{q,n} \quad \text{and} \quad \eta K^\eta_{p,n} = \Phi_{p,n}(\eta)
$$

where $\Phi_{p,n}$ stands for the Feynman-Kac semigroup defined in Section 12.1.

12.3.2 A particle absorption interpretation

In the following proposition, we provide a more explicit representation of the Markov transitions $K^\eta_{p,n}$ associated with the McKean transitions (12.21) in terms of the semigroup $Q_{p,n}$ of the unnormalized measures $\gamma_n = \gamma_p Q_{p,n}$ and the semigroup $\Phi_{p,n}$ of the normalized distributions $\eta_n = \Phi_{p,n}(\eta_p)$.

Proposition 12.3.3 *For any $0 \leq p \leq n$ and any probability measure η on E_p we have the following formulae*

$$
K^\eta_{p,n}(x_p, dx_n) = G_{p,n}(x_p)\, P_{p,n}(x_p, dx_n) + (1 - G_{p,n}(x_p))\, \Phi_{p,n}(\eta)(dx_n)
$$

Proof:
We prove the proposition using a backward induction on the parameter $p(\leq n)$. For $p = n$ or for $p = (n-1)$, the formulae result from the fact that

$K^{\eta}_{n,n} = Q_{n,n} = Id$, and $K^{\eta}_{n-1,n} = K_{n,\eta}$. We further assume that the formula has been proved for some $p(\leq n)$. Using the fact that

$$
\begin{aligned}
K^{\eta}_{p,n}(f)(x) &= Q_{p,n}(f)(x) + (1 - G_{p,n}(x))\, \Phi_{p,n}(\eta)(f) \\
&= \Phi_{p,n}(\eta)(f) + Q_{p,n}\left[f - \Phi_{p,n}(\eta)(f)\right](x)
\end{aligned}
$$

for any $x \in E_p$, we readily check that

$$
K^{\Phi_{p-1,p}(\eta)}_{p,n}(f) = \Phi_{p-1,n}(\eta)(f) + Q_{p,n}\left[f - \Phi_{p-1,n}(\eta)(f)\right]
$$

This implies that

$$
\begin{aligned}
K^{\eta}_{p-1,n}(f)(x) &= K_{p,\eta}\left(K^{\Phi_p(\eta)}_{p,n}(f)\right)(x) \\
&= \Phi_{p-1,n}(\eta)(f) + K_{p,\eta}\left\{Q_{p,n}\left[f - \Phi_{p-1,n}(\eta)(f)\right]\right\}(x)
\end{aligned}
$$

On the other hand, we have

$$
\Phi_{p-1,p}(\eta)\left\{Q_{p,n}\left[f - \Phi_{p-1,n}(\eta)(f)\right]\right\}
$$

$$
= \frac{1}{\eta(G_{p-1})}\, \eta Q_{p-1,n}\left[f - \Phi_{p-1,n}(\eta)(f)\right] = 0
$$

from which we conclude that

$$
K^{\eta}_{p-1,n}(f) = \Phi_{p-1,n}(\eta)(f) + Q_{p-1,n}\left[f - \Phi_{p-1,n}(\eta)(f)\right]
$$

This ends the proof of the proposition. ∎

In terms of the particle absorption models discussed in Section 3.3.2 and in Chapter 7, we have the following interpretation. The potential function $G_{p,n} := Q_{p,n}(1)$ is the probability of nonabsorption of the state at time n, starting at $x_p \in E_p$ at time p; and the Markov transition

$$
P_{p,n}(x_p, d_n) := \frac{Q_{p,n}(x_p, d_n)}{Q_{p,n}(1)(x_p)} = \Phi_{p,n}\left(\delta_{x_p}\right)(dx_n)
$$

represents the probability of a transition from a location x_p at time p to a state x_n at time n, given the fact that the particle has not been absorbed. In summary, we have the following interpretation

$$
K^{\eta}_{p,n}(x_p, dx_n)
$$

$$
= \underbrace{G_{p,n}(x_p)}_{\text{nonabsorption proba}} \underbrace{P_{p,n}(x_p, dx_n)}_{\text{nonabsorption transition}} + \underbrace{(1 - G_{p,n}(x_p))}_{\text{absorption proba}} \underbrace{\Phi_{p,n}(\eta)(dx_n)}_{\text{law of the selected state}}
$$

$$
\tag{12.22}
$$

12.3.3   Quantitative contraction estimates

In this section, we provide a brief discussion of the regularity properties of the semigroups $K_{p,n}^{\eta}$ of the nonlinear Markov chain models defined in Section 12.3.1.

Firstly, by the Definition (12.21) of K_{n+1,η_n}, we have

$$K_{n+1,\eta_n}(x_n,.) \geq (1 - \|G_n\|)\, \Psi_{G_n}(\eta_n) M_{n+1} \Rightarrow \beta\left(K_{n+1,\eta_n}\right) \leq \|G_n\|$$

This implies that for any $0 \leq p \leq n$ and any probability measure η on E

$$\sup_n \|G_n\| := \|G\| < 1 \Longrightarrow \beta\left(K_{p,n}^{\eta}\right) \leq \prod_{p \leq q < n} \|G_q\| \leq \|G\|^{(n-p)}$$

This shows that the nonlinear Markov chain $\overline{X}_n$ starting at time p with the distribution $\eta = \mathrm{Law}(\overline{X}_p)$ merges exponentially fast random states starting from two different locations, as soon as the absorption rate is sufficiently large. This result is not really surprising. At every time step $q \geq p$ the random state $\overline{X}_q$ has a chance to forget instantly its initial starting point $\overline{X}_p$ with a probability $(1 - G_q(\overline{X}_q))$, and to start in a new location randomly chosen with the distribution $\Phi_{p,q+1}(\eta)$.

This observation also indicates that the law of the random state $\overline{X}_n$ starting at time p on some location $\overline{X}_p = x_p$ randomly chosen with the distribution $\eta = \mathrm{Law}(\overline{X}_p)$ is very close to $\Phi_{p,n}(\eta)$.

Indeed, using the particle absorption interpretation formula presented in Proposition 12.3.3, we have

$$\left[K_{p,n}^{\eta}(x_p, dx_n) - \Phi_{p,n}(\eta)(dx_n)\right] = G_{p,n}(x_p)\left[P_{p,n}(x_p, dx_n) - \Phi_{p,n}(\eta)(dx_n)\right]$$

Using the fact that

$$\mu_{x_p} = \delta_{x_p} \Rightarrow P_{p,n}(x_p, dx_n) = \Phi_{p,n}(\mu_{x_p})(dx_n)$$

we find that

$$\left[K_{p,n}^{\eta}(x_p, dx_n) - \Phi_{p,n}(\eta)(y)\right] = G_{p,n}(x_p)\left[\Phi_{p,n}(\mu_{x_p}) - \Phi_{p,n}(\eta)\right](dx_n)$$

Under the regularity conditions stated in Theorem 12.2.5, we prove that

$$\left\|K_{p,n}^{\eta}(f) - \Phi_{p,n}(\eta)(f)\right\| \leq \rho_m\left((1 - \kappa_m)g\right)^{(n-p)/m}$$

for some parameters $\rho_m < \infty$ and $\kappa_m \in]0,1]$ satisfying (12.14).

Much more is true. Using (12.22), we also prove that

$$K_{p,n}^{\eta} = S_{G_{p,n},\eta} P_{p,n} \Longrightarrow \beta\left(K_{p,n}^{\eta}\right) \leq \beta\left(S_{G_{p,n},\eta}\right) \times \beta\left(P_{p,n}\right)$$

Recalling that

$$S_{G_{p,n},\eta}(x,y) \geq (1 - \|G_{p,n}\|)\, \Psi_{G_{p,n}}(\eta) \Rightarrow \beta\left(S_{G_{p,n},\eta}\right) \leq \|G_{p,n}\|$$

we readily prove the following theorem.

Theorem 12.3.4 *For any $0 \leq p \leq n$, we have the contraction estimate*

$$\beta\left(K_{p,n}^{\eta}\right) \leq \|G_{p,n}\| \times \beta\left(P_{p,n}\right)$$

and for any probability measures η and μ we have

$$\|\Phi_{p,n}(\eta) - \Phi_{p,n}(\mu)\|_{tv} \leq \frac{1}{\mu(G_{p,n}) \vee \eta(G_{p,n})} \; \beta\left(K_{p,n}^{\eta}\right)$$

The last assertion in the theorem comes from the decomposition

$$\begin{aligned}
\Phi_{p,n}(\eta) - \Phi_{p,n}(\mu) &= [\eta - \mu]K_{p,n}^{\eta} + \mu\left[K_{p,n}^{\eta} - K_{p,n}^{\mu}\right] \\
&= [\eta - \mu]K_{p,n}^{\eta} + (1 - \mu(G_{p,n}))\left[\Phi_{p,n}(\eta) - \Phi_{p,n}(\mu)\right]
\end{aligned}$$

This ends the proof of the theorem.　　　■

12.4　Backward Markovian semigroups

In this section, we further develop the semigroup analysis of the Backward Markov chain models introduced in Section 1.4.3.4.

In the first section, Section 12.4.1, we recall some basic facts about dual Markov transitions and time reversal Feynman-Kac formulae. In Section 12.4.1, we provide a dual Markovian interpretation of the semigroups $P_{n,p}$ and $R_{p+1}^{(n)}$ introduced in Section 12.1 and in Lemma 12.2.2.

In the second section, Section 12.4.2, we only discuss some more or less direct consequences of these Markov semigroup descriptions, deriving some new strong quantitative contraction estimates w.r.t. the total variation norm.

12.4.1　Dual Markov transition models

We further assume that the Markov transitions M_n are absolutely continuous with respect to some measures λ_n on E_n, and for any $(x, y) \in (E_{n-1} \times E_n)$ we have

$$H_n(x, y) := \frac{dM_n(x, \cdot)}{d\lambda_n}(y) > 0 \qquad (12.23)$$

Definition 12.4.1 *For any $n \geq 1$, we set*

$$H_{n-1,n}(x_{n-1}, x_n) = G_{n-1}(x_{n-1})H_n(x_{n-1}, x_n)$$

In this notation, we have the following backward formula

$$\mathbb{Q}_n(d(x_0, \ldots, x_n)) = \eta_n(dx_n) \prod_{q=1}^{n} \mathbb{M}_{q,\eta_{q-1}}(x_q, dx_{q-1})$$

with the collection of Markov transitions defined by

$$\mathbb{M}_{n+1,\eta_n}(x,dy) \propto H_{n,n+1}(y,x)\,\eta_n(dy)$$

Definition 12.4.2 *For any $p \leq n$, the $\mathbb{Q}_n$-conditional distribution of the p-th variable x_p given the n-th coordinate x_n is given by*

$$\mathbb{M}_{n,p,\eta_p}(x_n,dx_p) = \mathbb{M}_{n,\eta_{n-1}}\cdots\mathbb{M}_{p+1,\eta_p}(x_n,dx_p)$$

We notice that for any $0 \leq p \leq q \leq n$, these conditional distributions satisfy the semigroup property

$$\mathbb{M}_{n,p,\eta_p} = \mathbb{M}_{n,q,\eta_q}\mathbb{M}_{q,p,\eta_p}$$

Definition 12.4.3 *We denote by $\eta_{p|n}$ the p-th marginal of the measure $\mathbb{Q}_n$.*

By construction, we have that $\eta_{p|n} := \eta_n \mathbb{M}_{n,p,\eta_p}$. One simplification of Definition 12.4.1 comes from the following technical lemma.

Lemma 12.4.4 *For any $0 \leq p \leq n$, we have*

$$Q_{p,n}(x_p,dx_n) = H_{p,n}(x_p,x_n)\,\lambda_n(dx_n)$$

with a collection of functions $H_{p,n}$ on $(E_p \times E_n)$ satisfying for any $p \leq q \leq n$ the following recursive equation

$$
\begin{aligned}
H_{p,n}(x_p,x_n) &= \int Q_{p,q}(x_p,dx_q)H_{q,n}(x_q,x_n)\\
&= \int H_{p,q}(x_p,x_q)H_{q,n}(x_q,x_n)\,\lambda_q(dx_q) \qquad (12.24)
\end{aligned}
$$

The result is a direct consequence of the semigroup properties of $Q_{p,n}$; thus it is omitted. The next lemma provides an interpretation of the measures $\eta_{p|n}$ and $\mathbb{M}_{n,p,\eta_p}$ in terms of the functions $G_{p,n}$ and $H_{p,n}$.

Lemma 12.4.5 *For any $0 \leq p \leq q \leq n$, we have*

$$\eta_{q|n} = \Psi_{G_{q,n}}(\eta_q) = \eta_{p|n}R_{p+1}^{(n)}\cdots R_q^{(n)}$$

with the transitions $R_{p+1}^{(n)}$ given in Lemma 12.2.2 In addition,

$$\mathbb{M}_{n,p,\eta_p}(x_n,dx_p) \propto \eta_p(dx_p)H_{p,n}(x_p,x_n)$$

Proof:
Notice that the $\mathbb{Q}_n$ distribution of the pair (p,n)-coordinates (x_p,x_n) is given by the formula

$$
\begin{aligned}
\frac{\mathcal{Z}_p}{\mathcal{Z}_n}\,\eta_p(dx_p)\,Q_{p,n}(x_p,dx_n) &:= \frac{\mathcal{Z}_p}{\mathcal{Z}_n}\,\eta_p(dx_p)H_{p,n}(x_p,x_n)\,\lambda_n(dx_n)\\
&= \eta_n(dx_n)\left(\mathbb{M}_{n,\eta_{n-1}}\cdots\mathbb{M}_{p+1,\eta_p}\right)(x_n,dx_p)
\end{aligned}
$$

Thus, the law of the last n-th coordinate x_n is given by

$$\frac{\mathcal{Z}_p}{\mathcal{Z}_n} \, \eta_p(H_{p,n}(.,x_n)) \, \lambda_n(dx_n) = \eta_n(dx_n)$$

We also notice that the p-th marginal is given by

$$\frac{1}{\mathcal{Z}_n/\mathcal{Z}_p} \, \eta_p(dx_p) \, G_{p,n}(x_p) = \eta_n \mathbb{M}_{n,\eta_{n-1}} \cdots \mathbb{M}_{p+1,\eta_p}(dx_p) = \Psi_{G_{p,n}}(\eta_p)(dx_p)$$

The r.h.s. in the above display comes from the fact that

$$\mathcal{Z}_n/\mathcal{Z}_p = \gamma_p(G_{p,n})/\gamma_p(1) = \eta_p(G_{p,n})$$

This ends the proof of the second assertion. Finally, using the fact that

$$G_{p,n} = Q_{p,n}(1) = Q_{p+1}(G_{p+1,n})$$

we prove that

$$\Psi_{G_{p,n}}(\eta_p)R_{p+1}^{(n)}(f) \quad = \quad \frac{\eta_p Q_{p+1}(G_{p+1,n}f)}{\eta_p Q_{p+1}(G_{p+1,n})} = \Psi_{G_{p+1,n}}(\eta_{p+1})$$

This implies that

$$\eta_{p|n} R_{p+1}^{(n)} = \eta_{p+1|n}$$

We end the proof of the first assertion using an induction on the parameter p. The proof of the lemma is now completed.

∎

12.4.2 Quantitative contraction estimates

We further assume that the regularity condition (12.23) is met. In this situation, we recall that $\eta_n \ll \lambda_n$.

Subsequently in this section, we let $(\mathbb{Q}'_n, \Gamma'_n, \eta'_n, \gamma'_n, \mathcal{Z}'_n)$ be the Feynman-Kac models defined as $(\mathbb{Q}_n, \Gamma_n, \eta_n, \gamma_n, \mathcal{Z}_n)$ by replacing the initial measure η_0 by some probability measure $\eta'_0 \ll \eta_0 \in \mathcal{P}(E_0)$. In this notation, we have

$$\eta'_0 \ll \eta_0 \Rightarrow \frac{1}{\mathcal{Z}'_n} \, \eta'_0(H_{0,n}(.,x_n)) \, \lambda_n(dx_n) = \eta'_n(dx_n) \ll \eta_n(dx_n)$$

Definition 12.4.6 *For any initial distributions $\eta'_0 \ll \eta_0 \in \mathcal{P}(E_0)$, and/or any $n \geq 0$ we set*

$$h_n := d\eta'_n/d\eta_n$$

Lemma 12.4.7 *For any initial distributions $\eta_0' \ll \eta_0 \in \mathcal{P}(E_0)$, and any $p \leq n$, we have the functional formula*

$$h_n = \mathbb{M}_{n,p,\eta_p}(h_p)/\eta_n(\mathbb{M}_{n,p,\eta_p}(h_p))$$

In particular, we have the following recursion

$$h_n = \mathbb{M}_{n,\eta_{n-1}}(h_{n-1})/\eta_n(\mathbb{M}_{n,\eta_{n-1}}(h_{n-1}))$$

Proof:
We use the fact that

$$
\begin{aligned}
\Gamma'_n(d(x_0,\dots,x_n)) &= \Gamma_n(d(x_0,\dots,x_n))\ h_0(x_0) \\
&= \gamma_n(dx_n)\left\{\prod_{1\leq p\leq n}\mathbb{M}_{p,\eta_{p-1}}(x_p,dx_{p-1})\right\}h_0(x_0)
\end{aligned}
$$

to prove that

$$\gamma'_n(dx_n) = \gamma_n(dx_n) \times \mathbb{M}_{n,0,\eta_0}(h_0)(x_n) \Leftrightarrow d\gamma'_n/d\gamma_n = \mathbb{M}_{n,0,\eta_0}(h_0)$$

We also observe that

$$h_n = \frac{d\eta'_n}{d\eta_n} = \frac{\mathcal{Z}_n}{\mathcal{Z}'_n} \times \frac{d\gamma'_n}{d\gamma_n} = \frac{1}{\eta_n(d\gamma'_n/d\gamma_n)} \times \frac{d\gamma'_n}{d\gamma_n}$$

We end the proof using the fact that

$$h_n = \frac{\mathbb{M}_{n,0,\eta_0}(h_0)}{\eta_n(\mathbb{M}_{n,0,\eta_0}(h_0))} = \frac{\mathbb{M}_{n,p,\eta_p}\mathbb{M}_{p,0,\eta_0}(h_0)}{\eta_n(\mathbb{M}_{n,p,\eta_p}\mathbb{M}_{p,0,\eta_0}(h_0))} = \frac{\mathbb{M}_{n,p,\eta_p}(h_p)}{\eta_n(\mathbb{M}_{n,p,\eta_p}(h_p))}$$

This ends the proof of the lemma. ∎

By construction, we have the decomposition

$$h_n - 1 = \frac{1}{\eta_{0|n}(h_0)}\left[\mathbb{M}_{n,\eta_{n-1}}\cdots\mathbb{M}_{1,\eta_0}(h_0) - \eta_n\mathbb{M}_{n,\eta_{n-1}}\cdots\mathbb{M}_{1,\eta_0}(h_0)\right]$$

This formula expresses the oscillations of $h_n - 1$ in terms of the oscillations of the the function $\mathbb{M}_{n,\eta_{n-1}}\cdots\mathbb{M}_{1,\eta_0}(h_0)$.

We are now in position to state and prove the following quantitative contraction estimate.

Theorem 12.4.8 *We assume that $\eta_0' \ll \eta_0$ and $d\eta_0'/d\eta_0 \in [1/\rho, \rho]$, for some finite $\rho \geq 1$. For any $n \geq 0$, we have the uniform contraction estimate*

$$\|1 - d\eta'_n/d\eta_n\| \leq (\rho^2 - 1)\ \beta(\mathbb{M}_{n,\eta_{n-1}}\cdots\mathbb{M}_{1,\eta_0})$$

In addition, assume that the following condition is satisfied for η_n-almost every $x \in E_n$ and any $n \geq 0$

$$\mathbf{H_m(Q)}: \qquad H_{n,n+m}(x,y) \leq \tau_m \ H_{n,n+m}(x,y')$$

for some finite constant $\tau_m < \infty$, and some integer $m \geq 1$. In this situation, we have

$$\|1 - d\eta'_n/d\eta_n\| \leq (\rho^2 - 1) \ \left(1 - \tau_m^{-2}\right)^{\lfloor n/m \rfloor}$$

Proof:

The first assertion comes from the definition of the Dobrushin contraction coefficient and the fact that

$$1/\rho \leq h_0 \leq \rho \Rightarrow \text{osc}(h_0) \leq \rho - 1/\rho$$

To prove the last assertion, we observe that

$$\begin{aligned}
\mathbb{M}_{p+m,p,\eta_p}(x_n, dx_p) &= \left[\mathbb{M}_{p+m,\eta_{p+m-1}} \cdots \mathbb{M}_{p+1,\eta_p}\right](x_{p+m}, dx_p) \\
&= \frac{\eta_p(dx_p)H_{p,p+m}(x_p, x_{p+m})}{\eta_p\left(H_{p,p+m}(., x_{p+m})\right)}
\end{aligned}$$

Under our assumption, we have

$$\mathbb{M}_{p+m,p,\eta_p}(x_{p+m}, x_p) \leq \tau_m^2 \ \mathbb{M}_{p+m,p,\eta_p}(x'_{p+m}, x_p) \Rightarrow \beta\left(\mathbb{M}_{p+m,p,\eta_p}\right) \leq 1 - \tau_m^{-2}$$

Finally, we use the decomposition

$$\mathbb{M}_{km,\eta_{km-1}} \cdots \mathbb{M}_{1,\eta_0} = \mathbb{M}_{km,(k-1)m,\eta_p} \cdots \mathbb{M}_{2m,m,\eta_m} \mathbb{M}_{m,0,\eta_0}$$

to check that

$$\beta\left(\mathbb{M}_{n,\eta_{n-1}} \cdots \mathbb{M}_{1,\eta_0}\right) \leq \left(1 - \tau_m^{-2}\right)^{\lfloor n/m \rfloor}$$

This ends the proof of the theorem. ∎

Chapter 13

Intensity measure semigroups

13.1 Spatial branching models

In this section we further develop the semigroup analysis of the intensity measures of spatial branching processes discussed in Section 6.1 We consider a sequence of nonnegative measures μ_n and a collection of nonnegative potential functions G_n on some measurable state spaces E_n. We also let M_n be some Markov transitions from E_{n-1} into E_n. We associate with these objects the flow of nonnegative measures

$$\gamma_{n+1} = \gamma_n Q_{n+1} + \mu_{n+1} \tag{13.1}$$

with the initial condition $\gamma_0 = \mu_0$, and the integral operator Q_{n+1} from E_n into E_{n+1} is defined by

$$Q_{n+1}(x_n, dx_{n+1}) := G_n(x_n) \ M_{n+1}(x_n, dx_{n+1}) \tag{13.2}$$

These distributions can be interpreted as the intensity measures of a spatial branching process. We refer the reader to Chapter 6 for a detailed discussion on these spatial point processes.

Before concluding this section, we note that

$$\mu_n = 0 \Longrightarrow \gamma_{n+1} = \gamma_n Q_{n+1} \tag{13.3}$$

In this particular situation, the solution of the equation (6.3) is given by the following Feynman-Kac path integral formulae

$$\gamma_n(f) = \gamma_0(1) \ \mathbb{E} \left(f(X_n) \prod_{0 \leq p < n} G_p(X_p) \right) \tag{13.4}$$

where X_n stands for a Markov chain taking values in the state spaces E_n with initial distribution $\eta_0 = \gamma_0/\gamma_0(1)$ and Markov transitions M_n.

For a detailed discussion on these models, we refer to Chapter 12, dedicated to Feynman-Kac models, and the regularity properties of their semigroups.

In this specific situation, we recall that the distributions γ_n are given, for any $p \leq n$, by

$$\gamma_n = \gamma_p Q_{p,n} \quad \text{with} \quad Q_{p,n} = Q_{p+1} \ldots Q_{n-1} Q_n \tag{13.5}$$

For $p = n$, we use the convention $Q_{n,n} = Id$. We recall that the nonlinear semigroup associated to this sequence of distributions is given by

$$\eta_n = \Phi_{p,n}(\eta_p) := \Psi_{G_{p,n}}(\eta_p)P_{p,n} \qquad (13.6)$$

with the potential functions $G_{p,n}$ and Markov kernel $P_{p,n}$ defined by

$$G_{p,n} := Q_{p,n}(1) \quad \text{and} \quad P_{p,n}(x_p, dx_n) = Q_{p,n}(x_p, dx_n)/Q_{p,n}(x_p, E_n)$$

We also consider the parameters:

$$g_{p,n} = \sup_{x,y} \frac{G_{p,n}(x)}{G_{p,n}(y)} \quad \text{and} \quad \beta(P_{p,n}) = \sup_{x,y \in E_p} \|P_{p,n}(x, .) - P_{p,n}(y, .)\|_{\mathrm{tv}} \quad (13.7)$$

as well as the pair of parameters $(g_-(n), g_+(n))$ defined below

$$g_-(n) = \inf_{0 \leq p < n} \inf_{E_p} G_p \leq \sup_{0 \leq p < n} \sup_{E_p} G_p = g_+(n)$$

We also write $g_{-/+}(n)$ to refer to both parameters.

Definition 13.1.1 *We denote by I_n the compact intervals defined by*

$$I_n := [m_-(n), m_+(n)] \quad \text{where} \quad m_{-/+}(n) := \sum_{p=0}^{n} \mu_p(1)g_{-/+}(n)^{(n-p)} \quad (13.8)$$

13.1.1   Extended Feynman-Kac semigroups

In this section, we further develop the regularity properties of the evolution equations discussed in Section 6.1.3. For the convenience of the reader, we recall some basic notation. We denote by $\eta_n \in \mathcal{P}(E_n)$ the normalized distributions associated with the measures γ_n defined in (13.1); that is, for any $f_n \in \mathcal{B}_b(E_n)$, we have

$$\eta_n(f_n) = \gamma_n(f_n)/\gamma_n(1)$$

The following lemma collects some important properties of the sequence of intensity measures γ_n.

Lemma 13.1.2 *For any $0 \leq p \leq n$, we have the semigroup decomposition*

$$\gamma_n = \gamma_p Q_{p,n} + \sum_{p < q \leq n} \mu_q Q_{q,n} \quad \text{and} \quad \gamma_n = \sum_{0 \leq p \leq n} \mu_p Q_{p,n} \quad (13.9)$$

In addition, we also have the following formula

$$\gamma_n(1) = \sum_{p=0}^{n} \mu_p(1) \prod_{p \leq q < n} \eta_q(G_q) \quad (13.10)$$

In addition, we have $\gamma_n(1) \in I_n$, with the compact interval I_n defined in (13.8).

Proof:

The first couple of formulae are easily proved using a simple induction, and recalling that $\gamma_0 = \mu_0$. To prove the last assertion, we use an induction on the parameter $n \geq 0$. The result is obvious for $n = 0$. We also have by (6.3)

$$\gamma_{n+1}(1) = \gamma_n Q_{n+1}(1) + \mu_{n+1}(1) = \gamma_n(G_n) + \mu_{n+1}(1)$$

$$\Rightarrow \gamma_{n+1}(1) = \gamma_n(1) \, \eta_n(G_n) + \mu_{n+1}(1)$$

$$= \gamma_0(1) \prod_{p=0}^{n} \eta_p(G_p) + \sum_{p=1}^{n+1} \mu_p(1) \prod_{p \leq q \leq n} \eta_q(G_q)$$

Recalling that $\gamma_0(dx_0) = \mu_0(dx_0)$, we prove (13.10). The last assertion is immediate, and thus it is skipped. This ends the proof of the lemma. ∎

We let

$$\Lambda_{n+1} : (\mathbb{R}_+ \times \mathcal{P}(E_n)) \mapsto (\mathbb{R}_+ \times \mathcal{P}(E_{n+1}))$$

be the one step mapping associated with the flow

$$(\gamma_{n+1}(1), \eta_{n+1}) = \Lambda_{n+1} (\gamma_n(1), \eta_n) = \left(\Lambda^1_{n+1} (\gamma_n(1), \eta_n), \Lambda^2_{n+1} (\gamma_n(1), \eta_n) \right)$$

We also denote by $(\Lambda_{p,n})_{0 \leq p \leq n}$ the corresponding semigroup defined by

$$\forall 0 \leq p \leq n \qquad \Lambda_{p,n} = \Lambda_{p+1,n} \Lambda_{p+1} = \Lambda_n \Lambda_{n-1} \dots \Lambda_{p+1} \quad \text{with} \ \Lambda_{n,n} = Id$$

Using Lemma 13.1.2, one proves that the semigroup $\Lambda_{p,n}$ satisfies the pair of formulae described below:

Proposition 13.1.3 *For any $0 \leq p \leq n$, we have*

$$\Lambda^1_{p,n}(m, \eta) = m \, \eta Q_{p,n}(1) + \sum_{p < q \leq n} \mu_q Q_{q,n}(1) \tag{13.11}$$

$$\Lambda^2_{p,n}(m, \eta) = \alpha_{p,n} (m, \eta) \, \Phi_{p,n}(\eta)$$
$$+ (1 - \alpha_{p,n} (m, \eta)) \sum_{p < q \leq n} \frac{c_{q,n}}{\sum_{p < r \leq n} c_{r,n}} \Phi_{q,n}(\overline{\mu}_q) \tag{13.12}$$

with the collection of parameters $c_{p,n} := \mu_p Q_{p,n}(1)$, and the $[0,1]$-valued parameters $\alpha_{p,n} (m, \eta)$, defined below

$$\alpha_{p,n} (m, \eta) = \frac{m \eta Q_{p,n}(1)}{m \eta Q_{p,n}(1) + \sum_{p < q \leq n} c_{q,n}}$$

$$\leq \alpha^\star_{p,n}(m) := 1 \wedge \left[m \left\| \frac{Q_{p,n}(1)}{\sum_{p < q \leq n} c_{q,n}} \right\| \right] \tag{13.13}$$

We return to the spatial branching interpretation of the measures γ_n discussed in Section 6.1.1. In this context, one central question is the long time behavior of the total mass process $\gamma_n(1)$.

Notice that $\gamma_n(1) = \mathbb{E}(\mathcal{X}_n(1))$ is the expected size of the n-th generation. For time homogeneous models with null spontaneous branching $\mu_n = \mu = 0$, the exponential growth of these quantities is related to the logarithmic Lyapunov exponents of the Feynman-Kac semigroup $Q_{p,n}$.

The prototype of these models is the Galton-Watson branching process. In this context three typical situations may occur:

1) $\gamma_n(1)$ remains constant and equal to the initial mean number of individuals; 2) $\gamma_n(1)$ goes exponentially fast to 0; 3) $\gamma_n(1)$ grows exponentially fast to infinity.

The analysis of spatial branching point processes with $\mu_n = \mu \neq 0$ is slightly more involved. Loosely speaking, in the first situation discussed above the total mass process is generally strictly increasing; while in the second situation the additional mass injected in the system stabilizes the process.

13.1.2 Stability and Lipschitz regularity properties

In this section, we present a framework that allows transfering the regularity properties of the Feynman-Kac semigroups $\Phi_{p,n}$ to the semigroup $\Lambda_{p,n}$.

Definition 13.1.4 *For any $p \leq n$ and $\eta \in \mathcal{P}(E_p)$, we denote by $\mathcal{D}_{p,n,\eta}$ the integral operators from $\mathrm{Osc}(E_n)$ into $\mathrm{Osc}(E_p)$ defined by*

$$\mathcal{D}_{p,n,\eta}(f_n) = \frac{1}{2\|G_{p,n}\|\beta(P_{p,n})} \, G_{p,n} \, P_{p,n} \, (f_n - \Phi_{p,n}(\eta)(f_n)) \in \mathrm{Osc}(E_p) \tag{13.14}$$

By definition of $\Phi_{p,n}$, for any $f_n \in \mathrm{Osc}(E_n)$ and $p \leq n$ we have the decomposition

$$[\Phi_{p,n}(\mu) - \Phi_{p,n}(\eta)] (f) := \frac{2\|G_{p,n}\|\beta(P_{p,n})}{\eta(G_{p,n})} \, (\mu - \eta) \, \mathcal{D}_{p,n,\eta}(f_n)$$

This yields the following lemma.

Lemma 13.1.5 *For any $0 \leq p \leq n$, any $\eta, \mu \in \mathcal{P}(E_p)$, and any $f \in \mathrm{Osc}_1(E_n)$, we have*

$$|[\Phi_{p,n}(\mu) - \Phi_{p,n}(\eta)] (f)| \leq 2 \, g_{p,n} \, \beta(P_{p,n}) \, |(\mu - \eta)\mathcal{D}_{p,n,\eta}(f)| \tag{13.15}$$

Proposition 13.1.6 *For any $0 \leq p \leq n$, any $\eta, \eta' \in \mathcal{P}(E_p)$, and any $f \in \mathrm{Osc}(E_n)$, there exists a collection of functions $\mathcal{D}_{p,n,\eta'}(f) \in \mathrm{Osc}_1(E_p)$ whose values only depend on the parameters (p, n, η) and such that, for any $m \in I_p$, we have*

$$\left|\left[\Lambda_{p,n}^2(m, \eta) - \Lambda_{p,n}^2(m, \eta')\right] (f)\right|$$

$$\tag{13.16}$$

$$\leq 2 \, \alpha_{p,n}^{\star} \, g_{p,n} \, [\beta(P_{p,n}) \, |(\eta - \eta')\mathcal{D}_{p,n,\eta'}(f)| + \beta_{p,n} \, |(\eta - \eta')h_{p,n,\eta'}|]$$

with the collection of functions

$$h_{p,n,\eta'} = \frac{1}{2g_{p,n}} \frac{Q_{p,n}(1)}{\eta' Q_{p,n}(1)} \in \mathrm{Osc}(E_p)$$

and the sequence of parameters $\alpha^\star_{p,n}$ and $\beta_{p,n}$ defined below

$$\alpha^\star_{p,n} := \alpha^\star_{p,n}(m_+(p)) \quad and \quad \beta_{p,n} := \sum_{p<q\leq n} \frac{c_{q,n}}{\sum_{p<r\leq n} c_{r,n}} \beta(P_{q,n}) \qquad (13.17)$$

In the above display, $\alpha^\star_{p,n}(m)$, and resp. $m_+(p)$, stands for the parameter defined in (13.13), and resp. (13.8).

Proof:

First, we observe that

$$\Lambda^2_{p,n}(m,\eta) - \Lambda^2_{p,n}(m',\eta')$$

$$= \alpha_{p,n}(m,\eta) \left[\Phi_{p,n}(\eta) - \sum_{p<q\leq n} \frac{c_{q,n}}{\sum_{p<r\leq n} c_{r,n}} \Phi_{q,n}(\overline{\mu}_q) \right]$$

$$- \alpha_{p,n}(m',\eta') \left[\Phi_{p,n}(\eta') - \sum_{p<q\leq n} \frac{c_{q,n}}{\sum_{p<r\leq n} c_{r,n}} \Phi_{q,n}(\overline{\mu}_q) \right]$$

Using the following decomposition

$$ab - a'b' = a'(b - b') + (a - a')b' + (a - a')(b - b') \qquad (13.18)$$

which is valid for any $a, a', b, b' \in \mathbb{R}$, we prove that

$$\Lambda^2_{p,n}(m,\eta) - \Lambda^2_{p,n}(m',\eta')$$

$$= \alpha_{p,n}(m',\eta') [\Phi_{p,n}(\eta) - \Phi_{p,n}(\eta')]$$

$$+ \left[\Phi_{p,n}(\eta') - \sum_{p<q\leq n} \frac{c_{q,n}}{\sum_{p<r\leq n} c_{r,n}} \Phi_{q,n}(\overline{\mu}_q) \right] [\alpha_{p,n}(m,\eta) - \alpha_{p,n}(m',\eta')]$$

$$+ [\alpha_{p,n}(m,\eta) - \alpha_{p,n}(m',\eta')] [\Phi_{p,n}(\eta) - \Phi_{p,n}(\eta')]$$

$$(13.19)$$

For $m = m'$, using (13.19) we find that

$$\Lambda^2_{p,n}(m,\eta) - \Lambda^2_{p,n}(m,\eta')$$

$$= \alpha_{p,n}(m,\eta) \ [\Phi_{p,n}(\eta) - \Phi_{p,n}(\eta')]$$

$$+ \left[\Phi_{p,n}(\eta') - \sum_{p<q\leq n} \frac{c_{q,n}}{\sum_{p<r\leq n} c_{r,n}} \Phi_{q,n}(\overline{\mu}_q) \right] [\alpha_{p,n}(m,\eta) - \alpha_{p,n}(m,\eta')]$$

We also notice that

$$\alpha_{p,n}(m,\eta) = \frac{1}{1 + \mu_{p,n}/[m\eta Q_{p,n}(1)]}$$

from which we easily prove that

$$\alpha_{p,n}(m,\eta) - \alpha_{p,n}(m',\eta')$$

$$= \frac{\mu_{p,n}}{\mu_{p,n} + m\eta Q_{p,n}(1)} \frac{1}{\mu_{p,n} + m'\eta'Q_{p,n}(1)} [m\eta Q_{p,n}(1) - m'\eta'Q_{p,n}(1)]$$

and therefore

$$\alpha_{p,n}(m,\eta) - \alpha_{p,n}(m,\eta')$$

$$= (\alpha_{p,n}(m,\eta')(1 - \alpha_{p,n}(m,\eta)))\ [\eta - \eta']\left(\frac{Q_{p,n}(1)}{\eta'Q_{p,n}(1)}\right)$$

The proof of $\alpha_{p,n}(m,\eta) \leq \alpha_{p,n}^{\star}(m)$ is elementary. From the above decomposition, we prove the following upper bounds

$$|\alpha_{p,n}(m,\eta) - \alpha_{p,n}(m,\eta')| \leq \alpha_{p,n}^{\star}(m)\left|[\eta - \eta']\left(\frac{Q_{p,n}(1)}{\eta'Q_{p,n}(1)}\right)\right|$$

and

$$\left|\left[\Lambda_{p,n}^2(m,\eta) - \Lambda_{p,n}^2(m,\eta')\right](f)\right|$$

$$\leq \alpha_{p,n}^{\star}(m)\left[|[\Phi_{p,n}(\eta) - \Phi_{p,n}(\eta')](f)| + \left|[\eta - \eta']\left(\frac{Q_{p,n}(1)}{\eta'Q_{p,n}(1)}\right)\right|\right.$$

$$\left.\times\left|\sum_{p<q\leq n}\frac{c_{q,n}}{\sum_{p<r\leq n}c_{r,n}}\left[\Phi_{q,n}(\overline{\mu}_q) - \Phi_{q,n}(\Phi_{p,q}(\eta'))\right](f)\right|\right]$$

This yields

$$\left|\left[\Lambda_{p,n}^2(m,\eta) - \Lambda_{p,n}^2(m,\eta')\right](f)\right|$$

$$\leq \alpha_{p,n}^{\star}(m)\left[|[\Phi_{p,n}(\eta) - \Phi_{p,n}(\eta')](f)| + \beta_{p,n}\left|[\eta - \eta']\left(\frac{Q_{p,n}(1)}{\eta'Q_{p,n}(1)}\right)\right|\right]$$

The end of the proof is now a direct consequence of (13.15). This ends the proof of the proposition. ∎

13.1.3   Stability properties

Further on in this section, we will illustrate the stability properties of the sequence of probability distributions (γ_n, η_n) for time homogeneous $(G_n, M_n, \mu_n) = (G, M, \mu)$ in three typical scenarios.

$$1)\quad G = g_{-/+} = 1 \qquad 2)\quad g_+ < 1 \quad \text{and} \quad 3)\quad g_- > 1 \qquad (13.20)$$

arising in time homogeneous models

$$(E_n, G_n, M_n, \mu_n, g_-(n), g_+(n)) = (E, G, M, \mu, g_-, g_+) \qquad (13.21)$$

In the context of the spatial branching models discussed in Section 6.1.1, these three scenarios correspond to the case where, *independently from the additional spontaneous births*, the existing particle die, or survive, and spawn in such a way that either their number remains constant $(G = g_{-/+} = 1)$, decreases $(g_+ < 1)$, or increases $(g_- > 1)$.

13.1.3.1 Constant mass processes

We examine the first situation, where $G(x) = 1$ for any $x \in E$.

Proposition 13.1.7 *For any $n \geq 0$, we have*

$$\gamma_n(1) = \gamma_0(1) + \mu(1) \, n \quad and \quad \|\eta_n - \eta_\infty\|_{tv} = O\left(\frac{1}{n}\right)$$

as soon as M is chosen so that

$$\sum_{n \geq 0} \sup_{x \in E} \|M^n(x, .) - \eta_\infty\|_{tv} < \infty \quad for some invariant measure \, \eta_\infty = \eta_\infty M.$$

$$(13.22)$$

Proof:
We prove this proposition using the fact that the total mass process $\gamma_n(1)$ grows linearly w.r.t. the time parameter; that is, we have that

$$\gamma_n(1) = m_-(n) = m_+(n) = \gamma_0(1) + \mu(1) \, n \qquad (13.23)$$

Note that the estimates in (13.13) take the following form

$$\alpha_{p,n}(\gamma_p(1), \eta_p) \leq \alpha_{p,n}^\star(\gamma_p(1)) := 1 \wedge \frac{\gamma_0(1) + \mu(1) \, p}{\mu(1) \, (n-p)} \to_{(n-p) \to \infty} 0$$

In this particular situation, the time-inhomogeneous Markov transitions $M_{n,(\gamma_{n-1}(1), \eta_{n-1})} := \overline{M}_n$ introduced in (6.5) are given by

$$\overline{M}_n(x, dy) = \left(1 - \frac{\mu(1)}{\gamma_0(1) + n\mu(1)}\right) M(x, dy) + \frac{\mu(1)}{\gamma_0(1) + n\mu(1)} \, \overline{\mu}(dy)$$

In this situation, the mapping $\Lambda_{0,n}^2$ is given by

$$\Lambda_{0,n}^2(\gamma_0(1), \eta_0) := \frac{\gamma_0(1)}{\gamma_0(1) + n\mu(1)} \, \eta_0 M^n + \frac{n\mu(1)}{\gamma_0(1) + n\mu(1)} \, \frac{1}{n} \sum_{0 \leq p < n} \overline{\mu} M^p$$

The above formula shows that for a large time horizon n, the normalized distribution flow η_n is almost equal to $\frac{1}{n} \sum_{0 \leq p < n} \overline{\mu} M^p$. Let us assume that the Markov kernel M is chosen so that $\beta(M^m) < 1$, for some $m \geq 1$. In this situation, it is well known that there exists a single invariant measure $\eta_\infty = \eta_\infty M$ such that

$$\tau_n = \sup_{x \in E} \|M^n(x, .) - \eta_\infty\|_{tv} \leq \beta(M^m)^{\lfloor n/m \rfloor} \to_{n \uparrow \infty} 0$$

In this case, for any starting measure γ_0, we have

$$\|\eta_n - \eta_\infty\|_{\mathrm{tv}} \leq \frac{\gamma_0(1)}{\gamma_0(1) + n\mu(1)} \tau_n + \frac{n\mu(1)}{\gamma_0(1) + n\mu(1)} \frac{1}{n} \sum_{0 \leq p < n} \tau_p = O\left(\frac{1}{n}\right)$$

This ends the proof of the proposition. ∎

Next, we illustrate some consequences of the weak functional inequalities stated in Proposition 13.1.6. Notice that if $G = 1$, then we have

$$\Phi_{p,n}(\eta) = \eta M^{(n-p)}, \quad h_{p,n,\eta'} = 1/2 \quad c_{p,n} = \mu(1) \quad g_{p,n} = 1 \quad \alpha^\star_{p,n} \leq 1$$

with the functions $h_{p,n,\eta'}$ and the parameters $\alpha^\star_{p,n}$ defined in Proposition 13.1.6. Using (13.16) we readily prove the following corollary.

Corollary 13.1.8 *Let us assume that there exist $a < \infty$ and $0 < \lambda < \infty$ such that $\beta(M^n) \leq ae^{-\lambda n}$ for any $n \geq 0$. In this situation, we have*

$$\left| \left[\Lambda^2_{p,n}(m, \eta) - \Lambda^2_{p,n}(m, \eta') \right](f) \right| \leq 2ae^{-\lambda(n-p)} \left| (\mu - \eta) \mathcal{D}_{p,n,\eta'}(f) \right|$$

13.1.3.2 Decreasing mass processes

We assume that $g_+ < 1$, and the regularity condition $\mathbf{H}(\mathbf{G}, \mathbf{P})$ stated on page 371, is satisfied, for some finite parameters (v_1, v_2). In addition, we have

$$\beta(P_{0,n}) \leq a \, e^{-\lambda \, n}$$

for some finite positive parameters a and λ.

Proposition 13.1.9 *There exists a constant $c < \infty$ such that*

$$\forall f \in \mathcal{B}_n(E), \qquad |\gamma_n(f) - \gamma_\infty(f)| \vee |\eta_n(f) - \eta_\infty(f)| \leq c \, g^n_+ \, \|f\|$$

with the limiting measures

$$\gamma_\infty(f) := \sum_{n \geq 0} \mu Q^n(f) \quad and \quad \eta_\infty(f) := \gamma_\infty(f)/\gamma_\infty(1) \qquad (13.24)$$

Proof:
When $g_+ < 1$, the total mass process $\gamma_n(1)$ is uniformly bounded w.r.t. the time parameter. More precisely, we have that

$$m_{-/+}(n) = g^n_{-/+} \gamma_0(1) + \left(1 - g^n_{-/+}\right) \frac{\mu(1)}{1 - g_{-/+}}$$

This yields the rather crude estimates

$$m_- := \gamma_0(1) \wedge \frac{\mu(1)}{1 - g_-} \leq \gamma_n(1) \leq \gamma_0(1) \vee \frac{\mu(1)}{1 - g_+} := m_+ \qquad (13.25)$$

Furthermore, for any $f \in \mathcal{B}_b(E)$ with $\|f\| \leq 1$, we have the estimates

$$|\gamma_n(f) - \gamma_\infty(f)| \leq \gamma_0(1) \, \eta_0 Q^n(1) + \sum_{p \geq n} \mu Q^p(1)$$

$$\leq g_+^n \, [\gamma_0(1) + \mu(1)/(1 - g_+)] \longrightarrow_{n \to \infty} 0$$

In addition, using the fact that $\gamma_n(1) \geq \mu(1)$, we find that for any $f \in \mathrm{Osc}(E)$

$$|\eta_n(f) - \eta_\infty(f)| \leq \frac{1}{\gamma_n(1)} \, |\gamma_n[f - \eta_\infty(f)] - \gamma_\infty[f - \eta_\infty(f)]|$$

$$\leq g_+^n \, [\gamma_0(1)/\mu(1) + 1/(1 - g_+)] \longrightarrow_{n \to \infty} 0$$

This ends the proof of the proposition. ∎

Notice that the measure γ_∞ is the solution of the Poisson equation given by the formula

$$\gamma_\infty(Id - Q) = \mu$$

It is instructive to connect these mathematical objects with their continuous time version. To this end, we consider a potential function $G = e^{-V\Delta t}$, and a Markov transition $M = Id + L \, \Delta t$ associated with a discrete time approximation of continuous time Feynman-Kac model with time step Δt. In this context, the discrete time Feynman-Kac measures γ_n is a time discretization of the Feynman-Kac measures

$$\gamma_t(f) = \int_0^t \mathbb{E}_\mu \left(f(X_s) \exp\left(-\int_0^s V(X_r)dr\right) \right) \, ds$$

where X_t stands for a continuous time Markov process with infinitesimal generator L. When $t \uparrow \infty$, the measures γ_t tends to the solution

$$\gamma_\infty(f) = \int_0^\infty \mathbb{E}_\mu \left(f(X_s) \exp\left(-\int_0^s V(X_r)dr\right) \right) \, ds$$

of the Poisson equation $\gamma_\infty L^V = \mu$ with $L^V = L + V$.

Next, we estimate the parameter $\alpha_{p,n}(m)$ given in (13.13), when the regularity condition $\mathbf{H(G, P)}$, stated on page 371, is satisfied for some finite constants (υ_1, υ_2).

Proposition 13.1.10 *Assume the regularity condition* $\mathbf{H(G, P)}$ *stated on page 371 is satisfied, and we have* $g_+ < 1$. *In this situation, for any* $n > p$ *and any* $m \in I_p$, *we have*

$$\alpha_{p,n}^\star(m) \leq 1 \wedge \left[m \, a' \, e^{-\lambda'(n-p)} \right] \to_{(n-p) \uparrow \infty} 0 \tag{13.26}$$

for some finite positive constants (a', λ').

Proof:

We recall that $c_{p,n} := \mu Q_{p,n}(1)$ and

$$\alpha_{p,n}^{\star}(m) := 1 \wedge \left[m \left\| \frac{Q_{p,n}(1)}{\sum_{p<q\leq n} \mu Q_{q,n}(1)} \right\| \right]$$

On the other hand, using Proposition 12.2.4 we easily check that

$$\frac{1}{\sum_{p<q\leq n} \mu Q_{q,n}(1)/Q_{p,q}Q_{q,n}(1)}$$

$$\leq \mu(1)^{-1} e^{\upsilon_1} / \sum_{p<q\leq n} \left(\frac{1}{g_+}\right)^{q-p} = \frac{e^{\upsilon_1}}{\mu(1)(1-g_+)} \left(\left(\frac{1}{g_+}\right)^{n-p} - 1 \right)^{-1}$$

This ends the proof of the proposition. ∎

Proposition 13.1.11 *For any $f \in \mathrm{Osc}(E)$ and any $p \leq n$ we have*

$$\left| \left[\Lambda_{p,n}^2(m, \eta) - \Lambda_{p,n}^2(m, \eta') \right] (f) \right|$$

$$\leq 2 \left[1 \wedge \left(a'm_+ \ e^{-\lambda'(n-p)} \right) \right]$$

$$\times e^{\upsilon_1} \left[a \ e^{-\lambda(n-p)} \ |(\mu - \eta)\mathcal{D}_{p,n,\eta'}(f)| + |(\mu - \eta)h_{p,n,\eta'}| \right]$$

for some finite positive constants (a', λ'), the constant m_+ defined in (13.25), and the functions $(h_{p,n,\eta'}, \mathcal{D}_{p,n,\eta}(f)) \in \mathrm{Osc}(E)^2$ defined in Proposition 13.1.6.

Proof:

Using Proposition 13.1.10, we have

$$\sup_{m\in I_p} \alpha_{p,n}^{\star}(m) \leq 1 \wedge \left(a'm_+ \ e^{-\lambda'(n-p)} \right)$$

On the other hand, using Proposition 12.2.4 we also have that

$$g_{p,n} \leq e^{\upsilon_1} \qquad \beta_{p,n} \leq 1 \quad \text{and} \quad \beta(P_{p,n}) \leq a \ e^{-\lambda \ (n-p)}$$

The end of the proof is now a direct consequence of the estimate (13.16). This ends the proof of the proposition. ∎

13.1.3.3 Increasing mass processes

We assume that $g_- > 1$ and the regularity condition $\mathbf{H}(\mathbf{G}, \mathbf{P})$ stated on page 371 is satisfied for some finite parameters (υ_1, υ_2), with

$$\beta(P_{0,n}) \leq a \ e^{-\lambda \ n}$$

for some finite positive parameters a and λ.

Proposition 13.1.12 *The one step Feynman-Kac mapping $\Phi(\eta) = \Psi_G(\eta)M$ has a unique fixed point $\eta_\infty = \Phi(\eta_\infty)$. Furthermore, for any $n \geq 0$ we have*

$$\|\eta_n - \eta_\infty\|_{\mathrm{tv}} \leq a' \, e^{-\lambda' n} \quad and \quad \left| \frac{1}{n} \log \gamma_n(1) - \log \eta_\infty(G) \right| \leq b'/n \quad (13.27)$$

for some finite positive parameters (a', b', λ').

Proof:

When $g_- > 1$, the total mass process $\gamma_n(1)$ grows exponentially fast w.r.t. the time parameter and we can easily show that

$$g_- > 1 \Longrightarrow \gamma_n(1) \geq m_-(n) = \gamma_0(1) \, g_-^n + \mu(1) \, \frac{g_-^n - 1}{g_- - 1} \quad (13.28)$$

Under our assumptions, it is well known that for any initial distribution η_0, we have

$$\|\Phi_{0,n}(\eta_0) - \eta_\infty\|_{\mathrm{tv}} \quad \leq \quad a \, e^{-\lambda \, n} \quad (13.29)$$

$$\sup_{\eta \in \mathcal{P}(E)} \left| \frac{1}{n} \log \eta Q^n(1) - \log \eta_\infty(G) \right| \quad \leq \quad b/n \quad (13.30)$$

for some finite constant $b < \infty$. For a more thorough discussion on the stability properties of the semigroup $\Phi_{0,n}$, and the limiting measures η_∞, we refer the reader to [163], as well as to Chapter 12 in the present book, dedicated to Feynman-Kac semigroups.

The proof of the r.h.s. assertion in (13.27) is based on the decomposition

$$\frac{1}{n} \log \gamma_n(1) = \frac{1}{n} \log \gamma_0 Q^n(1) + \frac{1}{n} \log \left(1 + \sum_{1 \leq k \leq n} \mu Q^k(1)/\gamma_0 Q^k Q^{n-k}(1) \right)$$

The above formula is a direct consequence of the decomposition (13.9) stated in Lemma 13.1.2.

Using Proposition 12.2.4, we prove that

$$\sum_{1 \leq k \leq n} \mu Q^k(1)/\gamma_0 Q^k Q^{n-k}(1) \leq (\mu(1)/\gamma_0(1)) \exp(v_1) \sum_{1 \leq k \leq n} (1/g_-)^{n-k}$$

The end of the proof follows elementary calculations.

Now, we come to the proof of the l.h.s. assertion in (13.27). We simplify the notation and we set $\alpha_n := \alpha_{0,n}(\gamma_0(1), \eta_0)$ and $c_n := c_{0,n}$. Using (13.12), we find that for any $n > 1$

$$a^{-1} \|\eta_n - \eta_\infty\|_{\mathrm{tv}} \leq \alpha_n \, e^{-\lambda n} + (1 - \alpha_n) \sum_{0 \leq p < n} \frac{c_p}{\sum_{0 \leq q < n} c_q} \, e^{-\lambda p}$$

Recalling that
$$\mu(1) \, g_-^p \le c_p = \mu Q^p(1) \le \mu(1) \, g_+^p$$
we also obtain that
$$\sum_{0 \le p < n} \frac{c_p}{\sum_{1 \le q < n} c_q} \, e^{-\lambda p} \le \frac{1}{\left[\sum_{0 \le q < n} c_q\right]^{1/r}} \left[\sum_{0 \le p < n} c_p e^{-\lambda p r}\right]^{1/r}$$

$$\le \frac{1}{\left[\sum_{0 \le q < n} g_-^q\right]^{1/r}} \left[\sum_{0 \le p < n} (e^{-\lambda r} g_+)^p\right]^{1/r} \qquad (13.31)$$

for any $r \ge 1$. We conclude that
$$r > \frac{1}{\lambda} \, \log g_+ \implies \sum_{0 \le p < n} \frac{c_p}{\sum_{0 \le q < n} c_q} \, e^{-\lambda p} \le g_-^{-(n-1)/r}/(1 - e^{-\lambda r} g_+)^{1/r}$$

and therefore
$$a^{-1} \, \|\eta_m - \eta_\infty\|_{\mathrm{tv}} \le e^{-\lambda n} \vee \left(g_-^{-(n-1)/r} /(1 - e^{-\lambda r} g_+)^{1/r} \right) \to_{n \to \infty} 0$$

This ends the proof of the proposition. ∎

Proposition 13.1.13 *For any $f \in \mathrm{Osc}(E)$ and any $p \le n$ we have*
$$\left| \left[\Lambda_{p,n}^2(m, \eta) - \Lambda_{p,n}^2(m, \eta') \right](f) \right|$$

$$\le a_0 \, e^{-\lambda_0(n-p)} \, |(\mu - \eta)\mathcal{D}_{p,n,\eta'}(f)| + a_1 \, e^{-\lambda_1(n-p)} |(\mu - \eta)h_{p,n,\eta'}|$$

for some finite positive constants (a_0, λ_0) and (a_1, λ_1), with the functions $h_{p,n,\eta'}$ and $\mathcal{D}_{p,n,\eta}(f_n) \in \mathrm{Osc}(E)$ defined in Proposition 13.1.6.

Proof:
Using Proposition 12.2.4, we have
$$\alpha_{p,n}^\star \le 1 \qquad g_{p,n} \le e^{\upsilon_1} \quad \text{and} \quad \beta(P_{p,n}) \le a \, e^{-\lambda(n-p)}$$

Arguing as in (13.31), we prove that for any $r > \frac{1}{\lambda} \, \log g_+$
$$\beta_{p,n} \le g_-^{-(n-p-1)/r}/(1 - e^{-\lambda r} g_+)^{1/r}$$

from which we conclude that
$$\left| \left[\Lambda_{p,n}^2(m, \eta) - \Lambda_{p,n}^2(m, \eta') \right](f) \right|$$

$$\le a_0 \, e^{-\lambda_0(n-p)} \, |(\mu - \eta)\mathcal{D}_{p,n,\eta'}(f)| + a_1 \, e^{-\lambda_1(n-p)} |(\mu - \eta)h_{p,n,\eta'}|$$

with
$$a_0 = 2ae^{\upsilon_1} \quad a_1 = 2g_-^r e^{\upsilon_1}/(1 - e^{-\lambda r} g_+)^{1/r} \quad \lambda_0 = \lambda \quad \text{and} \quad \lambda_1 = \log(g_-)$$

This ends the proof of the proposition. ∎

13.2 Measure valued nonlinear equations

13.2.1 Regularity conditions

This section is concerned with the semigroup analysis of the measure valued processes discussed in Section 6.2. We denote by

$$\Lambda_{p,n} = \Lambda_n \circ \Lambda_{n-1} \circ \ldots \circ \Lambda_{p+1}$$

with $p \leq n$, the evolution semigroup of the process $(\gamma_n(1), \eta_n) \in (\mathbb{R}_+ \times \mathcal{P}(E_n))$ defined in (6.12).

We use the convention $\Lambda_{n,n} = Id$, the identity mapping for $p = 0$, and we denote by $\Lambda_{p,n}^1$ and $\Lambda_{p,n}^2$ the first and the second components of $\Lambda_{p,n}$. We consider the following regularity conditions.

(H_1) : *There exists a sequence of compact sets $I_n \subset (0, \infty)$, and some positive nondecreasing functions $\theta_{+/-,n}$ on $\mathbb{R}_+$ s.t. $\gamma_0(1) \in I_0$, and for any $(m, \eta) \in (I_n \times \mathcal{P}(E_n))$, we have*

$$\theta_{-,n}(m) \leq \eta\left(G_{n,m\eta}\right) \leq \theta_{+,n}(m)$$

In the above display, $G_{n,m\eta}$ stands for the collection of potential functions defined in (6.13).

The main implication of condition (H_1) comes from the fact that the total mass process $\gamma_n(1)$ evolves in compact sets

$$I_n \subset [m_n^-, m_n^+] \subset (0, \infty)$$

with the sequence of parameters $m_n^{+/-}$ defined by the recursive equations

$$m_{n+1}^- = m_n^- \times \theta_{-,n}(m_n^-) \quad \text{and} \quad m_{n+1}^+ = m_n^+ \times \theta_{+,n}(m_n^+)$$

with the initial conditions $m_0^- = m_0^+ = \gamma_0(1)$.

We also mention the (PHD) model presented in Section 6.3.2 satisfies condition (H_1), as long as the functions $(s_n, b_n, g_n(., y_n))$ are uniformly bounded, and $\mu_n(1) > 0$. It is also met when $\mu_n(1) = 0$, as long as $r_n = (s_n + b_n)$ is uniformly lower bounded, and $\mathcal{Y}_n \neq 0$, or $d_n < 1$.

(H_2) : *For any $n \geq 1$, $f \in \text{Osc}(E_n)$, and any (m, η), $(m', \eta') \in (I_n \times \mathcal{P}(E_n))$, the one step mappings $\Lambda_n = (\Lambda_n^1, \Lambda_n^2)$ defined in (6.14) satisfy the* *following Lipschitz type inequalities:*

$$\left|\Lambda_n^1(m,\eta) - \Lambda_n^1(m',\eta')\right| \leq c(n) \, |m - m'|$$
$$+ \int \, |[\eta - \eta'](\varphi)| \, \Sigma_{n,(m',\eta')}^1(d\varphi)$$

$$(13.32)$$

$$\left|\left[\Lambda_n^2(m,\eta) - \Lambda_n^2(m',\eta')\right](f)\right| \leq c(n) \, |m - m'|$$
$$+ \int \, |[\eta - \eta'](\varphi)| \, \Sigma_{n,(m',\eta')}^2(f,d\varphi)$$

$$(13.33)$$

for some finite constants $c(n) < \infty$, and some collection of bounded measures $\Sigma_{n,(m',\eta')}^1$ and $\Sigma_{n,(m',\eta')}^2(f,\cdot)$ on $\mathcal{B}_b(E_n)$ such that

$$\int \, osc(\varphi) \, \Sigma_{n,(m,\eta)}^1(d\varphi) \leq \delta\left(\Sigma_n^1\right) \quad and \quad \int \, osc(\varphi) \, \Sigma_{n,(m,\eta)}^2(f,d\varphi) \leq \delta\left(\Sigma_n^2\right)$$

for some $\delta\left(\Sigma_n^i\right) < \infty$ that do not depend on (m,η,f), with $i = 1,2$.

We mention that condition (H_2) can be replaced by the following stronger regularity condition:

(H_2') : *For any $n \geq 1$, $f \in Osc(E_n)$, and $(m,\eta), (m',\eta') \in (I_n \times \mathcal{P}(E_n))$, the integral operators $Q_{n,m\eta}$ defined in (6.11) satisfy the following Lipschitz type inequalities:*

$$\|Q_{n,m\eta}(f) - Q_{n,m'\eta'}(f)\|$$

$$(13.34)$$

$$\leq c(n) \, |m - m'| + \int \, |[\eta - \eta'](\varphi)| \, \Sigma_{n,(m',\eta')}(f,d\varphi)$$

for some collection of bounded measures $\Sigma_{n,(m',\eta')}(f,\cdot)$ on $\mathcal{B}_b(E_n)$ such that

$$\int \, osc(\varphi) \, \Sigma_{n,(m,\eta)}(f,d\varphi) \leq \delta\left(\Sigma_n\right)$$

for some $\delta\left(\Sigma_n\right) < \infty$ that does not depend on (m,η,f).

We prove $(H_2') \Rightarrow (13.32)$ using the decompositions

$$m\eta Q_{n,m\eta} - m'\eta' Q_{n,m'\eta'} = m\eta \left[Q_{n,m\eta} - Q_{n,m'\eta'}\right] + [m\eta - m'\eta'] Q_{n,m'\eta'}$$

and of course $[m\eta - m'\eta'] = [m - m']\eta + m'[\eta - \eta']$.

To prove $(H_2') \Rightarrow (13.33)$, we let $\gamma = m\eta$ and $\gamma' = m'\eta'$ and we use the decomposition

$$\left[\Lambda_n^2(m,\eta) - \Lambda_n^2(m',\eta')\right](f) = \frac{1}{\gamma Q_{n,\gamma}(1)} \, [\gamma Q_{n,\gamma} - \gamma' Q_{n,\gamma'}] \, (f - \Lambda_n^2(m',\eta')(f))$$

13.2.2 Some illustrations

Condition (H_2) is a rather basic weak type continuity property. It states that the one step transformations of the flow (6.12) are weakly Lipschitz, in the sense that the mass variations, as well as the integral differences w.r.t. functions f, can be controlled by the differences between the initial masses, and the initial measures, w.r.t. a collection of integrals, on a possibly infinite number of test functions.

This condition is clearly satisfied for a rather large class of one step transformations Λ_n. For instance, the (PHD) evolution equations discussed in Section 6.3 are associated with the collection of integral operators defined in (6.23). This class of models satisfies (H_2'), as long as the functions $h_n(y) + g'_{n,y}$, with $g'_{n,y} := d_n g_n(.,y)$, are uniformly bounded above and below. To prove this claim, we simply use the fact that

$$\|\widehat{g}_{n,\gamma} - \widehat{g}_{n,\gamma'}\| \leq c_n \left[|m' - m| + \int \mathcal{Y}_n(dy) \; |[\eta' - \eta](g'_{n,y})| \right]$$

with the functions $\widehat{g}_{n,\gamma}$ defined in Proposition 6.3.1.

After some elementary manipulations, this estimate is a direct consequence of the following decomposition

$$\widehat{g}_{n,\gamma}(x) - \widehat{g}_{n,\gamma'}(x) = \int \mathcal{Y}_n(dy) \; \frac{g'_{n,y}(x)}{h_n(y) + \gamma(g'_{n,y})} \frac{[\gamma' - \gamma] (g'_{n,y})}{h_n(y) + \gamma'(g'_{n,y})}$$

We end this section with some comments on the association tree based models discussed in Section 6.4. We recall that the flow of unnormalized measures $(B_n)_{n \geq 0}$ given in (6.43) satisfies the same type of equation as in (6.11); that is, we have that

$$B_{n+1} = B_n \mathcal{Q}_{n+1,B_n}$$

for some integral operator $\mathcal{Q}_{n+1,B_n}$ defined in (6.42). In this context, for any $a \in \mathbf{A_n}$, and any function $F \in \mathcal{B}_b(\mathbf{A_n} \times \mathbf{A}_{n+1})$, we have

$$\mathcal{Q}_{n+1,B_n}(F)(a) = \int \nu_{n+1}(db) \; \mathcal{G}_{B_n(1),\overline{B}_n}(a, b) \; F(a, b)$$

with the normalized association measure $\overline{B}_n = B_n / B_n(1) \in \mathcal{P}(\mathbf{A_n})$.

For any $B = mA \in \mathcal{M}_+(\mathbf{A_n})$, and $B' = m'A' \in \mathcal{M}_+(\mathbf{A_n})$, this implies that

$$[\mathcal{Q}_{n+1,B}(F) - \mathcal{Q}_{n+1,B'}(F)](a)$$

$$= \int \nu_{n+1}(db) \; [\mathcal{G}_{m,A}(a, b) - \mathcal{G}_{m',A'}(a, b)] \; F(a, b)$$

Thus, condition (H_2') is met as long as

$$|\mathcal{G}_{m,A}(a, b) - \mathcal{G}_{m',A'}(a, b)| \leq c(n) \; |m - m'| + \int |[A - A'](\varphi)| \; \Sigma_{n,B'}^{(b)}(d\varphi)$$

for some collection of bounded measures $\Sigma_{n,B'}^{(b)}$ on $\mathcal{B}_b(\mathcal{A}_n)$ such that $\int \mathrm{osc}(\varphi)\, \Sigma_{n,B'}^{(b)} \leq \delta\left(\Sigma_n^{(b)}\right)$, for some finite constant $\delta\left(\Sigma_n^{(b)}\right) < \infty$, whose values do not depend on the parameters $(m, A) \in (I_n \times \mathcal{P}(\mathcal{A}_n))$. Under the assumptions (6.37), we have

$$B := mA \in \mathcal{M}_+(\mathcal{A}_n) \quad \Rightarrow \quad G_{n,m\eta^{(A)}}^{(b)} = \alpha_n^{(b)}(B)\, G_n^{(b)}$$

$$\Rightarrow \quad \mathcal{G}_{m,A}(a,b) = \alpha_n^{(b)}(B)\, \eta_n^{(a)}(G_n^{(b)})$$

for some collection of parameters $\alpha_n^{(b)}(B)$ satisfying

$$\left|\alpha_n^{(b)}(B) - \alpha_n^{(b)}(B')\right| \leq c(n)\, |m - m'| + \int |[A - A'](\varphi)|\, \Sigma_{n,B'}^{(b)}(d\varphi)$$

with $B = mA$, and $B' = m'A'$.

This condition is clearly satisfied for the PHD model discussed in (6.38), as long as the functions $h_n(y_n) + d_n g_n(., y_n)$ are uniformly bounded from above and below. For instance, for $b = y_n \notin \{c, c'\}$ we have

$$\alpha_n^{(b)}(B) = \left[h_n(b) + \int B(da)\, \eta_n^{(a)}(d_n g_n(., b))\right]^{-1}$$

In this case, we can check that

$$\left|\alpha_n^{(b)}(B) - \alpha_n^{(b)}(B')\right| \leq c(n)\left|[B - B'](\varphi_n^{(b)})\right|$$

with the function $\varphi_n^{(b)}(a) := \eta_n^{(a)}(d_n g_n(., b))$. In the same way, we show that the condition (H_1) is also met.

13.3 Weak Lipschitz properties of semigroups

13.3.1 Nonlinear semigroups

In this section, we present an essential weak Lipschitz type regularity property of the semigroup $(\Lambda_{p,n})_{0 \leq p \leq n}$ associated with the one step transformations of the flow (6.14).

Proposition 13.3.1 *We assume that conditions (H_1) and (H_2) are satisfied. Then, for any $0 \leq p \leq n$, $f \in \mathrm{Osc}(E_n)$, and any $(m, \eta), (m', \eta') \in (I_p \times \mathcal{P}(E_p))$, we have the following Lipschitz type inequalities:*

$$\left|\Lambda_{p,n}^1(m,\eta) - \Lambda_{p,n}^1(m',\eta')\right|$$

$$\leq c_p(n)\, |m - m'| + \int |[\eta - \eta'](\varphi)|\, \Sigma_{p,n,(m',\eta')}^1(d\varphi)$$

and

$$\left| \left[\Lambda_{p,n}^2(m, \eta) - \Lambda_{p,n}^2(m', \eta') \right](f) \right|$$

$$\leq c_p(n) \, |m - m'| + \int \, \|[\eta - \eta'](\varphi)| \, \Sigma_{p,n,(m',\eta')}^2(f, d\varphi)$$

for some finite constants $c_p(n) < \infty$, *and some collection of bounded measures* $\Sigma_{p,n,(m',\eta')}^1$ *and* $\Sigma_{p,n,(m',\eta')}^2(f, \cdot)$ *on* $\mathcal{B}_b(E_p)$ *such that*

$$\begin{aligned}
\int \, \mathrm{osc}(\varphi) \, \Sigma_{p,n,(m,\eta)}^1(d\varphi) &\leq \delta\left(\Sigma_{p,n}^1\right) \\
\int \, \mathrm{osc}(\varphi) \, \Sigma_{p,n,(m,\eta)}^2(f, d\varphi) &\leq \delta\left(\Sigma_{p,n}^2\right)
\end{aligned} \tag{13.35}$$

for some $\delta\left(\Sigma_{p,n}^i\right) < \infty$ *that does not depend on* (m, η, f), *with* $i = 1, 2$.

Proof:
To prove this proposition, we use a backward induction on the parameter $1 \leq p \leq n$. For $p = (n-1)$, we have $\Lambda_{n-1,n}^i = \Lambda_n^i$, with $i = 1, 2$, so that the desired result is satisfied for $p = (n-1)$. We further assume that the estimates hold at a given rank $p < n$. To prove the estimates at rank $(p-1)$, we recall that

$$\Lambda_{p-1,n}(m, \eta) = \Lambda_{p,n}\left(\Lambda_p(m, \eta)\right) \Rightarrow \Lambda_{p-1,n}^i(m, \eta) = \Lambda_{p,n}^i\left(\Lambda_p(m, \eta)\right)$$

for any $i \in \{1, 2\}$. Under the induction hypothesis, we have

$$\left| \Lambda_{p-1,n}^1(m, \eta) - \Lambda_{p-1,n}^1(m', \eta') \right|$$

$$\leq c_p(n) \, |\Lambda_p^1(m, \eta) - \Lambda_p^1(m', \eta')|$$

$$+ \int \, \left| \left[\Lambda_p^2(m, \eta) - \Lambda_p^2(m', \eta') \right](\varphi) \right| \, \Sigma_{p,n,\Lambda_p(m',\eta')}^1(d\varphi)$$

On the other hand, we have

$$|\Lambda_p^1(m, \eta) - \Lambda_p^1(m', \eta')| \leq c(p) \, |m - m'| + \int \, \|[\eta - \eta'](\varphi)| \, \Sigma_{p,(m',\eta')}^1(d\varphi)$$

and

$$\left| \left[\Lambda_p^2(m, \eta) - \Lambda_p^2(m', \eta') \right](\varphi) \right|$$

$$\leq c(p) \, |m - m'| + \int \, \|[\eta - \eta'](\psi)| \, \Sigma_{p,(m',\eta')}^2(\varphi, d\psi)$$

The end of the proof is now clear. The analysis of $\Lambda_{p-1,n}^2$ follows the same reasoning; thus it is omitted. This ends the proof of the proposition. ∎

13.3.2   Contraction inequalities

This section is concerned with the long time behavior of nonlinear measure-valued processes of the form (6.12). The complexity of these models comes from the interaction function between the flow of masses $\gamma_n(1)$ and the flow of probability measures $\eta_n = \gamma_n/\gamma_n(1)$.

One natural way to start the analysis of these models is to study the stability properties of the measure-valued semigroup associated with a fixed flow of masses and the one associated with a fixed flow of probability measures. To simplify the presentation, we introduce the following notation

$$\boldsymbol{I_n} := \prod_{n \geq 0} I_n \quad \text{and} \quad \boldsymbol{\mathcal{P}(E_n)} := \prod_{n \geq 0} \mathcal{P}(E_n)$$

Definition 13.3.2 *We associate with* $m = (m_n)_{n \geq 0} \in \boldsymbol{I_n}$, *and* $\nu := (\nu_n)_{n \geq 0} \in \boldsymbol{\mathcal{P}(E_n)}$, *the semigroups*

$$\Phi^1_{p,n,\nu} := \Phi^1_{n,\nu_{n-1}} \circ \ldots \circ \Phi^1_{1,\nu_0} \quad \text{and} \quad \Phi^2_{p,n,m} := \Phi^2_{n,m_{n-1}} \circ \ldots \circ \Phi^2_{1,m_0} \quad (13.36)$$

with $0 \leq p \leq n$. *In the above display,* $\Phi^1_{n,\nu_{n-1}}$, *and* $\Phi^2_{n,m_{n-1}}$, *stands for the one step transformations*

$$\Phi^1_{n,\nu_{n-1}} \quad : \quad u \in I_{n-1} \quad \mapsto \Phi^1_{n,\nu_{n-1}}(u) := \Lambda^1_n(u,\nu_{n-1}) \in I_n$$

$$\Phi^2_{n,m_{n-1}} \quad : \quad \eta \in \mathcal{P}(E_{n-1}) \mapsto \Phi^2_{n,m_{n-1}}(\eta) := \Lambda^2_n(m_{n-1},\eta) \in \mathcal{P}(E_n)$$

By induction w.r.t. the time parameter n, we find that

$$\forall n \geq 1 \quad m_n = \Phi^1_{n,\nu_{n-1}}(m_{n-1}) \quad \text{and} \quad \nu_n = \Phi^2_{n,m_{n-1}}(\nu_{n-1})$$

$$\Updownarrow$$

$$\forall n \geq 0 \quad (m_n, \nu_n) = (\gamma_n(1), \eta_n)$$

as soon as $(m_0, \nu_0) = (\gamma_0(1), \eta_0)$.

In the cases that are of particular interest in multiple object tracking problems, the semigroups $\Phi^1_{p,n,\nu}$ and $\Phi^2_{p,n,m}$ will have a Feynman-Kac representation.

These models are rather well understood. A brief review on their contraction properties is provided in Section 12.1 and in Section 12.2. Further details can also be found in the monograph [163].

The first basic regularity property of these models which are needed is the following weak Lipschitz type property:

$(Lip(\Phi))$ *For any* $p \leq n$, $u, u' \in I_p$, $\eta, \eta' \in \mathcal{P}(E_p)$, *and* $f \in Osc_1(E_n)$, *we have the following Lipschitz inequalities*

$$\left| \Phi^1_{p,n,\nu}(u) - \Phi^1_{p,n,\nu}(u') \right| \quad \leq \quad a^1_{p,n} \, |u - u'| \tag{13.37}$$

$$\left| \left[\Phi^2_{p,n,m}(\eta) - \Phi^2_{p,n,m}(\eta') \right](f) \right| \quad \leq \quad a^2_{p,n} \int \, |[\eta - \eta'](\varphi)| \, \Omega^2_{p,n,\eta'}(f, d\varphi) \tag{13.38}$$

for some finite constants $a_{p,n}^i < \infty$, *with* $i = 1, 2$, *and some collection of Markov transitions* $\Omega_{p,n,\eta'}^2$ *from* $Osc_1(E_n)$ *into* $Osc_1(E_p)$, *with* $p \le n$, *whose values only depend on the parameters* p, n, *resp.* p, n *and* η'.

The semigroups $\Phi_{p,n,\nu}^1$, and $\Phi_{p,n,m}^2$ may, or may not, be asymptotically stable, depending on whether $a_{p,n}^i$ tends to 0, as $(n - p) \to \infty$.

The second step in the study of the stability properties of the semigroups associated with the flow (6.14) is the following continuity property:

$(Cont(\Phi))$ *For any* $n \ge 1$, $u, u' \in I_{n-1}$, $(\eta, \eta') \in \mathcal{P}(E_{n-1})^2$, *and any function* $f \in Osc_1(E_n)$, *we have*

$$\left| \Phi_{n,\eta}^1(u) - \Phi_{n,\eta'}^1(u) \right| \le \tau_n^1 \int |[\eta - \eta'](\varphi)| \; \Omega_{n,\eta'}^1(d\varphi) \quad (13.39)$$

$$\left| \left[\Phi_{n,u}^2(\eta) - \Phi_{n,u'}^2(\eta) \right](f) \right| \le \tau_n^2 |u - u'| \quad (13.40)$$

for some finite constants $\tau_n^i < \infty$, *with* $i = 1, 2$, *and resp. some collection probability measures* $\Omega_{n,\nu'}^1$ *on* $Osc_1(E_{n-1})$, *whose values only depend on the parameters* n, *resp.* (n, ν').

We are now in position to state the main result of this section.

Theorem 13.3.3 *Assume that the semigroups* $\Phi_{p,n,\nu}^1$ *and* $\Phi_{p,n,m}^2$ *satisfy conditions* $(Lip(\Phi))$ *and* $(Cont(\Phi))$ *with some parameters* $(a_{p,n}^i, \tau_n^i)_{i=1,2}$ *such that*

$$\tau^i := \sup_{n \ge 1} \tau_n^i < \infty \quad and \quad a_{p,n}^i \le c_i \; e^{-\lambda_i(n-p)}$$

for any $p \le n$, *and some finite parameters* $c_i < \infty$ *and* $\lambda_i > 0$, *with* $i = 1, 2$. *We further assume that the following condition is satisfied*

$$\lambda_1 \ne \lambda_2 \quad and \quad c_1 c_2 \; \tau^1 \tau^2 \le \left(1 - e^{-(\lambda_1 \wedge \lambda_2)} \right) \left(e^{-(\lambda_1 \wedge \lambda_2)} - e^{-(\lambda_1 \vee \lambda_2)} \right)$$

In this situation, for any $p \le n$, $u, u' \in I_p$, $\eta, \eta' \in \mathcal{P}(E_p)$, *and any* $f \in Osc_1(E_n)$, *we have the following Lipschitz inequalities*

$$\left| \Lambda_{p,n}^1(u', \eta') - \Lambda_{p,n}^1(u, \eta) \right|$$

$$\le c^{1,1} \; e^{-\lambda(n-p)} \, |u - u'| + c^{1,2} \; e^{-\lambda(n-p)} \int |[\eta - \eta'](\varphi)| \; \Sigma_{p,n,u',\eta'}^1(d\varphi)$$

and

$$\left| \Lambda_{p,n}^2(u', \eta')(f) - \Lambda_{p,n}^2(u, \eta)(f) \right|$$

$$\le c^{2,1} \; e^{-\lambda(n-p)} \, |u - u'| + c^{2,2} \; e^{-\lambda(n-p)} \int |[\eta - \eta'](\varphi)| \; \Sigma_{p,n,u'\eta'}^2(f, d\varphi)$$

In the above display, $\Sigma^1_{p,n,u',\eta'}$ and $\Sigma^2_{p,n,m'\eta'}(f,.)$ stand for some probability measures on $\mathrm{Osc}(E_p)$, and

$$\lambda = (\lambda_1 \wedge \lambda_2) - \log\left(1 + c_1 c_2 \; \tau^1 \tau^2 \frac{e^{(\lambda_1 \wedge \lambda_2)}}{e^{-(\lambda_1 \wedge \lambda_2)} - e^{-(\lambda_1 \vee \lambda_2)}}\right) > 0$$

In addition, the parameters $c^{i,j}$ are given by $c^{2,2} = c_2$, and

$$
\begin{aligned}
c^{2,1} &= c_1 c_2 \tau^2 / \left(e^{-(\lambda_1 \wedge \lambda_2)} - e^{-(\lambda_1 \vee \lambda_2)}\right) \\
c^{1,1} &= c_1 \left(1 + c^{2,1} \tau^1 / (e^{-\lambda} - e^{-\lambda_1})\right) \quad \text{and} \quad c^{1,2} = c_1 c_2 \tau^1 / (e^{-\lambda} - e^{-\lambda_1})
\end{aligned}
$$

The rest of this section is mainly concerned with the proof of this theorem.

We mention that the second elementary continuity condition $(\mathrm{Cont}(\Phi))$ allows us to enter the contraction properties of the semigroups $\Phi^1_{p,n,\nu}$ and $\Phi^2_{p,n,m}$ in the stability analysis of the flow of measures (6.14).

The functional contraction inequalities developed in the further development of this section are described in terms of the following collection of parameters.

Definition 13.3.4 *When the couple of conditions $(\mathrm{Lip}(\Phi))$ and $(\mathrm{Cont}(\Phi))$ stated above are satisfied, for any $i = 1, 2$ and $p \leq n$ we set*

$$\overline{a}^i_{p,n} = \tau^i_{p+1} \, a^i_{p+1,n} \qquad b_{p,n} = \sum_{p<q<n} \overline{a}^1_{p,q} \, \overline{a}^2_{q,n} \quad \text{and} \quad b'_{p,n} = \sum_{p\leq q<n} a^1_{p,q} \, \overline{a}^2_{q,n} \tag{13.41}$$

Our first objective is to quantify the weak Lipschitz continuity properties of the mappings

$$(m, \nu) \in (I_n \times \mathcal{P}(E_n)) \mapsto \left(\Phi^1_{p,n,\nu}(u), \Phi^2_{p,n,m}(\eta)\right)$$

Lemma 13.3.5 *We assume that the regularity conditions $(\mathrm{Lip}(\Phi))$ and $(\mathrm{Cont}(\Phi))$ are satisfied. In this situation, for any $p \leq n$, $u, u' \in I_p$, $\eta, \eta' \in \mathcal{P}(E_p)$, and $f \in \mathrm{Osc}_1(E_n)$, any $m = (m_n)_{n\geq 0}$, $m' = (m'_n)_{n\geq 0} \in I_n$, and any $\nu = (\nu_n)_{n\geq 0}$, $\nu' = (\nu'_n)_{n\geq 0} \in \mathcal{P}(E_n)$, we have the following estimates*

$$\left|\Phi^1_{p,n,\nu'}(u') - \Phi^1_{p,n,\nu}(u)\right|$$

$$\leq a^1_{p,n} |u - u'| + \sum_{p\leq q<n} \overline{a}^1_{q,n} \int |[\nu_q - \nu'_q](\varphi)| \; \Omega^1_{q+1,\nu'_q}(d\varphi)$$

and

$$\left|\Phi^2_{p,n,m'}(\eta')(f) - \Phi^2_{p,n,m}(\eta)(f)\right|$$

$$\leq a^2_{p,n} \int |[\eta - \eta'](\varphi)| \, \Omega^2_{p,n,\eta'}(f, d\varphi) + \sum_{p\leq q<n} \overline{a}^2_{q,n} \, |m_q - m'_q|$$

with the collection of parameters $\overline{a}^i_{p,n}$, $i = 1, 2$, defined in (13.41).

Proof:

The proof is based on the decomposition

$$\Phi^1_{p,n,\nu'}(u') - \Phi^1_{p,n,\nu}(u)$$

$$= \Phi^1_{p,n,\nu}(u') - \Phi^1_{p,n,\nu}(u)$$

$$+ \sum_{p<q\leq n} \left[\Phi^1_{q,n,\nu}(\Phi^1_{p,q,\nu'}(u')) - \Phi^1_{q-1,n,\nu}(\Phi^1_{p,q-1,\nu'}(u')) \right]$$

We also observe that

$$\Phi^1_{q-1,n,\nu}(\Phi^1_{p,q-1,\nu'}(u')) = \Phi^1_{q,n,\nu}\left(\Phi^1_{q-1,q,\nu}\left[\Phi^1_{p,q-1,\nu'}(u')\right]\right)$$
$$\Phi^1_{q,n,\nu}(\Phi^1_{p,q,\nu'}(u')) = \Phi^1_{q,n,\nu}\left(\Phi^1_{q-1,q,\nu'}\left[\Phi^1_{p,q-1,\nu'}(u')\right]\right)$$

Under our assumptions, we have

$$\left| \Phi^1_{p,n,\nu}(u') - \Phi^1_{p,n,\nu}(u) \right| \leq a^1_{p,n} \, |u - u'|$$

and

$$\left| \Phi^1_{q,n,\nu}(\Phi^1_{p,q,\nu'}(u')) - \Phi^1_{q-1,n,\nu}(\Phi^1_{p,q-1,\nu'}(u')) \right|$$

$$\leq a^1_{q,n} \, \left| \Phi^1_{q,\nu_{q-1}}\left[\Phi^1_{p,q-1,\nu'}(u')\right] - \Phi^1_{q,\nu'_{q-1}}\left[\Phi^1_{p,q-1,\nu'}(u')\right] \right|$$

$$\leq \bar{a}^1_{q-1,n} \int \left| [\nu_{q-1} - \nu'_{q-1}](\varphi) \right| \, \Omega_{q,\nu'_{q-1}}(d\varphi)$$

From these observations, we prove that

$$\left| \Phi^1_{p,n,\nu'}(u') - \Phi^1_{p,n,\nu}(u) \right|$$

$$\leq a^1_{p,n} \, |u - u'| + \sum_{p<q\leq n} \bar{a}^1_{q-1,n} \int \left| [\nu_{q-1} - \nu'_{q-1}](\varphi) \right| \, \Omega^1_{q,\nu'_{q-1}}(d\varphi)$$

This ends the proof of the first assertion.

In much the same way, we use the decomposition

$$\left[\Phi^2_{p,n,m'}(\eta') - \Phi^2_{p,n,m}(\eta) \right]$$

$$= \left[\Phi^2_{p,n,m}(\eta') - \Phi^2_{p,n,m}(\eta) \right]$$

$$+ \sum_{p<q\leq n} \left[\Phi^2_{q,n,m}(\Phi^2_{p,q,m'}(\eta')) - \Phi^2_{q-1,n,m}(\Phi^2_{p,q-1,m'}(\eta')) \right]$$

and the fact that

$$\Phi^2_{q-1,n,m}(\Phi^2_{p,q-1,m'}(\eta')) = \Phi^2_{q,n,m}\left(\Phi^2_{q-1,q,m}\left[\Phi^2_{p,q-1,m'}(\eta')\right]\right)$$
$$\Phi^2_{q,n,m}(\Phi^2_{p,q,m'}(\eta')) = \Phi^2_{q,n,m}\left(\Phi^2_{q-1,q,m'}\left[\Phi^2_{p,q-1,m'}(\eta')\right]\right)$$

and

$$\left| \Phi^2_{p,n,m}(\eta')(f) - \Phi^2_{p,n,m}(\eta)(f) \right| \leq a^2_{p,n} \int \left| [\eta - \eta'](\varphi) \right| \, \Omega^2_{p,n,\eta'}(f, d\varphi)$$

to show that

$$\left| \Phi^2_{q,n,m}(\Phi^2_{p,q,m'}(\eta')) - \Phi^2_{q-1,n,m}(\Phi^2_{p,q-1,m'}(\eta')) \right|$$

$$\leq a^2_{q,n} \int \left| [\Phi^2_{q,m_{q-1}} [\Phi^2_{p,q-1,m'}(\eta')] \right.$$

$$\left. - \Phi^2_{q,m'_{q-1}} [\Phi^2_{p,q-1,m'}(\eta')]](\varphi) \right| \; \Omega^2_{q,n,\Phi^2_{p,q,m'}(\eta')}(f, d\varphi)$$

$$\leq \bar{a}^2_{q-1,n} \left| m_{q-1} - m'_{q-1} \right|$$

Using these estimates we conclude that

$$\left| \left[\Phi^2_{p,n,m'}(\eta') - \Phi^2_{p,n,m}(\eta) \right] (f) \right|$$

$$\leq a^2_{p,n} \int |[\eta - \eta'](\varphi)| \; \Omega^2_{p,n,\eta}(f, d\varphi) + \sum_{p<q\leq n} \bar{a}^2_{q-1,n} \left| m_{q-1} - m'_{q-1} \right|$$

This ends the proof of the lemma.   ∎

The main technical result of this section is the following proposition.

Proposition 13.3.6 *Assume that conditions* $(Lip(\Phi))$ *and* $(Cont(\Phi))$ *are satisfied. In this situation, the estimates presented in Theorem 13.3.3 are met by replacing the parameters* $c^{i,j} e^{-\lambda(n-p)}$ *with the collection of parameters* $c^{i,j}_{p,n}$ *defined below*

$$c^{1,1}_{p,n} = a^1_{p,n} + \sum_{p\leq q<n} c^{2,1}_{p,q} \bar{a}^1_{q,n} \quad \text{and} \quad c^{1,2}_{p,n} = \sum_{p\leq q<n} c^{2,2}_{p,q} \bar{a}^1_{q,n}$$

$$c^{2,1}_{p,n} = b'_{p,n} + \sum_{l=1}^{n-p} \sum_{p\leq r_1<...r_l<n} b'_{p,r_1} \prod_{1\leq k\leq l} b_{r_k, r_{k+1}}$$

$$c^{2,2}_{p,n} = a^2_{p,n} + \sum_{l=1}^{n-p} \sum_{p\leq r_1<...r_l<n} a^2_{p,r_1} \prod_{1\leq k\leq l} b_{r_k, r_{k+1}}$$

with the convention $r_{l+1} = n$. *In particular, the collection of parameters* $\delta \left(\Sigma^i_{p,n} \right)_{i=1,2}$, $p \leq n$ *introduced in (13.35) are such that*

$$\delta \left(\Sigma^1_{p,n} \right) \leq c^{1,2}_{p,n} \quad \text{and} \quad \delta \left(\Sigma^2_{p,n} \right) \leq c^{2,2}_{p,n}$$

Proof:

We fix a parameter $p \geq 0$, and we let $(m_n)_{n\geq p}, (m'_n)_{n\geq p} \in \prod_{n\geq p} I_n$, and $(\nu_n)_{n\geq p}, (\nu'_n)_{n\geq p} \in \prod_{n\geq p} \mathcal{P}(E_n)$, be defined by

$$\forall q > p \quad m'_q = \Phi^1_{q,\nu'_{q-1}}(m'_{q-1}) \quad \text{and} \quad \nu'_q = \Phi^2_{q,m'_{q-1}}(\nu'_{q-1})$$

$$\forall q > p \quad m_q = \Phi^1_{q,\nu_{q-1}}(m_{q-1}) \quad \text{and} \quad \nu_q = \Phi^2_{q,m_{q-1}}(\nu_{q-1})$$

with the initial condition for $q = p$

$$(\nu_p, \nu_p') = (\eta, \eta') \quad \text{and} \quad (m_p, m_p') = (u, u')$$

By construction, we have

$$
\begin{aligned}
\nu_q' &= \Phi_{p,q,m'}^2(\eta') \quad \text{and} \quad \nu_q = \Phi_{p,q,m}^2(\eta) \\
m_q' &= \Phi_{p,q,\nu'}^1(u') \quad \text{and} \quad m_q = \Phi_{p,q,\nu}^1(u)
\end{aligned}
$$

In this case, using Lemma 13.3.5 it follows that

$$\left| \left[\Lambda_{p,n}^2(m', \eta') - \Lambda_{p,n}^2(m, \eta) \right] (f) \right|$$

$$\leq a_{p,n}^2 \int \left| [\eta - \eta'](\varphi) \right| \, \Omega_{p,n,\eta'}^2(f, d\varphi)$$

$$+ \sum_{p \leq q < n} \overline{a}_{q,n}^2 \left| \Lambda_{p,q}^1(m', \eta') - \Lambda_{p,q}^1(m, \eta) \right|$$

as well as

$$\left| \Lambda_{p,n}^1(m', \eta') - \Lambda_{p,n}^1(m, \eta) \right|$$

$$\leq a_{p,n}^1 \left| m - m' \right|$$

$$+ \sum_{p \leq q < n} \overline{a}_{q,n}^1 \int \left| [\Lambda_{p,q}^2(m', \eta') - \Lambda_{p,q}^2(m, \eta)](\varphi) \right| \, \overline{\Omega}_{p,q,m',\eta'}^1(d\varphi)$$

with the probability measure $\overline{\Omega}_{p,q,m',\eta'}^1 = \Omega_{q+1,\Lambda_{p,q}^2(m',\eta')}^1$.

These two estimates yield the following inequality

$$\left| \left[\Lambda_{p,n}^2(m', \eta') - \Lambda_{p,n}^2(m, \eta) \right] (f) \right|$$

$$\leq a_{p,n}^2 \int \left| [\eta - \eta'](\varphi) \right| \, \Omega_{p,n,\eta'}^2(f, d\varphi) + \left[\sum_{p \leq q < n} a_{p,q}^1 \, \overline{a}_{q,n}^2 \right] \left| m - m' \right|$$

$$+ \sum_{p \leq r < q < n} \overline{a}_{r,q}^1 \, \overline{a}_{q,n}^2 \int \left| [\Lambda_{p,r}^2(m', \eta') - \Lambda_{p,r}^2(m, \eta)](\varphi) \right| \, \overline{\Omega}_{p,r,m',\eta'}^1(d\varphi)$$

This implies that

$$\left| \left[\Lambda_{p,n}^2(m', \eta') - \Lambda_{p,n}^2(m, \eta) \right] (f) \right|$$

$$\leq b_{p,n}' \left| m - m' \right| + a_{p,n}^2 \int \left| [\eta - \eta'](\varphi) \right| \, \Omega_{p,n,\eta'}^2(f, d\varphi)$$

$$+ \sum_{p \leq r_1 < n} b_{r_1,n} \int \left| [\Lambda_{p,r_1}^2(m', \eta') - \Lambda_{p,r_1}^2(m, \eta)](\varphi) \right| \, \overline{\Omega}_{p,r_1,m',\eta'}^1(d\varphi)$$

Our next objective is to show that

$$\left| \left[\Lambda_{p,n}^2(m',\eta') - \Lambda_{p,n}^2(m,\eta) \right](f) \right|$$

$$\leq \alpha_{p,n}^k \, |m - m'| + \beta_{p,n}^k \int \, |[\eta - \eta'](\varphi)| \; \Theta_{p,n,\eta'}^k(f, d\varphi)$$

$$+ \sum_{p \leq r_1 < r_2 < \ldots < r_k < n} b_{r_1,r_2} \, \cdots \, b_{r_k,n}$$

$$\times \int \, \left| [\Lambda_{p,r_1}^2(m',\eta') - \Lambda_{p,r_1}^2(m,\eta)](\varphi) \right| \; \overline{\Omega}_{p,r_1,m',\eta'}^1(d\varphi)$$

for any $k \leq (n - p)$ for some Markov transitions $\Theta_{p,n,m'\eta'}^k(f, d\varphi)$ and the parameters

$$\alpha_{p,n}^k \;=\; b_{p,n}' + \sum_{l=1}^{k-1} \sum_{p \leq r_1 < \ldots r_l < n} b_{p,r_1}' \, b_{r_1,r_2} \ldots b_{r_l,n}$$

$$\beta_{p,n}^k \;=\; a_{p,n}^2 + \sum_{l=1}^{k-1} \sum_{p \leq r_1 < \ldots r_l < n} a_{p,r_1}^2 \, b_{r_1,r_2} \ldots b_{r_l,n}$$

We proceed by induction on the parameter k. Firstly, we observe that the result is satisfied for $k = 1$ with

$$(\alpha_{p,n}^1, \beta_{p,n}^1) = (b_{p,n}', a_{p,n}^2) \quad \text{and} \quad \Theta_{p,n,\eta'}^1 = \Omega_{p,n,\eta'}^2$$

We further assume that the result is satisfied at rank k. In this situation, using the fact that

$$\left| \left[\Lambda_{p,r_1}^2(m',\eta') - \Lambda_{p,r_1}^2(m,\eta) \right](\varphi) \right|$$

$$\leq b_{p,r_1}' \, |m - m'| + a_{p,r_1}^2 \int \, |[\eta - \eta'](\varphi')| \; \Omega_{p,r_1,\eta'}^2(\varphi, d\varphi')$$

$$+ \sum_{p \leq r_0 < r_1} b_{r_0,r_1} \int \, \left| [\Lambda_{p,r_0}^2(m',\eta') - \Lambda_{p,r_0}^2(m,\eta)](\varphi) \right| \; \overline{\Omega}_{p,r_0,m',\eta'}^1(d\varphi)$$

we conclude that

$$\left| \left[\Lambda_{p,n}^2(m',\eta') - \Lambda_{p,n}^2(m,\eta) \right](f) \right|$$

$$\leq \alpha_{p,n}^{k+1} \, |m - m'| + \beta_{p,n}^{k+1} \int \, |[\eta - \eta'](\varphi)| \; \Theta_{p,n,m'\eta'}^{k+1}(f, d\varphi)$$

$$+ \sum_{p \leq r_0 < r_1 < r_2 < \ldots < r_k < n} b_{r_0,r_1} \, b_{r_1,r_2} \, \cdots \, b_{r_k,n}$$

$$\int \, \left| [\Lambda_{p,r_0}^2(m',\eta') - \Lambda_{p,r_0}^2(m,\eta)](\varphi) \right| \; \overline{\Omega}_{p,r_0,m',\eta'}^1(d\varphi)$$

with

$$\alpha_{p,n}^{k+1} = \alpha_{p,n}^{k} + \sum_{p \leq r_1 < r_2 < \ldots < r_k < n} b_{p,r_1}' \, b_{r_1,r_2} \cdots b_{r_k,n}$$

$$\beta_{p,n}^{k+1} = \beta_{p,n}^{k} + \sum_{p \leq r_1 < r_2 < \ldots < r_k < n} a_{p,r_1}^2 \, b_{r_1,r_2} \cdots b_{r_k,n}$$

and the Markov transitions

$$\beta_{p,n}^{k+1} \, \Theta_{p,n,m'\eta'}^{k+1}(f, d\varphi)$$

$$= \beta_{p,n}^{k} \, \Theta_{p,n,\eta'}^{k}(f, d\varphi)$$
$$+ \sum_{p \leq r_1 < r_2 < \ldots < r_k < n} a_{p,r_1}^2 \, b_{r_1,r_2} \cdots b_{r_k,n} \left(\overline{\Omega}_{p,r_1,m',\eta'}^1, \Omega_{p,r_1,\eta'}^2 \right)(d\varphi)$$

We end the proof of the proposition using the fact that

$$\left| \Lambda_{p,n}^1(m', \eta') - \Lambda_{p,n}^1(m, \eta) \right|$$

$$\leq \left[a_{p,n}^1 + \sum_{p \leq q < n} c_{p,q}^{2,1} \, \overline{a}_{q,n}^1 \right] |m - m'|$$

$$+ \sum_{p \leq q < n} \overline{a}_{q,n}^1 \, c_{p,q}^{2,2} \int |[\eta - \eta'](\varphi')| \left[\overline{\Omega}_{p,q,m',\eta'}^1 \Theta_{p,q,\eta'} \right](d\varphi')$$

This proof of the proposition is now completed. ∎

We are now in position to prove Theorem 13.3.3.
Proof of Theorem 13.3.3:
Using Proposition 13.3.6, we readily check the following estimates

$$b_{p,n} \leq c\tau \sum_{p < q < n} e^{-\lambda_1(q-(p+1))} \, e^{-\lambda_2(n-(q+1))}$$

$$b_{p,n}' \leq c\tau^2 \sum_{p \leq q < n} e^{-\lambda_1(q-p)} \, e^{-\lambda_2(n-(q+1))}$$

with $c = c_1 c_2$ and $\tau = \tau^1 \tau^2$. We further assume that $\lambda_1 > \lambda_2$ and we set $\Delta = |\lambda_1 - \lambda_2|$. In this notation, we have that

$$b_{p,n} \leq c\tau e^{-\lambda_2((n-1)-(p+1))} \sum_{p < q < n} e^{-\Delta(q-(p+1))} \leq c\tau e^{-\lambda_2((n-1)-(p+1))} / (1 - e^{-\Delta})$$

In much the same way, if $\lambda_2 > \lambda_1$ we have

$$b_{p,n} \leq c\tau e^{-\lambda_1((n-1)-(p+1))} \sum_{p < q < n} e^{-\Delta(n-(q+1))} \leq c\tau e^{-\lambda_1((n-1)-(p+1))} / (1 - e^{-\Delta})$$

This implies that

$$b_{p,n} \leq c\tau e^{-(\lambda_1 \wedge \lambda_2)((n-1)-(p+1))} / (1 - e^{-\Delta})$$

In much the same way, it can be shown that

$$b'_{p,n} = \leq c\tau^2 e^{-(\lambda_1 \wedge \lambda_2)((n-1)-p)} / (1 - e^{-\Delta}) \tag{13.42}$$

We are now in a position to estimate the parameters $c_{p,n}^{i,j}$. We observe that

$$c_{p,n}^{2,2} \leq c_2 \, e^{-\lambda_2(n-p)}$$

$$+ c_2 \sum_{l=1}^{n-p} \left(\frac{c\tau^1 \tau^2 e^{2(\lambda_1 \wedge \lambda_2)}}{1 - e^{-\Delta}} \right)^l \sum_{p \leq r_1 < \ldots r_l < n} e^{-\lambda_2(r_1 - p)} e^{-(\lambda_1 \wedge \lambda_2)(n-r_1)}$$

When $\lambda_1 > \lambda_2$, we find that

$$c_{p,n}^{2,2} \leq c_2 \, e^{-\lambda_2(n-p)} \sum_{l=0}^{n-p} \left(\frac{c\tau e^{2\lambda_2}}{1 - e^{-\Delta}} \right)^l \binom{n-p}{l}$$

and therefore

$$c_{p,n}^{2,2} \leq c_2 \, e^{-\lambda_2(n-p)} \left(1 + c\tau \frac{e^{2\lambda_2}}{1 - e^{-\Delta}} \right)^{n-p} \Rightarrow c_{p,n}^{2,2} = c_2 \, e^{-\lambda(n-p)}$$

with

$$\lambda = \lambda_2 - \log \left(1 + c\tau \frac{e^{\lambda_2}}{e^{-\lambda_2} - e^{-\lambda_1}} \right) > 0$$

as long as

$$c\tau \leq \left(1 - e^{-\lambda_2} \right) \left(e^{-\lambda_2} - e^{-\lambda_1} \right)$$

When $\lambda_2 > \lambda_1$ we have $\lambda_2 = \lambda_1 + \Delta$, and we find that

$$c_{p,n}^{2,2} \leq c_2 \, e^{-\lambda_2(n-p)}$$

$$+ c_2 e^{-\lambda_1(n-p)} \sum_{l=1}^{n-p} \left(\frac{c\tau e^{2\lambda_1}}{1 - e^{-\Delta}} \right)^l \sum_{p \leq r_1 < \ldots r_l < n} e^{-\Delta(r_1 - p)}$$

from which it follows that

$$c_{p,n}^{2,2} \leq c_2 \, e^{-\lambda_1(n-p)} \left(1 + c\tau \frac{e^{2\lambda_1}}{1 - e^{-\Delta}} \right)^{n-p}$$

Using a similar line of argument as above, we have $c_{p,n}^{2,2} \leq c_2 \, e^{-\lambda(n-p)}$, with

$$\lambda = \lambda_1 - \log \left(1 + c\tau \frac{e^{\lambda_1}}{e^{-\lambda_1} - e^{-\lambda_2}} \right) > 0$$

as long as

$$c\tau \le \left(1 - e^{-\lambda_1}\right) \left(e^{-\lambda_1} - e^{-\lambda_2}\right)$$

We conclude that $c_{p,n}^{2,2} \le c_2 \, e^{-\lambda(n-p)}$, with

$$\lambda = (\lambda_1 \wedge \lambda_2) - \log\left(1 + c\tau \frac{e^{(\lambda_1 \wedge \lambda_2)}}{e^{-(\lambda_1 \wedge \lambda_2)} - e^{-(\lambda_1 \vee \lambda_2)}}\right) > 0$$

as long as

$$c\tau \le \left(1 - e^{-(\lambda_1 \wedge \lambda_2)}\right) \left(e^{-(\lambda_1 \wedge \lambda_2)} - e^{-(\lambda_1 \vee \lambda_2)}\right)$$

Using (13.42) we also show that

$$c_{p,n}^{2,1} \le c^{2,1} \, e^{-\lambda(n-p)} \quad \text{with} \quad c^{2,1} = c\tau^2 \, \frac{1}{e^{-(\lambda_1 \wedge \lambda_2)} - e^{-(\lambda_1 \vee \lambda_2)}}$$

Using these estimates, we find that

$$
\begin{aligned}
c_{p,n}^{1,1} &= c_1 \, e^{-\lambda_1(n-p)} + \sum_{p \le q < n} c_{p,q}^{2,1} \, c_1 \tau^1 \, e^{-\lambda_1(n-(q+1))} \\
&\le c_1 \, e^{-\lambda_1(n-p)} + c^{2,1} c_1 \tau^1 \sum_{p \le q < n} e^{-\lambda(q-p)} \, e^{-\lambda_1(n-(q+1))}
\end{aligned}
$$

Since $\lambda_1 > \lambda$, we find that

$$c_{p,n}^{1,1} \le c_1 \, e^{-\lambda_1(n-p)} + c^{2,1} c_1 \tau^1 \, e^{-\lambda((n-1)-p)}/(1 - e^{-\Delta'})$$

with $\Delta' = \lambda_1 - \lambda > 0$. This yields

$$c_{p,n}^{1,1} \le c^{1,1} \, e^{-\lambda(n-p)} \quad \text{with} \quad c^{1,1} := c_1 \left(1 + c^{2,1}\tau^1/(e^{-\lambda} - e^{-\lambda_1})\right)$$

Finally, we observe that

$$c_{p,n}^{1,2} = c\tau^1 \sum_{p \le q < n} e^{-\lambda(q-p)} \, e^{-\lambda_1(n-(q+1))} \le c\tau^1 \, e^{-\lambda((n-1)-p)}/(1 - e^{-\Delta'})$$

which implies that $c_{p,n}^{1,2} \le c^{1,2} \, e^{-\lambda(n-p)}$, with $c^{1,2} := c\tau^1/(e^{-\lambda} - e^{-\lambda_1})$. This ends the proof of the theorem. ∎

13.4 Stability properties of PHD models

13.4.1 An exponential stability theorem

This section is concerned with the contraction properties of the semigroups $\Phi_{p,n,\nu}^1$ and $\Phi_{p,n,m}^2$ associated with the PHD filter discussed in Section 6.3.2.

To simplify the analysis, we further assume that the clutter intensity, the detectability, as well as the survival and the spawning rates, introduced in Section 6.3.1, are time homogeneous and constants functions

$$(b_n, h_n, s_n, r_n) = (b, h, s, r)$$

To simplify the presentation, we also assume that the state spaces, the Markov transitions of the targets, and the likelihood functions, as well as the spontaneous birth rates, are time homogeneous; that is, we have that

$$(E_n, E_n^Y, M_n, g_n(x, y), \mu_{n+1}) = (E, E^Y, M, g(x, y), \mu)$$

Without further mention, we suppose that $r(1 - d) < 1$, $\mu(1) > 0$, $r > 0$, and for any $y \in E^Y$ we have

$$0 \le g^-(y) := \inf_{x \in E} g(x, y) \le \sup_{x \in E} g(x, y) := g^+(y) < \infty$$

Given a mapping θ from E^Y into $\mathbb{R}$, we set

$$\mathcal{Y}^-(\theta) := \inf_n \mathcal{Y}_n(\theta) \quad \text{and} \quad \mathcal{Y}^+(\theta) := \sup_n \mathcal{Y}_n(\theta)$$

The following theorem shows that PHD filter presented in Section 6.3.2 is exponentially stable for small clutter intensities, sufficiently high detection probability, and high spontaneous birth rates.

Theorem 13.4.1 *We assume that $\mathcal{Y}^+(g^+/g^-)$ and $\mathcal{Y}^+(g^+/(g^-)^2) < \infty$. In this situation, there exists some parameters $0 < \kappa_0 \le 1$, $\kappa_1 < \infty$, and $\kappa_2 > 0$ such that for any $d \ge \kappa_0$, $\mu(1) \ge \kappa_1$, and $h \le \kappa_2$, the semigroups $\Phi^1_{p,n,\nu}$ and $\Phi^2_{p,n,m}$ satisfy the pair of conditions $(Lip(\Phi))$ and $(Cont(\Phi))$, with some parameters $(a^i_{p,n}, \tau^i_n)_{i=1,2,p \le n}$, satisfying the stability conditions presented in Theorem 13.3.3.*

The proof of this stability theorem is provided on page 423.

13.4.2 Evolution equations

We recall from (6.23) that the PDH filter is defined by the evolution equation $\gamma_{n+1} = \gamma_n Q_{n+1,\gamma_n}$, with the integral operator

$$Q_{n+1,\gamma_n}(x_n, dx_{n+1}) = g_{n,\gamma_n}(x_n) M(x_n, dx_{n+1}) + \gamma_n(1)^{-1} \mu(dx_{n+1})$$

and the function g_{γ_n} defined below

$$g_{n,\gamma_n}(x) = r(1 - d) + rd \int \mathcal{Y}_n(dy) \frac{g(x, y)}{h + d\gamma_n(g(., y))}$$

We also notice that the total mass process is given by the following equation

$$
\gamma_{n+1}(1) \;=\; \Phi^1_{n+1,\eta_n}(\gamma_n(1))
$$
$$
=\; \gamma_n(1)\, r(1-d) + \int \mathcal{Y}_n(dy)\, w_{\gamma_n(1)}(\eta_n, y) + \mu(1)
$$

The evolution equation of normalized distribution flow is also given by

$$
\eta_{n+1}
$$

$$
= \Phi^2_{n+1,\gamma_n(1)}(\eta_n)
$$

$$
\propto \gamma_n(1)\, r(1-d)\, \eta_n M + \int \mathcal{Y}_n(dy)\, w_{\gamma_n(1)}(\eta_n, y)\, \Psi_{g(.,y)}(\eta_n) M + \mu(1)\, \overline{\mu}
$$

with the probability measure $\overline{\mu}$ and weight functions w defined below

$$
\overline{\mu}(dx) = \mu(dx)/\mu(1) \quad \text{and} \quad w_u(\eta, y) := r\left(1 - \frac{h}{h + du\eta(g(.,y))}\right)
$$

For null clutter parameter $h = 0$, we already observe that the total mass transformation Φ^1_{n+1,η_n} does not depend on the flow of probability measures η_n and it is simply given by

$$
\Phi^1_{n+1,\eta_n}(\gamma_n(1)) = \gamma_n(1)\, r(1-d) + r\, \mathcal{Y}_n(1) + \mu(1)
$$

In this particular situation, we have

$$
\gamma_n(1) = (r(1-d))^n \gamma_0(1) + \sum_{0 \le k < n} (r(1-d))^{n-1-k}(r\, \mathcal{Y}_k(1) + \mu(1))
$$

Now, we easily show that the pair of conditions (13.37) and (13.39) are satisfied with the parameters $a^1_{p,n} = (r(1-d))^{n-p}$ and $\tau^1_n = 0$.

13.4.3 Lipschitz regularity properties

In general, the total mass process is not explicitly known. Some useful estimates are provided by the following lemma.

Lemma 13.4.2 *We assume that the number of observations is uniformly bounded; that is, we have that $\mathcal{Y}^+(1) < \infty$. In this situation, the total mass process $\gamma_n(1)$ takes values in a sequence of compact sets $I_n \subset [m^-, m^+]$ with*

$$
m^- \;:=\; \frac{\mu(1)}{1 - r(1-d)}\left(1 + rd\, \mathcal{Y}^-\left(\frac{g^-}{h + d\mu(1)g^-}\right)\right)
$$
$$
m^+ \;:=\; \gamma_0(1) + \frac{r\mathcal{Y}^+(1) + \mu(1)}{1 - r(1-d)}
$$

Proof:

Using the fact that $\gamma_n(1) \geq \mu(1)$ we prove that

$$r\left(1 - \frac{h}{h + d\mu(1)\ g^-(y)}\right) \leq w_{\gamma_n(1)}(\eta_n, y) \leq r$$

from which we conclude that

$$\gamma_n(1)r(1-d)+r\ \mathcal{Y}_{h,n}(1)+\mu(1) \leq \Phi^1_{n+1,\eta_n}(\gamma_n(1)) \leq \gamma_n(1)\ r(1-d)+r\ \mathcal{Y}_n(1)+\mu(1)$$

with the random measures

$$\mathcal{Y}_{h,n}(dy) := \mathcal{Y}_n(dy)\ \frac{d\mu(1)\ g^-(y)}{h + d\mu(1)\ g^-(y)}$$

For any sequence of probability measures $\nu := (\nu_n)_{n\geq 0} \in \mathcal{P}(E)^{\mathbb{N}}$, and any starting mass $u \in [0, \infty[$, one concludes that

$$(r(1-d))^n\ u + \frac{r\mathcal{Y}_h^-(1) + \mu(1)}{1 - r(1-d)} \leq \Phi^1_{0,n,\nu}(u) \leq (r(1-d))^n\ u + \frac{r\mathcal{Y}^+(1) + \mu(1)}{1 - r(1-d)}$$

This implies that $\gamma_n(1), \gamma_n^N(1) \in I_n \subset [m^-, m^+]$ with

$$m^- := \frac{r\mathcal{Y}_h^-(1) + \mu(1)}{1 - r(1-d)} = \frac{\mu(1)}{1 - r(1-d)}\left(1 + rd\ \mathcal{Y}^-\left(\frac{g^-}{h + d\mu(1)g^-}\right)\right)$$

The end of the proof of the lemma is now completed. ∎

We are now in position to state the main result of this section.

Theorem 13.4.3 *We assume that the number of observations is uniformly bounded; that is, we have that $\mathcal{Y}^+(1) < \infty$. In this situation, the condition $(Lip(\Phi))$ is met with the Lipschitz constants*

$$a^i_{p,n} \leq \prod_{p \leq k < n} a^i_{k,k+1}$$

with $i = 1, 2$, and the sequence of parameters $\left(a^i_{n,n+1}\right)_{n\geq 0}$, $i = 1, 2$, defined below

$$a^1_{n,n+1} \leq r(1-d) + rdh\ \mathcal{Y}_n\left(\frac{g^+}{[h + dm^-g^-]^2}\right)$$

and

$$a^2_{n,n+1} \leq m^+ \frac{\beta(M)\left[(1-d) + d\ \mathcal{Y}_n\left(\frac{g^+}{h+dm^+g^+}\ \frac{g^+}{g^-}\right)\right] + hd\mathcal{Y}_n\left(\frac{g^+-g^-}{(h+dm^-g^-)^2}\right)}{(1-d)\ m^- + dm^-\mathcal{Y}_n\left(\frac{g^-}{h+dm^-g^-}\right) + \mu(1)/r}$$

In addition, condition $(Cont(\Phi))$ is met with the sequence of parameters

$$\tau^1_{n+1} \leq rdhm^+\ \mathcal{Y}_n\left(\frac{g^+ - g^-}{[h + dm^-g^-]^2}\right)$$

and

$$\tau_{n+1}^2 \le \frac{(1-d) + hd \; \mathcal{Y}_n \left(\frac{g^+}{(h+dm^- g^-)^2} \right)}{(1-d) \; m^- + dm^- \mathcal{Y}_n \left(\frac{g^-}{h+dm^- g^-} \right) + \mu(1)/r}$$

Proof:

For any $\eta \in \mathcal{P}(E)$, and any $(u, u') \in I_n^2$, we have

$$\left| \Phi_{n+1,\eta}^1(u) - \Phi_{n+1,\eta}^1(u') \right|$$

$$= |u - u'| \left[r(1-d) + rdh \int \mathcal{Y}_n(dy) \; \frac{\eta(g(\cdot, y))}{[h + du\eta(g(\cdot, y))][h + du'\eta(g(\cdot, y))]} \right]$$

$$\le |u - u'| \left[r(1-d) + rdh \; \mathcal{Y}_n \left(\frac{g^+}{[h + dm^- g^-)]^2} \right) \right]$$

This implies that condition (13.37) is satisfied with

$$a_{n,n+1}^1 \le r(1-d) + rdh \; \mathcal{Y}_n \left(\frac{g^+}{[h + dm^- g^-)]^2} \right)$$

In the same way, for any $\eta, \eta' \in \mathcal{P}(E)$ and any $u \in I_n$, we have

$$\Phi_{n+1,\eta}^1(u) - \Phi_{n+1,\eta'}^1(u)$$

$$= rdhu \int \mathcal{Y}_n(dy) \; \frac{1}{[h + du\eta(g(\cdot, y))][h + du\eta'(g(\cdot, y))]} \; (\eta - \eta') \, (g(\cdot, y))$$

We conclude that (13.39) is met with the parameter

$$\tau_{n+1}^1 \le rdhm^+ \; \mathcal{Y}_n \left(\frac{g^+ - g^-}{[h + dm^- g^-]^2} \right)$$

and the collection of probability measures

$$\Omega_{n,\eta'}^1(d\varphi) \propto \int \mathcal{Y}_n(dy) \; \frac{g^+(y) - g^-(y)}{[h + dm^- g^-(y)]^2} \; \delta_{\frac{g(\cdot, y)}{g^+(y) - g^-(y)}}(d\varphi)$$

Now, we come to the analysis of the mappings

$$\Phi_{n+1,u}^2(\eta) \propto r(1-d)u \; \eta M + \int \mathcal{Y}_n(dy) \; w_u(\eta, y) \; \Psi_{g(\cdot, y)}(\eta) M + \mu(1) \; \overline{\mu}$$

with the weight functions

$$w_u(\eta, y) := \frac{rdu\eta(g(\cdot, y))}{h + du\eta(g(\cdot, y))} = r \left(1 - \frac{h}{h + du\eta(g(\cdot, y))} \right)$$

Notice that

$$w^-(y) := \frac{rdm^- g^-(y)}{h + dm^- g^-(y)} \le w_u(\eta, y) \le w^+(y) := \frac{rdm^+ g^+(y)}{h + dm^+ g^+(y)}$$

To have a more synthetic formula, we extend the observation state space with two auxiliary points c_1, c_2 and we set

$$\mathcal{Y}_n^c = \mathcal{Y}_n + \delta_{c_1} + \delta_{c_2}$$

We extend the likelihood and the weight functions by setting $g(x, c_1) = g(x, c_2) = 1$, and

$$w^-(c_1) \quad := \quad r(1-d)m^- \leq w_u(\eta, c_1) := r(1-d)u \leq w^+(c_1) := r(1-d)m^+$$
$$w_u(\eta, c_2) \quad = \quad w^+(c_2) = w^-(c_2) := \mu(1)$$

In this notation, we find that

$$\Phi_{n+1,u}^2(\eta) \propto \int \mathcal{Y}_n^c(dy)\ w_u(\eta, y)\ \Psi_{g(.,y)}(\eta)M_y$$

with the collection of Markov transitions M_y defined below

$$\forall y \notin \{c_2\} \qquad M_y = M \quad \text{and} \quad M_{c_2} = \overline{\mu}$$

Notice that the normalizing constants $\mathcal{Y}_n^c(w_u(\eta, .))$ satisfy the following lower bounds

$$\mathcal{Y}_n^c(w_u(\eta, .)) \quad \geq \quad \mathcal{Y}_n^c(w^-) = r(1-d)\ m^- + \mathcal{Y}_n\left(w^-\right) + \mu(1)$$

We analyze the Lipschitz properties of the mappings $\Phi_{n+1,u}^2$ using the following decomposition

$$\Phi_{n+1,u}^2(\eta) - \Phi_{n+1,u}^2(\eta') = \Delta_{n+1,u}(\eta, \eta') + \Delta'_{n+1,u}(\eta, \eta')$$

with the signed measures

$$\Delta_{n+1,u}(\eta, \eta') = \int \mathcal{Y}_n^c(dy)\ \frac{w_u(\eta, y)}{\mathcal{Y}_n^c(w_u(\eta, .))}\ \left[\Psi_{g(.,y)}(\eta)M_y - \Psi_{g(.,y)}(\eta')M_y\right]$$

and

$$\Delta'_{n+1,u}(\eta, \eta')$$
$$= \frac{1}{\mathcal{Y}_n^c(w_u(\eta, .))} \int \mathcal{Y}_n^c(dy)\ [w_u(\eta, y) - w_u(\eta', y)]\ \left(\Psi_{g(.,y)}(\eta')M_y - \Phi_{n+1,u}^2(\eta')\right)$$

On the other hand, it is not difficult to check that

$$|\Delta_{n+1,u}(\eta, \eta')(f)|$$
$$\leq \frac{1}{\mathcal{Y}_n^c(w^-)}\ (r(1-d)m^+\ |(\eta - \eta')(M(f))|$$
$$+ \int \mathcal{Y}_n(dy)\ w^+(y)\ \frac{g^+(y)}{g^-(y)}\ \left|(\eta - \eta')(S_{\eta'}^y M(f))\right|)$$

for some collection of Markov transitions $S_{\eta'}^y$ from E into itself.

It is also readily checked that

$$|\Delta'_{n+1,u}(\eta,\eta')(f)| \le \frac{hrdm^+}{\mathcal{Y}_n^c(w^-)} \int \mathcal{Y}_n(dy) \frac{1}{(h+m^-dg^-(y))^2} |(\eta-\eta')(g(.,y))|$$

This clearly implies that condition (13.38) is satisfied with

$$a_{n,n+1}^2 \le \frac{1}{\mathcal{Y}_n^c(w^-)}$$

$$\times \left(\beta(M)\left[r(1-d)m^+ + \mathcal{Y}_n\left(\frac{w^+g^+}{g^-}\right)\right] + hrdm^+\mathcal{Y}_n\left(\frac{g^+-g^-}{(h+m^-dg^-)^2}\right)\right)$$

We analyze the continuity properties of the mappings $u \mapsto \Phi_{n+1,u}^2(\eta)$ using the following decomposition

$$\Phi_{n+1,u}^2(\eta) - \Phi_{n+1,u'}^2(\eta)$$

$$= \frac{1}{\mathcal{Y}_n^c(w_u(\eta,.))} \int \mathcal{Y}_n^c(dy) \left[w_u(\eta,y) - w_{u'}(\eta,y)\right] \left(\Psi_{g(.,y)}(\eta)M_y - \Phi_{n+1,u'}^2(\eta)\right)$$

This implies that

$$\left|\left[\Phi_{n+1,u}^2(\eta) - \Phi_{n+1,u'}^2(\eta)\right](f)\right|$$

$$\le \frac{1}{\mathcal{Y}_n^c(w^-)}\left[r(1-d) + hrd\,\mathcal{Y}_n\left(\frac{g^+}{(h+dm^-g^-)^2}\right)\right] |u-u'|$$

This shows that condition (13.40) is satisfied with

$$\tau_{n+1}^2 \le \frac{1}{\mathcal{Y}_n^c(w^-)}\left[r(1-d) + hrd\,\mathcal{Y}_n\left(\frac{g^+}{(h+dm^-g^-)^2}\right)\right]$$

This ends the proof of the theorem. ∎

13.4.4 Proof of the stability theorem

This section is concerned with the proof of Theorem 13.4.1.

There is no loss of generality to assume that $r(1-d) < 1/2 \le d$ and $\mu(1) \ge 1 \ge h$. Recalling that $m^- \ge \mu(1)$, one readily proves that

$$\frac{m^+}{\mu(1)} = \frac{\gamma_0(1)}{\mu(1)} + \frac{1}{1-r(1-d)}\left(1 + \frac{r}{\mu(1)}\mathcal{Y}^+(1)\right) \le 2 + \gamma_0(1) + 2r\mathcal{Y}^+(1) := \rho$$

If we set $\delta(g) := \rho \vee \mathcal{Y}^+\left(\frac{g^+}{g^-}\right) \vee \mathcal{Y}^+\left(\frac{g^+}{(g^-)^2}\right)$, then we find the rather crude estimates

$$a_{n,n+1}^1/r \le (1-d) + \frac{2h}{\mu(1)^2}\delta(g)$$

and

$$a^2_{n,n+1}/r \le \left[\beta(M)(1-d) + \frac{2h + \beta(M)}{\mu(1)} \right] \delta(g)$$

as well as

$$\tau^1_{n+1}/r \le \frac{2h}{\mu(1)} \, \delta(g)^2 \quad \text{and} \quad \tau^2_{n+1}/r \le \frac{1}{\mu(1)} \left[(1-d) + \frac{2h}{\mu(1)^2} \, \delta(g) \right]$$

from which we find that

$$\tau^1 \tau^2 \le \frac{2hr^2}{\mu(1)^2} \left[(1-d) + \frac{2h}{\mu(1)^2} \, \delta(g) \right] \delta(g)^2 \tag{13.43}$$

Thus, there exists some $0 < \kappa_0 \le 1$ and some $\kappa_1 < \infty$ so that for any $d \ge \kappa_0$ and any $\mu(1) \ge \kappa_1$ we have

$$
\begin{aligned}
a^1_{n,n+1} &\le r \left[(1-d) + \frac{2}{\mu(1)^2} \right] \delta(g) \doteq e^{-\lambda_1} < 1 \\
a^2_{n,n+1} &\le r \left[(1-d) + \frac{3}{\mu(1)} \right] \delta(g) \doteq e^{-\lambda_2} < 1 \quad \text{with} \quad 0 < \lambda_2 < \lambda_1
\end{aligned}
$$

Finally, using (13.43) we find some $\kappa_2 > 0$ such that for any $h \le \kappa_2$, we have that $\tau^1 \tau^2 \le \left(1 - e^{-\lambda_2}\right) \left(e^{-\lambda_2} - e^{-\lambda_1}\right)$. This ends the proof of the corollary. ∎

Chapter 14

Particle density profiles

In this chapter, we further develop the analysis of the different classes of mean field models presented in Section 1.5 and in Chapter 2.

The first sections are concerned with the mean field Feynman-Kac models discussed in Chapter 9. The fluctuation, and the concentration analysis of the general class of mean field models, discussed in Chapter 10, is further developed in Section 14.6.

The last two sections are dedicated to the convergence analysis of the branching intensity models presented in Section 2.3.3 and chapter 6 and the mean field approximations of positive measure valued equations discussed in Section 2.4 and in chapter 6. The chapter is centered around the fluctuation of particle density profiles of the McKean particle measures introduced in (1.54), around their limiting values.

14.1 Stochastic perturbation analysis

14.1.1 Local perturbation random fields

We consider a Feynman-Kac measure valued equation of the form (9.4). The mean field particle model ξ_n associated with these equations is defined in (9.15). We let $\mathcal{F}_n^N$ be the σ-field generated by ξ_p, with $p \leq n$.

In Section 2.1.1, we have seen that the random perturbation terms associated with these mean field models are encapsulated into the sequence of local random sampling errors $(V_n^N)_{n \geq 0}$, defined by the perturbation equations

$$\eta_n^N = \Phi_n^N\left(\eta_{n-1}^N\right) := \Phi_n\left(\eta_{n-1}^N\right) + \frac{1}{\sqrt{N}} V_n^N \qquad (14.1)$$

with the one step mapping Φ_n defined in (9.4). In the above display, the random $\mathcal{F}_n^N$-measurable mapping Φ_n^N encapsulates the randomness induced by the mean field sampling (9.15) of every particles

$$\xi_{n-1}^i \rightsquigarrow \xi_n^i \sim K_{n,\eta_{n-1}^N}\left(\xi_{n-1}^i, dx_n\right)$$

for every $1 \leq i \leq N$. In this notation, for any $0 \leq p \leq n$ we have the semigroup formulae

$$\eta_n^N = \Phi_{p,n}^N \left(\eta_p^N\right) \quad \text{with} \quad \Phi_{p,n}^N = \Phi_{p+1}^N \circ \Phi_{p+2}^N \circ \ldots \circ \Phi_n^N$$

with the convention $\Phi_{n,n}^N = Id$, the identity mapping, for $p = 0$.

We recall that the choice of the Markov transitions $(K_{n,\eta})_{n \geq 1, \eta \in \mathcal{P}(E_{n-1})}$ is far from being unique. Three possible different classes of transitions are provided in Section 4.4.5.

In Section 4.1.1 dedicated to genetic type branching particle models, we also presented a rather general branching model to approximate the selection transition of the genetic particle model, as soon as some of the local variance condition (4.4) is satisfied. For a more thorough discussion on these branching strategies, we refer the interested reader to [144, 161, 163].

In the context of mean field particle models, Theorem 9.5.3 ensures that for any fixed time horizon $n \geq 0$, the sequence of random fields V_n^N converges in law, in the sense of convergence of finite dimensional distributions, as the number of particles N tends to infinity, to a sequence of independent, Gaussian and centered random fields V_n, with a covariance function defined for any $f \in \mathcal{B}_b(E_n)$, and any $n \geq 0$, by the formula

$$\mathbb{E}(V_n(f)^2) = \eta_{n-1} K_{n,\eta_{n-1}}([f - K_{n,\eta_{n-1}}(f)]^2) \leq \sigma_n^2$$

with the uniform local variance parameters σ_n defined in (9.31).

Roughly speaking, these asymptotic results indicate that the random mappings Φ_n^N defined in (14.1) behave as the limiting one step transformation Φ_n, up to some Gaussian type perturbation random field, with the traditional Monte Carlo precision order $1/\sqrt{N}$.

To get some useful nonasymptotic quantitative estimates, we can look to conditional $\mathbb{L}_m$-mean error bounds. In this connection, we have seen in (11.3) that the local sampling random fields models V_n^N belong to the class of empirical processes discussed in Section 11.1. Thus, the stochastic analysis developed in Chapter 11 applies directly to these models. For instance, Proposition 9.5.5 provides the following almost sure estimate of the amplitude of the stochastic perturbations

$$\sup_{f \in \mathrm{Osc}(E_n)} \mathbb{E}\left(\left|V_n^N(f)\right|^m \mid \mathcal{F}_{n-1}^N\right)^{1/m} \leq b(m) \ (1 \wedge c(m)) \tag{14.2}$$

for any $m \geq 1$, with the constant $b(m)$ defined in (0.6) and $c(m)$ given by

$$c(m) = 6b(m) \ \max\left(\sqrt{2}\sigma_n, \left[\frac{2\sigma_n^2}{N^{\frac{m'}{2}-1}}\right]^{1/m'}\right)$$

In the above display, m' stands for the smallest even integer $m' \geq m$.

Suppose that N is chosen so that $2\sigma_n^2 N \geq 1$. In this situation,

$$c(m) = 6\sqrt{2}b(m)\sigma_n \max\left(1, \left[2\sigma_n^2 N\right]^{-(1/2-1/m')}\right) = 6\sqrt{2}b(m)\sigma_n$$

and therefore

$$\sup_{f\in\mathrm{Osc}(E_n)} \mathbb{E}\left(\left|V_n^N(f)\right|^m \mid \mathcal{F}_{n-1}^N\right)^{1/m} \leq b(m)\left(1 \wedge \left[6\sqrt{2}b(m)\sigma_n\right]\right) \quad (14.3)$$

Most of the stochastic perturbation analysis developed in this chapter is based on these $\mathbb{L}_m$-mean error bounds. Thus, these results apply directly without further work to any branching type numerical scheme of Feynman-Kac flows, as soon as the local approximation satisfies some $\mathbb{L}_m$-mean error bounds of the form (14.2).

14.1.2   Backward semigroup analysis

To get one step further in our discussion, we need to introduce the "global error" associated with the fluctuation of the particle density profiles η_n^N around their limiting values η_n.

Definition 14.1.1 *We let $W_n^{\eta,N}$ be the empirical random fields on $\mathcal{B}_b(E_n)$ defined for any $n \geq 1$ by*

$$\eta_n^N = \eta_n + \frac{1}{\sqrt{N}}\, W_n^{\eta,N} \quad (14.4)$$

One natural way to control the fluctuations and the concentration properties of the particle measures η_n^N around their limiting values η_n is to express the random fields $W_n^{\eta,N}$ in terms of the empirical random fields $(V_n^N)_{n\geq 0}$.

In the following picture, we have illustrated the random evolution of the N-mean field particle model described in Section 4.4 and in Section 9.4.

$$
\begin{array}{ccccccc}
\eta_0 & \to & \eta_1 = \Phi_1(\eta_0) & \to & \eta_2 = \Phi_{0,2}(\eta_0) & \to & \cdots & \to & \Phi_{0,n}(\eta_0) \\
\Downarrow & & & & & & \\
\eta_0^N & \to & \Phi_1(\eta_0^N) & \to & \Phi_{0,2}(\eta_0^N) & \to & \cdots & \to & \Phi_{0,n}(\eta_0^N) \\
& & \Downarrow & & & & \\
& & \eta_1^N & \to & \Phi_2(\eta_1^N) & \to & \cdots & \to & \Phi_{1,n}(\eta_1^N) \\
& & & & \Downarrow & & \\
& & & & \eta_2^N & \to & \cdots & \to & \Phi_{2,n}(\eta_2^N) \\
& & & & & & \Downarrow & & \vdots \\
& & & & & & \eta_{n-1}^N & \to & \Phi_n(\eta_{n-1}^N) \\
& & & & & & & & \Downarrow \\
& & & & & & & & \eta_n^N
\end{array}
$$

This picture gives a sound basis to the stochastic perturbation analysis developed later on in this chapter.

Intuitively, we first observe that the sampling error random fields models V_n^N (represented by the implication sign "$\Downarrow$") do not propagate but stabilize as soon as the semigroup $\Phi_{p,n}$ of the Feynman-Kac model is sufficiently stable.

This intuitive and natural idea is made clear by the pivotal formula

$$\eta_n^N - \eta_n \;=\; \sum_{q=0}^{n} \left[\Phi_{q,n}(\eta_q^N) - \Phi_{q,n}(\Phi_q(\eta_{q-1}^N)) \right]$$

or equivalently in terms of the local random fields

$$W_n^{\eta,N} = \sum_{q=0}^{n} \sqrt{N} \left[\Phi_{q,n}\left(\Phi_q(\eta_{q-1}^N) + \frac{1}{\sqrt{N}} \, V_q^N \right) - \Phi_{q,n}\left(\Phi_q(\eta_{q-1}^N) \right) \right] \quad (14.5)$$

with the convention $\Phi_0(\eta_{-1}^N) = \eta_0$ for $p = 0$.

In other words, we use the interpolating sequence of measures

$$0 \le p \le n \mapsto \Phi_{p,n}\left(\Phi_{0,p}^N(\eta_0^N) \right) = \Phi_{p,n}\left(\eta_p^N \right)$$

from the initial distribution

$$\Phi_{0,n}\left(\Phi_{0,0}^N(\eta_0^N) \right) = \Phi_{0,n}\left(\eta_0^N \right) \quad \text{to the measure} \quad \Phi_{n,n}\left(\Phi_{0,n}^N(\eta_0^N) \right) = \eta_n^N$$

In this notation, the decomposition (14.5) can be rewritten in the following form

$$
\begin{aligned}
W_n^{\eta,N} \;=\;\; & \Phi_{0,n}\left(\eta_0^N \right) - \Phi_{0,n}\left(\eta_0 \right) \\
& + \sum_{q=1}^{n} \sqrt{N} \left[\Phi_{q,n}\left(\Phi_{0,q}^N(\eta_0^N) \right) - \Phi_{q-1,n}\left(\Phi_{0,q-1}^N(\eta_0^N) \right) \right]
\end{aligned}
$$

These key decompositions of the random fields $W_n^{\eta,N}$ using the backward semigroup decomposition (14.5) first appear in the proof of theorem 1 in the article [152].

Note that each term on the r.h.s. represents the propagation of the p-th sampling local error

$$\Phi_p(\eta_{p-1}^N) \Rightarrow \eta_p^N = \Phi_p(\eta_{p-1}^N) + \frac{1}{\sqrt{N}} \, V_p^N$$

This observation emphasizes that the numerical analysis of the mean field particle algorithm or the one of any numerical approximation model based on local approximations is intimately related to the stability property of the nonlinear semigroup of the limiting model. These questions will be made clear later on in this chapter.

14.2   First order expansions

14.2.1   Feynman-Kac semigroup expansions

By Lemma 12.1.6, we first observe that the Feynman-Kac semigroup $\Phi_{p,n}$ is expressed in terms of a Boltzmann-Gibbs transformation and a Markov transport equation. More precisely, for any $0 \le p \le n$, we have

$$\Phi_{p,n}(\eta_p) = \Psi_{G_{p,n}}(\eta_p) P_{p,n} \tag{14.6}$$

This shows that the decomposition (14.5) can be rewritten in the following form

$$W_n^{\eta,N} = \sum_{p=0}^{n} \Delta W_{p,n}^{\eta,N} \tag{14.7}$$

with the random fields sequence $\Delta W_{p,n}^{\eta,N}$ defined by

$$\Delta W_{p,n}^{\eta,N} := \sqrt{N} \left[\Psi_{G_{p,n}} \left(\Phi_p(\eta_{p-1}^N) + \frac{1}{\sqrt{N}} V_q^N \right) - \Psi_{G_{p,n}} \left(\Phi_p(\eta_{p-1}^N) \right) \right] P_{p,n} \tag{14.8}$$

Our next objective is to derive a first order expansion of $\Delta W_{p,n}^{\eta,N}$ in terms of the local random fields V_q^N. Before proceeding, we use Lemma 3.1.5 to prove for any $p \le n$, and any couple of measures $\mu, \nu \in \mathcal{P}(E_p)$, the following first order decompositions

$$\left[\Psi_{G_{p,n}}(\mu) - \Psi_{G_{p,n}}(\nu) \right](f)$$

$$= \frac{1}{\mu(G_{p,n,\nu})} (\mu - \nu)(d_\nu \Psi_{G_{p,n}}(f))$$

$$= \frac{1}{\mu(G'_{p,n})} (\mu - \nu)(d'_\nu \Psi_{G_{p,n}}(f))$$

$$= (\mu - \nu)(d_\nu \Psi_{G_{p,n}}(f)) - \frac{1}{\mu(G_{p,n,\nu})} (\mu - \nu)^{\otimes 2} \left[G_{p,n,\nu} \otimes d_\nu \Psi_{G_{p,n}}(f) \right]$$

with the functions $G_{p,n,\nu}$ and $G_{p,n}$ on E_p, and the integral operators $d_\nu \Psi_{G_{p,n}}$ and $d'_\nu \Psi_{G_{p,n}}$ from $\mathcal{B}_b(E_n)$ into $\mathcal{B}_b(E_p)$ defined by

$$d_\nu \Psi_{G_{p,n}}(f) \quad := \quad G_{p,n,\nu} \left(f - \Psi_{G_{p,n}}(\nu)(f) \right) \quad \text{and} \quad G_{p,n,\nu} := G_{p,n}/\nu(G_{p,n})$$

$$d'_\nu \Psi_{G_{p,n}}(f) \quad := \quad G'_{p,n} \left(f - \Psi_{G_{p,n}}(\nu)(f) \right) \quad \text{and} \quad G'_{p,n} := G_{p,n}/\|G_{p,n}\| \tag{14.9}$$

We use these first order decompositions to derive Taylor's type expansions of $\Delta W_{p,n}^{\eta,N}$ in terms of the local random fields V_q^N. These decompositions are expressed in terms of the integral operators and the functions defined below.

Definition 14.2.1 *For any $0 \le p \le n$, and $N \ge 1$, we denote by $G_{p,n}^N$ the $\mathcal{G}_{p-1}^N$-measurable random functions on E_p given by*

$$G_{p,n}^N := \frac{G_{p,n}}{\Phi_p(\eta_{p-1}^N)(G_{p,n})} \le g_{p,n} \quad \text{and} \quad G'_{p,n} = G_{p,n}/\|G_{p,n}\| (\le 1)$$

Definition 14.2.2 *For any $0 \le p \le n$, and $N \ge 1$, we denote by $d_{p,n}^N$ and $d_{p,n}'^N$ the $\mathcal{G}_{p-1}^N$-measurable random integral operators defined by*

$$d_{p,n}^N := d_{\Phi_p(\eta_{p-1}^N)} \Phi_{p,n} \quad and \quad d_{p,n}'^N := d'_{\Phi_p(\eta_{p-1}^N)} \Phi_{p,n}$$

We also denote by $d_{p,n}$ and $d_{p,n}'$ the integral operators defined by

$$d_{p,n} := d_{\eta_p} \Phi_{p,n} \quad and \quad d_{p,n}' := d'_{\eta_p} \Phi_{p,n} \tag{14.10}$$

Using the differential rule stated in Lemma 3.1.9, we prove the following lemma.

Lemma 14.2.3 *For any $0 \le p \le q \le n$ we have the semigroup property*

$$d_{p,q} d_{q,n} = d_{p,n} \tag{14.11}$$

with $d_{n,n} = Id$, the identity operator, for $p = n$.

Proof:
We check this claim using the following decompositions

$$
\begin{aligned}
d_{p,q} d_{q,n} &= d_{\eta_p} \Phi_{p,q} \, d_{\eta_q} \Phi_{q,n} \\
&= d_{\eta_p} \Phi_{p,q} \, d_{\Phi_{p,q}(\eta_q)} \Phi_{q,n} = d_{\eta_p} \left(\Phi_{q,n} \circ \Phi_{p,q} \right) = d_{\eta_p} \Phi_{p,n} = d_{p,n}
\end{aligned}
$$

This ends the proof of the lemma. ∎

Using the decomposition (14.6), we find the following formulae

$$d_{p,n}^N := d_{\Phi_p(\eta_{p-1}^N)} \Psi_{G_{p,n}} P_{p,n} = d_{\Phi_p(\eta_{p-1}^N)} \Phi_{p,n}$$

and in much the same way we have

$$
\begin{aligned}
d_{p,n}'^N &:= d'_{\Phi_p(\eta_{p-1}^N)} \Psi_{G_{p,n}} P_{p,n} \\
&= \Phi_p(\eta_{p-1}^N)(G_{p,n}') \times d_{\Phi_p(\eta_{p-1}^N)} \Phi_{p,n}
\end{aligned}
$$

We also observe that

$$d_{p,n} := d_{\eta_p} \Psi_{G_{p,n}} P_{p,n} \quad and \quad d_{p,n}' := d'_{\eta_p} \Psi_{G_{p,n}} P_{p,n}$$

We end this section with some useful upper bounds

$$\|d_{p,n}(f)\| \vee \|d_{p,n}^N(f)\| \le g_{p,n}\, \beta(P_{p,n}) \quad and \quad \|G_{p,n}^N\| \le g_{p,n} \tag{14.12}$$

as well as

$$\|d_{p,n}'(f)\| \vee \|d_{p,n}'^N(f)\| \le \beta(P_{p,n}) \tag{14.13}$$

14.2.2 Decomposition theorems

This section is concerned with fluctuation decompositions of the particle profiles η_n^N around their limiting values η_n. The first theorem is a consequence of the backward semigroup analysis presented in Section 14.1.2 and in Section 14.2.1.

Theorem 14.2.4 *For any $0 \le p \le n$, $N \ge 1$, and any $f_n \in \mathcal{B}_b(E_n)$, we have the following first and second order decompositions*

$$W_n^{\eta,N}$$

$$= \sum_{p=0}^n \frac{1}{\eta_p^N(G_{p,n}^N)}\, V_p^N\left(d_{p,n}^N(f)\right)$$

$$= \sum_{p=0}^n \frac{1}{\eta_p^N(G_{p,n}'^N)}\, V_p^N\left(d_{p,n}'^N(f)\right)$$

$$= \sum_{p=0}^n V_p^N\left[d_{p,n}^N(f)\right] - \sum_{p=0}^n \frac{1}{\sqrt{N}}\, \frac{1}{\eta_p^N(G_{p,n}^N)}\, V_p^N(G_{p,n}^N)\, V_p^N\left[d_{p,n}^N(f)\right]$$

$$(14.14)$$

Proof:
By (14.7) and (14.8), to prove (14.14) it clearly suffices to prove that

$$\sqrt{N}\left[\Phi_{p,n}\left(\Phi_p(\eta_{p-1}^N) + \tfrac{1}{\sqrt{N}}\, V_p^N\right) - \Phi_{p,n}\left(\Phi_p(\eta_{p-1}^N)\right)\right](f)$$

$$= \frac{1}{\eta_p^N(G_{p,n}^N)}\, V_p^N\left(d_{p,n}^N(f)\right)$$

$$= \frac{1}{\eta_p^N(G_{p,n}'^N)}\, V_p^N\left(d_{p,n}'^N(f)\right)$$

$$= V_p^N\left[d_{p,n}^N(f)\right] - \frac{1}{\sqrt{N}}\, \frac{1}{\eta_p^N(G_{p,n}^N)}\, V_p^N(G_{p,n}^N)\, V_p^N\left[d_{p,n}^N(f)\right]$$

These decompositions are direct consequences of the formulae (14.9) applied to the following measures and functions

$$\nu = \Phi_p(\eta_{p-1}^N) \qquad \mu = \Phi_p(\eta_{p-1}^N) + \frac{1}{\sqrt{N}}\, V_p^N \quad \text{and} \quad f = P_{p,n}(f_n)$$

for some $f_n \in \mathcal{B}_b(E_n)$ and $N \ge 1$. This ends the proof of (14.14).
The end of the proof of the theorem is now completed. ∎

The second main result of this section is a fluctuation decomposition formula in terms of the end point of a martingale and a second order predictable random field.

Theorem 14.2.5 *For any $n \geq 0$, $N \geq 1$, and any $f_n \in \mathcal{B}_b(E_n)$, we have the decomposition*

$$W_n^{\eta,N}(f) = \sum_{p=0}^{n} V_p^N \left[d_{p,n}(f) \right] + \frac{1}{\sqrt{N}} R_n^N(f) \tag{14.15}$$

with the second order remainder term

$$R_n^N(f) := -\sum_{p=0}^{n-1} \frac{1}{\eta_p^N(\overline{G}_p)} W_p^{\eta,N}(\overline{G}_p) W_p^{\eta,N} \left[d_{p,n}(f) \right]$$

Before entering into the proof of the theorem, we make a couple of remarks. Firstly, using the differential rule (14.11) if we replace f by $d_{n,m}(f)$ for some $m \geq n$ we find that

$$W_n^{\eta,N} \left(d_{n,m}(f) \right) = M_n^{(N,m)}(f) + \frac{1}{\sqrt{N}} R_n^{(N,m)}(f)$$

with the martingale sequence $\left(M_n^{(N,m)} \right)_{0 \leq n \leq m}$ defined by

$$M_n^{(N,m)}(f) := \sum_{p=0}^{n} V_p^N \left[d_{p,m}(f) \right]$$

and the predictable second order term $\left(R_n^{(N,m)} \right)_{0 \leq n \leq m}$ given by

$$R_n^{(N,m)}(f) = -\sum_{p=0}^{n-1} \frac{1}{\eta_p^N(\overline{G}_p)} W_p^{\eta,N}(\overline{G}_p) W_p^{\eta,N} \left[d_{p,m}(f) \right]$$

On the other hand, the decomposition (14.14) stated in Theorem 14.2.4 already implies that

$$W_n^{\eta,N}(f) = \sum_{p=0}^{n} V_p^N \left[d_{p,n}(f) \right] + \frac{1}{\sqrt{N}} R_n^N(f)$$

with a remainder term

$$R_n^N(f) := \sum_{p=0}^{n} V_p^N \left[\widetilde{d}_{p,n}^N(f) \right] - \sum_{p=0}^{n} \frac{1}{\eta_p^N(G_{p,n}^N)} V_p^N(G_{p,n}^N) V_p^N \left[d_{p,n}^N(f) \right]$$

with the operator

$$\widetilde{d}_{p,n}^N := \sqrt{N} \left[d_{p,n}^N - d_{p,n} \right] \tag{14.16}$$

Some estimates of the weak $\mathbb{L}_m$-norm of $\widetilde{d}_{p,n}^N$ are provided in Lemma 14.3.4.

Now, we come to the proof of the theorem.

Proof of theorem 14.2.5:

The proof is based on the telescoping sum decomposition

$$\eta_n^N - \eta_n = \sum_{p=0}^{n} \left[\eta_p^N \overline{Q}_{p,n} - \eta_{p-1}^N \overline{Q}_{p-1,n} \right]$$

with the convention $\eta_{-1}^N \overline{Q}_{-1,n} = \eta_0 \overline{Q}_{0,n} = \eta_n$, for $p = 0$. Using the fact that

$$\eta_{p-1}^N \overline{Q}_{p-1,n}(f) = \eta_{p-1}^N (\overline{G}_{p-1}) \times \Phi_p \left(\eta_{p-1}^N \right) \overline{Q}_{p,n}(f)$$

we prove that

$$\left[\eta_n^N - \eta_n \right](f) = \sum_{p=0}^{n} \left[\eta_p^N - \Phi_p \left(\eta_{p-1}^N \right) \right] \overline{Q}_{p,n}(f) + R_n^N(f)$$

with the second order remainder term

$$R_n^N(f) := \sum_{p=1}^{n} \left(1 - \eta_{p-1}^N (\overline{G}_{p-1}) \right) \times \Phi_p \left(\eta_{p-1}^N \right) \overline{Q}_{p,n}(f)$$

Replacing f by the centered function $(f - \eta_n(f))$, and using the fact that $1 = \eta_{p-1}(\overline{G}_{p-1})$ and $\eta_p \left[d_{p,n}(f - \eta_n(f)) \right] = 0$, we conclude that

$$\left[\eta_n^N - \eta_n \right](f) = \sum_{p=0}^{n} \left[\eta_p^N - \Phi_p \left(\eta_{p-1}^N \right) \right] (d_{p,n}(f)) + \overline{R}_n^N(f)$$

with the second order remainder term

$$
\begin{aligned}
\overline{R}_n^N(f) &:= \sum_{p=1}^{n} \left[\eta_{p-1} - \eta_{p-1}^N \right] (\overline{G}_{p-1}) \\
&\qquad \times \left[\Psi_{G_{p-1}} \left(\eta_{p-1}^N \right) - \Psi_{G_{p-1}} (\eta_{p-1}) \right] (M_p (d_{p,n}(f))) \\
&= -\frac{1}{N} \sum_{p=1}^{n} W_{p-1}^{\eta,N} (\overline{G}_{p-1}) \\
&\qquad \times \frac{1}{\eta_{p-1}^N (\overline{G}_{p-1})} \, W_{p-1}^{\eta,N} \left(d_{\eta_{p-1}} \Psi_{G_{p-1}} (M_p (d_{p,n}(f))) \right)
\end{aligned}
$$

Finally, we observe that

$$
\begin{aligned}
d_{\eta_{p-1}} \Psi_{G_{p-1}} (M_p (d_{p,n}(f))) &= \frac{G_{p-1}}{\eta_{p-1}(G_{p-1})} (M_p(d_{p,n}(f)) - \eta_p(d_{p,n}(f))) \\
&= \frac{G_{p-1}}{\eta_{p-1}(G_{p-1})} M_p(d_{p,n}(f)) \\
&= \overline{Q}_{p-1,p}(d_{p,n}(f)) = \overline{Q}_{p-1,n}(f - \eta_n(f)) \\
&= d_{p-1,n}(f)
\end{aligned}
$$

This ends the proof of (14.15).

The end of the proof of the theorem is now completed. ∎

14.3   Some nonasymptotic theorems

14.3.1   Orlicz norm and $\mathbb{L}_m$-mean error estimates

In this section we analyze some more or less direct consequences of the pivotal first order decompositions developed in Section 14.2.2. Further on in this section $\tau_{k,l}(n)$, $b(m)$, and $c(m)$ stand for the parameters defined, respectively, in (12.4), (0.6), and in (9.33); and π_ψ is the Orlicz norm defined in (11.4).

The first main result is a nonasymptotic theorem on $\mathbb{L}_m$-mean error bounds and Orlicz norm estimates of the random field $W_n^{\eta,N}$ defined in (14.4). These random fields are related to the fluctuations of the particle profiles η_n^N around their limiting values η_n.

Theorem 14.3.1 *For any $m \geq 1$, $N \geq 1$, $n \geq 0$, and any $f \in \mathrm{Osc}(E_n)$, we have $\mathbb{L}_m$-mean error estimates*

$$\mathbb{E}\left(\left|W_n^{\eta,N}(f)\right|^m\right)^{1/m} \leq 2\, b(m)\,\, (1 \wedge c(m))\, \tau_{1,1}(n) \qquad (14.17)$$

In addition, we have the Orlicz norm upper bounds

$$\pi_\psi\left(\left|W_n^{\eta,N}(f)\right|\right) \leq 4\sqrt{2/3}\,\, \tau_{1,1}(n) \qquad (14.18)$$

$$\pi_\psi\left(\sup_{0 \leq p \leq n}\left|W_p^{\eta,N}(f)\right|\right) \leq 8\,\sqrt{\log\,(9+n)}\,\, \sup_{0 \leq p \leq n}\tau_{1,1}(p) \quad (14.19)$$

Proof:
Using the second decomposition in (14.14) we check that

$$\mathbb{E}\left(\left|W_n^{\eta,N}(f)\right|^m\right)^{1/m} \leq 2\sum_{p=0}^{n} g_{p,n}\,\beta(P_{p,n})\,\mathbb{E}\left(\left|V_p^N\left(f_{p,n}^N\right)\right|^m\right)^{1/m}$$

for any $m \geq 1$ and the random $\mathcal{G}_{p-1}^N$-measurable functions

$$f_{p,n}^N := d_{p,n}'^N(f)/(2\beta(P_{p,n})) \in \mathrm{Osc}(E_p)$$

In addition, we have the estimates

$$\pi_\psi\left(\left|W_n^{\eta,N}(f)\right|\right) \leq 2\sum_{p=0}^{n} g_{p,n}\,\beta(P_{p,n})\pi_\psi\left(\left|V_p^N\left(f_{p,n}^N\right)\right|\right)$$

On the other hand, combining (11.8) and (11.19) we have

$$\pi_\psi\left(\left|V_p^N\left(f_{p,n}^N\right)\right|\right) \leq \sqrt{8/3}$$

We end the proof using Proposition 9.5.5. The estimate (14.19) is a direct consequence of Lemma 11.3.6. This ends the proof of the theorem. ∎

The next corollary is now a direct consequence of Lemma 11.3.2, applied to the Orlicz norm upper bounds (14.18) and (14.19).

Corollary 14.3.2 *For any* $n \geq 0$, $N \geq 1$, $f \in \mathrm{Osc}(E_n)$, *and any* $x \geq 0$, *the probability of any of the following events*

$$\left| [\eta_n^N - \eta_n](f) \right| \leq 4 \sqrt{\frac{2(x + \log 2)}{3N}} \; \tau_{1,1}(n)$$

and

$$\sup_{0 \leq p \leq n} \left| [\eta_p^N - \eta_p](f) \right| \leq 8 \sqrt{(x + \log 2)} \sqrt{\frac{\log(9 + n)}{N}} \sup_{0 \leq p \leq n} \tau_{1,1}(p)$$

is greater than $1 - e^{-x}$.

Using Corollary 12.2.7, we also readily check the following uniform concentration inequalities.

Corollary 14.3.3 *We assume the regularity condition* $\mathbf{H}(\mathbf{G}, \mathbf{P})$ *is satisfied. In this situation, for any* $n \geq 0$, $N \geq 1$, $f \in \mathrm{Osc}(E_n)$, *and any* $x \geq 0$, *the probability of any of the following events*

$$\left| [\eta_n^N - \eta_n](f) \right| \leq 4 \sqrt{\frac{2(x + \log 2)}{3N}} \; \overline{\tau}_{1,1}$$

and

$$\sup_{0 \leq p \leq n} \left| [\eta_p^N - \eta_p](f) \right| \leq 8 \sqrt{(x + \log 2)} \sqrt{\frac{\log(9 + n)}{N}} \; \overline{\tau}_{1,1}$$

is greater than $1 - e^{-x}$, *with the parameters* $\overline{\tau}_{1,1}$ *defined in (12.10).*

In particular, when the regularity condition $\mathbf{H}_\mathbf{m}(\mathbf{G}, \mathbf{M})$ *is met for some* $m \geq 0$, *the above concentration inequalities are satisfied by replacing* $\overline{\tau}_{1,1}$ *by the finite parameters* $\overline{\tau}_{1,1}(m)$ *defined in (12.15).*

We end this section with a lemma that quantifies the difference between the integral operators $d_{p,n}^N$ and their limiting values $d_{p,n}$. The estimates are expressed in terms of the parameters $\tau_{k,l}(n)$, $b(m)$, and $c(m)$ defined, respectively, in (12.4), (0.6), and in (9.33). Notice that $d_{0,n} = d_{0,n}^N$, for $p = 0$.

Lemma 14.3.4 *We consider the integral operators* $\widetilde{d}_{p,n}^N$ *defined in (14.16). For any* $N \geq 1$, $r \geq 1$, *and any* $1 \leq p \leq n$ *and* $f \in \mathrm{Osc}(E)$, *we have*

$$\mathbb{E}\left[\left\| \widetilde{d}_{p,n}^N(f) \right\|^r \right]^{1/r} \leq b(r)\,(1 \wedge c(r))\; e_{p,n} \tag{14.20}$$

with some constant

$$e_{p,n} = 6 g_{p-1} g_{p,n}^2 \; \beta(P_{p,n}) \beta(M_p) \, \tau_{1,1}(p - 1)$$

Proof:
We set
$$\overline{d}_{p,n}^N := d_{p,n}^N - d_{p,n}$$

By definition of the integral operators, we use (3.17) to check that

$$\left\| \overline{d}_{p,n}^N(f) \right\| \leq g_{p,n} \; \beta(P_{p,n}) \; \left(\left| (\Phi_p(\eta_{p-1}) - \Phi_p(\eta_{p-1}^N))(\overline{G}_{p,n}) \right| \right.$$

$$\left. + \left| (\Phi_p(\eta_{p-1}) - \Phi_p(\eta_{p-1}^N))(\overline{d}_{p,n}(f)) \right| \right)$$

with the couple of functions

$$\overline{d}_{p,n}(f) := d_{p,n}(f)/\beta(P_{p,n}) \quad \text{and} \quad \overline{G}_{p,n} = G_{p,n}/\eta_p(G_{p,n})$$

On the other hand, by (3.24) we have

$$\left| (\Phi_p(\eta_{p-1}^N) - \Phi_p(\eta_{p-1}))(f) \right| \leq g_{p-1} \; \left| (\eta_{p-1}^N - \eta_{p-1})(d'_{p-1,p}(f)) \right|$$

and osc $\left(d'_{p-1,p}(f) \right) \leq \beta(M_p)$. Using Theorem 14.3.1, for any $f \in \mathrm{Osc}(E)$, we also have the $\mathbb{L}_r$-mean error estimates

$$\sqrt{N} \; \mathbb{E}\left[\left| (\Phi_p(\eta_{p-1}^N) - \Phi_p(\eta_{p-1}))(f) \right|^r \right]^{1/r} \leq b(r)\,(1 \wedge c(r)) \; e(p)$$

with the constant
$$e(p) := 2 \; g_{p-1} \; \beta(M_p) \; \tau_{1,1}(p-1)$$

From these estimates, we prove that (14.20). This ends the proof of the lemma.
∎

14.3.2 A nonasymptotic variance theorem

The main objective of this section is to prove the following nonasymptotic bias and variance estimates.

Theorem 14.3.5 *We assume that the local variance parameters σ_n defined in (9.31) are uniformly bounded, and we set $\sigma := \sup_{n \geq 0} \sigma_n < \infty$. In this situation, for any $n \geq 0$, $N \geq 1$, and any function $f \in \mathrm{Osc}(E)$, we have the bias estimates*

$$\left| \mathbb{E}\left(\eta_n^N(f) \right) - \eta_n(f) \right| \leq 4 \; \sigma^2 \; \tau_{2,1}(n)/N$$

and the variance estimates

$$\mathbb{E}\left(\left[\eta_n^N(f) - \eta_n(f) \right]^2 \right) \leq 4 \; \sigma^2 \; \tau_{1,1}(n)^2/N$$

Before getting into the proof of the theorem, we present some direct consequences of these variance estimates.

Corollary 14.3.6 *Assume that the regularity conditions of Theorem 14.3.5 are satisfied. When the regularity condition* $\mathbf{H}(\mathbf{G}, \mathbf{P})$ *is satisfied we have the uniform bias estimates*

$$\sup_{n \geq 0} \left| \mathbb{E} \left(\eta_n^N(f) \right) - \eta_n(f) \right| \leq 4 \, \sigma^2 \, \overline{\tau}_{2,1}/N$$

and

$$\sup_{n \geq 0} \mathbb{E} \left(\left[\eta_n^N(f) - \eta_n(f) \right]^2 \right) \leq 4 \, \sigma^2 \, \overline{\tau}_{1,1}^2/N$$

with $\sigma := \sup_{n \geq 0} \sigma_n$, *and the parameters* $\overline{\tau}_{k,l}$ *defined in (12.10).*

When one of the conditions $\mathbf{H_m}(\mathbf{G}, \mathbf{M})$ *stated in Theorem 12.2.5 and Theorem 12.2.6 is met for some* $m \geq 0$, *the above estimates are satisfied by replacing* $\overline{\tau}_{k,l}$ *by the finite parameters* $\overline{\tau}_{k,l}(m)$ *defined in (12.15).*

Now we come to the proof of the theorem.
Proof of Theorem 14.3.5:
Using the decomposition (14.14), we have

$$\sqrt{N} \, \mathbb{E} \left(W_n^{\eta,N}(f) \right) = - \sum_{p=0}^n \mathbb{E} \left(\frac{1}{\eta_p^N(G_{p,n}^N)} \, V_p^N(G_{p,n}^N) \, V_p^N \left[d_{p,n}^N(f) \right] \right)$$

On the other hand, we have

$$\mathbb{E} \left(V_p^N \left[d_{p,n}^N(f) \right]^2 \right) \leq \sigma^2 \, (2g_{p,n} \, \beta(P_{p,n}))^2$$

and

$$\begin{aligned}
\mathbb{E} \left(\frac{1}{\eta_p^N(G_{p,n}^N)^2} \, V_p^N(G_{p,n}^N)^2 \right) &\leq \mathbb{E} \left(\frac{1}{(\inf G_{p,n}^N)^2} \, \mathbb{E} \left(V_p^N(G_{p,n}^N)^2 \, | \, \mathcal{G}_{p-1}^N \right) \right) \\
&\leq \sigma^2 (2g_{p,n})^2
\end{aligned}$$

A simple application of Cauchy-Schwartz inequality yields the desired bias estimate.

In the same vein, using the decomposition (14.14), we prove that

$$\mathbb{E} \left(W_n^{\eta,N}(f)^2 \right)^{1/2} \leq \sum_{p=0}^n \mathbb{E} \left(\frac{1}{(\inf G_{p,n}^N)^2} \, \mathbb{E} \left(V_p^N(d_{p,n}^N(f))^2 \, | \, \mathcal{G}_{p-1}^N \right) \right)^{1/2}$$

Arguing as above, the proof of the variance estimate is now clear. This ends the proof of the theorem. ■

14.3.3   Discrete and continuous time models

We illustrate the estimates stated in Theorem 14.3.5 by applying them to the continuous time discretization models discussed in Section 12.2.3.1.

By (12.17), we have the estimates

$$\overline{\tau}_{k,1}(0) \leq \frac{1}{\alpha h} \exp\left(\alpha h + k(v/\alpha) e^{(v+\alpha)h}\right)$$

On the other hand, by (14.21) we have

$$\sigma^2 \leq \sigma^2(h) := 2 h (\lambda + \|V\|) + h^2 c(V,\lambda)$$

Using Theorem 14.3.5, we readily prove the following uniform bias estimates

$$N \sup_{h \in [0,1]} \sup_{n \geq 0} \left| \mathbb{E}\left(\eta_n^N(f)\right) - \eta_n(f) \right| \leq C(\lambda, \alpha, V)$$

with some constant

$$C(\lambda, \alpha, V) \leq \frac{4}{\alpha} \left((\lambda + \|V\|) + c(V,\lambda)\right) \exp\left(\alpha + 2(v/\alpha) e^{(v+\alpha)}\right) \quad (14.21)$$

For this class of models, it is difficult to obtain more general uniform estimates w.r.t. the parameter h. The main reason comes from the fact that the estimates of the Dobrushin contraction coefficient $\beta(P_{p,n})$ tend to 1, when h tends to 0. Next, we present some strategy to obtain some uniform estimates.

For instance, using the decomposition (16.2), we have

$$W_n^{\eta,N}(f) = M_n^N(f) + \frac{1}{\sqrt{N}} R_n^N(f)$$

with the first order term

$$M_n^N(f) := \sum_{p=0}^{n} V_p^N \left[d_{p,n}^N(f)\right]$$

and the second order remainder term

$$R_n^N(f) = -\sum_{p=0}^{n} \frac{1}{\eta_p^N(G_{p,n}^N)} V_p^N(G_{p,n}^N) V_p^N \left[d_{p,n}^N(f)\right]$$

Using the fact that $ab \leq (a^2 + b^2)/2$, we prove that

$$\left| R_n^N(f) \right| \leq \sum_{p=0}^{n} a_{p,n} \left(V_p^N \left[f_{p,n}^{(1)}\right]^2 + V_p^N \left[f_{p,n}^{(2)}\right]^2\right)$$

with $a_{p,n} = 2g_{p,n}^2 \beta(P_{p,n})$

$$f_{p,n}^{(1)} := G_{p,n}^N / \mathrm{osc}(G_{p,n}^N) \quad f_{p,n}^{(2)} := d_{p,n}^N(f) / \mathrm{osc}(d_{p,n}^N(f))$$

Using (14.3), for any $hN \geq 1/(4(\lambda + \|V\|))$, and any $m \geq 1$ we have

$$\mathbb{E}\left(|R_n^N(f)|^m\right)^{1/m} \leq c(m) \; \sigma^2(h) \; \bar{\tau}_{2,1}(0)$$

for some finite constant $c(m)$ whose values only depend on the parameter m and the constant $\sigma^2(h)$ defined in (14.21).

In the same vein, by the Burkholder-Davis-Gundy inequalities, for any $m \geq 2$ we have

$$\mathbb{E}\left(|M_n^N(f)|^m\right)^{1/m} \leq m \; \mathbb{E}\left(\left(\sum_{p=0}^n V_p^N \left[d_{p,n}^N(f)\right]^2\right)^{m/2}\right)^{1/m}$$

Now, we use the decomposition

$$\sum_{p=0}^n V_p^N \left[d_{p,n}^N(f)\right]^2 = \sum_{p=0}^n a_{p,n} V_p^N \left[f_{p,n}\right]^2$$

with

$$a_{p,n} = (2g_{p,n}\beta(P_{p,n}))^2 \quad \text{and} \quad f_{p,n} = d_{p,n}^N(f)/\mathrm{osc}(d_{p,n}^N(f))$$

to prove that for even integers m, we have

$$\mathbb{E}\left(|M_n^N(f)|^m\right)^{1/m} \leq c(m) \; \sigma(h) \; \bar{\tau}_{2,2}(0)^{1/2}$$

for some finite constant $c(m)$ whose values only depend on the parameter m.

Notice that

$$\sigma^2(h) \; \bar{\tau}_{2,1}(0) \leq [2 \; (\lambda + \|V\|) + \; c(V,\lambda)] \frac{1}{\alpha} \; \exp\left(\alpha + 2(v/\alpha) \; e^{(v+\alpha)}\right)$$

and

$$\sigma^2(h) \; \bar{\tau}_{2,2}(0) \leq [2 \; (\lambda + \|V\|) + \; c(V,\lambda)] \frac{1}{2\alpha} \; \exp\left(2\alpha + 2(v/\alpha) \; e^{(v+\alpha)}\right)$$

From these estimates, we readily prove the following theorem.

Theorem 14.3.7 *For any $m \geq 2$, any function $f \in \mathrm{Osc}(E)$, and any (h, N) s.t. $hN \geq 1/(4(\lambda + \|V\|))$, we have*

$$\sqrt{N} \; \sup_{n \geq 0} \mathbb{E}\left(|\eta_n^N(f) - \eta_n(f)|^m\right)^{1/m} \leq c_1(m) \; c_2(\lambda, \alpha, V)$$

for some finite constant $c_1(m)$ whose values only depend on the parameter m, and some finite constant $c_2(\lambda, \alpha, V)$ whose values only depend on the parameters $(\lambda, \alpha, \|V\|)$.

14.4   Fluctuation analysis

14.4.1   Central limit theorems

The aim of this section is to analyze the fluctuations of the particle density profiles η_n^N around their limiting values. We start our analysis with two decomposition theorems that express $W_n^{\eta,N}$ in terms of a linear combination of the local fluctuation random fields V_p^N, with $p \leq n$, up to a second order remainder term.

The first result is a consequence of the decomposition Theorem 14.2.4 and the $\mathbb{L}_p$-mean error bounds (14.12).

Theorem 14.4.1 *For any $n \geq 0$, $N \geq 1$, and any $f_n \in \mathcal{B}_b(E_n)$, we have the first order decomposition*

$$W_n^{\eta,N}(f) = \sum_{p=0}^{n} V_p^N \left[d_{p,n}^N(f)\right] + \frac{1}{\sqrt{N}}\, R_n^N(f) \qquad (14.22)$$

The second order remainder term $R_n^N(f)$ in the above display is such that

$$\sup_{f \in \mathrm{Osc}(E_n)} \mathbb{E}\left[\left|R_n^N(f)\right|^m\right]^{1/m} \leq b(2m)^2\, r(n) \qquad (14.23)$$

with some finite constant $r(n)$ such that

$$r(n) \leq 2 \sum_{p=0}^{n} (g_{p,n} - 1)\, g_{p,n}\, \beta(P_{p,n}) = 2\, (\tau_{2,1}(n) - \tau_{1,1}(n)) \qquad (14.24)$$

with the parameter $\tau_{2,1}(n)$ defined in (12.4).

Proof:

Using (14.14), we have (14.22) with

$$R_n^N(f) = -\sum_{p=0}^{n} \frac{1}{\eta_p^N(G_{p,n}^N)}\, V_p^N(G_{p,n}^N)\, V_p^N\left[d_{p,n}^N(f)\right]$$

On the other hand, we have

$$\mathbb{E}\left(\left[V_p^N(G_{p,n})\right]^{2m}\right)^{\frac{1}{2m}} \leq b(2m)\, \mathrm{osc}(G_{p,n})$$

Furthermore, using the upper bounds (14.12) we have that

$$\mathbb{E}\left(\left[V_p^N(d_{p,n}^N(f))\right]^{2m}\right)^{\frac{1}{2m}} \leq 2\, b(2m)\, g_{p,n}\, \beta(P_{p,n})$$

These bounds, in conjunction with Cauchy-Schwartz inequality, yield

$$\sup_{f \in \mathrm{Osc}(E_n)} \mathbb{E}\left[|R_n^N(f)|^m\right]^{1/m} \leq 2\, b(2m)^2 \sum_{p=0}^{n} (g_{p,n}-1)\, g_{p,n}\, \beta(P_{p,n})$$

This ends the proof of the theorem. ∎

The next decomposition theorem is a consequence of Theorem 14.2.5 and the $\mathbb{L}_m$-bounds presented in (14.17).

Theorem 14.4.2 *For any $n \geq 0$, $N \geq 1$, and any $f_n \in \mathrm{Osc}(E_n)$, we have the first order decomposition*

$$W_n^{\eta,N}(f) = \sum_{p=0}^{n} V_p^N\left[d_{p,n}(f_n)\right] + \frac{1}{\sqrt{N}}\, R_n^N(f_n) \tag{14.25}$$

The second order remainder term $R_n^N(f_n)$ in the above display is such that

$$\sup_{f_n \in \mathrm{Osc}(E_n)} \mathbb{E}\left[|R_n^N(f_n)|^m\right]^{1/m} \leq b(2m)^2\, r'(n) \tag{14.26}$$

for any $m \geq 1$, with some finite constant $r'(n)$ such that

$$r'(n) \leq (2\tau_{1,1}(n))^3 \sum_{0 \leq p < n} (g_p - 1) g_{p,n}\, \beta(P_{p,n})$$

with the parameter $\tau_{1,1}(n)$ defined in (12.4).

Proof:

We prove (14.25) using the decomposition (14.15) with the second order remainder term

$$R_n^N(f) := -\sum_{p=0}^{n-1} \frac{1}{\eta_p^N(\overline{G}_p)}\, W_p^{\eta,N}(\overline{G}_p)\, W_p^{\eta,N}[d_{p,n}(f)]$$

On the other hand, using the $\mathbb{L}_m$-bounds presented in (14.17) we have

$$\mathbb{E}\left[|W_p^{\eta,N}(G_p)|^{2m}\right]^{\frac{1}{2m}} \leq 2\, \mathrm{osc}(G_p)\, b(2m)\, \tau_{1,1}(n)$$

and using the upper bound (14.12)

$$\mathbb{E}\left[|W_p^{\eta,N}[d_{p,n}(f)]|^{2m}\right]^{\frac{1}{2m}} \leq 2\, \mathrm{osc}(d_{p,n}(f))\, b(2m)\, \tau_{1,1}(n)$$
$$\leq 4 g_{p,n}\, \beta(P_{p,n})\, b(2m)\, \tau_{1,1}(n)$$

The last estimate is proved using (14.12). These bounds, in conjunction with the Cauchy-Schwartz inequality, yield the estimates

$$\mathbb{E}\left[\left|R_n^N(f)\right|^m\right]^{1/m} \leq 8\, b(2m)^2\, \tau_{1,1}(n)^3 \sum_{0 \leq p < n} (g_p - 1)g_{p,n}\, \beta(P_{p,n})$$

This ends the proof of the theorem. ∎

Combining (14.25) with Theorem 9.5.3, the Slutsky's lemma and the continuous mapping theorem yield the following multivariate central limit theorem.

For genetic type IPS particle models associated with selection type transition (0.4) that depends on the $\eta_n^N - ess\sup G_n$, we need to ensure that $\eta_n^N - ess\sup G_n$ converge to $\eta_n - ess\sup G_n$, in probability, as $N \to \infty$. This property has been checked in Section 9.5.2.

Theorem 14.4.3 *The sequence of random fields $W_n^{\eta,N}$ converges in law in the sense of convergence of finite dimensional distributions, as the number of particles N tends to infinity, to a sequence of Gaussian, and centered random fields W_n^{η} defined by*

$$W_n^{\eta}(f) \quad := \quad \sum_{p=0}^{n} V_p\left[d_{p,n}(f)\right] = V_n(f) + W_{n-1}^{\eta}(d_{n-1,n}(f))$$

14.4.2   Time homogeneous models

It is instructive to apply the fluctuation Theorem 14.4.3 to the quasi-invariant measures computation problems arising in the analysis of particle absorption models discussed in Chapter 7. We return to the time homogeneous Feynman-Kac models $(G_n, M_n) = (G, M)$ presented in Section 7.6 and in Section 7.6.2. We let K_η be a McKean transition associated with this time homogeneous Feynman-Kac model. In this situation, we have

$$d_{p,n}(f) = \overline{Q}^{(n-p)}(1)\left(\frac{Q^{(n-p)}(f)}{Q^{(n-p)}(1)} - \eta_n(f)\right)$$

with the semigroup

$$\overline{Q}^{(n-p)}(f) := \frac{Q^{(n-p)}(f)}{\eta_p(Q^{(n-p)}(1))}$$

We recall that in this context, for any $q \geq 1$ we have that

$$Q^q(f) = \lambda^q\, h\, (M^h)^q(f/h)$$

as well as

$$\eta_\infty(Q^q(1)) = \lambda^q \quad \text{and} \quad \eta_\infty = \Psi_{1/h}(\eta_\infty^h)$$

Thus, if we choose $\eta_0 = \eta_\infty$, then we have $\eta_n = \eta_\infty$, for any $n \geq 0$, and for any p and $q \geq 0$

$$d_{p,p+q}(f) = d_q(f) := h \ (M^h)^q(1/h) \left(\frac{(M^h)^q(f/h)}{(M^h)^q(1/h)} - \Psi_{1/h}(\eta_\infty^h)(f) \right)$$

When the regularity condition (7.15) is met for some finite $\rho < \infty$ (which holds when condition $\mathbf{H_m(G, M)}$ stated on page 373 is satisfied), the contraction estimate (3.10) in conjunction with the fact that

$$\frac{(M^h)^q(f/h)(x)}{(M^h)^q(1/h)(x)} - \Psi_{1/h}(\eta_\infty^h)(f) = \left[\Psi_{1/h}(\delta_x(M^h)^q) - \Psi_{1/h}(\eta_\infty^h) \right](f)$$

yields $\|d_q(f)\| \leq \rho^2 \ \beta((M^h)^q)$.

Notice that in this situation, the random fields V_n are independent copies of the centered Gaussian random field V with covariance function

$$\mathbb{E}\left(V(f)^2\right) = \eta_\infty \ K_{\eta_\infty} \left([f - K_{\eta_\infty}(f)]^2 \right)$$

This shows

$$\mathbb{E}\left(W_n^\eta(f)^2\right) \ := \ \sum_{p=0}^{n} \eta_\infty \ K_{\eta_\infty} \left([d_p(f) - K_{\eta_\infty}(d_p(f))]^2 \right)$$

$$\leq \ 4 \ \rho^4 \ \sigma^2 \ \sum_{p=0}^{n} \beta((M^h)^q)^2$$

with the local variance parameter σ defined in (9.31).

14.5 Concentration inequalities

In this section, we investigate the concentration properties of the particle density profiles η_n^N around their limiting values η_n. The results are expressed in terms of the contraction parameters $\tau_{k,l}(n)$ and $\overline{\tau}_{k,l}(m)$ introduced in Definition 12.1.5, and in Corollary 12.2.7, and the functions (ϵ_0, ϵ_1) on $\mathbb{R}_+$ defined by

$$\epsilon_0(\lambda) = \frac{1}{2} \left(\lambda - \log(1 + \lambda) \right) \quad \text{and} \quad \epsilon_1(\lambda) = (1 + \lambda) \log(1 + \lambda) - \lambda \quad (14.27)$$

To have some more explicit estimates, we also use the following rather crude upper bounds provided in Section 11.1

$$\epsilon_0^{-1}(x) \leq 2(x + \sqrt{x}) \quad \text{and} \quad \epsilon_1^{-1}(x) \leq \frac{x}{3} + \sqrt{2x} \quad (14.28)$$

The first Section, Section 14.5.1, is concerned with the concentration properties of finite marginal models. More precisely, given some $f \in \mathrm{Osc}(E_n)$, we are interested in the deviation of empirical averages $\eta_n^N(f)$ around their limiting values $\eta_n(f)$.

Our first main result is a concentration inequality in terms of the parameters $g_{p,n}$ and $\beta(P_{p,n})$ introduced in Definition 12.1.4 and the uniform local variance parameter σ_n^2 defined in (9.31).

When the Feynman-Kac models satisfy the regularity condition $\mathbf{H}(\mathbf{G}, \mathbf{P})$ presented in Section 12.2.1, we also provide some uniform concentration estimates w.r.t. the time parameter.

Section 14.5.2 is concerned with the concentration properties of the empirical processes

$$W_n^{\eta,N} \ : \ f \in \mathcal{F}_n \longrightarrow W_n^{\eta,N}(f)$$

associated with a separable collection of functions $\mathcal{F}_n \subset \mathrm{Osc}(E_n)$.

Our main results are Olicz norm estimates of the supremum of the empirical averages $W_n^{\eta,N}(f)$ w.r.t. classes of functions $\mathcal{F}_n$ with finite entropy. Uniform concentration estimates w.r.t. the time parameter are also provided, as soon as the Feynman-Kac models satisfy the regularity condition $\mathbf{H}(\mathbf{G}, \mathbf{P})$ presented in Section 12.2.1.

14.5.1   Finite marginal models

In Section 4.4.4, we have seen the local fluctuation random fields V_n^N share the same statistical properties as the empirical process models analyzed in Section 11.8. Furthermore, the couple of first order decompositions stated in Theorem 14.4.1 and in Theorem 14.4.2 have the same form as the class of interacting processes introduced in (11.38).

Thus, by a rather direct application of Theorem 11.8.6, we prove the following exponential concentration property.

Theorem 14.5.1 *For any $n \geq 0$, any $f \in \mathrm{Osc}(E_n)$, and any $N \geq 1$, the probability of the event*

$$\left[\eta_n^N - \eta_n\right](f) \leq \frac{1}{N} \ \overline{r}(n) \ \left(1 + \epsilon_0^{-1}(x)\right) + 2b_n \ \overline{\sigma}_n^2 \ \epsilon_1^{-1}\left(\frac{x}{N\overline{\sigma}_n^2}\right)$$

is greater than $1 - e^{-x}$, for any $x \geq 0$, with the parameters

$$\overline{r}(n) := r(n) \wedge r'(n) \quad and \quad \overline{\sigma}_n^2 := \frac{1}{b_n^2} \sum_{0 \leq p \leq n} g_{p,n}^2 \ \beta(P_{p,n})^2 \ \sigma_p^2 \qquad (14.29)$$

for any choice of $b_n \geq \kappa(n)$.

In the above display, σ_n, $\kappa(n)$, $r(n)$, $r'(n)$ are the constants defined, respectively, in (9.31), (12.4), (14.24), and (14.23).

Proof:

Firstly, we note that the first term in the l.h.s. of the decomposition formula (14.25) can be rewritten in the following form.

$$\sum_{p=0}^{n} V_p^N \left[d_{p,n}(f)\right] = \sum_{p=0}^{n} a_p \, V_p^N \left[\delta_{p,n}(f)\right] \tag{14.30}$$

for any finite constants

$$a_p \geq 2 \sup_{0 \leq p \leq n} \left(g_{p,n}\beta(P_{p,n})\right) := 2\kappa(n) \tag{14.31}$$

with the integral operators $\delta_{p,n}$ from $\text{Osc}(E_n)$ into $\text{Osc}(E_p)$ defined for any $f \in \text{Osc}(E_n)$ by the normalized equation

$$\delta_{p,n}(f) = d_{p,n}(f)/a_p \in \text{Osc}(E_p)$$

We fix the final time horizon n, and with some slightly abusive notation we set

$$X_p = \xi_p = \left(\xi_p^i\right)_{1 \leq i \leq N} \quad \alpha_p = \sup_{0 \leq q \leq n} a_q = a_n^\star \quad \text{and} \quad f_p = d_{p,n}(f)/a_n^\star$$

In this notation, the linear combination equation (14.30) has the same form as (11.36); that is, we have that

$$\sum_{p=0}^{n} V_p^N \left[d_{p,n}(f)\right] = \sum_{p=0}^{n} \alpha_p \, V(X_p)\,[f_p] := V_n(X)(f)$$

Using the first order decomposition stated in Theorem 14.4.2, we conclude that

$$W_n^{\eta,N}(f) = V_n(X)(f) + \frac{1}{\sqrt{N}} \, R_n(X)(f)$$

with a second order remainder term $R_n(X)(f)$ such that

$$\mathbb{E}\left(|R_n(X)(f)|^m\right)^{1/m} \leq b(2m)^2 \, r'(n)$$

On the other hand, we also have the following almost sure local variance estimates

$$\mathbb{E}\left(V_p^N \left[f_p\right]^2 \big| \mathcal{G}_{p-1}^N\right)^{1/2} \leq \sigma_p \, \text{osc}(d_{p,n}(f))/a_n^\star \leq \sigma_p \, \|d_{p,n}(f)\|/b_n$$

with $b_n = a_n^\star/2 \geq \kappa(n)$, and the uniform local variance parameters σ_p defined in (9.31).

This implies that

$$\mathbb{E}\left(V_p^N \left[f_p\right]^2 \big| \mathcal{G}_{p-1}^N\right)^{1/2} \leq \sigma_p \, g_{p,n} \, \beta(P_{p,n})/b_n \tag{14.32}$$

with the uniform local variance parameters σ_p^2 defined in (9.31). It is also important to notice that the estimates (14.31) and (14.32) are also satisfied if we replace the functions $d_{p,n}(f)$ by the random functions $d_{p,n}^N(f)$.

The above discussion shows that the regularity condition stated in (11.37) is met by replacing the variance parameters in formula (11.37) by the constants $2\sigma_p g_{p,n}\beta(P_{p,n})/a_n^\star$. The end of the proof is now a direct consequence of Theorem 11.8.6. This ends the proof of the theorem.

∎

Using the estimates (14.28), we readily check the following corollary.

Corollary 14.5.2 *For any $n \geq 0$, any $f \in \mathrm{Osc}(E_n)$, and any $N \geq 1$, the probability of the event*

$$\left[\eta_n^N - \eta_n\right](f) \leq \frac{1}{N}\, p_n(x) + \frac{1}{\sqrt{N}}\, q_n(x)$$

is greater than $1 - e^{-x}$, for any $x \geq 0$, with the functions

$$p_n(x) \;=\; \bar{r}(n)\,\left(1 + 2(x + \sqrt{x})\right) + 2x\,\kappa(n)/3$$

$$q_n(x) \;=\; 2\sqrt{2x}\,\sqrt{\sum_{0 \leq p \leq n} g_{p,n}^2\,\beta(P_{p,n})^2\,\sigma_p^2}$$

In the above display, σ_n, $\kappa(n)$, and $\bar{r}(n)$ stand for the parameters defined in (9.31), (12.4), and (14.29).

We illustrate these exponential concentration inequalities with the Feynman-Kac models satisfying the regularity condition $\mathbf{H}(\mathbf{G},\mathbf{P})$ presented in Section 12.2.1. In this situation, by Proposition 12.2.4 we have

$$g := \sup_{n \geq 0} g_n < \infty \quad \text{and} \quad \sigma := \sup_{n \geq 1} \sigma_n < \infty$$

as well as

$$\bar{\tau}_{k,l} := \sup_{n \geq 0} \tau_{k,l}(n) < \infty \quad \text{and} \quad \bar{\kappa} := \sup_{n \geq 0} \kappa(n) < \infty$$

with the parameters σ_n^2, g_n, $\tau_{k,l}(n)$, and $\kappa(n)$ introduced, respectively, in (9.31), in Definition 12.1.4, and in (12.4).

Corollary 14.5.3 *We assume that condition $\mathbf{H}(\mathbf{G},\mathbf{P})$ is satisfied. In this situation, for any $n \geq 0$, any $f \in \mathrm{Osc}(E_n)$, and any $N \geq 1$, the probability of the event*

$$\left[\eta_n^N - \eta_n\right](f) \leq \frac{1}{N}\, p(x) + \frac{1}{\sqrt{N}}\, q(x)$$

is greater than $1 - e^{-x}$, for any $x \geq 0$, with the functions

$$
\begin{aligned}
p(x) &= \bar{r}\left(1 + 2(x + \sqrt{x})\right) + 2\bar{\kappa}x/3 \\
q(x) &= 2\sqrt{2\sigma^2 \bar{\tau}_{2,2}}\,\sqrt{x}
\end{aligned}
$$

and some parameter $\bar{r}$ such that

$$
\bar{r} \leq \left(2\;(\bar{\tau}_{2,1} - \bar{\tau}_{1,1})\right) \wedge \left((2\bar{\tau}_{1,1})^4(g-1)\right)
$$

If one of the regularity conditions $\mathbf{H_m(G, M)}$ stated in Section 12.2.1 is met for some $m \geq 0$, the above concentration inequalities are satisfied by replacing the parameters $\bar{\tau}_{k,l}$ and $\bar{\kappa}$ by the parameters $\bar{\tau}_{k,l}(m)$ and $\bar{\kappa}(m)$ defined in Corollary 12.2.7.

14.5.2 Empirical processes

The main aim of this section is to derive concentration inequalities for the particle density empirical processes

$$
W_n^{\eta,N} : f \in \mathrm{Osc}(E_n) \longrightarrow W_n^{\eta,N}(f) \in \mathbb{R}
$$

Given a separable collection of functions $\mathcal{F}_n \subset \mathrm{Osc}(E_n)$, we consider the Zolotarev seminorm on $\mathcal{M}(E_n)$ defined for any $\mu \in \mathcal{M}(E_n)$ by

$$
\|\mu\|_{\mathcal{F}_n} = \sup\left\{|\mu(f)| \; ; \; f \in \mathcal{F}_n\right\}
$$

Theorem 14.5.4 *We let $\mathcal{F}_n$ be a separable collection of measurable functions f_n on E_n, such that $\|f_n\| \leq 1$, $\mathrm{osc}(f_n) \leq 1$, with finite entropy $I(\mathcal{F}_n) < \infty$. We have the Orlicz norm estimates*

$$
\pi_\psi\left(\left\|W_n^{\eta,N}\right\|_{\mathcal{F}_n}\right) \leq c_{\mathcal{F}_n}\,\tau_{1,1}(n) \tag{14.33}
$$

and

$$
\pi_\psi\left(\sup_{0 \leq p \leq n}\left\|W_p^{\eta,N}\right\|_{\mathcal{F}_p}\right) \leq \sqrt{6\log(9+n)}\,\sup_{0 \leq p \leq n}\left(c_{\mathcal{F}_p}\,\tau_{1,1}(p)\right) \tag{14.34}
$$

with the parameter $\tau_{1,1}(n)$ defined in (12.4) and the constants

$$
c_{\mathcal{F}_n} \leq 24^2 \int_0^1 \sqrt{\log\left(8 + \mathcal{N}(\mathcal{F}_n, \epsilon)^2\right)}\,d\epsilon \tag{14.35}
$$

Proof:
Firstly, by Lemma 11.3.6 we observe that (14.33) implies (14.34). Now we come to the proof of (14.33). Using (14.14), for any function $f_n \in \mathrm{Osc}(E_n)$ we have the estimate

$$
\left|W_n^{\eta,N}(f_n)\right| \leq 2\sum_{p=0}^{n} g_{p,n}\beta(P_{p,n})\left|V_p^N(\delta_{p,n}^N(f_n))\right|
$$

with the $\mathcal{G}_{p-1}^N$-measurable random functions $\delta_{p,n}^N(f_n)$ on E_p defined by

$$\delta_{p,n}^N(f_n) = \frac{1}{2\beta(P_{p,n})} \, d_{p,n}'^N(f_n)$$

By construction, we have

$$\left\| \delta_{p,n}^N(f_n) \right\| \leq 1/2 \quad \text{and} \quad \delta_{p,n}^N(f_n) \in \mathrm{Osc}(E_p)$$

Using the uniform estimate (11.40), if we set

$$\mathcal{G}_{p,n}^N := \delta_{p,n}^N(\mathcal{F}_n) = \left\{ \delta_{p,n}^N(f) \; : \; f \in \mathcal{F}_n \right\}$$

then we also prove the almost sure upper bound

$$\sup_{N \geq 1} \mathcal{N}\left[\mathcal{G}_{p,n}^N, \epsilon \right] \leq \mathcal{N}(\mathcal{F}_n, \epsilon/2)$$

The end of the proof is now a direct consequence of Theorem 11.8.8. This ends the proof of the theorem. ∎

Next, we examine some consequences of the above theorem. The first result is a direct consequence of Lemma 11.3.2.

Corollary 14.5.5 *We assume that the conditions stated in Theorem 14.5.4 are satisfied. In this situation, for any $n \geq 0$, and any $N \geq 1$, the probability of any of the following events*

$$\left\| \eta_n^N - \eta_n \right\|_{\mathcal{F}_n} \leq \frac{c_{\mathcal{F}_n}}{\sqrt{N}} \, \tau_{1,1}(n) \, \sqrt{x + \log 2}$$

and

$$\sup_{0 \leq p \leq n} \left\| \eta_p^N - \eta_p \right\|_{\mathcal{F}_p} \leq c_{\mathcal{F}}(n) \, \sqrt{\frac{6 \log(9 + n)(x + \log 2)}{N}}$$

is greater than $1 - e^{-x}$, for any $x \geq 0$, with some constant

$$c_{\mathcal{F}}(n) \leq \max_{0 \leq p \leq n} \left(c_{\mathcal{F}_p} \, \tau_{1,1}(p) \right)$$

The above results can be turned into uniform concentration inequalities, under the regularity properties (12.7).

Corollary 14.5.6 *We consider time homogeneous Feynman-Kac models on some common measurable state space $E_n = E$. We also let $\mathcal{F}$ be a separable collection of measurable functions f on E, such that $\|f\| \leq 1$, $\mathrm{osc}(f) \leq 1$, with finite entropy $I(\mathcal{F}) < \infty$. We also assume that one of the regularity conditions $H(G, P)$ stated in Section 12.2.1 is met. In this situation, for any $n \geq 0$, and any $N \geq 1$, the probability of any of the following events*

$$\left\| \eta_n^N - \eta_n \right\|_{\mathcal{F}} \leq \frac{c_{\mathcal{F}}}{\sqrt{N}} \, \bar{\tau}_{1,1} \, \sqrt{x + \log 2}$$

and

$$\sup_{0 \le p \le n} \left\| \eta_p^N - \eta_p \right\|_{\mathcal{F}} \le c_{\mathcal{F}} \, \overline{\tau}_{1,1} \sqrt{\frac{6 \log (9 + n)}{N}} \sqrt{x + \log 2}$$

is greater than $1 - e^{-x}$, *for any* $x \ge 0$, *with the finite parameters* $\overline{\tau}_{1,1}$ *introduced in (12.4) and a constant* $c_{\mathcal{F}}$ *such that*

$$c_{\mathcal{F}} \le 24^2 \int_0^1 \sqrt{\log \left(8 + \mathcal{N}(\mathcal{F}, \epsilon)^2 \right)} \, d\epsilon$$

The last result is based on the covering number analysis presented in (11.16). This result can be applied to obtain uniform estimates w.r.t. the time horizon for the empirical repartition function associated with the particle density profiles.

Corollary 14.5.7 *We assume that the conditions stated in Corollary 14.5.6 are satisfied. When* $\mathcal{F}$ *stands for the indicator functions (11.15) of cells in* $E = \mathbb{R}^d$, *for some* $d \ge 1$, *the probability of any of the following events*

$$\left\| \eta_n^N - \eta_n \right\|_{\mathcal{F}} \le c \, \overline{\tau}_{1,1} \sqrt{\frac{d}{N} \, (x + 1)}$$

and

$$\sup_{0 \le p \le n} \left\| \eta_p^N - \eta_p \right\|_{\mathcal{F}} \le c \, \overline{\tau}_{1,1} \sqrt{\frac{d \log (n + e)}{N} \, (x + 1)}$$

is greater than $1 - e^{-x}$, *for any* $x \ge 0$, *for some universal constant* $c < \infty$ *that does not depend on the dimension.*

14.6   A general class of mean field particle models

14.6.1   First order decomposition theorems

This section is concerned with the concentration analysis of the general class of IPS models introduced in Chapter 10. Our first result applies to mean field models associated with a collection McKean transitions $K_{n,\eta}$ satisfying the weak Lipschitz type condition stated in (10.10).

Proposition 14.6.1 *Under the assumptions stated above, for any* $m \ge 1$, *any* $n \ge 1$, *and for any function* $f \in \mathrm{Osc}(E_n)$ *we have*

$$\mathbb{E} \left(\left| W_n^{\eta, N}(f) \right|^m \right)^{1/m} \le b(m) \, (1 \wedge c(m)) \, \sum_{p=0}^n \delta \left(T^{\Phi_{p,n}} \right)$$

with the constants $(b(m), c(m))$ *defined in (0.6) and (9.33). In addition, for* *any $x \geq 0$ the probability of the following event is greater than $1 - e^{-x}$*

$$\left|W_n^{\eta,N}(f)\right| \leq \sqrt{8/3} \ \sqrt{x + \log 2} \ \sum_{p=0}^{n} \delta\left(T^{\Phi_{p,n}}\right) \qquad (14.36)$$

Proof:
According to the remark stated on page 303 at the end of Section 10.2, the semigroup $\Phi_{p,n}$ satisfy the condition $(\Phi_{p,n})$ given in (14.37) for some collection of bounded integral operators $T_\eta^{\Phi_{p,n}}$. In this situation, we have

$$\left|[\Phi_{p,n}(\mu) - \Phi_{p,n}(\eta)](f)\right| \leq \int \left|(\mu - \eta)(h)\right| \ T_\eta^{\Phi_{p,n}}(f, dh)$$

for some integral operators $T_\eta^{\Phi_{p,n}}$ from $\mathcal{B}(E_n)$ into the set $\mathrm{Osc}(E_p)$ such that

$$\int \mathrm{osc}(h) \ T_\eta^{\Phi_{p,n}}(f, dh) \leq T_\eta^{\Phi_{p,n}}(1)(f) \leq \mathrm{osc}(f) \ \delta\left(T^{\Phi_{p,n}}\right) \qquad (14.37)$$

for some $\delta\left(T^{\Phi_{p,n}}\right) < \infty$. Using the decomposition (2.5), we find that

$$\left|W_n^{\eta,N}(f)\right| \leq \sum_{p=0}^{n} \int \left|V_p^N(h)\right| \ T_{\Phi_p(\eta_{p-1}^N)}^{\Phi_{p,n}}(f, dh)$$

from which we prove that

$$\mathbb{E}\left(\left|W_n^{\eta,N}(f)\right|^m\right)^{1/m}$$

$$\leq \sum_{p=0}^{n} \mathbb{E}\left(\left[\int \mathbb{E}\left(\left|V_p^N(h)\right|^m \mid \mathcal{F}_{p-1}^{(N)}\right)^{1/m} \ T_{\Phi_p(\eta_{p-1}^N)}^{\Phi_{p,n}}(f, dh)\right]^m\right)^{1/m}$$

The proof of the $\mathbb{L}_m$-mean error bound is now a direct consequence of proposition 9.5.5. The concentration inequality is a consequence of of the comparison lemma 11.3.2, and the estimate (11.22). This ends the proof of the proposition. ∎

To get some more refined estimate, we assume that the nonlinear mappings Φ_n and the McKean transitions $K_{n,\eta}$ satisfy the first order regularity properties presented in Section 10.2.

To simplify the presentation, we also set

$$d_{p,n}^{(N)} := d_{\Phi_p(\eta_{p-1}^N)}\Phi_{p,n} \quad \text{and} \quad \mathcal{R}_{p,n} = \mathcal{R}^{\Phi_{p,n}}$$

with the collection of first order operators $d_{\Phi_p(\eta_{p-1}^N)}\Phi_{p,n}$ and the second order integral operators $\mathcal{R}^{\Phi_{p,n}}$ defined in (10.3) and in (10.5).

Theorem 14.6.2 *For any $n \geq 0$, $N \geq 1$, and any $f_n \in \mathcal{B}_b(E_n)$, we have the first order decomposition*

$$W_n^{\eta,N}(f) = \sum_{p=0}^{n} V_p^N \left[d_{p,n}^N(f)\right] + \frac{1}{\sqrt{N}} R_n^N(f) \qquad (14.38)$$

The second order remainder term $R_n^N(f)$ in the above display is such that

$$\sup_{f \in \text{Osc}(E_n)} \mathbb{E}\left[\left|R_n^N(f)\right|^m\right]^{1/m} \leq b(2m)^2 \; r(n) \qquad (14.39)$$

with some constant

$$r(n) \leq \sum_{p=0}^{n} \delta(\mathcal{R}_{p,n})$$

In addition, we have the first order decomposition

$$W_n^{\eta,N}(f) = \sum_{p=0}^{n} V_p^N \left[d_{p,n}(f_n)\right] + \frac{1}{\sqrt{N}} R_n^N(f_n) \qquad (14.40)$$

The second order remainder term $R_n^N(f_n)$ in the above display is such that

$$\sup_{f_n \in \text{Osc}(E_n)} \mathbb{E}\left[\left|R_n^N(f_n)\right|^m\right]^{1/m} \leq b(2m)^2 \; r(n) \qquad (14.41)$$

with some constant

$$r'(n) \leq \sum_{p=0}^{n-1} \beta(d_{p+1,n}) \left(\sum_{q=0}^{p} \delta(T^{\Phi_{q,p}})\right)^2 \delta\left(R^{\Phi_{p+1}}\right)$$

Proof:
Under our regularity conditions, we have the almost sure estimates

$$\sup_{N \geq 1} \beta\left(d_{p,n}^{(N)}\right) \leq \beta\left(d\Phi_{p,n}\right) := \sup_{\eta \in \mathcal{P}(E_p)} \beta\left(d_\eta \Phi_{p,n}\right)$$

Using the first order expansion of the semigroup $\Phi_{p,n}$, we prove the following decomposition

$$
\begin{aligned}
W_n^{\eta,N} \; &:= \; \sqrt{N} \left[\eta_n^N - \eta_n\right] \\
&= \; \sqrt{N} \sum_{p=0}^{n} \left[\Phi_{p,n}(\eta_p^N) - \Phi_{p,n}\left(\Phi_p(\eta_{p-1}^N)\right)\right] = I_n^N + J_n^N
\end{aligned}
$$

with the pair of random measures (I_n^N, J_n^N) given by

$$I_n^N \; := \; \sum_{p=0}^{n} V_p^N d_{p,n}^{(N)} \quad \text{and} \quad J_n^N := \sqrt{N} \sum_{p=0}^{n} \mathcal{R}_{p,n}\left(\eta_p^N, \Phi_p(\eta_{p-1}^N)\right)$$

Combining the $\mathbb{L}_m$-mean error estimates stated in Proposition 9.5.5 with the generalized Minkowski integral inequality (cf. for instance [192]) we find that

$$N \, \mathbb{E} \left(\left| \mathcal{R}_{p,n} \left(\eta_p^N, \Phi_p(\eta_{p-1}^N) \right) (f_n) \right|^m \, \middle| \, \mathcal{F}_{p-1}^{(N)} \right)^{\frac{1}{m}} \leq b(2m)^2 \, \delta(\mathcal{R}_{p,n})$$

for any $f_n \in \mathrm{Osc}(E_n)$. This implies that

$$
\begin{aligned}
\mathbb{E} \left(\left| \sqrt{N} J_n^N(f_n) \right|^m \right)^{\frac{1}{m}} &= N \, \mathbb{E} \left(\left| \sum_{p=0}^{n} \mathcal{R}_{p,n} \left(\eta_p^N, \Phi_p(\eta_{p-1}^N) \right) (f_n) \right|^m \right)^{\frac{1}{m}} \\
&\leq b(2m)^2 \sum_{p=0}^{n} \delta(\mathcal{R}_{p,n})
\end{aligned}
$$

This ends the proof of the first assertion. The second assertion is based on the first order decompositions developed in the proof of Theorem 10.6.2. This ends the proof of the theorem.

∎

14.6.2   Concentration inequalitites

Theorem 10.6.2 shows that the fluctuations of η_n^N around the limiting measure η_n are precisely dictated by first order differential type operators $d_{\eta_p} \Phi_{p,n}$ of the semigroup $\Phi_{p,n}$ around the flow of measures η_p, with $p \leq n$. Furthermore, for any $f_n \in \mathrm{Osc}(E_n)$, one observes that

$$\mathbb{E}(W_n^\eta(f_n)^2) = \sum_{p=0}^{n} \mathbb{E} \left(\left(V_p \left[d_{\eta_p} \Phi_{p,n}(f_n) \right] \right)^2 \right) \tag{14.42}$$

$$\leq \sum_{p=0}^{n} \sigma_p^2 \, \beta(d\Phi_{p,n})^2 := \overline{\sigma}_n^2 \tag{14.43}$$

with the uniform local variance parameters:

$$\sigma_n^2 := \sup_{f_n \in \mathrm{Osc}(E_n)} \sup_{\mu \in \mathcal{P}(E_{n-1})} \left| \mu \left(K_{n,\mu} \left[f_n - K_{n,\mu}(f_n) \right]^2 \right) \right|$$

and the contraction parameter

$$\beta(d\Phi_{p,n}) := \sup_{\eta \in \mathcal{P}(E_p)} \beta(d_\eta \Phi_{p,n})$$

The exponential deviation events developed in this section are described in terms of the parameters

$$\overline{\sigma}_n^2 \leq \beta_n^2 := \sum_{p=0}^{n} \beta(d\Phi_{p,n})^2 \quad \text{and} \quad b_n^\star := \sup_{0 \leq p \leq n} \beta(d\Phi_{p,n})$$

Arguing as in Section 14.5.1, by a direct application of Theorem 11.8.6 we prove the following exponential concentration theorem.

Theorem 14.6.3 *For any $N \geq 1$, $n \geq 0$, $f_n \in \mathrm{Osc}(E_n)$, and any $x \geq 0$ the probability of each of the following pair of events is greater than $1 - e^{-x}$*

$$W_n^{\eta,N}(f_n) \leq \frac{r_n}{\sqrt{N}} \left(1 + \epsilon_0^{-1}(x)\right) + \sqrt{N} \, \bar{\sigma}_n^2 \, b_n^\star \, \epsilon_1^{-1} \left(\frac{x}{N\bar{\sigma}_n^2}\right)$$

and

$$W_n^{\eta,N}(f_n) \leq \frac{r_n}{\sqrt{N}} \left(1 + \epsilon_0^{-1}(x)\right) + \sqrt{2x} \, \beta_n$$

In the above display $r(n)$ stands for the parameter defined in (14.39), and (ϵ_0, ϵ_1) are the functions defined in (14.5).

The first and second order terms can be estimated more explicitly using the upper bounds

$$\epsilon_0^{-1}(x) \leq 2x + \log(1 + 2x + 2\sqrt{x}) + \frac{\log(1 + 2x + 2\sqrt{x}) - 2\sqrt{x}}{2x + 2\sqrt{x}} \leq 2(x + \sqrt{x})$$

and

$$\epsilon_1^{-1}(x) \leq \frac{\sqrt{2x} + (4x/3) - \log(1 + (x/3) + \sqrt{2x})}{\log(1 + (x/3) + \sqrt{2x})} \leq (x/3) + \sqrt{2x}$$

A detailed proof of these estimates is provided in the article [166]. A direct application of these estimates yields the following corollary.

Corollary 14.6.4 *For any $N \geq 1$, $n \geq 0$, $f_n \in \mathrm{Osc}(E_n)$, and any $x \geq 0$ the probability of each of the following events is greater than $1 - e^{-x}$*

$$[\eta_n^N - \eta_n](f_n) \leq \frac{1}{N} \left(r_n \left(1 + 2(x + \sqrt{x}) + b_n^\star \frac{x}{3}\right)\right) + b_n^\star \, \bar{\sigma}_n \sqrt{\frac{2x}{N}}$$

The second rough estimate in the r.h.s. of the above displayed formulae also leads to Bernstein type concentration inequalities.

Corollary 14.6.5 *For any $N \geq 1$ and any $n \geq 0$, we have the following Bernstein type concentration inequalities*

$$\frac{1}{N} \, \log \mathbb{P} \left([\eta_n^N - \eta_n](f_n) \geq \frac{r_n}{N} + \lambda\right)$$

$$\geq \frac{\lambda^2}{2} \left(\left(b_n^\star \, \bar{\sigma}_n + \frac{\sqrt{2}r_n}{\sqrt{N}}\right)^2 + \lambda \left(2r_n + \frac{b_n^\star}{3}\right)\right)^{-1}$$

and

$$-\frac{1}{N} \, \log \mathbb{P} \left([\eta_n^N - \eta_n](f_n) \geq \frac{r_n}{N} + \lambda\right) \geq \frac{\lambda^2}{2} \left(\left(\beta_n + \frac{\sqrt{2}r_n}{\sqrt{N}}\right)^2 + 2r_n\lambda\right)^{-1}$$

In terms of the random fields $W_n^{\eta,N}$, the first concentration inequality stated in Corollary 14.6.5 takes the following form

$$- \log \mathbb{P} \left(W_n^{\eta,N}(f_n) \geq \frac{r_n}{\sqrt{N}} + \lambda \right)$$

$$\geq \frac{\lambda^2}{2} \left(\left(b_n^\star \bar{\sigma}_n + \frac{\sqrt{2} r_n}{\sqrt{N}} \right)^2 + \frac{\lambda}{\sqrt{N}} \left(2 r_n + \frac{b_n^\star}{3} \right) \right)^{-1} \xrightarrow{N \to \infty} \frac{\lambda^2}{2 \left(b_n^\star \bar{\sigma}_n \right)^2}$$

14.7 Particle branching intensity measures

14.7.1 Nonlinear evolution equations

We consider a sequence of measurable state space $(E_n)_{n \geq 0}$, a collection of of Markov transitions M_n from E_{n-1} into E_n, a sequence of positive potential functions G_n on E_n, and a sequence of nonnegative measures μ_n on E_n. We let Q_n be the integral operator from $\mathcal{B}_b(E_n)$ into $\mathcal{B}_b(E_{n-1})$ defined by

$$Q_n(x_{n-1}, dx_n) = G_{n-1}(x_{n-1}) \, M_n(x_{n-1}, dx_n)$$

and we denote by $Q_{p,n}$, $p \leq n$, the corresponding semigroup of operators

$$Q_{p,n} = Q_{p+1} \cdots Q_{n-1} Q_n$$

We let $(G_{p,n}, P_{p,n})$ be the potential functions and the Markov transitions defined by

$$G_{p,n} := Q_{p,n}(1) \quad \text{and} \quad P_{p,n}(x_p, dx_n) = Q_{p,n}(x_p, dx_n)/Q_{p,n}(x_p, E_n)$$

We also consider the parameters:

$$g_{p,n} = \sup_{x,y} \frac{G_{p,n}(x)}{G_{p,n}(y)} \quad \text{and} \quad \beta(P_{p,n}) = \sup_{x,y \in E_p} \| P_{p,n}(x, .) - P_{p,n}(y, .) \|_{\mathrm{tv}}$$

We also consider the pair of parameters $(g_-(n), g_+(n))$ defined below

$$g_-(n) = \inf_{0 \leq p < n} \inf_{E_p} G_p \leq \sup_{0 \leq p < n} \sup_{E_p} G_p = g_+(n)$$

We also write $g_{-/+}(n)$ to refer to both parameters.

For any $n \geq 0$, we denote by $\bar{\mu}_n$ the normalized probability measures on E_n given for any function $f_n \in \mathcal{B}_b(E_n)$ by

$$\bar{\mu}_n(f_n) = \mu_n(f_n)/\mu_n(1)$$

We also let $S_{n,\eta}$ be some Markov transition satisfying the compatibility condition

$$\forall \eta \in \mathcal{P}(E_n) \qquad \Psi_{G_n}(\eta) = \eta S_{n,\eta}$$

Several examples of transitions S_{n,η_n} are discussed in Section 4.4.5.

We consider the extended Feynman-Kac path measures (γ_n, η_n) associated with the triplet (M_n, G_n, μ_n) defined in Section 6.1.2. We recall that these measures are defined for any $n \geq 0$ and any function $f_n \in \mathcal{B}_b(E_n)$ by the following evolution equations

$$\gamma_n = \gamma_{n-1}Q_n + \mu_n \quad \text{and} \quad \eta_n(f_n) = \gamma_n(f_n)/\gamma_n(1) \tag{14.44}$$

with the initial condition $\gamma_0 = \mu_0$. In Section 6.1.4 we have shown that these flows of measures satisfy the following measure valued equations

$$\gamma_{n+1}(1) = \gamma_n(1)\, \eta_n(G_n) + \mu_{n+1}(1) \tag{14.45}$$

$$\eta_{n+1} = \eta_n K_{n+1,(\gamma_n(1),\eta_n)} \tag{14.46}$$

with the Markov transitions

$$K_{n+1,(\gamma_n(1),\eta_n)} = S_{n,\eta_n} M_{n+1,(\gamma_n(1),\eta_n)}$$

$$M_{n+1,((\gamma_n(1),\eta_n))}(x,dy) := \alpha_n\left((\gamma_n(1),\eta_n)\right) M_{n+1}(x,dy)$$

$$+ \left(1 - \alpha_n\left((\gamma_n(1),\eta_n)\right)\right) \overline{\mu}_{n+1}(dy)$$

and the collection of $[0,1]$-parameters $\alpha_n\,(m,\eta)$ defined below

$$\alpha_n\,(\gamma_n(1),\eta_n) = \frac{\gamma_n(1)\, \eta_n(G_n)}{\gamma_n(1)\, \eta_n(G_n) + \mu_{n+1}(1)}$$

We let $\Lambda_{p,n}$ be the semigroup of the flow $(\gamma_n(1),\eta_n)$ given by

$$(\gamma_n(1),\eta_n) = \Lambda_{p,n}(\gamma_p(1),\eta_p)$$
$$= \left(\Lambda_{p,n}^1(\gamma_p(1),\eta_p), \Lambda_{p,n}^2(\gamma_p(1),\eta_p)\right) \in (\mathbb{R}_+ \times \mathcal{P}(E_n)) \tag{14.47}$$

For $p = n - 1$, sometimes we write Λ_n instead of $\Lambda_{n-1,n}$.

14.7.2 Mean field particle models

The mean field particle interpretation of these evolution equations is an E_n^N-valued sequence $\xi_n = \left(\xi_n^i\right)_{1 \leq i \leq N}$ defined as

$$\begin{cases} \gamma_{n+1}^N(1) = \gamma_n^N(1)\, \eta_n^N(G_n) + \mu_{n+1}(1) \\[2mm] \mathbb{P}\left(\xi_{n+1} \in dx \mid \gamma_n^N(1),\, \eta_n^N\right) = \prod_{i=1}^N K_{n+1,(\gamma_n^N(1),\eta_n^N)}(\xi_n^i, dx^i) \end{cases} \tag{14.48}$$

with the pair of measures (γ_n^N, η_n^N) defined for any $f_n \in \mathcal{B}_b(E_n)$ by

$$\eta_n^N := \frac{1}{N} \sum_{i=1}^{N} \delta_{\xi_n^i} \quad \text{and} \quad \gamma_n^N(f_n) := \gamma_n^N(1)\, \eta_n^N(f_n)$$

For any $n \geq 0$, and any $N \geq 1$, we have $\gamma_n(1)$ and $\gamma_n^N(1) \in I_n$ with the compact interval I_n defined below

$$I_n := [m_-(n), m_+(n)] \quad \text{where} \quad m_{-/+}(n) := \sum_{p=0}^{n} \mu_p(1) g_{-/+}(n)^{(n-p)} \quad (14.49)$$

In this situation, we have that

$$\gamma_n(1) = \sum_{p=0}^{n} \mu_p(1) \prod_{p \leq q < n} \eta_q(G_q) \in I_n$$

$$\gamma_n^N(1) = \sum_{p=0}^{n} \mu_p(1) \prod_{p \leq q < n} \eta_q^N(G_q) \in I_n$$

14.7.3 Particle intensity measures

We start this section with a simple unbiasedness property. We denote by $\mathcal{F}_n^{(N)}$ the σ-field generated by the random sequence $(\xi_p)_{0 \leq p \leq n}$.

Definition 14.7.1 *We let V_n^N be the centered random fields defined by the following formula*

$$\eta_n^N = \eta_{n-1}^N K_{n,(\gamma_n^N(1),\eta_{n-1}^N)} + \frac{1}{\sqrt{N}} V_n^N \qquad (14.50)$$

For $n = 0$, we use the convention $V_0^N = \sqrt{N}[\eta_0^N - \eta_0]$.

Definition 14.7.2 *We consider the pair of random fields*

$$W_n^{\eta,N} := \sqrt{N}[\eta_n^N - \eta_n] \quad \text{and} \quad W_n^{\gamma,N} := \sqrt{N}[\gamma_n^N - \gamma_n]$$

Proposition 14.7.3 *For any $0 \leq p \leq n$, and any $f \in \mathcal{B}(E_n)$, we have*

$$\mathbb{E}\left(\gamma_{n+1}^N(f) \,\Big|\, \mathcal{F}_p^{(N)}\right) = \gamma_p^N Q_{p,n+1}(f) + \sum_{p<q \leq n+1} \mu_q Q_{q,n+1}(f) \qquad (14.51)$$

In particular, we have the unbiasedness property:

$$\mathbb{E}\left(\gamma_n^N(f)\right) = \gamma_n(f)$$

Proof:

By construction of the particle model, for any $f \in \mathcal{B}(E_n)$ we have

$$\mathbb{E}\left(\eta_{n+1}^N(f) \mid \mathcal{F}_n^{(N)}\right) = \eta_n^N K_{n+1,(\gamma_n^N(1),\eta_n^N)}(f) = \Lambda_{n+1}^2\left(\gamma_n^N(1),\eta_n^N\right)(f)$$

with the second component Λ_{n+1}^2 of the transformation Λ_{n+1} introduced in 14.47. Using the fact that

$$\begin{aligned}
\Lambda_{n+1}^2\left(\gamma_n^N(1),\eta_n^N\right)(f) &= \frac{\gamma_n^N(1)\,\eta_n^N(Q_{n+1}(f)) + \mu_{n+1}(f)}{\gamma_n^N(1)\,\eta_n^N(Q_{n+1}(1)) + \mu_{n+1}(1)} \\
&= \frac{\gamma_n^N(Q_{n+1}(f)) + \mu_{n+1}(f)}{\gamma_n^N(Q_{n+1}(1)) + \mu_{n+1}(1)}
\end{aligned}$$

and

$$\gamma_{n+1}^N(1) = \gamma_n^N(1)\,\eta_n^N(G_n) + \mu_{n+1}(1) = \gamma_n^N(Q_{n+1}(1)) + \mu_{n+1}(1)$$

we prove that

$$\begin{aligned}
\mathbb{E}\left(\gamma_{n+1}^N(f) \mid \mathcal{F}_n^{(N)}\right) &= \mathbb{E}\left(\gamma_{n+1}^N(1)\,\eta_{n+1}^N(f) \mid \mathcal{F}_n^{(N)}\right) \\
&= \gamma_{n+1}^N(1)\,\mathbb{E}\left(\eta_{n+1}^N(f) \mid \mathcal{F}_n^{(N)}\right) \\
&= \gamma_n^N(Q_{n+1}(f)) + \mu_{n+1}(f)
\end{aligned}$$

This also implies that

$$\begin{aligned}
\mathbb{E}\left(\gamma_{n+1}^N(f) \mid \mathcal{F}_{n-1}^{(N)}\right) &= \mathbb{E}\left(\gamma_n^N(Q_{n+1}(f)) \mid \mathcal{F}_{n-1}^{(N)}\right) + \mu_{n+1}(f) \\
&= \gamma_{n-1}^N(Q_n Q_{n+1}(f)) + \mu_n(Q_{n+1}(f)) + \mu_{n+1}(f)
\end{aligned}$$

Iterating the argument one proves (14.51). The end of the proof is now clear. ∎

The next theorems provide some key martingale decomposition and a rather crude nonasymptotic variance estimate.

Theorem 14.7.4 *For any $n \geq 0$ and any function $f_n \in \mathcal{B}_b(E_n)$, we have the decomposition*

$$W_n^{\gamma,N}(f_n) = \sum_{p=0}^n \gamma_p^N(1)\,V_p^N(Q_{p,n}(f_n)) \qquad (14.52)$$

Proof:

We use the decomposition:

$$\gamma_{n+1}^N(f) - \gamma_{n+1}(f)$$

$$= \left[\gamma_{n+1}^N(f) - \mathbb{E}\left(\gamma_{n+1}^N(f) \mid \mathcal{F}_n^{(N)}\right)\right] + \left[\mathbb{E}\left(\gamma_{n+1}^N(f) \mid \mathcal{F}_n^{(N)}\right) - \gamma_{n+1}(f)\right]$$

By (14.51), we find that

$$\gamma_{n+1}^N(f) - \mathbb{E}\left(\gamma_{n+1}^N(f)\mid \mathcal{F}_n^{(N)}\right) = \gamma_{n+1}^N(f) - \left[\gamma_n^N(Q_{n+1}(f)) + \mu_{n+1}(f)\right]$$

Since we have

$$\gamma_n^N(Q_{n+1}(1)) + \mu_{n+1}(1) = \gamma_n^N(G_n) + \mu_{n+1}(1)$$
$$= \gamma_n^N(1)\,\eta_n^N(G_n) + \mu_{n+1}(1) = \gamma_{n+1}^N(1)$$

this implies that

$$\gamma_{n+1}^N(f) - \left[\gamma_n^N(Q_{n+1}(f)) + \mu_{n+1}(f)\right]$$

$$= \gamma_{n+1}^N(1)\left[\eta_{n+1}^N(f) - \frac{\left[\gamma_n^N(Q_{n+1}(f)) + \mu_{n+1}(f)\right]}{\left[\gamma_n^N(Q_{n+1}(1)) + \mu_{n+1}(1)\right]}\right]$$

$$= \gamma_{n+1}^N(1)\left[\eta_{n+1}^N(f) - \eta_n^N K_{n+1,(\gamma_n^N(1),\eta_n^N)}(f)\right]$$

and therefore

$$\gamma_{n+1}^N(f) - \mathbb{E}\left(\gamma_{n+1}^N(f)\mid \mathcal{F}_n^{(N)}\right) = \gamma_{n+1}^N(1)\left[\eta_{n+1}^N(f) - \eta_n^N K_{n+1,(\gamma_n^N(1),\eta_n^N)}(f)\right]$$

Finally, we observe that

$$\mathbb{E}\left(\gamma_{n+1}^N(f)\mid \mathcal{F}_n^{(N)}\right) - \gamma_{n+1}(f) = \gamma_n^N(Q_{n+1}(f)) - \gamma_n(Q_{n+1}(f))$$

from which we find the recursive formula

$$\left[\gamma_{n+1}^N - \gamma_{n+1}\right](f)$$

$$= \gamma_{n+1}^N(1)\left[\eta_{n+1}^N - \eta_n^N K_{n+1,(\gamma_n^N(1),\eta_n^N)}\right](f) + \left[\gamma_n^N - \gamma_n\right](Q_{n+1}(f))$$

The end of the proof of (14.52) is now obtained by a simple induction on the parameter n. This ends the proof of the theorem. ∎

Theorem 14.7.5 *When the regularity condition* $\mathbf{H}(\mathbf{G}, \mathbf{P})$ *stated on page 371 is satisfied for some finite parameters* (v_1, v_2), *then we have for any* $N > 1$ *and any* $n \geq 1$

$$\mathbb{E}\left(\left[\frac{\gamma_n^N(1)}{\gamma_n(1)} - 1\right]^2\right) \leq \frac{n+1}{N-1}\,\exp(2v_1)\left(1 + \frac{\exp(2v_1)}{(N-1)}\right)^n \qquad (14.53)$$

Proof:
Using the fact that

$$\mathbb{E}\left(\gamma_p^N(1)V_p^N(f^{(1)})\,\gamma_q^N(1)V_q^N(f^{(2)})\right)$$

$$= \mathbb{E}\left(\gamma_p^N(1)\gamma_q^N(1)V_p^N(f^{(1)})\,\mathbb{E}\left(V_q^N(f^{(2)})\mid \mathcal{F}_{q-1}^N\right)\right) = 0$$

for any $0 \le p < q \le n$, and any $f^{(1)} \in \mathcal{B}(E_p)$, and $f^{(2)} \in \mathcal{B}(E_q)$, we prove that

$$N \, \mathbb{E} \left(\left[\gamma_n^N(1) - \gamma_n(1) \right]^2 \right) = \sum_{p=0}^{n} \mathbb{E} \left(\gamma_p^N(1)^2 \, \mathbb{E} \left(V_p^N(Q_{p,n}(1))^2 | \mathcal{F}_{p-1}^N \right) \right)$$

Notice that

$$\frac{1}{\gamma_n(1)^2} = \frac{1}{\gamma_p(1)^2} \frac{1}{\eta_p(Q_{p,n}(1))^2} \left(\frac{\gamma_p(Q_{p,n}(1))}{\gamma_n(1)} \right)^2$$

$$\le \alpha_{p,n}^\star(\gamma_p(1))^2 \, \frac{1}{\gamma_p(1)^2} \frac{1}{\eta_p(Q_{p,n}(1))^2}$$

The r.h.s. estimate comes from the fact that

$$\frac{\gamma_p(Q_{p,n}(1))}{\gamma_n(1)} = \frac{\gamma_p(1) \, \eta_p(Q_{p,n}(1))}{\gamma_p(1) \, \eta_p(Q_{p,n}(1)) + \sum_{p<q\le n} \mu_q Q_{q,n}(1)}$$

$$= \alpha_{p,n}(\gamma_p(1), \eta_p) \le \alpha_{p,n}^\star(\gamma_p(1))$$

with the function $m \mapsto \alpha_{p,n}^\star(m)$ defined in (13.13).

Using the above decompositions, we readily prove that

$$N \, \mathbb{E} \left(\left[\frac{\gamma_n^N(1)}{\gamma_n(1)} - 1 \right]^2 \right)$$

$$\le \sum_{p=0}^{n} \alpha_{p,n}^\star(\gamma_p(1))^2 \, \mathbb{E} \left(\left(\frac{\gamma_p^N(1)}{\gamma_p(1)} \right)^2 \mathbb{E} \left(W_p^N(\overline{Q}_{p,n}(1))^2 | \mathcal{F}_{p-1}^N \right) \right)$$

with

$$\overline{Q}_{p,n}(1) = \overline{Q}_{p,n}(1)/\eta_p(Q_{p,n}(1)) \le g_{p,n}$$

If we set

$$U_n^N := \mathbb{E} \left(\left[\frac{\gamma_n^N(1)}{\gamma_n(1)} - 1 \right]^2 \right)$$

then we find that

$$N \, U_n^N \le a_n + \sum_{p=0}^{n} b_{p,n} \, U_p^N$$

with the parameters

$$a_n := \sum_{p=0}^{n} \left(g_{p,n} \alpha_{p,n}^\star(\gamma_p(1)) \right)^2 \quad \text{and} \quad b_{p,n} := \left(g_{p,n} \alpha_{p,n}^\star(\gamma_p(1)) \right)^2$$

Using the fact that $b_{n,n} \le 1$, we prove the following recursive equation

$$U_n^N \le a_n^N + \sum_{0 \le p < n} b_{p,n}^N \, U_p^N \quad \text{with} \quad a_n^N := \frac{a_n}{N-1} \quad \text{and} \quad b_{p,n}^N := \frac{b_{p,n}}{N-1}$$

Using an elementary proof by induction on the time horizon n, we prove the following inequality:

$$U_n^N \leq \left[\sum_{p=1}^n a_p^N \sum_{e \in \langle p,n \rangle} b^N(e) \right] + \left[\sum_{e \in \langle 0,n \rangle} b^N(e) \right] U_0^N$$

In the above display, $\langle p,n \rangle$ stands for the set of all integer valued paths $e = (e(l))_{0 \leq l \leq k}$ of a given length k from p to n

$$e_0 = p < e_1 < \ldots < e_{k-1} < e_k = n \quad \text{and} \quad b^N(e) = \prod_{1 \leq l \leq k} b_{e(l-1),e(l)}^N$$

We have also used the convention $b^N(\emptyset) = \prod_\emptyset = 1$ and $\langle n,n \rangle = \{\emptyset\}$, for $p = n$. Recalling that $\gamma_0^N = \gamma_0$, we conclude that

$$U_n^N \leq \sum_{p=1}^n a_p^N \sum_{e \in \langle p,n \rangle} b^N(e)$$

Under our regularity condition, using Proposition 12.2.4 we have that

$$\alpha_{p,n}^\star \leq 1 \quad \text{and} \quad g_{p,n} \leq \exp \upsilon_1$$

to prove that

$$\sup_{0 \leq p \leq n} a_p^N \leq (n+1) \, \exp\left(2\upsilon_1\right)/(N-1) \quad \text{and} \quad \sup_{0 \leq p \leq n} b_{p,n}^N \leq \exp\left(2\upsilon_1\right)/(N-1)$$

Using these rather crude estimates, we find that

$$U_n^N \leq a_n^N + \sum_{0 < p < n} a_p^N \sum_{l=1}^{(n-p)} \binom{n-p-1}{l-1} \left(\frac{\exp\left(2\upsilon_1\right)}{(N-1)} \right)^l$$

and therefore

$$U_n^N \leq \frac{(n+1)}{(N-1)} \, \exp\left(2\upsilon_1\right) \left(1 + \frac{\exp\left(2\upsilon_1\right)}{(N-1)} \sum_{0<p<n} \left(1 + \left(\frac{\exp\left(2\upsilon_1\right)}{(N-1)} \right) \right)^{n-p-1} \right)$$

$$= \frac{(n+1)}{(N-1)} \, \exp\left(2\upsilon_1\right) \left(1 + \frac{\exp\left(2\upsilon_1\right)}{(N-1)} \right)^n$$

This ends the proof of the theorem. ∎

14.7.4 Probability distributions

In the further development of this section $b(r)$ and $c(r)$ stand for the parameters defined, respectively, in (0.6) and in (9.33); and π_ψ is the Orlicz norm defined in (11.4).

Definition 14.7.6 *We consider the collection of parameters $a_p(n)$ given by*

$$a_p(n) := 2 \; (1 \wedge m_{p,n}) \; g_{p,n} \left[\beta(P_{p,n}) + \sum_{p<q\leq n} \frac{c_{q,n}}{\sum_{p<r\leq n} c_{r,n}} \; \beta(P_{q,n}) \right] \tag{14.54}$$

and the parameters $m_{p,n}$ given by

$$m_{p,n} = m_+(p)\|Q_{p,n}(1)\|/ \sum_{p<q\leq n} c_{q,n} \quad and \quad c_{p,n} := \mu_p Q_{p,n}(1)$$

with $m_+(p)$ defined in (14.49).

Theorem 14.7.7 *For any $n \geq 0$, $N \geq 1$, any $r \geq 1$, $f \in \mathrm{Osc}(E_n)$ we have*

$$\mathbb{E} \left(\left| W_n^{\eta,N}(f_n) \right|^r \right)^{\frac{1}{r}} \leq b(r) \; (1 \wedge c(r)) \; a(n) \tag{14.55}$$

and the Orlicz norm upper bounds

$$\pi_\psi \left(\left| W_n^{\eta,N}(f_n) \right| \right) \;\; \leq \;\; 2\sqrt{2/3} \; a(n)$$

$$\pi_\psi \left(\sup_{0\leq p\leq n} \left| W_p^{\eta,N}(f_p) \right| \right) \;\; \leq \;\; 2 \; \sqrt{\log(9+n)} \; \sup_{0\leq p\leq n} a(p)$$

with some finite constant such that $a(n) \leq \sum_{p=0}^{n} a_p(n)$. In the three scenarios discussed in Section 13.1.3, the constants $a(n)$ can be chosen so that $a = \sup_{n\geq 0} a_n < \infty$.

Proof:
We use the decomposition

$$\left(\gamma_n^N(1), \eta_n^N \right) - \left(\gamma_n(1), \eta_n \right) = \left[\Lambda_{0,n} \left(\gamma_0^N(1), \eta_0^N \right) - \Lambda_{0,n} \left(\gamma_0(1), \eta_0 \right) \right]$$

$$+ \sum_{p=1}^{n} \left[\Lambda_{p,n} \left(\gamma_p^N(1), \eta_p^N \right) - \Lambda_{p-1,n} \left(\gamma_{p-1}^N(1), \eta_{p-1}^N \right) \right] \tag{14.56}$$

to prove that

$$\eta_n^N - \eta_n = \left[\Lambda_{0,n}^2 \left(\gamma_0^N(1), \eta_0^N \right) - \Lambda_{0,n}^2 \left(\gamma_0(1), \eta_0 \right) \right]$$

$$+ \sum_{p=1}^{n} \left[\Lambda_{p,n}^2 \left(\gamma_p^N(1), \eta_p^N \right) - \Lambda_{p-1,n}^2 \left(\gamma_{p-1}^N(1), \eta_{p-1}^N \right) \right]$$

Using the fact that

$$\Lambda_{p-1,n}(m,\eta) = \Lambda_{p,n}\left(\Lambda_p(m,\eta)\right) \Rightarrow \Lambda_{p-1,n}^2(m,\eta) = \Lambda_{p,n}^2\left(\Lambda_p(m,\eta)\right)$$

we readily check that

$$\Lambda_p\left(\gamma_{p-1}^N(1), \eta_{p-1}^N\right)$$

$$= \left(\gamma_{p-1}^N(1)\eta_{p-1}^N(G_{p-1}) + \mu_p(1), \quad \Psi_{G_{p-1}}\left(\eta_{p-1}^N\right) M_{p,(\gamma_{p-1}^N(1),\eta_{p-1}^N)}\right)$$

$$= \left(\gamma_p^N(1), \quad \eta_{p-1}^N K_{p,(\gamma_{p-1}^N(1),\eta_{p-1}^N)}\right)$$

Since we have $\gamma_0^N(1) = \mu_0(1) = \gamma_0(1)$, one concludes that

$$\eta_n^N - \eta_n = \left[\Lambda_{0,n}^2\left(\gamma_0(1), \eta_0^N\right) - \Lambda_{0,n}^2\left(\gamma_0(1), \eta_0\right)\right]$$

$$+ \sum_{p=1}^{n}\left[\Lambda_{p,n}^2\left(\gamma_p^N(1), \eta_p^N\right) - \Lambda_{p,n}^2\left(\gamma_p^N(1), \eta_{p-1}^N K_{p,(\gamma_{p-1}^N(1),\eta_{p-1}^N)}\right)\right]$$

Using Proposition 13.1.6, for any $f_n \in \mathrm{Osc}(E_n)$ we prove that

$$W_n^{\eta,N}(f_n)$$

$$\leq 2\sum_{p=0}^{n}\left\{\alpha_{p,n}^\star \, g_{p,n}\left[\beta(P_{p,n})\, |V_p^N(\mathcal{D}_{p,n}^N(f))| + \beta_{p,n}\, |V_p^N(h_{p,n}^N)|\right]\right\}$$

with some $\mathcal{G}_{p-1}^N$-measurable random functions $\mathcal{D}_{p,n}^N(f)$ and $h_{p,n}^N \in \mathrm{Osc}(E_p)$.

Using the fact that $\gamma_p^N(1) \in I_p$, for any $p \geq 0$, the end of the proof is a direct consequence of Lemma 13.1.6 and Kintchine's type inequalities. The proof of the Orlicz norm estimates follows the same arguments as the ones we used in the proof of Theorem 14.3.1; thus it is omitted. The last assertion is a direct consequence of stability properties developed in Section 13.1.3. This ends the proof of the theorem. ∎

A direct consequence of this theorem is that it implies the almost sure convergence results:

$$\lim_{N\to\infty} \eta_n^N(f_n) = \eta_n(f_n) \quad \text{and} \quad \lim_{N\to\infty} \gamma_n^N(f) = \gamma_n(f_n)$$

for any bounded function $f_n \in \mathcal{B}_b(E_n)$.

The following exponential concentration estimates are direct consequences of Lemma 11.3.2, applied to the Orlicz norm upper bounds (14.18) and (14.19).

Corollary 14.7.8 *For any $n \geq 0$, $N \geq 1$, $f \in \mathrm{Osc}(E_n)$, and any $x \geq 0$ the probability of any of the following events*

$$\left|[\eta_n^N - \eta_n](f)\right| \leq 2\sqrt{\frac{2(x + \log 2)}{3N}}\, a(n)$$

and

$$\sup_{0 \le p \le n} \left| [\eta_p^N - \eta_p](f) \right| \le 4 \; \sqrt{(x + \log 2)} \; \sqrt{\frac{\log (9 + n)}{N}} \; \sup_{0 \le p \le n} a(p)$$

is greater than $1 - e^{-x}$. *In the three scenarios discussed in Section 13.1.3, the constants* $a(n)$ *can be chosen so that* $a = \sup_{n \ge 0} a_n < \infty$.

We end this section with the fluctuation properties of the N-particle approximation measures γ_n^N and η_n^N around their limiting values. Using the type of arguments as those used in the proof of the functional central limit theorems, Theorem 9.5.3, we prove the following theorem.

Theorem 14.7.9 *The sequence of random fields* $(V_n^N)_{n \ge 0}$ *converges in law, as* N *tends to infinity, to the sequence of* n *independent, Gaussian, and centered random fields* $(V_n)_{n \ge 0}$ *with a covariance function given for any* $f, g \in \mathcal{B}_n(E_n)$ *and* $n \ge 0$ *by*

$$\mathbb{E}(V_n(f)V_n(g))$$

$$= \eta_{n-1} K_{n,(\gamma_{n-1}(1),\eta_{n-1})} \left([f - K_{n,(\gamma_{n-1}(1),\eta_{n-1})}(f)][g - K_{n,(\gamma_{n-1}(1),\eta_{n-1})}(g)] \right) .$$

In addition, the pair of random fields $W_n^{\gamma,N}$ *and* $W_n^{\eta,N}$ *converge in law as* $N \to \infty$ *to a pair of centered Gaussian fields* W_n^γ *and* W_n^η *defined by*

$$W_n^\gamma(f) := \sum_{p=0}^{n} \gamma_p(1) \; V_p(Q_{p,n}(f)) \quad and \quad W_n^\eta(f) := W_n^\gamma \left(\frac{1}{\gamma_n(1)} (f - \eta_n(f)) \right)$$

14.8   Positive measure particle equations

14.8.1   Nonlinear evolution models

We consider a sequence of measurable state space $(E_n)_{n \ge 0}$, and a collection of bounded integral operators $Q_{n,\gamma}$ from $\mathcal{B}_b(E_n)$ into $\mathcal{B}_b(E_{n-1})$, indexed by the time parameter $n \ge 1$ and the set of measures $\gamma \in \mathcal{M}_+(E_n)$.

We associate with these objects the measure-valued dynamical systems $\gamma_n \in \mathcal{M}_+(E_n)$ defined by the following nonlinear evolution equations

$$\gamma_n = \gamma_{n-1} Q_{n,\gamma_{n-1}}$$

with initial measure $\gamma_0 \in \mathcal{M}_+(E_0)$. We let $\eta_n \in \mathcal{P}(E_n)$ be the normalized distributions given for any $f \in \mathcal{B}_b(E_n)$ by

$$\eta_n(f) = \gamma_n(f)/\gamma_n(1) \quad and \; we \; set \quad G_{n,\gamma_n} := Q_{n+1,\gamma_n}(1)$$

In Section 6.2.1 we have shown that these flows of measures satisfy the following measure valued equations

$$\begin{aligned}
\gamma_{n+1}(1) &= \eta_n(G_{n,\gamma_n})\,\gamma_n(1) \\
\eta_{n+1} &= \Psi_{G_{n,\gamma_n}}(\eta_n)\,M_{n+1,\gamma_n}
\end{aligned} \tag{14.57}$$

with the Markov transitions M_{n+1,γ_n} defined for any $f \in \mathcal{B}_b(E_{n+1})$ by

$$M_{n+1,\gamma_n}(f) := Q_{n+1,\gamma_n}(f)/Q_{n+1,\gamma_n}(1)$$

14.8.2 Mean field particle models

We consider a Markov transition S_{n,γ_n} from E_n into itself satisfying the following compatibility condition

$$\Psi_{G_{n,\gamma_n}}(\eta_n) = \eta_n S_{n,\gamma_n}$$

Several examples of transitions S_{n,γ_n} are discussed in Section 4.4.5. In this situation, the evolution Equations (14.57) can be rewritten in the following form

$$\left\{ \begin{aligned}
\gamma_{n+1}(1) &= \eta_n(G_{n,\gamma_n})\,\gamma_n(1) \\
\eta_{n+1} &= \eta_n K_{n+1,\gamma_n} \quad \text{with} \quad K_{n+1,\gamma_n} := S_{n,\gamma_n} M_{n+1,\gamma_n}
\end{aligned} \right. \tag{14.58}$$

The mean field type interacting particle system associated with these models is the E_n^N-valued Markov chain $\xi_n = \left(\xi_n^i\right)_{1 \le i \le N}$, with elementary transitions defined as

$$\gamma_{n+1}^N(1) = \gamma_n^N(1)\,\eta_n^N(G_{n,\gamma_n^N}) \tag{14.59}$$

$$\mathbb{P}\left(\xi_{n+1} \in dx \mid \xi_n\right) = \prod_{i=1}^{N} K_{n+1,\gamma_n^N}(\xi_n^i, dx^i) \tag{14.60}$$

with the pair of measures

$$\left(\gamma_n^N, \eta_n^N\right) \in \left(\mathcal{M}_+(E_n), \mathcal{P}(E_n)\right)$$

defined for any $f_n \in \mathcal{B}_b(E_n)$ by

$$\eta_n^N := \frac{1}{N}\sum_{j=1}^{N} \delta_{\xi_n^j} \quad \text{and} \quad \gamma_n^N(f_n) := \gamma_n^N(1) \times \eta_n^N(f_n)$$

In the above displayed formula, $dx = dx^1 \times \ldots \times dx^N$ stands for an infinitesimal neighborhood of a point $x = (x^1, \ldots, x^N) \in E_n^N$. The initial system $\xi_0^{(N)}$ consists of N independent and identically distributed random variables with common law η_0.

Definition 14.8.1 *We consider the sequence of local random sampling errors* $(V_n^N)_{n\geq 0}$ *defined by the perturbation equations*

$$\eta_n^N = \eta_{n-1}^N K_{n,\gamma_{n-1}^N} + \frac{1}{\sqrt{N}}\, V_n^N$$

We also let $W_n^{\gamma,N}$ *and* $W_n^{\eta,N}$ *be the empirical random fields on* $\mathcal{B}_b(E_n)$ *defined for any* $n \geq 1$ *by*

$$\gamma_n^N = \gamma_n + \frac{1}{\sqrt{N}}\, W_n^{\gamma,N} \quad \text{and} \quad \eta_n^N = \eta_n + \frac{1}{\sqrt{N}}\, W_n^{\eta,N}$$

We emphasize that the local sampling random fields V_n^N have similar properties as the ones discussed in Section 4.4.4. For instance, using Proposition 9.5.5 we readily prove the following $\mathbb{L}_m$-mean error estimates.

$$\sup_{f\in \mathrm{Osc}(E_n)} \mathbb{E}\left(\left|V_n^N(f)\right|^m \big| \mathcal{F}_{n-1}^N\right)^{1/m} \leq b(m)\,(1 \wedge c(m)) \qquad (14.61)$$

with the parameters $b(m)$ and $c(m)$ defined, respectively, in (0.6) and in (9.33) (with an appropriate version of the uniform local variance parameters given in (9.31)).

14.8.3 A nonasymptotic theorem

We further assume that the regularity conditions (H_1) and (H_2) presented in Section 13.2.1 are satisfied for some pair of positive functions $\theta_{+/-,n}$, some finite constants $c(n) < \infty$, and some collection of bounded measures $\Sigma_{n,(m',\eta')}^1$ and $\Sigma_{n,(m',\eta')}^2(f,.)$ Under the first condition, we recall that the total mass process $\gamma_n(1)$ and its N-approximation model $\gamma_n^N(1)$ are finite and they evolve at every time n in a series of compact sets

$$I_n \subset [m_n^-, m_n^+] \subset (0, \infty)$$

In the above display, the sequence of parameters $m_n^{+/-}$ is defined by the recursive equations

$$m_{n+1}^{+/-} = m_n^{+/-} \times \theta_{+/-,n}(m_n^{+/-}) \quad \text{with} \quad m_0^- = m_0^+ = \gamma_0(1)$$

In addition, by Proposition 13.3.1 we have the following weak Lipschitz type inequalities:

$$\left|\Lambda_{p,n}^1(m,\eta) - \Lambda_{p,n}^1(m',\eta')\right|$$
$$\leq c_p(n)\,|m - m'| + \int |[\eta - \eta'](\varphi)|\,\Sigma_{p,n,(m',\eta')}^1(d\varphi) \qquad (14.62)$$

and

$$\left| \left[\Lambda^2_{p,n}(m, \eta) - \Lambda^2_{p,n}(m', \eta') \right](f) \right|$$

$$\leq c_p(n) \, |m - m'| + \int \, |[\eta - \eta'](\varphi)| \, \Sigma^2_{p,n,(m',\eta')}(f, d\varphi) \tag{14.63}$$

for some finite constants $c_p(n) < \infty$, and some collection of bounded measures $\Sigma^1_{p,n,(m',\eta')}$ and $\Sigma^2_{p,n,(m',\eta')}(f, .)$ on $\mathcal{B}_b(E_p)$ such that

$$\begin{aligned}
\int \operatorname{osc}(\varphi) \, \Sigma^1_{p,n,(m,\eta)}(d\varphi) &\leq \delta\left(\Sigma^1_{p,n}\right) \\
\int \operatorname{osc}(\varphi) \, \Sigma^2_{p,n,(m,\eta)}(f, d\varphi) &\leq \delta\left(\Sigma^2_{p,n}\right)
\end{aligned} \tag{14.64}$$

for some finite constant $\delta\left(\Sigma^i_{p,n}\right) < \infty$, $i = 1, 2$, whose values do not depend on the parameters $(m, \eta) \in (I_p \times \mathcal{P}(E_p))$ and $f \in \operatorname{Osc}(E_n)$.

We are now in position to state and to prove the main result of this section.

Theorem 14.8.2 *Assume that the regularity conditions (H_1) and (H_2) presented in Section 13.2.1 are satisfied. In this situation, for any $N \geq 1$, $n \geq 0$, $m \geq 1$, and any $f_n \in \mathcal{B}_b(E_n)$, we have the estimates:*

$$\mathbb{E}\left(\left| W^{\gamma,N}_n(1) \right|^m \right)^{\frac{1}{m}} \leq b(m) \, (1 \wedge c(m)) \, \sum_{p=0}^{n} \delta\left(\Sigma^1_{p,n}\right) \tag{14.65}$$

and

$$\mathbb{E}\left(\left| W^{\eta,N}_n(f) \right|^m \right)^{\frac{1}{m}} \leq b(m) \, (1 \wedge c(m)) \, \sum_{p=0}^{n} \delta\left(\Sigma^2_{p,n}\right) \tag{14.66}$$

with the constant $b(m)$ and $c(m)$ defined in (14.61).

In addition, when the regularity conditions of Theorem 13.3.3 are satisfied, we have

$$\sup_{n \geq 0} \sum_{p=0}^{n} \delta\left(\Sigma^1_{p,n}\right) \leq c^{1,2}/\left(1 - e^{-\lambda}\right) \quad and \quad \sup_{n \geq 0} \sum_{p=0}^{n} \delta\left(\Sigma^2_{p,n}\right) \leq c^{2,2}/\left(1 - e^{-\lambda}\right)$$

with the parameters $\lambda > 0$ and $c^{i,j}$ defined in Theorem 13.3.3.

Proof:

We use the decomposition

$$\left(\gamma^N_n(1), \eta^N_n\right) - \left(\gamma_n(1), \eta_n\right)$$

$$= \left[\Lambda_{0,n}\left(\gamma^N_0(1), \eta^N_0\right) - \Lambda_{0,n}\left(\gamma_0(1), \eta_0\right)\right]$$

$$+ \sum_{p=1}^{n}\left[\Lambda_{p,n}\left(\gamma^N_p(1), \eta^N_p\right) - \Lambda_{p-1,n}\left(\gamma^N_{p-1}(1), \eta^N_{p-1}\right)\right]$$

Using the fact that

$$\Lambda_{p-1,p}\left(\gamma_{p-1}^N(1), \eta_{p-1}^N\right) = \left(\gamma_p^N(1), \Lambda_{p-1,p}^2\left(\gamma_{p-1}^N(1), \eta_{p-1}^N\right)\right)$$

we readily prove the following decomposition

$$\gamma_n^N(1) - \gamma_n(1)$$

$$= \left[\Lambda_{0,n}^1\left(\gamma_0^N(1), \eta_0^N\right) - \Lambda_{0,n}^1\left(\gamma_0(1), \eta_0\right)\right]$$

$$+ \sum_{p=1}^n \left[\Lambda_{p,n}^1\left(\gamma_p^N(1), \eta_p^N\right) - \Lambda_{p,n}^1\left(\gamma_p^N(1), \Lambda_{p-1,p}^2\left(\gamma_{p-1}^N(1), \eta_{p-1}^N\right)\right)\right]$$

Recalling that $\gamma_0^N(1) = \gamma_0(1)$, using (14.62), we find that

$$\left|W_n^{\gamma,N}(1)\right| \le \sum_{p=0}^n \int \left|[V_p^N(\varphi)]\right| \Sigma_{p,n}^{(N,1)}(d\varphi)$$

with the predictable measure

$$\Sigma_{p,n}^{(N,1)} = \Sigma_{p,n,(m,\eta)}^1$$

associated with the parameters

$$(m, \eta) = \left(\gamma_p^N(1), \Lambda_{p-1,p}^2\left(\gamma_{p-1}^N(1), \eta_{p-1}^N\right)\right)$$

with $0 < p \le n$; and for $p = 0$, we set $\Sigma_{0,n}^{(N,1)} = \Sigma_{0,n,(\gamma_0(1),\eta_0)}$.

Combining the generalized Minkowski's inequality with the $\mathbb{L}_m$-mean error bounds (14.61) we prove that

$$\mathbb{E}\left(\left|\int |W_p^N(\varphi)| \Sigma_{p,n}^{(N,1)}(d\varphi)\right|^m \left|\mathcal{F}_{p-1}^{(N)}\right.\right)^{\frac{1}{m}} \le b(m) \ (1 \wedge c(m)) \ \delta\left(\Sigma_{p,n}^1\right)$$

This clearly implies that

$$\mathbb{E}\left(\left|W_n^{\gamma,N}(1)\right|^m\right)^{\frac{1}{m}} \le b(m) \ (1 \wedge c(m)) \sum_{p=0}^n \delta\left(\Sigma_{p,n}^1\right)$$

The normalized occupation measures can be analyzed in the same way using the decomposition given below:

$$\eta_n^N - \eta_n$$

$$= \left[\Lambda_{0,n}^2\left(\gamma_0^N(1), \eta_0^N\right) - \Lambda_{0,n}^2\left(\gamma_0(1), \eta_0\right)\right]$$

$$+ \sum_{p=1}^n \left[\Lambda_{p,n}^2\left(\gamma_p^N(1), \eta_p^N\right) - \Lambda_{p,n}^2\left(\gamma_p^N(1), \eta_{p-1}^N K_{p,(\gamma_{p-1}^N(1),\eta_{p-1}^N)}\right)\right]$$

This ends the proof of the theorem.   ∎

Using the same arguments as the ones we used in the proof of Theorem 14.7.7 and Theorem 14.3.1, we prove the following corollary.

Corollary 14.8.3 *For any $N \geq 1$, and any $n \geq 0$, we have the Orlicz norm upper bounds*

$$\pi_\psi \left(\left| W_n^{\gamma, N}(1) \right| \right) \leq 2\sqrt{2/3} \, \delta_n^1$$

$$\pi_\psi \left(\sup_{0 \leq p \leq n} \left| W_p^{\gamma, N}(1) \right| \right) \leq 2 \sqrt{\log(9 + n)} \sup_{0 \leq q \leq p} \delta_p^1$$

and

$$\pi_\psi \left(\left| W_n^{\eta, N}(f_n) \right| \right) \leq 2\sqrt{2/3} \, \delta_n^2$$

$$\pi_\psi \left(\sup_{0 \leq p \leq n} \left| W_p^{\eta, N}(f_p) \right| \right) \leq 2 \sqrt{\log(9 + n)} \sup_{0 \leq q \leq p} \delta_p^2$$

with some finite constants

$$\forall i \in \{1, 2\} \qquad \delta_n^i \leq \sum_{p=0}^{n} \delta \left(\Sigma_{p,n}^i \right)$$

Corollary 14.8.4 *Assume that the regularity conditions of Theorem 13.3.3 are satisfied. In this situation, for any $N \geq 1$, and any $n \geq 0$, we have the Orlicz norm upper bounds*

$$\pi_\psi \left(\left| W_n^{\gamma, N}(1) \right| \right) \leq 2\sqrt{2/3} \, c^{1,2} / \left(1 - e^{-\lambda} \right)$$

$$\pi_\psi \left(\sup_{0 \leq p \leq n} \left| W_p^{\gamma, N}(1) \right| \right) \leq 2 \sqrt{\log(9 + n)} \, c^{1,2} / \left(1 - e^{-\lambda} \right)$$

as well as

$$\pi_\psi \left(\left| W_n^{\eta, N}(f_n) \right| \right) \leq 2\sqrt{2/3} \, c^{2,2} / \left(1 - e^{-\lambda} \right)$$

$$\pi_\psi \left(\sup_{0 \leq p \leq n} \left| W_p^{\eta, N}(f_p) \right| \right) \leq 2 \sqrt{\log(9 + n)} \, c^{2,2} / \left(1 - e^{-\lambda} \right)$$

with the parameters $\lambda > 0$ and $c^{i,j}$ defined in Theorem 13.3.3.

Chapter 15

Genealogical tree models

15.1 Some equivalence principles

Feynman-Kac models on path spaces and their genealogical tree based particle approximations have been discussed in several places in earlier chapters of the book. Roughly speaking, up to natural state space enlargement, we have seen that the Feynman-Kac measures $(\Gamma_n, \mathbb{Q}_n)$ defined in (1.38) and (1.37) can also be interpreted as the n-th time marginals of a Feynman-Kac model with a reference Markov chain associated with historical processes. In this interpretation, the genealogical tree based approximations of the measures $(\Gamma_n, \mathbb{Q}_n)$ can be interpreted as the mean field Feynman-Kac particle model associated with an historical process. In this interpretation, Feynman-Kac models on path spaces and their genealogical tree based approximations are mathematically equivalent to the particle density profiles developed in Chapter 14.

In this short chapter, we review some key equivalence principles that allows applying without further work most of the results presented in Chapter 14 dedicated to the stochastic analysis of particle density profiles.

15.1.1 Feynman-Kac semigroups on path spaces

In this section, we provide a brief reminder on Feynman-Kac semigroups associated with an historical process. We also derive a series of equivalence principles, including some useful estimates of the path space versions of the regularity parameters and the first order integral operators introduced in (12.4) and in Definition 14.2.2.

Feynman-Kac models associated with an historical process are discussed in Section 3.4. These models are defined as follows. We consider a Markov chain X_n with elementary transitions M_n from some measurable state space E_{n-1} into a possibly different measurable state space E_n. We also consider a collection of nonnegative potential functions G_n on E_n. The historical process $\mathbf{X}_n$ associated with X_n is the Markov chain on path space given by the sequence of random variables

$$\mathbf{X_n} := (X_0, \ldots, X_n) \in \boldsymbol{E_n} := (E_0 \times \ldots \times E_n)$$

We let $\mathbf{G}_n$ be a sequence of positive and bounded potential functions on $\boldsymbol{E_n}$

defined by

$$\mathbf{G_n} \; : \; \mathbf{x_n} = (x_0, \ldots, x_n) \in \mathbf{E_n} \mapsto \mathbf{G_n}(\mathbf{x_n}) = G_n(x_n)$$

Definition 15.1.1 *The Feynman-Kac model* $(\boldsymbol{\gamma_n}, \boldsymbol{\eta_n})$ *associated with the pair* $(\mathbf{G}_n, \mathbf{M}_n)$ *on the path spaces* $\mathbf{E_n}$ *is given for any functions* $\boldsymbol{f_n} \in \mathcal{B}_b(\mathbf{E_n})$, *by the formulae*

$$\boldsymbol{\eta_n}(\mathbf{f_n}) := \boldsymbol{\gamma_n}(\mathbf{f_n})/\boldsymbol{\gamma_n}(1) \quad \text{with} \quad \boldsymbol{\gamma_n}(\boldsymbol{f_n}) = \mathbb{E}\left(\boldsymbol{f_n}(\mathbf{X_n}) \prod_{0 \leq p < n} \mathbf{G}_p(\mathbf{X_p})\right)$$

Definition 15.1.2 *We consider the Feynman-Kac semigroups* $\mathbf{Q_{p,n}}$, $\mathbf{P_{p,n}}$, *and* $\boldsymbol{\Phi_{p,n}}$ *defined as the semigroups* $Q_{p,n}$, $P_{p,n}$, *and* $\Phi_{p,n}$, *introduced in Section 12.1, by replacing* (G_n, M_n) *by* $(\mathbf{G_n}, \mathbf{M_n})$.

Definition 15.1.3 *For any* $0 \leq p \leq n$, *we also denote by* $\mathbf{G_{p,n}}$ *the nonnegative function on* $\mathbf{E_p}$ *defined for any* $\mathbf{x_p} \in \mathbf{E_p}$ *by*

$$\mathbf{G_{p,n}}(\mathbf{x_p}) := \mathbf{Q_{p,n}}(1)(\mathbf{x_p})$$

and we set

$$\mathbf{g_{p,n}} := \sup_{\mathbf{x_p}, \mathbf{y_p}} (\mathbf{G_{p,n}}(\mathbf{x_p})/\mathbf{G_{p,n}}(\mathbf{y_p}))$$

Definition 15.1.4 *For any* $k, l \geq 0$, *we denote by* $\boldsymbol{\tau_{k,l}(n)}$ *and* $\boldsymbol{\kappa(n)}$ *the parameters defined as the parameters* $\tau_{k,l}(n)$ *and* $\kappa(n)$ *introduced in (12.4) by replacing* $(g_{p,n}, P_{p,n})$ *by* $(\boldsymbol{g_{p,n}}, \mathbf{P_{p,n}})$.

By construction, we recall that for any $\boldsymbol{f_n} \in \mathcal{B}_b(\mathbf{E_n})$, we have that

$$\boldsymbol{\gamma_n} = \boldsymbol{\gamma_p} \mathbf{Q_{p,n}}$$

and

$$\boldsymbol{\eta_n}(\boldsymbol{f_n}) = \frac{\boldsymbol{\eta_p}(\mathbf{Q_{p,n}}(\boldsymbol{f_n}))}{\boldsymbol{\eta_p}(\mathbf{Q_{p,n}}(1))} = \Psi_{\mathbf{G_{p,n}}}(\boldsymbol{\eta_p})\mathbf{P_{p,n}}(\boldsymbol{f_n})$$

with the Markov transition $\mathbf{P_{p,n}}$ from $\mathbf{E_p}$ into $\mathbf{E_n}$ defined by

$$\mathbf{P_{p,n}}(\boldsymbol{f_n}) \quad := \quad \mathbf{Q_{p,n}}(\boldsymbol{f_n})/\mathbf{Q_{p,n}}(1)$$

We also notice that the Feynman-Kac semigroup $\mathbf{Q_{p,n}}$ in the above display are defined by

$$\mathbf{Q_{p,n}}(\boldsymbol{f_n})(\mathbf{x_p})$$

$$:= \mathbb{E}\left(\boldsymbol{f_n}(\mathbf{X_n}) \prod_{p \leq q < n} \mathbf{G_q}(\mathbf{X_q}) \mid \mathbf{X_p} = \mathbf{x_p}\right)$$

$$= \mathbb{E}\left(\boldsymbol{f_n}(\mathbf{x_p}, X_{p+1}, \ldots, X_n) \prod_{p \leq q < n} G_q(X_q) \mid X_p = x_p\right)$$

for any $f_n \in \mathcal{B}_b(E_n)$, any $p \leq n$, and any $\mathbf{x_p} = (x_0, \ldots, x_p) \in \mathbf{E_p}$.

In the next proposition we summarize some equivalence principles discussed in Section 4.1.2

Proposition 15.1.5   • *The measures $(\gamma_\mathbf{n}, \eta_\mathbf{n})$ introduced in Definition 15.1.1 coincide with the couple of measures $(\Gamma_n, \mathbb{Q}_n)$ defined in (1.38) and (1.37); that is we have that*

$$\eta_\mathbf{n}(d\mathbf{x_n}) = \mathbb{Q}_n(d(x_0, \ldots, x_n)) \quad and \quad \gamma_\mathbf{n}(d\mathbf{x_n}) = \Gamma_n(d(x_0, \ldots, x_n))$$

where $d\mathbf{x_n}$ and $d(x_0, \ldots, x_n)$ stand for an infinitesimal neighborhood of the point $\mathbf{x_n} \in (x_0, \ldots, x_n) \in \mathbf{E_n} = (E_0 \times \ldots \times E_n)$.

• *For any functions $f_n \in \mathcal{B}_b(E_n)$ of the form*

$$f_n(\mathbf{x_n}) = f_n(x_n)$$

for any $\mathbf{x_n} = (x_0, \ldots, x_n) \in \mathbf{E_n}$, for some $f_n \in \mathcal{B}_b(E_n)$, we have

$$\mathbf{Q_{p,n}}(f_n)(\mathbf{x_p}) = Q_{p,n}(f_n)(x_p)$$

and

$$\mathbf{P_{p,n}}(f_n)(\mathbf{x_p}) = P_{p,n}(f_n)(x_p)$$

for any $p \leq n$, and any $\mathbf{x_p} = (x_0, \ldots, x_p) \in \mathbf{E_p}$, as well as

$$\gamma_\mathbf{n}(f_n) = \gamma_n(f_n) \quad and \quad \eta_\mathbf{n}(f_n) = \eta_n(f_n)$$

• *For any $0 \leq p \leq n$ and any $\mathbf{x_p} \in \mathbf{E_p}$ we have that*

$$\mathbf{G_{p,n}}(\mathbf{x_p}) = \mathbf{Q_{p,n}}(1)(\mathbf{x_p}) = Q_{p,n}(1)(x_p) = G_{p,n}(x_p)$$

and

$$\mathbf{g}_{p,n} := \sup_{\mathbf{x_p}, \mathbf{y_p}} \frac{\mathbf{G_{p,n}}(\mathbf{x_p})}{\mathbf{G_{p,n}}(\mathbf{y_p})} = \sup_{x_p, y_p} \frac{G_{p,n}(x_p)}{G_{p,n}(y_p)} = g_{p,n}$$

• *We further assume that the reference Markov chain X_n satisfies the condition $\mathbf{H(G, P)}$, stated on page 371, for some finite constants υ_1, υ_2. In this situation, we have*

$$g_{p,n} \leq \exp \upsilon_1 \quad and \quad \beta(\mathbf{P}_{p,n}) \leq 1 \tag{15.1}$$

as well as

$$\forall k, l \geq 0 \quad \tau_{k,l}(n) \leq (n+1) \exp(k\upsilon_1) \quad and \quad \kappa(n) \leq \exp \upsilon_1 \tag{15.2}$$

The last assertion in the above proposition is proved using the estimates presented in Proposition 12.2.9.

Proposition 15.1.6 *We consider a function $\boldsymbol{f_n} \in \mathcal{B}_b(\boldsymbol{E_n})$ that only depends on the p-th coordinate*

$$\boldsymbol{f_n}(\boldsymbol{x_n}) = \varphi_p(x_p)$$

for some function $\varphi_p \in \mathcal{B}_b(E_p)$, with $p \leq n$. In this situation, we have

$$\mathbf{P_{q,n}}(\boldsymbol{f_n})(\mathbf{x_q}) = \begin{cases} R_{q,p}^{(n)}(\varphi_p)(x_q) & if \quad q < p \\ \varphi_p(x_p) & if \quad q \geq p \end{cases}$$

with the Markov transitions $R_{q,p}^{(n)}$ defined in Lemma 12.2.2.

Proof:
We have

$$\mathbf{Q_{q,n}}(\boldsymbol{f_n})(\mathbf{x_q}) = \mathbb{E}\left(\varphi_p(X_p) \prod_{q \leq r < n} G_r(X_r) \mid X_0 = x_0, \ldots, X_q = x_q \right)$$

for any $\boldsymbol{x_q} = (x_0, \ldots, x_q)$. For $q \leq p$, this implies that

$$\mathbf{Q_{q,n}}(\boldsymbol{f_n})(\mathbf{x_q}) = \varphi_p(x_p)\, Q_{q,n}(1)(x_q) \quad \text{and} \quad \mathbf{P_{q,n}}(\boldsymbol{f_n})(\mathbf{x_q}) = \varphi_p(x_p)$$

For $q < p$ we have

$$\mathbf{Q_{q,n}}(\boldsymbol{f_n})(\mathbf{x_q})$$

$$= \mathbb{E}\left(\varphi_p(X_p) \left\{ \prod_{q \leq r < p} G_r(X_r) \right\} \times \prod_{p \leq r < n} G_r(X_r) \mid X_q = x_q \right)$$

$$= \mathbb{E}\left(\varphi_p(X_p) \left\{ \prod_{q \leq r < p} G_r(X_r) \right\} \times \mathbb{E}\left(\prod_{p \leq r < n} G_r(X_r) \mid X_p \right) \mid X_q = x_q \right)$$

$$= Q_{q,p}\left(\varphi_p\, Q_{p,n}(1) \right)(x_q)$$

In this situation, using Lemma 12.2.2 we conclude that

$$\begin{aligned} \mathbf{P_{q,n}}(\boldsymbol{f_n})(\mathbf{x_q}) &= \frac{Q_{q,p}\left(\varphi_p\, Q_{p,n}(1) \right)(x_q)}{Q_{q,p}\left(Q_{p,n}(1) \right)(x_q)} \\ &= \frac{P_{q,p}\left(\varphi_p\, Q_{p,n}(1) \right)(x_q)}{P_{q,p}\left(Q_{p,n}(1) \right)(x_q)} = R_{q,p}^{(n)}(\varphi_p)(x_q) \end{aligned}$$

This ends the proof of the proposition. ∎

15.1.2   Genealogical tree evolution models

Definition 15.1.7 *We let $\boldsymbol{\xi_n} = (\boldsymbol{\xi_n^i})_{1 \leq i \leq N} \in \mathbf{E_n^N}$ be a mean field N-particle model (9.15) associated with the flow of Feynman-Kac measures $\boldsymbol{\eta_n} \in \mathcal{P}(\boldsymbol{E_n})$ introduced in Definition 15.1.1. For any $1 \leq i \leq N$, we set*

$$\boldsymbol{\xi_n^i} = \left(\xi_{0,n}^i, \xi_{1,n}^i, \ldots, \xi_{n,n}^i \right) \in \mathbf{E_n^N} = (E_0 \times \ldots \times E_n)^N$$

We also denote by η_n^N the corresponding particle density profiles

$$\eta_n^N := \frac{1}{N} \sum_{i=1}^N \delta_{\xi_n^i} = \frac{1}{N} \sum_{i=1}^N \delta_{(\xi_{0,n}^i, \xi_{1,n}^i, \ldots, \xi_{n,n}^i)} \in \mathcal{P}(E_n)$$

Definition 15.1.8 *We consider the local random fields models*

$$\mathbf{V}_n^N := \sqrt{N} \left(\eta_n^N - \Phi_n \left(\eta_{n-1}^N \right) \right)$$

with the convention

$$\Phi_0 \left(\eta_{-1}^N \right) = \eta_0^N = \eta_0$$

for $n = 0$, so that $\mathbf{V}_0^N = V_0^N$. We also denote by $\mathbf{W}_n^{\eta,N}$ the sequence of random fields

$$\mathbf{W}_n^{\eta,N} = \sqrt{N} \left[\eta_n^N - \eta_n \right] = \sqrt{N} \left[\eta_n^N - \mathbb{Q}_n \right]$$

In Section 9.4.3, we have seen that these particle models in path space represent the ancestral lines of the particle interpretation of the n-th marginal measures η_n of the measures $\mathbb{Q}_n$ defined in (1.37).

Definition 15.1.9 *For any $0 \le p \le n$, and $N \ge 1$, we denote by*

$$\left(G_{p,n}^N, d_{p,n}^N, d_{p,n}'^N, d_{p,n}, d_{p,n}' \right)$$

the mathematical objects defined as the objects

$$\left(G_{p,n}^N, d_{p,n}^N, d_{p,n}'^N, d_{p,n}, d_{p,n}' \right)$$

introduced in Definition 14.2.2, by replacing $(\Phi_{p,n}, \eta_n, \eta_n^N)$ by $(\boldsymbol{\Phi}_{p,n}, \boldsymbol{\eta}_n, \boldsymbol{\eta}_n^N)$.

Using (14.12) and (14.13), it is readily checked that for any function $f_n \in \mathcal{B}_b(E_n)$ and any $p \le n$, we have

$$\|d_{p,n}(f_n)\| \vee \|d_{p,n}^N(f_n)\| \vee \|G_{p,n}^N\| \le g_{p,n}$$

and

$$\left\| d_{p,n}'(f_n) \right\| \vee \left\| d_{p,n}'^N(f_n) \right\| \le 1 \quad \text{and} \quad \text{osc} \left(d_{p,n}'^N(f_n) \right) \le 2$$

Using (15.1) we prove the following technical lemma.

Lemma 15.1.10 *Assume that the reference Markov chain X_n satisfies the condition $\mathbf{H}(\mathbf{G}, \mathbf{P})$, stated on page 371, for some finite constants υ_1, υ_2. In this situation, we have the uniform bounds*

$$\sup_{n \ge 0} \sup_{0 \le p \le n} \left[\|d_{p,n}(f_n)\| \vee \|d_{p,n}^N(f_n)\| \vee \|G_{p,n}^N\| \vee g_{p,n} \right] \le \exp \upsilon_1$$

15.2 Some nonasymptotic theorems

In this section, we present some nonasymptotic theorems, including Orlicz norms and $\mathbb{L}_p$-mean error estimates. In Section 15.2.2 we present a natural strategy to analyze the $\mathbb{L}_m$-norm of the first order martingale terms in the first order decompositions (14.14) and (14.15) using Burkholder-Davis-Gundy type inequalities.

15.2.1 Orlicz norms and $\mathbb{L}_p$-mean error estimates

In the further development of this section $b(m)$ and $c(m)$ stand for the parameters defined, respectively, in (0.6) and in (9.33); and π_ψ is the Orlicz norm defined in (11.4).

Using the equivalence principles presented in Section 15.1, we readily check the path space version of Theorem 14.3.1.

Theorem 15.2.1 *For any $m \geq 1$, $N \geq 1$, $n \geq 0$, and any $\boldsymbol{f_n} \in \mathrm{Osc}(\boldsymbol{E_n})$, we have $\mathbb{L}_m$-mean error estimates*

$$\mathbb{E}\left(\left|\mathbf{W}_{\mathbf{n}}^{\eta,\mathbf{N}}(\boldsymbol{f_n})\right|^m\right)^{1/m} \leq 2\, b(m)\ (1 \wedge c(m))\, \boldsymbol{\tau_{1,1}(n)} \tag{15.3}$$

In addition, we have the Orlicz norm upper bounds

$$\pi_\psi\left(\left|\mathbf{W}_{\mathbf{n}}^{\eta,\mathbf{N}}(\boldsymbol{f_n})\right|\right) \leq 4\sqrt{2/3}\ \boldsymbol{\tau_{1,1}(n)}$$

$$\pi_\psi\left(\sup_{0 \leq p \leq n}\left|\mathbf{W}_{\mathbf{p}}^{\eta,\mathbf{N}}(\mathbf{f_p})\right|\right) \leq 8\ \sqrt{\log(9+n)}\ \sup_{0 \leq p \leq n}\boldsymbol{\tau_{1,1}(p)}$$

Using (15.2) we prove the following corollary.

Corollary 15.2.2 *Assume that the reference Markov chain X_n satisfies the condition $\mathbf{H(G,P)}$, stated on page 371, for some finite constants υ_1, υ_2. In this situation, for any $m \geq 1$, $N \geq 1$, $n \geq 0$, and any collection of functions $\boldsymbol{f_n} \in \mathrm{Osc}(\boldsymbol{E_n})$, we have $\mathbb{L}_m$-mean error estimates*

$$\mathbb{E}\left(\left|\mathbf{W}_{\mathbf{n}}^{\eta,\mathbf{N}}(\boldsymbol{f_n})\right|^m\right)^{1/m} \leq 2\exp(\upsilon_1)\, b(m)\ (1 \wedge c(m))\,(n+1)$$

as well as the Orlicz norm upper bounds

$$\pi_\psi\left(\left|\mathbf{W}_{\mathbf{n}}^{\eta,\mathbf{N}}(\boldsymbol{f_n})\right|\right) \leq 4\sqrt{2/3}\exp(\upsilon_1)(n+1)$$

$$\pi_\psi\left(\sup_{0 \leq p \leq n}\left|\mathbf{W}_{\mathbf{p}}^{\eta,\mathbf{N}}(\mathbf{f_p})\right|\right) \leq 8\ \exp(\upsilon_1)\,(n+1)\sqrt{\log(9+n)}$$

We end this section with a direct consequence of Lemma 11.3.2.

Corollary 15.2.3 *Assume that the reference Markov chain X_n satisfies the condition* $\mathbf{H}(\mathbf{G}, \mathbf{P})$, *stated on page 371, for some finite constants* v_1, v_2. *In this situation, for any $x \geq 0$, $n \geq 0$, $N \geq 1$, and any collection of functions $\mathbf{f_n} \in \mathrm{Osc}(\mathbf{E_n})$, the probability of any of the events*

$$\left| \mathbf{W}_{\mathbf{n}}^{\eta, \mathbf{N}}(\mathbf{f_n}) \right| \leq 4\sqrt{2/3} \, \exp(v_1) \, (n+1) \, \sqrt{x + \log 2}$$

and

$$\sup_{0 \leq p \leq n} \left| \mathbf{W}_{\mathbf{p}}^{\eta, \mathbf{N}}(\mathbf{f_p}) \right| \leq 8 \exp(v_1) \, (n+1) \sqrt{\log(9+n)} \, \sqrt{x + \log 2}$$

is greater than $1 - e^{-x}$.

15.2.2 First order Burkholder-Davis-Gundy type estimates

The error bounds presented in Section 15.2 can be improved significantly using the path space version of the first order decompositions (14.14). The first order terms in the decompositions (14.14) and (14.15) are associated with triangular arrays of martingale sequences. One natural way to estimate their $\mathbb{L}_m$-norm is to use Burkholder-Davis-Gundy type inequalities.

For instance, the path space version of the decomposition (14.14) is given by

$$\mathbf{W}_{\mathbf{n}}^{\eta, \mathbf{N}}(\mathbf{f_n}) = \sum_{p=0}^{n} V_p^N(d_{p,n}^N(\mathbf{f_n})) + \frac{1}{\sqrt{N}} \, R_n^N(\mathbf{f_n})$$

with the second order term

$$R_n^N := -\sum_{p=0}^{n} \frac{1}{\eta_p^N(G_{p,n}^N)} \, V_p^N(G_{p,n}^N) \, V_p^N \left[d_{p,n}^N(\mathbf{f_n}) \right]$$

By Theorem 14.22, we readily check that

$$\sup_{\mathbf{f_n} \in \mathrm{Osc}(\mathbf{E_n})} \mathbb{E}\left[\left| R_n^N(\mathbf{f_n}) \right|^m \right]^{1/m} \leq b(2m)^2 \, \mathbf{r}(n)$$

with some finite constant

$$\mathbf{r}(n) \leq 2 \, (\tau_{2,1}(n) - \tau_{1,1}(n))$$

On the other hand, we observe that the collection of random variables $M_q^{(N,n)}(\mathbf{f_n}))$, $0 \leq q \leq n$, defined by

$$M_q^{(N,n)}(\mathbf{f_n})) := \sum_{p=0}^{q} V_p^N(d_{p,n}^N(\mathbf{f_n}))$$

forms a triangular array of martingale sequences. Using Burkholder-Davis-Gundy inequalities, we have

$$\mathbb{E}\left(\left|[M_n^{(N,n)}(f_n)]\right|^m\right)^{1/m}$$

$$\leq \alpha(m)\sqrt{(n+1)}\ \mathbb{E}\left(\left[\frac{1}{(n+1)}\sum_{p=0}^{n}\left(V_p^N(d_{p,n}^N(f_n))\right)^2\right]^{m/2}\right)^{1/m}$$

with $\alpha(m) = \max(m-1, 1/m-1)$.

For $m \geq 2$, we find that

$$\mathbb{E}\left(\left|[M_n^{(N,n)}(f_n)]\right|^m\right)^{1/m}$$

$$\leq (m-1)\sqrt{(n+1)}\ \left(\frac{1}{(n+1)}\sum_{p=0}^{n}\mathbb{E}\left(\left|V_p^N(d_{p,n}^N(f_n))\right|^m\right)\right)^{1/m}$$

$$\leq (m-1)(n+1)^{1/2}\ \left(\frac{1}{n+1}\sum_{p=0}^{n}(2g_{p,n})^m\right)^{1/m}$$

Combining these estimates, we prove the following proposition.

Proposition 15.2.4 *For any $m \geq 2$, $N \geq 1$, $n \geq 0$, and any collection of functions $f_n \in \mathrm{Osc}(E_n)$, we have $\mathbb{L}_m$-mean error estimates*

$$\mathbb{E}\left(\left|[\eta_n^N - \mathbb{Q}_n](f_n)\right|^m\right)^{1/m} \leq 2(m-1)\ \sqrt{\frac{(n+1)}{N}}\ g_n^\star + \frac{1}{N}\ b(2m)^2\ \mathbf{r(n)}$$

with the constants

$$g_n^\star \leq \sup_{0 \leq p \leq n}\ g_{p,n} \quad and \quad \mathbf{r(n)} \leq 2\ (\tau_{2,1}(n) - \tau_{1,1}(n))$$

Using (15.2) and Lemma 15.1.10, we also check the following corollary.

Corollary 15.2.5 *Assume that the reference Markov chain X_n satisfies the condition $\mathbf{H(G,P)}$, stated on page 371, for some finite constants v_1, v_2. In this situation, for any $m \geq 2$, $N \geq 1$, $n \geq 0$, and any collection of functions $f_n \in \mathrm{Osc}(E_n)$, we have $\mathbb{L}_m$-mean error estimates*

$$\mathbb{E}\left(\left|[\eta_n^N - \mathbb{Q}_n](f_n)\right|^m\right)^{1/m}$$

$$\leq 2(m-1)\ \sqrt{\frac{(n+1)}{N}}\ e^{v_1} + 2b(2m)^2\ \frac{(n+1)}{N}\ (e^{2v_1} - e^{v_1})$$

15.3 Ancestral tree occupation measures

In this section, we provide some nonasymptotic theorems for the occupation measures of the ancestral trees at some level.

Our first result is a nonasymptotic result for the p-th time marginals of the measures η_n^N. These distributions represent the occupation measures of the ancestral individuals at level p.

Proposition 15.3.1 *We consider a function $f_n \in \mathrm{Osc}(E_n)$ that only depends on the p-th coordinate*

$$f_n(x_n) = \varphi_p(x_p)$$

for some function $\varphi_p \in \mathrm{Osc}(E_p)$ and some integer $p \le n$. We also assume that the reference Markov chain X_n satisfies the condition $\mathbf{H}(\mathbf{G}, \mathbf{P})$, stated on page 371, for some finite constants v_1, v_2. In this situation, we have the bias estimate

$$\left| \mathbb{E}\left(\eta_n^N(f_n) \right) - \mathbb{Q}_n(f_n) \right| \le \frac{2}{N} \left(\exp v_1 - 1 \right) \left(\exp v_1 \right) \left(v_2 \exp v_1 + (n - p) \right)$$

In addition, for any $m \ge 2$ we have the $\mathbb{L}_m$-mean error bounds

$$\mathbb{E}\left(\left| [\eta_n^N - \mathbb{Q}_n](f_n) \right|^m \right)^{1/m}$$

$$\le (m - 1) \sqrt{\frac{(n + 1)}{N}} \, (2 \exp v_1) \left(\frac{v_2}{p + 1} + \frac{n - p}{n + 1} \right)^{1/m}$$

$$+ \frac{2}{N} (\exp v_1 - 1) \, (\exp v_1) \, (v_2 \exp v_1 + (n - p))$$

Before getting into the proof of this proposition, we note that for $m = 2$ the $\mathbb{L}_m$-mean error estimate stated in the proposition gives the following nonasymptotic variance estimate.

Corollary 15.3.2 *Under the assumptions of Proposition 15.3.1 we have the variance upper bounds*

$$\mathbb{E}\left(\left| [\eta_n^N - \mathbb{Q}_n](f_n) \right|^2 \right)$$

$$\le \frac{8}{N} \exp(2v_1) \left[\left(v_2 \frac{n + 1}{p + 1} + (n - p) \right) \right.$$

$$\left. + \frac{1}{N} (\exp v_1 - 1)^2 \, (v_2 \exp v_1 + (n - p))^2 \right]$$

Now, we come to the proof of the proposition.

Proof of Proposition 15.3.1:

We use the same arguments and the notation of Section 15.2.2. Using Proposition 15.1.6 we have

$$\operatorname{osc}\left(P_{q,n}(f_n)\right) \le \begin{cases} \beta(R_{q,p}^{(n)}) & \text{if} \quad q < p \\ 1 & \text{if} \quad q \ge p \end{cases}$$

On the other hand, we observe that

$$d_{q,n}^N(f_n)(x_q) = \frac{G_{q,n}(x_q)}{\Phi_q(\eta_q^N)(G_{q,n})} \left[P_{q,n}(f_n)(x_q) - \Psi_{G_{q,n}}(\eta_{q-1}^N)P_{q,n}(f_n)\right]$$

Using (15.1), this implies that

$$\left\|d_{q,n}^N(f_n)\right\| \le g_{q,n} \operatorname{osc}\left(P_{q,n}(f_n)\right) \le \begin{cases} \exp\upsilon_1 \beta(R_{q,p}^{(n)}) & \text{if} \quad q < p \\ \exp\upsilon_1 & \text{if} \quad q \ge p \end{cases}$$

Using the same arguments as in the proof of Theorem 14.4.1, for $m \ge 2$, we find that

$$\mathbb{E}\left(|[M_n^{(N,n)}(f_n))|^m\right)^{1/m}$$

$$\le (m-1)(n+1)^{1/2}\left(2\exp\upsilon_1\right)\left(\left(\frac{1}{p+1}\sum_{0\le q<p}\beta(R_{q,p}^{(n)})^m + \frac{n-p}{n+1}\right)\right)^{1/m}$$

Using (3.6), we also have that

$$\beta(R_{q,p}^{(n)}) \le \exp\upsilon_1\ \beta(P_{q,p})$$

This yields

$$\mathbb{E}\left(|[M_n^{(N,n)}(f_n))|^m\right)^{1/m}$$

$$\le (m-1)(n+1)^{1/2}\left(2\exp\upsilon_1\right)\left(\left(\frac{1}{p+1}\sum_{0\le q<p}\beta(P_{q,p})^m + \frac{n-p}{n+1}\right)\right)^{1/m}$$

$$\le (m-1)(n+1)^{1/2}\left(2\exp\upsilon_1\right)\left(\left(\frac{\upsilon_2}{p+1} + \frac{n-p}{n+1}\right)\right)^{1/m}$$

with the remainder term estimate

$$\mathbb{E}\left[|R_n^N(f_n)|^m\right]^{1/m} \le b(2m)^2\ \mathbf{r(n)}$$

with

$$\begin{aligned} \mathbf{r(n)} \quad &\le \quad 2(\exp\upsilon_1 - 1)\left(\exp\upsilon_1\right)\left(\left(\exp\upsilon_1\right)\sum_{q=0}^{p}\beta(P_{q,p}) + (n-p)\right) \\ &\le \quad 2(\exp\upsilon_1 - 1)\left(\exp\upsilon_1\right)\left(\upsilon_2\exp\upsilon_1 + (n-p)\right) \end{aligned}$$

This ends the proof of the proposition. ∎

We end this section with the path space version of Theorem 9.5.7.

Theorem 15.3.3 *For any $n \geq 0$, and any measurable function V_n on the path space E_n s.t. $\mathbb{Q}_n - \text{ess sup } V_n < \infty$, we have the convergence in probability*

$$\eta_n^N - \text{ess sup } V_n = \max_{1 \leq i \leq N} V_n(\xi_{0,n}^i, \ldots, \xi_{n,n}^i) \longrightarrow_{N \to \infty} \mathbb{Q}_n - \text{ess sup } V_n$$

15.4 Central limit theorems

This section is concerned with the path space version of the fluctuation analysis developed in Section 14.4. Using the equivalence principles presented in Section 15.1, we have

$$\eta_n = \Phi_n(\eta_{n-1}) = \eta_{n-1} K_{n,\eta_{n-1}}$$

for any collection of Markov transitions $K_{n,\eta_{n-1}}$ from E_{n-1} into E_n satisfying the compatibility condition

$$\eta_{n-1} K_{n,\eta_{n-1}} = \Phi_n(\eta_{n-1})$$

The local sampling random fields induced by the mean field particle model in path space are defined by the following equation

$$\eta_n^N = \Phi_n\left(\eta_{n-1}^N\right) + \frac{1}{\sqrt{N}} V_n^N$$

In this context, the path space version of Theorem 9.5.3 ensures that for any fixed time horizon $n \geq 0$, the sequence of random fields V_n^N converges in law, in the sense of convergence of finite dimensional distributions, as the number of particles N tends to infinity, to a sequence of independent, Gaussian, and centered random fields V_n, with a covariance function defined for any $f_n \in \text{Osc}(E_n)$, and any $n \geq 0$, by the formula

$$\mathbb{E}(V_n(f_n)^2) = \eta_{n-1} K_{n,\eta_{n-1}}([f_n - K_{n,\eta_{n-1}}(f_n)]^2) \leq 1$$

On the other hand, the path space version of Theorem 14.4.2 takes the following form.

Theorem 15.4.1 *For any $n \geq 0$, $N \geq 1$, and any $f_n \in \mathcal{B}_b(E_n)$, we have the first order decomposition*

$$W_n^{\eta,N}(f_n) = \sum_{p=0}^{n} V_p^N[d_{p,n}(f_n)] + \frac{1}{\sqrt{N}} R_n^N(f_n) \qquad (15.4)$$

The second order remainder term $R_n^N(f_n)$ in the above display is such that

$$\sup_{f_n \in \mathrm{Osc}(E_n)} \mathbb{E}\left[\left|R_n^N(f_n)\right|^m\right]^{1/m} \le b(2m)^2 \, r(n) \tag{15.5}$$

with some finite constant $r(n)$ such that

$$r(n) \le (2\tau_{1,1}(n))^3 \sum_{0 \le p < n} (g_p - 1) g_{p,n}$$

with the parameter $\tau_{1,1}(n)$ defined in (15.1.4).

Using (15.4), the Slutsky's lemma and the continuous mapping theorem yield the following multivariate central limit theorem.

Theorem 15.4.2 *The sequence of random fields $W_n^{\eta,N}$ converges in law in the sense of convergence of finite dimensional distributions, as the number of particles N tends to infinity, to a sequence of Gaussian, and centered random fields W_n^η defined by*

$$W_n^\eta(f_n) \; := \; \sum_{p=0}^{n} V_p\left[d_{p,n}(f_n)\right] = V_n(f_n) + W_{n-1}^\eta(d_{n-1,n}(f_n))$$

We end this section with an interpretation of the covariance functional of W_n^η in terms of the Feynman-Kac distributions $\mathbb{Q}_n$.

If we choose

$$K_{n,\eta_{n-1}}(x_{n-1}, dx_n) = \Phi_n(\eta_{n-1})(dx_n) \quad \text{and} \quad \eta_n(f_n) = 0$$

then we find that

$$\eta_p(d_{p,n}(f_n)) = \eta_p\left(\frac{G_{p,n}}{\eta_p(G_{p,n})} \, P_{p,n}\left(f_n - \Phi_{p,n}(\eta_p)(f_n)\right)\right) = 0$$

and therefore

$$\mathbb{E}\left(W_n^\eta(f_n)^2\right) \;=\; \sum_{p=0}^{n} \eta_p\left(d_{p,n}(f_n)^2\right)$$

$$=\; \sum_{p=0}^{n} \eta_p\left(\frac{G_{p,n}^2}{\eta_p(G_{p,n})^2} \, P_{p,n}(f_n)^2\right)$$

By construction, we have

$$P_{p,n}(f_n)(x_p) = \mathbb{E}_{\mathbb{Q}_n}\left(f_n(X_p, X_{p+1}, \ldots, X_n) \mid X_p = x_p\right)$$

By Lemma 12.4.5, we also have that

$$\Psi_{G_{p,n}}(\eta_p)(dx_p) = \frac{1}{\eta_p(G_{p,n})} \, G_{p,n}(x_p) \, \mathbb{Q}_p(dx_p) = \mathbb{Q}_{p|n}(dx_p)$$

where $\mathbb{Q}_{p|n}$ stands for the marginal of $\mathbb{Q}_n$ with respect to the first $p+1$ coordinates $(x_0, \ldots, x_p) = \mathbf{x_p}$. This implies that

$$\eta_p \left(\frac{G_{p,n}^2}{\eta_p(G_{p,n})^2} \, P_{p,n}(f_n)^2 \right) = \mathbb{E}_{\mathbb{Q}_n} \left(\frac{d\mathbb{Q}_{p|n}}{d\mathbb{Q}_p}(\mathbf{X_p}) \, \mathbb{E}_{\mathbb{Q}_n} \left(f_n(\mathbf{X_n}) \mid \mathbf{X_p} \right)^2 \right)$$

We conclude that

$$\mathbb{E} \left(W_n^\eta(f_n)^2 \right) = \mathbb{E}_{\mathbb{Q}_n} \left(\sum_{p=0}^{n} \frac{d\mathbb{Q}_{p|n}}{d\mathbb{Q}_p}(\mathbf{X_p}) \, \mathbb{E}_{\mathbb{Q}_n} \left(f_n(\mathbf{X_n}) \mid \mathbf{X_p} \right)^2 \right)$$

15.5 Concentration inequalities

The exponential concentration inequalities presented in Section 15.2 can be improved using the concentration analysis of interacting processes developed in Section 11.8. For instance, using the equivalence principles presented in Section 15.1, we readily check the path space version of Corollary 14.5.2.

Theorem 15.5.1 *For any $n \geq 0$, any $f_n \in \mathrm{Osc}(E_n)$, and any $N \geq 1$, the probability of the event*

$$\left[\eta_n^N - \eta_n \right] (f_n) \leq \frac{1}{N} \, p_n(x) + \frac{1}{\sqrt{N}} \, q_n(x)$$

is greater than $1 - e^{-x}$, for any $x \geq 0$, with the functions

$$p_n(x) = 2 \left(\tau_{2,1}(n) - \tau_{1,1}(n) \right) \left(1 + 2(x + \sqrt{x}) \right) + 2x \, \kappa(n)/3$$

$$q_n(x) = 2\sqrt{2x} \left(\sum_{0 \leq p \leq n} g_{p,n}^2 \right)^{1/2}$$

Using (15.2) and Lemma 15.1.10, we also check the following corollary.

Corollary 15.5.2 *Assume that the reference Markov chain X_n satisfies the condition $\mathbf{H(G, P)}$, stated on page 371, for some finite constants v_1, v_2. In this situation, for any $n \geq 0$, $f_n \in \mathrm{Osc}(E_n)$, and any $N \geq 1$, the probability of the event*

$$\left[\eta_n^N - \eta_n \right] (f_n) \leq \frac{1}{N} \, p_n(x) + \frac{1}{\sqrt{N}} \, q_n(x)$$

is greater than $1 - e^{-x}$, for any $x \geq 0$, with the functions

$$p_n(x) = 2 \left(\exp(2v_1) - \exp v_1 \right) (n+1) \left(1 + 2(x + \sqrt{x}) \right) + 2x \exp(v_1)/3$$

$$q_n(x) = 2\sqrt{2x} \, \exp(v_1) \sqrt{(n+1)}$$

Chapter 16

Particle normalizing constants

16.1 Unnormalized particle measures

16.1.1 Description of the models

We consider a Feynman-Kac model (1.37) associated with a sequence of Markov transitions M_n and potential functions G_n on some measurable state spaces E_n. The unnormalized Feynman-Kac measures $\gamma_n \in \mathcal{M}(E_n)$ are given for any $f_n \in \mathcal{B}_b(E_n)$ by the following formulae

$$\gamma_n(f_n) = \mathbb{E}\left(f_n(X_n) \prod_{0 \leq p < n} G_p(X_p)\right) = \eta_n(f_n) \prod_{0 \leq p < n} \eta_p(G_p)$$

The product formulae expressing γ_n in terms of the flow of normalized distributions $(\eta_p)_{0 \leq p \leq n}$ are proved in (9.3).

We denote by $(\eta_n^N)_{n \geq 0}$ the particle density profiles associated with a mean field particle model (9.15). A stochastic perturbation analysis of these particle density profiles is given in Chapter 14.

The N-particle approximation of the measures γ_n is defined by the formula

$$\gamma_n^N(f_n) := \eta_n^N(f_n) \prod_{0 \leq p < n} \eta_p^N(G_p)$$

These unnormalized particle models γ_n^N have a particularly simple form. They are defined in terms of product of empirical mean values $\eta_p^N(G_p)$ of the potential functions G_p w.r.t. the flow of normalized particle measures η_p^N after the p-th mutation stages, with $p < n$.

Thus, the deviation properties of γ_n^N around their limiting values γ_n should be related in some way to one of the interacting processes η_n^N developed in Section 14.5.

Definition 16.1.1 *We denote by $\overline{\gamma}_n^N$ the normalized models defined by the following formulae*

$$\overline{\gamma}_n^N(f) = \gamma_n^N(f)/\gamma_n(1) = \eta_n^N(f) \prod_{0 \leq p < n} \eta_p^N(\overline{G}_p)$$

with the normalized potential functions $\overline{G}_n$ defined in (12.1).

16.1.2   Fluctuation random fields models

The aim of this chapter is to analyze the deviations of the N-particle unnormalized measures γ_n^N around the limiting measures γ_n. Our analysis is expressed in terms of the random fields models defined below.

Definition 16.1.2 *We let $W_n^{\gamma,N}$ and $\overline{W}_n^{\gamma,N}$ be the random fields particle models defined by*

$$W_n^{\gamma,N} = \sqrt{N} \ \left[\gamma_n^N - \gamma_n\right]$$

and

$$\overline{W}_n^{\gamma,N} := W_n^{\gamma,N}(f)/\gamma_n(1)$$

In this notation, we observe that

$$\overline{\gamma}_n^N(1) := \gamma_n^N(1)/\gamma_n(1) = \prod_{0 \leq p < n} \eta_p^N(\overline{G}_p) = 1 + \frac{1}{\sqrt{N}} \ \overline{W}_n^{\gamma,N}(1)$$

For more general functions we also observe that for any function f on E_n, s.t. $\eta_n(f) = 1$, we have the decompositions

$$\overline{W}_n^{\gamma,N}(f)$$

$$= \sqrt{N} \ \left[\left(1 + \tfrac{1}{\sqrt{N}} \ \overline{W}_n^{\gamma,N}(1)\right) \left(\eta_n(f) + \tfrac{1}{\sqrt{N}} \ W_n^{\eta,N}(f)\right) - \eta_n(f)\right]$$

and

$$\overline{W}_n^{\gamma,N}(f)$$

$$= \sqrt{N} \ \left[\left(1 + \tfrac{1}{\sqrt{N}} \ \overline{W}_n^{\gamma,N}(1)\right) \left(1 + \tfrac{1}{\sqrt{N}} \ W_n^{\eta,N}(f)\right) - 1\right]$$

We readily deduce the following second order decompositions of the fluctuation errors

$$\overline{W}_n^{\gamma,N}(f) = \left[\overline{W}_n^{\gamma,N}(1) + W_n^{\eta,N}(f)\right] + \frac{1}{\sqrt{N}} \ \left(\overline{W}_n^{\gamma,N}(1) \ W_n^{\eta,N}(f)\right)$$

This decomposition allows reduction of the concentration properties of $\overline{W}_n^{\gamma,N}(f)$ to the ones of $W_n^{\eta,N}(f)$ and $\overline{W}_n^{\gamma,N}(1)$.

This chapter is divided into two parts: In the first part, we provide some key decompositions of $W_n^{\gamma,N}$ in terms of the local sampling errors V_n^N, as well as a pivotal exponential formula connecting the fluctuations of the particle free energies in terms of the fluctuations of the potential empirical mean values.

Then, we derive first order expansions and logarithmic concentration inequalities for particle free energy ratios $\overline{\gamma}_n^N(1) = \gamma_n^N(1)/\gamma_n(1)$.

16.2 Some key decompositions

This section is mainly concerned with the proof of two decomposition theorems.

Theorem 16.2.1 *For any $0 \leq p \leq n$, $N \geq 1$, and any function f on E_n, we have the decompositions*

$$W_n^{\gamma,N}(f) = \sum_{p=0}^{n} \gamma_p^N(1) \, V_p^N(Q_{p,n}(f)) \tag{16.1}$$

$$\overline{W}_n^{\gamma,N}(f) = \sum_{p=0}^{n} \overline{\gamma}_p^N(1) \, V_p^N(\overline{Q}_{p,n}(f)) \tag{16.2}$$

with the normalized Feynman-Kac semigroup $\overline{Q}_{p,n}$ defined in (12.2).

Proof:
We use the telescoping sum decomposition

$$\gamma_n^N - \gamma_n = \sum_{p=0}^{n} \left(\gamma_p^N Q_{p,n} - \gamma_{p-1}^N Q_{p-1,n} \right)$$

with the conventions $Q_{n,n} = Id$, for $p = n$; and $\gamma_{-1}^N Q_{-1,n} = \gamma_0 Q_{0,n}$, for $p = 0$. Using the fact that

$$\gamma_p^N(1) = \gamma_{p-1}^N(G_{p-1}) \quad \text{and} \quad \gamma_{p-1}^N Q_{p-1,n}(f) = \gamma_{p-1}^N(G_{p-1} M_p(Q_{p,n}(f)))$$

we prove that

$$\gamma_{p-1}^N Q_{p-1,n} = \gamma_p^N(1) \, \Phi_p \left(\eta_{p-1}^N \right) Q_{p,n}$$

The end of the proof of the first decomposition is now easily completed. We prove (16.2) using the following formulae

$$\begin{aligned}
\overline{Q}_{p,n}(f)(x) &= \frac{\gamma_p(1)}{\gamma_n(1)} \, Q_{p,n}(f)(x) \\
&= Q_{p,n}(f)(x) \prod_{p \leq q < n} \eta_q(G_q)^{-1} = \frac{Q_{p,n}(f)(x)}{\eta_p Q_{p,n}(1)}
\end{aligned}$$

This ends the proof of the theorem. ∎

Theorem 16.2.2 *For any $0 \leq p \leq n$, $N \geq 1$, we have the exponential formulae*

$$\overline{W}_n^{\gamma,N}(1)$$

$$= \sqrt{N} \left(\exp \left\{ \frac{1}{\sqrt{N}} \int_0^1 \sum_{0 \leq p < n} \frac{W_p^{\eta,N}(\overline{G}_p)}{1 + \frac{t}{\sqrt{N}} W_p^{\eta,N}(\overline{G}_p)} \, dt \right\} - 1 \right) \tag{16.3}$$

Proof:
The proof of (16.3) is based on the fact that

$$\log y - \log x = \int_0^1 \frac{(y-x)}{x + t(y-x)} \, dt$$

for any positive numbers x, y. Indeed, we have the formula

$$\begin{aligned}
\log\left(\gamma_n^N(1)/\gamma_n(1)\right) &= \log\left(1 + \frac{1}{\sqrt{N}} \frac{W_n^{\gamma,N}(1)}{\gamma_n(1)}\right) \\
&= \sum_{0 \le p < n} \left(\log \eta_p^N(G_p) - \log \eta_p(G_p)\right) \\
&= \frac{1}{\sqrt{N}} \sum_{0 \le p < n} \int_0^1 \frac{W_p^{\eta,N}(G_p)}{\eta_p(G_p) + \frac{t}{\sqrt{N}} W_p^{\eta,N}(G_p)} \, dt
\end{aligned}$$

∎

16.3   Fluctuation theorems

This section is concerned with the fluctuation analysis of the random fields $W_n^{\gamma,N}$ and the random variable $\sqrt{N} \, \log \overline{\gamma}_n^N(1)$. We start with a corollary of Theorem 16.2.2.

Corollary 16.3.1 *For any $0 \le p \le n$, $N \ge 1$, we have the decomposition*

$$\log \overline{\gamma}_n^N(1) = \frac{1}{\sqrt{N}} \sum_{0 \le q < n} V_q^N[h_{q,n}] + \frac{1}{N} R_n^{(N)}$$

with the first order functions

$$h_{q,n} := \sum_{q \le p < n} d_{q,p}(\overline{G}_p) = \overline{Q}_{q,n}(1) - 1 \tag{16.4}$$

with the semigroups $\overline{Q}_{p,n}$ and $d_{p,n}$ defined in (12.2) and (14.10), and with a second order remainder term $R_n^{(N)}$ such that

$$\mathbb{E}\left[\left|R_n^{(N)}\right|^m\right]^{1/m} \le b(2m)^2 \, r(n)$$

for any $m \ge 1$, for some finite constant $r(n) < \infty$.

Proof:

The proof is based on the following decomposition

$$\sqrt{N} \log \overline{\gamma}_n^N(1) = \sum_{0 \le p < n} W_p^{\eta,N}(\overline{G}_p) \int_0^1 \frac{dt}{1 + \frac{t}{\sqrt{N}} W_p^{\eta,N}(\overline{G}_p)}$$

$$= \sum_{0 \le p < n} W_p^{\eta,N}(\overline{G}_p) + \frac{1}{\sqrt{N}} R_n^{(N,1)}$$

with

$$R_n^{(N,1)} := - \sum_{0 \le p < n} W_p^{\eta,N}(\overline{G}_p)^2 \int_0^1 \frac{t}{1 + \frac{t}{\sqrt{N}} W_p^{\eta,N}(\overline{G}_p)} \, dt$$

Using the fact that

$$1 + \frac{t}{\sqrt{N}} W_p^{\eta,N}(\overline{G}_p) = (1 - t) \, \eta_p(\overline{G}_p) + t \, \eta_p^N(\overline{G}_p) \in [g_p^{-1}, g_p]$$

for any $m \ge 1$, we prove that

$$\mathbb{E}\left[\left|R_n^{(N,1)}\right|^m\right]^{1/m} \le \sum_{0 \le p < n} g_p \, \mathbb{E}\left(W_p^{\eta,N}(\overline{G}_p)^{2m}\right)^{1/m}$$

Using Theorem 14.3.1, we conclude that

$$\mathbb{E}\left[\left|R_n^{(N,1)}\right|^m\right]^{1/m} \le b(2m)^2 \, r_1(n)$$

with some constant

$$r_1(n) \le 16 \sum_{0 \le p < n} g_p^3 \, \tau_{1,1}(p)^2$$

On the other hand, using the decomposition Theorem 14.4.2, we have

$$W_p^{\eta,N}(\overline{G}_p) = \sum_{q=0}^p V_q^N \left[d_{q,p}(\overline{G}_p)\right] + \frac{1}{\sqrt{N}} R_p^{(N,2)}$$

with the second order remainder term $R_p^{(N,2)}$ such that

$$\mathbb{E}\left[\left|R_p^{(N,2)}\right|^m\right]^{1/m} \le b(2m)^2 \, r_2(p)$$

with some constant

$$r_2(p) \le (2g_p) \, (2\tau_{1,1}(p))^3 \sum_{0 \le q < p} (g_q - 1) g_{q,p} \, \beta(P_{q,p})$$

In summary, we have proved that

$$\sqrt{N} \log \overline{\gamma}_n^N(1) = \sum_{0 \le q < n} V_q^N \left[h_{q,n}\right] + \frac{1}{\sqrt{N}} R_n^{(N)}$$

with

$$R_n^{(N)} := R_n^{(N,1)} + \sum_{0 \le p < n} R_p^{(N,2)}$$

On the other hand, for any $f \in \mathcal{B}_b(E_p)$ we have

$$d_{q,p}(f) = \overline{Q}_{q,p}(f) - \overline{Q}_{q,p}(1)\,\eta_p(f) \quad \text{and} \quad \overline{G}_p = \overline{Q}_{p,p+1}(1)$$

and therefore

$$d_{q,p}(\overline{G}_p) = \overline{Q}_{q,p+1}(1) - \overline{Q}_{q,p}(1)$$

The proof of (16.4) is now easily completed. This ends the proof of the corollary. ∎

Combining (16.1) with Theorem 9.5.3, Slutsky's lemma, and the continuous mapping theorem, we easily prove the following theorem.

Theorem 16.3.2 *The sequence of random fields $W_n^{\gamma,N}$ converges in law in the sense of convergence of finite dimensional distributions, as the number of particles N tends to infinity, to a sequence of Gaussian, and centered random fields W_n^{γ} defined for any $f_n \in \mathcal{B}_b(E_n)$ by*

$$W_n^{\gamma}(f_n) := \sum_{p=0}^{n} \gamma_p(1)\, V_p(Q_{p,n}(f_n)) = V_n(f_n) + W_{n-1}^{\gamma}(Q_n(f_n))$$

In addition, we have the convergence in law

$$\lim_{N \to \infty} \sqrt{N} \, \log \overline{\gamma}_n^N(1) = L_n := \sum_{0 \le q < n} V_q\,[h_{q,n}]$$

with the functions $h_{q,n}$ defined in (16.4).

We end this section with some estimates of the asymptotic variances of the limiting Gaussian variables presented in Theorem 16.3.2.

Proposition 16.3.3 *We consider time homogeneous Feynman-Kac models $(G_n, M_n) = (G, M)$, and we assume that condition $\mathbf{H(G,P)}$ stated on page 371 is satisfied for some finite constants υ_1, υ_2. In this situation, we have the uniform estimates*

$$\sup_{n \ge 0} \mathbb{E}\left(\overline{W}_n^{\gamma}(f_n)^2\right) < \infty \quad \text{and} \quad \sup_{n \ge 0} \mathbb{E}\left(\left[L_n/\sqrt{(n+1)}\right]^2\right) \le (2\exp(2\upsilon_1)\,\upsilon_2)^2$$

Using the estimates (14.12), we readily check that

$$\mathrm{osc}\,(h_{q,n}) \le \sum_{q \le p < n} \mathrm{osc}\,(d_{q,p}(\overline{G}_p)) \le 2g_p \sum_{q \le p < n} g_{q,p}\beta(P_{q,p})$$

Using Proposition 12.2.4 we have that

$$\mathrm{osc}\,(h_{q,n}) \le 2\exp\,(2\upsilon_1)\,\upsilon_2$$

from which we conclude that
and therefore

$$\mathbb{E}\left(\left(\sum_{0\le q<n} V_q\,[h_{q,n}]\right)^2\right) = \sum_{0\le q<n} \mathbb{E}\left(V_q\,[h_{q,n}]^2\right)$$

$$\le (2\exp\,(2\upsilon_1)\,\upsilon_2)^2\,(n+1)$$

In much the same way, as a direct consequence of Theorem 16.3.2, the $\overline{W}_n^{\gamma,N}$ converges in law in the sense of convergence of finite dimensional distributions, as the number of particles N tends to infinity, to a sequence of of Gaussian, and centered random fields $\overline{W}_n^\gamma$ defined for any $f_n \in \mathrm{Osc}(E_n)$ by

$$\overline{W}_n^\gamma(f_n) := \sum_{p=0}^{n} V_p(\overline{Q}_{p,n}(f_n))$$

Recalling that

$$\overline{Q}_{p,n}(f_n) = \frac{Q_{p,n}(f_n)}{\eta_p Q_{p,n}(1)} = \frac{G_{p,n}}{\eta_p(G_{p,n})}\,P_{p,n}(f_n)$$

we find that

$$\mathrm{osc}\,(\overline{Q}_{p,n}(f_n)) \le 2g_{p,n}\,\beta(P_{p,n})$$

This implies that

$$\mathbb{E}\left(\overline{W}_n^\gamma(f_n)^2\right) = \sum_{p=0}^{n} \mathbb{E}\left(V_p(\overline{Q}_{p,n}(f_n))^2\right) \le 4\tau_{2,2}(n)$$

with the parameters $\tau_{2,2}(n)$ defined in (12.4). The end of the proof is now a direct consequence of (12.7). This ends the proof of the proposition. ∎

16.4   A nonasymptotic variance theorem

The main object of this section is to obtain some useful nonasymptotic variance estimate for the particle mass process $\gamma_n^N(1)$. Firstly, we observe that

$$\overline{W}_n^{\gamma,N}(1) = \sqrt{N}\left(\prod_{0\le p<n} \eta_p^N(\overline{G}_p) - 1\right) \quad \text{with} \quad \overline{G}_p = G_p/\eta_p(G_p)$$

In this notation, we have

$$\overline{\gamma}_n^N(1) := \gamma_n^N(1)/\gamma_n(1) = \prod_{0 \le p < n} \eta_p^N(\overline{G}_p) = 1 + \frac{1}{\sqrt{N}} \, \overline{W}_n^{\gamma,N}(1)$$

Combining (16.1) with the formula

$$\gamma_n(1)/\gamma_p(1) = \eta_p(G_{p,n}) \quad \text{with} \quad G_{p,n} := Q_{p,n}(1)$$

for any $0 \le p \le n$, we prove the following recursive formulae

$$\overline{W}_n^{\gamma,N}(1) = \sum_{0 \le p < n} \left(1 + \frac{1}{\sqrt{N}} \, \overline{W}_p^{\gamma,N}(1)\right) V_p^N(\overline{G}_{p,n})$$

$$= \sum_{0 \le p < n} V_p^N(\overline{G}_{p,n}) + \frac{1}{\sqrt{N}} \sum_{0 \le p < n} \overline{W}_p^{\gamma,N}(1) \, V_p^N(\overline{G}_{p,n})$$

with the normalized functions

$$\overline{G}_{p,n} := G_{p,n}/\eta_p(G_{p,n}) \le g_{p,n}$$

Notice that

$$\mathrm{osc}\left(\overline{G}_{p,n}\right) = \frac{1}{\eta_p(G_{p,n})} \, \mathrm{osc}\left(G_{p,n}\right) \le g_{p,n} - 1$$

In addition, when the stability condition $\mathbf{H(G, P)}$ stated on page 371 is satisfied for some finite constants v_1, v_2, by Proposition 12.2.4 we have the uniform estimates

$$\mathrm{osc}\left(\overline{G}_{p,n}\right) \le \exp v_1 - 1 \tag{16.5}$$

From the above recursions, we also easily prove the following decompositions that reflect in some sense the complexity of unnormalized particle measures

$$\overline{W}_n^{\gamma,N}(1) = \sum_{k=0}^{n} \frac{1}{N^{k/2}} \sum_{0 \le p_0 < \ldots < p_k < n} V_{p_0}^N(\overline{G}_{p_0,p_1}) \cdots V_{p_k}^N(\overline{G}_{p_k,n})$$

The main result of this section is the following theorem.

Theorem 16.4.1 *We assume that the local variance parameters σ_n defined in (9.31) are uniformly bounded, and we set $\sigma := \sup_{n \ge 0} \sigma_n < \infty$. For any $N \ge 1$, and for any $n \ge 0$, we have*

$$N \, \mathbb{E}\left(\left[\frac{\gamma_n^N(1)}{\gamma_n(1)} - 1\right]^2\right) \le \sum_{0 \le k \le n} \frac{\sigma^{2(k+1)}}{N^k} \, d_{k,n}$$

with

$$d_{k,n} = \sum_{0 \le p_0 < \ldots < p_k < n} \left(\mathrm{osc}(\overline{G}_{p_0,p_1}) \ldots \mathrm{osc}(\overline{G}_{p_k,n})\right)^2$$

In addition, when the stability condition $\mathbf{H}(\mathbf{G}, \mathbf{P})$ *stated on page 371 is satisfied for some finite constants* υ_1, υ_2, *we have the uniform variance estimates*

$$N \, \mathbb{E}\left(\left[\frac{\gamma_n^N(1)}{\gamma_n(1)} - 1\right]^2\right) \leq c \, n \, (1 + c/N)^n$$

with some constant $c \leq 4\sigma^2 \left[\exp \upsilon_1 - 1\right]^2$.

Using the fact that $\log(1 + x) \leq x$, for any $x \geq 0$, and $e^x - 1 \leq x(1 + e)$, for any $x \in [0, 1]$, we prove that

$$N \geq n \, c \Rightarrow (1 + c/N)^n \leq e^{nc/N} \leq 1 + n \, c \, (1 + e)/N \leq 2 + e$$

from which we conclude that

$$N \geq n \, c \Rightarrow N \, \mathbb{E}\left(\left[\frac{\gamma_n^N(1)}{\gamma_n(1)} - 1\right]^2\right) \leq c \, n \, (1 + c \, (1 + e) \, n/N)$$

and therefore

$$N \geq n \, c \Rightarrow N \, \mathbb{E}\left(\left[\frac{\gamma_n^N(1)}{\gamma_n(1)} - 1\right]^2\right) \leq c \, (2 + e) \, n \qquad (16.6)$$

The proof of the theorem is based on the following technical lemma. This result is rather well known, and its proof by induction on the parameter n is rather elementary; thus it is omitted.

Lemma 16.4.2 *We consider a triangular array of nonnegative parameters* $(b_{p,n})_{0 \leq p \leq n}$, *with* $n \geq 0$ *and a sequence of nonnegative parameters* $(a_n)_{n \geq 0}$. *We let* $(I_n)_{n \geq 0}$ *be a sequence of nonnegative numbers satisfying for any* $n \geq 0$ *the following recursions*

$$I_n \leq a_n + \sum_{0 \leq p < n} I_p \, b_{p,n}$$

In this situation, for any $n \geq 0$ *we have*

$$I_n \leq a_n + \sum_{1 \leq k \leq n} \sum_{0 \leq p_1 < \ldots < p_k < n} a_{p_1} b_{p_1, p_2} b_{p_2, p_3} \ldots b_{p_k, n}$$

For parameters $a_p \leq a_n$ *and* $b_{p,n} \leq b_n$, *for some* b_n *and any* $p \leq q \leq n$, *we have*

$$I_n \leq a_n \, (1 + b_n)^n$$

Now we come to the proof of the theorem.
Proof of Theorem 16.4.1:
Using the fact that $\left(1 + \frac{1}{\sqrt{N}} \, \overline{W}_p^{\gamma, N}(1)\right)$ are measurable w.r.t. $\mathcal{G}_{p-1}^N$ and

$$\mathbb{E}\left(V_p^N(f_1) V_q^N(f_2) \mid \mathcal{F}_{q-1}^N\right) = V_p^N(f_1) \, \mathbb{E}\left(V_q^N(f_2) \mid \mathcal{F}_{q-1}^N\right) = 0$$

for any $0 \leq p < q \leq n$ and any functions f_1, f_2 on E we prove that

$$
\mathbb{E}\left(\overline{W}_n^{\gamma,N}(1)^2\right) = \sum_{0 \leq p < n} \mathbb{E}\left(\left(1 + \frac{1}{\sqrt{N}}\,\overline{W}_p^{\gamma,N}(1)\right)^2 V_p^N(\overline{G}_{p,n})^2\right)
$$

$$
\leq \sigma^2 \sum_{0 \leq p < n} \mathrm{osc}(\overline{G}_{p,n})^2\,\mathbb{E}\left(\left(1 + \frac{1}{\sqrt{N}}\,\overline{W}_p^{\gamma,N}(1)\right)^2\right)
$$

If we set

$$
I_n^N := \mathbb{E}\left(\overline{W}_n^{\gamma,N}(1)^2\right)
$$

then we find that

$$
I_n^N \leq \sigma^2 \sum_{0 \leq p < n} \mathrm{osc}(\overline{G}_{p,n})^2\left(1 + \frac{1}{N}\,I_p^N\right) = a_n + \sum_{0 \leq p < n} I_p\,b_{p,n}^N
$$

with

$$
a_n := \sigma^2 \sum_{0 \leq p < n} \mathrm{osc}(\overline{G}_{p,n})^2 \quad \text{and} \quad b_{p,n}^N := \sigma^2\,\mathrm{osc}(\overline{G}_{p,n})^2/N
$$

The proof of the first assertion is now a consequence of Lemma 16.4.2. When the condition $\mathbf{H(G,P)}$ is met the same equations hold true with

$$
a_n = (2\,[\exp \upsilon_1 - 1]\,\sigma)^2 n \quad \text{and} \quad b_{p,n}^N := (2\,[\exp \upsilon_1 - 1]\,\sigma)^2/N
$$

The end of the proof of the theorem is again a direct consequence of Lemma 16.4.2. This ends the proof of the theorem. ∎

16.5 $\mathbb{L}_p$-mean error estimates

The main objective of this section is to extend the variance theorem to nonasymptotic $\mathbb{L}_m$-mean errors, for any $m \geq 2$. We have not really succeeded since our quantitative estimates do not depend on the variance parameter σ.

Theorem 16.5.1 *For any $n \geq 0$ and any $r \geq 2$, we have*

$$
\mathbb{E}\left(\left[\frac{\gamma_n^N(1)}{\gamma_n(1)} - 1\right]^r\right)^{1/r} \leq c(r)\,\left(\frac{n}{N}\right)^{1/2}\left(\frac{1}{n}\sum_{0 \leq k \leq n}\left(\frac{c(r)}{N^{1/2}}\right)^{rk} d_{k,n}^{(r)}\right)^{1/r}
$$

with the constant

$$
c(r) = 2^{1-1/r}\,(r-1)b(r)
$$

and some sequence of parameters

$$d_{k,n}^{(r)} \le \sum_{0 \le p_0 < \ldots < p_k < n} (p_1 p_2 \ldots p_k)^{r/2-1} \operatorname{osc}(\overline{G}_{p_0,p_1})^r \cdots \operatorname{osc}(\overline{G}_{p_k,n})^r$$

In addition, when the stability condition $\mathbf{H(G,P)}$ *stated on page 371 is satisfied for some finite constants* υ_1, υ_2, *we have*

$$\mathbb{E}\left(\left[\frac{\gamma_n^N(1)}{\gamma_n(1)} - 1\right]^r\right)^{1/r} \le c'(r) \left(\frac{n}{N}\right)^{1/2} \left(1 + \frac{1}{n}\left(\frac{n}{N}\right)^{r/2} c'(r)^r\right)^{n/r}$$

with some constant $c'(r) \le c(r)[\exp \upsilon_1 - 1]$.

Proof:
We use the decomposition

$$\overline{W}_n^{\gamma,N}(1) = M_{n-1}^{N,1} + \frac{1}{\sqrt{N}} M_{n-1}^{N,2}$$

with the sequence of random variables $(M_q^{N,i})_{0 \le q < n}$, $i = 1, 2$, defined by

$$M_q^{N,1} := \sum_{0 \le p \le q} V_p^N(\overline{G}_{p,n}) \quad \text{and} \quad M_q^{N,2} := \sum_{0 \le p \le q} \overline{W}_p^{\gamma,N}(1) \, V_p^N(\overline{G}_{p,n})$$

Notice that

$$\mathbb{E}\left(M_q^{N,1} | \mathcal{F}_{q-1}^N\right) = M_{q-1}^{N,1} + \mathbb{E}\left(V_q^N(\overline{G}_{q,n}) | \mathcal{F}_{q-1}^N\right) = M_{q-1}^{N,1}$$

and

$$\mathbb{E}\left(M_q^{N,2} | \mathcal{F}_{q-1}^N\right) = M_{q-1}^{N,2} + \overline{W}_q^{\gamma,N}(1) \, \mathbb{E}\left(V_q^N(\overline{G}_{q,n}) | \mathcal{F}_{q-1}^N\right) = M_{q-1}^{N,2}$$

This implies that $(M_q^{N,i})_{0 \le q < n}$, $i = 1, 2$, are martingales w.r.t. the filtration $(\mathcal{F}_q^N)_{0 \le q < n}$.

By the Burkholder-Davis-Gundy inequalities, for any $r > 1$ we have

$$\mathbb{E}\left(|M_q^{N,1}|^r\right)^{1/r} \le \alpha(r) \, \mathbb{E}\left(\left(\sum_{p=0}^q V_p^N(\overline{G}_{p,n})^2\right)^{r/2}\right)^{1/r}$$

with $\alpha(r) = \max(r-1, 1/r-1)$ (see for instance [458]).

On the other hand, for $r \ge 2$ we have $\alpha(r) = (r-1)$ and

$$\mathbb{E}\left[\left(\frac{1}{q+1}\sum_{p=0}^q V_p^N(\overline{G}_{p,n})^2\right)^{r/2}\right] \le \frac{1}{q+1}\sum_{p=0}^q \mathbb{E}\left(|V_p^N(\overline{G}_{p,n})|^r\right)$$

$$\le b(r)^r \frac{1}{q+1}\sum_{p=0}^q \operatorname{osc}(\overline{G}_{p,n})^r$$

from which we conclude that

$$\mathbb{E}\left(\left|M_{n-1}^{N,1}\right|^r\right)^{1/r} \leq b(r)(r-1)n^{1/2}\left[\frac{1}{n}\sum_{0\leq q<n}\mathrm{osc}(\overline{G}_{p,n})^r\right]^{1/r}$$

In much the same way, for any $r \geq 2$ we have

$$\mathbb{E}\left(\left|M_q^{N,2}\right|^r\right)^{1/r} \leq (r-1)\,\mathbb{E}\left(\left(\sum_{p=0}^{q}V_p^N(\overline{G}_{p,n})^2\overline{W}_p^{\gamma,N}(1)^2\right)^{r/2}\right)^{1/r}$$

and

$$\mathbb{E}\left[\left(\frac{1}{q+1}\sum_{p=0}^{q}V_p^N(\overline{G}_{p,n})^2\overline{W}_p^{\gamma,N}(1)\right)^{r/2}\right]$$

$$\leq \frac{1}{q+1}\sum_{p=0}^{q}\mathbb{E}\left(\left|\overline{W}_p^{\gamma,N}(1)\right|^r\mathbb{E}\left(\left|V_p^N(\overline{G}_{p,n})\right|^r\,\left|\mathcal{G}_{p-1}^N\right.\right)\right)$$

$$\leq b(r)^r\,\frac{1}{q+1}\sum_{p=0}^{q}\mathrm{osc}(\overline{G}_{p,n})^r\mathbb{E}\left(\left|\overline{W}_p^{\gamma,N}(1)\right|^r\right)$$

from which we conclude that

$$\mathbb{E}\left(\left|M_{n-1}^{N,2}\right|^r\right)^{1/r} \leq b(r)(r-1)n^{1/2}\left[\frac{1}{n}\sum_{0\leq q<n}\mathrm{osc}(\overline{G}_{q,n})^r\mathbb{E}\left(\left|\overline{W}_q^{\gamma,N}(1)\right|^r\right)\right]^{1/r}$$

Using the fact $((a+b)/2)^r \leq (a^r + b^r)/2$, for any $a,b \geq 0$ and $r \geq 1$, we find that

$$I_n^N := \mathbb{E}\left(\left|\overline{W}_n^{\gamma,N}(1)\right|^r\right) \leq 2^{r-1}\left(\mathbb{E}\left(\left|M_{n-1}^{N,1}\right|^r\right) + \frac{1}{N^{r/2}}\,\mathbb{E}\left(\left|M_{n-1}^{N,2}\right|^r\right)\right)$$

and therefore

$$I_n^N \leq a_n + \sum_{0\leq p<n}b_{p,n}^N\,I_p^N$$

for any $r \geq 2$ with

$$a_n := c(r)^r n^{r/2-1}\sum_{0\leq q<n}\mathrm{osc}(\overline{G}_{p,n})^r$$

and

$$b_{p,n}^N := c(r)^r n^{r/2-1}\frac{1}{N^{r/2}}\,\mathrm{osc}(\overline{G}_{p,n})^r$$

The end of the proof of the first estimate is easily completed. Under condition $\mathbf{H}(\mathbf{G},\mathbf{P})$, using (16.5) we have

$$a_n \leq c(r)^r n^{r/2}(\exp v_1 - 1)^r$$

and for any $p \leq q \leq n$

$$b_{p,q}^N := c(r)^r n^{r/2-1} \frac{1}{N^{r/2}} (\exp \upsilon_1 - 1)^r$$

This implies that

$$I_n^N \leq n^{r/2}(c(r) \exp \upsilon_1 - 1)^r \left(1 + \frac{1}{n} \left(\frac{n}{N}\right)^{r/2} (c(r) \exp \upsilon_1 - 1)^r\right)^n$$

This ends the proof of the theorem. ∎

16.6 Concentration analysis

This section is concerned with the exponential deviation properties of the particle free energy models introduced in Section 16.1.1. We follow the same line of arguments as the ones we used in Section 14.5.1 dedicated to the concentration properties of particle density profiles. Our results are expressed in terms of the contraction parameters $\tau_{k,l}(n)$ and $\overline{\tau}_{k,l}(m)$ introduced in Definition 12.1.5, and in Corollary 12.2.7, and the functions (ϵ_0, ϵ_1) defined in (14.27). In Section 16.6.1 we present some preliminary first order decompositions. The main result of this section is stated in Section 16.6.

16.6.1 A preliminary first order decomposition

Combining the exponential formulae (16.3) with the expansions (16.2), we derive first order decompositions for the random sequence $\sqrt{N} \log \overline{\gamma}_n^N(1)$. These expansions will be expressed in terms of the random predictable functions defined below.

Definition 16.6.1 *We let $h_{q,n}^N$ be the random $\mathcal{G}_{q-1}^N$-measurable functions given by*

$$h_{q,n}^N := \sum_{q \leq p < n} d_{q,p}^N(\overline{G}_p)$$

with the functions $d_{q,p}^N(\overline{G}_p)$ given in Definition 14.2.1.

Lemma 16.6.2 *For any $n \geq 0$ and any $N \geq 1$, we have*

$$\sqrt{N} \log \overline{\gamma}_n^N(1) = \sum_{0 \leq q < n} V_q^N \left(h_{q,n}^N\right) + \frac{1}{\sqrt{N}} R_n^N \qquad (16.7)$$

with a second remainder order term R_n^N such that

$$\mathbb{E}\left(\left|R_n^N\right|^m\right)^{1/m} \leq b(2m)^2 r(n)$$

for any $m \geq 1$, with some constant

$$r(n) \leq 8 \sum_{0 \leq p < n} g_p \left(2g_p \tau_{1,1}(p)^2 + \tau_{3,1}(p)\right) \tag{16.8}$$

Proof:

Using the exponential formulae (16.3), we have

$$\sqrt{N}\,\log\overline{\gamma}_n^N(1) = \sqrt{N}\log\left(1 + \frac{1}{\sqrt{N}}\overline{W}_n^{\gamma,N}(1)\right)$$

$$= \sum_{0 \leq p < n}\int_0^1 \frac{W_p^{\eta,N}(\overline{G}_p)}{1 + \frac{t}{\sqrt{N}}\,W_p^{\eta,N}(\overline{G}_p)}\,dt$$

This implies that

$$\sqrt{N}\,\log\overline{\gamma}_n^N(1) = \sum_{0 \leq p < n} W_p^{\eta,N}(\overline{G}_p) + \frac{1}{\sqrt{N}}R_n^{N,1}$$

with the (negative) second order remainder term

$$R_n^{N,1} = -\sum_{0 \leq p < n}\int_0^1 t\,\frac{W_p^{\eta,N}(\overline{G}_p)^2}{1 + \frac{t}{\sqrt{N}}\,W_p^{\eta,N}(\overline{G}_p)}\,dt$$

On the other hand, using (16.2) we have

$$\sum_{0 \leq p < n} W_p^{\eta,N}(\overline{G}_p) = \sum_{0 \leq q < n} V_q^N\left(h_{q,n}^N\right) + \frac{1}{\sqrt{N}}R_n^{N,2}$$

with the second order remainder order term

$$R_n^{N,2} := -\sum_{0 \leq q \leq p < n}\frac{1}{\eta_q^N(G_{q,p}^N)}\,V_q^N(G_{q,p}^N)\,V_q^N\left[d_{q,p}^N(\overline{G}_p)\right]$$

This gives the decomposition (16.7), with the second remainder order term

$$R_n^N := R_n^{N,1} + R_n^{N,2}$$

Using the fact that

$$1 + \frac{t}{\sqrt{N}}\,W_p^{\eta,N}(\overline{G}_p) = t\,\eta_p^N(\overline{G}_p) + (1-t) \geq t\,g_p^-$$

for any $t \in]0,1]$, with $g_p^- := \inf_x \overline{G}_p(x)$, we find that

$$|R_n^{N,1}| \leq \sum_{0 \leq p < n} \frac{1}{g_p^-} W_p^{\eta,N}(\overline{G}_p)^2$$

Using (14.17), we prove that

$$\mathbb{E}\left(|R_n^{N,1}|^r\right)^{1/r} \leq 4b(2r)^2 \sum_{0 \leq p < n} \frac{1}{g_p^-} \mathrm{osc}\left(\overline{G}_p\right)^2 \tau_{1,1}(p)^2$$

from which we conclude that

$$\mathbb{E}\left(|R_n^{N,1}|^r\right)^{1/r} \leq (4b(2r))^2 \sum_{0 \leq p < n} g_p^2 \tau_{1,1}(p)^2$$

In much the same way, we have

$$\mathbb{E}\left(|R_n^{N,2}|^r\right)^{1/r}$$

$$\leq \sum_{0 \leq q \leq p < n} g_{q,p} \; \mathbb{E}\left(\left|V_q^N(G_{q,p}^N)\right|^{2r}\right)^{1/2r} \mathbb{E}\left(\left|V_q^N\left[d_{q,p}^N(\overline{G}_p)\right]\right|^{2r}\right)^{1/2r}$$

and using (14.2), we prove that

$$\mathbb{E}\left(|R_n^{N,2}|^r\right)^{1/r} \leq 8b(2r)^2 \sum_{0 \leq q \leq p < n} g_p \, g_{q,p}^3 \, \beta(P_{q,p})$$

This ends the proof of the lemma. ∎

16.6.2 A concentration theorem

Using the first order decomposition presented in Section 16.6.1, we are now in position to state and to prove the following concentration theorem.

Theorem 16.6.3 *For any $N \geq 1$, $\epsilon \in \{+1, -1\}$, $n \geq 0$, and for any*

$$\varsigma_n^\star \geq \sup_{0 \leq q \leq n} \varsigma_{q,n} \quad \text{with} \quad \varsigma_{q,n} := \frac{4}{n} \sum_{q \leq p < n} g_{q,p} g_p \, \beta(P_{q,p})$$

the probability of the following events

$$\frac{\epsilon}{n} \log \overline{\gamma}_n^N(1) \leq \frac{1}{N} \, \overline{r}(n) \, \left(1 + \epsilon_0^{-1}(x)\right) + \varsigma_n^\star \, \overline{\sigma}_n^2 \, \epsilon_1^{-1}\left(\frac{x}{N\overline{\sigma}_n^2}\right) \qquad (16.9)$$

is greater than $1 - e^{-x}$, for any $x \geq 0$, with the parameters

$$\overline{\sigma}_n^2 := \sum_{0 \leq q < n} \sigma_q^2 \, (\varsigma_{q,n}/\varsigma_n^\star)^2 \quad \text{and} \quad \overline{r}(n) = r(n)/n$$

In the above display, $r(n)$ stands for the parameter defined in (16.8), and (ϵ_0, ϵ_1) the functions defined in (14.27).

Before getting into the proof of the theorem, we present simple arguments to derive exponential concentration inequalities for the quantities $\left|\overline{\gamma}_n^N(1) - 1\right|$. Suppose that for any $\epsilon \in \{+1, -1\}$, the probability of events

$$\frac{\epsilon}{n} \log \overline{\gamma}_n^N(1) \leq \rho_n^N(x)$$

is greater than $1 - e^{-x}$, for any $x \geq 0$, for some function ρ_n^N such that

$$\rho_n^N(x) \to_{N \to \infty} 0$$

In this case, the probability of event

$$-\left(1 - e^{-n\rho_n^N(x)}\right) \leq \overline{\gamma}_n^N(1) - 1 \leq e^{n\rho_n^N(x)} - 1$$

is greater than $1 - 2e^{-x}$, for any $x \geq 0$. Choosing N large enough so that $\rho_n^N(x) \leq 1/n$ we have

$$-2n\rho_n^N(x) \leq -\left(1 - e^{-n\rho_n^N(x)}\right) \quad \text{and} \quad e^{n\rho_n^N(x)} - 1 \leq 2n\rho_n^N(x)$$

from which we conclude that the probability of event

$$\mathbb{P}\left(\left|\overline{\gamma}_n^N(1) - 1\right| \leq 2n \, \rho_n^N(x)\right) \geq 1 - 2e^{-x}$$

Now, we come to the proof of the theorem.
Proof of Theorem 16.6.3:
We use the same reasoning as in Section 14.5.1. Firstly, we observe that

$$
\begin{aligned}
\left\|h_{q,n}^N\right\| &\leq \sum_{q \leq p < n} \left\|d_{q,p}^N(\overline{G}_p)\right\| \\
&\leq \sum_{q \leq p < n} g_{q,p} \, \mathrm{osc}(P_{q,p}(\overline{G}_p)) \leq 2 \sum_{q \leq p < n} g_{q,p} g_p \, \beta(P_{q,p}) = c_{q,n}/2
\end{aligned}
$$

and $\mathrm{osc}(h_{q,n}^N) \leq c_{q,n}$. Now, we use the following decompositions

$$\sum_{0 \leq q < n} V_q^N\left(h_{q,n}^N\right) = a_n^\star \sum_{0 \leq q < n} V_q^N\left(\delta_{q,n}^N\right)$$

with the $\mathcal{G}_{q-1}^N$-measurable functions

$$\delta_{q,n}^N = h_{q,n}^N/a_n^\star \in \mathrm{Osc}(E_q) \cap \mathcal{B}_1(E_q)$$

and for any constant $a_n^\star \geq \sup_{0 \leq q \leq n} c_{q,n}$.
On the other hand, we have the almost sure variance estimate

$$\mathbb{E}\left(V_q^N\left[\delta_{q,n}^N\right]^2 \big| \mathcal{G}_{q-1}^N\right) \leq \sigma_q^2 \, \mathrm{osc}(h_{q,n}^N)^2/a_n^{\star 2} \leq \sigma_q^2 c_{q,n}^2/a_n^{\star 2}$$

from which we conclude that

$$\mathbb{E}\left(V_q^N \left[\delta_{q,n}^N\right]^2\right) \le \sigma_q^2 c_{q,n}^2 / a_n^{\star 2}$$

This shows that the regularity condition stated in (11.37) is met by replacing the parameters σ_q in the variance formula (11.37) by the constants $\sigma_q c_{q,n}/a_n^\star$, with the uniform local variance parameters σ_p defined in (9.31).

The end of the proof is now a direct consequence of Theorem 11.8.6. This ends the proof of the theorem. ∎

Corollary 16.6.4 *Assume that the uniform local variance parameters defined in (9.31) are chosen so that $\sigma := \sup_{n\ge0}(\sigma_n)$ and the condition $\mathbf{H}(\mathbf{G},\mathbf{P})$ stated on page 371 is satisfied for some finite constants v_1, v_2. We set*

$$p(x) := \overline{r}\,\left(1 + 2(x + \sqrt{x})\right) + \varsigma^\star\, x/3 \quad \text{and} \quad q(x) = \sqrt{2\sigma^2 \overline{\tau}_{2,2} x}$$

with the parameters

$$\overline{r} := \varsigma^\star\, \overline{\tau}_{1,1}^2 + 2\sqrt{\varsigma^\star}\, \overline{\tau}_{3,1} \quad \text{and} \quad \varsigma^\star := (4 \exp v_1)^2$$

which stands for the finite parameters $\overline{\tau}_{k,l} < \infty$ introduced in Definition 12.1.5. In this situation, for any $N \ge 1$, and any $\epsilon \in \{+1, -1\}$, the probability of any of the following events

$$\frac{\epsilon}{n} \log \overline{\gamma}_n^N(1) \le \frac{1}{N}\, p(x) + \frac{1}{\sqrt{N}}\, q(x)$$

is greater than $1 - e^{-x}$, for any $x \ge 0$.

Proof:
Under condition $\mathbf{H}(\mathbf{G},\mathbf{P})$, using (12.10) we have $\overline{r}(n) = r(n)/n \le \overline{r}$, and for any $p < n$

$$\varsigma_{p,n}^2 = \left(\frac{4 \exp v_1}{n}\right)^2 (n-p)^2 \left(\frac{1}{n-p} \sum_{p \le q < n} g_{p,q}\, \beta(P_{p,q})\right)^2$$

$$\le \frac{(4 \exp v_1)^2 (n-p)}{n} \frac{}{n} \sum_{p \le q < n} g_{p,q}^2\, \beta(P_{p,q})^2$$

This implies that

$$\sum_{0 \le p < n} \varsigma_{p,n}^2 \le \frac{(4 \exp v_1)^2}{n} \sum_{0 \le q < n} \tau_{2,2}(q) \le (4 \exp v_1)^2 \overline{\tau}_{2,2}$$

and

$$\varsigma_{p,n} \le \varsigma^\star := (4 \exp v_1)^2$$

Using the upper bounds (14.28) we prove that the r.h.s. of (16.9) is bounded by

$$\frac{1}{N} \left(\overline{r} \ (1 + 2(x + \sqrt{x})) + \frac{x}{3} \ \varsigma^{\star} \right) + \frac{1}{\sqrt{N}} \ \sqrt{2\overline{\sigma}^2 x}$$

with

$$\overline{\sigma}_n^2 := \sigma^2 \left(\sum_{0 \le q < n} \varsigma_{q,n}^2 \right) / (\varsigma_n^{\star 2})^2 \le \sigma^2 \overline{\tau}_{2,2} := \overline{\sigma}^2$$

The end of the proof is now a consequence of the estimates (11.26) and (11.27). This ends the proof of the corollary.   ∎

Chapter 17

Backward particle Markov models

17.1 Description of the models

We consider a Feynman-Kac measure $\mathbb{Q}_n$ on path space associated with some collection of potential functions and Markov transitions (G_n, M_n) on some state measurable state spaces E_n. A more precise description of these models is provided in (1.37).

In the further development of this chapter, we further assume that the Markov transitions M_n are absolutely continuous with respect to some measures λ_n on E_n, and for any $(x, y) \in (E_{n-1} \times E_n)$ we have

$$H_n(x, y) := \frac{dM_n(x, \cdot)}{d\lambda_n}(y) > 0 \tag{17.1}$$

In this situation, we have

$$Q_n(x, dy) := G_{n-1}(x)\ M_n(x, dy) \ll \lambda_n(dy)$$

with the Radon-Nikodym derivative

$$\mathbb{H}_n(x, y) := \frac{dQ_n(x, \cdot)}{d\lambda_n}(y) = G_{n-1}(x)\ H_n(x, y)$$

We recall from Section 1.4.3.4 the following backward formula

$$\mathbb{Q}_n(d(x_0, \ldots, x_n)) = \eta_n(dx_n)\ \prod_{q=1}^{n} \mathbb{M}_{q, \eta_{q-1}}(x_q, dx_{q-1}) \tag{17.2}$$

with the collection of Markov transitions defined by

$$\mathbb{M}_{n+1, \eta_n}(x, dy) = \frac{\eta_n(dy)\ \mathbb{H}_{n+1}(y, x)}{\eta_n\left(\mathbb{H}_{n+1}(\cdot, x)\right)} \tag{17.3}$$

We consider the flow of particle density profiles $(\eta_n^N)_{n \geq 0}$ associated with a given McKean particle model. A detailed description of these particle models can be found in Section 4.4.3 and in Section 4.4.5.

We recall that the backward particle approximation of the distributions

$\mathbb{Q}_n$ associated with the flow of particle measures $(\eta_n^N)_{n \geq 0}$ is defined by the following formula

$$\mathbb{Q}_n^N(d(x_0, \ldots, x_n)) = \eta_n^N(dx_n) \prod_{q=1}^{n} \mathbb{M}_{q, \eta_{q-1}^N}(x_q, dx_{q-1}) \qquad (17.4)$$

Definition 17.1.1 *We consider the unnormalized particle measures*

$$\Gamma_n^N = \mathcal{Z}_n^N \times \mathbb{Q}_n^N \quad with \quad \mathcal{Z}_n^N = \prod_{0 \leq p < n} \eta_p^N(G_p)$$

$$\overline{\Gamma}_n^N = \frac{1}{\mathcal{Z}_n} \Gamma_n^N = \overline{\mathcal{Z}}_n^N \times \mathbb{Q}_n^N \quad with \quad \overline{\mathcal{Z}}_n^N = \frac{\mathcal{Z}_n^N}{\mathcal{Z}_n} = \prod_{0 \leq p < n} \eta_p^N(\overline{G}_p)$$

and the normalized potential functions $\overline{G}_n = G_n / \eta_n(G_n)$. *For any* $0 \leq p < q \leq n$, *we set*

$$E_{[p,n]} = (E_p \times \ldots \times E_n)$$

Sometimes we write $\boldsymbol{E_n}$ *instead of* $E_{[0,n]}$.

This section is concerned with the stochastic analysis of the backward Markov particle measures defined in (17.4). We consider the fluctuation random fields defined below.

Definition 17.1.2 *We let* $(W_n^{\mathbb{Q},N}, W_n^{\Gamma,N}, \overline{W}_n^{\Gamma,N})$ *be the sequence of random fields models defined by*

$$W_n^{\mathbb{Q},N} = \sqrt{N} \left(\mathbb{Q}_n^N - \mathbb{Q}_n \right)$$

$$W_n^{\Gamma,N} = \sqrt{N} \left(\Gamma_n^N - \Gamma_n \right) \quad and \quad \overline{W}_n^{\Gamma,N} = \sqrt{N} \left(\overline{\Gamma}_n^N - \mathbb{Q}_n \right)$$

The random measures $(\Gamma_n^N, \mathbb{Q}_n^N)$ can be interpreted as distributions on the complete genealogical tree model associated with the genetic type interpretation of the flow of measures $(\eta_n^N)_{n \geq 0}$. A more thorough discussion on these genealogical tree models is provided in Section 4.1.6.

The analysis of the fluctuation random fields of backward particle models $(\Gamma_n^N, \mathbb{Q}_n^N)$ is slightly more involved than the one of the genealogical tree particle models developed in Chapter 15. Here, the main difficulty is to deal with the nonlinear dependency of these backward particle Markov chain models with the flow of particle density profiles η_n^N.

In Section 17.2, we provide some preliminary key backward conditioning principles. We also introduce some predictable integral operators involved in the first order expansions of the fluctuation random fields discussed in Section 17.5. In Section 17.4, we illustrate these models in the context of additive functional models. In Section 17.9, we put together the semigroup techniques developed in earlier sections to derive a series of quantitative concentration inequalities.

17.2 Conditioning principles

Definition 17.2.1 *For any $p \leq n$, we denote by $\Gamma_{n|p}$ the integral operator from E_p into $E_{[p+1,n]}$ defined by*

$$\Gamma_{n|p}(x_p, d(x_{p+1}, \ldots, x_n)) = \prod_{p < q \leq n} Q_q(x_{q-1}, dx_q)$$

By construction, we have the backward recursive formula

$$\Gamma_{n|p}(x_p, d(y_{p+1}, \ldots, y_n)) \tag{17.5}$$
$$= Q_{p+1}(x_p, dy_{p+1}) \ \Gamma_{n|p+1}(x_{p+1}, d(y_{p+2}, \ldots, y_n))$$

Proposition 17.2.2 *For any $0 \leq p \leq n$, the $\mathbb{Q}_n$-conditional distribution of $(X_{p+1}, \ldots, X_n)$ given the random state $X_p = x_p$ is defined by the following integral operator*

$$\overline{\Gamma}_{n|p}(x_p, d(x_{p+1}, \ldots, x_n)) = \frac{1}{\Gamma_{n|p}(1)(x_p)} \ \Gamma_{n|p}(x_p, d(x_{p+1}, \ldots, x_n))$$

Proof:
By definition of the unnormalized Feynman-Kac measures Γ_n, we have

$$\Gamma_n(d(x_0, \ldots, x_n)) = \Gamma_p(d(x_0, \ldots, x_p)) \ \Gamma_{n|p}(x_p, d(x_{p+1}, \ldots, x_n))$$

This implies that

$$\mathbb{Q}_n(d(x_0, \ldots, x_n)) = \mathbb{Q}_{n,p}(d(x_0, \ldots, x_p)) \times \overline{\Gamma}_{n|p}(x_p, d(x_{p+1}, \ldots, x_n))$$

with the $\mathbb{Q}_n$-distribution of the random states $(X_0, \ldots, X_p)$

$$\mathbb{Q}_{n,p}(d(x_0, \ldots, x_p)) := \frac{1}{\eta_p(G_{p,n})} \ \mathbb{Q}_p(d(x_0, \ldots, x_p)) \ G_{p,n}(x_p)$$

This ends the proof of the proposition. ∎

Now, we discuss some backward conditioning principles.

Definition 17.2.3 *For any $n \geq 0$ and $N \geq 1$, we let $\mathbb{Q}_{n|n}$, respectively $\mathbb{Q}_{n|n}^N$, be the $\mathbb{Q}_n$, resp. $\mathbb{Q}_n^N$, conditional distribution of the variables $(x_0, \ldots, x_{n-1})$ given the terminal random state x_n, and defined by the backward Markov transitions*

$$\mathbb{Q}_{n|n}(x_n, d(x_0, \ldots, x_{n-1})) := \prod_{q=1}^{n} \mathbb{M}_{q, \eta_{q-1}}(x_q, dx_{q-1})$$

and

$$\mathbb{Q}_{n|n}^N(x_n, d(x_0, \ldots, x_{n-1})) := \prod_{q=1}^{n} \mathbb{M}_{q, \eta_{q-1}^N}(x_q, dx_{q-1}) \tag{17.6}$$

Using the backward Markov chain formulation (17.2), we have

$$\mathbb{Q}_n(d(x_0,\ldots,x_n)) = \eta_n(dx_n)\,\mathbb{Q}_{n|n}(x_n,d(x_0,\ldots,x_{n-1}))$$

and

$$\mathbb{Q}_n^N(d(x_0,\ldots,x_n)) = \eta_n^N(dx_n)\,\mathbb{Q}_{n|n}^N(x_n,d(x_0,\ldots,x_{n-1}))$$

We end this section with some properties of backward Markov transitions associated with a given initial probability measure that may differ from the one associated with the Feynman-Kac measures. These mathematical objects appear in a natural way in the analysis of the N-particle approximation transitions $\mathbb{Q}_n^N$.

Definition 17.2.4 *For any $0 \leq p \leq n$ and any probability measure $\eta \in \mathcal{P}(E_p)$, we denote by $\mathbb{M}_{n+1,p,\eta}$ the Markov transition from E_{n+1} into $E_{[p,n]}$ defined by*

$$\mathbb{M}_{n+1,p,\eta}\,(x_{n+1},d(x_p,\ldots,x_n)) := \prod_{p\leq q\leq n} \mathbb{M}_{q+1,\Phi_{p,q}(\eta)}(x_{q+1},dx_q)$$

For $\eta = \eta_p$, sometimes we write $\mathbb{M}_{n+1,p}$ instead of $\mathbb{M}_{n+1,p,\eta_p}$ the Markov transition

$$\mathbb{M}_{n+1,p}(x_{n+1},d(x_p,\ldots,x_n)) := \prod_{p\leq q\leq n} \mathbb{M}_{q+1,\eta_q}(x_{q+1},dx_q)$$

We observe that $\mathbb{M}_{n+1,p,\eta}$ can alternatively be defined by the pair of recursions

$$\mathbb{M}_{n+1,p,\eta}\,(x_{n+1},d(x_p,\ldots,x_n))$$

$$= \mathbb{M}_{n+1,p+1,\Phi_{p+1}(\eta)}\,(x_{n+1},d(x_{p+1},\ldots,x_n)) \times \mathbb{M}_{p+1,\eta}(x_{p+1},dx_p) \qquad (17.7)$$

$$= \mathbb{M}_{n+1,\Phi_{p,n}(\eta)}(x_{n+1},dx_n)\,\mathbb{M}_{n,p,\eta}\,(x_n,d(x_p,\ldots,x_{n-1}))$$

For $p = 0$, we also notice that

$$\mathbb{M}_{n,0,\eta}\,(x_n,d(x_0,\ldots,x_{n-1})) := \prod_{1\leq q\leq n} \mathbb{M}_{q,\Phi_{0,q-1}(\eta)}(x_q,dx_{q-1})$$

so that

$$\eta = \eta_0 \implies \mathbb{M}_{n,0,\eta_0} = \mathbb{Q}_{n|n} \quad \text{and} \quad \Phi_{0,n}(\eta_0) \otimes \mathbb{M}_{n,0,\eta_0} = \eta_n \otimes \mathbb{Q}_{n|n} = \mathbb{Q}_n$$

For more general initial distribution, recalling that

$$G_{0,n} = Q_{0,n}(1) \Rightarrow \eta Q_{0,n}(1) = \eta(G_{0,n})$$

we have that

$$\Phi_{0,n}(\eta)(dx_n) \, \mathbb{M}_{n,0,\eta}(x_n, d(x_0, \dots, x_{n-1}))$$

$$= \frac{1}{\eta Q_{0,n}(1)} \, \eta(dx_0) \, \Gamma_{n|0}(x_0, d(x_1, \dots, x_n))$$

$$= \Psi_{G_{0,n}}(\eta)(dx_0) \, \overline{\Gamma}_{n|0}(x_0, d(x_1, \dots, x_n))$$

The next lemma provides an alternative representation of the integral operators $\mathbb{M}_{n,p,\eta}$.

Lemma 17.2.5 *For any $0 \leq p < n$ and any probability measure $\eta \in \mathcal{P}(E_p)$, we have*

$$\eta Q_{p,n}(dx_n) \, \mathbb{M}_{n,p,\eta}(x_n, d(x_p, \dots, x_{n-1})) = \eta(dx_p) \, \Gamma_{n|p}(x_p, d(x_{p+1}, \dots, x_n))$$

In other words, we have

$$\Phi_{p,n}(\eta)(dx_n) \, \mathbb{M}_{n,p,\eta}(x_n, d(x_p, \dots, x_{n-1}))$$

$$= \Psi_{G_{p,n}}(\eta)(dx_p) \, \overline{\Gamma}_{n|p}(x_p, d(x_{p+1}, \dots, x_n))$$

as well as

$$\mathbb{M}_{n,p,\eta}(x_n, d(x_p, \dots, x_{n-1}))$$

$$= \frac{(\eta \times \Gamma_{n-1|p})(d(x_p, \dots, x_{n-1})) \, \mathbb{H}_n(x_{n-1}, x_n)}{(\eta Q_{p,n-1})(\, \mathbb{H}_n(\cdot, x_n))} \qquad (17.8)$$

$$\propto (\eta \times \overline{\Gamma}_{n-1|p})(d(x_p, \dots, x_{n-1})) \, Q_{p,n-1}(1)(x_{n-1}) \, \mathbb{H}_n(x_{n-1}, x_n)$$

Proof:
We prove the lemma by induction on the parameter $n(> p)$. For $n = p+1$, we have

$$\mathbb{M}_{p+1,p,\eta}(x_{p+1}, dx_p) = \mathbb{M}_{p+1,\eta}(x_{p+1}, dx_p)$$

and

$$\Gamma_{p+1|p}(x_p, dx_{p+1}) = Q_{p+1}(x_p, dx_{p+1})$$

By definition of the transitions $\mathbb{M}_{p+1,\eta}$, we have

$$\eta Q_{p+1}(dx_{p+1}) \, \mathbb{M}_{p+1,p,\eta}(x_{p+1}, dx_p) = \eta(dx_p) \, \Gamma_{p+1|p}(x_p, dx_{p+1})$$

We suppose that the result has been proved at rank n. In this situation, we notice that

$$\eta(dx_p) \, \Gamma_{n+1|p}(x_p, d(x_{p+1}, \dots, x_{n+1}))$$

$$= \eta(dx_p) \, \Gamma_{n|p}(x_p, d(x_{p+1}, \dots, x_n)) Q_{n+1}(x_n, dx_{n+1})$$

$$= \eta Q_{p,n}(dx_n) \, Q_{n+1}(x_n, dx_{n+1}) \, \mathbb{M}_{n,p,\eta}(x_n, d(x_p, \dots, x_{n-1}))$$

$$= \eta Q_{p,n}(1) \, \Phi_{p,n}(\eta)(dx_n) \, Q_{n+1}(x_n, dx_{n+1}) \, \mathbb{M}_{n,p,\eta}(x_n, d(x_p, \dots, x_{n-1}))$$

Using the fact that

$$\Phi_{p,n}(\eta)(dx_n)\, Q_{n+1}(x_n, dx_{n+1}) = \Phi_{p,n}(\eta)Q_{n+1}(dx_{n+1})\, \mathbb{M}_{n+1,\Phi_{p,n}(\eta)}(x_{n+1}, dx_n)$$

and

$$\eta Q_{p,n}(1)\, \Phi_{p,n}(\eta)Q_{n+1}(dx_{n+1}) = \eta Q_{p,n+1}(dx_{n+1})$$

we conclude that

$$\eta(dx_p)\, \Gamma_{n+1|p}(x_p, d(x_{p+1}, \ldots, x_{n+1}))$$

$$= \eta Q_{p,n+1}(dx_{n+1})\, \mathbb{M}_{n+1,\Phi_{p,n}(\eta)}(x_{n+1}, dx_n)\, \mathbb{M}_{n,p,\eta}(x_n, d(x_p, \ldots, x_{n-1}))$$

$$= \eta Q_{p,n+1}(dx_{n+1})\, \mathbb{M}_{n+1,p,\eta}(x_{n+1}, d(x_p, \ldots, x_n))$$

This ends the proof of the lemma. ∎

17.3 Integral transport properties

Definition 17.3.1 *For any $0 \le p \le n$ and $N \ge 1$, we denote by $D_{p,n}^N$ and $L_{p,n}^N$ the $\mathcal{G}_{p-1}^N$-measurable integral operators from $\mathcal{B}_b(\boldsymbol{E_n})$ into $\mathcal{B}_b(E_p)$ defined by*

$$D_{p,n}^N(x_p, d(y_0, \ldots, y_n))$$

$$\tag{17.9}$$

$$:= \mathbb{Q}_{p|p}^N(x_p, d(y_0, \ldots, y_{p-1}))\, \delta_{x_p}(dy_p)\, \Gamma_{n|p}(x_p, d(y_{p+1}, \ldots, y_n))$$

and the normalized operators

$$L_{p,n}^N(x_p, d(y_0, \ldots, y_n))$$

$$:= \mathbb{Q}_{p|p}^N(x_p, d(y_0, \ldots, y_{p-1}))\, \delta_{x_p}(dy_p)\, \overline{\Gamma}_{n|p}(x_p, d(y_{p+1}, \ldots, y_n))$$

For $p \in \{0, n\}$, we use the conventions

$$D_{n,n}^N(x_n, d(y_0, \ldots, y_n)) = L_{n,n}^N(x_n, d(y_0, \ldots, y_n))$$
$$= \mathbb{Q}_{n|n}^N(x_n, d(y_0, \ldots, y_{n-1}))\, \delta_{x_n}(dy_n)$$

and

$$D_{0,n}^N(x_0, d(y_0, \ldots, y_n)) = \delta_{x_0}(dy_0)\, \Gamma_{n|0}(x_0, d(y_1, \ldots, y_n))$$
$$L_{0,n}^N(x_0, d(y_0, \ldots, y_n)) = \delta_{x_0}(dy_0)\, \overline{\Gamma}_{n|0}(x_0, d(y_1, \ldots, y_n))$$

By construction, for any $p \le n$ we notice that

$$D_{p,n}^N(1) = Q_{p,n}(1) = G_{p,n} \quad \text{and} \quad L_{p,n}^N(1) = 1 \tag{17.10}$$

Definition 17.3.2 *By construction, for any bounded function $\boldsymbol{f_n}$ on the path space $\boldsymbol{E_n}$, we have some almost sure estimate*

$$\sup_{N \geq 1} \operatorname{osc}\left(L_{p,n}^N(\boldsymbol{f_n})\right) \leq l_{p,n}(\boldsymbol{f_n}) \leq \operatorname{osc}(\boldsymbol{f_n}) \tag{17.11}$$

for some finite constant $l_{p,n}(\boldsymbol{f_n})$, whose values only depend on the time parameters $p \leq n$ and on the function $\boldsymbol{f_n}$. We also consider the parameters

$$\mathbf{a(n)} := \sum_{p=0}^{n} g_{p,n}\, l_{p,n} \quad \text{with} \quad l_{p,n} := \sup_{\boldsymbol{f_n} \in \operatorname{Osc}(\boldsymbol{E_n})} l_{p,n}(\boldsymbol{f_n}) \tag{17.12}$$

and the path space $\boldsymbol{E_n} := (E_0 \times \ldots \times E_n)$.

The main reason for introducing these integral operators comes from the following integral transport properties.

Lemma 17.3.3 *For any $0 \leq p \leq n$, and any $N \geq 1$, and any function $\boldsymbol{f_n}$ on the path space $\mathbf{E}_n$, we have the almost sure formulae*

$$\eta_p^N D_{p,n}^N = \left[\eta_p^N Q_{p+1}\right] D_{p+1,n}^N = \eta_p^N(G_p) \times \Phi_{p+1}(\eta_p^N) D_{p+1,n}^N \tag{17.13}$$

and

$$\frac{\eta_p^N D_{p,n}^N(\boldsymbol{f_n})}{\eta_p^N D_{p,n}^N(1)} = \Psi_{G_{p,n}}\left(\eta_p^N\right) L_{p,n}^N(\boldsymbol{f_n}) \tag{17.14}$$

$$= \Psi_{G_{p+1,n}}\left(\Phi_{p+1}(\eta_p^N)\right) L_{p+1,n}^N(\boldsymbol{f_n}) \tag{17.15}$$

Proof:

We check (17.13) using (17.5) and

$$\eta_p^N(dx_p)Q_{p+1}(x_p, dy_{p+1}) = \left[\eta_p^N Q_{p+1}\right](dy_{p+1}) \times \mathbb{M}_{p+1,\eta_p^N}(y_{p+1}, dx_p) \tag{17.16}$$

More precisely, we have

$$\eta_p^N(dx_p)D_{p,n}^N(x_p, d(y_0, \ldots, y_n))$$

$$:= \eta_p^N(dx_p)Q_{p+1}(x_p, dy_{p+1})\mathbb{Q}_{p|p}^N(x_p, d(y_0, \ldots, y_{p-1}))$$

$$\times \delta_{x_p}(dy_p)\ \Gamma_{n|p+1}(y_{p+1}, d(y_{p+2}, \ldots, y_n))$$

Using (17.16), this implies that

$$\eta_p^N(dx_p)D_{p,n}^N(x_p, d(y_0, \ldots, y_n))$$

$$:= \eta_p^N Q_{p+1}(dy_{p+1})\ \mathbb{M}_{p+1,\eta_p^N}(y_{p+1}, dx_p)\mathbb{Q}_{p|p}^N(x_p, d(y_0, \ldots, y_{p-1}))$$

$$\times \delta_{x_p}(dy_p)\ \Gamma_{n|p+1}(y_{p+1}, d(y_{p+2}, \ldots, y_n))$$

from which we conclude that

$$\eta_p^N D_{p,n}^N(\boldsymbol{f_n})$$

$$:= \int \; [\eta_p^N Q_{p+1}] \, (dy_{p+1}) \; \mathbb{Q}_{p+1|p+1}^N(y_{p+1}, d(y_0, \dots, y_p))$$

$$\times \; \Gamma_{n|p+1}(y_{p+1}, d(y_{p+2}, \dots, y_n)) \; \boldsymbol{f_n}\,(y_0, \dots, y_n)$$

$$= \left([\eta_p^N Q_{p+1}] \, D_{p+1,n}^N\right)(\boldsymbol{f_n})$$

This ends the proof of the first assertion. Now, using (17.13) we have

$$\frac{\eta_p^N D_{p,n}^N(\boldsymbol{f_n})}{\eta_p^N D_{p,n}^N(1)} = \frac{\Phi_{p+1}\left(\eta_p^N\right) D_{p+1,n}^N(\boldsymbol{f_n})}{\Phi_{p+1}\left(\eta_p^N\right) D_{p+1,n}^N(1)}$$

Using (17.10), we readily prove (17.14) and (17.15). This ends the proof of the lemma. ∎

We end this section with a key decomposition lemma and an estimate of the deviation of the functions $D_{p,n}^N$ around their limiting values $D_{p,n}$.

Lemma 17.3.4 *For any $N \geq 1$ and any $n \geq 0$, we have the decomposition*

$$\left[\mathbb{Q}_{n|n}^N - \mathbb{Q}_{n|n}\right](x_n, d(x_0, \dots, x_{n-1}))$$

$$= \sum_{0 \leq p \leq n} \left[\mathbb{M}_{n,p,\eta_p^N} - \mathbb{M}_{n,p,\Phi_p(\eta_{p-1}^N)}\right](x_n, d(x_p, \dots, x_{n-1}))$$

$$\times \mathbb{Q}_{p|p}^N(x_p, d(x_0, \dots, x_{p-1}))$$

Proof:

Using the recursions (17.7), we prove that

$$\mathbb{M}_{n+1,p,\eta_p^N}\,(x_{n+1}, d(x_p, \dots, x_n))$$

$$= \mathbb{M}_{n+1,p+1,\Phi_{p+1}(\eta_p^N)}\,(x_{n+1}, d(x_{p+1}, \dots, x_n)) \times \mathbb{M}_{p+1,\eta_p^N}\,(x_{p+1}, dx_p)$$

On the other hand, we also have

$$\mathbb{Q}_{p+1|p+1}^N(x_{p+1}, d(x_0, \dots, x_p)) = \mathbb{M}_{p+1,\eta_p^N}\,(x_{p+1}, dx_p) \, \mathbb{Q}_{p|p}^N(x_p, d(x_0, \dots, x_{p-1}))$$

from which we conclude that

$$\mathbb{M}_{n+1,p+1,\Phi_{p+1}(\eta_p^N)}\,(x_{n+1}, d(x_{p+1}, \dots, x_n)) \, \mathbb{Q}_{p+1|p+1}^N(x_{p+1}, d(x_0, \dots, x_p))$$

$$= \mathbb{M}_{n+1,p,\eta_p^N}\,(x_{n+1}, d(x_p, \dots, x_n)) \, \mathbb{Q}_{p|p}^N(x_p, d(x_0, \dots, x_{p-1}))$$

The end of the proof is now a direct consequence of the following decomposition

$$\mathbb{Q}_{n|n}^N(x_n, d(x_0, \ldots, x_{n-1})) - \mathbb{Q}_{n|n}(x_n, d(x_0, \ldots, x_{n-1}))$$

$$= \sum_{1 \le p \le n} \Big[\mathbb{M}_{n,p,\eta_p^N}(x_n, d(x_p, \ldots, x_{n-1})) \, \mathbb{Q}_{p|p}^N(x_p, d(x_0, \ldots, x_{p-1}))$$

$$-\mathbb{M}_{n,p-1,\eta_{p-1}^N}(x_n, d(x_{p-1}, \ldots, x_{n-1})) \, \mathbb{Q}_{p-1|p-1}^N(x_{p-1}, d(x_0, \ldots, x_{p-2})) \Big]$$

$$+\mathbb{M}_{n,0,\eta_0^N}(x_n, d(x_0, \ldots, x_{n-1})) - \mathbb{M}_{n,0,\eta_0}(x_n, d(x_0, \ldots, x_{n-1}))$$

with the conventions

$$\mathbb{M}_{n,0,\eta_0^N}(x_n, d(x_0, \ldots, x_{n-1})) \, \mathbb{Q}_{0|0}^N(x_0, d(x_0, \ldots, x_1))$$

$$= \mathbb{M}_{n,0,\eta_0^N}(x_n, d(x_0, \ldots, x_{n-1}))$$

for $p = 0$, and for $p = n$

$$\mathbb{M}_{n,n,\eta_n^N}(x_n, d(x_n, \ldots, x_{n-1})) \, \mathbb{Q}_{n|n}^N(x_n, d(x_0, \ldots, x_{n-1}))$$

$$= \mathbb{Q}_{n|n}^N(x_n, d(x_0, \ldots, x_{n-1}))$$

This ends the proof of the lemma. ∎

Theorem 17.3.5 *We assume that the following regularity condition is met for any $n \ge 1$ and for any pair of transition state $(x, y) \in E_{[n-1,n]}$*

$$(H_0^+) \qquad 0 < h_n^-(y) \le H_n(x, y) \le h_n^+(y) \qquad (17.17)$$

for some positive functions (h_n^-, h^+) on E_n.

In this situation, for any $N \ge 1$, $0 \le p \le n$, $x_n \in E_n$, $m \ge 1$, and any $\mathbf{f}_{n-1} \in \mathrm{Osc}(\mathbf{E}_{n-1})$, we have

$$\sqrt{N} \, \mathbb{E} \left(\left| \left[\mathbb{Q}_{n|n}^N - \mathbb{Q}_{n|n} \right] (\mathbf{f}_{n-1})(x_n) \right|^m \right)^{\frac{1}{m}}$$

$$(17.18)$$

$$\le b(m) \left(\sum_{0 \le p < n} g_{p,n-1} \right) (h_n^+/h_n^-)(x_n)$$

with the parameters $b(m)$ defined in (0.6).

Proof:
Using Lemma 17.3.4, for any function $\mathbf{f}_{n-1} \in \mathcal{B}_b(\mathbf{E}_{n-1})$ and any $x_n \in E_n$ we have

$$\left[\mathbb{Q}_{n|n}^N - \mathbb{Q}_{n|n} \right] (\mathbf{f}_{n-1})(x_n) = \sum_{0 \le p \le n} \left[\mathbb{M}_{n,p,\eta_p^N} - \mathbb{M}_{n,p,\Phi_p(\eta_{p-1}^N)} \right] \left(T_{p,n}^N(\mathbf{f}_{n-1}) \right)(x_n)$$

with the random $\mathcal{G}_{p-1}^N$-measurable functions

$$T_{p,n}^N(\mathbf{f_{n-1}})(x_p, \dots, x_{n-1})$$

$$= \int \mathbb{Q}_{p|p}^N(x_p, d(x_0, \dots, x_{p-1})) \, \mathbf{f_{n-1}}((x_0, \dots, x_{p-1}), (x_p \dots, x_{n-1}))$$

The end of the proof of (17.18) is a direct consequence of formula (17.8). This ends the proof of the theorem. ∎

Using the generalized Minkowski inequality (cf. for instance [192]), we prove the following corollary.

Corollary 17.3.6 *We assume that the regularity assumptions of Theorem 17.3.5 are satisfied with $(h_n^+/h^-) \in \mathbb{L}_1(\lambda_n)$. In this situation, for any $N \geq 1$, $0 \leq p \leq n$, $m \geq 1$, we have*

$$\sqrt{N} \, \sup_{\mathbf{f_{n-1}} \in \mathrm{Osc}(\mathbf{E_{n-1}})} \mathbb{E}\left(\left[\left\|\left[\mathbb{Q}_{n|n}^N - \mathbb{Q}_{n|n}\right](\mathbf{f_{n-1}})\right\|_{\mathbb{L}_1(\lambda_n)}\right]^m\right)^{1/m}$$

$$\leq b(m) \, \sigma(n)$$

with the parameters $b(m)$ defined in (0.6), and some constant

$$\sigma(n) \leq \left(\sum_{0 \leq p < n} g_{p,n-1}\right) \lambda_n(|h_n^+/h^-|)$$

The comparison lemma, Lemma 11.3.1 combined with (0.6), (11.5), and (11.21), yields the following exponential concentration inequality.

Corollary 17.3.7 *We assume that the regularity assumptions of Theorem 17.3.5 are satisfied with $(h_n^+/h^-) \in \mathbb{L}_1(\lambda_n)$. In this situation, for any $N \geq 1$, we have*

$$\pi_\psi\left(\sqrt{\frac{N}{\sigma^2(n)}} \left\|\left[\mathbb{Q}_{n|n}^N - \mathbb{Q}_{n|n}\right](\mathbf{f_{n-1}})\right\|_{\mathbb{L}_1(\lambda_n)}\right) \leq \sqrt{8/3}$$

In particular, for any $x \geq 0$ the probability of the event

$$\left\|\left[\mathbb{Q}_{n|n}^N - \mathbb{Q}_{n|n}\right](\mathbf{f_{n-1}})\right\|_{\mathbb{L}_1(\lambda_n)} \leq 2 \, \sigma(n) \sqrt{\frac{2(x + \log 2)}{3N}}$$

is greater than $1 - e^{-x}$.

17.4 Additive functional models

In this section we provide a brief discussion on the action of the operators $D_{p,n}^N$ and $L_{p,n}^N$ on additive linear functionals associated with some collection of functions $f_n \in \mathrm{Osc}(E_n)$ and given by

$$f_n(x_0, \ldots, x_n) = \sum_{p=0}^{n} f_p(x_p) \qquad (17.19)$$

For any additive functional f_n of the form (17.19), sometimes we denote by $\overline{f}_n$ the normalized additive functional

$$\overline{f}_n(x_0, \ldots, x_n) = \frac{1}{n+1} \sum_{p=0}^{n} f_p(x_p) \qquad (17.20)$$

By construction, we have

$$D_{p,n}^N(f_n) = Q_{p,n}(1) \left[\sum_{0 \le q < p} \left[\mathbb{M}_{p,\eta_{p-1}^N} \cdots \mathbb{M}_{q+1,\eta_q^N} \right] (f_q) + \sum_{p \le q \le n} R_{p,q}^{(n)}(f_q) \right]$$

with triangular array of Markov transitions $R_{p,q}^{(n)}$ introduced in Definition 12.2.2.

In much the same way by definition of normalized operators $L_{p,n}^N$, we also have that

$$L_{p,n}^N(f_n) = \sum_{0 \le q < p} \left[\mathbb{M}_{p,\eta_{p-1}^N} \cdots \mathbb{M}_{q+1,\eta_q^N} \right] (f_q) + \sum_{p \le q \le n} R_{p,q}^{(n)}(f_q)$$

Using the estimates (12.9), we prove the following upper bounds

$$\mathrm{osc}(L_{p,n}^N(f_n))$$

$$\le \sum_{0 \le q < p} \beta \left(\mathbb{M}_{p,\eta_{p-1}^N} \cdots \mathbb{M}_{q+1,\eta_q^N} \right) + \sum_{p \le q \le n} g_{q,n} \times \beta (P_{p,q})$$

There are many ways to control the Dobrushin operator norm of the product of the random matrices defined in (4.14). For instance, we can use the multiplicative formulae

$$\beta \left(\mathbb{M}_{p,\eta_{p-1}^N} \cdots \mathbb{M}_{q+1,\eta_q^N} \right) \le \prod_{p < k \le q} \beta \left(\mathbb{M}_{k+1,\eta_k^N} \right)$$

One of the simplest but rather crude ways to proceed is to assume that

$$H_n(x, y) \le \tau \, H_n(x, y') \qquad (17.21)$$

for any x, y, y' and for some finite constant $\tau < \infty$. In this situation, we find that

$$\mathbb{M}_{k+1,\eta_k^N}(y, dx) \leq \tau^2 \, \mathbb{M}_{k+1,\eta_k^N}(y', dx)$$

from which we conclude that

$$\beta\left(\mathbb{M}_{k+1,\eta_k^N}\right) \leq 1 - \tau^{-2}$$

We further assume that the condition $\mathbf{H}(\mathbf{G}, \mathbf{P})$ stated in Section 12.2.1 is met for some parameters (υ_1, υ_2). In this situation, using Proposition 12.2.4 we prove that

$$\mathrm{osc}\left(L_{p,n}^N(\boldsymbol{f_n})\right) \leq \sum_{0 \leq q < p} \left(1 - \tau^{-2}\right)^{(p-q)} + \upsilon_2 \, \exp \upsilon_1$$

from which we prove the following uniform estimates

$$\sup_{0 \leq p \leq n} \mathrm{osc}\left(L_{p,n}^N(\boldsymbol{f_n})\right) \leq \tau^2 + \upsilon_2 \, \exp \upsilon_1 \tag{17.22}$$

17.5   A stochastic perturbation analysis

As in Section 14.1.2, we develop a stochastic perturbation analysis that allows expressing $W_n^{\Gamma,N}$ and $W_n^{\mathbb{Q},N}$ in terms of the local sampling random fields $(V_p^N)_{0 \leq p \leq n}$.

These first order expansions presented will be expressed in terms of the first order functions $d_\nu \Psi_G(f)$ introduced in (3.14) and the random $\mathcal{G}_{p-1}^N$-measurable functions $G_{p,n}^N$ introduced in Definition 14.2.1.

Definition 17.5.1 *For any $N \geq 1$, any $0 \leq p \leq n$, and any bounded measurable function $\boldsymbol{f_n}$ on the path space $\boldsymbol{E_n}$, we let $\mathbf{d}_{\mathbf{p,n}}^{\mathbf{N}}(\boldsymbol{f_n})$ be the $\mathcal{G}_{p-1}^N$-measurable functions*

$$\mathbf{d}_{\mathbf{p,n}}^{\mathbf{N}}(\boldsymbol{f_n}) = d_{\Phi_p(\eta_{p-1}^N)} \Psi_{G_{p,n}} \left(L_{p,n}^N(\boldsymbol{f_n})\right)$$

We also denote by $\mathbf{d}_{\mathbf{p,n}}(\boldsymbol{f_n})$ the function

$$\mathbf{d}_{\mathbf{p,n}}(\boldsymbol{f_n}) = d_{\eta_p} \Psi_{G_{p,n}} \left(L_{p,n}(\boldsymbol{f_n})\right)$$

with the first order operators $d_{\eta_p} \Psi_{G_{p,n}}$ and $d_{\Phi_p(\eta_{p-1}^N)} \Psi_{G_{p,n}}$ from $\mathcal{B}(E_p)$ into itself defined in (3.14)

Using (17.11), we readily prove the following upper bounds

$$\left\| \mathbf{d}_{\mathbf{p,n}}^{\mathbf{N}}(\boldsymbol{f_n}) \right\| \leq \frac{\|G_{p,n}\|}{\Phi_p(\eta_{p-1}^N)(G_{p,n})} \, \mathrm{osc}\left(L_{p,n}^N(\boldsymbol{f_n})\right) \leq g_{p,n} \, l_{p,n}(\boldsymbol{f_n})$$

In addition, using Proposition 12.2.4, for any additive functional $\boldsymbol{f_n}$ of the form (17.19), we have

$$\left\|d_{p,n}^{\mathbf{N}}(\boldsymbol{f_n})\right\| \leq \exp(v_1)\left(\tau^2 + v_2 \exp v_1\right)$$

and

$$\mathbf{a(n)} \leq \exp(v_1)\left(\tau^2 + v_2 \exp v_1\right)(n+1)$$

as soon as the regularity conditions (17.21) and the condition $\mathbf{H(G,P)}$ stated in Section 12.2.1 are met for some finite parameters τ and (v_1, v_2).

We are now in position to state and to prove the following decomposition theorem.

Theorem 17.5.2 *For any $0 \leq p \leq n$, and any function $\boldsymbol{f_n}$ on the path space E^{n+1}, we have*

$$\mathbb{E}\left(\Gamma_n^N(\boldsymbol{f_n})\,|\mathcal{G}_p^N\right) = \gamma_p^N\left(D_{p,n}^N(\boldsymbol{f_n})\right) \tag{17.23}$$

In addition, we have

$$W_n^{\Gamma,N}(\boldsymbol{f_n}) = \sum_{p=0}^{n} \gamma_p^N(1)\, V_p^N\left(D_{p,n}^N(\boldsymbol{f_n})\right) \tag{17.24}$$

$$W_n^{Q,N}(\boldsymbol{f_n}) = \sum_{p=0}^{n} \frac{1}{\eta_p^N(G_{p,n}^N)} V_p^N\left(d_{p,n}^{\mathbf{N}}(\boldsymbol{f_n})\right) \tag{17.25}$$

$$= \sum_{p=0}^{n} V_p^N\left(d_{p,n}^{\mathbf{N}}(\boldsymbol{f_n})\right)$$

$$- \sum_{p=0}^{n} \frac{1}{\eta_p^N(G_{p,n}^N)} \frac{1}{\sqrt{N}}\, V_p^N(G_{p,n}^N) \times V_p^N\left(d_{p,n}^{\mathbf{N}}(\boldsymbol{f_n})\right) \tag{17.26}$$

Proof:

To prove the first assertion, we use a backward induction on the parameter p. For $p = n$, the result is immediate since we have

$$\Gamma_n^N(\boldsymbol{f_n}) = \gamma_n^N(1)\, \eta_n^N\left(D_{n,n}^N(\boldsymbol{f_n})\right)$$

We suppose that the formula is valid at a given rank $p \leq n$. In this situation, using the fact that $D_{p,n}^N(\boldsymbol{f_n})$ is a $\mathcal{G}_{p-1}^N$-measurable function, we prove that

$$\mathbb{E}\left(\Gamma_n^N(\boldsymbol{f_n})\,|\,\mathcal{G}_{p-1}^N\right)$$

$$= \mathbb{E}\left(\gamma_p^N\left(D_{p,n}^N(\boldsymbol{f_n})\right)\,|\,\mathcal{G}_{p-1}^N\right) = (\gamma_{p-1}^N Q_p)D_{p,n}^N(\boldsymbol{f_n}) \tag{17.27}$$

Applying (17.13), we also have that

$$\gamma_{p-1}^N Q_p D_{p,n}^N = \gamma_{p-1}^N D_{p-1,n}^N$$

from which we conclude that the desired formula is satisfied at rank $(p-1)$. This ends the proof of the first assertion.

Now, combining Lemma 17.3.3 and (17.23), the proof of the second assertion is simply based on the following decomposition

$$\left(\Gamma_n^N - \Gamma_n\right)(\boldsymbol{f_n})$$

$$= \sum_{p=0}^n \left[\mathbb{E}\left(\Gamma_n^N(\boldsymbol{f_n}) \,\middle|\, \mathcal{G}_p^N\right) - \mathbb{E}\left(\Gamma_n^N(\boldsymbol{f_n}) \,\middle|\, \mathcal{G}_{p-1}^N\right)\right]$$

$$= \sum_{p=0}^n \gamma_p^N(1) \left(\eta_p^N\left(D_{p,n}^N(\boldsymbol{f_n})\right) - \frac{1}{\eta_{p-1}^N(G_{p-1})} \, \eta_{p-1}^N\left(D_{p-1,n}^N(\boldsymbol{f_n})\right)\right)$$

To prove the final decomposition, we use the fact that

$$[\mathbb{Q}_n^N - \mathbb{Q}_n](\boldsymbol{f_n}) = \sum_{0 \le p \le n} \left(\frac{\eta_p^N D_{p,n}^N(\boldsymbol{f_n})}{\eta_p^N D_{p,n}^N(1)} - \frac{\eta_{p-1}^N D_{p-1,n}^N(\boldsymbol{f_n})}{\eta_{p-1}^N D_{p-1,n}^N(1)}\right)$$

with the conventions $\eta_{-1}^N D_{-1,n}^N = \eta_0 \Gamma_{n|0}$, for $p = 0$.

Finally, we use (17.14) and (17.15) to check that

$$[\mathbb{Q}_n^N - \mathbb{Q}_n] = \sum_{0 \le p \le n} \left(\Psi_{G_{p,n}}\left(\eta_p^N\right) - \Psi_{G_{p,n}}\left(\Phi_p(\eta_{p-1}^N)\right)\right) L_{p,n}^N$$

We end the proof using the first order expansions of the Boltzmann-Gibbs transformation developed in Section 3.16

$$\sqrt{N} \left(\Psi_{G_{p,n}}\left(\eta_p^N\right) - \Psi_{G_{p,n}}\left(\Phi_p(\eta_{p-1}^N)\right)\right) L_{p,n}^N(\boldsymbol{f_n})$$

$$= \frac{1}{\eta_p^N(G_{p,n}^N)} \, V_p^N\left(\mathbf{d_{p,n}^N}(\boldsymbol{f_n})\right)$$

$$= V_p^N\left(\mathbf{d_{p,n}^N}(\boldsymbol{f_n})\right) - \frac{1}{\eta_p^N(G_{p,n}^N)} \, \frac{1}{\sqrt{N}} \, V_p^N(G_{p,n}^N) \times V_p^N\left(\mathbf{d_{p,n}^N}(\boldsymbol{f_n})\right)$$

This ends the proof of the theorem. ∎

Lemma 17.5.3 *For any $N \ge 1$, $n \ge 0$, and any bounded function $\boldsymbol{f_n}$ on the path space $\boldsymbol{E_n}$, we have the first order decomposition*

$$W_n^{\mathbb{Q},N}(\boldsymbol{f_n}) = \sum_{p=0}^n V_p^N\left(\mathbf{d_{p,n}^N}(\boldsymbol{f_n})\right) + \frac{1}{\sqrt{N}} \, R_n^N(\boldsymbol{f_n}) \qquad (17.28)$$

with a second order remainder term $R_n^N(\boldsymbol{f_n})$ such that

$$\mathbb{E}\left(\left|R_n^N(\boldsymbol{f_n})\right|^m\right)^{1/m} \le b(2m)^2 \, r_n(\boldsymbol{f_n})$$

for any $m \geq 1$, with some finite constant

$$r_n(\boldsymbol{f_n}) \leq 4 \sum_{0 \leq p \leq n} g_{p,n}^2 \, l_{p,n}(\boldsymbol{f_n}) \tag{17.29}$$

Proof:

Using (17.26), we find the decomposition (17.28) with the second order remainder term

$$R_n^N(\boldsymbol{f_n}) = -\sum_{p=0}^{n} R_{p,n}^N(\boldsymbol{f_n})$$

with

$$R_{p,n}^N(\boldsymbol{f_n}) := \frac{1}{\eta_p^N(G_{p,n}^N)} \, V_p^N(G_{p,n}^N) \, V_p^N\left(\mathbf{d}_{\mathbf{p,n}}^{\mathbf{N}}(\boldsymbol{f_n})\right)$$

On the other hand, we have

$$\mathbb{E}\left(\left|R_{p,n}^N(\boldsymbol{f_n})\right|^m \left| \mathcal{G}_{p-1}^N \right.\right)^{1/m}$$

$$\leq g_{p,n} \, \mathbb{E}\left(\left|V_p^N(G_{p,n}/\|G_{p,n}\|)\right|^{2m} \left| \mathcal{G}_{p-1}^N \right.\right)^{1/(2m)}$$

$$\times \, \mathbb{E}\left(\left|V_p^N\left(d_{p,n}^N(\boldsymbol{f_n})\right)\right|^{2m} \left| \mathcal{G}_{p-1}^N \right.\right)^{1/(2m)}$$

Using (14.2), we prove that

$$\mathbb{E}\left(\left|R_{p,n}^N(\boldsymbol{f_n})\right|^m \left| \mathcal{G}_{p-1}^N \right.\right)^{1/m} \leq 4b(2m)^2 \, g_{p,n}^2 \, l_{p,n}(\boldsymbol{f_n})$$

The end of the proof is now clear. This ends the proof of the lemma. ∎

17.6 Orlicz norm and $\mathbb{L}_m$-mean error estimates

In this section, we present some direct consequences of the decompositions stated in Theorem 17.5.2. Firstly, using the same arguments as the ones we used in the proof of Theorem 14.3.1, we prove the following corollary.

Corollary 17.6.1 *For any $N \geq 1$, $n \geq 0$, and any collection of functions $\boldsymbol{f_n} \in \mathrm{Osc}(\boldsymbol{E_n})$, we have the Orlicz norm upper bounds*

$$\pi_\psi\left(W_n^{\mathbb{Q},N}(\boldsymbol{f_n})\right) \leq 4\sqrt{2/3} \, \mathbf{a(n)}$$

and

$$\pi_\psi \left(\sup_{0 \le p \le n} |W_n^{Q,N}(\mathbf{f_p})| \right) \le 8 \sqrt{\log(9+n)} \sup_{0 \le p \le n} \mathbf{a(p)}$$

with the finite constants $\mathbf{a(n)}$ defined in (17.12).

Using Lemma 11.3.2, we also have the following concentration inequalities.

Corollary 17.6.2 *For any $N \ge 1$, $n \ge 0$, any collection of functions $\mathbf{f_n} \in \mathrm{Osc}(\mathbf{E_n})$, and any $x \ge 0$, the probability of any of the following events*

$$\left| [Q_n^N - Q_n](\mathbf{f_n}) \right| \le 4 \sqrt{\frac{2(x + \log 2)}{3N}} \, \mathbf{a(n)}$$

and

$$\sup_{0 \le p \le n} \left| [Q_p^N - Q_p](\mathbf{f_p}) \right| \le 8 \sqrt{(x + \log 2)} \sqrt{\frac{\log(9+n)}{N}} \sup_{0 \le p \le n} \mathbf{a(p)}$$

is greater than $1 - e^{-x}$, with the finite constants $\mathbf{a(n)}$ defined in (17.12).

Next, we examine an important unbiasedness property. The decomposition (17.24) implies that Γ_n^N are unbiased; that is, for any $\mathbf{f_n} \in \mathrm{Osc}(\mathbf{E_n})$ we have that

$$\mathbb{E}\left(\Gamma_n^N(\mathbf{f_n}) \right) = \Gamma_n(\mathbf{f_n})$$

On the other hand, using the decomposition (17.26) we prove that

$$\mathbb{E}\left(W_n^{Q,N}(\mathbf{f_n}) \right) = -\frac{1}{\sqrt{N}} \sum_{p=0}^{n} \mathbb{E}\left[\frac{1}{\eta_p^N(G_{p,n}^N)} \, V_p^N(G_{p,n}^N) \times V_p^N\left(\mathbf{d_{p,n}^N(f_n)} \right) \right]$$

This implies that

$$\left| \mathbb{E}\left(W_n^{Q,N}(F_n) \right) \right|$$

$$\le \frac{1}{\sqrt{N}} \sum_{p=0}^{n} g_{p,n} \, \mathbb{E}\left[V_p^N(G_{p,n}/\|G_{p,n}\|)^2 \right]^{1/2} \mathbb{E}\left[V_p^N\left(\mathbf{d_{p,n}^N}(F_n) \right)^2 \right]^{1/2}$$

Using (17.22) and Proposition 12.2.4, we summarize the discussion above with the following corollary.

Corollary 17.6.3 *For any $0 \le p \le n$, and any function $\mathbf{f_n}$ on the path space $\mathbf{E_n}$, we have*

$$\mathbb{E}\left(\Gamma_n^N(\mathbf{f_n}) \right) = \Gamma_n(\mathbf{f_n})$$

and

$$\left| \mathbb{E}\left(W_n^{Q,N}(\mathbf{f_n}) \right) \right| \le \frac{4}{\sqrt{N}} \sum_{p=0}^{n} g_{p,n}^2 \, l_{p,n}(\mathbf{f_n})$$

In addition, when (17.21) and the regularity condition $\mathbf{H}(\mathbf{G}, \mathbf{P})$ *stated on page 371 are satisfied for some finite constants* υ_1, υ_2, *then for normalized additive functionals* $\overline{\boldsymbol{f}}_n$ *of the form (17.20) we have the uniform unbiasedness property*

$$\sup_{n \geq 0} \left| \mathbb{E}\left(\mathbb{Q}_n^N(\overline{\boldsymbol{f}}_n)\right) - \mathbb{Q}_n(\overline{\boldsymbol{f}}_n) \right| \leq \frac{4}{N} \, \exp\left(2\upsilon_1\right) \, \left(\tau^2 + \upsilon_2 \, \exp \upsilon_1\right)$$

Corollary 17.6.4 *For any* $0 \leq p \leq n$, *and any function* $\boldsymbol{f}_n \in \mathcal{B}_b(\boldsymbol{E}_n)$, *and any* $m \geq 2$ *we have*

$$\mathbb{E}\left(\left|W_n^{\mathbb{Q},N}(\boldsymbol{f}_n)\right|^m\right)^{1/m}$$

$$\leq 2(m-1) \, b(m) \, (1 \wedge c(m)) \, \sqrt{(n+1)} \, d_{n,m}(\boldsymbol{f}_n) + \frac{1}{\sqrt{N}} \, b(2m)^2 r_n(\boldsymbol{f}_n)$$

with some constants $(b(m), c(m))$ *defined in (0.6) and (9.33),* $r_n(\boldsymbol{f}_n)$ *defined in (17.29), and some constants* $d_{n,m}(\boldsymbol{f}_n)$ *such that*

$$d_{n,m}^m(\boldsymbol{f}_n) \quad \leq \quad \frac{1}{n+1} \, \sum_{p=0}^{n} g_{p,n}^m \, l_{p,n}^m(\boldsymbol{f}_n)$$

Proof:
Notice that the sequence of random variables $(M_q^{(N,n)})_{0 \leq q \leq n}$

$$M_q^{(N,n)} := \sum_{p=0}^{q} V_p^N \left(\boldsymbol{d}_{p,n}^N(\boldsymbol{f}_n)\right) \quad \text{with} \quad \Delta M_p^{(N,n)} = V_p^N \left(\boldsymbol{d}_{p,n}^N(F_n)\right)$$

is a martingale w.r.t. the filtration $\mathcal{G}_q^N = \sigma(\xi_p, \quad p \leq q)$ generated by the random variables ξ_p, with $p \leq q$. Arguing as in the proof of Theorem 16.5.1, by the Burkholder-Davis-Gundy inequalities (see for instance [458]), for any $rm > 1$ we have

$$\mathbb{E}\left(\left|M_q^{(N,n)}\right|^m\right)^{1/m} \leq \max\left(m-1, 1/m - 1\right) \mathbb{E}\left(\left(\sum_{p=0}^{q} \left(\Delta M_p^{(N,n)}\right)^2\right)^{m/2}\right)^{1/m}$$

On the other hand, for $m \geq 2$ we have

$$\mathbb{E}\left[\left(\frac{1}{q+1} \sum_{p=0}^{q} \left(\Delta M_p^{(N,n)}\right)^2\right)^{m/2}\right] \leq \frac{1}{q+1} \sum_{p=0}^{q} \mathbb{E}\left(\left|\Delta M_p^{(N,n)}\right|^m\right)$$

and by Proposition 9.5.5

$$\mathbb{E}\left(\left|\Delta M_p^{(N,n)}\right|^m\right)^{1/m} \leq 2 \, b(m) \, (1 \wedge c(m)) \, g_{p,n} \, l_{p,n}(\boldsymbol{f}_n)$$

with the constants $b(m)$, $c(m)$, and $l_{p,n}(\boldsymbol{f_n})$ defined in (0.6), (9.33), and in (17.11). We conclude that

$$\mathbb{E}\left(\left|M_q^{(N,n)}\right|^m\right)^{1/m}$$

$$\leq 2b(m)\ (1 \wedge c(m))\ (m-1)\ (q+1)^{1/2}\left[\tfrac{1}{q+1}\sum_{p=0}^q g_{p,n}^m\ l_{p,n}^m(\boldsymbol{f_n})\right]^{1/m}$$

The end of the proof is now a direct consequence of the first order decomposition (17.28). This ends the proof of the corollary. ■

The following nonasymptotic estimates for normalized additive functionals are direct consequences of Corollary 17.6.4.

Corollary 17.6.5 *We assume that (17.21) and the regularity condition* $\mathbf{H}(\mathbf{G}, \mathbf{P})$ *stated on page 371 are satisfied for some finite constants* v_1, v_2, τ. *In this situation, for any normalized additive functionals* $\overline{\boldsymbol{f}}_{\boldsymbol{n}}$ *of the form (17.20), and any* $m \geq 2$, *we have*

$$\mathbb{E}\left(\left|[\mathbb{Q}_n^N - \mathbb{Q}_n]\,(\overline{\boldsymbol{f}}_{\boldsymbol{n}})\right|^m\right)^{1/m} \leq \left[\frac{(m-1)b(m)(1 \wedge c(m))}{\sqrt{N(n+1)}} + \frac{b(2m)^2}{N}\right]d$$

with some constants $(b(m), c(m))$ *defined in (0.6) and (9.33), and some finite constant* d *such that*

$$d \leq 4\exp\left(2v_1\right)\left(\tau^2 + v_2\ \exp v_1\right) \tag{17.30}$$

17.7 Some nonasymptotic variance estimates

Several consequences of the decomposition Theorem 17.5.2 are now emphasized. We further assume that the local variance parameters σ_n defined in (9.31) are uniformly bounded, and we set $\sigma := \sup_{n \geq 0} \sigma_n < \infty$.

Using the fact that

$$\frac{\gamma_p(1)}{\gamma_n(1)} = \frac{\gamma_p(1)}{\gamma_p Q_{p,n}(1)} = \frac{1}{\eta_p Q_{p,n}(1)}$$

we prove the following decomposition

$$\overline{W}_n^{\Gamma,N}(\boldsymbol{f_n}) = \sqrt{N}\left(\overline{\gamma}_n^N(1)\,\mathbb{Q}_n^N - \mathbb{Q}_n\right)(\boldsymbol{f_n}) = \sum_{p=0}^n \overline{\gamma}_p^N(1)\,V_p^N\left(\overline{D}_{p,n}^N(\boldsymbol{f_n})\right)$$

$$\tag{17.31}$$

with the pair of parameters $\left(\overline{\gamma}_n^N(1), \overline{D}_{p,n}^N\right)$ defined below

$$\overline{\gamma}_n^N(1) := \frac{\gamma_n^N(1)}{\gamma_n(1)} \quad \text{and} \quad \overline{D}_{p,n}^N(\boldsymbol{f_n}) = \frac{D_{p,n}^N(\boldsymbol{f_n})}{\eta_p Q_{p,n}(1)} \qquad (17.32)$$

Using the fact that the random fields V_n^N are centered given $\mathcal{F}_{n-1}^N$, we have

$$\mathbb{E}\left(\overline{W}_n^{\Gamma,N}(\boldsymbol{f_n})^2\right) = \sum_{p=0}^n \mathbb{E}\left(\overline{\gamma}_p^N(1)^2 \; \mathbb{E}\left[V_p^N\left(\overline{D}_{p,n}^N(\boldsymbol{f_n})\right)^2 \mid \mathcal{F}_{p-1}^N\right]\right)$$

Using the estimates

$$\|D_{p,n}^N(\boldsymbol{f_n})\| \leq \|Q_{p,n}(1)\| \; \|\boldsymbol{f_n}\|$$
$$\|\overline{D}_{p,n}^N(\boldsymbol{f_n})\| \leq \|\overline{Q}_{p,n}(1)\| \; \|\boldsymbol{f_n}\| \quad \text{with} \quad \overline{Q}_{p,n}(1) = \frac{Q_{p,n}(1)}{\eta_p Q_{p,n}(1)} (17.33)$$

we prove the nonasymptotic variance estimate

$$\mathbb{E}\left(\overline{W}_n^{\Gamma,N}(\boldsymbol{f_n})^2\right) \leq \sum_{p=0}^n \mathbb{E}\left(\overline{\gamma}_p^N(1)^2\right) \; \|\overline{Q}_{p,n}(1)\|^2$$

$$= \sum_{p=0}^n \left[1 + \mathbb{E}\left([\overline{\gamma}_p^N(1) - 1]^2\right)\right] \; \|\overline{Q}_{p,n}(1)\|^2$$

for any function $\boldsymbol{f_n}$ such that $\|\boldsymbol{f_n}\| \leq 1$. When the stability condition $\mathbf{H(G,P)}$ stated on page 371 is satisfied for some finite constants υ_1, υ_2, using Theorem 16.4.1 and Proposition 12.2.4, we prove that

$$\mathbb{E}\left(\overline{W}_n^{\Gamma,N}(\boldsymbol{f_n})^2\right) \leq \exp(2\upsilon_1) \sum_{p=0}^n \left[1 + \frac{c\,p}{N}\left(1 + \frac{c}{N}\right)^p\right]$$

with some constant $c \leq 4\sigma^2 \left[\exp \upsilon_1 - 1\right]^2$. For $N \geq n\,c$, using the same arguments as on page 490, we find that

$$\mathbb{E}\left(\overline{W}_n^{\Gamma,N}(\boldsymbol{f_n})^2\right) \leq \exp(2\upsilon_1) \left[3 + e\right] \; (n+1)$$

On the other hand, using the decomposition

$$\left(\overline{\gamma}_n^N(1)\,\mathbb{Q}_n^N - \mathbb{Q}_n\right) = \left[\overline{\gamma}_n^N(1) - 1\right]\,\mathbb{Q}_n^N + \left(\mathbb{Q}_n^N - \mathbb{Q}_n\right)$$

we prove that

$$\mathbb{E}\left(\left[\mathbb{Q}_n^N(\boldsymbol{f_n}) - \mathbb{Q}_n(\boldsymbol{f_n})\right]^2\right)^{1/2} \leq \frac{1}{\sqrt{N}}\,\mathbb{E}\left(\overline{W}_n^{\Gamma,N}(\boldsymbol{f_n})^2\right)^{1/2} + \mathbb{E}\left(\left[\overline{\gamma}_n^N(1) - 1\right]^2\right)^{1/2}$$

When the stability condition $\mathbf{H}(\mathbf{G}, \mathbf{P})$ stated on page 371 is satisfied for some finite constants υ_1, υ_2, we find that

$$\mathbb{E}\left(\left[\mathbb{Q}_n^N(\boldsymbol{f_n}) - \mathbb{Q}_n(\boldsymbol{f_n})\right]^2\right)^{1/2} \leq \sqrt{\frac{(n+1)}{N}}\left(\exp\left(\upsilon_1\right) + \sqrt{[3+e]}\right)$$

Some interesting bias estimates can also be obtained using the fact that

$$\mathbb{E}\left(\mathbb{Q}_n^N(\boldsymbol{f_n})\right) - \mathbb{Q}_n(\boldsymbol{f_n}) = \mathbb{E}\left(\left[1 - \overline{\gamma}_n^N(1)\right]\ \left[\mathbb{Q}_n^N(\boldsymbol{f_n}) - \mathbb{Q}_n(\boldsymbol{f_n})\right]\right)$$

and the following easily proved upper bound

$$\left|\mathbb{E}\left(\mathbb{Q}_n^N(\boldsymbol{f_n})\right) - \mathbb{Q}_n(\boldsymbol{f_n})\right| \leq \mathbb{E}\left(\left[1 - \overline{\gamma}_n^N(1)\right]^2\right)^{1/2} \mathbb{E}\left(\left[\mathbb{Q}_n^N(\boldsymbol{f_n}) - \mathbb{Q}_n(\boldsymbol{f_n})\right]^2\right)^{1/2}$$

Using (16.6), we find that

$$N\ \left|\mathbb{E}\left(\mathbb{Q}_n^N(\boldsymbol{f_n})\right) - \mathbb{Q}_n(\boldsymbol{f_n})\right| \leq a\ (n+1)$$

with some constant

$$a \leq 2\sigma\left[\exp\upsilon_1 - 1\right]\sqrt{(2+e)} \times \left(\exp\left(\upsilon_1\right) + \sqrt{[3+e]}\right)$$

as soon as

$$N \geq 4n\sigma^2\left[\exp\upsilon_1 - 1\right]^2$$

From these estimates, we readily prove the following theorem.

Theorem 17.7.1 *Assume that the local variance parameters σ_n defined in (9.31) are uniformly bounded, and we set $\sigma := \sup_{n \geq 0} \sigma_n < \infty$. We also assume that condition $\mathbf{H}(\mathbf{G}, \mathbf{P})$ stated on page 371 is satisfied for some finite constants υ_1, υ_2. In this situation, for any $n \geq 0$, $N \geq 1$, and any function $\boldsymbol{f_n}$ such that $\|\boldsymbol{f_n}\| \leq 1$, we have the nonasymptotic variance estimate*

$$N\ \mathbb{E}\left(\left[\mathbb{Q}_n^N(\boldsymbol{f_n}) - \mathbb{Q}_n(\boldsymbol{f_n})\right]^2\right) \leq c\ (n+1)$$

with some constant

$$c \leq \left(\exp\left(\upsilon_1\right) + \sqrt{[3+e]}\right)^2$$

In addition, we have

$$N\ \left|\mathbb{E}\left(\mathbb{Q}_n^N(\boldsymbol{f_n})\right) - \mathbb{Q}_n(\boldsymbol{f_n})\right| \leq a\ (n+1)$$

with some constant

$$a \leq 2\sigma\left[\exp\upsilon_1 - 1\right]\sqrt{(2+e)c} \quad \text{as soon as} \quad N \geq 4n\sigma^2\left[\exp\upsilon_1 - 1\right]^2$$

17.8   Fluctuation analysis

In the further development of this section we assume that the following regularity condition is met for any $n \geq 1$ and for any pair of states $(x, y) \in (E_{n-1}, E_n)$

$$(H^+) \quad h_n^-(y) \leq H_n(x, y) \leq h_n^+(y) \text{ with } (h_n^+/h_n^-) \in \mathbb{L}_2(\eta_n) \text{ and } h_n^+ \in \mathbb{L}_1(\lambda_n) \tag{17.34}$$

Definition 17.8.1 *For any $0 \leq p \leq n$ we denote by $D_{p,n}$ the integral operators from $\mathcal{B}_b(E_n)$ into $\mathcal{B}_b(E_p)$ defined by*

$$D_{p,n}(x_p, d(y_0, \dots, y_n))$$

$$:= Q_{p|p}(x_p, d(y_0, \dots, y_{p-1})) \; \delta_{x_p}(dy_p) \; \Gamma_{n|p}(x_p, d(y_{p+1}, \dots, y_n)) \tag{17.35}$$

and the normalized operators

$$L_{p,n}(x_p, d(y_0, \dots, y_n))$$

$$:= Q_{p|p}(x_p, d(y_0, \dots, y_{p-1})) \; \delta_{x_p}(dy_p) \; \overline{\Gamma}_{n|p}(x_p, d(y_{p+1}, \dots, y_n))$$

We let $\overline{D}_{p,n}$ be the normalized operator from $\mathcal{B}_b(E_n)$ into $\mathcal{B}_b(E_p)$ defined for any $f_n \in \mathcal{B}_b(E_n)$ by

$$\overline{D}_{p,n}(f_n) := \frac{D_{p,n}(f_n)}{\eta_p D_{p,n}(1)} = \frac{D_{p,n}(f_n)}{\eta_p Q_{p,n}(1)}$$

By construction, we have that

$$\eta_p D_{p,n}(f_n)$$

$$= \int Q_p(d(y_0, \dots, y_{p-1}, y_p)) Q_{p+1}(y_p, dy_{p+1}) \dots Q_n(y_{n-1}, dy_n) f_n(y_0, \dots, y_n)$$

$$= \eta_p(Q_{p,n}(1)) \times Q_n(f_n)$$

from which we prove that

$$\Psi_{G_{p,n}}(\eta_p) L_{p,n} = \frac{1}{\eta_p(G_{p,n})} \; \eta_p D_{p,n} = Q_n$$

and

$$\begin{aligned} \mathbf{d}_{p,n}(f_n) &= \frac{G_{p,n}}{\eta_p(G_{p,n})} \; \left[L_{p,n}(f_n) - \Psi_{G_{p,n}}(\eta_p) \left(L_{p,n}(f_n) \right) \right] \\ &= \frac{G_{p,n}}{\eta_p(G_{p,n})} \; L_{p,n} \left[f_n - Q_n(f_n) \right] \end{aligned} \tag{17.36}$$

Theorem 17.8.2 *For any* $N \geq 1$, $0 \leq p \leq n$, $x_p \in E_p$, $m \geq 1$, *and* $\boldsymbol{f_n} \in \mathcal{B}_b(\boldsymbol{E_n})$ *such that* $\|\boldsymbol{f_n}\| \leq 1$, *we have*

$$\sqrt{N} \ \mathbb{E}\left(\left|\left[D_{p,n}^N - D_{p,n}\right](\boldsymbol{f_n})(x_p)\right|^m\right)^{\frac{1}{m}} \leq a(n) \ b(m) \ \left(h_p^+/h_p^-\right)(x_p) \quad (17.37)$$

for some finite constants $a(n) < \infty$, *resp.* $b(m) < \infty$, *whose values only depend on the parameters* n, *resp. on* m.

Proof:
Notice that

$$\left[D_{p,n}^N - D_{p,n}\right](x_p, d(y_0, \ldots, y_n))$$

$$= \left[\mathbb{Q}_{p|p}^N - \mathbb{Q}_{p|p}\right](x_p, d(y_0, \ldots, y_{p-1})) \ \delta_{x_p}(dy_p) \ \Gamma_{n|p}(x_p, d(y_{p+1}, \ldots, y_n))$$

Using Lemma 17.3.4, we find that

$$\left[D_{p,n}^N - D_{p,n}\right](x_p, d(y_0, \ldots, y_n))$$

$$= \left[\sum_{0 \leq q \leq p} \left[\mathbb{M}_{p,q,\eta_q^N} - \mathbb{M}_{p,q,\Phi_q\left(\eta_{q-1}^N\right)}\right](x_p, d(y_q, \ldots, y_{p-1}))\right.$$

$$\left. \times \ \delta_{x_p}(dy_p) \ \mathbb{Q}_{q|q}^N(x_q, d(y_0, \ldots, y_{q-1}))\right] \times \Gamma_{n|p}(x_p, d(y_{p+1}, \ldots, y_n))$$

This yields

$$\left[D_{p,n}^N - D_{p,n}\right](\boldsymbol{f_n})(x_p) = \sum_{0 \leq q \leq p} \left[\mathcal{M}_{p,q,\eta_q^N} - \mathcal{M}_{p,q,\Phi_q\left(\eta_{q-1}^N\right)}\right]\left(T_{p,q,n,x_p}^N(\boldsymbol{f_n})\right)(x_p)$$

with the random function $T_{p,q,n,x_p}^N(\boldsymbol{f_n})$ defined below

$$T_{p,q,n,x_p}^N(\boldsymbol{f_n})(x_q, \ldots, x_{p-1})$$

$$:= \int \Gamma_{n|p}(x_p, d(x_{p+1}, \ldots, x_n)) \ \mathbb{Q}_{q|q}^N(x_q, d(x_0, \ldots, x_{q-1})) \ \boldsymbol{f_n}(x_0, \ldots, x_n)$$

Using formula (17.8), we prove that for any $m \geq 1$ and any function F on $E_{[q,p-1]}$

$$\sqrt{N} \ \mathbb{E}\left(\left|\left[\mathcal{M}_{p,q,\eta_q^N} - \mathcal{M}_{p,q,\Phi_q\left(\eta_{q-1}^N\right)}\right](F)(x_p)\right|^m \ \Big| \ \mathcal{F}_{q-1}^N\right)^{\frac{1}{m}}$$

$$\leq a(n) \ b(m) \ \|F\| \ \left(h_p^+/h_p^-(x_p)\right)$$

for some finite constants $a(n) < \infty$ and $b(m) < \infty$ whose values only depend on the parameters n and m. Using these almost sure estimates, we easily prove (17.37). This ends the proof of the theorem. ∎

Theorem 17.8.3 *We suppose that the following regularity condition (17.34) is satisfied. In this situation, the sequence of random fields $W_n^{\Gamma,N}$, resp. $\overline{W}_n^{\Gamma,N}$ and $W_n^{\mathbb{Q},N}$, converges in law, as $N \to \infty$, to the centered Gaussian fields W_n^{Γ}, resp. $W_n^{\mathbb{Q}}$, defined for any $f_n \in \mathcal{B}_b(E_n)$ by*

$$W_n^{\Gamma}(f_n) = \sum_{p=0}^{n} \gamma_p(1) \, V_p\left(D_{p,n}(f_n)\right)$$

$$\overline{W}_n^{\Gamma}(f_n) = \sum_{p=0}^{n} V_p\left(\overline{D}_{p,n}(f_n)\right) = \gamma_n(1)^{-1} \, W_n^{\Gamma}(f_n)$$

$$W_n^{\mathbb{Q}}(f_n) = \sum_{p=0}^{n} V_p\left(G_{p,n} \, L_{p,n}(f_n - \mathbb{Q}_n(f_n))\right) = \overline{W}_n^{\Gamma}(f_n - \mathbb{Q}_n(f_n))$$

Proof:

Using Theorem 17.5.2, we have the decomposition

$$W_n^{\Gamma,N}(f_n) = \sum_{p=0}^{n} \gamma_p^N(1) \, V_p^N\left(D_{p,n}(f_n)\right) + R_n^{\Gamma,N}(f_n)$$

with the second order remainder term

$$R_n^{\Gamma,N}(f_n) := \sum_{p=0}^{n} \gamma_p^N(1) \, V_p^N\left(f_{p,n}^N\right) \quad \text{with} \quad f_{p,n}^N := [D_{p,n}^N - D_{p,n}](f_n)$$

Using the estimates (17.33), we notice that

$$\|f_{p,n}^N\| \le \|D_{p,n}(f_n)\| + \|D_{p,n}^N(f_n)\| \le 2 \, \|G_{p,n}\| \, \|f_n\|$$

By Slutsky's lemma and by the continuous mapping theorem it clearly suffices to check that $R_n^{\Gamma,N}(F_n)$ converge to 0, in probability, as $N \to \infty$. To prove this claim, we notice that

$$\mathbb{E}\left(V_p^N\left(f_{p,n}^N\right)^2 \mid \mathcal{F}_{p-1}^N\right) \le \Phi_p\left(\eta_{p-1}^N\right)\left(\left(f_{p,n}^N\right)^2\right)$$

On the other hand, we have

$$\Phi_p\left(\eta_{p-1}^N\right)\left(\left(f_{p,n}^N\right)^2\right) = \int \lambda_p(dx_p) \, \Psi_{G_{p-1}}\left(\eta_{p-1}^N\right)(H_p(.,x_p)) \, f_{p,n}^N(x_p)^2$$

from which we prove that

$$\Phi_p\left(\eta_{p-1}^N\right)\left(\left(f_{p,n}^N\right)^2\right) \le \eta_p\left(\left(f_{p,n}^N\right)^2\right)$$

$$+ \int \lambda_p(dx_p) \, \left|[\Psi_{G_{p-1}}\left(\eta_{p-1}^N\right) - \Psi_{G_{p-1}}(\eta_{p-1})](H_p(.,x_p))\right| \, f_{p,n}^N(x_p)^2$$

This yields the rather crude estimate

$$\Phi_p\left(\eta_{p-1}^N\right)\left(\left(\boldsymbol{f_{p,n}^N}\right)^2\right) \leq \eta_p\left(\left(\boldsymbol{f_{p,n}^N}\right)^2\right)$$

$$+4\|G_{p,n}\|^2 \int \lambda_p(dx_p)\,\left|\left[\Psi_{G_{p-1}}\left(\eta_{p-1}^N\right) - \Psi_{G_{p-1}}\left(\eta_{p-1}\right)\right](H_p(.,x_p))\right|$$

from which we conclude that

$$\mathbb{E}\left(V_p^N\left(\boldsymbol{f_{p,n}^N}\right)^2\right) \leq \int \eta_p(dx_p)\,\mathbb{E}\left[\left(\boldsymbol{f_{p,n}^N}(x_p)\right)^2\right]$$

$$+4\|G_{p,n}\|^2 \int \lambda_p(dx_p)\,\mathbb{E}\left(\left|\left[\Psi_{G_{p-1}}\left(\eta_{p-1}^N\right) - \Psi_{G_{p-1}}\left(\eta_{p-1}\right)\right](H_p(.,x_p))\right|\right)$$

Using Proposition 9.5.6, and the weak regularity properties of Boltzmann-Gibbs transformations presented in Lemma 3.1.5, we prove that

$$\sqrt{N}\,\mathbb{E}\left(\left|\left[\Psi_{G_{p-1}}\left(\eta_{p-1}^N\right) - \Psi_{G_{p-1}}\left(\eta_{p-1}\right)\right](H_p(.,x_p))\right|\right) \leq a(p)\,h_p^+(x_p)$$

for some finite constant $a(p) < \infty$. Using Theorem 17.37, we prove that

$$\sqrt{N}\,\mathbb{E}\left(V_p^N\left(\boldsymbol{f_{p,n}^N}\right)^2\right) \leq a(n)\left(\frac{1}{\sqrt{N}}\,\eta_p\left(\left(\frac{h_p^+}{h_p^-}\right)^2\right) + \lambda_p(h_p^+)\right)$$

for some finite constant $a(n) < \infty$. The end of the proof of the first assertion now follows standard computations. To prove the second assertion, we use the following decomposition

$$\sqrt{N}\,[\mathbb{Q}_n^N - \mathbb{Q}_n](\boldsymbol{f_n}) = \frac{1}{\overline{\gamma}_n^N(1)}\,\overline{W}_n^{\Gamma,N}(\boldsymbol{f_n} - \mathbb{Q}_n(\boldsymbol{f_n}))$$

with the random fields $\overline{W}_n^{\Gamma,N}$ defined in (17.31). We complete the proof using the fact that $\overline{\gamma}_n^N(1)$ tends to 1, almost surely, as $N \to \infty$. This ends the proof of the theorem.   ∎

We end this section with some comments on the asymptotic variance associated to the Gaussian fields $W_n^{\mathbb{Q}}$. Using (17.36), we prove that

$$\mathbf{d_{p,n}}(\boldsymbol{f_n})(x_p)$$

$$= \frac{G_{p,n}(x_p)}{\eta_p(G_{p,n})}\,\int \left[L_{p,n}(\boldsymbol{f_n})(x_p) - L_{p,n}(\boldsymbol{f_n})(y_p)\right]\,\Psi_{G_{p,n}}(\eta_p)(dy_p)$$

and

$$\|\mathrm{osc}\left(\mathbf{d_{p,n}}(\boldsymbol{f_n})\right)\| \leq 2\,g_{p,n}\,\mathrm{osc}(L_{p,n}(\boldsymbol{f_n}))$$

Using the same arguments as the ones we used in Section 17.4, for normalized additive functionals $\overline{f}_n$ of the form (17.20) we prove that

$$(n+1) \left\| \operatorname{osc} \left(\mathbf{d}_{\mathbf{p},\mathbf{n}}(\overline{f}_n) \right) \right\|$$

$$\leq 2 \, g_{p,n} \left[\sum_{0 \leq q < p} \beta \left(\mathbb{M}_{p,\eta_{p-1}} \cdots \mathbb{M}_{q+1,\eta_q} \right) + \sum_{p \leq q \leq n} g_{q,n} \beta \left(P_{p,q} \right) \right]$$

When the regularity conditions (17.21) and $\mathbf{H}(\mathbf{G},\mathbf{P})$ stated on page 371 are satisfied for some finite constants v_1, v_2, τ, we conclude that

$$\mathbb{E} \left(W_n^{\mathbb{Q}}(\overline{f}_n)^2 \right)^{1/2} \leq 2 \exp(v_1) \left(\tau^2 + v_2 \, \exp v_1 \right)$$

17.9 Concentration inequalities

17.9.1 Finite marginal models

In this section, we investigate the concentration properties of the backward particle measures $\mathbb{Q}_n^N$ around their limiting values $\mathbb{Q}_n$.

Theorem 17.9.1 *For any $N \geq 1$, $n \geq 0$, and any bounded function f_n on the path space E_n, the probability of the events*

$$\left[\mathbb{Q}_n^N - \mathbb{Q}_n \right] (f_n) \leq \frac{r_n(f_n)}{N} \left(1 + (L_0^\star)^{-1}(x) \right) + 2b_n \, \overline{\sigma}_n^2 \, (L_1^\star)^{-1} \left(\frac{x}{N\overline{\sigma}_n^2} \right)$$

is greater than $1 - e^{-x}$, for any $x \geq 0$, with

$$\overline{\sigma}_n^2 := \frac{1}{b_n^2} \sum_{0 \leq p \leq n} g_{p,n}^2 \, l_{p,n}(f_n)^2 \sigma_p^2 \leq \sum_{0 \leq p \leq n} \sigma_p^2$$

and for any choice of $b_n \geq \sup_{0 \leq p \leq n} g_{p,n} \, l_{p,n}(f_n)$. In the above displayed formulae, σ_n are the uniform local variance parameters defined in (9.31) and $l_{p,n}(f_n)$ and $r_n(f_n)$ are the parameters defined, respectively, in (17.11) and (17.29).

Before getting into the proof of the theorem, we present some direct consequences of these concentration inequalities for normalized additive functionals (we use the estimates (11.26) and (11.27)).

Corollary 17.9.2 *We assume that (17.21) and the regularity condition $\mathbf{H}(\mathbf{G},\mathbf{P})$ stated on page 371 are satisfied for some finite constants v_1, v_2, τ. We also suppose that the parameters σ_n defined in (9.31) are uniformly bounded*

$\sigma = \sup_{n \geq 0} \sigma_n < \infty$. In this notation, for any $N \geq 1$, $n \geq 0$, and any normalized additive functional $\overline{f}_n$ on the path space E_n, the probability of the events

$$\left[\mathbb{Q}_n^N - \mathbb{Q}_n\right](\overline{f}_n)$$

$$\leq \frac{d}{N}\left(1 + (L_0^\star)^{-1}(x)\right) + \exp(v_1)\,\sigma^2\,(L_1^\star)^{-1}\left(\frac{x}{N(n+1)\sigma^2}\right)$$

is greater than $1 - e^{-x}$, for any $x \geq 0$, with a finite constant d satisfying the upperbound (17.30).

Using the estimates (14.28), we readily check the following corollary.

Corollary 17.9.3 *We assume that the assumptions of Corollary 17.9.2 are satisfied. Given some finite constant d satisfying the upperbound (17.30), we set*

$$p_n(x) = d\left(1 + 2(x + \sqrt{x})\right) + \frac{\exp(v_1)}{3(n+1)}\,x$$

$$q_n(x) = \exp(v_1)\,\sqrt{\frac{2x\sigma^2}{(n+1)}}$$

In this situation, for any $x \geq 0$, the probability of the following event is greater than $1 - e^{-x}$

$$\left[\mathbb{Q}_n^N - \mathbb{Q}_n\right](\overline{f}_n) \leq \frac{1}{N}\,p_n(x) + \frac{1}{\sqrt{N}}\,q_n(x)$$

Proof of Theorem 17.9.1:

We use the same line of arguments as the ones we used in Section 14.5.1. Firstly, we notice that

$$\left\|\mathbf{d}_{\mathbf{p,n}}^{\mathbf{N}}(\boldsymbol{f_n})\right\| \leq g_{p,n}\,l_{p,n}(\boldsymbol{f_n})$$

This yields the decompositions

$$\sum_{p=0}^{n} V_p^N\left(\mathbf{d}_{\mathbf{p,n}}^{\mathbf{N}}(\boldsymbol{f_n})\right) = \sum_{p=0}^{n} a_p V_p^N\left(\delta_{\mathbf{p,n}}^{\mathbf{N}}(\boldsymbol{f_n})\right)$$

with the functions $\delta_{\mathbf{p,n}}^{\mathbf{N}}(\boldsymbol{f_n}) = \mathbf{d}_{\mathbf{p,n}}^{\mathbf{N}}(\boldsymbol{f_n})/a_p \in \mathrm{Osc}(E_p)$, and for any finite constants $a_p \geq 2\sup_{0 \leq p \leq n} g_{p,n}\,l_{p,n}(\boldsymbol{f_n})$. On the other hand, we also have that

$$\mathbb{E}\left(V_p^N\left(\delta_{\mathbf{p,n}}^{\mathbf{N}}(\boldsymbol{f_n})\right)^2 \big| \mathcal{G}_{p-1}^N\right) \leq \frac{4}{(a_n^\star)^2}\,\sigma_p^2\,g_{p,n}^2\,l_{p,n}(\boldsymbol{f_n})^2$$

with $a^\star := \sup_{0 \leq p \leq n} a_p$, and the uniform local variance parameters σ_p^2 defined in (9.31).

This shows that the regularity condition stated in (11.37) is met by re-placing the parameters σ_p in the variance formula (11.37) by the constants $2\sigma_p g_{p,n} l_{p,n}(f_n)/a_n^\star$, with the uniform local variance parameters σ_p defined in (9.31). Using Theorem 11.8.6, we easily prove the desired concentration prop-erty. This ends the proof of the theorem. ∎

17.9.2 Empirical processes

Using the same reasoning as in Section 14.5.2, we prove the following concentration inequality.

Theorem 17.9.4 *We let $\mathcal{F}_n$ be a separable collection of measurable functions f_n on E_n, such that $\|f_n\| \leq 1$, $\mathrm{osc}(f_n) \leq 1$, with finite entropy $I(\mathcal{F}_n) < \infty$.*

$$\pi_\psi\left(\left\|W_n^{\mathbb{Q},N}\right\|_{\mathcal{F}_n}\right) \leq c_{\mathcal{F}_n} \sum_{p=0}^{n} g_{p,n} \|l_{p,n}\|_{\mathcal{F}_n}$$

with the functional $f_n \in \mathcal{F}_n \mapsto l_{p,n}(f_n)$ defined in (17.11) and

$$c_{\mathcal{F}_n} \leq 24^2 \int_0^1 \sqrt{\log\left(8 + \mathcal{N}(\mathcal{F}_n, \epsilon)^2\right)} \, d\epsilon$$

In particular, for any $n \geq 0$, $x \geq 0$, and any $N \geq 1$, the probability of the following event is greater than $1 - e^{-x}$

$$\sup_{f_n \in \mathcal{F}_n} \left|\mathbb{Q}_n^N(f_n) - \mathbb{Q}_n(f_n)\right| \leq \frac{c_{\mathcal{F}_n}}{\sqrt{N}} \sum_{p=0}^{n} g_{p,n} \|l_{p,n}\|_{\mathcal{F}_n} \sqrt{x + \log 2}$$

Proof:
Using (17.25), for any function $f_n \in \mathrm{Osc}(E_n)$ we have the estimate

$$\left|W_n^{\mathbb{Q},N}(f_n)\right| \leq 2 \sum_{p=0}^{n} g_{p,n} \, l_{p,n}(f_n) \left|V_p^N(\delta_{p,n}^N(f_n))\right|$$

with the $\mathcal{G}_{p-1}^N$-measurable random functions $\delta_{p,n}^N(f_n)$ on E_p defined by

$$\delta_{p,n}^N(f_n) = \frac{1}{2l_{p,n}(f_n)} \frac{G_{p,n}}{\|G_{p,n}\|} \left[L_{p,n}^N(f_n) - \Psi_{G_{p,n}}\left(\Phi_p\left(\eta_{p-1}^N\right)\right) L_{p,n}^N(f_n)\right]$$

By construction, we have

$$\left\|\delta_{p,n}^N(f_n)\right\| \leq 1/2 \quad \text{and} \quad \delta_{p,n}^N(f_n) \in \mathrm{Osc}(E_p)$$

Using the uniform estimate (11.40), if we set

$$\mathcal{G}_{p,n}^N := \delta_{p,n}^N(\mathcal{F}_n) = \left\{\delta_{p,n}^N(f_n) \; : \; f_n \in \mathcal{F}_n\right\}$$

then we also prove the almost sure upper bound

$$\sup_{N \geq 1} \mathcal{N}\left[\mathcal{G}^N_{p,n}, \epsilon\right] \leq \mathcal{N}(\mathcal{F}_n, \epsilon/2)$$

The end of the proof is now a direct consequence of Theorem 11.8.8 and Theorem 11.3.2. This ends the proof of the theorem. ∎

Definition 17.9.5 *We let* $\mathcal{F} = (\mathcal{F}_n)_{n \geq 0}$ *be a sequence of separable collections* $\mathcal{F}_n$ *of measurable functions* f_n *on* E_n, *such that* $\|f_n\| \leq 1$, $\mathrm{osc}(f_n) \leq 1$, *and finite entropy* $I(\mathcal{F}_n) < \infty$.

For any $n \geq 0$, *we set*

$$J_n(\mathcal{F}) := 24^2 \sup_{0 \leq q \leq n} \int_0^1 \sqrt{\log\left(8 + \mathcal{N}(\mathcal{F}_q, \epsilon)^2\right)}\, d\epsilon$$

We also denote by $\Sigma_n(\mathcal{F})$ *the collection of additive functionals defined by*

$$\Sigma_n(\mathcal{F}) = \{\boldsymbol{f_n} \in \mathcal{B}(\mathbf{E}_n) \text{ such that } \quad \forall \ \mathbf{x_n} = (x_0, \ldots, x_n) \in \mathbf{E}_n$$

$$\boldsymbol{f_n}(\mathbf{x_n}) = \sum_{p=0}^n f_p(x_p) \quad \text{with} \quad f_p \in \mathcal{F}_p, \text{ for } 0 \leq p \leq n\}$$

and

$$\overline{\Sigma}_n(\mathcal{F}) = \{\boldsymbol{f_n}/(n+1) \ : \ \boldsymbol{f_n} \in \Sigma_n(\mathcal{F})\}$$

Theorem 17.9.6 *For any* $N \geq 1$, *and any* $n \geq 0$, *we have*

$$\pi_\psi\left(\left\|W_n^{\mathbb{Q},N}\right\|_{\overline{\Sigma}_n(\mathcal{F})}\right) \leq a_n\, J_n(\mathcal{F}) \sum_{p=0}^n g_{p,n}$$

with some constant

$$a_n \leq \sum_{0 \leq q < p} \beta_{p,q} + \sum_{p \leq q \leq n} \beta\left(R_{p,q}^{(n)}\right)$$

and for any collection of $[0,1]$*-valued parameters* $\beta_{p,q}$ *such that*

$$\forall 0 \leq q \leq p \qquad \sup_{N \geq 1} \beta\left(\mathbb{M}_{p,\eta_{p-1}^N} \ldots \mathbb{M}_{q+1,\eta_q^N}\right) \leq \beta_{p,q}$$

Proof:
By definition of the operator $\mathbf{d}^{\mathbf{N}}_{\mathbf{p,n}}$ given in Definition 17.5.1, for additive functionals $\boldsymbol{f_n}$ of the form (17.19), we find that

$$\begin{aligned}
\mathbf{d}'^{\mathbf{N}}_{\mathbf{p,n}}(\boldsymbol{f_n}) &:= \Phi_p(\eta^N_{p-1})(G_{p,n}) \times \mathbf{d}^N_{p,n}(\boldsymbol{f_n}) \\
&= G_{p,n}\left[Id - \Psi_{G_{p,n}}\left(\Phi_p(\eta^N_{p-1})\right)\right] L^N_{p,n}(\boldsymbol{f_n}) \\
&= \sum_{0 \leq q < p} \mathbf{d}^{(\mathbf{N},1)}_{\mathbf{p,q,n}}(f_q) + \sum_{p \leq q \leq n} \mathbf{d}^{(\mathbf{N},2)}_{\mathbf{p,q,n}}(f_q)
\end{aligned}$$

with

$$\mathbf{d}_{\mathbf{p,q,n}}^{(\mathbf{N,1})}(f_q) = G_{p,n} \left[Id - \Psi_{G_{p,n}} \left(\Phi_p(\eta_{p-1}^N) \right) \right] \left(\mathbb{M}_{p,q}^{(N)}(f_q) \right)$$

$$\mathbf{d}_{\mathbf{p,q,n}}^{(\mathbf{N,2})}(f_q) = G_{p,n} \left[Id - \Psi_{G_{p,n}} \left(\Phi_p(\eta_{p-1}^N) \right) \right] \left(R_{p,q}^{(n)}(f_q) \right)$$

and

$$\mathbb{M}_{p,q}^{(N)} := \mathbb{M}_{p,\eta_{p-1}^N} \cdots \mathbb{M}_{q+1,\eta_q^N}$$

This implies that

$$\left| V_p^N \left(\mathbf{d}_{\mathbf{p,n}}^{\prime \mathbf{N}}(f_n) \right) \right|$$

$$\leq 2 \| G_{p,n} \| \left(\sum_{0 \leq q < p} \beta_{p,q} \, V_p^N \left(\delta_{\mathbf{p,q,n}}^{(\mathbf{N,1})}(f_q) \right) \right. \tag{17.38}$$

$$\left. + \sum_{p \leq q \leq n} \beta \left(R_{p,q}^{(n)} \right) \, V_p^N \left(\delta_{\mathbf{p,q,n}}^{(\mathbf{N,2})}(f_q) \right) \right)$$

with

$$\delta_{\mathbf{p,q,n}}^{(\mathbf{N,1})}(f_q)$$

$$= \frac{1}{2\beta \left(\mathbb{M}_{p,q}^{(N)} \right)} \frac{G_{p,n}}{\| G_{p,n} \|} \, G_{p,n} \left[Id - \Psi_{G_{p,n}} \left(\Phi_p(\eta_{p-1}^N) \right) \right] \mathbb{M}_{p,q}^{(N)}(f_q)$$

and

$$\delta_{\mathbf{p,q,n}}^{(\mathbf{N,2})}(f_q)$$

$$= \frac{1}{2\beta \left(R_{p,q}^{(n)} \right)} \frac{G_{p,n}}{\| G_{p,n} \|} \, G_{p,n} \left[Id - \Psi_{G_{p,n}} \left(\Phi_p(\eta_{p-1}^N) \right) \right] R_{p,q}^{(n)}(f_q)$$

By construction, we have that

$$\left\| \delta_{\mathbf{p,q,n}}^{(\mathbf{N,i})}(f_q) \right\| \leq 1/2 \quad \text{and} \quad \text{osc} \left(\delta_{\mathbf{p,q,n}}^{(\mathbf{N,i})}(f_q) \right) \leq 1$$

for any $i \in \{1,2\}$. We set

$$\mathcal{G}_{p,q,n}^{(N,i)} := \delta_{\mathbf{p,q,n}}^{(\mathbf{N,i})}(\mathcal{F}_q) = \left\{ \delta_{\mathbf{p,q,n}}^{(\mathbf{N,i})}(f_q) \, : \, f_q \in \mathcal{F}_q \right\}$$

Using the uniform estimate (11.40), we also prove the almost sure upper bound

$$\sup_{N \geq 1} \mathcal{N} \left[\mathcal{G}_{p,q,n}^{(N,i)}, \epsilon \right] \leq \mathcal{N}(\mathcal{F}_q, \epsilon/2)$$

Using Theorem 11.8.8, we prove that

$$\pi_\psi \left(\sup_{f_n \in \Sigma_n(\mathcal{F})} \left| V_p^N \left(\mathbf{d}_{\mathbf{p,n}}^{\prime \mathbf{N}}(f_n) \right) \right| \right) \leq a_n \, \| G_{p,n} \| \, J_n(\mathcal{F})$$

The end of the proof of the theorem is now a direct consequence of the decomposition (17.25). This ends the proof of the theorem. ∎

Corollary 17.9.7 *We assume that (17.21) and the regularity condition* **H(G, P)** *stated on page 371 are satisfied for some finite constants v_1, v_2, τ, and we have $J(\mathcal{F}) := \sup_{n \geq 0} J_n(\mathcal{F}) < \infty$. In this situation, we have the uniform estimates*

$$\sup_{n \geq 0} \pi_\psi \left(\left\| W_n^{\mathbb{Q}, N} \right\|_{\overline{\Sigma}_n(\mathcal{F})} \right) \leq c_\mathcal{F}$$

and

$$\pi_\psi \left(\sup_{0 \leq p \leq n} \left\| W_n^{\mathbb{Q}, N} \right\|_{\overline{\Sigma}_n(\mathcal{F})} \right) \leq c_\mathcal{F} \sqrt{6 \log (9 + n)}$$

with some constant

$$c_\mathcal{F} \leq \exp(v_1) \left(\tau^2 + v_2 \exp v_1 \right) J(\mathcal{F})$$

In particular, for any time horizon $n \geq 0$, and any $N \geq 1$, the probability of any of the following events

$$\sqrt{N} \left\| \mathbb{Q}_n^N - \mathbb{Q}_n \right\|_{\overline{\Sigma}_n(\mathcal{F})} \leq c_\mathcal{F} \sqrt{x + \log 2}$$

and

$$\sqrt{N} \sup_{0 \leq p \leq n} \left\| \mathbb{Q}_p^N - \mathbb{Q}_p \right\|_{\overline{\Sigma}_p(\mathcal{F})} \leq c_\mathcal{F} \sqrt{6 \log (9 + n)} \sqrt{(x + \log 2)}$$

is greater than $1 - e^{-x}$, for any $x \geq 0$.

Proof:
Using Proposition 12.2.2, and Proposition 12.2.4, we prove that

$$\sum_{0 \leq q < p} \beta_{p,q} + \sum_{p \leq q \leq n} \beta \left(R_{p,q}^{(n)} \right) \leq \tau^2 + v_2 \exp v_1$$

This ends the proof of the theorem. ∎

We end this section with a direct consequence of (11.16).

Corollary 17.9.8 *We assume that (17.21) and the regularity condition* **H(G, P)** *stated on page 371 are satisfied. We let $\mathcal{F}_n$ be the set of indicator product functions of cells in the path space $E_n = \mathbb{R}^d$, for some $d \geq 1$, $p \geq 0$.*

In this situation, for any time horizon $n \geq 0$, and any $N \geq 1$, the probability of any of the following events

$$\left\| \mathbb{Q}_n^N - \mathbb{Q}_n \right\|_{\overline{\Sigma}_n(\mathcal{F})} \leq c \sqrt{\frac{d}{N}(x + 1)}$$

and

$$\sqrt{N} \sup_{0 \leq p \leq n} \left\| \mathbb{Q}_p^N - \mathbb{Q}_p \right\|_{\overline{\Sigma}_p(\mathcal{F})} \leq c \sqrt{\frac{d}{N}(x + 1) \log (n + 1)}$$

is greater than $1 - e^{-x}$, for any $x \geq 0$, with some finite constant $c < \infty$.

Bibliography

[1] E. H. L. Aarts and J. H. M. Korst. *Simulated Annealing and Boltzmann Machines*. Wiley (1989).

[2] E. H. L. Aarts and J. K. Lenstra. *Local Search in Combinatorial Optimization*. Wiley (1989).

[3] Y. Achdou and O. Pironneau. Computational methods for option pricing. *SIAM, Frontiers in Applied Mathematics series* (2005).

[4] Y. Achdou, F. Camilli, and I. Capuzzo Dolcetta. Mean field games: convergence of a finite difference method. ArXiv:1207.2982v1 (2012).

[5] E. Aïdékon, B. Jaffuel. Survival of branching random walks with absorption. *Stochastic Processes and Their Applications*, vol. 121, pp. 1901–1937 (2011).

[6] R. Adamczak. A tail inequality for suprema of unbounded empirical processes with applications to Markov chains. *EJP*, vol. 13, no. 34, pp. 1000–1034 (2008).

[7] D. Aldous and U. Vazirani. Go with the Winners Algorithms. In Proc. 35th Symp. Foundations of Computer Sci., pp. 492–501 (1994).

[8] K. Allen, C. Yau, and J. A. Noble. A recursive, stochastic vessel segmentation framework that robustly handles bifurcations. In Proceedings, Medical Image Understanding and Analysis, pp. 19–23 (2008).

[9] D. Andersson, B. Djehiche. A maximum principle for SDEs of mean-field type. *Applied Mathematics and Optimization*, vol. 63, no. 3, pp. 341–356 (2010).

[10] J. L. Anderson and S. L. Anderson. A Monte Carlo implementation of the nonlinear filtering problem to produce ensemble assimilations and forecasts. *Monthly Weather Review*, vol. 127, no. 12, pp. 2741–2758 (1999).

[11] C. Andrieu, A. Doucet, and R. Holenstein. Particle Markov chain Monte Carlo methods. *Journal Royal Statistical Society B.*, vol. 72, no. 3, pp. 269–342 (2010).

[12] C. Andrieu and A. Doucet. Particle filtering for partially observed Gaussian state space models. *J. Royal Statistical Society* B, vol. 64, no. 4, pp. 827–836 (2002).

[13] F. Antonelli and A. Kohatsu-Higa. Rate of convergence of a particle method to the solution of the McKean–Vlasov equation. *The Annals of Applied Probability*, vol. 12, no. 2, pp. 423–476 (2002).

[14] L. Arnold, L. Demetrius, and V.M. Gundlach. Evolutionary formalism for products of positive random matrices. *Ann. Appl. Probab.*, vol. 4, 859–901, (1994).

[15] D. Assaf, L. Goldstein, and E. Samuel-Cahn. An unexpected connection between branching processes and optimal stopping. *J. Appl. Prob.*, vol. 37, pp. 613–626 (2000).

[16] R. Assaraf, M. Caffarel, and A. Khelif. Diffusion Monte Carlo methods with a fixed number of walkers. *Phys. Rev. E*, vol. 61, pp. 4566–4575 (2000).

[17] S. Asmussen, and H. Hering. *Branching Processes*. Birkhäuser, Boston (1983).

[18] S. Asmussen, P.W. Glynn. *Stochastic Simulation. Algorithms and Analysis*. Springer Series: Stochastic modelling and applied probability, vol. 57 (2007).

[19] K. B. Athreya and P.E. Ney. *Branching Processes*, Springer-Verlag, New York (1972).

[20] S. K. Au and J.L. Beck. Estimation of small failure probabilities in high dimensions by subset simulation. *Probabilistic Engineering Mechanics*, vol. 16, no. 4, pp. 263–277 (2001).

[21] S. K. Au and J. L. Beck. Subset simulation and its application to seismic risk based on dynamic analysis. *Journal of Engineering Mechanics ASCE*, pp. 901–917 (2003).

[22] S. K. Au, J. Ching, J. L. Beck. Application of subset simulation methods to reliability benchmark problems. *Structural Safety*, vol. 29, no. 3, pp.183–193 (2007).

[23] R. J. Aumann Markets with a continuum of traders. *Econometrica*, vol. 32, No. 1-2, pp. 39–50 (1964).

[24] A. Baddeley, P. Gregori, J. Mateu, R. Stoica, and D. Stoyan. *Case Studies in Spatial Point Process Modeling*, Springer Lecture Notes in Statistics 185, Springer-Verlag, New York (2006).

[25] C. Baehr, C. Beigbeder, F. Couvreux, A. Dabas, and B. Piguet. Retrieval of the turbulent and backscattering properties atmospheric properties using a nonlinear filtering technique applied to Doppler Lidar observations. Submitted to *Journal of Atmospheric and Oceanic Technology* (2012).

[26] C. Baehr. Nonlinear filtering for observations on a random vector field along a random path. M2AN - ESAIM, vol. 44, no. 5, pp. 921–945 (2010).

[27] C. Baehr. Probabilistic models for atmospheric turbulence to filter measurements with particle approximations, Ph.D. thesis, Paul Sabatier University, Toulouse, Speciality: Applied Mathematics, Option Probability (2008).

[28] C. Baehr and O. Pannekoucke. Some issues and results on the EnKF and particle filters for meteorological models in *Chaotic Systems: Theory and Applications.*, C. H. Skiadas and I. Dimotikalis, Eds. World Scientific 27-34. DOI No: 10.1142/9789814299725-0004 (2010).

[29] C. Baehr. Stochastic modeling and filtering of discrete measurements for a turbulent field. application to measurements of atmospheric wind. Int. Journal Modern Phys. B, vol. 23, no. 28-29, pp. 5424–5433 (2009).

[30] A. Bain, D. Crisan. *Fundamentals of Stochastic Filtering*, Springer Series: Stochastic Modelling and Applied Probability, vol. 60, Springer Verlag (2008).

[31] E. Balkovsky, G. Falkovich, and A. Fouxon. Intermittent distribution of inertial particles in turbulent flows, *Phys. Rev. Lett.*, vol. 86, 2790–2793 (2001).

[32] V. Bally and D. Talay. The law of the Euler scheme for stochastic differential equations. *Probability Theory and Related Fields*, vol. 104, no. 1, pp. 43–60 (1996).

[33] A. L. Barabàsi, R. Albert, and H. Jeong. Mean-field theory for scale-free random networks. *Physica A: Statistical Mechanics and its Applications*, vol. 272, no. 1-2, pp. 173–187 (1999)

[34] O. Bardou, S. Bouthemy, and G. Pagès. Optimal quantization for the pricing of swing options. *Applied Mathematical Finance*, vol. 16, Issue 2, pp. 183–217 (2009).

[35] J. Barral, R. Rhodes, V. Vargas. Limiting laws of supercritical branching random walks. *Comptes Rendus de l'Académie des Sciences - Series I - Mathematics*, vol. 350, no 9-10, pp. 535–538 (2012).

[36] S. Barthelmé and N. Chopin. ABC-EP: Expectation Propagation for Likelihood-free Bayesian Computation, ICML 2011 (Proceedings of the 28th International Conference on Machine Learning), L. Getoor and T. Scheffer (Eds.), 289–296 (2011).

[37] N. Bartoli and P. Del Moral. *Simulation et aux Algorithmes Stochastiques.* Cépaduès Édition (2001).

[38] L. E. Baum and T. Petrie. Statistical inference for probabilistic functions of finite state Markov Chains. *The Annals of Mathematical Statistics*, vol. 37, no. 6, pp. 1554–1563 (1966).

[39] L. E. Baum and J. A. Eagon. An inequality with applications to statistical estimation for probabilistic functions of Markov processes and to a model for ecology. *Bulletin of the American Mathematical Society*, vol. 73, no. 3, pp. 360–363 (1967).

[40] L. E. Baum and G. R. Sell. Growth transformations for functions on manifolds. *Pacific Journal of Mathematics*, vol. 27, no. 2, pp. 211–227 (1968).

[41] L. E. Baum, T. Petrie, G. Soules, and N. Weiss. A maximization technique occurring in the statistical analysis of probabilistic functions of Markov Chains. *The Annals of Mathematical Statistics*, vol. 41, no. 1, pp. 164–171 (1970).

[42] L. E. Baum. An inequality and associated maximization technique in statistical estimation of probabilistic functions of a Markov process. *Inequalities*, vol. 3, pp. 1–8. (1972).

[43] E. R. Beadle and P. Djuric. A fast-weighted Bayesian bootstrap filter for nonlinear model state estimation. *IEEE Transactions on Aerospace and Electronic Systems.* vol. 33, no. 1, pp. 338–343 (1997).

[44] N. Bellomo, M. Pulvirenti. *Modeling in Applied Sciences: A kinetic theory approach.* Birkhaüser Boston (2000).

[45] A. Bellouquid and M. Delitala. Mathematical modeling of complex biological systems. A kinetic theory approach. Birkhaüser, Boston (2006).

[46] C. Bender and J. Steiner. Least-squares Monte Carlo for backward SDEs. in *Numerical Methods in Finance*. (2011).

[47] S. Ben Hamida and R. Cont. Recovering volatility from option prices by evolutionary optimization. *Journal of Computational Finance*, Vol. 8, No. 4 (2005).

[48] A. Bensoussan and J. Frehse. Stochastic games for N players. *Journal of Optimization Theory and Applications*, vol. 105, no. 3, pp. 543–565 (2000).

[49] J. Berestycki, N. Berestycki, and J. Schweinsberg. Critical branching Brownian motion with absorption: survival probability. *ArXiv:1212.3821v2 [math.PR]* (2012)

[50] J. O. Berger. *Statistical Decision Theory and Bayesian Analysis.* Springer Series: Statistics (2nd ed.). Springer-Verlag (1985).

[51] J. M. Bernardo and A.F.M. Smith. *Bayesian Theory.* Wiley (1994).

[52] J. D. Biggins. Uniform convergence of martingales in the branching random walk. *Ann. Probab.* vol. 20, pp. 137–151 (1992).

[53] P. Billingsley. *Probability and Measure.* John Wiley & Sons, New York (1986).

[54] K. Binder. *Monte Carlo and Molecular Dynamics Simulations in Polymer Science.* Oxford University Press, New York (1995).

[55] T. Björk, M. H. A. Davis, and C. Landèn. Optimal investment under partial information. *Mathematical Methods of Operations Research* (2010).

[56] S. Boi, V. Capasso, and D. Morale. Modeling the aggregative behavior of ants of the species Polyergus rufescens. *Nonlinear Analysis: Real World Applications*, vol. 1, pp. 163–176 (2000).

[57] F. Bolley, A. Guillin, and F. Malrieu. Trend to equilibrium and particle approximation for a weakly self consistent Vlasov-Fokker-Planck Equation *ESAIM Mathematical Modelling and Numerical Analysis*, vol. 44, no. 5, 867–884 (2010).

[58] L. Boltzmann. *Leçons sur la Théorie des Gaz.* Gauthiers-Villars, tome I, (1902), tome II (1905), Réédition, Jean Gabay, Paris (1987).

[59] V. S. Borkar. Controlled diffusion processes. *Probability Surveys*, vol. 2, pp. 213-244 (2005).

[60] L. Bornn, P. Jacob, P. Del Moral, and A. Doucet. An adaptive interacting Wang-Landau algorithm for automatic density exploration. arXiv:1109.3829v2 [stat.CO] ArXiv April. 2012 [33p] . To appear in *Journal of Computational and Graphical Statistics* (2013).

[61] M. Bossy and D. Talay. A stochastic particle method for some one-dimensional nonlinear PDE. *Mathematics and Computer in Simulation*, vol. 38, pp. 43–50 (1995).

[62] M. Bossy and D. Talay. Convergence rate for the approximation of the limit law of weakly interacting particles: application to the Burgers equation. *Ann. Appl.Probab.*, vol. 6, pp. 818–861 (1996).

[63] M. Bossy and D. Talay. A stochastic particle method for the McKean-Vlasov and the Burgers equation. *Mathematics of Computation.* vol. 66, no. 217, pp. 157–192 (1997).

[64] B. Bouchard and X. Warin. Monte-Carlo valorisation of American options: facts and new algorithms to improve existing methods. *Numerical Methods in Finance*, Springer (2011).

[65] A. Bouchard-Côté, and S. Sankararaman, M. I. Jordan. Phylogenetic Inference via Sequential Monte Carlo. *Systematic Biology*, vol. 61, no. 4, pp. 579–593 (2012).

[66] B. Bouchard and N. Touzi. Discrete time approximation and Monte Carlo simulation of backward stochastic differential equations. *Stochastic Process Appl.* 111, pp. 175–206 (2004).

[67] J. P. Bouchaud and M. Potters. Back to basics: historical option pricing revisited. *Philosophical Transactions: Mathematical, Physical & Engineering Sciences*, 357(1758), pp. 2019–2028 (2002).

[68] P. Bougerol and J. Lacroix. *Products of Random Matrices with Applications to Schrödinger Operators*. Birkhäuser, Boston (1985).

[69] N. Bou-Rabee and M. Hairer. Nonasymptotic mixing of the MALA algorithm. *IMA Journal of Numerical Analysis* (2012).

[70] R. D. Bourgin and R. Cogburn On determining absorption probabilities for Markov chains in random environments. *Advances in Applied Probability*, vol. 13, no. 2, pp. 369–387 (1981).

[71] P. P. Boyle and D. Emanuel. Discretely adjusted option hedges. *Journal of Financial Economics*, vol. 8, pp. 259–282 (1980).

[72] R. N. Bradt, S. M. Johnson, and S. Karlin. On sequential designs for maximizing the sum of n observations. *Ann. Math. Statist.*, vol. 27, pp. 1060–1070 (1956).

[73] M. D. Bramson. Maximal displacement of branching Brownian motion. *Comm. Pure Appl. Math.*, vol 31, pp. 531–581 (1978).

[74] L. Breiman. *Probability*. Original edition published by Addison-Wesley, 1968; reprinted by Society for Industrial and Applied Mathematics (1992).

[75] J. Bretagnolle. A new large deviation inequality for U-statistics of order 2. *ESAIM: Probability and Statistics*. Vol. 3, pp. 151–162 (1999).

[76] D. Brigo, T. Bielecki, and Fr. Patras. *Credit Risk Frontiers*. Wiley–Bloomberg Press (2011).

[77] S. Brooks, A. Gelman, G. Jones, and X.-L. Meng. *Handbook of Markov Chain Monte Carlo*. Chapman and Hall/CRC Press, col. 25, iss. 1 (2012).

[78] J. K. Brooks. Representations of weak and strong integrals in Banach spaces. *Proc. Nat. Acad. Sci. U.S.A.*, vol. 63, pp. 266–270 (1969).

[79] M. Broadie and P. Glasserman. Estimating security prices using simulation. *Management Science*, 42, pp. 269–285 (1996).

[80] M. Broadie and P. Glasserman. A stochastic mesh method for pricing high-dimensional American options. *Journal of Computational Finance*, vol. 7, pp. 35–72, (2004).

[81] F. T. Bruss. The art of a right decision: Why decision makers want to know the odds-algorithm. *Newsletter of the European Mathematical Society*, Issue 62, pp. 14–20 (2006).

[82] A. Budhiraja, P. Del Moral, and S. Rubenthaler. Discrete time Markovian agents interacting through a potential. *Journal ESAIM Probability & Statistics* (2012).

[83] R. Buckdahn, B. Djehiche, J. Li, and S. Peng. Mean-field backward stochastic differential equations: a limit approach. *The Annals of Probability*, vol. 37, no. 4, pp. 15241–565 (2009).

[84] R. Buckdahn, B. Djehiche, J. Li, and S. Peng. Mean-field backward stochastic differential equations and related partial differential equations. *Stochastic Processes and Their Applications*, vol. 119, no. 10, pp. 3133–3154 (2007).

[85] J. A. Bucklew. *Large Deviation Techniques in Decision, Simulation, and Estimation*. John Wiley & Sons, New York (1990).

[86] J. A. Bucklew. *Introduction to Rare Event Simulation*. Springer-Verlag (2004).

[87] J. A. Bucklew. *Introduction to Rare Event Simulation*. Springer-Verlag, New York (2004).

[88] G. Burgers, P. Jan van Leeuwen, and G. Evensen. Analysis scheme in the ensemble Kalman filter. *Monthly Weather Review*, vol. 126, no. 6, pp. 1719–1724 (1998).

[89] F. Campillo and V. Rossi. Convolution particle filtering for parameter estimation in general state-space models. In *Proceedings of the 45th IEEE Conference on Decision and Control*, San Diego (USA), December (2006).

[90] F. Campillo and V. Rossi. Convolution filter based methods for parameter estimation in general state space models. *IEEE Transactions on Aerospace and Electronic Systems*, vol. 45, no. 3, pp. 1063–1071 (2009).

[91] E. Cancès, B. Jourdain, and T. Lelièvre. Quantum Monte Carlo simulations of fermions. A mathematical analysis of the fixed-node approximation, *ESAIM: M2AN*, vol. 16, no. 9, pp. 1403–1449, (2006).

[92] E. Cancès, F. Legoll, and G. Stoltz. Theoretical and numerical comparison of sampling methods for molecular dynamics. *ESAIM: M2AN*, vol. 41, no. 2, pp. 351–389 (2005).

[93] O. Cappé, E. Moulines, and T. Rydèn. *Inference in Hidden Markov Models*, Springer-Verlag (2005).

[94] O. Cappe, S. J. Godsill, and E. Moulines. An overview of existing methods and recent advances in SMC. *Proceedings of the IEEE*, vol. 95, no. 5, pp. 899–924 (2007)

[95] R. Carmona, and S. Crépey. Particle methods for the estimation of credit portfolio loss distributions. *International Journal of Theoretical and Applied Finance*, vol. 13, no. 4, pp. 577–602 (2010).

[96] R. Carmona, P. Del Moral, P. Hu, and N. Oudjane. An introduction to particle methods in finance. In *Numerical Methods in Finance*. Numerical Methods in Finance Series: Springer Proceedings in Mathematics, vol. 12, no. XVII, p. 3–49 (2012).

[97] R. Carmona, P. Del Moral, P. Hu, and N. Oudjane. Numerical Methods in Finance Series: Springer Proceedings in Mathematics, Bordeaux, June 2010 vol. 12, no. XVII (2012).

[98] R. Carmona, J.-P. Fouque, and D. Vestal. Interacting Particle systems for the computation of rare credit portfolio losses. *Finance and Stochastics*, vol. 13, no. 4, pp. 613–633 (2009).

[99] R. Carmona and N. Touzi. Optimal multiple stopping and valuation of swing options. *Mathematical Finance*, vol. 18, no. 2, pp. 239–268 (2008).

[100] F. Caron, P. Del Moral, A. Tantar, and E. Tantar. Simulation particulaire, Dimensionnement en conditions extrêmes. *Research contract no. 2010-IFREMER-01* with IFREMER, (92p) September (2010).

[101] F. Caron, P. Del Moral, A. Doucet, and M. Pace. On the conditional distributions of spatial point processes. *Adv. in Appl. Probab.*, vol. 43, no. 2, pp. 301–307. (2011).

[102] F. Caron, P. Del Moral, M. Pace, and B. N. Vo. On the stability and the approximation of branching distribution flows, with applications to nonlinear multiple target filtering stoch. *Analysis and Applications*, vol. 29, no. 6, p. 951–997 (2011).

[103] F. Caron, P. Del Moral, A. Doucet, and M. Pace. Particle approximations of a class of branching distribution flows arising in multi-target tracking *SIAM J. Control Optim.*, vol. 49, pp. 1766–1792 (2011).

[104] J. Carpenter, P. Clifford, and P. Fearnhead. An improved particle filter for nonlinear problems. *IEE Proceedings* F, vol. 146, pp. 2–7 (1999).

[105] J. F. Carrière. Valuation of the early-exercise price for options using simulations and nonparametric regression. *Insurance: Mathematics and Economics*, vol. 19, pp. 19–30 (1996).

[106] R. Casarin. Simulation methods for nonlinear and non-Gaussian models in finance. *Premio SIE, Rivista Italiana degli Economisti*, vol. 2, pp. 341–345 (2005).

[107] R. Casarin and C. Trecroci. Business cycle and stock market volatility: a particle filter approach. *Cahier du CEREMADE N. 0610*, University Paris Dauphine (2006).

[108] P. Cattiaux, P. Collet, A. Lambert, S. Martinez, S. Méléard, and J. S. Martin. Quasi-stationary distributions and diffusion models in population dynamics. *Ann. Probab.*, vol. 37, no. 5, pp. 1926–1969 (2009).

[109] P. Cattiaux, A. Guillin, and F. Malrieu. Probabilistic approach for granular media equations in the nonuniformly convex case. *Probability Theory and Related Fields*. vol. 140, no.1–2, pp. 19–40 (2008).

[110] J. A. Cavender. Quasi-stationary distributions of birth-and-death processes. *Adv. Appl. Probab.*, vol. 10, no. 3, pp. 570–586 (1978).

[111] C. Cercignani. *The Boltzmann Equation and Its Applications*. Springer-Verlag, New York (1988).

[112] C. Cercignani. *Rarefied Gas Dynamics*. Cambridge University Press, Cambridge (2000).

[113] C. Cercignani, R. Illner, and M. Pulvirenti. *The Mathematical Theory of Dilute Gases*. Springer (1994).

[114] A. Cerny. Dynamic programming and mean variance hedging in discrete time. *Tanaka Business School Discussion Papers* TBS/DP04/15 London: Tanaka Business School (2004).

[115] R. Cerf. Une théorie asymptotique des algorithmes génétiques, Ph.D., Université Montpellier II (1994).

[116] R. Cerf. Asymptotic convergence of a genetic algorithm, *C.R. Acad. Sci. Paris*, no. 319, Série I, pp. 271–276 (1994).

[117] R. Cerf. Asymptotic convergence of genetic algorithms. *Adv. Appl. Probab.*, vol. 30, pp. 521–550 (1998).

[118] R. Cerf. A new genetic algorithm, *Annals of Applied Probab*, vol. 3, pp. 778–817 (1996).

[119] R. Cerf. The dynamics of mutation-selection algorithms with large population sizes, *Ann. Inst. Henri Poincaré*, vol. 32 no.4, pp. 455–508 (1996).

[120] R. Cerf. An asymptotic theory for genetic algorithms, *Artificial Evolution*, Lecture Notes in Computer Science vol. 1063, pp. 37–53, Springer-Verlag (1996).

[121] F. Cerou, A. Guyader, R. Rubinstein, and R. Vaismana. Smoothed splitting method for counting. *Stochastic models* (2011), vol. 10, no. 27, pp. 626–650.

[122] F. Cérou, P. Del Moral, and A. Guyader. A nonasymptotic variance theorem for unnormalized Feynman-Kac particle models. Annales Institute Henri Poincoré, Proba. and Stat., vol. 47, no. 3, pp 629–669 (2011).

[123] F. Cerou, P. Del Moral, T. Furon, and A. Guyader. Sequential Monte Carlo for rare event estimation. *Statistics and Computing*, vol. 22, no. 3, pp. 795–808 (2012).

[124] F. Cérou, P. Del Moral, F. LeGland, and P. Lézaud. Limit theorems for multilevel splitting algorithms in the simulation of rare events. *Proceedings of the 2005 Winter Simulation Conference.* M. E. Kuhl, N. M. Steiger, F. B. Armstrong, and J. A. Joines, Eds. (2005).

[125] F. Cérou, P. Del Moral, F. Le Gland, and P. Lézaud. Genealogical models in entrance times rare event analysis. *Alea*, vol. 1, pp. 181–203 (2006).

[126] P. M. Chaikin, T. C. Lubensky. *Principles of Condensed Matter Physics.* Cambridge University Press, London (2007).

[127] M. Chaleyat-Maurel and D. Michel. Des résultats de nonexistence de filtres de dimension finie. *C. R. Acad. Sc. de Paris Série I Math.*, no. 296, no. 22, pp. 933–936 (1983).

[128] S. Chapman and T. Cowling. *The Mathematical Theory of Non-Uniform Gases.* Cambridge University Press (1952).

[129] R. Chelli, S. Marsili, A. Barduci, and P. Procacci. Generalization of the Jarzynski and Crooks nonequilibrium work theorems in molecular dynamics simulations *Phys. Rev. E*, vol. 75, no. 5, 050101(R) (2007)

[130] L. Y. Chen. On the Crooks fluctuation theorem and the Jarzynski equality. *J Chem Phys.*, vol. 129, no. 9, 091101 (2008)

[131] N. Chen and P. Glasserman. Malliavin Greeks without Malliavin calculus. *Stochastic Processes and Their Applications*, vol. 117, pp. 1689–1723 (2007).

[132] R. Chen and J. Liu. Mixture Kalman filters. *J. Royal Statistical Society B*, vol. 62, pp. 493–508 (2000).

[133] N. Chopin. A sequential particle filter method for static models. *Biometrika*, vol. 89, pp. 539–552 (2002).

[134] N. Chopin, P. E. Jacob, and O. Papaspiliopoulos. SMC^2: An efficient algorithm for sequential analysis of state-space models. arXiv:1101.1528v3 (2012). *Journal of the Royal Statistical Society: Series B (Statistical Methodology)* (2013).

[135] N. Chopin, T. Lelièvre, and G. Stoltz. Free energy methods for efficient exploration of mixture posterior densities. *Statistics and Computing*, vol. 22, no. 4, pp. 897–916 (2012).

[136] Y. S. Chow, H. Robbins and D. Siegmund. *Great Expectations: The Theory of Optimal Stopping*, Houghton Mifflin, Boston (1971).

[137] D. E. Clark and J. Bell. Convergence results for the particle PHD filter, *Signal Processing, IEEE Transactions*, volume 54, issue 7, pp. 2652–2661, (2006).

[138] E. Cleland, G. H. Booth, and A. Alavi. Communications: survival of the fittest: Accelerating convergence in full configuration-interaction quantum Monte Carlo. *The Journal of Chemical Physics*, vol. 132, 041103 (2010).

[139] E. Clément, D. Lamberton, and P. Protter. An analysis of a least squares regression method for American option pricing. *Finance and Stochastics*, vol. 6, pp. 449–472, (2002).

[140] J. Cohen. Products of random matrices and related topics in mathematics and science: a bibliography. *Contemp. Math.*, vol. 50, pp. 337–358 (1986).

[141] F. Comets, T. Shiga, and N. Yoshida. Probabilistic analysis of directed polymers in a random environment: A review. *Adv. Stud. in Pure Math.* (2004).

[142] D. D. Creal. A survey of sequential Monte Carlo methods for economics and finance. *Econometric Reviews*, vol. 31, no. 3, pp. 245–296. (2012).

[143] D. Crisan and B. Rozovsky (Eds). *Handbook of Nonlinear Filtering*, Cambridge University Press (2009).

[144] D. Crisan, P. Del Moral, and T. Lyons. Discrete filtering using branching and interacting particle systems. *Markov Processes and Related Fields*, vol. 5, no. 3, pp. 293–318 (1999).

[145] D. Crisan, P. Del Moral, and T. Lyons. Interacting particle systems approximations of the Kushner Stratonovitch equation. *Advances in Applied Probability*, vol. 31, no. 3, pp. 819–838 (1999).

[146] G. E. Crooks. Nonequilibrium measurements of free energy differences for microscopically reversible Markovian systems. *J. Stat. Phys.*, vol. 90, 1481 (1998).

[147] D. J. Daley and D. Vere-Jones. *An Introduction to the Theory of Point Processes*, Springer, New York (1988).

[148] D. A. Dawson. Critical dynamics and fluctuations for a mean field model of cooperative behavior. *J. Statistical Physics*, vol. 31, no. 1, pp. 29–85 (1983).

[149] D. A. Dawson, and P. Del Moral. Large deviations for interacting processes in the strong topology. *Statistical Modeling and Analysis for Complex Data Problem*. P. Duchesne and B. Rémillard, Eds. Springer, pp. 179–209 (2005).

[150] M. deBruijne, and M. Nielsen. Shape particle filtering for image segmentation. In Proceedings, Medical Image Computing and Computer-Assisted Intervention, pp.168–175 (2004).

[151] P. Del Moral. Nonlinear filtering: interacting particle solution. *Markov Processes and Related Fields*, vol. 2, no. 4, pp. 555–580 (1996).

[152] P. Del Moral. Measure valued processes and interacting particle systems. Application to nonlinear filtering problems. *Annals of Applied Probability*, vol. 8, no. 2, pp. 438–495 (1998).

[153] P. Del Moral. Fundamentals of stochastic filtering by Alan Bain and Dan Crisan Book Review [13p.], *Bulletin of the American Mathematical Society*, October (2010).

[154] P. Del Moral, J. Jacod, and P. Protter. The Monte-Carlo method for filtering with discrete-time observations. *Probability Theory and Related Fields*, vol. 120, pp. 346–368 (2001).

[155] P. Del Moral and J. Jacod. The Monte-Carlo Method for filtering with discrete time observations. Central Limit Theorems. *The Fields Institute Communications, Numerical Methods and Stochastics*, T.J. Lyons T.S. Salisbury, and Eds., American Mathematical Society (2002).

[156] P. Del Moral and J. Jacod. Interacting particle filtering with discrete observations. In *Sequential Monte Carlo Methods in Practice*, Statistics for Engineering and Information Science. Springer. A. Doucet, J. F. G. de Freitas, N. J. Gordon., Eds., pp. 43–77 (2001).

[157] P. Del Moral, L. Kallel, and J. Rowe. Modeling genetic algorithms with interacting particle systems. *Revista de Matematica, Teoria y aplicaciones*, vol. 8, no. 2, July (2001).

[158] P. Del Moral, A. Doucet, and A. Jasra. On adaptive resampling procedures for sequential Monte Carlo methods (HAL-INRIA RR-6700) [46p], (Oct. 2008). *Bernoulli*, vol. 18, no. 1, pp. 252–278 (2012).

[159] P. Del Moral, A. Doucet, and A. Jasra. An adaptive sequential Monte Carlo method for approximate Bayesian computation (DOI: 10.1007/s11222-011-9271-y). *Statistics and Computing*, vol. 1–12, August 03 (2011).

[160] P. Del Moral and M. Ledoux. On the convergence and the applications of empirical processes for interacting particle systems and nonlinear filtering. *Journal of Theoretical Probability*, vol. 13, no. 1, pp. 225–257 (2000).

[161] P. Del Moral and L. Miclo. *Branching and Interacting Particle Systems Approximations of Feynman-Kac Formulae with Applications to Non-Linear Filtering*. Sminaire de Probabilits XXXIV, J. Azma and M. Emery and M. Ledoux and M. Yor, Lecture Notes in Mathematics, Springer-Verlag, Berlin, vol. 1729, pp. 1–145 (2000).

[162] P. Del Moral and L. Miclo. On the convergence and the applications of the generalized simulated annealing. *SIAM Journal on Control and Optimization*, vol. 37, no. 4, pp. 1222–1250, (1999).

[163] P. Del Moral. *Feynman-Kac Formulae: Genealogical and Interacting Particle Systems with Applications*, Springer-Verlag, New York (2004).

[164] P. Del Moral and A. Doucet. *On a Class of Genealogical and Interacting Metropolis Models*. Séminaire de Probabilités XXXVII, J. Azéma and M. Emery and M. Ledoux and M. Yor, Eds., Lecture Notes in Mathematics 1832, Springer-Verlag, Berlin, pp. 415–446 (2003).

[165] P. Del Moral, A. Doucet, and A. Jasra. Sequential Monte Carlo samplers. *J. Royal Statist. Soc. B*, vol. 68, pp. 411–436 (2006).

[166] P. Del Moral and E. Rio. Concentration inequalities for mean field particle models. *Ann. Appl. Probab*, vol. 21, no. 3, pp. 1017–1052. (2011).

[167] P. Del Moral and L. Miclo. Annealed Feynman-Kac models. *Comm. Math. Phys.*, vol. 235, pp. 191–214 (2003).

[168] P. Del Moral, and A. Guionnet. Large deviations for interacting particle systems. Applications to nonlinear filtering problems. *Stochastic Processes and Their Applications*, vol. 78, pp. 69–95 (1998).

[169] P. Del Moral and A. Guionnet. On the stability of measure valued processes with applications to filtering. *C. R. Acad. Sci. Paris Sér. I Math.*, vol. 329, pp. 429–434 (1999).

[170] P. Del Moral and A. Guionnet. On the stability of interacting processes with applications to filtering and genetic algorithms. *Ann. Inst. Henri Poincaré*, vol. 37, no. 2, pp. 155–194 (2001).

[171] P. Del Moral and J. Garnier. Genealogical particle analysis of rare events. *Annals of Applied Probability*, vol. 15, no. 4, pp. 2496–2534 (2005).

[172] P. Del Moral and J. Garnier. Simulations of rare events in fiber optics by interacting particle systems. *Optics Communications* vol. 257, pp. 205–214 (2006).

[173] P. Del Moral and N. Hadjiconstantinou. An introduction to probabilistic methods, with applications. vol. 44, no. 5, pp. 805–830 M2AN (2010).

[174] P. Del Moral, S. Hu, and L.M. Wu. Moderate deviations for mean field particle models HAL-INRIA no. 00687827 [37p] (2012).

[175] P. Del Moral and A. Doucet. Particle motions in absorbing medium with hard and soft obstacles. *Stochastic Anal. Appl.*, vol. 22, pp. 1175–1207 (2004).

[176] P. Del Moral and L. Miclo. Particle approximations of Lyapunov exponents connected to Schrdinger operators and Feynman-Kac semigroups. *ESAIM: Probability and Statistics*, no. 7, pp. 171–208 (2003).

[177] P. Del Moral and L. Miclo. Genealogies and Increasing Propagations of Chaos for Feynman-Kac and Genetic Models. *Annals of Applied Probability*, vol. 11, no. 4, pp. 1166–1198 (2001).

[178] P. Del Moral and L. Miclo. Strong propagations of chaos in Moran's type particle interpretations of Feynman-Kac measures. *Stochastic Analysis and Applications*, vol. 25, no. 3, pp. 519–575 (2007).

[179] P. Del Moral and L. Miclo. A Moran particle system approximation of Feynman-Kac formulae. *Stochastic Processes and Their Applications*, vol. 86, pp. 193–216 (2000).

[180] P. Del Moral and L. Miclo. On the Stability of Non Linear Semigroup of Feynman-Kac Type. *Annales de la Faculté des Sciences de Toulouse*, vol. 11, no. 2, pp. 135–175 (2002)

[181] P. Del Moral and L. Miclo. Asymptotic Results for Genetic Algorithms with Applications to Non Linear Estimation. *Proceedings Second EvoNet Summer School on Theoretical Aspects of Evolutionary Computing.* Ed. B. Naudts, L. Kallel. Natural Computing Series, Springer (2000).

[182] P. Del Moral, M. Ledoux, and L. Miclo. Contraction properties of Markov kernels. *Probab. Theory and Related Fields*, vol. 126, pp. 395–420, (2003).

[183] P. Del Moral, L. Miclo, F. Patras, and S. Rubenthaler. The convergence to equilibrium of neutral genetic models. *Stochastic Analysis and Applications*. vol. 28, no. 1, pp. 123–143 (2009).

[184] P. Del Moral and P. Lezaud. Branching and interacting particle interpretation of rare event probabilities. *Stochastic Hybrid Systems: Theory and Safety Critical Applications*, H. Blom and J. Lygeros, Eds. Springer-Verlag, Heidelberg (2006).

[185] P. Del Moral, P. Jacob, A. Lee, L. Murray, and G. Peters. Feynman-Kac particle integration with geometric interacting jumps. arXiv:1211.7191 November (2012).

[186] P. Del Moral, P. Hu, and L.M. Wu. On the concentration properties of interacting particle processes. *Foundations and Trends in Machine Learning*, vol. 3, no. 3-4, pp. 225–389 (2012).

[187] P. Del Moral and Fr. Patras. Interacting path systems for credit risk. *Credit Risk Frontiers*. D. Brigo, T. Bielecki, and F. Patras, Eds. *Wiley–Bloomberg Press*, (2011), 649–674. *Short announcement available as Interacting path systems for credit portfolios risk analysis. INRIA:RR-7196 (2010)*.

[188] P. Del Moral, F. Patras., and S. Rubenthaler. Convergence of U-statistics for interacting particle systems. *Journal of Theoretical Probability*, vol. 24, no. 4, pp 1002–1027 December (2011).

[189] P. Del Moral, Fr. Patras., and S. Rubenthaler. Coalescent tree based functional representations for some Feynman-Kac particle models. *Annals of Applied Probability*, vol. 19, no. 2, pp. 1–50 (2009).

[190] P. Del Moral, Fr. Patras., and S. Rubenthaler. A mean field theory of nonlinear filtering. *Handbook of Nonlinear Filtering*, D. Crisan and B. Rozovsky, Eds. Cambridge University Press (2009).

[191] P. Del Moral, Br. Rémillard, and S. Rubenthaler. Monte Carlo approximations of American options that preserve monotonicity and convexity. In *Numerical Methods in Finance*. Numerical Methods in Finance Series: Springer Proceedings in Mathematics, vol. 12, no. XVII (2012).

[192] P. Del Moral and A. Doucet. Interacting Markov chain Monte Carlo methods for solving nonlinear measure-valued equations. *The Annals of Applied Probability*, vol. 20, no. 2, pp. 593–639 (2010).

[193] P. Del Moral, A. Doucet, and S. S. Singh. A backward particle interpretation of Feynman-Kac formulae *M2AN ESAIM*. vol. 44, no. 5, pp. 947–976 (Sept. 2010).

[194] P. Del Moral, A. Doucet, and S. S. Singh. Forward smoothing using sequential Monte Carlo. *Technical Report CUED/F-INFENG/TR 638. Cambridge University Engineering Department* (2009).

[195] P. Del Moral, A. Doucet, and S. S. Singh. Computing the filter derivative using Sequential Monte Carlo. *Technical Report. Cambridge University Engineering Department* (2011).

[196] P. Del Moral, P. Hu, and N. Oudjane. Snell envelope with small probability criteria. *Applied Mathematics and Optimization*, vol. 66, no. 3, pp. 309–330, December (2012).

[197] P. Del Moral, P. Hu, N. Oudjane, Br. Rémillard. On the robustness of the Snell envelope. *SIAM Journal on Financial Mathematics*, vol. 2, pp. 587–626 (2011).

[198] P. Del Moral, M.A. Kouritzin, and L. Miclo. On a class of discrete generation interacting particle systems. *Electronic Journal of Probability*, no. 16, no. 26 (2001).

[199] P. Del Moral, and S. Tindel. A Berry-Esseen theorem for Feynman-Kac and interacting particle models. *Annals of Applied Probability*, vol. 15, no. 1B, pp. 941–962 (2005).

[200] P. Del Moral, J. Tugaut. Uniform propagation of chaos for a class of inhomogeneous diffusions. *hal-00798813* (2013).

[201] B. E. Brunet, B. Derrida, A. H. Muller, and S. Munier. Effect of selection on ancestry: and exactly soluble case and its phenomenological generalization. *Phys. Review E*, vol. 76, 041104 (2007).

[202] B. Derrida, M. R. Evans, and E. R. Speer. Mean field theory of directed polymers with random complex weights. *Commun. Math. Phys.*, vol. 156, pp. 221–244 (1993).

[203] B. Derrida, and D. Simon. The survival probability of a branching random walk in the presence of an absorbing wall. *Europhys. Lett. EPL*, vol. 78, Art. 60006 (2007).

[204] B. Derrida, and D. Simon. Quasi-stationary regime of a branching random walk in presence of an absorbing wall. *J. Statist. Phys.*, vol. 131, pp. 203–233 (2008).

[205] B. Derrida, and H. Spohn. Polymers on disordered trees, spin glasses, and traveling waves. *Journal of Statistical Physics*, vol. 51, nos. 5/6, pp. 817–840 (1988).

[206] P. Diaconis. Some things we've learned (about Markov chain Monte Carlo). Preprint Stanford University (2012).

[207] P. Diaconis, G. Lebeau, and L. Michel. Gibbs/Metropolis algorithm on a convex polytope. *Math. Zeitschrift*, published online: Aug. (2011).

[208] P. Diaconis, G. Lebeau, and L. Michel. Geometric Analysis for the Metropolis Algorithm on Lipschitz Domains. *Invent. Math.*, vol. 185, no. 2, pp. 239–281 (2010).

[209] P. Diaconis and L. Miclo. On characterizations of Metropolis type algorithms in continuous time. *Alea*, vol. 6, pp. 199–238 (2009).

[210] P. Diaconis and G. Lebeau. Micro-local analysis for the Metropolis algorithm. *Mathematische Zeitschrift*, vol. 262, no. 2, pp. 441–447 (2009).

[211] P. Diaconis. The Markov Chain Monte Carlo revolution. *Bull. Amer. Math. Soc.*, November (2008).

[212] P. Diaconis and F. Bassetti. Examples comparing importance sampling and the Metropolis algorithm. *Illinois Journal of Mathematics*, vol. 50, no. 1–4, pp. 67–91 (2006).

[213] P. Diaconis J. Neuberger. Numerical results for the Metropolis algorithm. *Experimental Math.*, vol. 13, no. 2, pp. 207–214 (2004).

[214] P. Diaconis and L. Billera. A geometric interpretation of the Metropolis-Hastings algorithm. *Statis. Sci.*, vol. 16, no. 4, pp. 335–339 (2001).

[215] P. Diaconis and A. Ram. Analysis of systematic scan Metropolis Algorithms using Iwahori-Hecke algebra techniques. *Michigan Journal of Mathematics*, vol. 48, no. 1, pp. 157–190 (2000).

[216] P. Diaconis, L. Saloff-Coste. What do we know about the Metropolis algorithm? *Jour. Comp. System Sciences*, vol. 57, pp. 20–36 (1998).

[217] P. Diaconis and P. Hanlon. Eigen-analysis for some examples of the Metropolis algorithm. *Contemporary Math.*, vol. 138, pp. 99–117 (1992).

[218] P. J. Diggle. *Statistical Analysis of Spatial Point Patterns*, 2nd ed., Arnold (2003).

[219] R. L. Dobrushin. Description of a random field by means of conditional probabilities, and the conditions governing its regularity. *Theor. Proba. Appli.*, vol. 13, pp. 197–224 (1968).

[220] R. Douc, A. Garivier, E. Moulines, and J. Olsson. On the forward filtering backward smoothing particle approximations of the smoothing distribution in general state spaces models. *Annals of Applied Probab.*, vol. 21, no. 6, pp. 2109–2145 (2011).

[221] A. Doucet, B. Vo, and S. S. Singh. Sequential Monte Carlo methods for Bayesian multi-target filtering with random finite sets. *IEEE Trans. Aerospace Elec. Systems*, vol. 41, pp. 1224–1245 (2005).

[222] A. Doucet, J.F.G. de Freitas, and N.J. Gordon, Eds. *Sequential Monte Carlo Methods in Practice*. Springer-Verlag, New York (2001).

[223] A. Doucet and A.M. Johansen. A tutorial on particle filtering and smoothing: fifteen years later in *Handbook of Nonlinear Filtering*, D. Crisan and B. Rozovsky, Eds., Cambridge University Press (2009).

[224] A. Doucet, L. Montesano, and A. Jasra. Optimal filtering for partially observed point processes, *Proceedings ICASSP* (2006).

[225] A. Doucet, S.J. Godsill, and C. Andrieu. On sequential Monte Carlo sampling methods for Bayesian filtering. *Statistics and Computing*, vol. 10, pp. 197–208 (2000).

[226] C. Dubarry and S. Le Corff. Nonasymptotic deviation inequalities for smoothed additive functionals in nonlinear state-space models, Preprint (2012).

[227] C. Dubarry, P. Del Moral, and E. Moulines, Ch. Vergé. On the convergence of Island particle models, Preprint (2012).

[228] P. Dupuis and H. Wang. Importance sampling for Jackson networks. *Queueing Systems*, vol. 62, pp. 113–157 (2009).

[229] P. Dupuis, K. Leder, and H. Wang. Importance sampling for weighted-serve-the-longest-queue, *Math. of Operations Research*, vol. 34, pp. 642–660 (2009).

[230] P. Dupuis, K. Leder, and H. Wang. Large deviations and importance sampling for a tandem network with slow-down. *QUESTA*, vol. 57, pp. 71–83 (2007).

[231] P. Dupuis, K. Leder, and H. Wang. Importance sampling for sums of random variables with regularly varying tails. *ACM Trans. on Modelling and Computer Simulation*, vol. 17, pp. 1–21 (2007).

[232] P. Dupuis, D. Sezer, and H. Wang. Dynamic importance sampling for queueing networks. *Annals of Applied Probability*, vol. 17, pp. 1306–1346 (2007).

[233] P. Dupuis and H. Wang. Sub-solutions of an Isaacs equation and efficient schemes for importance sampling. *Mathematics of Operations Research*, vol. 32, pp. 1–35 (2007).

[234] P. Dupuis and H. Wang. Optimal stopping with random intervention times. *Adv. Applied Probability*, vol. 34, pp.1–17 (2002).

[235] P. Dupuis and H. Wang. On the convergence from discrete to continuous time in an optimal stopping problem. *Annals of Applied Probability*, vol. 15, pp. 1339–1366 (2005).

[236] P. Dybvig. Distributional analysis of portfolio choice. *The Journal of Business*, vol. 61, no. 3, pp. 369–393 (1988).

[237] A. Eberle and C. Marinelli. Quantitative approximations of evolving probability measures and sequential MCMC methods. *Probability Theory and Related Fields*, pp. 1–37 (2008).

[238] A. Eberle and C. Marinelli. Stability of nonlinear flows of probability measures related to sequential MCMC methods. *ESAIM: Proceedings*, vol. 19, pp. 22–31 (2007).

[239] A. Eberle and C. Marinelli. Lp estimates for Feynman-Kac propagators with time-dependent reference measures. *Journal of mathematical analysis and applications*, vol. 365, no. 1, pp. 120–134. (2010).

[240] A. Economou. Generalized product-form stationary distributions for Markov chains in random environments with queueing applications. *Adv. in Appl. Probab.*, vol. 37, no. 1, pp. 185–211 (2005).

[241] D. Egloff. Monte Carlo algorithms for optimal stopping and statistical learning. *Annals of Applied Probability*, 15, pp. 1–37 (2005).

[242] D. Egloff, M. Kohler, and N. Todorovic. A dynamic look-ahead Monte Carlo algorithm for pricing Bermudan options. *Ann. Appl. Probab.*, 17, pp. 1138–1171 (2007).

[243] N. El Karoui, S. Peng, and M. C. Quenez. Backward stochastic differential equations in finance. *Mathematical Finance*, vol. 7, no. 1, pp. 1–71 (1997).

[244] M. Ellouze, J.P. Gauchi, and J.C. Augustin. Global sensitivity analysis applied to a contamination assessment model of Listeria monocytogenes in cold smoked salmon at consumption. *Risk Anal.*, vol. 30, pp. 841–852 (2010).

[245] M. Ellouze, J.P. Gauchi, and J.C. Augustin. Use of global sensitivity analysis in quantitative microbial risk assessment: application to the evaluation of a biological time temperature integrator as a quality and safety indicator for cold smoked salmon. *Food Microbiol.*, vol. 28, no. 4, pp 755–769 (2011).

[246] M. El Makrini, B. Jourdain and T. Lelièvre. Diffusion Monte Carlo method: Numerical analysis in a simple case. *ESAIM: M2AN*, vol. 41, no. 2, pp. 189–213, (2007).

[247] G. Evensen. Sequential data assimilation with a nonlinear quasi-geostrophic model using Monte Carlo methods to forecast error statistics. *Journal of Geophysical Research (Oceans)*, vol. 99, no. C5, pp. 10143–10162 (1994).

[248] G. Evensen. Ensemble Kalman filter: Theoretical formulation and practical implementations. *Ocean Dynamics*, vol. 53, no. 4, pp. 343–367 (2003).

[249] G. Evensen. *Data Assimilation: The Ensemble Kalman Filter*. Springer-Verlag, Berlin (2006).

[250] G. Falkovich, K. Gawedzki, and M. Vergassola, Particles and fields in fluid turbulence. *Rev. Modern Phys.*, vol. 73, pp. 913–975 (2001).

[251] Y. E. Famao, S.U. Lin, L. Shukai, and L. Shengyang. Extraction of complex object contour by particle filtering. In *IEEE Int. Geoscience and Remote Sensing Symposium*, pp. 3711–3713 (2005).

[252] P. Fearnhead. Computational methods for complex stochastic systems: a review of some alternatives to MCMC. *Statistics and Computing*, vol. 18, pp. 151–171 (2008).

[253] P. Fearnhead, P. Clifford. On-line inference for hidden Markov models via particle filters. *Journal of the Royal Statistical Society*. Series B, vol. 65, no. 4, pp. 887–899 (2003).

[254] P. A. Ferrari, H. Kesten, S. Martinez, and P. Picco. Existence of quasi-stationary distributions. A renewal dynamical approach. *Ann. Probab.*, vol. 23, no. 2, pp. 501–521 (1995).

[255] P. A. Ferrari and N. Maric. Quasi stationary distributions and Fleming-Viot processes in countable spaces. *Electron. J. Probab.*, vol. 12, no. 24, pp. 684–702 (2007).

[256] G. Fishman. *Monte Carlo. Concepts, Algorithms and Applications*. Springer-Verlag (1996).

[257] J. J. Florentin. Optimal observability and optimal control. *J. of Electronics and Control*. vol. 13, pp. 263–279 (1962).

[258] I. Florescu and F. Viens. Stochastic volatility: option pricing using a multinomial recombining tree. *Applied Mathematical Finance*, vol. 15, no. 2, pp. 151–181 (2008).

[259] J. P. Fouque, G. Papanicolaou, and R. Sircar. *Derivatives in Financial Markets with Stochastic Volatility*. Cambridge University Press (2000).

[260] E. Fournié, J. M. Lasry, J. Lebuchoux, P. L. Lions, and N. Touzi. Applications of Malliavin calculus to Monte Carlo methods in finance, *Finance and Stochastics*, vol. 3, pp. 391–412 (1999).

[261] E. Fournié, J. M. Lasry, J. Lebuchoux, and P. L. Lions. Applications of Malliavin calculus to Monte Carlo methods in finance. II, *Finance and Stochastics*, vol. 5, pp. 201–236 (2001).

[262] H. Frauenkron, U. Bastolla, E. Gerstner, P. Grassberger, and W. Nadler. New Monte Carlo algorithm for protein folding. *Physical Review Letters*, vol. 80, no. 14, pp. 3149–3152 (1998).

[263] H. Frauenkron, M.S. Causo, and P. Grassberger. Two-dimensional self-avoiding walks on a cylinder. *Phys. Rev., E*, vol. 59, pp. 16–19 (1999).

[264] M. C. Fu, D. B. Madan, and T. Wang. Pricing continuous Asian options: A comparison of Monte Carlo and Laplace transform inversion methods. *Journal of Computational Finance*, vol. 2, pp. 49–74, (1998).

[265] J. Gärtner. On the McKean-Vlasov limit for interacting particles. *Math. Nachr.*, vol. 137, pp. 197–248 (1988).

[266] D. Gasbarra. Particle filters for counting process observations. http://www.rni.helsinki.fi/~dag/newpart2.ps, Research report Helsinki University, (2001).

[267] J. P. Gauchi, J. P. Vila, and L. Coroller. New prediction confidence intervals and bands in the nonlinear regression model: Application to the predictive modelling in food. *Communications in Statistics, Simulation and Computation*, vol. 39, no. 2, pp. 322–330 (2009).

[268] J. P. Gauchi, C. Bidot, J.C. Augustin, and J. P. Vila. Identification of complex microbiological dynamic system by nonlinear filtering. *6th Int. Conference on Predictive Modelling in Foods, Washington DC* (2009).

[269] H. Gea and S. Asgarpoorb. Parallel Monte Carlo simulation for reliability and cost evaluation of equipment and systems. *Electric Power Systems Research*, vol. 81, no. 2, pp. 347–356 (2011).

[270] S. Geman and D. Geman. Stochastic relaxation, Gibbs distributions, and the Bayesian restoration of images. *IEEE Transactions on Pattern Analysis and Machine Intelligence*, vol. 6, no. 6, pp. 721–741 (1984).

[271] V. Genon-Catalot, T. Jeantheau, and C. Laredo. Conditional likelihood estimators for hidden Markov models and stochastic volatility models. *Scandinavian Journal of Statistics*, vol. 30, pp. 297–316 (2003).

[272] J. Geweke and G. Durham. Massively Parallel Sequential Monte Carlo for Bayesian Inference. Available at SSRN: http://ssrn.com/abstract=1964731 or http://dx.doi.org/10.2139/ssrn.1964731 (2011).

[273] J. W. Gibbs. *Elementary Principles in Statistical Mechanics*. Yale University Press, New Haven, CT (1902).

[274] W. R. Gilks, S. Richardson, and D.J. Spiegelhalter. *Markov chain Monte Carlo in practice*. Chapman & Hall, London (1996).

[275] W. R. Gilks and C Berzuini. Following a moving target Monte Carlo inference for dynamic Bayesian models. *Journal of the Royal Statistical Society: Series B*, vol. 63, no. 1, pp. 127–146 (2002).

[276] W. R. Gilks, S. Richardson, and D.J. Spiegelhalter. *Markov chain Monte Carlo in practice*. Chapman & Hall (1996).

[277] E. Giné. *Lectures on some aspects of the bootstrap*. Lecture Notes in Maths, vol. 1665, Springer (1997).

[278] E. Giné, R. Latala, J. Zinn. Exponential and moment inequalities for *U*-statistics. *High Dimensional Proba. II*. Birkhäuser Boston, pp.13–38 (2000).

[279] F. Giraud, P. Del Moral. Nonasymptotic analysis of adaptive and annealed Feynman-Kac particle models. arXiv:1209.5654 (2012).

[280] F. Giraud, P. Minvielle, M. Sancandi, and P. Del Moral. Rao-Blackwellised interacting Markov chain Monte Carlo for electromagnetic scattering inversion. *Journal of Physics: Conference Series* 386 (1), 012008 (2012).

[281] M. Girolami, B. Calderhead. Riemann manifold Langevin and Hamiltonian Monte Carlo methods. *Journal of the Royal Statistical Society: Series B*, vol. 73, no. 2, pp. 123–214 (2011).

[282] P. Glasserman, P. Heidelberger, P. Shahabuddin, and T. Zajic. Multilevel splitting for estimating rare event probabilities. *Operations Research*, vol. 47, pp. 585–600 (1999).

[283] P. Glasserman and B. Yu. Number of paths versus number of basis functions in American option pricing. *Ann. Appl. Probab.*, 14, pp. 1–30 (2004).

[284] P. Glasserman. *Monte Carlo Methods in Financial Engineering.* Springer-Verlag (2004).

[285] E. Gobet, J. P. Lemor, and X. Warin. A regression-based Monte-Carlo method for backward stochastic differential equations. *Annals Applied Probability*, vol. 15, pp. 2172–2202 (2005).

[286] D. E. Goldberg. *Genetic Algorithms in Search, Optimization and Machine Learning.* Addison-Wesley, Reading, MA (1989).

[287] D. A. Gomes, J. Mohr, and R. Rigao Souza. Discrete time, finite state space mean field games. *Journal de Mathématiques Pures et Appliquées*, vol. 93, issue 3, pp. 308-328 (2010).

[288] N. J. Gordon, D. Salmond, and A. F. M. Smith. A novel approach to state estimation to nonlinear non-Gaussian state estimation. *IEE Proceedings* F, vol. 40, pp. 107–113 (1993).

[289] N. J. Gordon, D. Salmond, and E. Craig. Bayesian state estimation for tracking and guidance using the bootstrap filter. *Journal of Guidance, Control, and Dynamics*, vol. 18, no. 6, pp. 1434–1443 (1995).

[290] N. J. Gordon, A hybrid bootstrap filter for target tracking in clutter. *IEEE Transactions on Aerospace and Electronic Systems*, vol. 33, no. 1, pp. 353–358 (1997).

[291] F. Gosselin. Asymptotic behavior of absorbing Markov chains conditional on nonabsorption for applications in conservation biology. *Ann. Appl. Probab.*, vol. 11, pp. 261–284 (2001).

[292] J. Goutsias, R. Mahler, and H. Nguyen, Eds. *Random Sets Theory and Applications*, Springer-Verlag New York (1997).

[293] I. Goodman, R. Mahler, and H. Nguyen. *Mathematics of Data Fusion*, Kluwer Academic Publishers (1997).

[294] H. Grad. Principles of the kinetic theory of gases. In *Flugge's Handbuch des Physik*, Springer-Verlag, vol. XII. pp. 205–294 (1958).

[295] C. Graham and S. Méléard. Stochastic particle approximations for generalized Boltzmann models and convergence estimates *Ann. Probab.*, vol. 25, no. 1, pp. 115–132 (1997).

[296] C. Graham. A statistical physics approach to large networks. Probabilistic models for nonlinear partial differential equations (Montecatini Terme, 1995), Lecture Notes in Math., Springer Berlin, vol. 1627, pp.127–147 (1996).

[297] C. Graham and S. Méléard. Dynamic asymptotic results for a generalized star-shaped loss network. *Ann. Appl. Probab.*, vol. 5, no. 3, pp. 666–680 (1995).

[298] C. Graham. Homogenization and propagation of chaos to a nonlinear diffusion with sticky reflection. *Probab. Theory Related Fields*, vol. 101, no. 3, pp. 291–302 (1995).

[299] C. Graham and S. Méléard. Fluctuations for a fully connected loss network with alternate routing. *Stochastic Process. and Appl.*, vol. 53, no. 1, pp. 97–115 (1994).

[300] C. Graham and S. Méléard. Chaos hypothesis for a system interacting through shared resources. *Probab. Theory Related Fields*, vol. 100, no. 2, pp. 157–173 (1994).

[301] C. Graham and S. Méléard. Propagation of chaos for a fully connected loss network with alternate routing. *Stochastic Process. Appl.*, vol. 44, no. 1, pp. 159–180 (1993).

[302] C. Graham. Nonlinear diffusion with jumps. *Ann. Inst. H. Poincaré Probab. Statist.*, vol. 28, no. 3, pp. 393–402 (1992).

[303] C. Graham. McKean-Vlasov Ito-Skorohod equations, and nonlinear diffusions with discrete jump sets. *Stochastic Process. Appl.*, vol. 40, no. 1, pp. 69–82 (1992).

[304] C. Graham. Nonlinear limit for a system of diffusing particles which alternate between two states. *Appl. Math. Optim.*, vol. 22, no. 1, pp. 75–90 (1990).

[305] C. Graham and M. Métivier. System of interacting particles and nonlinear diffusion reflecting in a domain with sticky boundary. *Probab. Theory Related Fields*, vol. 82, no. 2, pp. 225–240 (1989).

[306] C. Graham. The martingale problem with sticky reflection conditions, and a system of particles interacting at the boundary. *Ann. Inst. H. Poincar Probab. Statist.*, vol. 24, no. 1, pp. 45–72 (1988).

[307] I. Grigorescu and M. Kang. Hydrodynamic limit for a Fleming-Viot type system. *Stochastic Process. Appl.*, vol. 110, no. 1, pp. 111–143 (2004).

[308] A. Guyader, N. Hengartner, and E. Matzner-Løber. Simulation and estimation of extreme quantiles and extreme probabilities. *Applied Mathematics & Optimization* (2011).

[309] P. Grassberger. Pruned-enriched Rosenbluth method: Simulations of θ polymers of chain length up to 1,000,000. *Phys. Rev. E*, pp. 3682–3693 (1997).

[310] P. Grassberger. Go with the winners: A general Monte Carlo strategy. *Computer Physics Communications*, vol. 147, no. 1, pp. 64–70 (2002).

[311] E. J. Green. Continuum and finite player noncooperative models of competition. *Econometrica*, vol. 52, no. 4, pp. 975–993 (1984).

[312] G. Grimmett, and D. Stirzaker. *Probability and Random Processes*, 2nd ed. Oxford University Press, Oxford (1992).

[313] D. Grünbaum. Translating stochastic density-dependent individual behavior with sensory constraints to an Eulerian model of animal swarming. *J. Math. Biology*, vol. 33, pp. 139–161 (1994).

[314] D. Grünbaum and A. Okubo. Modelling social animal aggregations. In *Frontiers of Theoretical Biology* (S. Levin, Ed.), Lectures Notes in Biomathematics, 100, Springer-Verlag, New York, pp. 296–325 (1994).

[315] S. Gueron, S.A. Levin, and D.I. Rubenstein. The dynamics of herds: from individuals to aggregations, *J. Theor. Biol.*, 182, pp. 85–98 (1996).

[316] N. M. Haan and S. J. Godsill. Sequential methods for DNA sequencing. Acoustics, Speech, and Signal Processing, 2001. *Proceedings (ICASSP'01). 2001 IEEE International Conference*, vol. 2. IEEE (2001).

[317] O. H. Hald. Convergence of random methods for a reaction diffusion equation. *SIAM J. Sci. Statist. Comput.*, vol. 2, pp. 85–94 (1981).

[318] D. L. Hanson and F. T. Wright. A bound on tail probabilities for quadratic forms in independent random variables. *Ann. Math. Stat.*, vol. 43, no. 2, pp. 1079–1083 (1971).

[319] T. E. Harris. *The Theory of Branching Processes*, Springer-Verlag, Berlin (1963).

[320] T. E. Harris. Branching processes. *Ann. Math. Statist.*, 19, pp. 474–494 (1948).

[321] J. W. Harris, and S.C Harris. Survival probability for branching Brownian motion with absorption. *Electron. Comm. Probab.*, vol. 12, pp. 89–100 (2007).

[322] T. E. Harris and H. Kahn. Estimation of particle transmission by random sampling. *Natl. Bur. Stand. Appl. Math. Ser.*, vol. 12, pp. 27–30 (1951).

[323] P. H. Haynes and J. Vanneste. What controls the decay rate of passive scalars in smooth random flows? *Phys. Fluids*, vol. 17, 097103 (2005).

[324] K. Heine. Unified framework for sampling/Importance resampling algorithms. *Proceedings of FUSION Conference*, Philadelphia (2005)

[325] A. W. Heemink, M.Verlaan, and A. J. Segers. Variance reduced ensemble Kalman filtering. *Monthly Weather Review*, vol. 129, no. 7, pp. 1718–1728 (2001).

[326] J. H. Hetherington. Observations on the statistical iteration of matrices. *Phys. Rev. A*, vol. 30, pp. 2713–2719 (1984).

[327] J. H. Holland. *Adaptation in Natural and Artificial Systems*. University of Michigan Press, Ann Arbor (1975).

[328] C. Houdré. P. Reynaud-Bouret. Exponential inequalities, with constants, for U-statistics of order two. *Stochastic inequalities and applications*, Progr. Probab., 56 Birkhäuser, Basel, pp. 55–69 (2003)

[329] M. Huang, R.P. Malhamé, and P.E. Caines. Large population stochastic dynamical games. Closed loop McKean Vlasov systems and the Nash certainty equivalence principle. *Communications in Information and Systems*, vol. 6, no. 3, pp. 221–252 (2006).

[330] M. Huang, R.P. Malhamé, and P.E. Caines. Individual and mass behaviour in large population stochastic wireless power control problems: Centralized and Nash equilibrium solutions. *Proceedings of the 42nd IEEE Conference on Decision and Control, Maui, Hawaii*, December 2003, pp. 98–103 (2003).

[331] M. Huang, R.P. Malhamé, and P.E. Caines. Large-population cost-coupled LQG problems: generalizations to nonuniform individuals. *Proceedings of the 43rd IEEE Conference on Decision and Control, Atlantis, Paradise Island, Bahamas*, December 2004, pp. 3453–3458 (2004).

[332] M. Huang, R.P. Malhamé, P.E. Caines. An invariance principle in large population stochastic dynamic games. *Journal of Systems Science & Complexity*, vol. 20, no. 2, pp. 162–172 (2007).

[333] M. Huang, R.P. Malhamé, and P.E. Caines. Large-population cost-coupled LQG problems with nonuniform agents: individual-mass behavior and decentralized ?-Nash equilibria. *IEEE Transactions on Automatic Control*, vol. 52, no. 9, pp. 1560–1571 (2007).

[334] M. Huang, P.E. Caines, and R.P. Malhamé. Social optima in mean field LQG control: centralized and decentralized strategies. *Proceedings of the 47th Annual Allerton Conference, Allerton House, UIUC, Illinois*, September 2009, pp. 322–329 (2009).

[335] M. Huang, P.E. Caines, and R.P. Malhamé, Social certainty equivalence in mean field LQG control: social, Nash and centralized strategies. *Proceedings of the 19th International Symposium on Mathematical Theory of Networks and Systems, Budapest, Hungary*, July 2010, pp. 1525–1532 (2010).

[336] Y. Iba. Population Monte Carlo algorithms. Transactions of the Japanese Society for Artificial Intelligence, vol. 16, no. 2, pp. 279–286 (2001).

[337] E. Ikonen, P. Del Moral, and K. Najim. A genealogical decision tree solution to optimal control problems. In *IFAC Workshop on Advanced Fuzzy/Neural Control, Oulu, Finland*, pp. 169–174 (2004).

[338] E. Ikonen, K. Najim, and P. Del Moral. Application of genealogical decision trees for open-loop tracking control. In *Proceedings of the 16th IFAC World Congress, Prague, Czech* (2005).

[339] M. Isard and A. Blake. Condensation conditional probability density propagation for visual tracking. *International Journal of Computer Vision*, vol. 29, pp. 5–28 (1998).

[340] J. Jacod and A. Shiryaev. *Limit Theorems for Stochastic Processes.* Springer, New York (1987).

[341] P. Jagers. *Branching Processes with Biological Applications.* John Wiley & Sons, London (1975).

[342] C. Jarzynski. Nonequilibrium equality for free energy differences. *Phys. Rev. Lett.*, vol. 78, 2690 (1997).

[343] C. Jarzynski. Equilibrium free-energy differences from nonequilibrium measurements: A master-equation approach. *Phys. Rev. E*, vol. 56, 5018 (1997).

[344] A. H. Jazwinski. *Stochastic Processes and Filtering Theory*. Academic Press, New York (1970).

[345] M. S. Johannes, N. G. Polson, and J. R. Stroud. Optimal filtering of jump diffusions: extracting latent states from asset prices. *Review of Financial Studies*, 22, issue. 7, pp. 2759–2799 (2009).

[346] A. M. Johansen, P. Del Moral, and A. Doucet. Sequential Monte Carlo samplers for rare events. In *Proceedings of 6th International Workshop on Rare Event Simulation, Bamberg, Germany* (2006).

[347] W. B. Johnson, G. Schechtman, and J. Zinn. Best constants in moment inequalities for linear combinations of independent and exchangeable random variables. *Ann. Probab.*, vol. 13, no. 1, pp. 234–253 (1985).

[348] B. Jourdain. Diffusion processes associated with nonlinear evolution equations for signed measures. *Methodology and Computing in Applied Probability*, vol. 2, no. 1, pp. 69–91 (2000).

[349] B. Jourdain and F. Malrieu. Propagation of chaos and Poincar inequalities for a system of particles interacting through their cdf. *Annals of Applied Probability*, vol. 18, no. 5, 1706–1736 (2008).

[350] M. Kac. On distributions of certain Wiener functionals, *Trans. American Math. Soc.*, vol. 61, no. 1, pp. 1–13, (1949).

[351] M. Kac. Foundations of kinetic theory. *Proceedings 3rd Berkeley Sympos. Math. Stat. and Proba.*, vol. 3, pp. 171-197 (1956).

[352] O. Kallenberg. *Foundations of Modern Probability*. Springer-Verlag, New York (1997).

[353] R. E. Kalman. A new approach to linear filtering and prediction problems. *Transactions of the ASME–Journal of Basic Engineering, Series D.*, vol. 82, pp. 25–45 (1960).

[354] I. Karatzas and S.E. Shreve. *Methods of Mathematical Finance*. Springer (1998).

[355] I. Karatzas and S. E. Shreve, *Brownian Motion and Stochastic Calculus*, Graduate Texts in Mathematics, Springer (2004).

[356] S. Karlin. *Mathematical Methods and Theory in Games, Programming and Economics*, in two volumes, Addison-Wesley, London (1959).

[357] S. Karlin. Stochastic models and optimal policy for selling an asset. In *Studies in Applied Probability and Management Science*. K. J. Arrow, S. Karlin, and H. Scarf, Eds., Stanford University Press, 148-158 (1962).

[358] H. Kesten. Branching Brownian motion with absorption. *Stochastic Process. and Appl.*, vol. 37, pp. 9–47 (1978).

[359] J. F. C. Kingman. The first birth problem for age dependent branching processes. *Ann. Probab.*, vol. 3, pp. 790–801 (1975).

[360] V. N. Kolokoltsov. *Nonlinear Markov Processes and Kinetic Equations.* Cambridge Univ. Press (2010).

[361] V. N. Kolokoltsov, J. Li, W. Yang. Mean field games and nonlinear Markov processes. arXiv:1112.3744v2 (2012).

[362] V. N. Kolokoltsov, and O. A. Malafeyev. *Analysis of Many Agent Systems of Competition and Cooperation. Game Theory for All* (in Russian). St. Petersburg University Press (2007).

[363] V. N. Kolokoltsov, and O.A. Malafeyev. *Understanding Game Theory. Introduction to the Analysis of Many Agent Systems of Competition and Cooperation.* World Scientific (2010).

[364] H. R. Künsch. State-space and hidden Markov models. In *Complex Stochastic Systems.* O. E. Barndorff-Nielsen, D. R. Cox, and C. Kluppelberg, Eds., CRC Press, pp. 109–173 (2001).

[365] J.G. Kemeny and J. L. Snell. *Finite Markov Chains.* Springer-Verlag, New York (1976).

[366] G. Kitagawa. Monte Carlo filter and smoother for non-Gaussian nonlinear state space models. *J. Comp. Graph. Statist.*, vol. 5, pp. 1–25 (1996).

[367] P. E. Kloeden, E. Platen, and N. Hofmann. Extrapolation methods for the weak approximation of Itô diffusions. *SIAM Journal on Numerical Analysis*, vol. 32, no. 5, pp. 1519–1534 (1995).

[368] P. E. Kloeden and E. Platen. *Numerical Solution of Stochastic Differential Equations*, vol. 23, Springer (2011).

[369] P. E. Kloeden and E. Platen. A survey of numerical methods for stochastic differential equations. *Stochastic Hydrology and Hydraulics*, vol. 3, no. 5, pp. 155-178 (1989).

[370] E. Platen and K. Kubilius. Rate of weak convergence of the Euler approximation for diffusion processes with jumps. School of Finance and Economics, University of Technology, Sydney (2001).

[371] A. Kohatsu-Higa and M. Montero. *Malliavin Calculus in Finance.* Handbook of Computational and Numerical Methods in Finance, Birkhäuser, pp. 111–174 (2004).

[372] A. Kohatsu-Higa and S. Ogawa. Weak rate of convergence for an Euler scheme of Nonlinear SDE's. In *Monte Carlo Methods and Its Applications*, vol. 3, pp. 327–345 (1997).

[373] V. N. Kolokoltsov and V. P. Maslov. *Idempotent Analysis and Its Applications*, vol. 401 of *Mathematics and its Applications* with an appendix by P. Del Moral on Maslov optimization theory. Kluwer Academic Publishers Group, Dordrecht (1997).

[374] K. Kremer and K. Binder. Monte Carlo simulation of lattice models for macromolecules. *Computer Physics Reports*, vol. 7, no. 6, pp. 259–310 (1988).

[375] S. N. Ethier and T. G. Kurtz. *Markov Processes: Characterization and Convergence*, Wiley Series Probability & Statistics (1986).

[376] A. Lachapelle, J. Salomon, and G. Turinici. Computation of mean field equilibria in economics. *Mathematical Models and Methods in Applied Sciences*, vol. 20, no. 4, pp. 567–588 (2010).

[377] T. Laffargue, K. D. Nguyen Thu Lam, J. Kurchan, and J. Tailleur. Large deviations of Lyapunov exponents. ArXiv 1302.6254 (2013).

[378] L. D. Landau, E. M. Lifshitz. Course of theoretical physics. *Physical kinetics*, E. M. Lifshitz and L. P. Pitaevskii eds. vol. 10, Oxford, Pergamon (1981).

[379] O.E. Lanford and D. Ruelle. Observables at infinity and states with short range correlations in statistical mechanics. *Comm. Math. Physics*, vol. 13, pp. 194–215 (1969).

[380] J. M. Laskry and P. L. Lions. Contrôle stochastique avec informations partielles et applications à la Finance. *Comptes Rendus de l'Académie des Sciences - Series I - Mathematics*, vol. 328, issue 11, pp. 1003–1010 (1999).

[381] J. M. Lasry and P.L. Lions. Mean field games. *Japanese J. Math*, vol. 2, no. 1, pp. 229–260 (2007).

[382] J. M. Lasry and P.L. Lions. Jeux à champ moyen. I. Le cas stationnaire. *C.R. Math. Acad. Sci. Paris.* 343(10), pp. 619-625 (2006).

[383] J. M. Lasry and P.L. Lions. Jeux à champ moyen. II. Horizon fini et contrôle optimal. *C.R. Math. Acad. Sci. Paris.* 343(10), pp. 679–684 (2006).

[384] J. M. Lasry and P.L. Lions. Mean field games. *Cahiers de la Chaire Finance et Développement Durable* (2) (2007).

[385] D. Lefèvre. An introduction to utility maximization with partial observation. *Finance*, vol. 23 (2002).

[386] F. Le Gland and N. Oudjane, A robustification approach to stability and to uniform particle approximation of nonlinear filters: the example of pseudo-mixing signals. *Stochastic Processes and Their Applications*, vol. 106, no. 2, pp. 279–316 (2003).

[387] F. Le Gland and N. Oudjane. Stability and uniform approximation of nonlinear filters using the Hilbert metric and application to particle filters. *Annals Applied Probability*, vol. 14, no. 1, 144–187 (2004).

[388] F. Le Gland, V. Monbet, and V. D. Tran. Large sample asymptotics for the ensemble Kalman filter. In *Handbook on Nonlinear Filtering*, D. Crisan and B. Rozovskii, Eds. Oxford University Press, Oxford, pp. 598–631 (2011).

[389] V. Tran, V. Monbet and F. Le Gland. Filtre de Kalman d'ensemble et filtres particulaires pour le modèle de Lorenz. Actes de la Manifestation des Jeunes Chercheurs STIC (MajecSTIC'06), Lorient, November 22-24 (2006).

[390] M. Lei and C. Baehr. Unscented and ensemble transform-based variational filter. Accepted to publication in *Physica D : Nonlinear Phenomena* (2012).

[391] M. Ledoux. The concentration of measure phenomenon. *AMS Monographs, Providence* (2001).

[392] T. Lelièvre, M. Rousset, and G. Stoltz. Computation of free energy differences through nonequilibrium stochastic dynamics: the reaction coordinate case. *J. Comp. Phys.*, vol. 222, no. 2, pp. 624–643 (2007).

[393] T. Lelièvre, M. Rousset, and G. Stoltz. *Free energy computations: A mathematical perspective.* 472 pp., Imperial College Press (2010).

[394] C. Léonard. Une loi des grands nombres pour des systèmes de diffusions avec interaction et à coéfficients non bornés. *Ann. Inst. Henri Poincaré*, vol. 22, no. 2, pp. 237–262 (1986).

[395] T. Li, J.F. Zhang. Asymptotically optimal decentralized control for large population stochastic multiagent systems. *IEEE Transactions on Automatic Control*, vol. 53, no. 7, pp. 1643–1660 (2008).

[396] L. Li, J. Bect, and E. Vazquez. Bayesian subset simulation: a kriging-based subset simulation algorithm for the estimation of small probabilities of failure. arXiv:1207.1963 v1 (2012).

[397] C. Liang, G. Cheng, D. L. Wixon. T. C. Balser. An absorbing Markov chain approach to understanding the microbial role in soil carbone stabilization. *Biogeochemistry*, vol. 106, pp. 303–309 (2011).

[398] A. M. L. Liekens. Evolution of Finite Populations in Dynamic Environments. Ph.D. thesis, Technische Universiteit Eindhoven (2005).

[399] M. Liebscher, S. Pannier, J. Sickert, and W. Graf. Efficiency improvement of stochastic simulations by means of subset sampling, *In Proceedings of the 5th German LS-DYNA Forum 2006. DYNAmore GmbH*, Ulm (2006).

[400] A. E. B. Lim. Mean variance hedging when there are jumps. *SIAM J. Control Optim.*, vol. 44, no. 5, pp. 1893–1922 (2005).

[401] P.L. Lions. Théorie des jeux à champ moyen et applications. Cours au Collège de France (2007-2008).

[402] P.L. Lions. *Mathematical Topics in Fluid Mechanics.* Oxford Science Publications. Clarendon Press. Oxford, vol. 1 (1996), vol 2 (1998).

[403] J.S. Liu. *Monte Carlo Strategies in Scientific Computing.* Springer Verlag, New York 2001.

[404] G. Liu and L. J. Hong. Revisit of stochastic mesh method for pricing American options. *Operations Research Letters*, vol. 37, no. 6, pp. 411–414 (2009).

[405] J.S. Liu and J.Z. Zhang. A new sequential importance sampling method and its applications to the 2-dimensional hydrophobichydrophilic model. *J. Chem. Phys.*, vol. 117, no. 7, pp. 3492–3498 (2002).

[406] J.S. Liu and R. Chen. Sequential Monte-Carlo methods for dynamic systems. *J. Am. Stat. Assoc.*, vol. 93, pp. 1032–1044 (1998).

[407] F. A. Longstaff and E. S. Schwartz. Valuing American options by simulation: a simple least-squares approach. *Review of Financial Studies*, vol. 14, pp. 113–147 (2001).

[408] H. F. Lopes, N. G. Polson, and C. M. Carvalho. Bayesian statistics with a smile: A resampling-sampling perspective. *Brazilian Journal of Probability and Statistics*, vol. 26, no. 4, pp. 358–371 (2012).

[409] S. McGinnity. Multiple model bootstrap filter for maneuvering target tracking. *IEEE Transactions on Aerospace and Electronic Systems*, vol. 36, no. 3, pp. 1006–1012 (2000).

[410] R. Mahler. A theoretical foundation for the Stein-Winter Probability Hypothesis Density (PHD) multi-target tracking approach, *Proc. MSS Nat'l Symp. on Sensor and Data Fusion*, vol. I, San Antonio TX (2000).

[411] R. Mahler. Multi-target Bayes filtering via first-order multi-target moments, *IEEE Trans. Aerospace & Electronic Systems*, vol. 39, no. 4, pp. 1152–1178 (2003).

[412] R. Mahler. Global integrated data fusion, *Proc. 7th Nat. Symp. on Sensor Fusion*, vol. 1, Sandia National Laboratories, Albuquerque, ERIM Ann Arbor, MI, pp. 187–199 (1994).

[413] R. Mahler. Random-set approach to data fusion, *SPIE*, vol. 2234 (1994).

[414] P. Major. Estimation of multiple random integrals and U-statistics. *Moscow Mathematical Journals*, vol. 10 no. 4, pp. 747–763 (2010).

[415] P. Major. Tail behaviour of multiple random integrals and U-statistics. *Probability Reviews Series*, vol. 2 pp. 448–505 (2005).

[416] P. Major. An estimate about multiple stochastic integrals with respect to a normalized empirical measure. *Studia Scientarum Mathematicarum Hungarica*, vol. 42, no. 3, pp. 295-341 (2005).

[417] P. Major. A multivariate generalization of Hoeffding's inequality. *Electronic Communication in Probability*, vol. 2, pp. 220–229 (2006).

[418] F. Malrieu. Convergence to equilibrium for granular media equations and their Euler schemes *Annals of Applied Probability*, vol. 13, no. 2, pp. 540–560 (2003).

[419] F. Malrieu. Inégalités de Sobolev logarithmiques pour des problèmes d'évolutions non linéaires. Ph.D. dissertation, Hal-00001287. *Université Paul Sabatier* (2001).

[420] S. Martinez, S. San Martin, and D. Villemonais. Existence and uniqueness of quasi- stationary distributions and fast return from infinity. Submitted to *Journal of Applied Probability*, 15 p. (2012).

[421] J. C. Mattingly, A. M. Stuart, and D. J. Higham. Ergodicity for SDEs and approximations: locally Lipschitz vector fields and degenerate noise. *Stochastic Process. Appl.*, vol. 101, no. 2, pp. 185–232 (2002).

[422] F. A. Matsen and J. Wakeley. Convergence to the island-model coalescent process in populations with restricted migration. *Genetics*, vol. 172, pp. 701–708 (2006).

[423] J. C. Maxwell. On the dynamical theory of gases. *Philos. Trans. Roy. Soc. London Ser. A*, vol. 157, pp. 49–88 (1867).

[424] J. C. Maxwell. On stresses in rarified gased arising from inequalities of temperatures. *Philos. Trans. Roy. Soc. London Ser. A*, vol. 170, pp. 231–256 (1879).

[425] J. C. Maxwell. *The Scientific Letters and Papers of James Clerk Maxwell.* vol. 2., pp. 1862-1873, Cambridge University Press (1995).

[426] J. Maynard Smith. *Evolution and the Theory of Games.* Cambridge University Press, Cambridge (1982).

[427] L. Mazliack. Approximation of a partially observable stochastic control problem. *Markov Processes and Related Fields*, vol. 5, pp. 477–486 (1999).

[428] H. P. McKean Jr. A class of Markov processes associated with nonlinear parabolic equations. *Proc. Nat. Acad. Sci. USA.* vol. 56, pp. 1907–1911 (1966)

[429] S. Méléard. Probabilistic interpretation and approximations of some Boltzmann equations. In Stochastic models. *Soc. Mat. Mexicana, Mexico*, pp. 1–64 (1998).

[430] S. Méléard. Asymptotic behaviour of some interacting particle systems; McKean-Vlasov and Boltzmann models. Probabilistic models for nonlinear partial differential equations. Lecture Notes in Mathematics, vol. 1627-1996, pp. 42–95 (1996).

[431] S. Méléard and S. Roelly-Coppoletta. A propagation of chaos result for a system of particles with moderate interaction. *Stochastic Process. Appl.*, vol. 26, pp. 317–332, (1987).

[432] S. Méléard. and D. Villemonais. Quasi-stationary distributions and population processes. Submitted to *Probability Surveys*, 65 p. (2011).

[433] V. Melik-Alaverdian and M.P. Nightingale. Quantum Monte Carlo methods in statistical mechanics, *Internat. J. of Modern Phys. C*, vol. 10, pp. 1409–1418 (1999).

[434] M. Métivier. Quelques problèmes liés aux systèmes infinis de particules et leurs limites. Lecture Notes in Math., vol. 1204, pp. 426–446 (1984).

[435] N. Metropolis and S. Ulam. The Monte Carlo method. *Journal of the American Statistical Association*, vol. 44, no. 247, pp. 335–341 (1949).

[436] N. Metropolis. The Beginning of the Monte Carlo method. *Los Alamos Science*, no. 15, pp. 125–130 (1987).

[437] N. Metropolis, A.W. Rosenbluth, M.N. Rosenbluth, A.H. Teller, and E. Teller. Equation of state calculations by fast computing machines. *Journal of Chemical Physics*, vol. 21, no. 6, pp. 1087–1092 (1953).

[438] S. P. Meyn and R.L. Tweedie. *Markov Chains and Stochastic Stability.* Springer-Verlag, London. Available at: probability.ca/MT (1993).

[439] S. P. Meyn. *Control Techniques for Complex Networks.* Cambridge University Press (2007).

[440] S. Mischler, C. Mouhot. Kac's program in kinetic theory. arXiv:1107.3251 (2011), to appear in *Inventiones mathematicae* (2013).

[441] S. Mischler, C. Mouhot, B. Wennberg. A new approach to quantitative propagation of chaos for drift, diffusion and jump processes. arXiv:1101.4727 (2011).

[442] D. Morale, V. Capasso, and K. Oeslschläger. An interacting particle system modeling aggregation behavior: from individual to populations. *J. Math. Bio.*, vol. 50, pp. 49–66 (2005).

[443] J. E. Moyal. The general theory of stochastic population processes. *Acta Mathematica*, vol. 108, pp. 1–31 (1962).

[444] K. P. Murphy, A. Doucet, N. De Freitas, and S. Russell. Rao-Blackwellised particle filtering for dynamic Bayesian networks. in *Proc. Uncertainty in Artificial Intelligence* (2000).

[445] T. Nagai and M. Mimura. Asymptotic behaviour for a nonlinear degenerate diffusion equation in population dynamics. *SIAM J. Appl. Math.*, 43, pp. 449–464 (1983).

[446] T. Nagai and M. Mimura. Some nonlinear degenerate diffusion equations related to population dynamics, *J. Math. Soc. Japan*, 35, pp. 539–561 (1983).

[447] K. Najim, E. Ikonen, and P. Del Moral. Open-loop regulation and tracking control based on a genealogical decision tree. *Neural Computing & Applications*, vol. 15, no. 3–4, pp. 339–349 (2006).

[448] J. Neveu. Multiplicative martingales for spatial branching processes. In *Seminar on Stochastic Processes, 1987.* E. Cinlar, K. L. Chung and R. K. Getoor, eds. *Prog. Probab. Statist.* vol.15, pp. 223–241. Birkhäuser, Boston (1988).

[449] M. Nourian, P.E. Caines, and R.P. Malhamé. Mean field analysis of controlled Cucker-Smale type flocking: linear analysis and perturbation equations. *Proceedings of the 18th IFAC World Congress, Milan*, August 2011, pp. 4471–4476 (2011).

[450] K. Oelschläger. A martingale approach to the law of large numbers for weakly interacting stochastic processes. *Ann. Probab.*, vol. 12, pp. 458–479 (1984).

[451] K. Oelschläger. On the derivation of reaction-diffusion equations as limit of dynamics of systems of moderately interacting stochastic processes. *Prob. Th. Rel. Fields*, vol. 82, pp. 565–586 (1989).

[452] K. Oelschläger. Large systems of interacting particles and porous medium equation, *J. Differential Equations*, vol. 88, pp. 294–346 (1990).

[453] B. K. Øksendal and A. Sulem. *Applied Stochastic Control of Jump Diffusions.* Springer, Berlin (2007).

[454] B. K. Øksendal. *Stochastic Differential Equations: An Introduction with Applications*, 6th ed. Springer, Berlin (2003).

[455] A. Okubo. Dynamical aspects of animal grouping: swarms, school, flocks and herds. *Adv. BioPhys.*, vol. 22, pp. 1–94 (1986).

[456] T.H. Otway. The Dirichlet problem for elliptic-hypebolic equations of Keldysh type. Lecture Notes in mathematics, Springer (2012).

[457] J. Olsson and T. Rydén. Rao-Blackwellisation of particle Markov chain Monte Carlo methods using forward filtering backward sampling. *IEEE Transactions on Signal Processing*, vol. 59, no. 10, pp. 4606–4619 (2011).

[458] A. Osekowski. Two inequalities for the first moments of a martingale, its square function and its maximal function *Bull. Polish Acad. Sci. Math.*, vol. 53, pp. 441–449 (2005).

[459] M. Pace, P. Del Moral, and F. Caron. Comparison of Implementations of Gaussian mixture PHD filters, *Proceedings of the 13th International Conference on Information Fusion* 26-29 July 2010 EICC Edinburgh, UK (2010).

[460] M. Pace. Comparison of PHD based filters for the tracking of 3D aerial and naval scenarios. 2010 *IEEE International Radar Conference Washington*, DC, on May 10–14 (2010).

[461] M. Pace and P. Del Moral. Mean-field PHD filters based on generalized Feynman-Kac flow. *IEEE Journal of Selected Topics in Signal Processing*, vol. 7, no. 2 (2013).

[462] G. Pagès and B. Wilbertz. Optimal quantization methods for pricing American style options. *Numerical Methods in Finance* (2011).

[463] G. Pagès and H. Pham. Optimal quantization methods for nonlinear filtering with discrete-time observations. *Bernoulli*, vol. 11, pp. 893–932 (2005).

[464] G. Pagès, H. Pham, and J. Printems. Optimal quantization methods and applications to numerical problems in finance. In *Handbook of Computational and Numerical Methods in Finance*. S. T. Rachev, Ed. Birkhäuser, Boston, pp. 253–297 (2004).

[465] G. Pagès. A space vector quantization method for numerical integration. *Journal of Comput. Appl. Math.*, 89, pp. 1–38 (1997).

[466] G. Pagès and J. Printems. Functional quantization for numerics with an application to option pricing. *Monte Carlo Methods and Appl.*, vol. 11, no. 4, pp. 407–446 (2005).

[467] N. Papageorgiou, B. Rémillard, and A. Hocquard. Replicating the properties of hedge fund returns. *Journal of Alternative Investments*, vol. 11, pp. 8–38 (2008).

[468] A. Papoulis. *Brownian Movement and Markov Processes. Probability, Random Variables, and Stochastic Processes*, 2nd ed. McGraw-Hill, New York, pp. 515–553 (1984).

[469] E. Pardoux, D. Talay. Discretization and simulation of stochastic differential equations. *Acta Applic. Math.*, vol. 3, no. 1, pp. 22–47 (1985).

[470] G. Peskir, and A. Shiryaev. *Optimal Stopping and Free-Boundary Problems*. Birkhäuser, Basel (2006).

[471] G. Pflug. *Stochastic Optimization*. Kluwer (1997).

[472] H. Pham, W. Runggaldier, and A. Sellami. Approximation by quantization of the filter process and applications to optimal stopping problems under partial observation. *Monte Carlo Methods and Applications*, vol. 11, no. 1, pp. 57–81 (2005).

[473] H. Pham. Mean-variance hedging under partial observation. *International Journal of Theoretical and Applied Finance*, vol. 4, pp. 263–284 (2001).

[474] H. Pham, and M. C. Quenez. Optimal portfolio in partially observed stochastic volatility models. *Annals of Applied Probability*, vol. 11, pp. 210–238 (2001).

[475] H. Pham, M. Corsi, and W. Runggaldier. Numerical approximation by quantization of control problems in finance under partial observations. In *Handbook of Numerical Analysis*, vol. 15, pp. 325–360 (2009).

[476] P. Pérez, A. Blake, and M. Gangnet. Jetstream: probabilistic contour extraction with particles. In *Proceedings, IEEE International Conference on Computer Vision*, pp. 524–531 (2001).

[477] B. Perthame. *Transport Equations in Biology. Frontiers in Mathematics*. Birkhaüser Verlag (2007).

[478] V. M. Pillai and M. P. Chandrasekharan. An absorbing Markov chain model for production systems with rework and scrapping. *Computers and Industrial Engineering*, vol. 55, no. 3, pp. 695–706 (2008).

[479] V. Plachouras, I. Ounis, and G. Amati. The static absorbing model for hyperlink analysis on the web. *Journal of Web Engineering*, vol. 4, no. 2, pp. 165–186 (2005).

[480] D. Pollard. *Empirical processes: Theory and Applications*. NSF-CBMS Regional Conference Series in Probability and Statistics. Publication of the Institute of Mathematical Statistics, Hayward, CA, vol. 2 (1990).

[481] S.T. Rachev. *Probability Metrics and the Stability of Stochastic Models.* Wiley, New York (1991).

[482] T. Rainsford and A. Bender. Markov approach to percolation theory based propagation in random media. *IEEE Transactions on Antennas and Propagation.* DOI:10.1109/TAP.2008.922626 (2008) vol 56, no. 3, pp. 1402–1412, May (2008).

[483] B. R. Rambharat and A. E. Brockwell. Sequential Monte Carlo pricing of American-style options under stochastic volatility models. *The Annals of Applied Statistics*, vol. 4, no. 1, pp. 222–265 (2010).

[484] C. R. Reeves and J. E. Rowe. *Genetic Algotihms: Principles and Perspectives: A Guide to GA Theory.* Kluwer Academic Publishers (2003).

[485] B. Rémillard, A. Hocquard, H. Langlois, and N. Papageorgiou. Optimal hedging of American options in discrete time. *Numerical Methods in Finance* (2011).

[486] B. Rémillard and S. Rubenthaler. Optimal hedging in discrete and continuous time. *Technical Report G-2009-77*, Cahiers du Gerad (2009).

[487] S. Rémy, O. Pannekoucke, T. Bergot, and C. Baehr. Adaptation of a particle filtering method for data assimilation in a 1D numerical model used for fog forecasting. *Quarterly Journal of Royal Meteorological Society.* vol. 138, no. 663, pp. 536–551, January (2012).

[488] D. Revuz and Marc Yor. *Continuous Martingales and Brownian Motion*, Springer-Verlag, New York (1991).

[489] C. Reyl, T. M. Antonsen, and E. Ott. Vorticity generation by instabilities in chaotic fluid flows, *Physica D*, vol. 111, pp. 202–226 (1998).

[490] G. Ridgeway and D. Madigan. A sequential Monte Carlo method for Bayesian analysis of massive datasets. *Data Mining and Knowledge Discovery*, vol. 7, no. 3, pp. 301–319 (2003).

[491] B. D. Ripley. *Stochastic Simulation.* Wiley & Sons (1987).

[492] E. Rio. Local invariance principles and their applications to density estimation. *Probability and Related Fields*, vol. 98, pp. 21–45 (1994).

[493] C.P. Robert and G. Casella. *Monte Carlo Statistical Methods*, 2nd ed. Springer-Verlag (2004).

[494] G. O. Roberts and R. L. Tweedie. Exponential convergence of Langevin distributions and their discrete approximations. *Bernoulli*, vol. 2, pp. 341–363 (1996).

[495] G. O. Roberts and R. L. Tweedie. Geometric convergence and central limit theorems for multi-dimensional hastings and metropolis algorithms. *Biometrika*, vol. 1, pp. 95–110 (1996).

[496] R. Rogerson, R. Shimer, and R. Wright. Search-theoretic models of the labor market: a survey. *Journal of Economic Literature*, 43, pp. 959–988 (2005).

[497] V. Rossi and J. P. Vila. Nonlinear filtering in discrete time: a particle convolution approach. *Ann. I.SU.P.*, vol. 50, no. 3, pp. 71–102 (2006).

[498] M. Rousset. On the control of an interacting particle approximation of Schroedinger ground states, *SIAM J. Math. Anal.*, vol. 38, no. 3, pp. 824–844, (2006).

[499] M. Rousset and G. Stoltz. Equilibrium sampling from nonequilibrium dynamics. *J. Stat. Phys.*, vol. 123, no. 6, pp. 1251–1272 (2006).

[500] R. Rubinstein. Entropy and cloning methods for combinatorial optimization, sampling and counting using the Gibbs sampler. *Information Theory and Statistical Learning*, pp. 385–434 (2009).

[501] R. Rubinstein. The Gibbs cloner for combinatorial optimization, counting and sampling. *Methodology and Computing in Applied Probability*, vol. 11, no. 4, pp. 491–549 (2009).

[502] R. Rubinstein. Randomized algorithms with splitting: why the classic randomized algorithms do not work and how to make them work. *Methodology and Computing in Applied Probability*, vol. 12, no. 1, pp. 1–50 (2010).

[503] R. Rubinstein. *Simulation and the Monte Carlo Method.* Wiley (1981).

[504] R. Rubinstein. *Monte Carlo Optimization Simulation and Sensitivity of Queueing Networks.* Wiley (1986).

[505] R. Rubinstein and D. P. Kroese. *The Cross-Entropy Method. A Unified Approach to Combinatorial Optimization, Simulation and Machine Learning.* Springer-Verlag (2005).

[506] R. Rubinstein, B. Melamed. Classical and Modern Simulation. Wiley (1998).

[507] W. Runggaldier and L. Stettner. *Approximations of Discrete Time Partially Observed Control Problems.* Applied Mathematics Monographs CNR, Giardini Editori, Pisa (1994).

[508] S. F. Schmidt. The Kalman filter: recognition and development for aerospace applications. *Journal of Guidance and Control*, vol. 4, no. 1, pp. 4–8 (1981).

[509] E. Seneta. *Nonnegative Matrices and Markov Chains*, 2nd rev. ed., XVI, p. 288, Springer Series in Statistics (1981).

[510] S.S. Singh, B.N. Vo, A. Baddeley, and S. Zuyev. Filters for Spatial Point Processes. *SIAM J. Control and Optimization*, vol. 47, no. 4, pp. 2275–2295 (2009).

[511] C. Schäfer and N. Chopin. Adaptive Monte Carlo on binary sampling spaces. *Statistics and Computing*. doi:10.1007/s11222-011-9299-z (2012).

[512] M. Schweizer. Variance-optimal hedging in discrete time. *Math. Oper. Res.*, vol. 20, no. 1, pp. 1–32 (1995).

[513] O. Schutze, C. A. Coello Coello, A.A. Tantar, E. Tantar, P. Bouvry, P. Del Moral, and P. Legrand. *Evolve - A Bridge between Probability, Set Oriented Numerics, and Evolutionary Computation II*. Advances in Intelligent Systems and Computing, Volume 175, Springer (2012).

[514] D. Sherrington and S. Kirkpatrick. Solvable model of a spin-glass, *Physics Review Letters*, vol. 35, no. 26, pp. 1792–1796 (1975).

[515] Z. Shi. Branching random walks. École d'été de Probabilités de Saint Flour (2012).

[516] T. Shiga and H. Tanaka. Central limit theorems for a system of Markovian particles with mean field interactions. *Z. Wahrscheinlichkeitstheorie, verw. Gebiete*, vol. 69, pp. 439–459 (1985).

[517] A. N. Shiryaev. On optimal methods in quickest detection problems. *Theory Prob. and Appl.*, vol. 8, pp. 22–46 (1963).

[518] A. N. Shiryaev. *Probability*. Graduate Texts in Mathematics, 2nd ed., vol. 95, Springer (1996).

[519] O. Skare, E. Bolviken, and L. Holden. Improved sampling-importance resampling and reduced bias importance sampling. *Scandinavian Journal of Statistics*, vol. 30, no. 4, pp. 719–737 (2003).

[520] Z. Skolicki and K. De Jong. The influence of migration sizes and intervals on island models. In *Proceedings of the Genetic and Evolutionary Computation Conference (GECCO-2005)*. ACM Press (2005).

[521] A. F. M. Smith and G. O. Roberts. Bayesian Computation Via the Gibbs Sampler and Related Markov Chain Monte Carlo Methods. *Journal of the Royal Statistical Society. Series B.*, vol. 55, no. 1, pp. 3–23 (1993).

[522] A. F. M. Smith and A. E. Gelfand. Bayesian statistics without tears: a sampling-resampling perspective. *The American Statistician*, vol. 46, no. 2, pp. 84–88 (1992).

[523] P. J. Smith, M. Shafi, and H. Gao. Quick simulation: A review of importance sampling techniques in communication systems. *IEEE J. Select. Areas Commun.*, vol. 15, pp. 597–613 (1997).

[524] L. J. Snell. Applications of martingale system theorems. *Trans. Amer. Math. Soc.*, vol. 73, pp. 293–312 (1952).

[525] R. Srinivasan. *Importance Sampling - Applications in Communications and Detection*, Springer-Verlag, Berlin (2002).

[526] H. E. Stanley. *Mean Field Theory of Magnetic Phase Transitions. Introduction to Phase Transitions and Critical Phenomena.* Oxford University Press (1971).

[527] N. Starr. How to win a war if you must: optimal stopping based on success runs. *Ann. Math. Statist.*, vol. 43, pp. 1884–1893, (1972).

[528] N. Starr. Optimal and adaptive stopping based on capture times. *J. Appl. Prob.*, vol. 11, pp. 294–301 (1974).

[529] N. Starr and M. Woodroofe. Gone fishin: Optimal stopping based on catch times. *Univ. Mich. Tech. Report*, no. 33, Dept. of Statistics (1974).

[530] N. Starr, R. Wardrop, and M. Woodroofe. Estimating a mean from delayed observations. *Z. fur Wahr.*, vol. 35, pp. 103–113 (1976).

[531] D. Steinsaltz and S. N. Evans. Markov mortality models: implications of quasi-stationarity and varying initial distributions. *Theor. Pop. Biol.*, vol. 65, pp. 319–337 (2004).

[532] D. Stoyan, W. Kendall, and J. Mecke, *Stochastic Geometry and Its Applications*, 2nd ed., Wiley (1995).

[533] R. L. Stratonovich. Conditional Markov processes. *Theory of probability and its applications*, vol. 5, pp. 156–178 (1960).

[534] R. L. Stratonovich. Optimum nonlinear systems which bring about a separation of a signal with constant parameters from noise. *Radiofizika*, vol. 2, no. 6, pp. 892–901 (1959).

[535] R. L. Stratonovich. On the theory of optimal nonlinear filtering of random functions. *Theory of Probability and Its Applications*, 4, pp. 223–225 (1959).

[536] R. L. Stratonovich. Application of the Markov processes theory to optimal filtering. *Radio Engineering and Electronic Physics*, vol. 5, no. 11, pp. 1–19 (1960).

[537] F. Suzat, C. Baehr, and A. Dabas. A fast atmospheric turbulent parameters estimation using particle filtering. Application to LIDAR observations. *J. Phys. Conf. Ser.*, vol. 318-072019 doi:10.1088/1742-6596/318/7/072019 (2011).

[538] A. S. Sznitman. Topics in propagation of chaos, course given at the Saint-Flour Probability Summer School, 1989. Lecture Notes in Math., 1464, Springer, Berlin, pp. 164–251 (1991).

[539] W. J. Stewart. *Introduction to the Numerical Solution of Markov Chains.* Princeton University Press, Princeton, NJ (1995).

[540] J. Tailleur, S. Tanese-Nicola, J. Kurchan. Kramers equations an supersymmetry. *Journal Stat. Phys.*, vol. 122, pp. 557–595 (2006).

[541] J. Tailleur, J. Kurchan. Probing rare physical trajectories with Lyapunov weighted dynamics. *Nature Physics*, vol.3 pp. 203–207 (2007).

[542] M. Talagrand. Concentration of measure and isoperimetric inequalities in product spaces. *Inst. Hautes Études Sci. Publ. Math.*, no. 81, pp. 73–205 (1995).

[543] M. Talagrand. A new look at independence. *Ann. Probab.*, vol. 24, pp. 1–34 (1996).

[544] M. Talagrand. New concentration inequalities in product spaces. *Invent. Math.*, vol. 125, pp. 505–563 (1996).

[545] H. Tanaka. Probabilistic treatment of the Boltzmann equation of Maxwellian molecules. *Z. Wahrsch. Verw. Gebiete*, vol. 46, no. 1, pp. 67–105 (1979).

[546] S. Tanese-Nicola, and J. Kurchan. Metastable states, transitions, basins and borders at finite temperature. *Journal Stat. Phys.*, vol. 116, no. 5–6, pp. 1202–1245 (2004).

[547] E. Tantar, A. Tantar, P. Bouvry, P. Del Moral, C. A. Coello Coello, P. Legrand, and O. Schutze. *Evolve: A bridge between Probability, Set Oriented Numerics and Evolutionary Computation I.* Springer Series: Studies in Computational Intelligence, vol. 447, no. XII, Springer (2013).

[548] H. Tembine, Q. Zhu, and T. Basar. Risk-sensitive mean-field stochastic differential games. *Proceedings of the 18th IFAC World Congress*, Milan, August 2011, pp. 3222–3227 (2011).

[549] J. L. Thorne, H. Kishino, and I.S. Painter. Estimating the rate of evolution of the rate of molecular evolution. *Molecular Biology and Evolution*, vol. 15, no. 12, pp. 1647–1657 (1998).

[550] S. T. Tokdar and R. E. Kass. Importance sampling: a review. *Wiley Interdisciplinary Reviews: Computational Statistics*, vol. 2, no. 1, pp. 54–60 (2010).

[551] V. D. Tran. Assimilation de données: les propriétés asymptotiques du filtre de Kalman d'ensemble. *Université de Bretagne Sud*, June 29 (2009).

[552] Y. K. Tsang, E. Ott, T. M. Antonsen, and P. N. Guzdar. Intermittency in two-dimensional turbulence with drag. *Phys. Rev. E*, vol. 71, 066313 (2005).

[553] J. N. Tsitsiklis and B. Van Roy. Regression methods for pricing complex American-style options. *IEEE Transactions on Neural Networks*, vol. 12, no. 4 (special issue on computational finance), pp. 694–703 (2001).

[554] G. E. Uhlenbeck, G. W. Ford. *Lectures in Statistical Mechanics*. American Mathematical Society, Providence, RI (1963).

[555] R. Van Handel. Uniform time average consistency of Monte Carlo particle filters. *Stoch. Proc. Appl.*, 119, pp. 3835–3861 (2009).

[556] J. Vanneste. Estimating generalized Lyapunov exponents for products of random matrices. *Phys. Rev. E*, vol. 81, 036701 (2010).

[557] F. Viens. Portfolio optimization under partially observed stochastic volatility. *Proceedings of the 8th International Conference on Advances in Communication and Control: Telecommunications / Signal Processing*, pp. 3–12, (2001).

[558] J. P. Vila and V. Rossi. Nonlinear filtering in discret time: a particle convolution approach. *Biostatistic group of Montpellier, Technical Report 04-03* (available at http://vrossi.free.fr/recherche.html) (2004).

[559] A. N. Van der Vaart and J.A. Wellner. *Weak Convergence and Empirical Processes with Applications to Statistics*. Springer Series in Statistics. Springer, New York (1996).

[560] P. Varaiya. N-players stochastic differential games. *SIAM J. Control Optim.*, vol. 14, pp. 538–545 (1976).

[561] D. Villemonais. Interacting particle systems and Yaglom limit approximation of diffusions with unbounded drift. *Electronic Journal of Probability*, vol. 16 (2011).

[562] D. Villemonais. Approximation of quasi-stationary distributions for 1-dimensional killed diffusions with unbounded drifts. arXiv:0905.3636 (2009).

[563] D. Villemonais. Interacting particle processes and approximation of Markov processes conditioned to not be killed. In revision for *ESAIM: Probability and Statistics*, p. 26 (2011).

[564] B. T. Vo, B. N. Vo, and A. Cantoni. The cardinalized probability hypothesis density filter for linear Gaussian multi-target models, *Proc. 40th Conf. on Info. Sciences & Systems*, Princeton, (2006).

[565] B. T. Vo, B. N. Vo, and A. Cantoni. Analytic implementations of the Cardinalized Probability Hypothesis Density Filter. *IEEE Trans. Signal Processing*, vol. 55, no. 7, part 2, pp. 3553–3567 (2007).

[566] B. N. Vo and W. K. Ma. The Gaussian mixture probability hypothesis density filter. *IEEE Trans. Signal Processing, IEEE Trans. Signal Processing*, vol. 54, no. 11, pp. 4091–4104 (2006).

[567] B. T. Vo. Random finite sets in multi-object filtering, Ph.D. thesis, University of Western Australia (2008).

[568] K. Wolny. Geometric ergodicity of heterogeneously scaled Metropolis-adjusted Langevin algorithms (MALA). Poster Seventh Workshop on Bayesian Inference in Stochastic Processes (BISP 2011), Madrid, Spain, 1-3.09.2011 (2011).

[569] M. D. Vose. *The Simple Genetic Algorithm: Foundations and Theory.* MIT Press (1999).

[570] A. Wald. *Sequential Analysis*, John Wiley & Sons, New York (1947).

[571] A. Wald. *Statistical Decision Functions*, John Wiley & Sons, New York (1950).

[572] D. Whitley, S. Rana, and R. B. Heckendorn. The island model genetic algorithm: on separability, population size and convergence. *Journal of Computing and Information Technology*, vol. 7, no. 1, pp. 33–47 (1999).

[573] A. H. Wright and M. D. Vose. Stability of vertex fixed points and applications. In *Foundations of Genetic Algorithms 3*, Morgan Kaufman Publishers, vol. 3 (1995).

[574] A. M. Yaglom. Certain limit theorems of the theory of branching random processes. *Doklady Akad. Nauk SSSR (N.S.)*, vol. 56, pp. 795–798 (1947).

[575] T. Yang., P.G. Metha, and S.P. Meyn. A mean-field control-oriented approach to particle filtering. *Proceedings of the American Control Conference (ACC)*, June 2011, pp. 2037–2043 (2011).

[576] L. Zhang and S. Dai. Application of Markov model to environment fate of phenanthrene in Lanzhou reach of Yellow river. *Chemosphere*, vol. 67, pp. 1296–1299 (2007).

Index